Optical production technology

Optical production technology

D. F. HORNE, M.B.E.

C.Eng., F.I.Prod.E., A.F.R.Ae.S.

Group Production Engineer, Rank Precision Industries Ltd, 1948–58
General Manager (G. B. Kalee Division), Rank Precision Industries Ltd, 1958–62
Group Production Adviser, Hilger and Watts Ltd, 1962–69
Optical Production Engineering Manager (Metrology Division), Rank Precision Industries Ltd, 1969–70
Lecturer, School of Production Studies, Cranfield Institute of Technology, 1971
Technology Tutor, The Open University, 1972

Adam Hilger : London

COPY B

Produced and Published in Great Britain by
Adam Hilger Ltd
Rank Precision Industries Ltd
31 Camden Road, London NW1

SBN 85274 187 1

Printed in Great Britain by
William Clowes & Sons, Limited
London, Beccles and Colchester

To
My Wife

AUTHOR'S PREFACE

On 11 March 1970, I received a letter from Mr Neville Goodman, Managing Director of Adam Hilger Ltd, inviting me to be 'the author of a new book on optical working—a book resembling *Prism and Lens Making* in its wealth of information, but entirely new and up to date and authoritative for a new generation of opticians'.

This was a request for a major task which, after some considerable thought, I accepted, and I hope that *Optical Production Technology* will approach the high standard set by Mr F. Twyman, F.R.S., some thirty years ago.

Inevitably, in this field there are some techniques which have not changed since then. The descriptions of these, together with some historical matter, have been extracted from *Prism and Lens Making*. This is noted at the relevant points in the text.

However, there have been many new developments in optical machines, equipment, and materials over the last twenty years, and to ensure technical accuracy I have included papers and articles by specialists who have taken part in the rapid progress during this period.

This book is addressed to students at Universities, Colleges, and Institutes of Technology. Also, I hope that managers, technologists, and technicians in industry will find the information useful and easy to understand. There are over 300 photographs (and as many charts and diagrams) to illustrate the optical machines, equipment, and processes discussed in the text.

I hope that I have not made any technical errors or significant omissions in the text, but if I have, I should like to be advised so that corrections or additions can be made in the next issue.

Any opinions which I have expressed are, of course, my responsibility and do not necessarily indicate the policy of Rank Precision Industries Ltd or any other Company which has given information to me.

It would have been difficult to write this book without the cooperation of the Directors of Rank Precision Industries Ltd, and I wish to thank them for their permission for me to describe processes, and take photographs of equipment, in each of the optical departments of the Group.

I have received help from a large number of people, but within Rank Precision Industries Ltd I particularly wish to thank R. Shipp, D. Gane, G. Wiseman, S. Mansfield, C. Rhodes, C. H. Hutt, S. C. Bottomley, C. Wright, A. Ashton, P. Allan, J. Reed, A. Monery, K. F. Rippon, R. Elam, J. Woolley, R. Bettell, D. J. Day, D. G. Monk, J. Eade, G. Cleaver, T. W. C. Fisher, H. W. Martin, J. R. Oliver, D. Demaine, and A. J. Bennett.

Information was also freely given by Mr A. J. Munro, General Manager, Vickers Ltd (Vickers Instruments), York, and by F. B. R. Hogg and F. Grimes on microscope production processes. Astronomical-telescope production information was provided by Mr G. E. Manville, General Manager, Sir Howard Grubb, Parsons & Company Ltd, Newcastle-upon-Tyne, as well as D. Brown and H. A. McLeod. Screen production information was provided by Mr T. Harkness and G. T. J. Knowles of Andrew Smith Harkness Ltd, Elstree.

In each of the factories mentioned I was allowed to take photographs, and permission for publication has been given to me. Mr A. A. S. Moore, Technical Director, United Kingdom Optical Co. Ltd, has very kindly provided the whole of Chapter 7, on the production of spectacle lenses, with the help of Mr O. P. Raphael on ophthalmic prescription work.

Permission has been given by the following Companies or Publishers for reproduction of material. The names are listed in the order in which material appears for the first time in the book.

Adam Hilger Ltd, *Prism and Lens Making*
SIRA Institute (Dr D. C. Cornish, B.Sc.)
I.C. Optical Systems Ltd (Mr H. W. Yates, Managing Director)
Autoflow Engineering Ltd (Mr H. C. Davis, Managing Director)

Abrafract Ltd (Mr R. A. Bingley, Technical Director)
The Roditi International Corporation Ltd (Mr D. Shamash)
The Carborundum Company (Miss M. Mitchell)
Industrial Diamond Information Bureau (Mr T. Marles)
Impregnated Diamond Products Ltd (Mr J. F. Finnimore)
Universal Grinding Wheel Company Ltd (Mr J. Woodiwiss)
Machine Shop Equipment Ltd (Mr E. J. C. Holland)
Securitas Non-Inflammable Products Company Ltd (Mr B. Baker)
G.S. Chemicals Ltd (Mr F. J. G. Graham, Managing Director)
The Penetone Company Ltd (Mr D. Hoggard)
Quadralene Chemical Products Ltd (Mr N. W. Vale, Technical Director)
Chance-Pilkington (Pilkington Optical Division) (Mr A. B. Scrivenor)
Pilkington Brothers Ltd (Dr W. M. Hampton)
Eastman Kodak Company, U.S.A. (Mr W. C. Colsman)
Jena[er] Glaswerk Schott and Gen, Mainz (Dr Carsten Eden)
Thermal Syndicate Ltd (Dr G. Hetherington, Technical Director)
Owens-Illinois Development Center, U.S.A. (Dr G. A. Simmons)
Combined Optical Industries Ltd (Mr M. Sigmund, Managing Director)
English Glass Company Limited (Mr T. J. Lawson)
South London Electrical Equipment Co. Ltd (Mr D. A. Rastall)
Wilhelm Loh, Wetzlar (Mr H. V. Scan)
R. Howard Strasbaugh Inc., U.S.A. (Mr R. H. Strasbaugh)
Bryant Symons & Co., Ltd (Mr R. J. R. Cully, Managing Director)
Machinery and Production Engineering (Mr P. A. Sidders, Chief Associate Editor)
Mr Leo H. Narodny, International Optical Ltd, Barbados
Bell & Howell Company, U.S.A. (Mr R. Hartmann)
Pilkington Brothers Ltd, Research and Development Laboratories (Mr D. W. Harper)
Solex (Gauges) Ltd (Mr C. J. Tanner, General Manager)
Moore Special Tool Co., U.S.A. (Mr R. W. Kuba)
National Physical Laboratory (Mr G. D. Drew)
Varian Associates Ltd U.S.A. (Mr C. W. Vince)
Ealing Beck Ltd (Mr P. S. Kenrick)
Edwards High Vacuum International (Dr L. Holland, Director of Research)
Rank Xerox Ltd (Mr F. Wickstead, Production Director, and Mr A. S. Pratt)
LogEtronics Inc., U.S.A. (Photomechanisms Division) (Mr William Kacin, Vice-President and General Manager)
Pilkington Brothers Ltd (Sir Alastair Pilkington)
Science Research Council (Mr R. J. Harris)
Australian and New Zealand Association for the Advancement of Science (Mr F. L. Freedman)
Mt Stromlo and Siding Spring Observatories, Australia
Edwards High Vacuum Plant Ltd (Mr R. H. Robinson, Director)
Sir Howard Grubb, Parsons & Company Ltd (Mr George Sisson)
Barr & Stroud Limited
Science Journal (Mr Robin Clarke, Editor)
Metals Research Limited (Mr D. J. Berry, Group Marketing Manager)
Royal Radar Establishment (Mr O. Jones and Mr G. W. Fynn)
Walter Bunter S.A.
Engis Ltd (Mr R. J. Swan, Director)

The frontispiece was photographed by Mr Walter Nurnberg, F.R.P.S., in the factory of Sir Howard Grubb, Parsons and Company Ltd, and is reproduced with permission of the editor of *Engineering*.

Photographs in §§ 11.8, 18.4, and 18.5, supplied by the National Physical Laboratory and the Royal Radar Establishment, are Crown Copyright reserved, and I wish to thank the Director of N.P.L. and the Controller of H.M.S.O. for permission to reproduce them.

I also acknowledge the efforts made by my wife in typing my manuscript and thank her and my family for their help to me during the time of writing this book. Indeed, I must thank them doubly; for they are even now granting me similar assistance in my collaboration with Mrs C. E. Arregger in the writing of a companion volume on *Dividing and Ruling* for the same publishers.

August 1972 D. F. HORNE

CONTENTS

1 Historical

1.1 Introduction

In the early part of the seventeenth century, the telescope and microscope were invented and the production of lenses therefore became an important factor in scientific development. The early lenses were produced with primitive equipment, but improved lathes for grinding and polishing lenses were developed by Ippolito Francini of Florence between 1623 and 1653, and these lathes provided some of the lenses for Galileo's instruments.

Further developments were made to manufacturing techniques by Carlo Antonio Manzini in 1648, followed by major improvements in production accuracy by Giuseppe Campani in 1660. In 1664, Campani published a book describing several telescopes he had made for observations of the planet Saturn and revealed his invention of a lathe for grinding and polishing lenses, which lathe enabled lens blanks, cut from sheet glass, to be ground without previously having been cast in a mould.

In 1766, Fougeroux de Bondaroy visited Bologna to see Campani's lathes and shop equipment. His purpose was to report to the Académie Royale des Sciences, and he stated that the choice of glass was important and that improved equipment with many minute details of workmanship all contributed to Campani's successful product. De Bondaroy commented that Campani's glass had few defects and was imported from Venice, while the workshop equipment included numerous metal patterns of all types and sizes which were used progressively in the grinding of the lenses.

At the same time, Christiaan Huygens of Holland and his brother Constantine developed their own techniques for the production of lenses for telescopes and, in 1661, Christiaan visited England to describe his methods to the Royal Society.

Although spherical aberration was understood, no aspherical lenses were made in the seventeenth century; for the defect was not critical owing to the extremely long focus used in the telescopes at that time.

Meanwhile, in England, the spectacle-makers turned their attention towards the making of lenses for optical instruments and, among them, John Marshall was an outstanding craftsman and the first to be approved by the Royal Society. By the end of the seventeenth century, the pioneering work in optical workshop practice had been completed and optical science was ready for the rapid progress of the eighteenth century.

From this point on, we cannot do better than to quote from Mr F. Twyman's account of historical development.

1.2 Ancient optical work

As with most of the useful arts, the development of lens-making has been in an order the reverse of what the scientific man feels to be logical. Instead of first studying the principles of the process, then putting the process into operation, and then finding a use for the product—a course followed by the science-born electrical industry—man first discovered some optical uses of accidentally produced lens-shaped bodies, and only then set himself deliberately to make lenses, leaving till quite recent times the study of the process of lens polishing.

It is a pleasing fancy that the possibility of using a lens as a burning glass may be related to the supposed ability of the priestly classes during the Nilotic and Mesopotamian civilizations to 'bring down fire from heaven' during religious ceremonies. The word *focus* (Latin) meant originally a hearth or burning place, and its etymology goes back to roots which suggest that originally it had associations with temple altars or places where sacrifices were burned (see the larger *Oxford Dictionary*). It may be noted that the modern French word 'foyer' is used for both 'hearth' and 'focus'.

Gunther (1923) points out that in England, the practice of kindling the new fire on Easter Eve by a burning glass was not uncommon in the Middle Ages; an entry to that effect occurs in the Inventory of the Vestry, Westminster Abbey, in 1388.

The references of Pliny and other ancient writers show quite clearly that burning glasses were known to them in the shape of glass spheres filled with water; and passages from Greek and Roman writers have been cited as showing that they knew of the magnifying properties of lenses, or at least of such glass spheres filled with water. The very thorough account of the subject by Wilde (1838–43) denies to the ancients all knowledge of spectacle lenses whether for short or long sight, or indeed of any kind of lenses, if we except the spheres of glass filled with water just mentioned; he maintains that the lens-shaped glasses or crystals which have been found from time to time among the relics of departed civilizations were made by polishers of jewels for purposes of ornament. Mach (1926), on the other hand, seems to tend, on the whole, to the opinion that a few archaeological objects were made, and intended to be used, as lenses.

Beck (1928) adduces further evidence to show that lens-shaped objects were used as magnifying and burning glasses from very early times. He points out that the usual varieties of glass have been continuously made from the time of the Eighth Egyptian Dynasty, whilst a piece of glass in the Ashmolean Museum is claimed to be First Dynasty, if not pre-dynastic. A large piece of blue glass from Abu Shahrein in Mesopotamia dates from about 3000 B.C. Beck continues:

> But whatever may have been the original date of the invention of glass, we know that by the fourteenth century B.C. there was a well-established centre of glass manufacture in Egypt, and a totally different one in the Aegean, where the technique was in use which did not penetrate into Egypt until a very much later date. Also, although the amount of transparent glass made in Egypt at that time was only a trifling proportion of the total output, much of the Aegean glass was transparent and a considerable amount colourless.
>
> The date of the first manufacture of colourless glass need not however, limit us in finding a possible date for early magnifiers, as crystal was always to hand and the earliest magnifiers known are in that material.
>
> The first reference to a lens that I know of in literature is in *The Comedy of the Clouds* by Aristophanes, which was performed in 434 B.C. In the second act comes the following passage:
>
> STREPSIADES: You have seen at the druggists that fine transparent stone with which fires are kindled?
> SOCRATES: You mean glass?
> STREPSIADES: Just so.
> SOCRATES: Well what will you do with that?
> STREPSIADES: When a summons is sent me, I will take this stone and placing myself in the sun I will, though at a distance, melt all the writing of the summons.
>
> *Note.*—The point of the remark of Strepsiades in the above passage is that the writing was a summons for debt. Such a writing would be traced on wax, and the suggestion is that, if he melted it with his burning glass, the record would be lost and he would thus be freed from his debts.
>
> . . . Lactantius in A.D. 303 says that a glass globe filled with water and held in the sun could light a fire even in the coldest weather.
>
> Now lenses sufficiently good to make burning glasses, would make magnifying glasses . . .

There are in the Egyptian department of the

British Museum two magnifying glasses which would make excellent burning glasses, except for the tarnish. They are about $2\frac{1}{2}$ in. diameter and about $3\frac{1}{2}$ in. focus and would magnify three diameters. They have been ground and are not merely cast. The flat surface has been ground against another flat surface with a rotary motion as at the present. For example; one of these glasses (22522 Egyptian Department) was found at Tanis, and definitely dated A.D. 150.

. . . but the most conclusive proof as to the early magnifying glasses is the discovery (1927) by Mr E. J. Forsdyke, in Crete, of two crystal magnifying lenses that date back at least as early as 1200 B.C., and probably 1600 B.C., as most of the small objects from the tombs where they were found are of that date.

About half-a-dozen other examples of objects which, whether intended to be used as magnifying glasses or not, were—and indeed, in some cases, are—*capable* of being so used are given in Chapter 3 of Greeff's book (1921). One glass is in the Volkermuseum in Berlin among the well-known objects excavated at Troy by Schliemann; it is supposed to date from the second half of the third century B.C.

It will be seen that the methods of polishing jewels for ornaments were available when men first felt the urge to polish lenses for use. Even then the first use was probably a ritual in connection with religious rather than secular purposes and possibly kept in close secrecy; a secrecy which still tends to linger in the 'mystery' of the trade.

The name of Alhazen is often mentioned in connection with the early history of Optics.

Alhazen (abu Ali Al-Hasan Ibu Alhasan) was born at Basra and died in Cairo in A.D. 1038. He solved the problem of finding the point on a convex mirror at which a ray coming from one given point shall be reflected to another. His treatise on optics was translated into Latin by Witelo (1270) and published by F. Risner in 1572 with the title *Optical thesaurus Alhazeni libri VII cum ejusdem libro de crepusculis et nubium ascensionibus* (*Enc. Brit.*, 11th Ed., 1910–11, 'Alhazen'). The only mention Alhazen made of lenses, however, appears to be his statement that if an object is placed at the base of the larger segment of a glass sphere, it will appear magnified.

1.3 The invention of spectacles

We must come to the end of the thirteenth century for the first authentic mention of the use of spectacles, which appears to be that of Meissner (1260–80), when he expressly states that old people derive advantage from spectacles (Bock, 1903). In the archives of the old Abbey of Saint-Bavon-le-Grand, the statement is found that Nicolas Bullet, a priest, in 1282 used spectacles in signing an agreement (Pansier, 1901). The first picture in which spectacles are known to have appeared is by Tomaso de Modena, in the Church of San Nicola in Treviso, and is of date 1362 (Oppenheimer, 1908).

Martin says:

> The invention of spectacles for the short and the long sighted is mentioned as a *quite recent* discovery in a manuscript of 1299 in Florence. Bernard Gordon, Professor at Montpelier, in his work *Lilium medicinae* begun in 1305 alludes to spectacles as a means of remedying visual defects. Giordano da Rivalto, in a sermon given on February 23rd, 1305, remarks that this invention *is not yet twenty years old.* The man who copied the sermon says he himself saw and conversed with the inventor. Thus it was about 1285—no earlier—that spectacles were invented. We know besides that the inventor was the Florentine, Salvino d'Armato degli Armati, who died in 1317. He hid his secret to keep it as a monopoly. But Brother Alessandro Spina of Pisa, who died in 1313, having seen spectacles made by Salvino d'Armati and having succeeded in making similar ones, hastened to make public the secret.
>
> Primitive spectacles consisted of two pieces of leather which were fastened on to a cap, worn low over the forehead.

Greeff (1921) has examined a great mass of data about the invention of spectacles. He points out that, although we can scarcely neglect the evidence of pictures, it would be naïve to conclude that any spectacles shown were of the epoch the picture represents. Even in depicting very ancient scenes, painters from the time of the van Eycks used to introduce a pair of spectacles to add verisimilitude when they wished to represent a person sunk in study or meditation. The author instances many cases of such anachronisms, for example, Moses is furnished with spectacles in a miniature painted in Heidelberg about 1456. The painter aimed not at historical accuracy in detail, but at representing things as their contemporaries saw and felt them; spectacles in their pictures were, therefore, of their own period.

In another chapter, Greeff examines the suggestions put forward from time to time that spectacles originated in India or in China, but he finds them groundless. He concludes, after citing many authorities, that there is absolutely no evidence that the Chinese had spectacles before they originated in Europe and shows good ground for supposing that they came in through Malacca in the early sixteenth century. Dr Greeff attributes, however, to Prof. Hirth of Columbia University, New York, the statement that the Chinese had mirrors both concave and convex, of bronze, in the first century B.C.

It may be accepted, from this and like evidence, that the use of spectacles dates from about A.D. 1280.

Brockwell (1948) says:

> As for the making of the lenses, Roger Bacon, in his *Opus Majus* of about 1266, and in his *Perspectivae Pars Tertia*, showed that 'by placing a segment of a sphere on a book with its plane side down, one can make small letters appear large'. He communicated his knowledge of optics to his friend Heinrich Goethals, who, travelling in Italy in 1285, handed on his information to Alessandro della Spina, a monk in Pisa, who 'could make anything he liked' [*operava di sua mano ogni cosa che volesse*], and who died in 1313.

Mr Twyman found no account of how spectacle lenses were made before William Bourne (*c.* 1585) who says:

> These sortes of glasses ys grounde upon a toole of Iron made of purpose, somewhat hallowe, or concave inwardes. And may be made of any kynde of glasse, but the clearer the better. And so the Glasse, after that yt ys full rounde, ys made fast with syman uppon a small block, and so ground by hand untill yt ys bothe smoothe and allso thynne, by the edges, or sydes, but thickest in the middle.

Nothing else is said by him of the materials, tools, or method of working.

Baptista Porta

Very different is the account given by Baptista Porta of Naples (1591) in his famous book on Natural Magic—a technical encyclopaedia embracing subjects as diverse as optics, magnetism, cosmetics, cooking, alchemy, pharmacy, and practical jokes. Among much that is trivial, debased, and revolting is also to be found much, like his description of optical polishing, which shows a keen quest after knowledge and accurate knowledge of a singularly wide range of subjects. The following extract is from Book 17, Chapter XXI in the English translation published in 1658, but the matter is identical with the Latin edition of 1591. The translation has, however, rendered the original 'pilae vitrae'—the phrase employed (as by Pliny) to describe a *hollow* glass ball—as 'Glass balls', and the reader must bear this in mind if he wishes to follow the description correctly.

> In Germany there are made Glass-balls, whose diameter is a foot long, or there abouts. The Ball is marked with the Emrilstone round and is so cut into many small circles, and they are brought to Venice. Here with a handle of wood are they glewed on, by Colophonia melted. And if you will make Convex Spectacles, you must have a hollow iron dish, that is a portion of a great sphaere, as you will have your spectacles more or less Convex; and the dish must be perfectly polished. But if we seek for concave spectacles, let there be an Iron ball, like to those we shoot with Gun-powder from the Great Brass Cannon; the superficies whereof is two, or three foot about. Upon the Dish or Ball, there is strewed whitesand, that comes from Vincentia, commonly called Saldame, and with water it is forcibly rubbed between our hands, and that so long until the superficies of that circle shall receive the form of the Dish, namely a Convex superficies or else a Concave

superficies upon the superficies of the Ball, that it may fit the superficies of it exactly. When that is done heat the handle at a soft fire, and take off the spectacle from it, and join the other side of it to the same handle with Colophonia, and work as you did before, that on both sides it may receive a Concave or Convex superficies, then rubbing it over again with the Power of Tripolis that it may be exactly polished; when it is perfectly polished, you shall make it perspicuous thus. They fasten a woollen-cloth upon wood; and upon this they sprinkle water of Depart, and powder of Tripolis; and by rubbing it diligently, you shall see it take a perfect glass. Thus are your great Lenticulars and spectacles made at Venice.

1.4 Telescope and microscope lenses

Cherubin d'Orléans

In 1671 appeared the well-known book by a Père Cherubin d'Orléans, which not only deals with optics, with telescopes (including binoculars) and microscopes, their theory and construction and use, but with lens-making and the various machines invented by himself to lessen the labour and increase the speed and accuracy of polishing lenses for telescopes. These descriptions are so good, and show such thoughtful personal knowledge of the subject, that they would be suitable to place in the hands of an optical apprentice today.

His materials were—for the tools, iron and brass; for the mallets (molettes) for holding the lenses, lead, tin or (which he prefers) copper. He gives full particulars for making the patterns for casting, making the moulds, and turning and grinding the tools. He also describes lathes which he invented for turning the tools.

Of the cement for attaching the lens to the mallet, he says that some make it of best pitch (which must not be burnt) and sifted ashes of vine cuttings; but that he prefers to add (to the pitch) a fourth part of good grape jelly and, in place of the ashes, finely ground ochre or whiting.

As a good material for grinding he recommends broken grindstones. These are graded by putting the powder into a large vessel full of water, agitating it well, letting it settle a little for the coarse particles to settle out, and then pouring off quickly into another vessel most of the liquid, which will carry with it the finer material. This one allows to settle entirely, gently pouring off the remaining liquid. The sediment (containing the useful grains) is treated in this way several times, and in this way one separates out the grains of several degrees of 'strength' which are kept separately for use according to the nature of the work. For polishing material he used Tripoli (preferably that of Germany) or 'potée d'estain' (putty powder).

Of the Tripoli he says that the lightest is the best. If of good quality it can be used in the lump, as Nature produces it, otherwise ground with brandy or (failing that) white wine, and kept in a closed jar of water to soften for four or five months. It can be sun-dried, and used in lumps, or used wet straight from the jar.

The putty powder he prefers is that made by calcining tin (he gives detailed instructions for its preparation). The glass was provided by broken Venice mirrors, and was thus in a form polished on both sides and suitable for examination.

He claims to have made improvement in the mode of polishing in that he stretched across the concave tool, which he had used for grinding, a soft thin piece of leather of uniform thickness, fixing it in position with a ring which just fitted the circumference of the tool. On this he rubbed his lens, pressing it down so that the leather was forced to fit the surface of the grinding tool. Another way he used was to coat the surface of the grinding tool with paper, the latter being pasted in position with many precautions to avoid wrinkling or other inequalities of the surface. This paper he moistened with Tripoli powder and so used for polishing. Of the machines he describes as invented by himself several may well have been the progenitors of some in use today. One of them may be mentioned in which the optical tool, mounted on a vertical spindle, is rotated by means of a rope and suspended weight so that the operator's hands are free. The tool is turned accurately to the desired radius by means of a radius arm pivoted at one end, and bearing a turning tool at the other, by means of which the tool is turned to the desired concavity. The same machine is used in grinding and polishing the glasses, the latter being held in the hand as in free-hand working.

Hooke

About this time Hooke was working on the microscope (Hooke 1667). He describes a way of making microscope objective lenses; he drew a piece of broken Venice glass in a lamp into a thin thread, then held the end of this thread in a flame till a globule of glass was formed. He then polished a flat surface on the thread side of the globule, first on a whetstone and then on a smooth metal plate with Tripoli; but these lenses being too small he used good plano-convex object glasses, and there is no indication that he made these himself.

Leeuwenhoek

The great Dutch microscopist, Leeuwenhoek (1719), made his own lenses, but left no account of his methods. He says in a letter to Leibnitz dated 28 September 1715:

> As to your idea of encouraging young men to polish glass—as it were to start a school of glass polishing—I do not myself see that would be of much use. Quite a number, who had time on their hands at Leyden, became keen on polishing glasses, owing to my discoveries; indeed there were three masters of that art in that town, who instructed students who were interested in such things. But what was the result of their labour? Nothing at all, so far as I have learnt.
>
> Now to every study the proposed object is this: to acquire wealth by knowledge, or celebrity by reputation for learning. But that is not to be gained either by polishing glasses or by discovering abstruse things. And then I am convinced hardly one in a thousand is properly fitted to take up this study; for much time is consumed in it and money is wasted, and if one is to make any progress in it one's mind must be for ever on the stretch, thinking and speculating. The majority of men are not sufficiently inflamed with the love of knowledge for that. Indeed, many whom it by no means becomes, do not hesitate to ask, What does it matter whether we know these things or not?

1.5 Newton and the reflecting telescope

Newton was the first to make a successful reflecting telescope. It is true that James Gregory had proposed his Gregorian telescope in 1663 and the next year came to London to commission the manufacture of such a telescope from Reive, a famous London optician. It was to be 6 feet focus but the figure proved so bad that the attempt was abandoned. Gregory was of the opinion that the failure was due to Reive's trying to polish the mirrors with cloth; Gregory must therefore have had in mind the possibility of a more perfect medium for a polisher. The passage from Newton (1721) is reputed to be the first *publication* of the use of pitch for a polisher.

Newton completed his first reflecting telescope in 1668. It had an aperture of 1 inch only and focal length of 6 inches.

The reasons why Newton adopted the reflecting form of telescope will be mentioned in Chapter 10.

In the list of contents of the *Philosophical Transactions* for 25 March 1672, one reads 'An accompt of a new kind of Telescope, invented by Mr Isaac Newton.' In the 'Accompt' it is stated that for metal he tried various mixtures of copper, tin, and arsenic. With one of them, consisting of copper 6 oz, tin 2 oz, arsenic 1 oz, a friend of his said he had 'polish't better than he did the other'. He used putty powder for polishing his metal mirror.

It will be seen that the metal he used was very like speculum metal, defined, in *Encyclopaedia Britannica* IXth Ed., Article 'Bronze', as consisting of two parts of copper to one part of tin. Newton (1721) makes some important remarks on polishing, which though referring to mirrors are also applicable to lenses, and appears to have been the first to use pitch for polishing, an innovation of the very greatest importance. Newton says in the reference cited:

> The Polish I used was in this manner. I had two round Copper Plates each five inches in diameter, the one convex, the other concave, ground very true to one another. On the convex I ground the Object-Metal or Concave which

was to be polished, till it had taken the figure of the Convex and was ready for the Polish. Then I pitched over the convex very thinly, by dropping melted Pitch upon it and warming it to keep the Pitch soft, whilst I ground it with the concave copper, wetted to make it spread evenly all over the convex. Thus by working it well I made it as thin as a Groat,* and after the convex was cold I ground it again to give it as true a figure as I could. Then I took Putty which I had made very fine by washing it from all its grosser particles, and lay a little of this upon the pitch, I ground it upon the Pitch with the concave Copper till it had done making a noise; and then upon the pitch I ground the Object-Metal with a brisk motion, for about two or three minutes of time, leaning hard upon it. Then I put fresh putty upon the Pitch and ground it again till it had done making a noise, and afterwards ground the Object-Metal upon it as before. And this work I repeated till the Metal was polished, grinding it the last time with all my strength for a good while together, and frequently breathing upon the pitch to keep it moist without laying on any more fresh Putty. The object-metal was two inches broad and about one third of an inch thick, to keep it from bending. I had two of these Metals, and when I had polished them both I tried which was best, and ground the other again to see if I could make it better than that which I kept. And thus by many trials I learn'd the way of polishing, till I made those two reflecting Perspectives I spake of above. For this Art of Polishing will be better learned by repeated Practice than by my description. Before I ground the Object-Metal on the Pitch, I always ground the Putty on it with the concave Copper till it had done making a noise, because if the Particles of the Putty were not by this means made to stick fast in the Pitch, they would by rolling up and down grate and fret the Object-Metal and fill it full of little holes.

But because metal is more difficult to polish than Glass and is afterwards very apt to be spoiled by tarnishing and reflects not so much Light as Glass quick-silvered over does: I propound to use instead of the Metal, a Glass ground concave on the foreside, and as much convex on the back-side, and quicksilvered over on the convex side. The Glass must be everywhere of the same thickness exactly. Otherwise it will make objects look coloured and indistinct. By such a Glass I tried about five or six Years ago to make a reflecting telescope of four Feet in length to magnify about 150 times, and I satisfied myself that there wants nothing but a good Artist to bring the design to perfection. For the glass being wrought by one of our London Artists after such a manner as they grind Glasses for Telescopes, tho' it seemed as well wrought as the object-glasses used to be, yet when it was quick-silvered, the Reflexion discovered innumerable Inequalities all over the Glass. And by reason of these Inequalities, Objects appeared indistinct in the Instrument. For the errors of reflected Rays caused by an Inequality of the Glass are about five times greater than the Errors of refracted rays caused by the like Inequalities. Yet by this Experiment I satisfied myself that the Reflexion on the concave side of the Glass, which I feared would disturb the Vision, did no sensible prejudice to it, and by consequence that nothing is wanting to perfect these Telescopes but good Workmen who can grind and polish Glasses truly spherical. An Object-Glass of a fourteen Foot Telescope made by an Artificer at London, I once mended considerably by grinding it on Pitch and Putty, and leaning very easily on it, in the grinding, lest the Putty should scratch it. Whether this way may not do well enough for polishing these reflecting Glasses, I have not yet tried. But he that shall try either this or any other way of polishing which he may think better, may do well to make his Glasses ready

* Mr Twyman expressed his indebtedness to Mr E. S. G. Robinson, Deputy Keeper of Coins and Medals of the British Museum, London, for the following information.

The groat in Newton's time was approximately half a millimetre thick. It was, however, not in wide currency, but only a Maundy piece, and the thickness of the earlier *currency* groats, which in his day would be considerably worn, would hardly have been more than one-third of a millimetre.

If, then, Newton intended his phrase to be taken literally and was not merely using the familiar phrase 'thin as a groat'—most unlikely in view of his usual precision of statement—he was using a thickness of pitch not more than one-third of what is customary today. There is much to be said for such practice, since it would make possible the use of a much softer pitch without distortion—thus obtaining less liability to scratch or sleek.

for polishing by grinding them without that violence wherewith our London workmen press their Glasses in grinding. For by such violent pressure, Glasses are apt to bend a little in the grinding, and such bending will certainly spoil their figure.

1.6 Early grinding and polishing machines

Some optical grinding and polishing machinery appears to have been made and used early in the seventeenth century. Dr C. A. Crommelin (1929) describes and illustrates machines made or proposed by Descartes, Huygens, Hooke, Hevelius, and others.

Herschel

Herschel (*Collected Papers*, 1912) in 1774 used a pitch polisher for polishing the speculum mirrors for his telescope. He mentions the polishing operation as having been carried out by ten men on one occasion. He gave an account of the polishing of a large speculum by a machine which he made to avoid the necessity of employing so many men, but published no details of his working methods, as he evidently looked upon these as trade secrets. He did, however, contribute a vague paper to the Royal Society in 1789, in which he stated that he had at last completed a polishing machine that really worked, whereas he had tried and failed six years before. I think we may take it that from 1789 onwards he used machines for polishing and figuring *all* his mirrors. The only details that have ever been published are incorporated in an article ('Telescope') written by J.F.W.H. for the 8th edition of the *Encyclopaedia Britannica*. Some extracts from this article are given by Dreger in p. XLIX of Vol. I of the *Collected Scientific Papers*. Here there is a diagram showing the principle employed in the movement and rotation of mirror and polisher at each stroke.

Fraunhofer

Fraunhofer, who made telescope lenses of great excellence, is said to have been first to use proof spheres for testing the accuracy of surfaces, but Dévé (1936) attributes their earliest use to the French firm of Laurent (now Messrs Jobin et Yvon).

Lord Rosse

Finally, in what we may still call the historical period, Lord Rosse (Parsons, 1926) (then bearing the title of Lord Oxmantown—he succeeded to the title of Earl of Rosse in 1841) described before the Royal Society a machine for polishing large specula.

In the course of the same papers he describes the method of preparation of his rouge (Parsons, 1926, p. 98) by calcination, at a dull red heat, of peroxide of iron produced as a precipitate with ammonia water from a dilute solution of iron sulphate. Lord Rosse (father of Sir Charles Parsons of steam turbine fame) used 'ammonia soap' with the water used in polishing. This probably retarded tarnishing; it is known that swabbing speculum mirrors with dilute ammonia will remove tarnish.

These steps in the development of methods of lens-making must suffice, but it should be added that surviving accounts very possibly leave unnamed the workers to whom the methods (probably originally derived from the ancient art of polishing stones) are due.

2 The Nature of Grinding and Polishing

The object of polishing is to produce regular transparent surfaces on a piece of glass or other clear substance. The surfaces are usually required to be flat or spherical, although occasionally departures from these shapes are needed in order to obtain some optical advantage not otherwise attainable. The process is divided into two: grinding and polishing. They are commonly held to be quite different in character, although this opinion is not universal.

Since 1950, all countries concerned with the polishing of glass have made considerable efforts to find out how a good polish is produced, so that effective improvements in method might be made to increase output and improve the quality of finish on optical components.

There are three main theories on how a glass surface is polished:

(1) Mechanical removal or wear (Newton, Rayleigh *et al.*)

(2) Athermic surface flow (Beilby *et al.*)

(3) Formation of a silica-gel surface by hydrolysis (Grebenshchikov *et al.*)

It would appear that all three theories are true to a greater or less extent but, although polishing times have been reduced by means of more powerful and stiffer machines capable of high speeds, there is no sign to date of any fully new type of polishing machine on the market.

Polishing speed, pressure, type of glass, surface finish from grinding, type of polishing material, and polishing compound, all have an important influence on what exactly happens during the process.

From 1962 to 1967, at Sira Institute, Chislehurst, Kent, Dr D. C. Cornish, B.Sc., with the help of I. M. Watt, M.Sc., and A. F. D. Verral, Grad. Inst. P., carried out a comprehensive series of experiments and research into previous work on this subject. His conclusions are as follows:

> Many of the early observations were of a qualitative nature and readily lent themselves to two or more interpretations and indeed this aspect of the problem is still with us today. However, with the methods then available it is doubtful whether further advances would have been made even with the most careful qualitative and quantitative experimenting. The phenomenon of polishing occurs beyond the scope of the optical microscope. It was only with the advent of the electronic miscroscope, phase-contrast techniques, the surface profilograph, etc., and perhaps more important, a new realisation of the significance of the chemical and physical properties of solid surfaces, that there occurred the great surge of work on the mechanism of glass polishing.
>
> However, what is to be concluded from this published work? It seems to be now accepted by most authorities that the hydrolysis of the glass surface plays an important part in the process. It is most likely that in the presence of water it is this hydrolysed layer itself that is removed by the polishing agent. How this is achieved is

still obscure. Physical adsorption and chemical combination are two processes that have been proposed and yet it is well known that individual particles leave polishing tracks behind them that are recorded not only in the surface layer, but also imprinted in the substrate without appearing visible on the surface. It would appear, therefore, that the physical action of simply ploughing off material cannot be ignored. The fact that the polishing tracks can remain hidden by the surface layer can also be used as evidence for surface flow although it must be borne in mind that this may simply be the result of propagation of strain into the substrate. However, it is undeniable that under certain conditions the surface of glass may be moved by means of plastic flow. This is mainly from the result of studies of single scratch marks made in the absence of water, but glass can be polished in media other than water and, indeed, is commercially polished in the dry state with the use of a lap and a polishing agent only. Under such conditions it seems that plasticity of the surface must play an important part.

Intimately connected with the removal of material from the surface is the problem of how the polish (or levelling) is achieved. It can be by any one or by any combination of the following processes; an athermic flow process, a redeposition process, or by regarding the hydrolysis as a self-levelling process. The contribution of each of these mechanisms to the whole process under various conditions remains to be determined.

Mr F. Twyman also carefully studied the work of others along with his own experiments, and the rest of this chapter is largely in his words.

The polished surface

A polished surface is characterized by the absence of cavities and projections having dimensions in excess of the wavelength of the reflected light, the present practical limits for the most accurately prepared glass surfaces being such that the deviations from a theoretical plane are reduced to the order of 20 angstrom units: approximately the same accuracy can be obtained in polishing metal surfaces. In the case of polished glass surfaces, this order of accuracy corresponds to projections of dimensions not exceeding the thickness of one or two molecular layers.

It is pointed out by Grebenshchikov (1931 and 1935) that, apart from the property of reflecting light, polished surfaces exhibit other characteristics which differ from those associated with the bulk material. Thus, polished surfaces usually have an elevated mechanical strength and an increased mechanical resistance, and polished materials have different heat and electrical surface conductivity, electro-magnetic characteristics, and crystalline structure, from those of the bright faces of a single crystal.

Lord Rayleigh (1903) states that the particles of emery in grinding glasses appear to act by pitting the glasses, i.e. by breaking out small fragments. He points out that surfaces may be ground so fine that a candle is seen reflected at an angle of incidence not exceeding 60° and, indeed, that at grazing incidence even coarsely ground surfaces behave as if polished. Wave theory shows that a regularly corrugated surface behaves as if absolutely plane, provided that the distance apart of the corrugations is less than a wavelength of light. He says:

> In view of these phenomena we recognize that it is something of an accident that polishing processes, as distinct from grinding, are needed at all; and we may be tempted to infer that there is no essential difference between the operations. This appears to have been the opinion of Herschel whom we may regard as one of the first authorities on such a subject. But, although perhaps no sure conclusion can be demonstrated, the balance of evidence appears to point in the opposite direction. It is true that the same powders may be employed in both cases. In one experiment a glass surface was polished with the same emery as had been used effectively a little earlier in the grinding. The difference is in the character of the backing. In grinding the emery is backed by a hard surface, e.g. of glass, while during the polishing the powder (mostly rouge in these experiments) is imbedded in a comparatively yielding substance, such as pitch. Under these conditions, which preclude more than a moderate pressure, it seems probable that no pits are formed by the breaking out of fragments, but that the material is worn away (at first, of course, on the eminences) almost molecularly.

The opinion of Herschel referred to by Lord Rayleigh is from *Enc. Met.*, Art. 'Light', p. 447, 1849. Herschel says:

> ... it may reasonably be asked, how any regular reflection can take place on a surface polished by art, when we recollect that the process of polishing is, in fact, nothing more than grinding down large asperities into smaller ones by the use of hard gritty powders which, whatever degree of mechanical comminution we may give them, are yet vast masses, in comparison with the ultimate molecules of matter, and their action can only be considered as an irregular tearing up by the roots of every projection that may occur in the surface. So that, in fact, a surface artificially polished must bear somewhat of the same kind of relation to the surface of a liquid, or a crystal, that a ploughed field does to the most delicately polished mirror, the work of human hands.

Lord Rayleigh continues:

> The progress of the operation is easily watched with a microscope, provided, say, with a $\frac{1}{4}$-inch object glass. The first few minutes suffice to effect a very visible change. Under the microscope it is seen that little facets, parallel to the general plane of the surface, have been formed on all the more prominent eminences. The facets, although at this stage but a very small fraction of the whole area, are adequate to give a sensible specular reflection, even at perpendicular incidence.... The polish of individual parts of the surface does not improve in the process. As soon as they can be observed at all, the facets appear absolutely structureless....
>
> ... Of course, the mere fact that no structure can be perceived does not of itself prove that pittings may not be taking place of a character too fine to be shown by a particular microscope or by any possible microscope. But so much discontinuity, as compared with the grinding action, has to be admitted in any case, that one is inevitably led to the conclusion that in all probability the operation is a molecular one, and that no coherent fragments containing a large number of molecules are broken out. If this be so, there would be much less difference than Herschel thought between the surfaces of a polished solid and of a liquid.

These passages still refer to polishing by the same emery as was used in the grinding.

The nature of the ground glass surfaces was studied very thoroughly by Preston (1922, 1926). It is not possible after reading his account or the earlier one of French (1916) to think of a ground glass surface as merely a number of intersecting cuts or grooves, a view which was formerly held by some. Preston, as a result of a number of careful observations, confirms that the process carried out preparatory to polishing the surface produces a great number of conchoidal fractures from which pieces of glass have been broken. Preston also observed another peculiarity of ground surfaces, namely that below the broken surface there is a region of small cracks which must be removed by polishing if the surface is to be perfectly clear.

The large part played by splintering in the grinding of glass at once marks it off from the other examples of polishing.

While splintering plays a very large part in the grinding of glass, ploughing plays only a small one, at any rate, till the later stages are reached. It has been shown that from the nature of the case only a very shallow furrow can be ploughed in so hard and brittle a material as glass, and there is great danger that splintering may occur even in the final stages of grinding.

A phenomenon observed by Twyman (*Proc. Optical Convention*, 1905, 52) when making plates for Michelson Echelons is described in his words:

> Supposing the plate to have been corrected to the required accuracy, will the cutting of it up cause distortion? The answer to this is; under certain circumstances, yes: but with due care, no. For instance, the plates must not, of course, be cut up with a diamond. Neither, in using the slitting wheel, must too great speed or force be used. As a matter of fact, we always use the old-fashioned hand saw fed with emery, that being the safest; and with this we find that neither on the proof plane nor by the interference test described above can any distortion be detected.
>
> A point worth mentioning is this. The edges of the plates are ground to a very fine matt surface. Now for some reason the region near a matt surface is in a state of strain. It seems probable that this is due to the grinding

material which, in crushing pieces out of the glass surface, subjects the part near the surface to a permanent strain from which it does not completely recover. This is easily detected by the use of polarized light, in the case of small pieces, say 1 mm thick. With the case of fairly massive pieces of glass, however, which do not permit the strain to be transferred inward by the bending of the whole piece, the strain becomes confined to an extremely thin skin of glass, and does not seem to have the slightest deleterious effect on the action of the echelon.

A strain such as mentioned above is entirely removed by polishing, which shows that it originates extremely near the surface.

One may reasonably suppose, in fact, that as the thickness removed by polishing in a particular case was within about 1/5000 of an inch, this is the thickness in which the strain originates, and an approximate calculation seems to indicate that the strain is *of the order* of the crushing strain of glass. If the glass is thin, say 1 or 2 mm this pressure over the skin bends the glass as a whole, and consequently is transferred inwards to a considerable depth; but if the glass is thick it does not bend by an appreciable amount and strain is extremely local.

Dr J. A. Anderson of the Mt Wilson Observatory found that some of the Mt Wilson's old lenses showed surface cracks which were extremely fine. Dr Anderson measured the depths of these hair cracks optically and found them to be about 30 wavelengths deep. Such hair cracks have also been observed on old telescope lenses in different parts of the world, and it is thought that the periodic (diurnal, especially) heating and cooling over a span of years might have caused them. This effect may possibly be ascribed to fatigue-failure of the surface layer if the layer was rendered weaker in the process of polishing.

Dr Anderson reproduced such surface cracks in the laboratory by heating ordinary lenses to 100°–120°C and subsequently chilling them on a cool metal plate. The depths of such cracks were also found to be of the same order as the natural cracks.

These observations suggest that the skin of a rouge-polished lens has physical properties different from the bulk of the lens body. This may be possible if the polishing takes place through thermal (softening) action at the superficial layer (*J.O.S.A.*, 1949, **39**, 92).

Attempts made by Rayleigh (1908) to discover whether the surface of polished glass is different in physical properties from the mass, resulted in the conclusion that, while grease and moisture on the surface (though extremely difficult to avoid) did not have much effect on the optical properties of the surface, yet even a recently polished surface is in a highly complicated condition.

Beilby (1921) brought fresh light on the problem of polish by his observations on metals, glass and Iceland Spar. According to Beilby, in polishing, molecules are set in gliding motion by the polisher so that they form an extremely thin film of fluid subject to surface tension, and this, he thinks, accounts for the smooth surface which is left by polishing.

It is worth quoting some passages selected from pp. 107–110 of his book *Aggregation and Flow of Solids* (Beilby, 1921, Macmillan, London):

> When glass is marked by the passage over it of a hard point, it may be either scratched, furrowed or cleft. A scratch is caused by the splintering of the glass along the track over which the hard body has moved. A furrow is ploughed when the tool is so formed and guided that its point or points lay hold of a layer only a few molecules in depth.
>
> A certain amount of the glass is shaved off, but the perfectly smooth coating of the groove which is left shows that the surface layer has passed through the mobile or liquid condition. A cleft results when a fine wedge-like point is drawn along the surface. The entry and passage of the thicker part of the wedge may result in furrowing or splintering at the outer surface, but these are not essential features of the true glass-cutter's cleaving scratch which, to be effective, must have forced the glass apart till a cleft is started.

Confirmation of Beilby's observations were obtained by a modified form of X-ray analysis, devised by P. B. Hirsch and J. N. Kellar (1948) of the Crystallographic Laboratory, the University of Cambridge, which can be used under certain conditions to measure the thickness of thin surface layers which differ in physical properties from the main crystal. This technique has been applied as a test case to measure the thickness of a ground and polished layer on the surface of calcite, of the order of one wave-length of red light in depth, and indirect confirmation obtained of the approximate correctness of the measurement.

French (1916–17) made a number of experiments which led him to the conclusion—to a great extent in harmony with the views of Beilby—that the surface of glass is converted in polishing into a form having properties materially different from the remainder. It can, for example, receive smooth-sided scratches ('sleeks') whereas scratches which are deep (called by the optician 'cuts') invariably consist of a series of conchoidal fractures. French actually went so far as to state that he believed the glass to become melted, a view which some regarded at the time as fantastic. In one illuminating passage (1917, p. 23) he draws a sharp distinction between two stages of wet and of dry polishing: 'The function of the first stage is to remove material; the function of the second is to fill up sleeks.'

Preston (1926) on the other hand concluded, after a careful examination of the views of previous workers, that the process of polishing is principally one of microscopic abrasion, although 'flow or fusion of some sort on a molecular scale may in fact be operative simultaneously with the more important phenomena of mechanical abrasion'.

The view that the temperature is sufficiently high for surface melting in polishing was adopted by Macaulay (1926, 1927, 1931). Macaulay was able to detect the products of thermal decomposition of the powder used for polishing glass plates. This evidence was questioned because many reactions may occur at a freshly exposed surface due to causes other than temperature. It seemed, however, that Macaulay was right in the view that he held, in the light of experiments of Bowden and Ridler (1936). Working at the Laboratory of Physical Chemistry, Cambridge, these authors deduced from their experiments the temperature of the surface layers of bodies during their sliding on one another, by using the rubbing contact of the two substances—actually two different metals—as a thermocouple and determining the electromotive force generated on sliding.

The behaviour of readily fusible metals confirmed that the temperature measured was a real one, for with metals of low melting point, such as gallium, Wood's metal, or lead, the measured temperature rose to a constant value which could not be exceeded and which corresponded numerically to the melting temperature of each metal. With less fusible metals the local surface temperature may exceed 1000°C.

The matter was carried further by Bowden and Hughes (1937). In this investigation the influence of the relative melting point of the polisher and the solid was determined. If high local temperatures really occur at the points of rubbing contact, as previous experiments (Bowden and Ridler, 1936) had proved, one would expect the relative melting point to be an important factor in the process, and the experiments showed that this is the case. Various

Table 2.1. The influence of relative melting points of polisher and solid

Polishing material	*Melting point of polishing material* (°C)	*Substance to be polished*	*Melting point of substance* (°C)	*Vickers hardness of substance*	*Results*
Camphor	178	Wood's alloy	69	25	polish
Camphor	178	Tin	232	4	no polish
Oxamide powder or camphor	417	Tin	232	4	polish
		Speculum metal	745	?	no polish
		Copper	1083	?	no polish
Lead oxide powder	888	Speculum metal	745	505	polish
		Nickel	1452	164	no polish
		Molybdenum	2470	234	no polish
Chromic oxide	1990	produced polish on all the metals tried			
Ferric oxide	1560				
Oxamide	417	Lead glass	469		polish
		Soda glass	600		slight polish
		Pyrex glass	815		barely perceptible polish
		Quartz glass	1710		no polish

materials were used as polishers, differing greatly in melting point. In some cases the polishing materials were massive substances such, for example, as camphor. In other cases they were in the more usual form of powders used on appropriate bases, camphor or otherwise. Table 2.1, compiled from the paper cited, shows some of the results. It will be observed that it is not the softness of the material that is the important factor in deciding whether it can be polished by a given polishing material, but whether its melting point is lower than that of the latter. For example, Wood's alloy is polished easily by camphor, although it is harder than tin which is not, the reason being that the Wood's alloy melts at a lower temperature, and tin at a higher temperature, than the camphor.

It will be seen that in the polishing of glass, the division is not so sharp; for example, all the glasses of course have a much higher melting point than the oxamide, but the nectart has a softening point almost the same; on the other hand, soda glass polishes slightly and the pyrex also takes a barely perceptible polish. It must be remembered, however, that glasses have a very long range over which they become of increasing softness with rising temperature, whereas the metals which have been considered have very definite melting points. When high melting oxides, such as chromium ferric-oxide, and zinc oxide are used they readily cause polish on quartz and all the glasses.

The paper also contains some interesting information on the loss of weight which accompanies polishing, showing that in the case of metals there is a definite removal of material. It is not the softest material which is necessarily removed most quickly but the one with the lowest melting point —for example, Wood's alloy polished on thick filter paper loses more weight than lead, although it is five times as hard on the Vicker's scale, the reason being that the lead has the higher melting point. Gallium, with a Vicker's hardness only slightly greater than lead, loses nearly ninety times as much weight.

More positive evidence of the nature of the surface layer in certain cases has been attained by electron diffraction and I am indebted to Professor G. I. Finch of the Department of Chemical Technology, Laboratory of Applied Physical Chemistry, Imperial College of Science and Technology, for the following interesting particulars of observations made in that laboratory.

Electron diffraction, at grazing incidence, with 50–60 kV electrons, gives the results in Table 2.2.

Professor Finch also remarks that, in electron-microscopic examination of the results of polishing glass scratched by diamond, the 'flowing over' of the scratches is 'remarkably evident'. For this study he uses moist rouge with a soft polishing pad. Finch, Quarrell and Roebuck (1934) also, in studying the photoelectric properties and structure of certain surfaces, obtained results that the authors considered 'confer an objective reality upon the Beilby layer which raises its existence from the realm of hypothesis to that of established fact'.

Hirsch and Kellar (1948) have measured the thickness of the surface layer resulting from the polishing of crystals, which in the case of ground and polished calcite (Iceland spar) was found to

Table 2.2. Electron diffraction study of surfaces

Surface	*E. D. Pattern*	*Conclusion*
Fresh glass fracture conchoidal Polished glass (microscope object glass) Polished plateglass (Pilkington's make) Polished silica glass Polished fused quartz	Two diffuse haloes and much diffuse background scattering	Amorphous
Polished fused quartz after heating during 2 hours at 400°C	Faint spots and fainter arcs	Unidentified crystal structure
Lead glass polished	Diffuse haloes and much background scattering	Amorphous
Lead glass, after prolonged weathering	Ring and arc pattern. (See also, Kamogawa, Phys. Rev., 58, 660, 1940)	Lead sulphide and lead sulphate crystals

be 7500Å ($7 \cdot 5 \times 10^{-4}$ mm). The method depends on measuring the variation of intensity of X-ray refraction from the polished surface of the crystals at various angles of incidence, and is being applied by the authors to the study of the ground and natural surfaces of single crystals and polycrystalline aggregates. It could be applied, they say, to measuring the thickness of any surface layer for which the value of the refraction coefficient differs from that of the substrate material, for example a layer of one metal electrode deposited on a different metal.

Further information on the subject will be found in Thompson (1930) and Browning (1944).

Mr Twyman made one or two personal observations of his own: Some glass is certainly removed in polishing. No one who has done any figuring by local retouching will doubt this. In correcting large plates for Michelson echelons he never found that retouching, which reduced the thickness of the area polished, caused any rise in the adjacent surface. Thus it is not usually just a matter of sweeping removed glass along the polished surface; the glass is taken up by the polisher or comes away with the rouge.

He used, in retouching, to keep account of the amount of rubbing and the quantity of glass removed. Counting the number of circular sweeps with a $1\frac{1}{4}$-inch diameter cloth polisher with the rouge fairly moist, and taking strokes of about 1 inch in diameter, he found that 100 strokes per inch of the area being polished removed about one Newton's ring (using Michelson's test)—that is about $\frac{1}{150000}$ inch. It follows that a single sweep with such a polisher would remove one hundredth of this, that is $\frac{1}{15} \times 10^{-6} = 7 \times 10^{-8}$ inch. Now if we consider a molecule of silica to have the dimensions of one lattice spacing of a silica crystal, the size of such a molecule will be about 10^{-7} inch. Since one can apparently continue this process of reduction to any extent merely by continuing the rubbing (he had himself carried the process to a depth of about a dozen Newton rings) one can only come to the conclusion either that one is removing the glass in portions of less than molecular dimensions, which is scarcely consistent with the picture of a flowing liquid, or that one is effecting a closer packing of the molecules.

On the other hand, although in ordinary figuring the glass is removed, either being carried away with the polisher or becoming mixed with the rouge, yet in certain exceptional circumstances one can get a transfer of the glass of a comparatively massive character, to which no other word seems applicable except 'flow'. In some small prisms, of which at one time Hilger's polished a considerable number, there was an obtuse angle, and if one surface was being polished singly by the optician, occasionally when the rouge was allowed to dry up pretty thoroughly the prism was found suddenly to develop, on the surface not being polished, a small bulge. Speaking from memory, Mr Twyman thought its height was something like $\frac{1}{4}$ mm. (French, in the paper cited, observed what may have been the same phenomenon, but he found the 'lump' to consist of a mixture of rouge and glass.)

Dr W. E. Williams told Mr Twyman that Dr Tillyer, of The American Optical Company, had demonstrated that optical polishing can be done either with a removal of glass and loss of weight or with progressive gain in weight dependent merely on the conditions of polishing. In the latter case, spot tests after etching the top layer with hydrofluoric acid, show the presence of Fe_2O_3 which must have been in the glass in a colourless transparent form. The probability is that the many different explanations of polishing arise, since it can be brought into being in different ways.

Finally, on this question of the removal of glass by polishing, Ray (1949) notes that the chemical examination of spent rouge shows the presence of silica (which must come out of the glass), but not enough to account for the total polish. He examined the polished facets of a half-polished glass surface on an electron microscope at magnifications of 8000 to 20 000 but was unable to detect any structure such as would arise from the rouge particles acting as tiny cutters. He also examined fresh and spent rouge samples under the electron microscope and found that the spent rouge seemed to be better dispersed and had more rounded corners than the fresh rouge. The particle sizes were mostly around $0 \cdot 25$ µm.

A paper by Lord Rayleigh describes experiments in which he proved that the reflecting power of a polished silica surface varies materially according to the treatment.

These variations in reflecting power were

measured by immersing the polished pieces in a liquid of the same refractive index as the interior of the material, and it was found that surfaces, polished by methods which do not quickly remove the material, may reflect in the liquid as much as 0·28 per cent of the incident light. A kind of burnishing seems to take place in these instances which modifies the surface and may bring its refractive index up from 1·461 (the ordinary value of fused silica) to as much as 1·6—quite as high as light flint glass, and much higher than any known variety of silica. On the other hand, surfaces polished by a process which removed material rapidly, or surfaces washed in hydrofluoric acid, do not reflect appreciably in the liquid. These effects are found in a less degree in ordinary glass, and, in a very much less degree, in crystalline quartz. In normal cases the reflected light changes in tint from red to blue as the refractive index of the immersion fluid is increased through the critical value for minimum reflection. The modified silica surface is anomalous in this respect, reflection being red on either side of the minimum value. An explanation is suggested for this in the paper. The thickness of the modified layer was measured as 0·06 λ where λ is the wavelength of green light in air.

These experiments explain the variable reflecting powers earlier found from the interfaces of fused silica or glass surfaces in optical contact (Rayleigh, 1936).

Contacted surfaces of crystal quartz give a reflection which is practically independent of the way in which the surfaces have been polished. The mean distance between the two crystals, when they are put in optical contact, is found to be about seven times the spacing of the layers of silicon atoms within the crystals.

In these earlier experiments (1936) Lord Rayleigh found that, using a power-driven pitch polisher very wet and pressing lightly, the reflecting power was increased; allowing it to become nearly dry, so that it dragged heavily and tended to squeak, the reflecting power was diminished. It must be remembered that the higher reflecting power was, in the circumstances of this experiment, an indication that the refractive index of the surface layer was different from that of the interior. Now the dragging and squeaking polisher approximates to the condition under which any commercial polishing is finished; heavy dragging of the polisher indicates, as is shown by the experiment, rapid removal of material and, therefore, effective working from the optician's point of view.

The practical optician, in polishing glass with rouge on pitch, finds that if the pitch is too hard he gets scratches. Further, he finds that when he requires to polish materials softer than glass, such as Iceland spar, he needs not only a soft pitch polisher, but a 'soft' polishing material, such as putty powder.

Two questions may therefore be involved from his point of view; the hardness of the polishing medium and the hardness of the polishing powder —or what is regarded as the hardness of the polishing powder.

Exactly on what ground he feels (as I suppose we all instinctively do) that putty powder is softer than rouge, I am unable to say, but the grounds for his doing so are undoubtedly shaky. For example, the following figures (*Rutley's Mineralogy*, H. H. Read, 24th Edition, Murby) for the hardness of minerals are very significant:

Haematite (Ferric Oxide Fe_2O_3)
Hardness 5·5 to 6·5 on Mohs' scale
Cassiterite (Stannic Oxide SnO_2)
Hardness 6 to 7 on Mohs' scale

Although inferences based on natural minerals are probably not valid it seems at least likely that putty powder is somewhat *harder* than rouge. Certainly, therefore, hardness is not the sole factor. The paper by Bowden and Hughes (1937, cited above) makes it clear beyond all doubt, that it is essential for the polishing material to be of a higher melting point than that of the substance to be polished. There are thus two factors in question for satisfactory polishing, the relative 'hardness' of the polisher and polished, and also the relative melting points of the polishing powder and the substance polished. Probably, in the avoidance of scratches, what is important is not the hardness of the individual crystals of the polishing material, but the force required to break down the aggregates which, in the case of certain samples of rouge, while liable in the initial stages of polishing to cause scratches, undoubtedly cause them to be faster polishing materials than the finer rouges.

Grebenschikov (1931, 1935), in reviewing the

position, points out that three basic theories of glass polishing have been proffered.

The first of these draws an analogy between polishing and grinding, according to which polishing is produced by breaking down the normal crystalline structure of the material by a mosaic of scratches so fine that the destruction of continuity produced by them becomes invisible.

The second theory assumes plastic deformation, flow and recrystallization of the glass, and views the polishing process essentially as a redistribution, as opposed to withdrawal of the surface layers.

Thirdly, it has been suggested that the upper layers of the polished article are fused under heat generated during the polishing process, with subsequent solidification to form an amorphous glassy substance on the surface.

Objections can be drawn to each of the above explanations, but it is probable that all three effects take place to a greater or less extent, according to the degree of polish attained and the type of machine used to effect the polish. However, of late, much attention has been given to chemical aspects of the polishing process, in an endeavour to explain some of the known effects of varying the physical and chemical properties of the material to be polished, the properties of the polishing medium, the chemical composition of liquids introduced in polishing processes, and the characteristics of the polishing base.

This leads, says the author, to a theory which suggests that the liquid medium (usually water) first reacts with the glass to form a thin protecting film, which tends to absorb particles of the grinding material. Thus if an appreciable amount of grinding material is absorbed simultaneously both by the glass surface and the surface of the polishing base, this will act as a form of binder between the polishing base and the glass—and the motion of the polishing base will tend to tear the film from the projections left after grinding.

Similarly, in considering the mechanism of glass-grinding, it has been suggested that the coolant enters into the grinding process proper by virtue of chemical reaction with the glass at the base of the cavities formed by the scratching action of the abrasive; the supposition being that the presence of the reaction-product sets up stresses which tend to assist in the scouring off of the glass contained between adjacent surface cavities and scratches previously made by the abrasive.

The foregoing summary of Grebenschikov's view was prepared by Mr T. H. Redding of the British Scientific Instrument Research Association (now Sira Institute).

Summary

It appears clear, from the facts cited, that:

(1) glass is to a certain extent removed in polishing;
(2) the surface sometimes undergoes a physical change and sometimes does not;
(3) in certain circumstances the surface is actually molten;
(4) the relative melting point of the polishing powder and material being polished is a factor of prime importance.

Whether appreciation of these facts will help the optician to polish faster or better may be a matter of doubt, but it will serve to occupy his mind while he carries out this useful, but, it must be admitted, often tedious process.

3 Single Surface Working

3.1 Introduction

Although machines, materials, and methods of production for medium and large quantities of optical components have changed considerably since 1950, there remains a demand—an increasing demand—for very small quantities needed in development departments and for sophisticated equipment, which will never be required as more than a few sets.

This problem of satisfying university and industrial research departments with small quantities—perhaps only one—will always be with us, and there is no alternative but to use skilled operators and, substantially, hand methods. For example, the cost of a trepanning cutter may be unwarranted if only one lens of a particular size is needed with no guarantee that more will be required. Manufacturing costs are therefore very high and economic prices equally difficult to justify to the customer, who is nearly always on a tight budget.

The methods in use today, and probably also for the future, do not differ much from those described by Mr F. Twyman in his book *Prism and Lens Making*, and it is sensible to reprint an extract from this book concerning methods that have not changed.

Mr H. W. Yates, who was Head of the Optical Workshop at the Hilger Division of Hilger & Watts Ltd and is now Managing Director of I.C. Optical Systems Ltd, Imperial College, London SW 7, has advised the author on the methods he uses for small-quantity work, and the remainder of this chapter will be extracts from the descriptions by Mr Twyman, modified by the author, where necessary, to suit present-day working conditions.

The chapter deals with methods which have been in use for several centuries as applied to the making by hand of a single prism or lens. For the benefit of laboratory artificers and amateur lens polishers, descriptions have been included of machines which, although not of the highest efficiency, involve only a minimum of equipment while still serving to reduce the tedium formerly associated with hand glass-working.

The processes and materials are subject to numerous modifications in the manufacture of prisms and lenses in quantity; nevertheless, the fundamental principles remain the same, and are directly descended from the earliest known methods of glass polishing. The manufacture of lenses and prisms in quantity is described in Chapters 6 and 8.

Just a word first of all about the accuracy of surface which has to be achieved. A piece of window glass about 1 foot square will usually have surfaces flat to within about one-hundredth part of an inch—to be precise, if you put it on a flat plate, there will be places where there is a separation between the two surfaces of perhaps one-hundredth of an inch. A piece of good plate glass of the same size would have errors up to something like one-thousandth of an inch. In a second-rate pair of binoculars, the prisms have a flatness to within one ten-thousandth of an inch; this also is about the accuracy to which spectacle lenses are made by a good firm. In binoculars of the best quality, the inaccuracies will not amount to more than one hundred-thousandth of an inch and, in work of the utmost accuracy required for certain scientific purposes, the surfaces must not depart from flatness by more than one-millionth of an inch: sometimes, indeed, even higher accuracy must be attained.

The manufacture of single optical components, whether prisms, lenses or parallel plates, consists of four main operations, which may be summarized as follows:

(*a*) Sawing the blank from the raw (slab) glass,
(*b*) Rough grinding to size and angle (or radius),
(*c*) Fine grinding while maintaining angles (or radii),
(*d*) Polishing and finishing the part.

3.2 Making a single prism

We will suppose that it is desired to make a glass prism for use in a spectroscope, and that a slab of suitable glass has been obtained from the glass-maker.

(*a*) *Sawing the blank*

Three methods of sawing glass are available, and in order of increasing efficiency they are:

(1) Remove the teeth from a hacksaw blade on a tool grinder, insert the blade in the usual frame, and feed with a mud of carborundum and water while cutting.

(2) Use carborundum cutting disks, as used by engineers for cutting off carbide tips. It is important that the grit and bond be suitable; the makers will supply the correct disk if told the purpose for which it is required.

(3) Use diamond-charged saws. These are obtainable commercially in several forms, or may be improvised with the help of a small power-driven lathe, using for the blade soft iron from 0·03 to 0·04 inch thick.

Regarding the respective merits of the three methods, it can be said that the hacksaw method, although slow and tedious, is for this very reason fairly safe in unskilled hands. Both the carborundum saw and the diamond blade need to be rotated at high speed to be effective, the actual speeds of the former being prescribed by the makers. If a small lathe is available, it can, by the addition of a suitable rest, be used for the turning, notching and charging of the blade, and also for the actual sawing.

Instead of by sawing, slabs can be divided by scoring and breaking, and a special breaking press makes this method available for very thick pieces of raw stock. The glass is first scored by a hard steel roller and placed, with the scored surface uppermost, on the table of the press. It is so positioned that the score lies parallel to, and vertically above, a knife-edge a little above the plane of the table. Pressure is then brought to bear on the upper surface of the glass by two more knife-edges, parallel to the first and roughly equidistant on either side of the score (the distance between them subtending an angle of about 60 degrees at the lower knife-edge). These two are free to assume positions in the plane of the upper surface of the slab to compensate for any lack of parallelism between the two glass surfaces: they are also lined with padding to prevent extreme pressure points and resultant local fractures. Their pressure breaks the glass cleanly along the line of the score and, in general, the fractured surfaces produced by this means are flat to within about 8 per cent of the thickness of the glass. The equipment can be used for breaking slabs of glass from $\frac{1}{4}$ inch up to 8 inches thickness.

(*b*) *Charging the blade*

In early times, the blade was charged with diamond in a crude but effective manner. A few largish diamonds of the type known as 'Industrial Bort' were crushed, say between a couple of flat irons, and then mixed with the fingers into a paste formed by moistening the ash of brown paper with paraffin oil. The diamond pieces were not more than about 0·007 inch in diameter.

This black paste was wiped on the edge of the blade with the fingers, while the blade was being used to cut a piece of hard glass or quartz. This procedure drove the diamonds into the edge of the saw, which was then ready for use.

(*c*) *Cutting the prisms to shape*

We will assume that the required shapes have been marked out on the slab with a grease pencil;

3.2.1. A slab of glass marked for cutting into eight prisms.

3.2.2. The angle of a prism being measured with a protractor. With care, an accuracy of 2 minutes of arc is possible.

the prism is now sawn out leaving one-eighth of an inch to be removed during the ensuing processes. (Fig. 3.2.1 shows a slab marked for cutting eight prisms.)

Three of its surfaces are to be polished, the fourth left grey. If the maker's melting number is removed during this stage, it should immediately be engraved with a marking diamond on the piece or pieces that are left. The melting number is the only accurate indication of the optical properties of the glass, and glass of unknown properties is of very little use.

(d) Rough grinding to angle

The rough grinding to angle can be carried out with the aid of a flat disk of iron screwed to a firm support, on which is spread the usual mud of carborundum and water, the grade of carborundum being 80 followed by 160. If the disk or tool is screwed to a powder-driven vertical shaft, and surrounded by a metal tank to catch the abrasive and water thrown off, it then becomes a conventional hand roughing machine and the rate of removal of material is very much greater. If it is not possible to power-drive the spindle, then a post should be erected around which the worker can walk, taking a small pace to one side after every half-dozen strokes; the wear on the tool surface is then distributed symmetrically round the centre, and is easier to correct.

Any one rectangular prism face is now ground square to the chosen base, using an engineer's square to test for squareness; with care an accuracy of 2 minutes of arc can be achieved. This surface, and the chosen base, are then marked with a grease pencil to avoid regrinding in error. A good protractor, such as that shown in Fig. 3.2.2, is then set to the required angle of the prism and a second rectangular face ground to this angle and perpendicular to the base. Mark this face also with a grease pencil and proceed to grind the third face, keeping it also perpendicular to the base and to angle.

As the exact amount of material removed during the fine grinding process which follows depends on the skill of the worker, it is customary to leave 1/100th of an inch full on each ground surface. With modern methods of sawing and grinding one can, however, treat the sawn surface in every respect as a trued one, and an allowance of about 5/1000ths of an inch is ample.

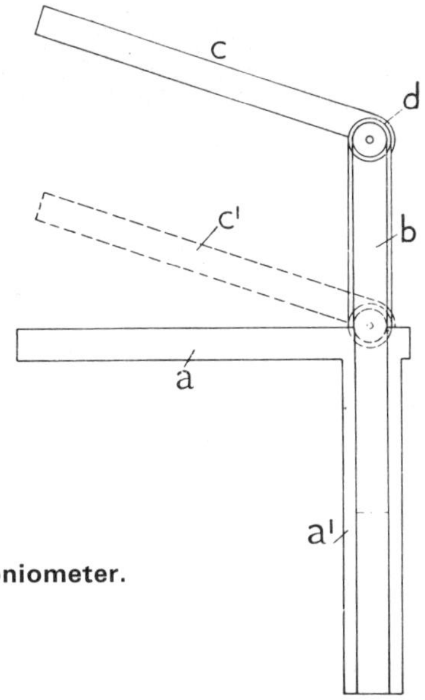

3.2.3. Hand goniometer.

Fig. 3.2.3 shows a hand goniometer which is useful for setting to angles in the measurement of prisms of acute angle when they are truncated at the pointed end.

(e) Truing, fine truing, smoothing, and testing the flatness of truing tools

The rough ground faces are now 'trued' and 'smoothed' with fine and finer abrasives, a suitable series having the following grain sizes:

Truing and fine truing	Aloxite 320 D followed by Aloxite 400
Smoothing	Aloxite 500 and 600

The first essential for these processes is a set of accurately flat tools.

The rate of removal of glass by pitch polishers is so slow that much time can be saved by smoothing the surface flat to nearly the same precision as is ultimately required in the polished state. For this reason the final smoothing of small single flat surfaces is done on the true tools. But the amount of smoothing on these should be limited to as few 'emeries' as possible, as each 'emery' destroys their state of flatness a little, entailing the reflattening of the true tools at more or less frequent intervals, depending on the skill of the worker to grind down all parts of the tool at the same rate. It is therefore advantageous to keep the truing tools in nearly as good condition as the true tools. A quick way of gauging the flatness of tools is as follows.

Select from plate glass (a $\frac{1}{4}$-inch thickness is suitable for a tool of 9-inch diameter) a piece 1 inch longer than the diameter and $\frac{1}{2}$ inch wider than the radius of the tool, i.e. a piece 10 inches long and 5 inches wide for a tool of 9-inch diameter. The only area of the plate which need be really flat is a strip about $\frac{3}{4}$ inch wide along one side of its length; the remaining area can be several wavelengths out of flat.

Having cleaned and dusted this side of the testplate and the surface of the tool, place the two together so that the flat strip on the testplate is over the centre of the tool and insert the end of a visiting card under the opposite edge of the testplate, thus forming a small wedge angle of air between tool and plate, tapering to zero along the flat edge of the testplate in contact with the centre of the tool.

Now take a swab of cotton wool wound on to the end of a dogwood peg, or a small camel-hair brush, load it with methylated spirit, and stroke it quickly along the contact edge of the testplate and tool. Capillarity will cause the liquid to run between testplate and tool. Repeat till the wedge of liquid from the zero edge is the same width as the flat strip on the testplate, i.e. $\frac{3}{4}$ inch. Allow to stand for 5 minutes.

Owing to surface tension the liquid will in this interval have adjusted itself to a uniform thickness along its thicker edge and the departure from flatness of the tool can be judged in the same way as though this thicker edge were a Newton's fringe between two glass surfaces inclined at this same angle; i.e. if the said liquid edge is straight the tool is flat; if its middle dips towards the zero end of the wedge, the tool is concave; if away from the edge the tool is convex. A hump at each end and one in the centre is a sign of a zonal depression, and so forth. Astigmatism is disclosed by slowly rotating the proof plane on the tool allowing rest intervals between rotation and reading times.

For truing tools of 9-inch diameter a visiting card is sufficiently thin and is convenient in that its thickness is about 0·008 inch, so that when inserted $\frac{1}{2}$ inch between glass and tool it gives a wedge angle of 100 wavelengths per inch, which is easy to remember. There is the further advantage that the methylated spirit evaporates more quickly than with a thinner separating piece. The assessment of sign and degree of curvature can be made in a few seconds by lowering the eye to nearly tool level and looking along the liquid wavefront; viewed thus, end-on, a departure from straightness of 0·02 inch (equal to an out-of-flatness of 2 wavelengths) can be detected. As a check on this the beginner can place a light-weight straight-edge on the testplate and measure the sag, but he must be careful not to press on the testplate, as this will cause the liquid to flow and it may take a few minutes for it to settle down to a state of equilibrium again.

To revert, now, to the truing and smoothing.

During this fine grinding process, the angles of the prisms are maintained to the required precision, and care must be taken to rub the prism in circular strokes around the centre and over the whole surface of the tool, especially the region near the extreme edge, so as to keep the tool a good shape. It must be remembered that more than one half of the total area of the tool is included in a marginal area only one sixth of the diameter of the tool in width. The face which is not to be polished is then trued. The final smoothing, in 500 and 600, of the faces which are to be polished must be performed on the flattest tool available, which should preferably be slightly convex rather than concave.

Throughout this book a 'grinding', or as the workshops would say, an 'emery', is taken to mean that the workpiece is ground with the specified abrasive grade until the emery has distinctly lost its 'bite' or cut. Using a tool of 9-inch diameter this may take from 5 to 15 minutes, depending on how much emery is applied in the first place.

If much trouble is encountered with scratches when smoothing with Aloxite 600, polishing can proceed straight from a worked-down smooth with 500, though polishing will naturally take a little longer.

Unless the prism faces are very flat in the grey, polishing them flat can be a very lengthy process. It is possible to get Newton's fringes by using a testplate directly on a smoothed surface, by observing at an oblique angle, but there are several objections to this procedure. First, there is a risk of scratching both the testplate and the smoothed surface. Secondly, the weight of the testplate may deform the surface being tested.

A useful method in which these disadvantages are minimized is to run a suitable ink—blue condensed ruling ink is to be favoured—between the surface to be tested and a testplate. In the regions of contact there is a white patch. This is surrounded by a pink area and this change of colour enables differences of flatness of as little as 2λ to be detected. Such departures from flatness are easily removed in polishing.

This ink possesses the advantage of being dichromatic. Increasing thickness is evidenced by the following sequence of colours: white, pink, blue and black with intermediate hues. To improve the sensitiveness a denser and more mobile liquid might be sought.

It is probable that this test could be used with advantage for metal surfaces which are bright enough to give a 20 per cent reflection even when they are so scored as to render the Newton's ring test out of the question—for example, ground stock, scraped surfaces, lathe slides, etc.

Fuller information about abrasives is given in Chapter 4.

When the faces have been smoothed, all the edges and corners are chamfered with 500 abrasive. The chamfering of edges and corners has a double utility in that it increases the resistance of the edge or corner to accidental damage, and also prevents the minute splinters in a ground corner from breaking out during the polishing process and causing scratches.

The prism is now ready for polishing, and after a thorough clean-up of the working space, the preparation of the polisher can be commenced.

(*f*) *Preparing the polisher*

Opticians usually keep plane tools in sets of four, a set comprising three accurately flat, and a fourth—which need not be free from pores or other surface defects—for use as a backing tool for a pitch polisher.

Pure Swedish wood pitch is heated in a thick saucepan over a low heat until it is of such a hardness that at room temperature it can be readily but not deeply indented with the thumbnail. A surprising degree of consistency can be

(a)

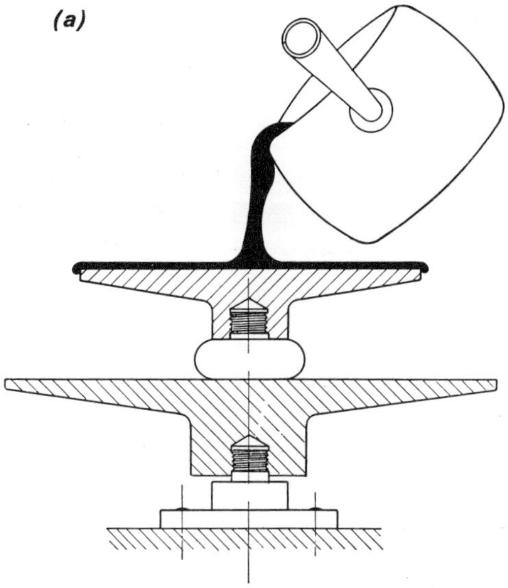

(b) *(c)*

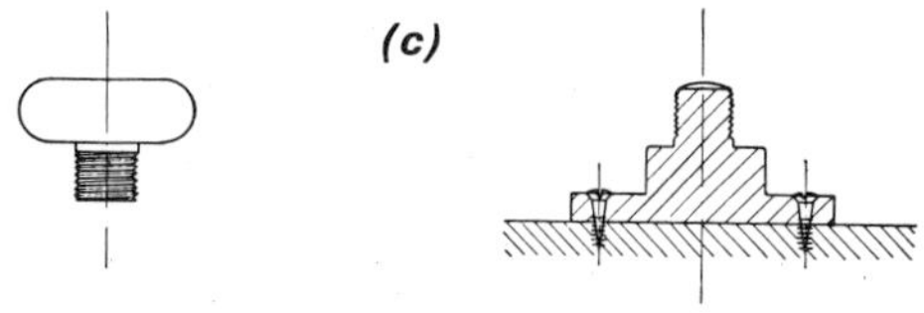

3.2.4. Making a flat polisher.
(*a*) Pouring out the pitch.
(*b*) Handle for the optical tool or block-holder.
(*c*) Nose for holding tools on the bench.

achieved with nothing more than the thumbnail test and a little experience.

While the pitch is being hardened, a small sample is taken every half-hour or so, cooled rapidly in cold water to room temperature, and tested for viscosity. When this is satisfactory, a wooden handle is screwed into the back of the polisher holder, which is then put on the gas ring and heated until it is just too hot to be comfortable to the hand. The hot tool is lifted from the gas ring, and the molten pitch, which should be of the consistency of a stiff treacle (molasses), is then poured over the hot tool so as to form an even layer $\frac{1}{4}$-inch thick. The polisher is then pressed face down on to one of the flat tools, which is itself screwed to the bench or post by means of a nose. (Fig. 3.2.4.)

If the polisher is inclined to stick to the cold tool, it can easily be dislodged by a smart tap with a hammer on the boss of the tool. Irregularities in the polisher surface can be removed by warming the polisher face downwards about a foot above a gas-ring; this will also help to remove air bubbles near the surface. After each warming it should be re-pressed on the flat tool which, if necessary, can be cooled with water as repeated pressings raise its temperature to near the point at which the pitch will readily adhere.

It is not out of place to remark here that although, with much experience, it may be possible to obtain very flat surfaces using tools that are themselves far from true, it is not to be recommended. Time spent in the flattening of working tools will be saved in the polishing stage many times over.

When the pressing is completed, and the polisher is smooth and flat all over, it is placed in a tank of cold water, where it should remain for a quarter of an hour; pitch is a poor conductor of heat.

Reticulations are now cut in the polisher surface, leaving facets of three-eighths to three quarters of an inch square. For making these reticulations, one uses a strip of brass about $\frac{1}{2}$ inch wide by 1/32nd inch thick. This is pressed firmly against the surface of the polisher with the fingers of the left hand and acts as a ruler to guide a graver, consisting of a three-square file sharpened to a triangular point, with which a series of grooves is made to form the desired checkered surface. If many polishers are to be made, then a grooving tool (Fig. 3.2.5) can be used just before the final cooling.

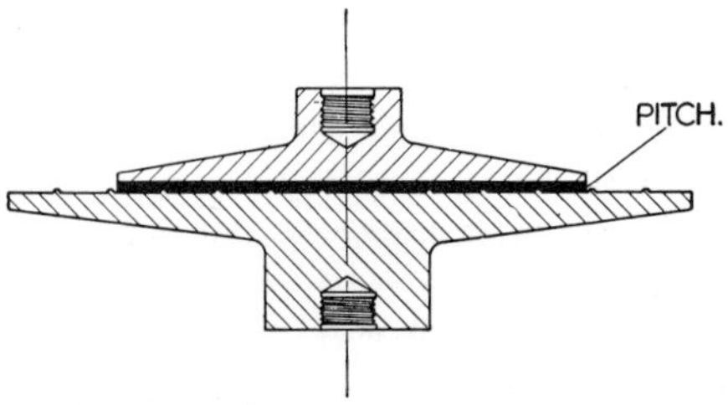

3.2.5. Grooving the surface of a polisher.

The polisher, while still hot, is pressed twice on the grooving tool, being turned through a right angle between the pressings.

After a final brief pressing on the flat tool the polisher is cooled in cold water.

(g) Stamping and rubbing up

Whilst the polisher is cooling off, a piece of Brussels netting (cotton net having about 10 meshes to the linear inch) or mosquito netting is soaked in water. The flat tool, on which the polisher was pressed, is heated over a low gas to about 80°–90°C; the polisher is then screwed to the post and painted with a fair quantity of wet rouge, the wooden handle being transferred to the hot flattening tool.

The wet Brussels netting, about an inch larger all round than the polisher, is now stretched evenly over the polisher surface, and the hot tool pressed firmly on top. If the tool temperature is about right the net will be forced into the pitch by a pressure of 20–30 lb applied for about $\frac{1}{2}$ minute.

If the tool is not hot enough, it must be further heated and painted with more wet oxide; if too hot it will stick to the polisher and will need cleaning, first with petrol (gasoline) and then with methylated spirit. Until some experience has been gained it will repay the novice to start with the flattening tool too cold rather than too hot.

When the netting is pressed right into the pitch, so that tiny flat facets are visible between the meshes, the tool is slid off and the netting raised simultaneously, when it should be found that the netting comes away from the polisher without any pitch adhering to it.

The tool is now cooled until only 'hand warm', painted again with rouge, and rubbed firmly over the surface of the polisher until the action is quite smooth and noiseless.

(h) Polishing the prism

Polishing of the prism can now be commenced.

Fig. 3.2.6 shows the way in which the prism is held for flattening the surface. The best practice is generally to hold the prism quite near the base, but it must be remembered that the process of flattening a surface is extremely complicated; although a better adjustment of pressure may be obtained by holding the prism very near the base, this is offset by the heat of the hand which, in that location, has a marked effect on the distortion of the surface by temperature. One has to consider the polishing of materials as different in heat conductivity as glass and quartz (crystalline quartz has approximately fifteen times the conductivity of glass). Another variable factor is the temperature of a man's fingers; they differ very much in this respect from time to time and from individual to individual.

3.2.6. Polishing a prism by hand on a spindle rotating at one revolution a minute, after smoothing on a flat tool.

A skilled optical polisher will instinctively be on his guard against this effect of temperature if the prism feels cold to his touch. A technique frequently adopted is to stick pieces of Perspex on each side of the prism with double-sided sticky tape, such as Evostick Twinstick. The Perspex will insulate the operator's hands from the prism. If the prism is made from a water-soluble crystal, the operator must use plastic gloves during polishing.

A long oval polishing stroke will be found the easiest to start with, and a pressure of about 1–2 lb/sq. inch should be applied. When hand polishing, it is desirable to have the polisher rotating slowly on a vertical shaft, say at 1 r.p.m., as this helps to keep the polisher a symmetrical shape and avoids astigmatism. If the rotating post is not available, good work can be done by erecting a post around which the worker can walk, taking a step to the side after every half-dozen strokes.

It is the general practice to apply wet cerium oxide to the surface of the polisher with a soft bristle brush from time to time.

Clean water should be added to the tool with a brush and the tool rubbed on the polisher for one or two minutes at the end of each 'wet'. A 'wet' is the time elapsing between the rubbing of the polisher and the point where the rouge begins to dry and considerable resistance to polishing becomes evident; it varies from two or three minutes to a quarter of an hour, depending on the amount of water added to the polisher when rubbing up.

Particular care must be taken during the end of a wet. Owing to the adhesion of the glass to the polisher it is necessary to use a stronger force to move the glass or polisher as the case may be. Consequently the pitch is liable to become a little soft and go out of shape. This effect can be minimized by a frequent change in the direction of movement of the polisher. If, however, the polisher does become non-uniform in surface there is no remedy but to clean the forming tool, warm the polisher and form it up again.

For the final polishing strokes after being assured that the forming tool is the correct shape and the surface of the prism free from grey, though probably not perfectly free from faint sleeks, warm the tool very slightly and wet the surface with the brush without cerium oxide. Rub the polisher with the forming tool and then rub the prism on the polisher.

Without delay proceed with the final polishing of the prism, working slowly over the polisher which ought to feel nice and smooth, but do not let it dry.

(*i*) *Maintaining tool flatness*

Optical workers who do a lot of hand polishing keep their tools flat in the following way.

If two disks are ground together, the upper one will tend to become concave, and the lower one convex. This principle is well known to amateur telescope makers, who continue the process until the desired radius of curvature is produced, and then use the other disk as a backing tool for the polisher.

If the tool which is used as the former is rubbed on top of the polisher it also will become more concave, until the stage is reached where flat surfaces can no longer be obtained, and the tool must be corrected by grinding with emery.

If, however, the polisher is rubbed on top of the tool the previous trend will be reversed; in fact it will even, in time, make the tool definitely convex. Thus by discriminating reversal of tool and polisher during forming, both can be kept flat for very long periods.

The simplest way to test a polished surface for flatness is by means of Newton's fringes, using a proof plate which is known to be flat; the fringes, by revealing the thickness of the air gap between the surface and the proof plane, show what is virtually a contour map of the surface. The thicknesses at points on two adjacent fringes differ by one-half the wavelength of light used, that is by about one hundred-thousandth of an inch. More precisely, for the green radiation from a mercury lamp the difference is 10·5 millionths of an inch per fringe.

The laying of a testplate on an iron tool to test the latter, even when it has acquired a degree of polish, is to be deprecated unless more than one testplate is available; testplates used regularly on tools become deeply and generally scratched to an extent that may endanger the glass surface on which they are subsequently used.

One strong objection to the use of proof planes is the difficulty of avoiding scratching the work. It is true that scratching can be avoided by meticulous cleaning, but this very often takes quite a time and is particularly difficult when the work is taken off the machine for examination. It is good practice, therefore, to use an interferoscope, the only difference being that the interference is observed between the surface to be tested and a proof plane with which the surface does not come into contact. This has a further advantage in that, since a gap of as much as $\frac{1}{4}$ inch can be used between the two surfaces, there is not that tendency to promote the harmful effects of placing together a proof plane and a piece of work which are not of identical temperature.

Primary testplates can be made without reference to a master flat by use of the principle generally attributed to Sir Joseph Whiteworth:

> In 1840 he attended the meeting of the British Association in Glasgow, and read a paper on the preparation and value of true planes, describing the method which he had successfully used for making them when at Maudsleys, and which depended on the principle that if any two of three surfaces

exactly fit each other, then all three must be true planes. The accuracy of workmanship thus indicated was far ahead of what was contemplated at the time as possible in mechanical engineering. (*Encyclopaedia Britannica*, 11th Edn, 28, 616.)

To make proof plates, three disks of plate glass, say 4 inches in diameter and $\frac{3}{4}$ inch thick, are ground together with Aloxite alumina powder 320, 500, and 600 until all are of a fine smooth finish and free from scratches. Each in turn is partly polished on the polisher, and when bright enough to see through they are laid together a pair at a time in monochromatic light.

The resulting appearance of each combination (*ab*, *bc*, and *ca*) should be recorded, calling a convex resultant positive and vice versa. Taking the simplest possible case, if all three combinations show two rings concave, then obviously all three plates must be one ring concave each, and if any one plate is worked to show with either of the others only one fringe concave then that plate will be nearly flat. A similar improvement is now carried out on the other two and so on until each combination shows uniform coloration over the whole surface—or as near to that perfection as circumstances require. The closeness of approach to contact may be easily estimated by use of tinfoil, which is available in leaves 0·0005 inch thick, of remarkable consistency. If a pair of tools swing about the middle, and three small pieces of foil placed symmetrically round the periphery of the lower tool are sufficient to annul the tendency to swing about the middle, then the tools are in contact to within this limit (0·0005 inch). If the tools bind at the edge, a small piece of tinfoil is placed in the centre of the lower tool, and if this makes them swing about the centre then again the tools are in contact to this limit.

If all three combinations of the three tools pass these tests with tinfoil 0·00075 inch thick, then we can be certain that all the tools are flat to 0·00075 inch. A 9-inch tool with a sagitta of 0·00075 inch has a radius of curvature of 1125 feet and a slope at one inch from the edge of the block of 54 seconds relative to a plane surface.

What error arises if the surfaces are not polished quite flat? The sagitta (or sag) of a spherical but nearly flat surface varies as the square of the diameter. Thus, if a 2-inch diameter proof plate applied to the block shows that the surface is $2\frac{1}{2}$ Newton's rings out of flat, the 9-inch diameter block will be 50 rings out of flat (=0·0005 inch). This indicates a radius of curvature of approximately 1300 feet and an inclination to the average surface in a position 1 inch from the edge of the block amounting to 40 seconds.

This error must be added to or subtracted from that due to lack of flatness of the blocking tool according to the sign of curvature of each. A testplate large enough to cover at least half the block is useful. If the wax is cleaned from a central area of the blocking tool, enabling a reflection to be obtained therefrom, the Angle Dekkor (§ 11.19) may be used to measure the (mean) error of parallelism to an accuracy of about 10 seconds.

Since the errors referred to above are those that can easily occur in ordinary working it will be seen that errors of parallelism may occur up to $54+40+10 = 104$ seconds $= 1'44''$ unless care is taken to get the tools flat to a higher accuracy, using a large testplate; and in that case a more severe mechanical test can be obtained by using thin plates of mica instead of tinfoil. If the measurement of parallelism be made, not by the Angle Dekkor, but by an interferoscope (§ 11.7), the parallelism can be measured and corrected to within 1 second of arc.

To sum up, in order to produce block work of a high order of parallelism, say 30 seconds, it is necessary not only to observe the most scrupulous care in sticking the plates down, but also to ensure that the blocking tool, smoothing tool or tools, and the polisher forming tool, are flat to a very high order of accuracy, and to provide an interferoscope for checking the parallelism.

Plates of glass can be laid down on a glass tool and fixed by wax, as described above, to an accuracy of about 4 seconds. The Angle Dekkor enables the top surface to be controlled to within about 10 seconds. It is therefore worth while devoting a good deal of care to getting the tools flat and in contact. If they are corrected, and the surface polished flat enough to result in a 35-second error, the procedure would assure a parallelism to within about 50 seconds, or if meticulous care were taken to get the tools in contact and flat, one might get down to 30 seconds accuracy straight from the block.

In order to obtain good fringes and few, optical

parts laid together for this test should be as clean and free from dust as possible. When it is realized that a piece of dust only a ten-thousandth of an inch in diameter will introduce 10 fringes of green light, the importance of scrupulous cleanliness will be appreciated.

(j) Finishing the prism

Reverting to the prism which we left in the final stages of polishing, the surfaces should now be tested with one of the testplates, the making of which was described in the last few paragraphs.

When laying down the testplate on a piece of finished or nearly-finished optic, the utmost care should be taken to avoid scratching the glass. Testplates once in position should not be moved about the surface by sliding; every piece of dust may scratch the testplate or prism.

When the prism surface shows one circular fringe or less, it will be good enough for all but the most exacting work.

All that remains to do now is to grind the grey bases and the unpolished back of the prism with a fine abrasive, and to chamfer the edges and corners. Needless to say, great care will be taken whilst so doing to avoid damaging the polished surfaces, which is all too easily done with dense flint glasses.

Before passing on to consider the making of a single lens, it is well to realize that to polish single surfaces of the highest quality is generally considered the most severe test of the skill of the optician. This art will not be acquired except by the expenditure of many hours of patient trial, nor is it the most vigorous worker who necessarily achieves success most quickly (Fig. 3.2.7). In the end, it is a question rather of patient thought than of manual dexterity. There is one advantage in hand polishing which is lost when a machine is used, namely that the optician can feel exactly whether the polisher is 'taking'—on the outside or in the middle—and at once alter his procedure accordingly. If it is the middle of the polisher which is operative, the piece being polished rotates freely as it is moved; if the reverse, it binds and offers more resistance to movement.

3.2.7. The late Mr S. W. Graves, B.E.M., of Hilger & Watts Ltd, polishing by hand a quartz plate 2 mm thick to a high order of flatness and parallelism.

3.3 Making a single lens

The process used for the making of a single lens is the same as that outlined for making a prism, with the appropriate modifications to take account of the fact that lens surfaces are usually very far from flat. The first stage is the same as before; a blank is sawn from the slab of glass about $\frac{1}{4}$ inch over size, and usually in the form of an octagon, or square with the corners removed. This is ground with coarse carborundum to a roughly circular shape, the thickness—in the case of a bi-convex lens—being left about 1/10 inch more than the finished size. If one or both surfaces are concave they must, of course, be roughed before the centre thickness can be ascertained.

Where the glass is in plate form, or has been reduced to plate form by slitting or grinding, an alternative to the method just described is 'shanking'. This is a very old-fashioned method, but it compares well in speed with modern methods where the number of disks to be made is small. In the use of the shanks it is customary to mark out in pencil the diameter of disk required, and then to break away around the margin, keeping just outside the circle (Fig. 3.3.1). Where quantity production is in question the quickest method is trepanning.

Let us assume that we are to make a biconvex lens, one surface being considerably deeper than the other. It will be necessary to provide a pair of 'true tools', accurately turned and lapped to the required radius, for each surface. In addition, it is highly desirable to provide a roughing tool and a polisher holder for each surface, and thus avoid using the true tool as support for the polisher and for roughing, which latter will tend to alter the shape of the tools. The preparation of curved optical tools is described in § 4.3.

If the depth of curve is not great, that is if the radius of curvature is five times the tool diameter or greater, the roughing tool and polisher holder can be of the same radii as the true tools, though not, of course, lapped together. If the radius of curvature is less than this the curves should be modified as shown in Fig. 4.3.2.

The lens blank is ground in the roughing tool with coarse carborundum until the flat surface has nearly disappeared, leaving a small circular patch or 'witness' in the centre about $\frac{1}{4}$ inch in diameter.

The other surface is ground in a similar manner on the other roughing tool, leaving a similar witness.

Now both surfaces are ground in turn with fine carborundum (160) until each witness just disappears; it can then be assumed that the pits of the coarse carborundum have been removed.

After thoroughly cleaning out the roughing tool and working space, grinding is carried on with 320 Aloxite, or other abrasive of like grain, still in the roughing tool, until the blank is about 0·02 inch over the finished centre thickness.

Two or three grindings are then carried out in the true tool, using the same abrasive as the last and taking care to work as evenly as possible over the whole tool surface.

The thickness of the lens should be measured at each stage either with a dial gauge or with a micrometer. If micrometers are used in the optical shop constantly, they should be tested and adjusted at least once a month; the presence of grinding material in the optical shop causes a good deal of wear.

The fine smoothing of the lens is performed in the same way, the tool being carefully cleaned each time the grade of emery is changed. One should avoid using more abrasive than necessary, particularly with the very fine grades, for if too much is used the work 'rides' on the surface, the pressure on each grain being insufficient to crush the glass.

When the fine smoothing process is complete, the lens should be of a fine, even grey, free from scratches, and within the allowed tolerance for centre thickness; the polishing removes very little material.

The importance of scrupulous cleanliness cannot be over-stated; when passing from one grade of abrasive to a finer one, the tool, hands, sponge, and bench must be thoroughly cleansed of all traces of the coarser grit, otherwise scratches are very likely to appear, for which the abrasive supplier may be wrongly blamed. It is indeed desirable to use a separate sponge for each grade of abrasive.

Before proceeding to the preparation of the polisher, the lens should be chamfered on both sides; if the blank still exhibits the irregular edges of the original lump, the chamfering should be carried on until the intersection of the surface and the chamfer is a full circle.

(a) Preparing curved polishers

The method of pressing the polisher is very similar to that detailed for the prism, except that the polisher holder is now of course curved to the radius of the surface. Into the concave tool, which is warmed for this purpose, is poured melted pitch. While this is still warm, the convex tool (used cold, and if necessary dipped into water once or twice to cool it during the process) is used to press the pitch over the surface of the concave tool till it forms a thin coating adhering to the latter, a little more than $\frac{1}{4}$ up to $\frac{3}{8}$ inch thick. The convex tool is moved about in the socket thus formed and

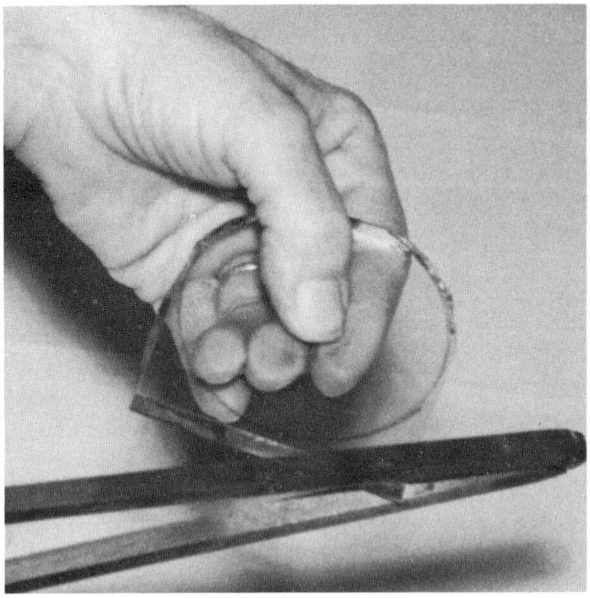

3.3.1. Shanking a disk of glass as a lens blank.

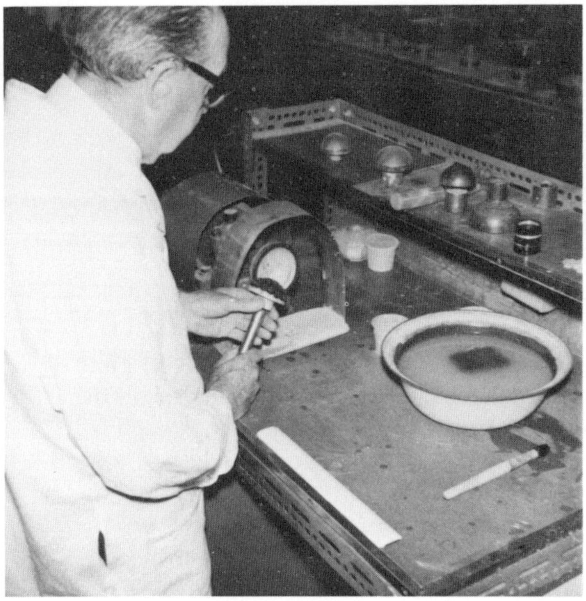

3.3.2. Smoothing a single block of lenses on a side lathe.

wetted occasionally with cold water to prevent its sticking, while the concave tool is allowed to cool down.

The reticulation is then cut in the surface of the pitch so that its surface is broken into squares of about $\frac{1}{2}$ inch side.

The grooves of the reticulation can be made in one of two ways, according to individual preference. If square reticulations are preferred, as on the flat polisher, then a more flexible straight edge than the normal steel rule is used to guide the cutting tool. This is pressed down on the curved surface of the polisher with the thumb and fingers of the left hand while the grooves are cut with a sharpened triangular file. Many workers prefer circular rings because they are much easier to renew when necessary. Ringed polishers are, however, not suitable for large lenses, as they are liable to stick; further they are liable to cause zoning, which although not usually severe enough to matter on a transmitting surface is not permissible on a mirror. To generate circular grooves, the polisher is rotated on a lathe or polishing machine at a speed of say 200 r.p.m. and the grooves cut as before. A little practice will ensure that the grooves are clean-edged and deep enough to remain through the stamping process.

(b) Polishing the lens

Polishing is carried out as for a prism. Some difficulty may be experienced in holding the lens whilst polishing, particularly if the curves are deep, and it is advisable in such case to stick the lens to a suitably shaped support ('mallet') by means of pitch. (Figs 3.3.2 and 4.3.2.)

For the best work it is necessary to re-form the polisher at frequent intervals by rubbing it on the true tool. If the weather be cold, or the pitch unduly hard, the tool may be slightly warmed to assist the polisher to take up the desired shape more quickly.

When the polishing of both surfaces has been completed, we have a biconvex lens which needs only to be edged.

(c) Centring and edging the lens

The method of edging to be described here is ancient and crude, but is nevertheless in fairly wide use even today, particularly in connection with optical elements of large size and high quality.

A piece of brass tubing is chucked in a power lathe, and turned true parallel, and to fit exactly into the same cell as the lens. The end of this tube is trued by chamfering the inside and outside at

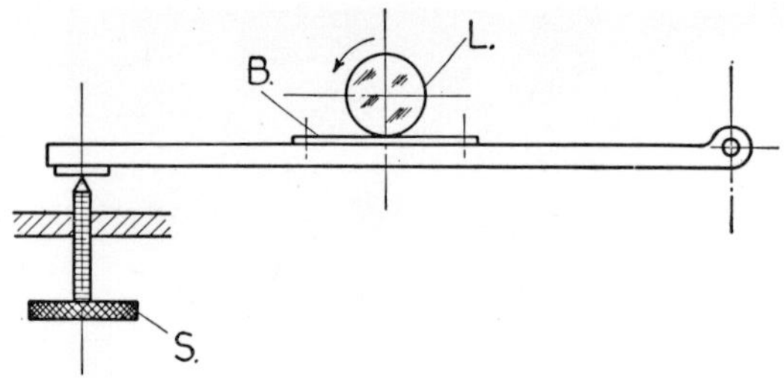

3.3.3. Edging a lens.

30° until the chamfers meet at an angle of 60°, and a very small flat, say 0·003 inch wide, made on the intersection of the chamfers. The tube is warmed with a Bunsen burner until pitch will adhere to it, and the lens, also warmed, is pressed on to the warm pitch. If the lens is correctly set on the chuck the brass rim of the latter will be uniformly visible through the lens. The lathe is now rotated and the reflections in the lens observed; if it happens to be rotating about its optical axis both the reflections will be stationary; if this is not the case the lens must be adjusted for position by pushing it sideways (warming the pitch again if necessary) until this condition is satisfied.

Cold water is then dripped from a sponge on the chuck, causing the pitch to become relatively hard. A hinged brass plate is then brought against the edge of the lens (Fig. 3.3.3) and 320 grade abrasive fed in until the lens is edged truly circular and almost down to the diameter of the chuck. Then, with a section of brass tube whose internal diameter is the same as the outside diameter of the chuck, the edging should be completed by hand, using truing emery and moving the tube to and fro axially to avoid grooving.

Before removing the lens from the tube, the exposed edge can be chamfered, by means of a deep concave optical tool, using the same abrasive. The lens is then warmed off the tube and allowed to cool. The pitch is cleaned off and the other edge can then be chamfered. The pitch is removed by soaking the lens in turpentine.

The main disadvantages of this method is that when edging a convex lens, the edge thickness increases as the edging proceeds. Thus, by the time the full width of the final lens edge is reached, the brass grinding plate, which itself has been ground away in the earlier stages, will produce a rounded edge. This shortcoming may be remedied either by frequently moving the piece of brass to a fresh place, or arranging to traverse it mechanically.

This simple method of hand polishing of prisms and lenses, although long discarded for quantity production, is still in use when work of high class has to be produced in small quantities, and is likely to become more widely used as the larger research organizations determine that provision should be made in their laboratories for making small optical elements which may at any time be required for some urgent project.

4 Optical Tools, Abrasives and Materials in General Use

4.1 Introduction

Since 1950, the development of optical tools and materials has been dramatic, and their contribution to increased productivity in the industry cannot be overemphasized. The mass production industries, making spectacle lenses and cameras, have undoubtedly provided the large market necessary for such developments.

Abrasives and materials can now be relied on to give consistent results, and so 'in-house' grading is no longer necessary (§ 4.11). The production of synthetic diamonds suitable for use in cutters designed to drill and mill hard metals, as well as rocks and ceramic materials, has made it possible for the optical industry to have an entirely new range of cutting tools at economical prices.

Polishing powders will now stay in suspension, instead of settling to the bottom of tanks and clogging up pipes, and this makes practicable the continuous feed of polishing slurry. A consequential gain is adequate cooling of the polishing pitch, and therefore higher speeds and pressures have become feasible without spoiling the shape of the tools.

Plastic polishing tools have been developed, with a better performance than pitch, where the highest accuracy of lens surface is not required. All the developments of optical machinery would have been of little use if the quality of the consumable materials had not been greatly improved in this same period of time.

4.2 Making optical tools

Materials

The material used for truing, smoothing, and polisher-forming tools is usually mild steel for 1-inch (25 mm) radius and under. For tools with radii greater than 25 mm the material is either gun-metal or 'Meehanite' cast iron. This type of cast iron has a very fine structure and is free of porosity, having been treated with calcium silicide when it was molten metal. Gun-metal has an advantage in that pitch does not stick to it as readily as to iron. Polisher holders or runners need to be as light as possible and invariably aluminium is the material used in their construction. (Fig. 4.2.1.)

Gauges

Gauges for the smaller optical tools may be prepared by turning a disk of brass about 2 mm thick to the required diameter as measured by a micrometer, the edge being reduced to about 0·5 mm. This is used as a gauge for the turning of the corresponding gauge for convex tools. For gauges

4.2.1. ▷
Aluminium runners on left and right of the photograph. Gunmetal true tools in the centre and at the back. Gauges for the initial turning on the shelf and in the foreground.

▽ 4.2.2. The Autoflow Precision Spherical Generator.
(*left*) Turning a large-radius aluminium runner.
(*right*) Turning a small-radius cast-iron smoothing tool. Note the Solex Spherometer on the front of the machine and the column indicator on the wall. (See § 11.11.)

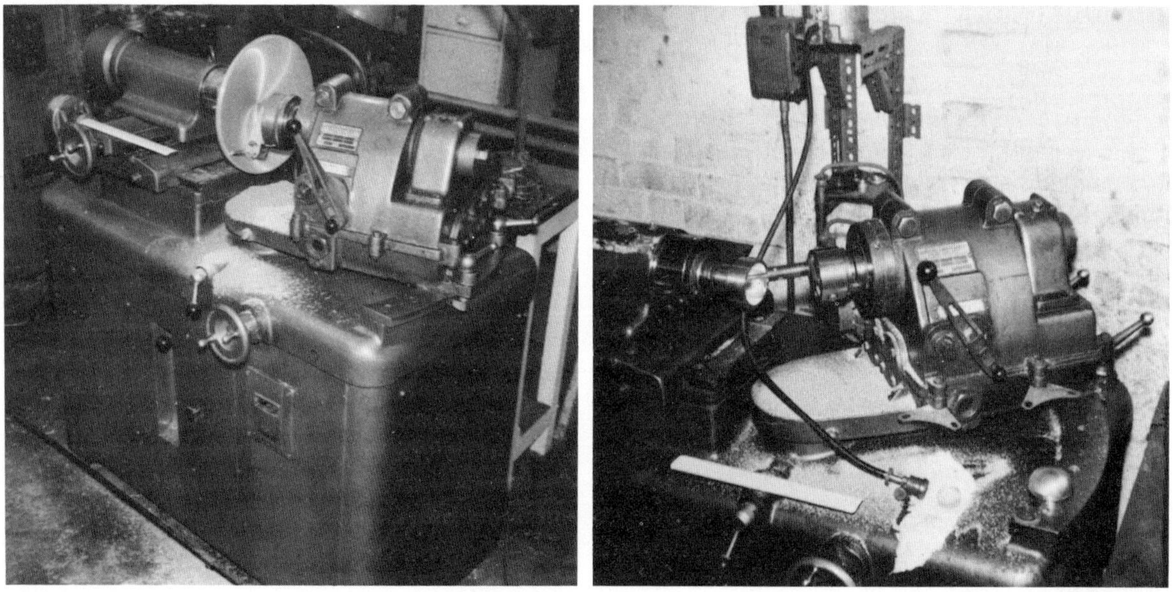

too great in radius to be turned on a lathe, unless a special attachment is used, it is far better to use a lap cutter.

Autoflow Precision Spherical Lap Cutter

The quickest and most accurate way of making optical tools is by radius generation on a lap cutter. (Fig. 4.2.2.) Control of the radius can be by means of a Solex Spherometer, contacting the radius generated and then comparing with an optical testplate for the curve required. By this means, the radius generated can be made within ten light bands of correct curve 0·0001 inch (0·0025 mm).

The Autoflow model 317 lapcutter has a range for plano tools up to 15-inch diameter, and for hemispherical tools up to 8-inch diameter, using a fully adjustable single-point cutter. The principles of head setting are exactly the same as for curve-generating lenses. (This is discussed in detail in § 6.4.)

Two conditions must be satisfied to produce true spherical surfaces:

(*a*) The cutting tool must pass exactly through the centre of the tool, i.e. the milling marks must intersect at a point.

(*b*) The plane of rotation of the vertical head must pass through the axis of rotation of the tool.

4.3 Making up sets of optical tools

When making up a set of three pairs of smoothing tools it is usual to allow a difference of 0·002 inch (0·05 mm) radius between them—the last pair being true to curve required. (Fig. 4.3.1.)

After curve generating, the pairs of tools must be rubbed together for a few minutes with fine emery, when the concave and convex curves should touch over nearly the whole surface. The tools must then be tested with a spherometer to ensure that the final curve is as intended.

If the radius of curvature is incorrect, then a carborundum cupped abrasive wheel of a diameter approximating the width of the high places should be pressed against the tool to grind away the elevations. After this grinding operation the tools must be lapped together again with truing emery, and checked with the Spherometer. When making a set of runners, such as tools for polisher or lens block, it is necessary to calculate the radii and so leave the correct gap for polishing pitch or lens and mallet.

Typical dimensions for this gap are as follows:

Radius of curvature	*Blocking tool gap for mallet pitch*	*Polishing tool gap for pitch*
0–1 inch (25 mm)	0·05 inch (1·25 mm)	0·05 inch (1·25 mm)
1–2 inch (50 mm)	0·06 inch (1·5 mm)	0·05 inch (1·25 mm)
2 inch (50 mm) upwards	0·08 inch (2 mm)	0·05 inch (1·25 mm)

In the case of blocking tools a gap must also be allowed for the thickness of the lens, and at this point the lens blank, disk, or moulding will be oversize to an extent of about 0·1 inch (2·5 mm) for lenses over 1 inch (25 mm) diameter, and 0·05 inch (1·25 mm) for lenses under 1 inch (25 mm)

4.3.1. True tools, smoothing tools, and runners.

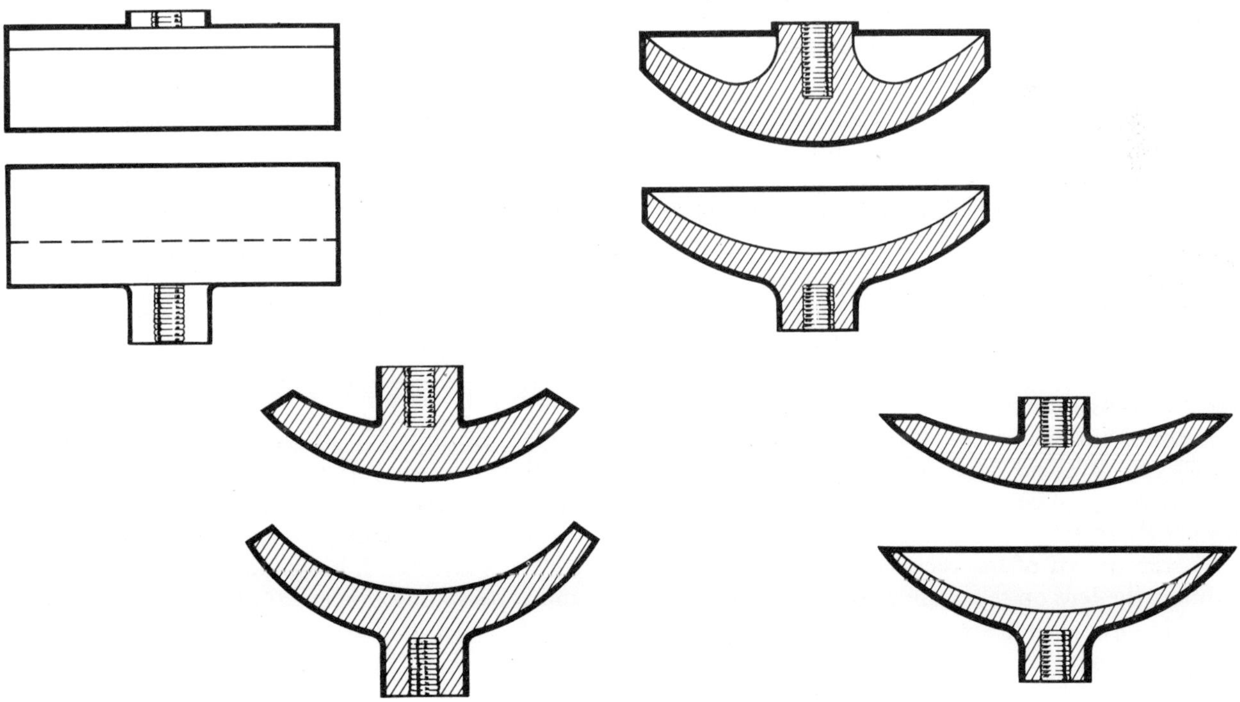

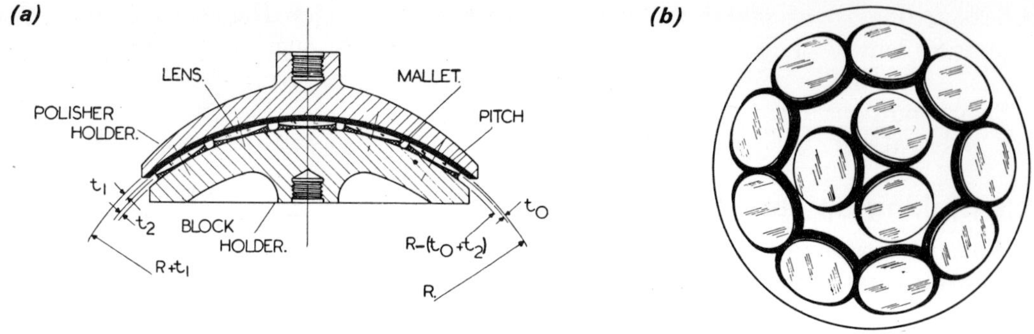

4.3.2. Making a block of lenses.
(*a*) Sectional elevation of lens block and working tool.
(*b*) Plan of completed block.

diameter. On this basis the lens thickness must be calculated. (Fig. 4.3.2.)

(*a*) To match a positive lens curve, the blocking tool radius must be:

lens radius − (*lens centre thickness* + *mallet pitch*)

(b) To match a negative lens curve the blocking tool radius must be:

lens radius + (*lens edge thickness* + *mallet pitch*)

If R is the radius of curvature of the lens surface polished, t_0 the thickness of the lens, t_1 that of the polisher and t_2 that of the mallet, then the radius of the polisher holder should equal $R+t_1$ and the radius of the blockholder should be equal to $R-(t_0+t_2)$. This ensures that the polishers and mallets are of uniform thickness.

Flat tools are made as a set of four, of which three are finishing tools. The ideal shape for a final smoothing tool is very slightly convex, say about four fringes in nine inches. The tool with which the polisher is rubbed up should be about three fringes convex in nine inches. (See §3.2)

4.4 Tools for lenses

Machined aluminium recessed tools

Recessed aluminium blocks for holding lenses are machined all over, and then the recesses are accurately bored on a machine such as the Deckel or Adcock & Shipley mill. (Fig. 4.4.1.) The recesses should be machined by a special cutter ground to the designed shape of the recess.

An accurate steel gauge is required for checking the recesses and a spherometer, located by the gauge, should be used to check the depth of each recess. To avoid marks on the lenses, they should be pushed into the recesses and not turned. The lenses are retained by a mixture of six parts of rosin to one part beeswax with a little black shellac for colour to aid examination of the glass surface.

In the design of the recesses it is necessary to support the lens on the outside edge only and also provide slots so that the lenses can be easily lifted out after polishing with a vacuum sucker. (See § 6.9.)

Epoxy resin recessed tools

For small quantities of lenses, where individually machined, metal recessed blocks would be too expensive, it is possible to use epoxy-resin moulded blocks. (Fig. 4.4.2.)

For lenses such as aspheric condenser lenses, where the spherical side requires optically polishing, epoxy runners are sufficiently accurate and the tooling can be made up very quickly. The epoxy resin which has been proved satisfactory is supplied by CIBA (ARL) Ltd., Duxford, Cambridge. Araldite CY 219 is a liquid epoxy resin of medium viscosity. It is used with Hardener HY 219 and Accelerator DY 219, both liquids of low viscosity, to provide a general purpose casting which, when cured, has good mechanical strength and is resistant to chemical attack. When used for recessed tools which are to be heated by an R.F. induction coil, the epoxy resin must be loaded with cast iron cuttings (from the lap cutter) to

(a)

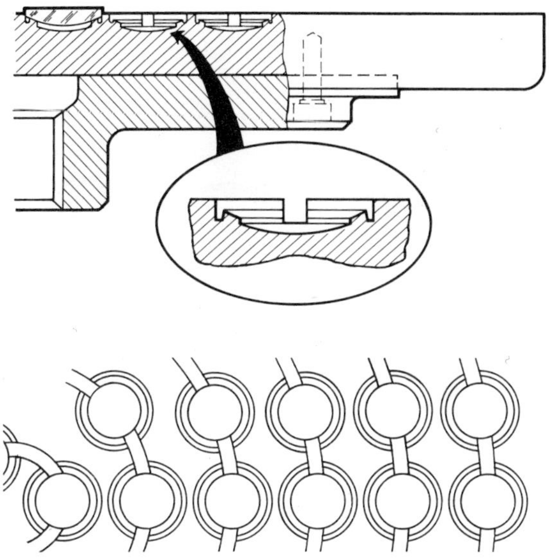

(b)

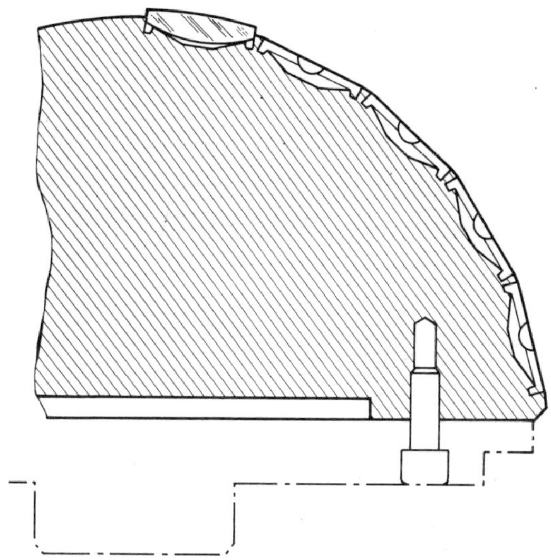

4.4.1. *(a)* Recessed block for plano lenses.
(b) Recessed block for convex lenses.
(c) Recessed block for concave lenses.

(c)

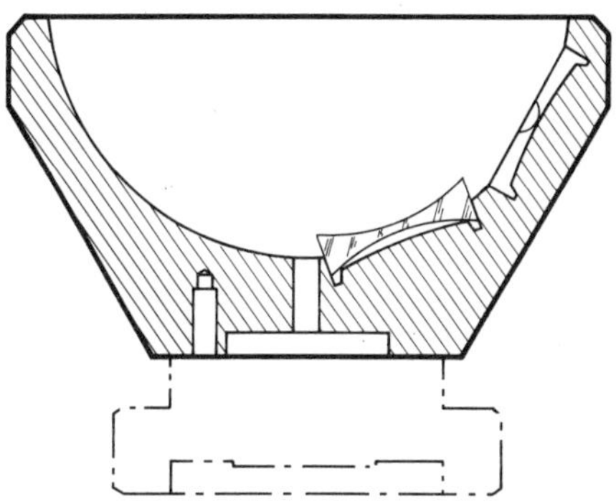

4.4.2. Overhead view of a fixture for moulding epoxy tools (recessed blocks or runners). On the left is a box of iron turnings for mixing with the resin before pouring it into the mould.

conduct heat throughout the tool. Because of the very low initial viscosity, this aids the thorough impregnation of the iron cuttings and at the same time ensures the reproduction of the finest mould features. Shrinkage during the cure and after release is very low, so that after years of working conditions, the dimensional accuracy is retained. The rate of cure can be adjusted by altering the quantity of accelerator DY 219 and the cure is completed without the application of heat or pressure. (See § 6.6.)

The polymerized resin is light in weight, easy to handle, and will withstand a considerable amount of abuse without effect on the accuracy of the tool. Because of the chemical resistance of the cured resin, the tools can be cleaned when necessary by normal cleaning methods.

It must be recognized that the accuracy of the epoxy tool depends on the accuracy of the form tool, inserts, and centring fixture. Initial preparation requires the application of a thin coat of wax to all surfaces which do not require bonding to the polymerized resin. These surfaces include the form tool, centring fixture, and hardened steel inserts (which have been previously magnetized). The steel shank is then placed in the centring fixture and locked in by means of a lock washer. The magnetized steel inserts are then located in the form tool equidistant from each other. The inserts represent the recesses in the finished block, and they are located in the hemisphere by scribe marks on the surface of the form tool.

The proportions of 100 parts of Araldite CY 219 to 50 parts Hardener HY 219 are mixed together at room temperature, stirring thoroughly, and 0·5 to 1·5 parts of Accelerator DY 219 are then added, again stirring thoroughly until a uniform colour is obtained. The small quantity of Accelerator is best measured out in a graduated pipette or measuring cylinder. About 100 per cent by weight of iron cuttings are then added and the mixing continued for 3 to 4 minutes.

The mixed epoxy resin must then be poured carefully into the form tool to about 2 mm below the lip level to allow for displacement. The centring fixture with locked shank is then inserted while taking care that centring is accurate. After curing at room temperature for about two to three days (possibly less time just to effect removal from the jigs), the recessed tool can be easily removed from the form tool by lightly tapping the shank. The lock washer is then removed, which permits removal of the centring fixture. The inserts are then easily removable by means of a jack screw in the centre of each insert. At a concentration of 0·5 to 1·5 parts of Accelerator, the usable life of unfilled mixture at 20°C is 3 to 6 hours, so that only sufficient resin should be mixed for the immediate work ahead. Similar epoxy resin materials are sold as RP 3261 REN by Red Products in the U.S.A.

4.5 Polishers

The function of a polisher is to provide an accurately shaped flat or spherical medium for applying the polishing material. Polishers are usually made of pitch, wax, plastic, Rexine cloth, or felt.

Swedish pitch polisher

Brown Swedish pitch is purchased from I. A. Hutchinson Ltd, 16 St John Lane, London EC1, or the White Sea & Baltic Co. Ltd, Hayne Street, Charterhouse Square, London EC1. The pitch is tested for viscosity and then boiled with or without linseed oil and tested until it is correct for the intended use. The pitch is then boiled and poured through a fine muslin strainer into cardboard containers. The following extract from *Prism and Lens Making* by F. Twyman illustrates a satisfactory device for measuring the viscosity of samples of pitch tested during the boiling operation. (Fig. 4.5.1.)

4.5.1. Pitch–viscosity testing apparatus.

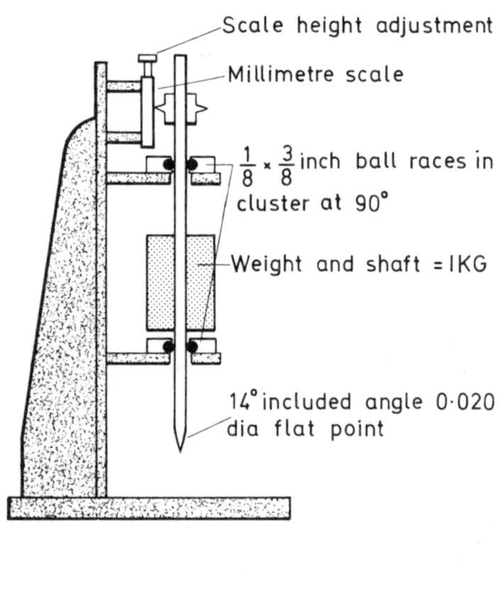

The illustration shows a simple but quite satisfactory device for measuring viscosity of samples of pitch removed from the cauldron from time to time during the hardening process. It consists of a piece of steel $\frac{1}{4}$-inch diameter, with a truncated conical point of 14° included angle, terminating in a $\frac{1}{2}$-mm diameter flattened point. Attached to this is a weight. The rod with its weight is held loosely

4.5.2. **Electric pitch pots under an extractor canopy. On the front left side of the bench is a pitch–viscosity testing apparatus. In the foreground is an electric pitch-mixing machine for polishing pitch.**

in a vertical position by the top of the rod passing through an eyelet in a wooden upright which itself is supported on a flat wooden stand and has a total weight of 1 kg. The point of the rod is allowed to bear on the top of the pitch, or other substance whose viscosity it is desired to control, which is immersed in water and thus kept at any desired temperature. The length of time taken for the rod to fall a given distance is determined.

It will be found that the most useful grades of pitch to stock in temperate climates are two, in which the rod falls $1\frac{1}{2}$ and 3 mm in 5 minutes respectively at a temperature of 70°F, the former being used mainly during the summer months and the latter in the winter.

The problem of maintaining the shape of a pitch polisher over a long period has led to loading it with dried wood flour. No. 80 mesh wood flour from I. A. Hutchinson Ltd is used by some opticians for this purpose. Some opticians introduce cerium oxide or zirconium oxide into the pitch.

A typical specification, in terms of parts by weight, for polishing large blocks of lenses and plano work would be:

Swedish pitch	36
Amber rosin	10
Wood flour	4
Zirconium oxide	46

With two or three annular rings cut in the pitch, the polisher is applied to the block whilst still warm and rubbed by hand for a short while before engaging the pin of the machine.

Boiling of pitch would normally be carried out in an electrically heated, thermostatically controlled pot with heated tap, installed under a metal hood for extracting the oil fumes. Mixing of pitch would normally be in an electrically powered mixing machine kept specially for a particular specification to avoid contamination. (Fig. 4.5.2.)

Burgundy pitch polisher

Specially prepared polishing pitch compounds are sold in the U.S.A. by Universal Shellac & Supply Co., 540 Irving Avenue, Brooklyn 27, N.Y., and imported into Great Britain by Abrafract Ltd, Owlerton, Sheffield 6. Mixtures of Burgundy pitch with cerium oxide or rouge and rosin are very satisfactory polishing materials.

Wax polishers

With wax polishers, as with pitch, a great variety of mixtures have been tried around beeswax, rosin and shellac, with fillers such as wood flour or zirconium oxide. Wax polishers are less apt to cause scratches and sleeks than pitch polishers, but they are more difficult to make flat. Wax mixtures do not flow to the same extent as pitch. They are harder and more suitable for small blocks of lenses. Wax polishers polish more slowly than pitch and do not so readily produce accurate surfaces.

Typical specifications for small polishing tools, in terms of parts by weight, would be:

(1)	Beeswax	3·5
	Rosin	2
(2)	Rosin	2·75
	Burgundy pitch	2
	Venetian turpentine	$\frac{1}{2}$ teaspoonful to 2 lb of pitch
(3)	Rosin	4
	Shellac	12
	Zirconium oxide	30
(4)	'$1\frac{1}{2}$ mm' pitch	16
	Rosin	2
	Beeswax	1
	Willow-wood flour	12

The addition of wood flour makes it possible to apply greater pressure without distortion of the surface of the polisher. The mixing of waxes should also be carried out in an electrically heated pot, with automatic powered mixing, kept specially for a particular specification. Always strain the ingredients thoroughly through a cheesecloth.

Plastic polisher

The use of plastic for high-speed polishing has been a development since 1962, when Rowland Products (British agents: Abrafract Ltd, Owlerton Sheffield 6) introduced a polyurethane foamed material containing zirconium oxide. The hardness and flexibility of the polishing pad can be varied to suit the conditions.

Plastic polishing has to a large extent replaced felt polishing and has also replaced pitch as a polishing medium for a substantial proportion of plano polishing. The quality of polish depends on the nature of slurry, the speed of polishing, and the pressure which is applied. Plastic polishers generally work best under low-speed high-pressure conditions, but this depends on the type of lens and glass being polished, together with the available machinery. (See § 8.9.)

The Rowland Products lens-polishing compound is supplied in three grades: LP-46 soft, LP-26 medium hard, and LP-30 hard. The materials soften in water and, once a polishing tool is made, the polishing material must not be allowed to dry or it will crack. The material works well with any polishing agent, but this agent must be continuously fed to avoid any dry spots which will cause glazing.

Self-adhesive polishing cloth

'Hyprocel-Pellon' is a foam-fabric material composed of more than 50 per cent chemicals with accurately controlled uniform thickness. Due to the excellent abrasive and lubricant retaining properties of the material a mirror finish can be obtained with the appropriate grades of Hyprez diamond compounds and abrasive slurries. (§§ 4.9 and 18.6.)

'Hyprocel-Pellon' is supplied with a pressure sensitive adhesive layer, protected with a sheet of backing paper, and when it is applied to the lapping plate a clean flat disk should be pressed on the Hyprocel-Pellon to smooth out any irregularities before work is commenced. 'Polimetal' and 'Microcloth' are also specially prepared materials suitable for use with Hyprez diamond compounds.

These materials can be obtained from Engis Ltd, Park Wood Trading Estate, Maidstone, Kent, who are the agents in England. 'Polimetal' is manufactured by Hyprez, 81 Rue de Vernaz, B.P. 15, Gaillard, France. 'Microcloth' is manufactured by Buehler Ltd, Evanston, Illinois, U.S.A. For further details, see Table 4.5.1.

4.6 Storage and control of optical tools

Testplates are made flat or spherical to a small fraction of a fringe, but in the case of spherical testplates the radii are known only to the accuracy of the test equipment used to measure them. (See § 11.9.) Testplates are usually made singly by skilled hand polishers and so are very expensive.

Most optical shops have a range of testplates for various radii which have been used in the past, and it is essential that designers, as far as is practicable, use existing gauge radii or realize the extra cost and inevitable delay if new radii are specified.

There will be a 'master' set of testplates for each curve in carefully controlled storage. A 'working set' of testplates, preferably made of quartz, will be kept in an individual box and stored in radius order. (Fig. 4.6.1.)

Table 4.5.1. Details of machine and processing

	Stage 1	*Stage 2*
(1.) Type of machine used	Engis Mark III lapping and polishing machine	
(2.) Size of lapping plate	12 inch diameter (30·48 cm)	
(3.) Speed of lapping plate	150 r.p.m.	165 r.p.m.
(4.) Stainless steel work holding block:		
Diameter	5 inch diameter (12·7 mm.)	
Weight	With weight disk—$9\frac{1}{4}$ lb (4·195 kg) Without weight disk—4 lb (1·814 kg)	
Carrying capacity	7—$1\frac{1}{4}$ inch (3·7 mm) diameter slices mounted with wax	
(5.) Speed of work holding blocks	375 r.p.m.	412 r.p.m.
(6.) Type of polishing cloth	Self-adhesive Microcloth	
(7.) Type of abrasive	1–L–05/36 Hyprez Formula L Diamond Compound (1 μ particle size)	$\frac{1}{4}$–L–05/38 Hyprez Formula L Diamond Compound ($\frac{1}{4}$ μ particle size)
(8.) Amount of Hyprez compound:		
First loading on polishing cloth	1·6 g	1·6 g
Subsequent loadings	1 g per 3 production runs	
(9.) Type of lubricating fluid	Hyprez WS Fluid	Hyprez ADWS Fluid
	(20 ml (approx.) per production run)	
(10.) Polishing time cycle	15 minutes with weight, 5 minutes without weight.	10 minutes with weight, 5 minutes without weight.
(11.) Average stock removal rate	40–60 micrometre (0·0016–0·0024 inch) per 20 minutes	$2\frac{1}{2}$–5 micrometre (0·0001–0·0002 inch) per 15 minutes

Each tool must be clearly marked with its radius and type according to its radius as follows:

Radius	*Type Pattern*
0–30 mm	A
30 mm–60 mm	B
60 mm–100 mm	C
100 mm upwards	D

Surfacers should also be marked according to their use as follows:

S1	First rougher	
S2	Second rougher	
S3	Smoother	302
S4	Smoother	$302\frac{1}{2}$
S5	Smoother	303
S6	Smoother	$303\frac{1}{2}$

Polishers and runners should have a letter identification such as L for lap. If some are specifically forming tools, then another letter, say F, should be added so that a tool marked LF20 would show that it was a tool 20 mm in radius which was a lap and had been made into a forming tool.

Stock control cards should be used to record movements of tools in and out of stores (Fig. 9.6.1.) and filed in type and radius order. The authority for issue of tools should be by tool requisition. (See § 9.6.)

4.6.1. Optical tool stores. True tools and runners on each side of the gangway. Glass test-plates in the boxes on the end wall.

4.7 Mallet pitch and waxes

The composition and properties of mallet pitch (pitch buttons in U.S.A.) are as variable as those of polishing pitch, but the mallet pitch should always be more viscous than polishing pitch. However, the mallet pitch must not be too hard or the lenses will be strained during the making of the block and, when they are knocked off after polishing, they will spring back and assume a bad figure. Mallet pitch is usually mixed and poured into cardboard tubes so that sticks are made and can be readily stored and used by the operators.

Typical specifications, in parts by weight, are as follows:

For holding lenses and prisms on jigs during grinding

	Beeswax	Rosin
Soft	1	1
Medium	1	4
Hard	1	8

As an alternative use No. 4 Blanchard Wax (yellow colour) from Universal Shellac & Supply Co., 540 Irving Avenue, Brooklyn 27, N.Y.

Graticule circle malleting

Beeswax	Rosin
1	20

Mallets for medium to small blocks
(Continuous-feed slurry)

Shellac	2
Swedish Pitch	4
Rosin	5

Mallets for O.G. lenses
For manually prepared mallets:

Gas pitch	1
Swedish pitch	1

For mallets prepared by machine: Glencom compound (Goulston Optical compounds). This compound is often too soft for normal precision optical purposes so it is then mixed with a harder pitch.

Another mixture suitable for machine or manually prepared mallets is as follows:

Blocking pitch (American Optical)	12
Plaster of Paris	6
Shellac flakes	1

As a blocking pitch for 'Hard on' flint lenses, Green Handi-stik bars of blocking pitch, from Universal Shellac & Supply Co., are convenient and suitable.

Edging wax

Suitable waxes for holding lenses on chucks during the centring and edging operations are as follows:

(1)	Rosin	12
	Beeswax	1
	Shellac flakes added to the mixture of rosin and beeswax	4
(2)	Rosin	4
	Beeswax (Hopkins & Williams Ltd)	3
	Shellac flakes	2

(3) No. 9 Black centring wax from Universal Shellac & Supply Co.

Stick pitch for dripping on small lenses

This pitch, after mixing, is poured out into steel moulds about 100 mm × 6 mm × 6 mm; when set, the sticks are ready for use. The mixture is:

Blocking pitch (American Optical)	12
Shellac flakes	1

Recess tool wax

This wax is used to hold lenses into recessed blocks during curve generating, grinding and polishing. It consists of:

Rosin	6
Beeswax	1
Black shellac	1

Stacking wax

For sticking prisms to glass or metal angular jigs before Blanchard grinding. Also used to hold graticules to plano during smoothing and polishing:

Rosin	1
Beeswax	1

For prisms prior to squaring and grinding:
$\frac{1}{4}$ lb Flaked Shellac with 2 teaspoons of Oil of Cassia.

Table 4.7.1. General properties of wax and pitch

Material	*Softening point (°C)*	*Melting point (°C)*	*Specific gravity*
Beeswax	—	62–66	0·959–0·967
Shellac	—	78–80	1·08 –1·13
Pitch soft	35–60	50–70	1·250–1·265
Pitch medium	50–80	70–90	1·265–1·285
Pitch hard	80–115	90–140	1·285–1·330
Rosin	60–80	100–140	1·07 –1·09

4.8 Polishing compounds

Materials in general use for polishing optical components are ferric oxide, cerium oxide, zirconium oxide for glass and putty powder for Iceland Spar and synthetic halide crystals.

The influence of pressure and rotation speed are important factors in the efficiency of polishing. Pressure has a linear relationship to glass removal and speed has an optimum value based on the volume and efficiency of the supply of polishing fluid.

Polishing efficiency is also influenced by the hardness of the glass, for example, borosilicate glass BK7 has about five times the wear resistance of dense flint glass SF2.

Ferric oxide Fe_2O_3 (rouge)

The rouge manufactured by Hopkins & Williams Ltd has been specially prepared for the best optical work.

Rouge is a red oxide of iron and should have a good red colour. It should contain little or no free sulphate; otherwise the powder will aggregate in balls which cause sleeks in the polished surface. To a large extent, rouge has been replaced by cerium or zirconium oxides, which are faster cutting materials and cause less soiling to hands, clothes, and surroundings.

Cerium oxide CeO_2

Cerium oxide is a flesh-coloured powder, insoluble in water, organic solvents, and most acids. It is easily removed from tools, glass, clothing, and hands by washing with water. Polishing times are about three times faster than ferric oxide, and as the cerium oxide can be suspension treated, this material is ideal for continuous feed slurry.

Slurry should be maintained between pH 5·5 and 6·5 for good polishing times. (See § 7.8.)

There are several manufacturers of reliable grades of cerium oxide, such as:

CEROX by Lindsay (Universal Polishing Compounds)
RAREOX by Davison Chemical Company (Universal Polishing Compounds)
CERIROUGE by Thorium Ltd (Autoflow Engineering Ltd)
REGIPOL 137 by Powergrind Ltd, Faroday Rd, Peterlee, Co. Durham.

The type of product available is illustrated by the following information, supplied by Thorium Ltd, on 'Cerirouge'.

In the glass-polishing industry, no single polishing powder has optimum physical, chemical, and economic characteristics for every application. The industry needs several different grades to cater for polishing a wide variety of products —spectacle lenses, precision optics, mirrors, plate glass, television tubes, camera and instrumental lenses, etc. The essential constituent of 'Cerirouge' polishing powders is cerium oxide, a chemical associated in nature with a group of elements which used to be referred to as rare earths, though nowadays they are often called lanthanides. At Thorium Ltd, where 'Cerirouge' is manufactured, the lanthanides are separated from bastnaesite which comes from Mountain Pass, in California.

'Cerirouge' has the following advantages:

Speed. Short polishing-cycle times increase output, reducing capital costs through more effective use of existing plant.

Life and consistency. The consistency is such that there is little fall-off in efficiency during continuous processing. This long life greatly aids production control and economic processing.

Special treatments. 'Cerirouge' is anti-foam treated, anti-cake treated, suspension treated, and pH controlled.

Cleanliness. 'Cerirouge' is absolutely clean in use, giving stain- and streak-free lenses. Personnel and plant alike benefit from cleaner working conditions.

Chemical and physical characteristics.

Grade	*O*	*90*	*H*	*E*
Total rare-earth oxides, %	80·3	99	99·2	99·2
Cerium oxide, %	39·3	89·1	47	90·3
Surface area, m^2/g	5·0	6·1	5·2	3·6
Particle size, % below 3·5 μ	80	85	90	90
Normal density, g/cm^3	1·28	1·20	1·0	1·10
pH (40 g/l slurry)	6·4	6·0	6·8	6·5

Grades of 'Cerirouge'. There are four grades, specially formulated for specific applications.

Plate glass. 'Cerirouge' O—this recently improved polisher is the least expensive cerium-based powder available, and is especially suitable for hand operations and scratch removal. It is of outstanding quality for its price, and is intended to replace other less efficient polishing powders, particularly in wasteful operations.

Ophthalmic. 'Cerirouge' 90—a newly developed product with a high cerium content. Careful laboratory and user tests indicate that 'Cerirouge' 90 combines the advantages of speed of action, life, and quality of finish at an economic cost. Where optimum speed of production is essential, 'Cerirouge' E is recommended.

Scientific and optical. 'Cerirouge' E—a polishing powder of the highest quality, containing a very large percentage of cerium oxide. 'Cerirouge' E polishes glass surfaces to the highest standards in short polishing-cycle times. A long polishing life ensures consistent results during the continuous operation of modern machinery.

White oxide—a powder that is suitable for special optics or stain removal.

'Cerirouge' H—a high-purity product with a lower cerium-oxide content than 'Cerirouge' E. It is a popular polishing powder, often used where material control is difficult, and gives surfaces comparable to those produced with 'Cerirouge' E on some special glasses.

Concentration. Nominally 4 ounces per pint (200 g/l), variable by experimentation to suit individual requirements.

Zirconium oxide ZrO_2

This material is cheaper than cerium oxide and is used more as a filler for polishing tools than as a slurry. Zirconium oxide is a fast polishing agent, but for high quality work has not been found to be as satisfactory as cerium oxide.

Putty powder

This is used for polishing soft materials, mirrors of soft metal alloys, and certain halide crystals. Putty powder is used wet in exactly the same way as rouge, but in order that it shall work well it must penetrate its support cloth or velvet uniformly. Best grades are 90 per cent tin oxide, but putty powder used for optical polishing is usually at least 50 per cent lead oxide.

The industrial use of putty powder is controlled by factory legislation in Great Britain and special extraction ventilation, protective clothing, and washing facilities are needed. In most cases Linde Alumina powder is a satisfactory replacement for putty powder and does not require special health precautions.

Linde Alumina powder

Linde powders are 99·9 per cent pure aluminium oxide of uniquely shaped fine particle sizes suitable for low scratch and high stock removal rates. This sapphire polishing powder is suitable for glass, crystals, plastics and metal where clear, scratch-free surfaces are required. It is made by the Roditi International Corporation Ltd.

Additives to abrasive and polishing compounds

One of the problems of grinding and polishing has been the packing and clogging of abrasive and polishing slurries. The compound 'Everflo' is a concentrated liquid additive which accelerates and enhances the action of abrasives by isolating the particles. This material is supplied by the Universal Shellac and Supply Company, 538, Irving Avenue, Brooklyn 27 N.Y., to whom the following information is due.

> In a slurry of abrasive and water, 'Everflo' shows great affinity for abrasive particles such as aluminium oxide, silicon carbide, boron carbide, cerium oxide, zirconium dioxide, etc. 'Everflo' encircles each abrasive particle with a microscopically thin film which isolates it from its neighbours. Thus while the abrasive particles

do not float in the water, the surrounding film prevents them from agglomerating and 'packing'. Gentle agitation is all that is necessary to bring the abrasive into a homogeneous mixture ready to be applied. The film-forming property of 'Everflo' gives an abrasive slurry the following desirable properties:

1. Pumps, lines, and hoses will not clog with caked abrasive even if left undisturbed for long periods.
2. Individual abrasive grains remain separate, preventing the formation of a sticky mud that is difficult to get into circulation.
3. Heavier abrasive grains do not precipitate out of the slurry.
4. The adherent 'Everflo' film prevents the abrasive from 'balling' up to cause unequal action during grinding and polishing.
5. Impellers and pumps will not be damaged because the slurry runs in a gentle yielding mass, not in a packed abrasive mud.

The microscopically thin film that surrounds each abrasive granule enables that particle to adhere to rotating and oscillating laps with great tenacity. Instead of being thrown off by centrifugal force, the adhering abrasive grains remain in contact with the work for longer periods of time. This results in more stock removal per unit time. Simultaneously, the abrasive grains begin to break down to progressively finer particles, resulting in a more highly refined surface than could be obtained with that particular size abrasive. A recent survey conducted among ophthalmic and precision optical shops has disclosed that on an average, there is a 20 per cent saving of fining time, thus resulting in 20 per cent more surfaces being produced when 'Everflo' was used as directed. In subsequent polishing operations, because of the highly refined surfaces obtained, a 10 per cent saving in polishing time was also achieved.

Abrasives do not actually 'grind' brittle substances such as glass, quartz, crystals, ceramics, silicon, etc. (See Chapter 2.) Instead the numerous sharp points of even the finest abrasive grains serve as minute pressure points which actually splinter the material away (called conchoidal fracturing). A fine ground surface appears smooth and uniform to the unaided eye, but seen under magnification it actually is rough and irregular. The surface consists of numerous criss-crossing shallow craters, deeper fissures, and much deeper extremely narrow cracks (called micro-cracks). The finer the abrasive, the greater will be the tendency of the particles to cling together (balling up) and the slurry to dry unevenly. The homogeneity of the slurry is thus altered and its resultant action will not be uniform. The craters, fissures, and micro-cracks vary greatly in width and depth. When 'Everflo' is added to an abrasive slurry, clumping of abrasive grains is prevented and uneven drying of the mixture is eliminated. The lubricant in 'Everflo' prevents harsh action of the abrasive on brittle substances. The resultant craters and fissures are shallower and narrower and consequently easier and faster to polish out. It is found, by the precision optical industry and in ophthalmic laboratories, that with 'Everflo' sleeks and scratches are all but eliminated with fewer fining operations, so saving polishing time. Also, large blocks of lenses and large diameter work do not dry out, because of its lubricity, and spindles can be left unattended for longer periods.

While the work is being 'ground' the metal lap is also being abraided away. The metal laps begin to lose their true radii because of the unequal pressure that is exerted on them as they rotate and oscillate. On toric laps, since there is no true rotation but only oscillation about a small central area, the pressure is so unevenly distributed that the lap wear is greatly accelerated at the centre. The convex lap becomes too 'weak' and the concave lap becomes too 'strong'. The uneven wear is so great that only the first three to five lenses will have the desired curvature that falls within standard tolerances. Naturally, to turn out lenses of the correct foci laps have to be retrued very often.

Here again the lubricity in 'Everflo' tends to minimize the frictional forces acting in the areas of greatest pressure. Lap wear with the 'Everflo' additive is materially reduced. Experience has shown that the 'Life' of a true curve can be extended by 20–40 per cent on spherical laps and from 30–100 per cent on toric laps.

In automatic polishing, the continuous action

of the impeller blades beat air into the polishing slurry, forming a layer of froth which floats on the surface and prevents the polishing particles from coming in contact with the lens and polisher, thus slowing the polishing cycle. 'Everflo', unlike other anti-foam agents, suppresses foaming without slowing polishing. Because of the lubricity of 'Everflo', a lens being polished cannot run too dry, and hence it remains cool and therefore undistorted.

'Everflo' is also bacteriostatic, and will prevent foul odours from building up in the tank of the polishing machine.

YMS (*Younger miracle suspension*) is made by Younger Medoptics Manufacturing, 3788 S. Broadway Place, Los Angeles, California 90007.

This compound is marketed in England by Autoflow Engineering Limited, Lawford Road, Rugby, and is supplied as a pink colour powder which is particularly good for use with 'Cerirouge' to give high-quality lens polishing. About 9 grams of the powder should be mixed with 4 pints of water and then shaken vigorously until all of the YMS is dissolved before adding another 2 pints of water to give 6 pints of solution.

To evaluate it for polishing, clean the bowl and pour in as much YMS (pink label) as you would water in your regular operation. Start the pump, add the normal amount of cerium or other polishing material, and after a few minutes of circulation put the lens to work.

YMS (green label) for abrasives may be evaluated in the same way, for emery, aluminium oxide, garnet powder, Aloxite, etc. With this agent you can use finer grades (micro-mesh) which will reduce polishing time. You will also find that this product will help prevent burning of the lens surface when using fine particles.

When you finally need to clean the bowl due to gross contamination with glass, you will find that none of the solids adhere to the bowl, tools, or pump mechanism. It can be readily cleaned by merely flushing with water.

'Hyprez' diamond compounds

For the lapping and polishing of semi-conductor and electro-optic wafers, laser rods, fibre optics embedded in resin, calcium fluoride, stainless steel mirrors, and many types of metal surfaces, diamond is a suitable material because it maintains sharp cutting points better than any other known abrasive. Diamond particles can be crushed and accurately graded into a range of sizes from 90 micrometer (0·0036 inch) to 0·1 micrometer (0·000004 inch). The graded particles can be dispersed in a chemical carrier which provides permanent suspension, cooling, and lubrication of the lapping operation.

Hyprez Formula L Compound is specially for semi-conductor slices where rapid stock removal and perfect surface finish are required. The compound can be thinned to any consistency with vegetable oils.

Hyprez Spray Diamond lapping compound in aerosol form is soluble in both water and oil, and, in conjunction with Hyprez Type W fluid, is

Table 4.8.1. Hyprez diamond compounds available

General application	*Micron sizes*	*Equivalent mean sizes in inches*	*Mesh size comparison*	*Compound colour*
Final polishing	$\frac{1}{10}$	0·000 004	Finer than *140 000	Grey
	$\frac{1}{4}$	0·000 01	* 25 000	
	1	0·000 04	* 14 000	Blue
	3	0·000 125	* 8 000	Green
Lapping and pre-polishing	6	0·000 25	* 3 000	Yellow
	8	0·000 33	* 1 800	Red

General application	*Micron sizes*	*Equivalent mean sizes in inches*	*Mesh size comparison*	*Compound colour*
Lapping and pre-polishing	14	0·000 56	*1 200	Brown
	25	0·001	600	Mahogany
Fast stock removal	45	0·001 8	300	Purple
	60	0·002 4	230	Orange
	90	0·003 6	170	White

*Estimated

suitable for lapping gallium phosphide wafers.

Hyprez fluids, available in aerosols and bulk containers, are essential for additional lubrication and will prevent the clogging of diamond particles during lapping and polishing operations.

Hyprez diamond compounds can be obtained from Engis Ltd, Park Wood Trading Estate, Maidstone, Kent, who are the agents in England for Hyprez, 81 Rue de Vernaz, B.P. 15, Gaillard, France.

Table 4.8.2. Hyprez Formula L Compound

Micron size	*Grade*	*Colour*	*Micron size*	*Grade*	*Colour*
1	1–L–00/37 1–L–00/35	blue	14	14–L–05/33 14–L–05/28	brown
3	3–L–00/37 3–L–05/33	green	25	25–L–00/30 25–L–00/25	mahogany
6	6–L–00/35 6–L–00/30	yellow	45	45–L–00/30 45–L–05/28 45–L–05/28	purple
8	8–L–00/35 8–L–00/30	red	60 90	60–L–00/25 90–L–00/25	orange white

4.9 Temperature control of polishing slurry

Although many polishing machines have electric heaters in the tanks containing the polishing compound there is some doubt whether they are really effective in maintaining constant working temperature.

The viscosity of pitch varies with temperature and the ideal position is to have the polishing tools at the same temperature, summer or winter, day or night. Complete air conditioning of all the polishing sections is the ideal condition but this may be impracticable, or uneconomic. (See § 9.8.) However, the temperature change is so great from summer to winter that different pitch mixtures have to be used, which upset times and procedures. A series of tests were carried out to prove if heating the slurry was worth while. Good polishing was achieved within the slurry temperature of 17°C to 32°C, but outside this range the pitch was too hard or too soft. However, with or without heaters, the temperature of the slurry followed closely the change in ambient temperature of the polishing shop during the working day. Results of the test suggest that, with good space heating, a moderate boost in the slurry temperature at the beginning of the day is all that is required.

4.10 Abrasives

For truing and smoothing, Aloxite corundum (aluminium oxide), in its various forms of emery and sapphire powders, is the most popular abrasive material, whereas Carborundum (silicon carbide) is the fastest material for rough grinding.

Typical grain sizes for various operations are given in Table 4.10.1.

The impure, natural forms of emery have been used from time immemorial and the natural corundums are found in U.S.A., Canada, Madagascar, and elsewhere.

Boron carbide is the hardest known material next to diamond, and is also suitable for truing and smoothing.

Table 4.10.1. Grain sizes suitable for various operations

Operation	*Abrasive*	*Grain size (1/100 mm)*
rough grinding singly	Carborundum	36 (36)
fine roughing, singly	Carborundum	20 (20)
truing, singly	Corundum	10 (10)
truing, one emery in the block	Corundum	10 (10)
smoothing in the block	Corundum	500 (2·5)
fine smoothing in the block	Corundum	600 (1·0)

Grading of emery

The usual nomenclature for distinguishing the emeries of different grades is to name them 1 minute, 2 minute, 40 minute, 60 minute, up to 240 minute (although emeries of more than 120 minutes are very little used). These designations indicate the duration of the decantation by means of which they have been selected by the following process. The following passage is freely translated from Dévé (1936, pp. 34–36):

> The process of grading is known as elutriation and is based on the time occupied by the grains in passing through a vessel of water 1 metre high and about 30 cm in diameter. The weight which causes the fall of a particular grain is proportional to its volume, that is to say to the cube of its dimensions. The force which resists that fall through the water is mainly a force of fluid friction proportional to the area, that is to say to the square of its dimensions. If then one doubles the linear dimensions the volume is multiplied by 8 and the surface by 4. If we imagine 8 little cubes of 1 mm side they will weigh as much as a single cube of the same material of 2 mm side. The surface of such a little cube will be 6 sq. mm, whence the 8 small cubes will together have an area of 48 sq. mm, while the surface of the large cube will be 24 sq. mm; thus when the large cube and the 8 little ones which weigh in total as much as the large one are thrown into the water together, the force which restricts the fall of the little cubes is double that for the large cube which will therefore arrive at the bottom well before the little ones. The procedure, then, is that after the emeries are crushed the mixture of grains of various sizes of about 10 litres volume is put into the vessel, which is then filled with water. One then stirs up emeries of all kinds thoroughly from the bottom of the vessel with a forked tool. Water is then added to make it overflow and carry out all the floating impurities. One then lets the vessel rest for 2 hours, at the end of which time the water will only contain the finer grains of emery. The water with the floating emery is decanted by a tap half-way up the vessel or by a syphon, and the water thus drawn off together with the floating grains is emptied into another well-polished and very clean vessel where one lets it remain for several days. At the end of that time one throws away the water and dries the deposit, which is the 120 minute emery. The process of stirring, filling up and decanting is repeated after 1 hour, the resulting emery being then called 60 minute emery.

Fused alumina

The following information has been extracted from *Facts about Fused Alumina* by permission of The Carborundum Company, Niagara Falls, New York.

Crystalline alumina in nature. Corundum, the naturally occurring crystalline form of aluminium oxide, or alumina (Al_2O_3) exists in nature in varying degrees of purity and under various names. The purest form, if of gem quality, is the white sapphire. Blue sapphires and rubies are gem quality corundum containing small amounts of mineral oxide colorants. Emery is an impure corundum containing iron oxides.

These forms were known to the ancients and there is some evidence that the Egyptians used corundum or emery as an abrasive for carving hieroglyphics on their stone monuments. India, where so many of the corundum gems were discovered, may also have used the lower grades for abrasive purposes at an early date, but the first deposit of any commercial importance was that near Cape Emeri on the island of Naxos in the Aegean Sea. Pliny describes this material. The terms Naxos emery and Grecian emery are used to refer to it. The name Turkish emery applies to the somewhat more impure and slightly softer material coming from deposits in Asia Minor, while American emery refers to a still softer grade in the United States, the first U.S. deposit being discovered in 1864. The importance of these deposits has diminished in recent years owing to the development of electric-furnace-fused alumina.

Corundum was recognized as the crystalline form of aluminium oxide by Charles Greville in 1798 and, in 1804, R. J. Hauy recognized that corundum, sapphire, and emery were three forms of the same material.

Development of fused alumina. As a natural material, the purity, and thus the abrasive properties of emery and corundum, are not readily subject to control. A deposit of acceptable quality may suddenly be depleted. Fortunately a means to improve on nature became available by the last quarter of the nineteenth century with increasing knowledge of electrochemistry.

Attention was focused on these techniques by the success which attended two developments of this period. In 1886, Charles Martin Hall produced aluminium metal by the electrolysis of the aluminous ore, bauxite, dissolved in cryolite. This improved method of production lowered the price of aluminium sufficiently so that by 1891 this metal began to be available for commercial applications. The value of electrochemistry was further demonstrated when Edward Goodrich Acheson, in 1891, discovered an electric furnace method of producing crystalline silicon carbide, which he trademarked 'Carborundum'. This was the first synthetic abrasive and, its sale, reaching significant volume by 1893, gave impetus to the application of electric furnace methods in the abrasive field.

The second synthetic abrasive was developed in 1900 when Charles B. Jacobs fused bauxite in the electric furnace and allowed the melt to cool slowly so as to produce alumina crystals of abrasive quality. It is this fused alumina which makes up the largest volume of abrasives used today, exceeding even silicon carbide in annual tonnage consumed. A long series of refinements of the original method and improvements in the furnace used have been developed in the succeeding years.

Commercial fused alumina grain is available today in a number of types: regular grain, high titania grain, grain having a more finely crystalline structure, and a pure white grain. Small variations in the amount of minor impurities contained in the alumina and differences in the raw materials or in the furnacing process account for the different types. The effect of these differences will be evident in the grinding action. Controlled commercial manufacture permits very close duplication of the physical characteristics of the various types so that the uniformity of any type is closer than that of the natural material. The petrology of commercial fused alumina abrasives and of fused cast high alumina refractories has been studied by H. N. Baumann, Jr, and their crystal habit described in detail.

The electric furnace. There are two major types of furnace in use today, the Higgins furnace and the Hutchins furnace. Both comprise a water-cooled steel shell containing the charge, into which carbon electrodes are positioned. In the Higgins type, the shell is a truncated cone, open at both ends, and with the smaller diameter up. This shell fits over a round carbon block base which is slightly hollowed out to form a hearth. At the end of the furnacing cycle, the shell can be lifted off, leaving the ingot resting on the carbon base.

In the Hutchins furnace, the shell is also a truncated cone, but closed on the smaller diameter base on which it rests. It is, in effect, a solid-bottomed bucket with lifting and dumping handles on the sides. The bottom of the bucket is covered with a layer of pitch and coke and the bucket rests on rods in a water bath to allow free circulation of water underneath it.

The electric furnace process. To reduce bauxite to fused alumina the furnace shell is filled about one quarter full with the charge, which consists of a mixture of bauxite, coke, and iron borings. The electrodes are lowered to just above the level of the mix and a line of coke is laid on the charge between them. This coke will carry the electric current until the bauxite fuses and the mix becomes conducting. As the charge fuses, additional mix is added at a rate to keep the bath temperature at about 2000°C. The electrodes are gradually consumed, and the level at which they are set is adjusted to keep pace with the filling furnace.

The process takes 16 to 36 hours depending on the type and size of furnace, amount of charge, and type of fused material desired. When furnacing is complete, the electrodes are raised and the mass is allowed to cool.

Slow cooling promotes crystal growth and the rate of cooling depends mainly on the size of the ingot or pig. The coarsely crystalline material will be produced by allowing the fusion to cool in the furnace shell in large pigs weighing about 6 tons each. After several days, the pig is removed from the shell and allowed to cool further. Cooling takes about one week.

To produce the material of finer crystalline structure, cooling must be more rapid. The pigs are smaller and may be made by casting the molten mass rather than by allowing it to cool in the furnace shell. This type is described in a patent issued to F. J. Tone in 1916.

Another type of fused alumina grain is made by

the electric furnace reduction of bauxite as before but with the addition of iron sulphide or other sulphides to the fusion. The fusion product is a mass of irregular single crystals of aluminum oxide in a water-soluble, glassy matrix. This mass is readily disintegrated and the single crystals may be classified in abrasive grain sizes.

For the production of the white fused alumina, a material purer than bauxite is used as the raw material. This is 'white alumina ore', produced from bauxite by the Bayer process which almost completely removes the oxides of iron, silicon, and titanium. No reduction to remove impurities is required in the furnacing operation, so neither coke nor iron is added to the furnace charge and the operation is the simple electric furnace fusion of the white alumina ore.

Raw materials. Fused alumina is produced from bauxite, coke, and iron, usually in the form of iron borings. If the fused alumina produced is to be of controlled quality, the raw materials must be carefully selected.

Fused alumina grain. When the large pig of fused alumina has cooled, a two-ton steel ball or 'skull crusher' is dropped on it from a travelling crane, breaking the pig into chunks. Any readily apparent large lumps of ferrosilicon or other impure portions not desired in the final product are removed by hand, being broken free by hand sledging if necessary. The chunks are then reduced in jaw crushers to about 6 inches and smaller. This material is passed along a belt where it is superficially hand picked on the way to gyratory crushers, which reduce it to 2 inches and finer.

For the next step in reduction, selection is made between hammer milling or roll crushing, the deciding factor being the shape desired in the grain. After this step the grain is a mixture of usable grit sizes.

The grain is group sized by passing through a series of screens. The various size lots are then passed through a magnetic separator to remove the smaller bits of ferrosilicon which have been broken free in the crushing operations.

At this point the grain may be treated by any one of a number of different processes to improve the final shape. Such processes include: impact crushing, air blasting, or pan milling. Any such treatment, generally called mulling in the abrasive industry, does not greatly alter grain size, but does knock off brittle corners. Roasting, washing, or chemical treatment with acid or alkali followed by washing, may be applied here, these treatments being designed to improve surface condition or toughness.

The grain is now ready for final sizing. For the screenable sizes, from 6 to 240 grit size, this is accomplished by passing through a series of screens. Either here, or just before final sizing, the grain may be again treated magnetically. For grain which is to be incorporated with a ceramic bond, a more intense magnetic treatment will be used than for grain intended for other purposes.

If the operation as described above, does not produce a large enough proportion of the finer grit sizes, before the mulling operation the excess coarse grit sizes will be ball or rod milled to reduce them further and increase the proportion of fines.

This operation is typical for the preparation of screenable sizes from 6 to 240 mesh but any number of minor variations are possible. To produce the finer grit sizes, the fines from these operations are further ball milled. They may also be given an acid treatment and washed. Classification of these sizes is accomplished by water sedimentation, hydraulic flotation, or air classification.

For screenable grit sizes, the grit size number refers to the number of openings to the linear inch of the wire or silk screen through which all the material will just pass. The distribution of material finer than that which will just pass through is controlled by the specifications given in the National Bureau of Standards Recommendation. Adherence to these specifications is on a voluntary basis, but the major producers and users of abrasive grains accept them as standards.

Alumina grain is available in the following screened sizes:

4	12	24	54	90	180
6	14	30	60	100	220
8	16	36	70	120	240
10	20	46	80	150	

in closely sized powders:

280	500	800
320	600	900
400	700	1000

and as broadly sized powders or flours, designated F, FF, FFF, and FFFF.

Another series of fine powders called optical powders is also available. These are essentially like the numerical series but are more closely controlled and sizes are designated by symbols denoting the average particle size in micrometres.

Abrasive grain is sold loose for use in lapping operations. Lapping with the finer grit sizes will produce a high polish such as is desired on metallographic specimens and for precision machine elements. Loose grain may also be used for glass grinding and polishing plate glass. As the name implies, optical powders are intended for finishing lenses of eye-glasses and optical instruments.

Fused alumina grain marketed by The Carborundum Company is identified by the trade mark Aloxite. The company's special lithographic abrasive grain is identified by the trade mark Lithograin and its pressure blasting abrasives by the trade mark Blastite.

Abrasive wheels. The variables which control the cutting action of the abrasive wheel are: the kind and type of abrasive, the grain size of the abrasive, the type and amount of bond, and the amount of pore space, or porosity of the wheel.

In general, fused alumina abrasive wheels are used for grinding materials of high tensile strength, while silicon carbide wheels are used on materials of low tensile strength. As noted in the section on the crude product, a variety of types is available within the class of fused alumina grain itself. Selection of the one best suited for the particular purpose must be made from the following types: regular grain, high titania grain, finely crystalline grain, and the pure white grain. Most grinding operations use the regular grain. The tougher high titania grain is used in snagging wheels for heavy stock removal and in coated products. The finely crystalline type is used in wheels for billet grinding and snagging. The more friable white type is desirable where a freer and cooler cutting action is required as in grinding hard, heat sensitive, high-alloy tool steels.

Selection of grain size depends on the material to be ground, the speed at which the wheel is to operate, and the surface finish required.

The amount of pore space in the wheel determines its structure. The more porous wheels are faster and cooler cutting, because the smaller amount of bond holds the abrasive grains less firmly and the grains break away more quickly, keeping the wheel always sharp. The more porous structure also allows more space for chip clearance, and the wheel does not load so quickly. The denser wheels are stronger, hold their shape more accurately and wear longer.

4.11 Commercially available abrasive grains

The chemical and physical properties of commercially available aluminium oxide, silicon, carbide and boron carbide grains are given in the following subsections.

The *chemical analyses* reported in these summaries have been obtained by traditional chemical methods. These analyses are important to the user because they affect the properties of the products made from these grains or powders. The toughness of aluminium oxide grain is, in part, a function of the amounts of other metal oxides present. Titania, silica, zirconia, and sodium oxide are of particular interest in this respect.

The *colour* of the aluminium oxide is also a function of the metallic oxides present and their oxidation states. Some of the principal oxides which affect the colour are chromium, iron, and titanium.

The *crystallographic descriptions* have been verified by X-ray diffraction techniques. For abrasive use it is important that silicon carbide be in the hexagonal system: alpha silicon carbide. Beta or cubic crystalline silicon carbide is considerably softer and less stable and is, in general, unsuitable for abrasive use. Silicon carbide grains are fragments of whole crystal or are single crystals themselves. Therefore, crystal-size measurements of silicon carbide grains would be meaningless.

Aluminium oxide grains, however, are normally composed of many small crystals cemented together by a matrix material, and in general, the finer the crystal size the tougher the abrasive grain. Petrographic thin sections of aluminium oxide grain, ground to 30-micrometre thickness when observed through a polarizing microscope, will show this crystal structure. Average crystal size is measured and the amount of matrix material is estimated from these thin sections.

Two different determinations of *porosity* are made from these thin sections. The individual pores are measured, to obtain an average pore size and the total volume of all the pores is estimated. As expected, a grain with a high pore volume and large pores, fractures more readily than a grain which is nearly solid.

The *particle specific gravity*, measured with a conventional pycnometer, is an indication of the amount of porosity. The closer this value approaches the absolute specific gravity for the material (SiC, sp. g. 3·23; Al_2O_3, sp. g. 3·99) the fewer pores or voids.

Hardness is measured on individual grains by the $Knoop_{100}$ microhardness method.

The *shape* of these particles is described as basically blocky in all cases. That is, the shape of the particle approaches that of a cube rather than that of a needle or plate. The edge sharpness, which is more variable, is controlled in manufacturing, depending on the intended end use of the grain. Bulk density is also a measure of shape. A sharp-edged grain will have a lower bulk density than a dull-edged grain of the same composition and particle size distribution.

Friability of grain is related to chemistry, edge sharpness, porosity, and crystallography. A ball mill attrition test has been developed which ranks these grains in the order of their friability. Even though this test does not provide an absolute measurement, it does provide an index from 0 to 100, very tough to very friable, for ranking these grains. Toughness is the complement of friability, and these grains may be ranked in order of toughness by subtracting the friability index numbers from 100.

Grain sizes are normally graded in accordance with the U.S. Department of Commerce Commercial Standard CS271–65 although the makers sometimes attach special grading curves. All of the grain sizes may be tested in accordance with the test procedure described in CS271–65. This test procedure is also available as an Electro Minerals Division publication, Test Specification No. M-T-170.

Powder size curves are determined by two different test methods. The most common method is the U.S. Sedimentometer method, Electro Minerals Division Test Specification No. M-T-96. This test procedure is also described in the U.S. Department of Commerce Commercial Standard No. CS217-59. This method is used with the numerical grades (240–1000), F grades (GF1–FFFF), optical powders, K5WT, and pressure blasting compound. All other powders, except the unfused buffing powders, have been analysed by the Whitby (M.S.A.) centrifuge, Electro Minerals Division.

Aluminium oxide, Type W

Aluminium oxide is produced by melting bauxite at extremely high temperatures in an arc-type electric furnace. Chemically, bauxite is a claylike aluminium hydroxide, but before use in the electric furnace, the bauxite is calcined and the water driven off. It is then mixed with iron particles and coke. In the electric arc, the coke reduces the impurities which combine with the iron and are precipitated to the bottom of the furnace. The heat melts and converts the bauxite into a hard substance consisting of approximately 95 per cent aluminium oxide. The furnace product is crushed and graded to produce abrasive grains.

Type W is a semi-friable, medium density, fused aluminium oxide abrasive. This general-purpose grain is widely used for lapping, polishing, and other finishing operations on materials such as glass, ceramics, brass, bronze, aluminium, granite, silicon, etc.

Typical chemical and physical properties of Type W aluminium oxide are as follows:

Chemistry (in percentages):

Al_2O_3 (by difference)	97·03
TiO_2	2·10
SiO_2	0·50
Fe_2O_3	0·20
Na_2O	0·02
ZrO_2	0·13
MgO–CaO	0·02

Crystallography: Alpha-alumina in the trigonal class of the hexagonal system. Average crystal size: 600 micrometres, low matrix.

Porosity: Total volume: 1.0 per cent. Average size: 10 micrometres.

Ball mill friability: 14 grit—50.

Particle specific gravity: 14 grit—3·93 g/cc.

Knoop$_{100}$ hardness number: 2090.

Shape: Blocky, sharp edged.

Grain sizes available: 10, 12, 14, 16, 20, 24, 30, 36, 46, 54, 60, 70, 80, 90, 100, 120, 150, 180, and 220 grit.

Powder sizes available: 240, 280, 320, 400, 500, 600, 700, 800, 900, and 1000 grit. Also CFl, F, FF, FFF Coarse, FFFF.

Silicon carbide, Type RA

Silicon carbide is made from pure glass sand and finely ground petroleum coke. The silicon of the sand combines with the coke, which is carbon. It is an extremely sharp abrasive and approaches the diamond in hardness; when crystals break, the broken surfaces present new cutting edges. Its extreme hardness and uniformity of shape result in a fast-cutting grain with remarkable resistance to breakdown.

Prepared especially for glass edging and optical lens grinding, silicon carbide is also used extensively for lapping and other industrial polishing for fast stock removal.

Type RA is a black, semi-friable, medium density, silicon carbide abrasive. It is used primarily for ring and pinion gear lapping, lapping metal seals, ceramics, blasting, glass grinding, ultrasonic machining of hard materials, and polishing granite.

Typical chemical and physical properties of Type RA are as follows:

Chemistry (in percentages):

SiC	98·7
SiO_2	0·48
Si	0·30
Fe	0·09
Al	0·10
C	0·30

Crystallography: Alpha silicon carbide in the hexagonal and rhombohedral classes of the hexagonal system.

Ball mill friability: 14 grit—64.

Particle specific gravity: 14 grit—3·20 g/cc.

Knoop$_{100}$ hardness number: 2480.

Shape: Blocky, sharp edged.

Grain sizes available: 6, 8, 10, 12, 14, 16, 20, 24, 30, 36, 46, 54, 60, 70, 80, 90, 100, 120, 150, 180, and 220 grit.

Powder sizes available: 240, 280, 320, 400, 500, and 600 grit. Also, CFl, F, FF, FFF coarse, FFF, and FFFF. Also, 800, 1000, and 1200 grit in limited quantities.

Boron carbide powder, abrasive grade

Boron carbide abrasive is the hardest material manufactured for commercial use. It is produced by fusing boric acid glass and petroleum coke in the electric furnace. Available in granular or powdered form where it is widely used as a lapping abrasive for ultrasonic machining.

It is a most satisfactory substitute for diamond powder for cutting and polishing materials having a Mohs hardness of 8 or more. In ultrasonic machining operations, its extreme hardness furnishes faster cutting rates and greater accuracies as against softer abrasives of the same grit size.

Boron carbide abrasive grade powder is an electric furnace crystalline product made from high purity anhydrous boric oxide and graphite. The raw material, furnacing, and final processing are carefully controlled to insure the highest purity.

The main applications of the abrasive grade of the powder are to lapping and ultrasonic machining.

Typical chemical and physical properties of the powder are as follows:

Chemistry (in percentages):

Total boron	76·0
Total boron + total carbon	98·0
Anhydrous boric oxide	0·3
All others	1·7

Crystallography: Rhombohedral.

Melting point: 2350°C.

Particle specific gravity: 2·45 g/cc.

Shape: Angular—sharp.

Particle size by MSA Whitby centrifuge analysis: Average 4 micrometre.

Range: 8–0·5 micrometre. Some agglomerates larger than 8 micrometres may be reported owing to dispersion problems.

4.12 Synthetic diamond cutters and tools

Although natural diamonds have been used for hundreds of years, little serious investigation into their properties was made until Sir Ernest Oppenheimer, then Chairman of De Beers, established the Diamond Research Laboratory in March 1947. Many attempts were made in the last century to synthesize diamonds, but these tended to be empirical in nature until physical theory had been sufficiently developed.

The situation is now very different as research into all that concerns diamond is being carried out in industrial laboratories and universities in many parts of the world. The stimulus for this research has come principally from the greatly increased use of diamond in the engineering industries.

While it is comparatively simple to define a good rough gem in terms of colour, size, and freedom from visual impurities or other blemishes, the criteria of a good industrial diamond and the knowledge and skill required to use it to its best advantage, either as a single stone, or in the shape of fragmented particles, are more complex and more directly related to the basic properties of the material.

The following description of diamond processing has been extracted from *Microscope on Performance* by the De Beers Diamond Research Laboratory with the permission of the Industrial Diamond Information Bureau.

Synthetic diamond

The discovery of a method of making diamonds suitable for industrial use was the climax of a long history of human interest and endeavour, many accidents and many failures. Of the earlier attempts, those by Henri Moissan, a French professor of chemistry around the turn of the century, probably came closest to success. Moissan dissolved graphite in molten iron, rapidly quenching the solution in water to cool it. Another experimenter, a Scottish chemist by the name of James Hannay, heated gaseous hydrocarbons under pressure in the belief that diamond would crystallize out. Both claimed to have made diamond.

From what is known today about the temperatures and pressures required to synthesize diamond, it is difficult to credit that either these two or their lesser known contemporaries succeeded. Some crystals reputedly made by Hannay, which were examined by the British Museum in 1943 and pronounced to be diamonds, are believed to be splinters of natural diamond which had been substituted for Hannay's material.

The first diamond. Within a month of the first announcement of successful synthesis in the United States, Sir Ernest Oppenheimer, then Chairman of De Beers Consolidated Mines, instructed the Diamond Research Laboratory to make independent efforts to produce synthetic diamonds in South Africa, and a new laboratory, called the Adamant Research Laboratory, was established. It opened its doors in October 1956. The new laboratory was able to draw on the knowledge and experience gained by scientists in the Diamond Research Laboratory, some of whom were seconded to join the new staff.

Initially, attention was devoted to designing a pressure chamber capable of withstanding great pressures and temperatures. A chamber was built and experiments began in May 1957. By September of the next year the necessary conditions had been attained and the first diamond was produced. It was a tiny particle measuring only 0·4 mm by 0·25 mm, but X-ray diffraction tests proved that it was indeed a diamond. (Fig. 4.12.1.)

It was found that improvements in the pressure chamber were necessary to give consistent results and to maintain the critical conditions of pressure and temperature required. The starting material was varied in attempts to improve the yields, and literally thousands of experiments were carried out until, on 16 September 1959, the essential conditions were at last achieved and the Adamant Research Laboratory was able to manufacture diamond reproducibly. The establishment of factories for commercial manufacture in Springs, South Africa, and Shannon, Ireland, followed in quick succession.

To understand the methods developed by the laboratory, which is now a division of the Diamond Research Laboratory, to win, or recover, the diamond, it is necessary to know something about the process. In brief, diamond synthesis begins with a precision-machined capsule made of pyrophyllite which has the virtues of not flowing easily under pressure and of being an electrical insulator. Graphite and a metal solvent are packed

△4.12.1. A lattice model ot the crystal structure of a diamond.

4.12.2.▷
Manufactured diamond, consisting of exceptionally strong single crystals of cubo-octagonal shape suitable for saw abrasive.

△4.12.3. Three 10 000-ton presses at Ultra High Pressure (Ireland) Ltd, Shannon, for diamond synthesis.

4.12.4. Compacted fragments of diamond, carbon, and metal after recovery from the ultra-high-temperature presses used for synthesis.

into the capsule which is then placed into the snugly fitting bore of a die and subjected, for a given period of time, to simultaneous heat and pressure inside a high-capacity hydraulic press. The time cycles of pressure, temperature, and subsequent cooling are varied according to the size and type of diamond required, for one of the wonders of diamond synthesis is the degree (nowadays) to which specialization is possible (Figs 4.12.2 and 4.12.3).

Chemical recovery. The compressed, hardly recognizable capsule is now removed from the press and recovery begins. The capsule is first subjected to light crushing. Identifiable pieces of pyrophyllite and other waste material are picked out by hand and the balance is sent to the chemical cleaning section. Here it is subjected to treatment in a variety of acid baths until excess metal and other non-diamond matter is dissolved. The residue of diamond is then cleaned, dried and sent to the sorting section for sizing and grading according to shape (Fig. 4.12.4.)

Separation into different sizes is done by sieving through a nest of six or seven sieves with the mesh perforations decreasing in size from the topmost to the bottom.

Sorting into shapes. Shape sorting is carried out on an ingenious machine developed in the Diamond Research Laboratory. Known as a vibrating or sorting table, it consists of an electrically vibrated metal 'table' set at an angle, with from twelve to fourteen receptacles along one of its sides. From a glass container on the opposite side are fed the diamond particles to be sorted. The vibrations cause the roundest or blockiest shapes to roll quickly into the receptacle at the bottom and the others to bounce their way into different receptacles higher up the table, the flattest and most needle-like shapes travelling to the highest.

This method, the best ever devised for sorting by shape, is also used to separate natural grit particles. Tabling, although now a production process, is still done continuously at the Laboratory as a control check and for experimental work connected with the development of new diamond products.

The diamonds of industry are many and various. In fact, it would be difficult to think of any branch of technology in which there has been greater specialization than in the application of diamond. There are more than a dozen major classifications of industrial diamonds to start with, ranging from the near-flawless single stones which, but for their colour, would rank as gems, to poor quality bort suitable only for crushing into small particles called grit. And when one gets down to the grit sizes, the De Beers Industrial Diamond Division alone markets fourteen different types, including the synthetic products from its presses. Practically all these types are further subdivided into ten and sometimes twelve different mesh sizes and, at the very bottom of the dimensional scale, there are the micrometre powders in nine sizes, the finest of which is finer than face powder.

Drilling achievement. Because grinding, sawing, and drilling collectively comprise 90 per cent of total industrial diamond utilization in the world today, it is with the smaller diamonds—the grit particles—that the Laboratory is principally concerned. Less research is done on whole stone drilling material, but it is pertinent to record one of the De Beers Diamond Research Laboratory's past achievements—the perfection of 'Hardcore', sold universally as the ideal processed diamond for drilling. Because of a shortage of sound drilling diamonds, producers began to lose sight of the empirical wisdom of the diamond driller of old who selected his stones by tilting the parcel he was offered and keeping those that rolled off. Constant laboratory and field testing proved that the best drilling diamond was indeed smooth and round.

It was found that, if the right sort of conditioning was applied, even rough and irregular-shaped stones could be made suitable for drilling. Equipment was developed which subjected the diamonds to repeated impact to eliminate, by fracture, those which had concealed flaws. Only the strongest survived to undergo the second process which rounded them in a centrifugal attritioning mill. Drilling performances immediately improved, but the Laboratory felt that friction resistance could be reduced still further if the mat surfaces were smoothed. A further process was developed to impart a high surface gloss to the diamond, and Hardcore was born.

Beginning of specialization. Turning to the smaller particles—the grit sizes used in sawing and grinding—attention was drawn earlier to the proliferation of types. The earliest differentiation was by size. Although there was a certain amount of overlapping, it was accepted in the industrial diamond trade that diamond below 50 micrometres (270 mesh) was the most suitable for polishing and lapping applications, that diamond between 40 (325 mesh) and 180 (80 mesh) was best for all forms of grinding and that the saw range was from 150 micrometres (100 mesh) to 1000 micrometres (18 mesh).

In the broad grinding range, crushed natural diamond was sold in parcels containing sizes which might vary from 80 to 325 mesh, a hair-raising situation in the light of today's insistence on accurate sizing. It was not until about 1954 that the first move was made towards a more sophisticated subdivision.

The De Beers Diamond Research Laboratory and a sister company, Diamond Abrasive Products, began experimenting with the separation of crushed natural diamond particles into different shapes. An arbitrary division into four shapes—blocky, more elongated, neither too blocky nor too flat, and perfect needles or flats—was decided upon and each shape was allocated a number. The term 'shape count' was coined. It indicated a measure of the average shape of a certain quantity of crushed diamond particles, and the Laboratory evolved a process to separate flats from the other shapes.

Grinding-wheel makers were encouraged to use diamond particles of a more blocky shape for metal bond wheels and flatter shapes for the less strong resin bonds. This improved the efficiency of both types of grinding wheel and the Laboratory's action was undoubtedly the precursor of true grit specialization as it is practised today.

The impact of synthetic diamond. The advent of factory-made diamond increased the tempo of specialization and brought the technical brains of the producers of both natural and synthetic diamonds to bear on the question of providing specific types of grit for particular applications.

By 1960 synthetic diamond had effectively penetrated the preserve of natural diamond in resin-bond grinding fields. The onus was now on the De Beers Diamond Research Laboratory to prove that the crushed natural product was as acceptable as manufactured diamond crystals. One of the first achievements of its scientists was the invention of the shape-sorting table. Shape separation was now a vital issue.

After two years of development the Laboratory was able to show that scientifically selected natural grit would perform as efficiently in resin-bond grinding wheels as its factory-made counterpart. The new product SND-RB (selected natural diamond for resin-bond applications) consisted of irregularly shaped grit with a large proportion of needles, able to break down at the same rate as the resin bond, and all the time presenting new cutting faces to the workpiece. For increased friability, when it was called for, De Beers scientists applied an additional heat treatment.

In the meantime research workers were busily testing the first De Beers synthesized grit from the presses against specifications which the Laboratory had laid down. The first of the synthetic products was a little disappointing, but the scientists, by modifying certain manufacturing and raw material parameters, were able to set improved standards; within a short time, a far superior grit, known as RDA (resin-bond diamond abrasive), was available.

No one at that stage had challenged the superiority of natural diamond for metal-bond grinding wheels, especially when the principle of shape selection was applied also to this type of diamond. To produce a uniform blocky particle, the Laboratory designed impact crushers. The particles were given a stricter tabling and specifications were drawn up for the new product, SND-MB (selected natural diamond for metal-bond applications).

Science was now beginning to play a bigger role in the setting of standards and a laboratory-designed unit, which was given the name of 'friatester', became standard equipment for the friability testing of grits at De Beers plants.

Next to engage the attention of the De Beers team was a synthetic abrasive which would be able to perform as well as natural diamond in the tougher metal-bond applications. The need for a synthetic substitute was magnified by threats to the supply of the natural product posed by civil war in the Congo, main source of bort.

Completely different manufacturing parameters had to be established and, after many tests, specifications were drawn up and the Laboratory's pilot plant produced its first consistently blocky particles. The new grit, MDA (metal-bond diamond abrasive), had to meet severe impact resistance tests and its strength specifications were the highest devised at that stage for a grit made by De Beers.

Metal cladding. The most dramatic diamond abrasive development in recent years was the introduction of metal-clad diamonds for resin-bond grinding wheels. The scientists reasoned that the jacketing of diamond particles with a metal coat would improve the keying of the particle in the resin bond, since metal adheres much better to resin than does diamond; that the metal coat would serve to retain the fractured particles of diamond, thereby prolonging wheel life; and that it would act as a heat sink by delaying too-rapid transmission of heat through the thermally highly conductive diamond, thus avoiding premature charring of the bond.

The De Beers research group worked at this project on a broad scale and was able to publish results which were both spectacular and technically extremely interesting. The effect of covering each individual particle of grit with a carefully graduated coating of metal was to double the efficiency of grinding wheels in wet grinding conditions. This product was designated RDA-MC (resin-bond diamond abrasive—metal clad).

A recent change in nomenclature led to the replacement in some cases, of the letters MC by symbols which more accurately describe the product, stipulating the basic material used in the coating and the percentage of cladding in terms of the total weight of the particle. RDA-MC is now marketed by De Beers under the name of RDA55N, indicating a 55 per cent coating of nickel (N). Similarly the grit perfected by the Diamond Research Laboratory for dry grinding applications (referred to later) is known as RDAR50C, denoting an RDA particle of reduced strength (R) clad with a composite coating (C) which accounts for 50 per cent of the total particle weight. The Laboratory then set its sights on a synthetic equivalent for EMB (engineered metal bond), a strong, blocky De Beers natural diamond product which had long been favoured for impregnation into saws to cut stone and concrete.

This type of work demanded a diamond grit substantially bigger than that which was suitable for grinding, and the growth of large artificial crystals represented a tremendous challenge to the scientists. Ultimately the use of bigger presses, variations in synthesis parameters and tighter control of growing conditions produced the desired grit, SDA (saw diamond abrasive), which, because it was expected to stand up to much severer conditions of tool manufacture and tool usage, had to satisfy a new concept in quality control standards. The most important of these was described as thermal stability, the ability of a grit to withstand high temperatures without impairment of its physical properties. The tests involved heating the particles to 1110°C in an inert atmosphere for 1 hour. After heating, they were re-sieved to ascertain the fraction which had undergone degradation. The impact strength of the underlying crystals was measured and its reduction as a result of the heating process was calculated.

So good was the quality of the new product that the opportunity was taken of upgrading specifications for MDA, since this was also metal-bond

4.12.5. (*left*) **SND-MB selected natural diamond metal–bond for grinding and drilling glass.**

4.12.6. (*right*) **MDA-S metal-bond diamond abrasive, super-strength for lens grinding.**

grit, but in the smaller sizes. This was called MDA-S (MDA-superstrength).

Selected Natural Diamond–Metal Bond

The diamond. De Beers SND-MB (selected natural diamond for metal-bond applications) is a multi-purpose metal-bond abrasive ideal for grinding and drilling hard metal alloys and ceramics. The grit is well-shaped and blocky with smooth regular surfaces and sharp cutting edges. It has outstanding thermal stability and resembles EMB in appearance, but is smaller in size. (Fig. 4.12.5.)

The tools. SND-MB is used in metal-bond grinding wheels, electrolytic grinding wheels, slitting saws (in the larger mesh sizes), electroplated saws, lapping tools and hones and dental drills. Its excellent thermal stability enables it to be set in tools subjected to a high temperature during manufacture.

Where it is used. The main applications for SND-MB are the grinding of tungsten carbide and other hard metals and alloys and the grinding of glass, hard plastics and ceramics. It is also used for drilling glass, ceramics and hard plastics, for honing engine cylinder linings and bearings and for lapping metal alloys to close tolerances and fine finishes.

Processing methods. Starting with the largest size, the diamond is subjected to a form of impact crushing which improves the shape of the strong particles without destroying the cutting edges. At the same time, weak and poorly-shaped particles are broken down into smaller, but stronger, particles. After each crushing cycle, the material is screened to remove the under-size which has been generated and table-sorted to remove suitable material, so that only reject particles are re-processed. The next largest size is then processed in an identical manner.

Quality control factors. With all natural grits the main criterion is particle shape. Only the strongest particles survive in the size being crushed, so that shape improvement goes hand-in-hand with increase in overall particle strength.

Highly trained operators carry out 'shape counts' whereby numerical values are ascribed to particles of various blockiness. An average shape count is determined from a representative sample of 100 particles, and this figure must satisfy specifications laid down for each size.

Metalbond Diamond Abrasive—Superstrength

The grit. Tougher, harder materials require stronger abrasives. This is a generalization, and like all generalizations can be faulted in particular cases; however, it is a very good rule to work to when grinding ceramics, stone and glass. For this reason, De Beers scientists and production engineers have developed MDA-S—Metalbond Diamond Abrasive-Superstrength. This is a new and much improved abrasive for service in all metal-bond applications where arduous conditions prevail.

The benefits of this new tough abrasive are displayed in its superior ability to deal with very hard workpieces and, where service conditions are less arduous, in enhanced stock removal rates and longer wheel life.

Its physical characteristics. MDA-S consists of well-formed strong, blocky single crystals, all exhibiting clean, well-defined cutting edges and points. The majority of the crystals have a cubo-octahedral shape, which is near the ideal, and are clear and free from harmful inclusions. (Fig. 4.12.6.)

Strength is probably the most important single property required of an abrasive grit of this type. No diamond is classified as MDA-S unless its resistance to impact fracture exceeds a specified high minimum value. This, together with rigid control of particle shape and crystallinity, ensures that MDA-S is the toughest abrasive available today in its field of application.

MDA-S must be able to withstand the temperatures encountered in tool formation without loss of strength or particle integrity. Like all other De Beers metal-bond products, it is therefore subjected to a thermal stability test before being accepted by the Quality Control Division. In this test the diamond grit is heated to 1110°C (2012°F) in an inert atmosphere for 1 hour. It is re-screened and, if more than a defined, very small percentage then passes the check screen as undersize, the material is rejected. If it passes this rigorous test, its impact strength is measured. The values before and after heating are compared and an index of thermal stability is computed. Any diamond not conforming to the defined standard of thermal stability is rejected as unacceptable.

Where it is used. MDA-S is an abrasive primarily designed for metal-bond application, and in particular for grinding the toughest workpiece materials. Typical examples include cylindrical and surface grinding of hard alumina and other tough refractories, the pencil edging of glass, lens grinding, stone sawing, and surfacing, etc. Where edge-breakage is a problem with brittle workpieces, or the grinding wheel appears to be grinding too hard, a lower concentration of diamond in the grinding wheel usually provides an acceptable solution; if not, the controlled friability of De Beers MDA diamond is recommended for this type of application.

Natural and synthetic micron powders

The diamond. De Beers SND and DA micron powders are especially prepared for all precision operations and particularly those which demand an ultra-fine finish.

Natural SND Micron. The Natural Micron is produced from virgin diamond subjected to a strictly controlled shape improvement procedure to ensure accurate grading, producing blocky-shaped characteristics. The particles have sharp cutting edges to ensure maximum performance.

Synthetic DA Micron. The Synthetic Micron Powders contain a high proportion of predominantly cubo-octahedral single crystals.

The synthesized powder has extremely high wear resistance.

Where micron powders are used. Micron powders are used in powder and paste form, in metal- and resin-bond wheels, and in hones, cloth laps, electroplated diamond tools, and slitting disks. They are essential in many applications, such as piercing, re-boring and polishing plastic forming and wire-drawing dies, polishing precious and semi-precious stones, and producing superfine finishes on metallic surfaces. Synthetic ruby for watch bearings, and other parts for precision instruments, demand extremely accurate surface finishes that can only be achieved with carefully graded diamond powders.

Quality control. All De Beers Micron Powders are produced under extremely strict laboratory conditions to ensure the continuity of a consistent product, accurately graded to very close tolerances by sedimentation and centrifuging techniques.

Final products are subjected to rigid checks for impurity content and size distribution. Particle counts are made on not less than 1000 particles from each production batch at up to 4000× magnification.

The maximum permitted undersize is 10 per cent of particles counted, while oversize particles are limited to 5 per cent. The largest oversize permitted is $1\frac{1}{2}\times$ the upper limit of the size range. Batches containing material falling outside these limits are rejected.

Availability. De Beers micron powders are available in a wide range of sizes, as follows:

SND micron

$0-\frac{1}{2}$	$\frac{1}{2}-1$	1–5	4–8	8–25	20–40
0–1	$\frac{1}{2}-3$	2–4	6–12	10–20	30–60
0–2	$1-2\frac{1}{4}$	2–6	8–15	15–30	

DA Micron

$0-\frac{1}{2}$	$\frac{1}{2}-3$	2–6	8–15	15–30	30–60
$0-\frac{1}{2}$	1–3	4–8	8–25	20–30	
0–2	1–5	6–10	10–20	20–40	
0–4	2–4	6–12	15–25	30–50	

4.13 Electroplated diamond wheels

An extract from an article titled 'The versatility of electro-plated diamond wheels' by John Prosser, of Impregnated Diamond Products Ltd, gives brief details of the manufacturing process for electrometallic diamond wheels.

The electrometallic wheel, as it is called, has tremendous versatility in its ability to meet new requirements and new demands. Perhaps this can be most easily understood by the simplicity and flexibility of its production and the variations which can be incorporated in the finished product.

Impregnated Diamond Products first started to manufacture electrometallic wheels at Gloucester in 1949 and the production department was considerably strengthened as a result of the merger with Universal Grinding Wheel Co. Ltd, Stafford, around 1960. In the few years following this amalgamation, the department expanded several times and the introduction of new techniques and developments has enabled I.D.P. to remain in the forefront of the field of electroplated diamond wheels.

What is an electrometallic wheel? Essentially it consists of a metal body which has diamond particles 'keyed' to the surface by an electrodeposited layer of metal. In most cases the diamond exists as a single layer of particles which, when viewed as a surface, represents a high concentration of cutting points. This high concentration gives the electrometallic wheel a number of distinctive properties, which are not always encountered in other types of bonded diamond wheels. (Fig. 4.13.1.)

Manufacturing process. (a) *Preparation of the basis metal.* The success of the electrometallic wheel depends largely on the adhesion of the diamond layer. Ordinary electroplating requires a fairly high standard of adhesion; but the diamond tool manufacturers' requirements are even greater, especially if one considers the shear forces to which the coating is subjected.

The key to success relies upon two factors: (1) choice of basis material and (2) the proper surface preparation.

It is generally accepted that the best adhesion is achieved when using mild steel and therefore where possible, this material should be selected.

Surface preparation involves a treatment which thoroughly degreases and cleans the surface, followed by an acid attack. The exact procedure employed will vary depending on the nature of the basis material; but a typical pre-treatment cycle consists of the following:

Solvent degrease
Electrolytic degrease in a hot alkaline solution
Water swill
Electrolytic etch in acid
Water swill
Electroplate

(b) *Diamond fixing.* Contrary to what might be expected no form of adhesive is used to fix the diamond in position. The tool is made cathodic in the electroplating bath and, after a short period of plating, the diamond is applied to the surfaces to be coated.

Those diamond particles in contact with the surface will be gripped as the depositing metal builds up and, once a layer of diamonds has been keyed on—as little as a few ten thousandths of an inch of deposit is enough to fix diamond particles as large as 0·020 inch—the excess diamond is removed and plating continued to secure and complete the bonding of the diamond particles.

A stopping-off paint is used to prevent plating and diamond pickups on unwanted areas.

4.13.1. A peripheral wheel with diamonds keyed to the surface, with a section through the diamond periphery.

4.13.2. A surface with high-concentration 18-mesh diamond particles keyed in position on graphite.

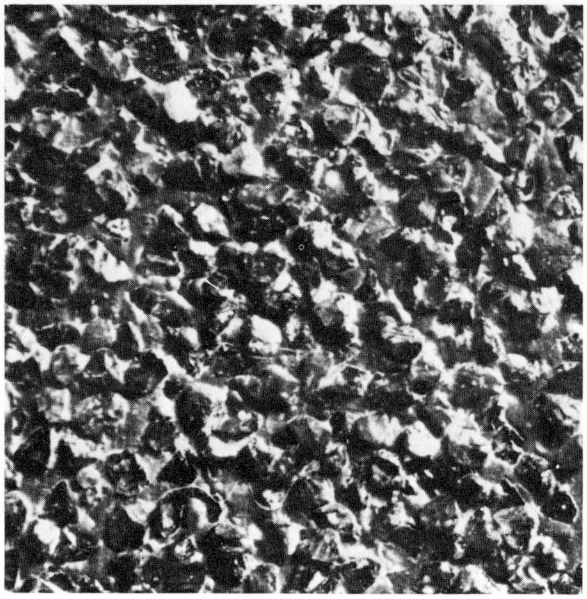

4.13.3. The same surface as in Fig. 4.13.2, with diamond particles secured and bonded by electro-deposited nickel.

During plating, the tools to be coated are the anode and cathode. The anode, which is made of nickel, is shaped to suit the contours of the components being coated to ensure an even thickness of plating. The plating solution is nickel sulphonate. The tools are slowly rotated in the solution during plating until the thickness of nickel is about one-half the thickness of the diamond particle size. The diamond powder is then sprinkled on the top of the solution over the revolving tools and as the diamond powder falls on the tools it is plated on the nickel.

Hollow drills are all nickel and are made in the following way. A steel mandrel is coated with graphite, and diamond powder is rolled into the graphite (Fig. 4.13.2). The mandrel is then placed in the plating solution and nickel is plated on the graphite. Diamond powder is then sprinkled on the surface of the solution over the mandrel and plating is continued. Once again diamond powder

4.13.4. Trepanning tools.

(*a*) 0·75-inch electromagnetic drill.

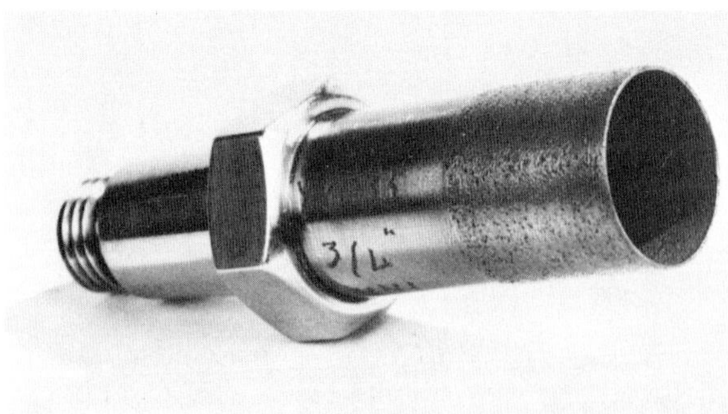

(*b*) Special-size tool for lens blanks.

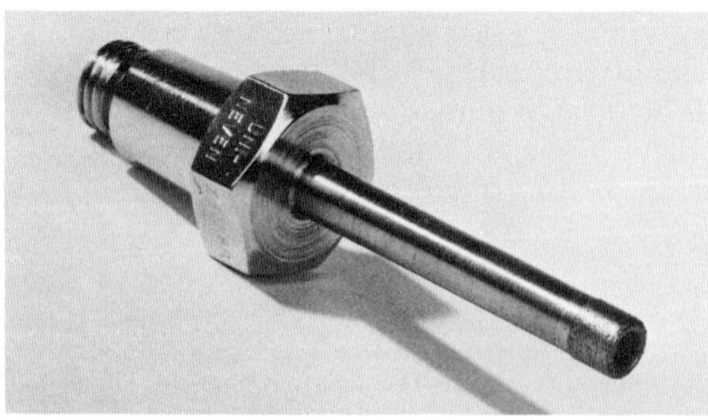

4.13.5. **Spec 100 electrometallic 0·5-inch hemispherical countersink.**

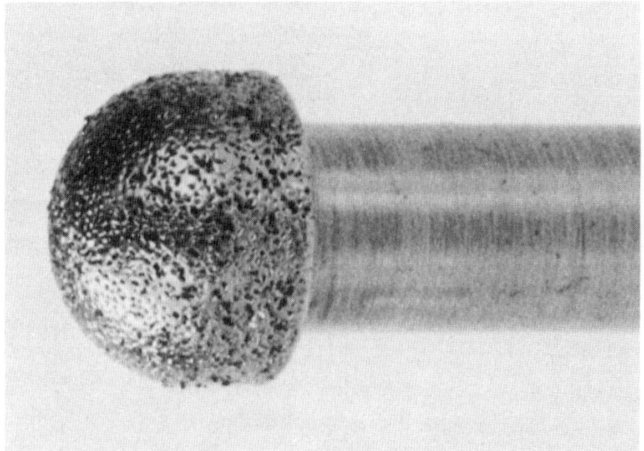

is sprinkled on the surface of the solution over the mandrel and plating is continued until the thickness is correct to specification. (Fig. 4.13.3.) The mandrel is then removed from the plating tank and the outside surface ground until it is a cylinder 0·010 inch (0·25 mm) oversize. Extract the mandrel and solder on an adaptor (Fig. 4.13.4). This process takes about 24 hours and is suitable for hollow drills from 2-mm to 100-mm diameter.

Hollow metal-bond tube drills are used in drilling machines converted for internal-coolant feed. (See § 6.3.) A number of plates of glass to the required specification are normally waxed together with a piece of scrap plate glass at the bottom. The drill is passed through the components into the scrap piece in order to minimize chipping on the lowest plate of optical glass. A speed range on the drilling machine of 2500 r.p.m. to 500 r.p.m. is necessary, with plenty of power available. For countersinking holes two types of tool are available, conical and hemispherical. (Fig. 4.13.5.) The conical type is for use in fixed-head drilling machines, but, for hand drills, the hemispherical electrometallic-bond tool is used with an oscillating motion to distribute wear. Coolant must be applied externally. Electrometallic tools are used for grinding plastic lenses because they are less apt to clog during continuous grinding than are metalbonded tools made by powder metallurgy.

4.14 Diamond [metal bond] grinding wheels

Diamond wheels are intended for the grinding of hard and brittle materials which produce short broken chips. The optimum performance of diamond wheels for the grinding of specific materials depends on the correct grit size, the diamond concentration, the depth of impregnation, the cutting speed, and the wear resistance of the selected bond. The grit size, concentration, and bond of diamond wheels are always interdependent to a certain extent. When ordering diamond wheels, give preference to the sturdiest type available and the largest admissible wheel diameter for the machine, bearing in mind the cutting speeds. The wheel speed should be in the region of 5000 and 6000 feet per minute, and the recommended speed must be maintained when the wheel is under load.

The type of coolant used bears a direct relation to the performance of the diamond wheel and must be varied to suit the material being ground. A copious supply of coolant should be directed at the point of cut.

Typical diamond grain sizes for milling tools to be used for curve-generating lenses are as follows:

	B.S. sieve fraction
Coarse grinding	85/100
Intermediate grinding	170/200
Fine grinding	240/300

Grains and powders available are listed on p. 63.

The classic Round Nose milling tool has an edge radius equal to half the wall thickness. Tools are specified by pitch circle, diameter (middle diameter), and wall thickness. Another design is the Bull Nose milling tool which gives maximum

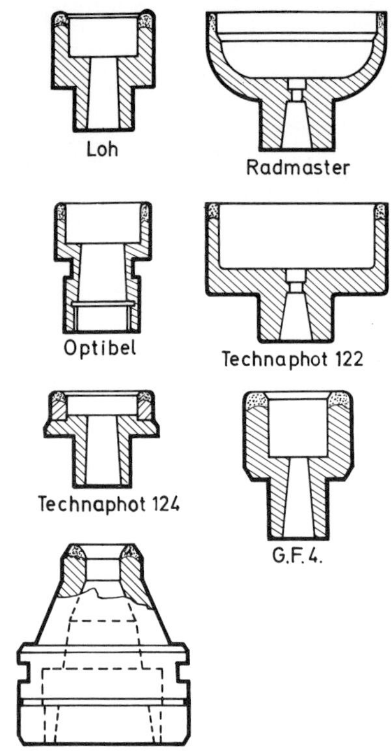

◁ **4.14.1. Standard milling tools for the machines indicated.**

▽ **4.14.2. Metal-bonded milling tool for curve-generating lenses on an Optibel machine.**

strength and stock removal properties. This type is suitable for interrupted cuts such as surfacing a large block of lenses. (Fig. 4.14.1.)

For the scientific optical trade the bonds used are either bronze, copper, or ferrous. Diamond concentration varies between 50 and 90 and, allied with dimensions and layer depth, fixes the diamond content and the price of the tool. Generally speaking, lower concentrations are supplied at greater layer depth than high concentrations in order to maintain an optimum life of the crown relative to the cost of the body threads and tapers. Milling tools rotating at high speed seem to work better with 90 concentration, whereas lower speeds suit 50 concentration. Autoflow and Optibel machines seem to prefer 50 concentration whereas LOH and Dama machines work better with 90 concentration. (Fig. 4.14.2.)

Lens-centring machines are basically cylindrical grinders and, for flat components, the grain size should be 240 or 280, whilst 320 is more suitable for chamfering. Concentrations are usually 90 for flat components and 140 for chamfering. High concentration assists in improving the surface finish on the lens but also protects the bond. This gives a slow wear rate and so preserves the accuracy of the lens edge-form. The use of lower concentrations to reduce price is uneconomic owing to the more frequent reconditioning necessary.

Cutting of optical glass by a cutting disk is an accurate and efficient method. The wheel speed must be between 5000 and 8000 s.f.p.m. and adequate coolant must be used. The grain sizes are usually 85 to 100 to give a minimum chip finish, whereas concentrations are low, between 15 and 20. The bonds are usually bronze in cutting disks.

4.15 Manufacture of metal-bonded diamond wheels

By courtesy of Impregnated Diamond Products Ltd, the following details are given of the methods used in making Uni-Neven bronze bonded wheels.

The diamond wheels are all made to special order, and each order on the shops specifies the weight of diamond and bond for each tool. The

Table 4.15.1. Diamond grain

New coding *Natural diamond*	*Synthetic diamond*	*B.S. sieve fraction B.S.S. 410–1962*	*Equivalent American sieve fraction ASTME 11*
18	—	18/22	20/25
22	—	22/25	25/30
25	—	25/30	30/35
30	—	30/36	35/40
36	36 S	36/44	40/45
44	44 S	44/52	45/50
52	52 S	52/60	50/60
60	60 S	60/72	60/70
72	72 S	72/85	70/80
85	85 S	85/100	80/100
100	100 S	100/120	100/120
100/120	—	100/120/150	—
120	120 S	120/150	120/140
150	150 S	150/170	140/170
170	170 S	170/200	170/200
170/200	—	170/200/240	—
200	200 S	200/240	200/230
240	240 S	240/300	230/270
280	280 S	300/350	270/325

Table 4.15.2. Diamond powder

I.D.P. code (Natural diamond only)	*Micron range*	*Median size*
320	20/40	30
400	14/28	21
500	9/18	13
600	6/12	9
800	4·5/9	6·5
1000	2·5/7	4·5
1200	1/5	3
1400	0·5/3	2
1600	0/2	1·5
2000	0/1	1

Also available in 0–¼ and 0–½ Micron sizes

Table 4.15.3. Recommended I.D.P. diamond size equivalents to the German standard DIN 848

DIN 848	*Micron range*	*I.D.P. Code No.*	*DIN 848*	*Micron range*	*I.D.P. Code No.*
D 30	20–40	320	D 150	120–200	85
D 50	40–60	240	D 250	200–300	60
D 70	60–80	170	D 350	300–400	44
D 100	80–120	120	D 500	400–600	25

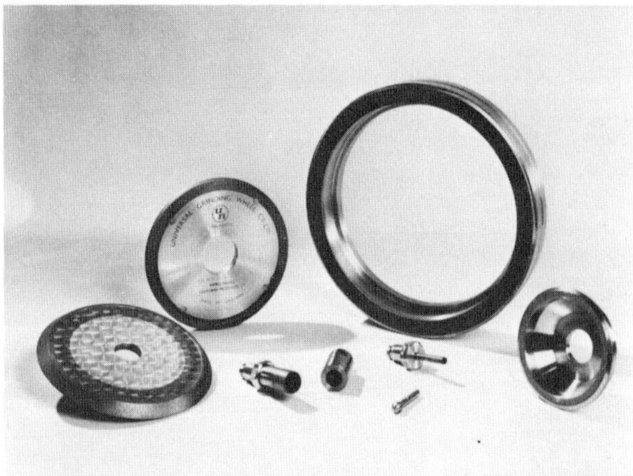

◁4.15.1. **Diamond wheels and drills.**
(*a*) **Brilliant cutting wheel.**
(*b*) **Peripheral wheel.**
(*c*) **Ring cup wheel for Blanchard mill.**
(*d*) **Cup wheel for surfacing glass.**
(*e*) **Two metal-bonded electrometallic drills.**
(*f*) **Hemispherical electrometallic countersink.**

▽4.15.2. **Metal-bonded drills.**

diamond grains and the 60/40 copper-tin powders are accurately weighed. The diamond grains and the bronze powder are then placed in a 4-oz glass jar with three flint pebbles, plus some oil to hold the grains together during subsequent processes. To avoid contamination, the glass jar is used only for the one size of diamond and bond. The jar is placed on a rolling machine and the materials are completely mixed in about 15 minutes. The Nimonic steel moulds are coated with graphite in the forming chambers to prevent scoring of the surface by the diamonds. After the mixture has been placed in the mould and evenly spaced, the mould is placed under a hydraulic press and compressed cold. The mould is then warmed up in the furnace, removed, placed under the press, and compressed again. The mould is then placed in the furnace again and brought up to a temperature of:

650°C	for bronze bond
750°C	for copper bond
875°C	for steel or cobalt bond

for a time of 60–90 minutes according to the size of the mould. It is then placed under the press and compressed again. After the mould has been removed from the press, it is allowed to cool. Then the diamond bonded tool is removed and machined before being fitted to its adaptor, to which it is retained by screws or Araldite adhesive. The tools are dimensionally checked before returning them to stores; for any change in the mould volume due to distortion will require a change in the quantity of mixture required to make up a tool. (Fig. 4.15.1. and 4.15.2.)

4.16 Impregnated pellets for diamond smoothing

Grain sizes vary from 20–40 to 1–5 micrometres (I.D.P. Code 1200), and the following finishes are obtainable:

Grain size	*IDP Code*	*CLA (micrometre)*
15–30	400	0·65
10–20	500	0·55
6–12	600	0·50
4–8	800	0·47
2–6	1000	0·45

Diamond grading

There are at least three different ways of defining the size of diamond particles.

Linear definition. The 'size' of the grain is its greatest dimension, so a rectilinear sliver 20 × 5 × 3 micrometres is a *20 micrometre particle.*

Projected area definition. The size of the particle is given as the diameter of the circle of similar area to the surface of the particle, so our grain size is *11 micrometres.*

Volume equivalent (Stokes' law) definition. Here it is accepted that particles of equal volume are of equal dimension, since they precipitate through a liquid at the same speed, irrespective of shape. The size is then given as the diameter of a sphere of the same volume as the rectilinear figure, so the $20 \times 5 \times 3$ micrometre sliver becomes a *grain of 8 micrometres.*

Both the German DIN 848 and American Coding CS.261–63 use this latter method of definition, so that when the upper limit of say 5–10 micrometres bulk range powder is given as 15 micrometre, slivers of up to 40 micrometre length could be present, although the numbers of such splinters are limited.

Impregnated Diamond Products check all powder used for pellet manufacture on the *linear definition.* The sizing given is the bulk range which indicates that 80 per cent of the grains are within the limits given, with 10 per cent above and 10 per cent below. This can be illustrated by considering 10–20 micrometre bulk range powder. This could be expected to have 80 per cent of the particles in the 10–20 range: 10 per cent would be approximately 5–10 micrometre: and 10 per cent, 20 to approximately 35 micrometre. The DIN 848 system refers to the medium size, i.e. D.15. I.D.P. code number is 500.

To describe a powder as of closer tolerance than those shown above is unrealistic and the production of this powder would be impractical.

Although diamond pellet smoothing is still in its infancy, I.D.P. has decided to standardize on 45 concentration, as low concentrations at fine grain sizes are seldom satisfactory.

In respect of pellet techniques, no concrete advantages have been found, to our knowledge, in going to higher concentrations. The bond used is bronze and this has been standardized at D11. The specification for all standard pellets is therefore 45D11.

Despite every effort and control to eliminate stray large grain diamonds, it is not realistic to say that these never occur in any manufacturer's product. Obviously, the quality of a batch of pellets depends to a large extent on the diamond grading and the absence of large particles. Nevertheless, if scratching is obtained, recourse must be made to magnifying glass and gouge. Pellets are usually fixed to their holders with Araldite and assistance on the best techniques can be obtained from C.I.B.A.

A coverage of between 25 and 50 per cent of the smoothing tool is normally used; with a greater coverage as the radius increases. (§ 6.7.)

4.17 Cutting fluids for grinding glass and crystal materials

In addition to possessing the normal functions of a metal-cutting fluid (such as heat transfer, swarf removal, and reduction of friction), a glass-cutting fluid, when grinding glass, must actively participate in the actual grinding process and have a significant effect upon the quality of the ground-glass surface including the depth of the layer fractured during grinding, and the efficiency and life of the grinding wheel. It is therefore very important to select a suitable cutting fluid composition for the diamond grinding of glass. In contrast to metals, glasses are generally hard brittle materials showing a marked tendency to break down under severe thermal or mechanical stresses. There are some exceptions, such as fused silica and Cer-Vit, which have excellent resistance to thermal shock.

A continuous flow of coolant is essential in grinding glass with diamond tools. Intermittent cooling is worse than dry grinding since the abrupt temperature changes can cause cracking of materials which are susceptible to thermal shock. In grinding glass, it is important to remember that the swarf is very abrasive. It follows that the slides of the machine must be kept clear of this debris, and the coolant must be able to support the abrasive particles and remove them to a suitable settling tank. Experience shows that non-aqueous liquids such as paraffin, turpentine, or proprietary oils such as Honilo, are best for this purpose. Water-soluble oils are not as efficient, but often have to be used because of other considerations.

The following proprietary cutting compounds are suitable for use with diamond tools in the applications mentioned.

For milling machines

(*a*) Castrol Cleeredge AA7050. Dilution with water 20:1. (Castrol Ltd, Lysander Road, Purley Way, Croydon.)

(*b*) Houghton Glassgrind 960. Dilution with water 60:1 (use ratio 10:1 for diamond smoothing). (Société des Produits, Houghton, 7 Rue Ampère, Puteaux (Seine), or from agents: Edgar Vaughan & Co. Ltd, Legge Street, Birmingham 4.)

(*c*) Supercut 3321. Dilution with water 60:1. (Supercut Diamond Tool Company, U.S.A.)

For edging machines with pressure chucks

Neat Esso Mentor 28. This has properties similar to paraffin (Kerosene) for cutting, but without the fire and health hazards.

For machining soluble crystals

Castrol Honilo 430 (W. B. Dick & Co. Ltd).

Coolants for glass grinding

For satisfactory results in grinding glass with diamond tools, and achieving the best possible finish without scratches, it is necessary to have the correct type of tool, a machine stiff enough to prevent vibrations, and a satisfactory, clean coolant at the optimum dilution ratio for the work. The coolant must equally be capable of reducing heat strain and providing lubrication between the tool and the work piece, so preventing excessive wear of tools.

Coolants should have the following desirable properties:

(*a*) A low viscosity, which improves the wetting action and gives better cooling and assists the settling of glass sludge.

(*b*) Chemical neutrality, being neither acid nor alkaline, but close to pH 7.

(*c*) Rust-proofing qualities to avoid damage to equipment.

(*d*) Uniform qualities from one batch to another.

(*e*) Non-foaming properties.

(*f*) Chemical stability: the coolant must not break down and become rancid, causing unpleasant odours and possibly breeding bacteria which would infect the operators.

(*g*) Coherence: the coolant must not separate or form gummy deposits on the side walls of tanks or machinery.

(*h*) Ability to wash away glass chips and influence the surface finish. A rich mixture results in a freer cutting action of the diamond tools and lower power consumption.

Coolant should be applied to the work by flooding the grinding zone at low pressure.

4.18. Centrifugal coolant clarifiers

A very fine finish on the workpiece can be marred by scratches from ground glass in the coolant carried to the grinding zone. Every effort must be made to remove the glass particles before the coolant reaches the diamond tool and workpiece.

If the solids content builds up to 2 per cent or higher, the diamond-tool face will load up with fine glass, the power consumption will increase, the tool will 'glaze' and the workpieces may break. Even if the workpiece does not break, the surface finish and form will be very poor. If the glass content builds up beyond a certain level, operators will develop dermatitis. (See § 4.24.)

Removal of the glass particles from the coolant is not an easy matter, as 60 per cent of the total volume may be less than 10 micrometres (0·0004 inch) particle size.

Normal filtration methods are not very satisfactory because of the small particle size. The two most effective methods are (*a*) allowing the glass to settle out and then decanting, which is a slow method, or (*b*) centrifuging, which is a high-speed continuous operation. Several companies make suitable centrifuging equipment, such as the American Tool & Machine Co., the De Laval Separator Company, the Bird Centrifuge Co., and Machine Shop Equipment Ltd.

A Centrifugal Coolant Clarifier (Fig. 4.18.1) made by Machine Shop Equipment Ltd (Model 5–19–20) has a 700-c.c. liner, and its rotor revolves at 6000 r.p.m., applying approximately 2600 g centrifugal force to the particles, which forces them through the coolant until they are trapped against the wall of the liner. The higher the viscosity of the coolant the longer it will take for the

4.18.1. **Centrifugal coolant clarifier. In the centre of the hinged lid is the rotatable coolant inlet. The rear unit is the coolant tank, which can be bolted to the centrifuge unit.**

glass particles to be forced through the coolant. The following details are a guide to the performance of this small but effective centrifuge:

Leads to better workpieces at less cost
Improves surface finish
Increases wheel or tool life
Increases machine life
Reduces frequency of wheel dressing
Allows machine to work to maximum efficiency all the time
Substantially reduces coolant oil costs by increasing oil life
All contaminants removed and coolants clarified
Solids collected in removable flexible liner
Speedy and easy sludge disposal—no mess
Minimum running costs
Adaptable design
Totally enclosed
Bacteria in solubles inhibited

Clarification principle

An open-topped rotor is driven at high speed. Into this is fitted a long-life removable flexible liner shaped to the inner contour of the rotor. The liner is secured to the rotor by a removable vane assembly consisting of an open-topped ring and a number of vertical paddle-type vanes.

Dirty coolant enters at the top of the rotor assembly and is deflected by the centre cone towards the bottom. Centrifugal force acts on the dirty coolant, packing the foreign matter into a 'cake' against the wall of the liner. The clarified coolant overflows through the hole in the top of the vane assembly and is returned to the machine. (Fig. 4.18.2.)

Clarification continues until the vane assembly is full.

MSE Clarifiers are compact self-contained units that stand alongside the machine tools they serve. As the dirty coolant flows from the machine tool under gravity it is necessary to ensure that the clarifier chosen has an appropriate inlet height. Generally, when the clarifier is positioned closely

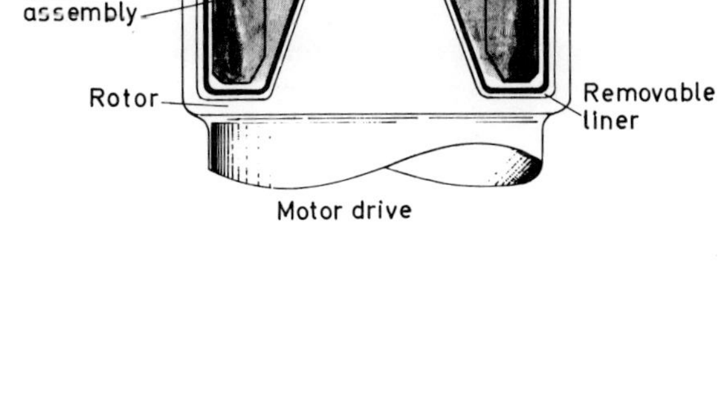

4.18.2. The rotor assembly of an MSE Clarifier. This recovers practically all foreign matter (some of which may be valuable) from the coolant.

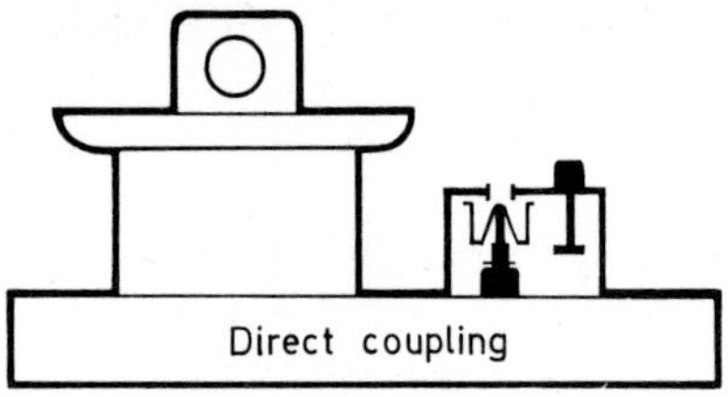

4.18.3. Simple connections of the coolant clarifier.

adjacent to the machine, about 2 inch fall is a satisfactory minimum. Only simple pipework and electrical connections are necessary. (Fig. 4.18.3.)

The flexible liner

For clean and speedy disposal of the foreign matter extracted from the coolant, all MSE Clarifiers are fitted with the long-life flexible liner described under 'Clarification Principle'.

Frequency of emptying the liner depends largely on stock removal rate of the machine tool. The procedure takes not longer than 2 minutes and can be shortened even further by the use of spare liners, allowing the full liner to be emptied at leisure.

These liners are the result of long investigations to find a mouldable material which withstands known coolants and which is strong and reliable over a long period of constant use.

4.19 Cleaning liquids

In addition to tap water, distilled water or deionized water for substances soluble therein, the following cleaning liquids are needed in an optical shop.

(*a*) For cleaning off shellac, Canada Balsam, waxes, and varnishes use methylated spirit and alcohol.

(*b*) For cleaning off pitch, rosin, beeswax, or mixtures of these, use benzene, turpentine, petrol, or paraffin.

(*c*) For polymerized HT cement, use acetone.

(*d*) For cleaning metals use nitric acid, aqua regia or strong sulphuric acid with chromic acid added.

The addition of a drop of nitric acid will often assist cleaning with other liquids. For example, the optical contacting of surfaces is difficult even when the surfaces appear quite clean; it may be successfully accomplished, however, if the surfaces are lightly wiped with cotton wool dipped in dilute nitric acid or ammonia. The surfaces should then be washed off with water, finishing with ammonia, and wiping the latter off with an old linen handkerchief which has been boiled in distilled water.

Moisture condensed from breath is a useful cleaning medium. Distilled or deionized water should always be used for cleaning, and not tap water, which is usually very far from being clean.

Carbon tetrachloride and chloroform are sometimes used as powerful solvents, but they are toxic and good ventilation for removing the vapour is essential. A non-inflammable cleaning spirit for dissolving pitch from lenses without removing shellac is named Securitine and supplied by the Securitas Non-Inflammable Products Co. Ltd. 12–14 Middle Street, London EC1.

Another good solvent for removing pitch, but not shellac, is G 700 solvent supplied by G. S. Chemicals Ltd, Old Forge Works, Chipstead, Sevenoaks, Kent. G 700 is a blend of aliphatic and chlorinated solvents with a flash point higher than 150°F (65°C) and neutral pH. This is a safe cleaning agent for optical components.

Dunkit

A very powerful paint stripper and decarbonizing fluid is named Dunkit, supplied by the Penetone Company Ltd, Bassington Industrial Estate, Cramlington, Northumberland.

When cleaning pitch and shellac lenses, Dunkit may leave a slight oil-like deposit, but this is easily removed by detergent in warm water or Formula 861 Solvent emulsion degreaser followed by water rinse.

The following information is due to the Penetone Company Ltd.

Dunkit is a heavy duty, highly concentrated liquid stripper and carbon remover for dip tank use. It is a single-phase stripper composed of 100 per cent active ingredients, including powerful solvents, emulsifiers, and wetting agents formulated to attack and strip a wide range of deposits, including carbon, heavy sludge, dried and oxidized oils, synthetic or natural rubber, paint, and all types of organic coatings, plastic films, etc. Dunkit is non-flammable and safe on metals, including zinc, aluminium and magnesium. It meets corrosion requirements of AAF 20043 and PC 111. Dunkit will not damage ceramics, glass, stone, or similar non-porous surfaces.

Unlike many solvents or strippers Dunkit is particularly advantageous owing to excellent rinsing qualities. Those parts which have been treated in a Dunkit solution and subsequently rinsed in water will have no residual smell or deposit which could cause problems on further treatment of the parts. Furthermore, there is no discomfort or toxicity to operators handling parts which have been treated in Dunkit and thoroughly rinsed.

Unlike many conventional cold strippers which are not supplied in concentrated form owing to the inclusion of a water seal, Dunkit is formulated as a concentrated stripper. Only water need be added when the tank is filled, and this then absorbs certain inhibitors from the concentrated Dunkit and acts as a 'blanket' on top of the concentrated solution to prevent excessive evaporation. Furthermore, parts are automatically pre-rinsed in the water seal on withdrawal from the tank. Thus the bulk of all active material drains back and is not lost in the carry-over.

For all tank operations, water is first added to the tank to a height of approximately a quarter of the required working level of the total liquid contents. Dunkit is then pumped or poured into the tank through the water seal until the working level has been reached. The solution should then be gently stirred and allowed to stand for at least 2 hours before use. This is important, since the water seal takes time to separate through the active Dunkit and to absorb the inhibitors.

To use the tank simply immerse the parts to be cleaned in Dunkit *below the water seal* until the contamination or deposit is softened. Soak time will vary from 30 seconds to overnight, depending on the type of film and thickness of coating to be removed. Most paints are removed within 15 minutes, hard carbon in an overnight soak, and lighter carbon deposits usually within 2 hours, as with rubber and plastics. Always use Dunkit at room temperature and install tanks in an area where there is adequate ventilation.

After components have been softened in the Dunkit tank, they should be removed and thoroughly rinsed in water, steam, or solvent before handling.

It is necessary to top up the tank with water since the seal will gradually be used owing to the drag-out etc. A minimum of 4 inches of water seal should always be present on top of the active Dunkit solution. As with all stripping compounds, contact with the skin should be avoided, and operators should wear protective clothes and goggles. The tank should be covered when not in use, and cleaned parts should be thoroughly rinsed before handling. In case of accidental spillage the affected part of the skin should be washed with copious amounts of fresh water.

Quadralene Laboratory Detergent

A good detergent suitable for a late stage in a cleaning process is Quadralene Laboratory Detergent (QLD) supplied by Quadralene Chemical Products Ltd, Liversage Works, Bateman Street, Derby.

The detergent consists of powerful wetting agents, sequestrating compounds and buffered alkalies. It is capable of giving results usually associated with more dangerous cleaning agents such as strong acids, oxidizing agents, or alkalies. QLD is normally used as a 2 per cent solution in water at temperatures of about 65°C with ultrasonic agitation (20 g of powder per litre of solution). Treated optical components must be rinsed in deionized water and dried immediately after removing from the cleaning solution unless a further treatment such as by vapour iso-propyl-alcohol is to follow. Cleaning times are of the order of 2 minutes.

4.20 Stains on glass

The usual protective coat for a polished surface during subsequent operations is given by shellac. However some glass, such as DEDF, easily stains, and a more suitable material in such cases is green-coloured Sealac, a vinyl-type of plastic which can be removed by its own remover—Sealac Remover. Acetone will also remove Sealac, which is supplied by W. Canning & Co., Ltd, PO Box 288, Great Hampton Street, Birmingham.

Pitch does not stick to Sealac as well as to shellac, but Sealac is better at preventing stain on the polished surface and is usually applied by dipping. Sealac can also be used to protect theodolite circle blanks or other graticule finished surfaces to prevent atmospheric contamination before the coating of resist required by the next process. Sealac, in its white colourless form, can be used as a protection against B.S.8 cement or other cements which may accidentally touch optical surfaces during the cementing operation.

Another protective coating is 'Airdrying spraying strippable coat' supplied by Carr's Paints Ltd, Birmingham (Reference 6074 A/600). This coating can be removed by soaking in acetone or ultrasonically in Dunkit.

A trichlorethylene-bound lacquer sold under the trade name Trilac is also used to protect polished surfaces and is applied by a spray gun. (See § 6.9.)

4.21 Optical cements (See also § 6.11.)

Optical cements in order to be satisfactory in use must flow freely to a thin layer during application and have a consistent refractive index. The cement must be transparent in the wavelength region where it will be used and be homogeneous. Lenses over about 5-inch (125-mm) diameter should not be cemented because of the problem of distortion.

Canada Balsam (refractive index 1.54)

This is a natural product. A pale yellow liquid of pine-like odour is applied in the cold, followed sometimes by a heat treatment for a period at 70°C. Since it is a solvent-loss cement, shrinkage is unavoidable. Optical components can be separated by dry heat at 120°C or by heating in a solvent such as benzene, trichlorethylene, or carbon tetrachloride. This cement is easy to use, but will not stand high working temperatures. It is quite satisfactory in O.G.s, cover-glasses, or other components not likely to meet high temperatures. The cement should be preserved in a small glass bottle provided with a ground-in stopper. The bottle should always be kept filled.

Ross No. 24 Optical Cement (refractive index 1·46)

This may be considered a plasticized H.T. cement made from *n*-butyl methacrylate, catalyst, inhibitor, and Canada Balsam. Although the cement is thermosetting, polymerization takes place slowly at room temperature or in sunlight, resulting in an increase in viscosity. Edged lenses up to 3-inch (75-mm) diameter are cemented at room temperature, and clamped in position on jigs, and then placed in an oven at 70°C and left overnight (16 hours). Lenses 4 inches (100 mm) or over should be left for about 40 hours. The cement should be kept in a refrigerator at −40°C when not in use. Components can be separated by heating in an organic solvent such as toluene or xylene.

Barr & Stroud No. 8 Optical Cement (refractive index 1·56)

This is a very stable cement but has three disadvantages.

(1) Its yellow colour ('cut-off' at 4000 angstroms).

(2) Moderately high viscosity.

(3) The difficulty of separating components cemented with it. (But, of course, correctly made joints do not require separation and this is not a serious disadvantage.)

The cement consists of three components:

	Parts by weight
Resin	10
Flexibilizer	5
Catalyst	1

The three components must be thoroughly mixed for 5 minutes followed by spinning in a centrifuge for 20 minutes at 2500 r.p.m. to remove air bubbles. After mixing, the shelf life of the cement is 4 to 5 hours at 20°C which may be extended by storing in a refrigerator. Curing time is three hours at 80°C, but baking overnight for 12 hours at 70°C is an alternative.

The surfaces to be joined must be clean, dry, and free from grease. The following cleaning process is recommended for work of high quality:

(1) Saturate surfaces of glass with 74 per cent meths by pouring.
(2) Rinse under running tap.
(3) Immerse in 10 per cent solution of Teepol and water for 5 minutes and swab surfaces with cotton wool.
(4) Rinse under running tap.
(5) Spray with 74 per cent meths.
(6) Rinse under running tap.
(7) Spray with distilled water from Millipore filtered flask.
(8) Spin dry.
(9) Store with face to be cemented downwards.

Cementing operations are best performed at room temperature, although components may be warmed a little in an oven before application of the cement. BS8 cement withstands tropical tests satisfactorily either when used to cement graticule cover-glasses or for trepanned glasses after cementing.

If it is necessary to separate joints cemented with BS8 cement there are two methods available:

(1) If the annealing temperature of the glass is above 400°C then the faulty components can be heated in a furnace at 400°C for 5 hours, followed by cooling down for 24 hours. The resulting carbon can be removed by careful use of a razor blade followed by a paint stripper, such as Solvo Strypa supplied by the Penetone Company.

(2) Alternatively, separation can be achieved by immersing in an oil bath on an electric hotplate at about 280°C using Midland Silicone oil MS 200/100. This process should be carried out in a fume cupboard and it is important that the heating and cooling be slow to avoid breaking the glass.

BS8 cement is supplied by Barr & Stroud Ltd, Anniesland, Glasgow, W3.

Shell Epicote ER 817 Cement (refractive index 1·57)

This cement is similar to BS8 but absorbs less at the blue end of the spectrum. It fulfils all the conditions required for service use. It is prepared by mixing the two component parts by weight:

Shell Epicote Resin	ER817	10 parts
Spikure Hardener	K61B	1 part

Mix together and remove bubbles by raising the temperature to not more than 50°C. After cementing, bake for 3 hours at 70°C. The shrinkage of this cement is $\frac{1}{2}$ to 1 per cent, whereas the old type H.T. cement (*n*-butyl methacrylate) had a shrinkage of 10 to 13 per cent.

Cellulose Caprate Optical Cement
(refractive index 1·47–1·49)

This is a very useful general purpose cement (supplied by Eastman Kodak) and the cementing techniques are similar to those of Canada Balsam.

The cement can be applied cold by cutting a slice from the end of the stick. The cement softens in the range 95°–100°C and is quite fluid at 120°C. Alternatively, the lens components, together with the syringe containing the cement heated in an oven until the cement becomes liquid. Cement is then applied to one of the surfaces to be mated and the two surfaces matched. After cementing, the components are allowed to cool slowly within the oven to avoid residual strain. Components can be separated by dry heat at 120°C or by warming in toluene. Residual cement may be removed with toluene.

A mixture of 4 parts by weight of cellulose caprate with 3 parts by weight of liquid paraffin is a satisfactory ultra-violet cement suitable for calcite polarizers and will transmit 71 per cent at 2500 angstroms wavelength. The mixture should be heated slowly in an oven to 100°C and stirred until homogeneous. It can then be treated as if it were Canada Balsam.

Laminac 4128 Optical Cement
(refractive index 1·54)

Laminac cement is supplied by B.I.P. Chemicals Ltd, Oldbury, Warley, Worcestershire.

It is a three-component mixture in the following proportions by weight:

Laminac Beetle resin	100
Catalyst	3·5
Accelerator	1·5

However, these proportions may be varied to change the setting time of the cement from 5 minutes to 2 hours. Laminac cement is often used when lenses have to be very accurately centred. The cement will withstand temperatures up to

70°C. Components can be separated by immersing in hot Midland silicone oil MS 200/100 in exactly the same process as for BS8 cement, but the decementing is easier with Laminac 4128. Final edging should be carried out after cementing.

Crystic 191 Optical Cement

This is a two-component mixture suitable for large high-accuracy lenses, curing at room temperature, with characteristics similar to Laminac 4128.

4.22 Adhesions for permanent fixing

Araldite epoxy resin

Araldite offers a number of advantages over conventional fastening processes, and joints of great strength can be produced. Bonding can be carried out at relatively low temperatures and hardening takes place by chemical reaction, without the emission of vapour, and with negligible shrinkage. Araldite joints are resistant to attack by fumes, moisture and most acids, alkalies, and solvents.

The general procedure for cementing glass is to degrease, abrade the surface with carborundum and water slurry, dry, and degrease. Warm the glass for $\frac{1}{2}$ hour at 100°C and apply the Araldite adhesive before the glass cools to room temperature. The curing temperature and time must be correct for successful bonding.

Araldite is supplied by CIBA (A.R.L.) Ltd., Duxford, Cambridge.

Resiweld epoxy adhesive

This has given satisfactory results in the joining together of optical components. The characteristics of Resiweld Adhesive No. 8, is that it is a two-part mixture in the ratio 100 parts of Part A to 6 parts of Part B, which must be stirred together before use. The pot life is about 4 hours at 25°C. Heat is required to cure the adhesive at 82°C for 3 hours. Strong solvents are useful for cleaning the epoxy resin from tools or equipment before it hardens. Resiweld epoxy adhesive is supplied by Fuller Adhesives International S.A., P.O. Box 4868, Nassau, Bahamas.

Evo-Stik Twinstik

A double-sided transfer tape; it is a very light but strong fabric tissue which has been impregnated with a special pressure-sensitive adhesive. It remains permanently tacky on both sides and is backed with a high-quality release paper. Twinstik is suitable for bonding light articles to metal, wood, and most plastics. It will also stick to glass. It is suitable for holding correction weights on free running-flat components during high-quality plano polishing. It is also suitable for holding small mirrors in position on instruments where other means of support are difficult to arrange and where vibration or violent movement are unlikely.

The tape is resistant to water and mild detergents, has excellent ageing properties and resists UV light. The tape must be applied to clean dry surfaces initially, at a temperature of 8–32°C. Evo-Stik Cleaner 33 is recommended. Evo-Stik is supplied by Evode Ltd, Common Road, Stafford.

The Minnesota Mining and Manufacturing Co. Ltd

This company also makes a range of adhesives, some of which are of interest to the optical industry.

4.23 Growth of fungus on lenses

Fungal growth on optical components in the tropics has been a problem for many years, and there is no evidence that any of the thin film coatings are a deterrent to the mycelia attacking the glass. Attempts to clean glass surfaces after they have been contaminated by mould will fail, as the glass itself will have been damaged and prevention of the mould growth is the only effective cure to the problem.

Most fungal damage arises during storage, and silica-gel bags will reduce the ambient humidity and so make attack less likely. Sealing of the optical parts is usually impossible owing to the need to maintain relative movements of parts for focusing or rotation, as in theodolites or other surveying instruments.

The general approach is to introduce a fungicide which is relatively non-volatile but still adequately lethal. Phenylmercuric acetate and Merthiosal (MTS) are used with some success. All fungicides at present in use:

(1) are toxic to humans,
(2) cause condensation on glass surfaces,
(3) have a deleterious effect on optical cements,
(4) tend to produce dermatitis,
(5) corrode metal parts, particularly aluminium alloys.

4.24 Dermatitis

A large number of the chemicals and materials used in the optical industry involve a risk to the skin, and occasionally dermatitis occurs as a result of contact with liquids. A barrier cream such as Kerodex (supplied by Scientific Pharmaceuticals Ltd, 1 Eden Street, London NW1) should always be available for operators likely to be in contact with toxic materials.

5 Dioptric Substances

5.1 Introduction

The first part of this chapter explains the elementary physics associated with optics and, because this does not change, the words of Mr F. Twyman have been used where his explanation cannot be improved upon.

A dioptric substance is a transparent material through which rays of light pass in straight lines. Although for certain purposes coloured materials may be used, yet in the majority of cases the glass, or whatever the substance may be, should be colourless, clear, free from bubbles or opaque pieces, and free from any defect which will cause the deviation of rays of light passing through it.

At the time of Newton the optician required no more of the glass than that it should conform to the above-mentioned qualities, and although two main types were known (crown and flint) neither Newton nor any opticians made use of their different properties. Glass as made in different countries and for different purposes varied in chemical composition; for example, window glass and mirror glasses would be soda glasses, Venetian and Bohemian were traditional potash glasses, glass from which the domestic glassware known as 'crystal' was made owed its brilliancy to the presence of lead, but there was no preference for one type of glass over another for *optical* reasons, nor any use made of their various properties before the discovery, in the middle of the eighteenth century, that lenses could be made free from 'colour' by use of two different kinds of glass (achromatic object glasses). The differing properties of crown and flint glasses which made them suitable for combination in this way were their *refractive indices* and their *dispersive powers*.

In the middle of the eighteenth century, flint glass, or 'English crystal' as it was once called, was principally made in England. In making it, sand was replaced by powdered flint—hence the name. The denomination 'crown' glass had its origin in the primitive method of its fabrication. With the aid of a tube about 2 yards long, about 20 lb was taken from the furnace and blown into a bulb. The tube was rotated to flatten the bulb, and then, by means of hot glass applied to the latter, an iron rod was attached to it. The tube being removed, the glass was put in the furnace again and a constant rotation given to it by means of the iron handle, whereby the centrifugal force caused the flattened sphere to open and finally to take the form of a crown, ∩, which was gradually extended more and more to form a round disk. It was from the supposed resemblance of the form ∩ to a crown that the glass gained the name of 'crown' glass.

A description of the method of manufacturing optical glass, together with much information concerning its faults and testing, will be found in Chapter 14. The plant and processes used in the manufacture of glass table-ware, bottles, tubing, rod, and sheet, are described in *Making Glass*, from the Glass Manufacturers Federation, 19 Portland Place, London, W1.

CHROMATIC ABERRATION OF LENSES AND ITS CORRECTION

Before considering the different properties of glass, and other glass-like substances, it is necessary, or at least desirable, to give those who are not acquainted with even the elements of lens calculations, an idea of the properties which are required to be known for such calculations.

If we consider a simple lens, such as that shown in Fig. 5.1.1, we see that the lens may be looked upon as being made up of a large number of truncated prisms. We see further, that a beam of light passing through any one of these prisms is deviated, and owing to what is known as the 'dispersion' of the glass, the violet rays are deviated more than the red ones; the blue, green, and orange rays occupying intermediate positions.

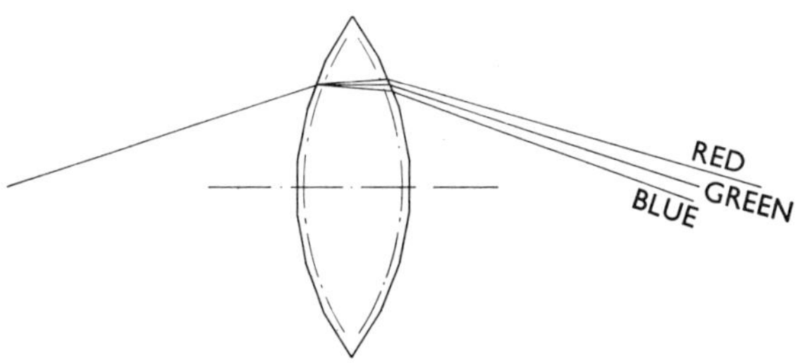

5.1.1. A simple lens represented as an assembly of small prisms.
The diagram explains the origin of chromatic aberration.

Such behaviour causes the rays of different colours to come to a focus at different distances from the lens. One will thus get either a reddish image with a blue surround, or a blue image with a ruddy surround according to the focusing of the telescope, and in either case the definition will be defective.

Sir Isaac Newton, deeming this defect unavoidable in refracting instruments, proceeded to develop the reflecting telescope.

It was left to a gentleman of Essex, Chester Moor Hall, in the early eighteenth century, to show that the chromatic aberration, as it is called, could be corrected by making use of two different kinds of glass. This idea was perfected in 1758 by the English optician, John Dollond, who is frequently credited with the original discovery of the technique.

It may be interesting to add that Hall based his discovery on the erroneous opinion that the eye is achromatic, that his first instrument was made as early as 1733, and that telescopes of his were certainly in use as late as 1827. (*Enc. Brit.*, 11th Ed., article 'Telescope.')

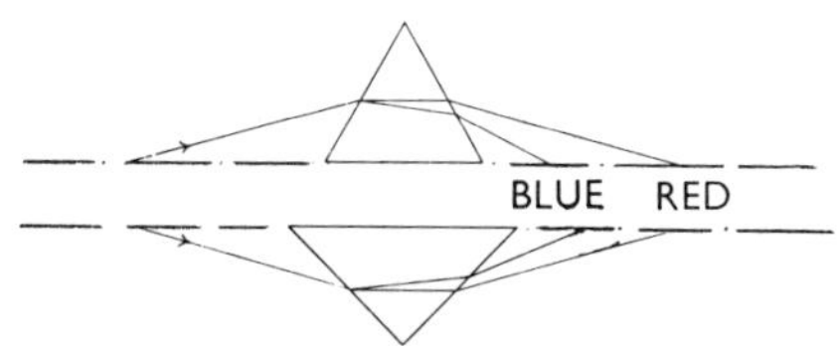

5.1.2. Deviation and dispersion.
The two prisms have equal deviation for the red, but unequal dispersion.

Optical glass may still be generally classified into two families, crown and flint. Crown glass has a dispersion considerably less than that of flint glass. If two prisms are taken, one crown and one flint, such that the *deviations* for the red light are the same, then the dispersions will differ, as shown in Fig. 5.1.2. Alternatively, if the flint prism is made weaker so that the *dispersions* are the same, then the deviations will be different (Fig. 5.1.3). It will be easily understood that by putting together two prisms of equal dispersion in opposition, the dispersions can be neutralized while yet leaving a certain amount of deviation.

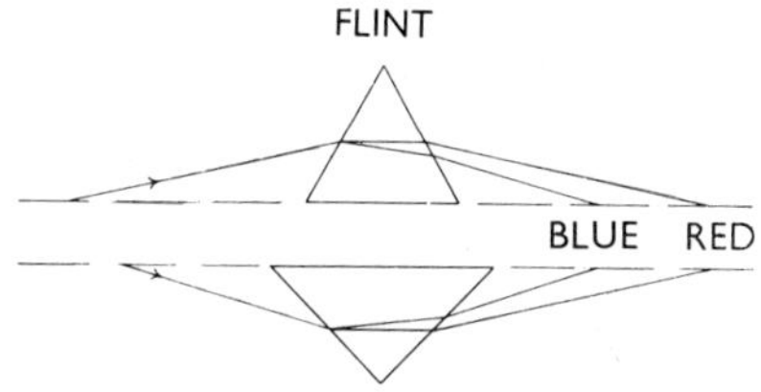

5.1.3. Deviation and dispersion.
The two prisms have equal dispersion, but unequal deviation.

Reverting to the consideration of a simple lens, it now becomes clear that an assembly of compound prisms, each of zero dispersion, would constitute a lens with zero dispersion also. A lens so formed is called achromatic, as the image it gives is, to a fair approximation, colourless.

Now the properties which will tell a lens-designer, at a single glance, whether two glasses are suitable for making a simple achromatic lens are n, the refractive index, and ν (the Greek letter pronounced *nu*), which is defined as the reciprocal of the dispersive power.

THE OPTICALLY IMPORTANT PROPERTIES OF DIOPTRIC SUBSTANCES

Refractive index

It was well known prior to 1617 that rays of light on passing to a different medium, say from air to water or from water to air, were deviated, but the exact law governing the degree of this refraction only became available to the world after its discovery by Snell shortly before 1617. It was first published and, indeed, first utilized by Descartes after the death of Snell.

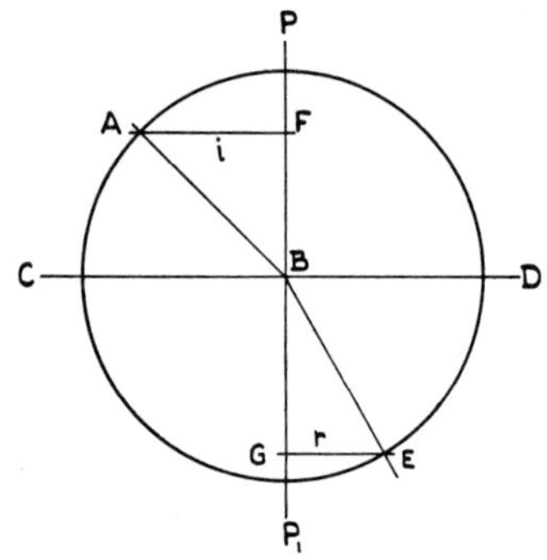

5.1.4. Snell's law.

Snell's law underlies the design of all optical instruments in which prisms or lenses are concerned. Referring to Fig. 5.1.4, if a ray of light AB falls on the surface CD of a piece of glass, or other transparent substance, it changes its direction to the course BE. Draw PP_1 perpendicular to the surface and passing through B and call i the angle ABP and r the angle P_1BE. Snell's law says that sin i/sin r is a constant, no matter at what angle the ray of light falls on the piece of glass. This constant is called the refractive index of the glass and is usually denoted by the letter n. The refractive index varies for different rays of the spectrum and it is usual to state the refractive index of a glass for the yellow sodium lines (present as dark lines in the solar spectrum) and to indicate it as n_D. There are two of these lines close together, known as D_1 and D_2, but the wavelength taken to define the refractive index is the mean between the two, and is represented by D.

The lens-computer requires to know not only n_D but what is known as the 'dispersive power'. In glass catalogues it is customary to give the inverse of this, and it is this which is usually designated by the Greek letter ν referred to above.

Spherical and chromatic aberrations

A lens with convex surfaces will form an image of a distant object, but a focus so formed, even with monochromatic light, can never be of good definition. In a positive lens, the rays which pass through the marginal part are brought to a shorter focus than those that go through the middle part of the lens (these are known as the paraxial rays), while in a negative lens the same is true of the virtual image (Fig. 5.1.5). This failure of a single lens with spherical surfaces to bring the rays from a point to a point focus is known as *spherical aberration*. As we have seen, there is another type of aberration, the *chromatic aberration*; since the refractive index of every dioptric substance is different for different colours, the rays which pass through the marginal part of a lens will not all be brought to a focus at the same point (Fig. 5.1.1), the blue rays being concentrated nearer to the lens than the red ones. Spherical as well as chromatic aberration can be corrected by combining a positive crown lens with a negative flint one.

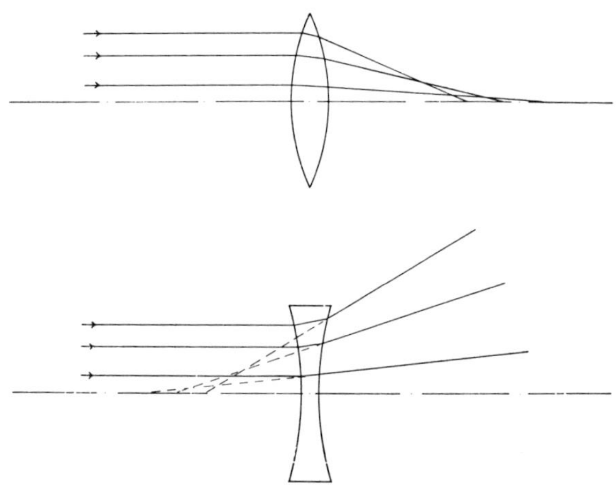

5.1.5. Spherical aberration.

The principle, first used to correct the *chromatic aberration* by Moor Hall and Dolland, is simple and has been outlined above. Since flint glass has a considerably greater dispersion than crown, it is possible by putting a crown and a flint lens together to annul the chromatic aberration while at the same time leaving the lens with sufficient refracting power to bring rays from a distant point to a focus. This invention was the starting point of the deliberate efforts to make glasses with different properties with a view to making more perfect lenses.

The two most important properties of a glass are, then, the refractive index and the ν value, where

$$\nu = \frac{n_D - 1}{n_F - n_C}$$

in which expression n_D, n_F, and n_C are the refractive indices for D, F, and C—three Fraunhofer lines in the solar spectrum in the yellow, bluish green, and red respectively. The refractive index gives an indication of the refracting power of a material, while ν is a measure of the dispersing properties of the glass, a large value indicating a small dispersion.

If the ν values of the crown and flint glasses are nearly the same, the curves of the lens will be extremely deep, and this is a disadvantage not only in manufacture (since deep curves are more expensive to make than shallow ones), but also because deep curves can introduce other types of aberrations.

The first aim, therefore (in making simple achromatic lenses), is to choose ν values as different as possible in the two glasses.

By combining the methods of correcting these two types of aberration, it is possible to correct the chromatic and spherical aberrations simultaneously and tables are available giving data for such lenses.

Absorption

The colour of a transparent substance is due to the absorption of some of the visible rays, which for practical purposes may be considered as extending from 4000 Å in the violet to 9000 Å in the red, although under very special circumstances it is possible to see radiations of considerably shorter wavelengths than 4000 Å. The transparency of ordinary glass extends only a little beyond this range both in the longer and shorter wavelengths, and since the ultra-violet and infrared regions of the spectrum are of great importance (the former for the analysis of metals and alloys and the latter for the analysis and estimation of organic substances), the optician must know how to work those substances which are transparent to these regions.

One of these, Iceland spar, has an importance of a character different from that of the others. Although it is transparent a good way into the ultra-violet, its importance in optical instruments depends not so much on that as on its having been the first known means of producing fully polarized light. It is true that nowadays we have Polaroid available in large sheets, but the perfection of polarization obtained with this material, although it has improved greatly in recent years, still falls short of that which can be obtained with Iceland spar, which therefore remains the only practical polarizing material for the manufacture of high-grade polarimeters.

Most of the other substances which are more transparent than glass in the ultra-violet and the infra-red were also originally obtained from natural sources, although they can now be produced in both larger and more perfect specimens in the laboratory.

Birefringence

For the production of polarized light of the highest purity, it is customary to insert into the light train a birefringent crystal whereby the incident beam is split into two beams which are polarized at right angles. All else being equal, a crystal having a higher birefringence than the others is to be preferred as wider fields of polarized light would become available and the optical path length of existing instruments could be reduced. In view of the comparatively recent progress made in the production of artificial crystals, it is to be expected that within a few years the monopoly held by Iceland spar (calcite) for well over a hundred years may be concluded and another crystal come into favour.

On looking through a cleft crystal of Iceland spar one cannot fail to see the double beams, one of which must be deleted and the other transmitted to give plane polarized light. Attempts to isolate one of the beams were not very successful until about 1830 when Nicol (1766–1851) made his famous discovery that, by taking a cleft crystal having a ratio of the long edge of the crystal to the short end face of about 3 to 1, and cutting it symmetrically from blunt corner to blunt corner, polishing the cut faces and cementing them together with Canada balsam, the ordinary ray was reflected from the balsam film and the extraordinary transmitted.

The field of light thus polarized does not extend over the entire aperture, as can be seen by tilting the prism when a blue band will be seen on the side, marking the limit of transmission of the extraordinary ray, and on the other side a sudden increase in intensity, marking the limit of reflection of the extraordinary ray and beyond which double images are seen. The internal field is about 14° and the external about 24°, but the middle line of the field is inclined to the longitudinal axis about 10° unsymmetrically and, for this reason, the end faces of the prisms are trimmed to various angles.

It was early realized that, where a smaller field would suffice, the length of prism could be reduced. Nicol himself in 1839 found that he might use shorter pieces, and in 1857 Foucault reduced the overall length to little more than the width by substituting a thin air film for the balsam film.

The chief disadvantage of the oblique-ended form of prism is that, on rotation of the prism, the transmitted beam moves around the axis of rotation. In conical light this is particularly objectionable and also introduces astigmatism, so a series of square ended prisms appeared such as Thompson's, the Glan-Foucault and the Soret, which latter have the advantage that they can be used over the region which would be cut out by a balsam film, viz. from 2950 Å to the limit of transmission of the spar at 2150 Å.

Iceland spar cleaves so easily that either nature or man has cleft it before it reaches the test room.

When cleft by nature the faces are sometimes covered with an opaque deposit which must be removed by further cleaving. This is done by scoring the four surfaces along the same cleavage plane or as nearly on the one plane as can be estimated with a sharp knife such as a safety razor blade set in a handle. This produces a strain on the cleavage plane around the crystal such that, with sufficiently small pieces, a continuation of the scoring may cause the obscure face to flake off leaving a transparent cleft face on the main block. If it does not part readily, then the knife is placed on and in a line with the scored line and given a smart tap with a hammer. If on looking through this face sufficient imperfections are seen, the crystal is rejected; if not, the opposite face is cleft in the same way and the crystal again inspected for imperfections. This process of cleaving off a face from a side is repeated till the crystal is rejected or passed.

The method of inspection is to project a concentrated light beam through the spar and look for reflections of this light from particles at all possible angles, using a magnifying lens if great clearness is desired, as in the polarizers of colorimeters. The apparatus consists of a box from which one side and one end has been removed. In the remaining end, a hole is made and a large angle condensing lens inserted. On the axis of this lens and outside the box is mounted a powerful light source, at such distance from the lens that the conjugate focus is within the box, and the divergent beam therefrom emerges through the space at the other end without infringing on the side of the box. The light which passes through the end of the box should be absorbed by black material, allowed to pass into another room, etc., so as to leave the room in darkness save for the unavoidable scatter of light from dust in the atmosphere and on the lens, which latter should be reduced by a mask between lens and observer's eye.

The spar is held in the hand in the narrowest part of the beam and tilted through all possible orientations while the eye scans the interior watching for any signs of a reflection.

It is worth while spending some time over this examination, especially if the spar is to be used for prisms in which the slightest blemish will be a cause for rejection (the half-shadow prism of a polarimeter for example). There is an optimum distance for the eye from the crystal since, on the one hand, the multiple reflections within the crystal flash intermittent beams into the eye which cause the pupil to close and, on the other hand, the 8-mm diameter of the pupil has to scan a very large part if not the whole of the envelope of a sphere of radius equal to the distance between crystal and eye, and the eye must be sighted more or less on that part of the crystal in which the defect exists.

DURABILITY

Durability is a measure of the resistance of a polished surface of glass to atmospheric attack. The varying humidity of the atmosphere results in corresponding variations in the water content of the glass surface, accompanied by alternations of deposition and evaporation of moisture. These alternations result in the separation of soluble constituents from the glass and the production of a surface film. In obtaining desired optical properties, either by developing a new type of glass or modifying an existing one, the final choice of batch should be decided by the durability.

SPECIFICATION CONTROL

The following information has, with permission, been extracted from the Chance-Pilkington Catalogue.

All optical glass produced by the alloy pot process is checked for refractive index and dispersions by measurement of samples from each pot, and each pot identified with a melt number.

All control measurements are made with a Hilger-Chance Precision Refractometer (§ 11.2).

A data sheet is despatched with each consignment of glass and the values quoted on this sheet, unless otherwise stated, refer to specimens cooled at the rate of 20°C per day in the sensitive region. The refractive properties of glass depend on its thermal history during annealing and it follows that small differences may occur between various specimens of the same melting, especially if these are annealed on different occasions.

The values for V [appearing as ν in earlier pages of this chapter] in the catalogue tables are quoted to the second decimal place: the same system is used for the melt data sheets but accuracy in the second decimal place is not implied where the latter are concerned. This will be understood if the liability to error of $\pm 0{\cdot}00002$ for the routine measurement of the mean dispersion is borne in mind.

The catalogue types can generally be reproduced having refractive indices within $\pm 0{\cdot}0010$ of the standard values.

The V values of repetitive meltings of a given type do not usually vary by more than plus or minus 0·5 per cent of the values given in the catalogue.

The refractive index of a glass is dependent upon its temperature and, for a number of glasses, values of the temperature coefficients of refractive index have been determined at certain wavelengths over the temperature range $20°\mathrm{C} \pm 5°\mathrm{C}$.

Light transmission

The purity of the raw materials used for optical-glass manufacture is strictly controlled to minimize absorption of light by the product. Special attention is also given to the purity of the refractory materials used in either tank construction or pot manufacture. One of the chief reasons for the use of alloy pots for melting special types containing rare earths and certain corrosive glasses of high barium content is that contamination by solution of the container is eliminated.

Transmission data expressed in terms of the extinction coefficients, alternatively known as absorption coefficient, per centimetre path length of all types for the ultra-violet, visible and infra-red portions of the spectrum are given in the data sheets.

Extinction coefficient K which is used in the catalogue, and does not include reflection losses, is defined by the expression $T - \mathrm{e}^{-kd}$, where T is the transmittance of a sample d cm thick.

The Thermodyne Test* was devised to give as close a simulation of the conditions of natural exposure as is possible for a test of reasonably short duration. This makes use of the most important features of atmospheric exposure, namely variation in temperature and humidity, and the conditions are capable of accurate control. The method does not give a numerical value but it is possible to group glasses by visual inspection.

Results obtained by the Thermodyne Test have been compared with the results of exposing glass samples to tropical atmospheres for prolonged periods, although it should be emphasized that the conditions of a natural exposure test are incapable of close repetition. On the basis of the Thermodyne Test the probable behaviour of a particular glass type can be stated under the worst conditions likely to be encountered in practice.

Refractive properties

In view of the development of modern optical system designing, the optical data quoted for each glass type has been considerably extended in comparison with previous optical catalogue issues. The index of refraction in air, n, and the

* See W. M. Hampton, *Proc. Phys. Soc.* (London) 1942, **54**, 400.

constringence value, V, for each type is duplicated with reference to two sets of spectral lines. The original basis is retained using the lines C, d, and F identified in the table below, and on this basis,

$$V_d = \frac{n_d - 1}{n_F - n_C}.$$

In addition, a new basis has been introduced, employing the lines C′, e and F′, also identified in the same table, and in this case

$$V_e = \frac{n_e - 1}{n_F{}' - n_C{}'}.$$

These refractive properties have been based on measurements made by the National Physical Laboratory, Teddington, covering ultra-violet, visible and infra-red transmitting ranges for each glass type using approximately thirty spectral lines. Those in the visible region are identified below with their wavelengths expressed in Ångstrom Units (1 A.U. $\equiv 10^{-1}$ millimicrons $\equiv 10^{-4}$ micrometers $\equiv 10^{-8}$ cm).

Colour	*Hydrogen*	*Helium*	*Cadmium*	*Mercury*
Red	C 6563	—	C′ 6439	—
Yellow	—	d 5876	—	—
Green	—	—	—	e 5461
Blue	F 4861	—	F′ 4800	g 4358
Violet	—	—	—	h 4047

The control of physical properties is determined by the method of manufacture.

All optical glass produced by the continuous process is checked at frequent intervals for refractive index and V value and periods segregated so that the refractive index does not vary more than $\pm 0{\cdot}0002$ and the V value not more than $\pm 0{\cdot}1$ of the values quoted on the data sheet. Glass made during these periods is released on an optical release number. It is possible, therefore, that within a batch of glass supplied in a particular release number, the index and V value can vary by the amount quoted.

Glass types

The glass types are arranged in groups and described as 'crowns' or 'flints' with a generally accepted, but arbitrary, division at $V_d = 55$ for indices (n_d) below 1·6 and $V_d = 50$ for higher indices. The type reference number on each data sheet consists of six digits; the first three indicate the excess of the refractive index (n_d) over 1 and the remaining figures define the V_d value.

Types made in alloy pots are identified in the tables by letter A. Types made by the continuous process are identified in the tables by the letter T.

Resistance to acid attack

The ability of a glass to withstand chemical attack in grinding and polishing operations is more reliably assessed from an acid durability test than from the Thermodyne Test. To get a quantitative measurement is extremely difficult and is the subject of international research. A visually qualitative grading of durability has been based on the exposure of polished samples to 0·5N nitric acid. The nature and extent of the surface corrosion is graded using a three-digit code.

The first digit indicates the time required for signs of attack to be noticeable. The second digit indicates the severity of attack, whilst the third digit or term describes the particular form of attack:

First term

1 No signs of attack observed in 100 hours.
2 Signs of attack observed between 10 hours and 100 hours.
3 Signs of attack observed between 1 hour and 10 hours.
4 Signs of attack observed between 0·1 hour and 1 hour.
5 Signs of attack observed between 0·01 hour and 0·1 hour.
6 Signs of attack observed between 0·001 hour and 0·01 hour.
7 Signs of attack observed in less than 0·001 hour.

Second term

·1 Mild attack
·2 Light attack
·3 Medium attack
·4 Heavy attack
·5 Very heavy attack.

Third term

a	Stain	Stain apparent when viewed by reflection.
b	Acid polish	Material removed uniformly leaving a high degree of polish.
c	Acid etch	Material removed unevenly leaving a rough pitted surface.

Example. A glass developing an acid etched surface in 6 minutes with heavy attack would be rated as 5.4.c. Experience of optical processing indicates that a glass can be satisfactorily processed if its first term rating is 6 or better.

Thermal expansion

Values of the coefficient of linear thermal expansion, α, for the temperature range 10°–100°C are given in tables for most types of glass.

Resistance to atmospheric attack

Several methods have been devised for testing glass surfaces for resistance to atmospheric attack. Some of these are open to criticism because the conditions of test bear little relation to conditions of natural exposure, examples being the Autoclave Test using super-heated steam and the Solubility Test using a powdered sample in contact with various chemical reagents. Two other tests, the Iodoeosin Test and the Electrical Conductivity Test, give numerical classifications, but they serve only for those glasses whose resistance depends on their alkali content. Other glasses, notably those containing a high proportion of barium, would appear from these tests to be more resistant to atmospheric attack than is found by the experience of actual use.

5.2 Glass compositions

The art of glass manufacture depends on melting together a number of substances which, on cooling rapidly, will set to give a clear transparent solid. During cooling, glass viscosity increases continuously and there is no definite temperature at which the glass can be said to solidify. Also, on heating from room temperature, it will soften more and more without any break in the temperature curve, and there is no sharp change from a solid state. The glassy state is termed 'vitreous' and glasses are often called supercooled liquids.

The most important glass former is silica; others are germanium dioxide, beryllium fluoride, and phosphorus pentoxide. What makes a substance a glass former is a question of relative atomic size of the atoms making up the substance. Firstly the cation (silicon, phosphorus, etc.) must be large enough to space around itself a number of anions (oxygen, fluorine, etc.) to permit the formation of a three-dimensional lattice structure: and secondly, the anion must be small enough in relation to the cation to fit into such a structure.

The high stability of silica glasses is due to the ratio of the number of silicon and oxygen atoms. Each silicon atom can group around itself four oxygen atoms in the shape of a regular tetrahedron, while each oxygen atom can hold two silicon atoms.

Before 1939, sand supplies for the manufacture of British optical glass came almost exclusively from Lippe & Hohenbockaer, in Germany, but since 1941, sand with the essential high purity as regards iron content for colourless glass, has been obtained from deposits in the Western Highlands at Loch Aline, although some is imported from Belgium. The details of typical glass compositions in Table 5.2.1 (p. 83) have been supplied by Chance-Pilkington.

The barium crown and barium flint glasses (D.B.C., S.B.F., S.B.C., range) move very rapidly from the liquid to the 'solid' state and cause handling problems. These glasses have compositions not suited to the usual continuous process (Fig. 5.2.1), where glass is melted in a furnace and passes down a platinum tube and is extruded, or gobs of glass are cut off for automatic moulding.

The barium glasses react with refractories and cause damage, also the temperature–viscosity characteristics are incompatible, so a special electric-induction melting (E.M.) process has been developed for their separate production.

Because of the particular structure of glass, care must be taken in cooling from the high temperature essential to making the glass. Any temperature gradients that exist are 'frozen' into the glass, which, instead of being optically homogeneous and isotropic, becomes optically anisotropic—behaving

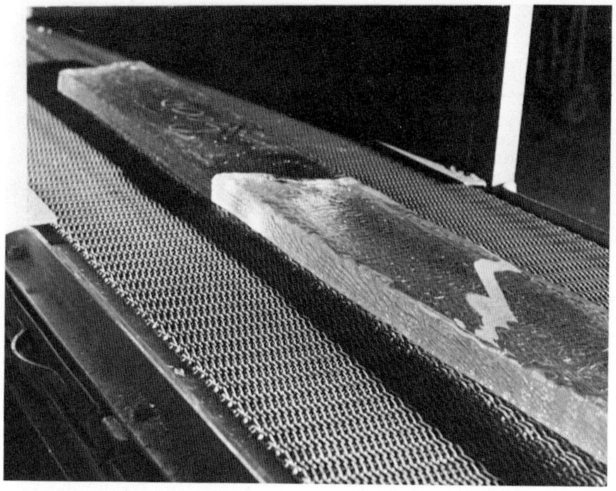

△ **5.2.1. Optical glass as extruded in a continuous process. Passing through the lehr.**

5.2.2. ▷
A pan loaded for the annealing kiln.

5.2.3. ▷
The pan being placed in the annealing kiln.

as a uniaxial crystal, with different optical qualities in different directions.

Transparent substances which possess the same optical properties in all directions are said to be isotropic. Transparent substances which are not isotropic, like the majority of crystals, are said to be anisotropic. Birefringence is the property of doubling the rays of light possessed by all anisotropic substances. Light does not travel at exactly the same speed in the doubled rays and these can separate from one another, and so give rise to two images.

Annealing is therefore an essential process after moulding, as well as after initial forming of slabs from the furnace. (Figs 5.2.2 and 5.2.3.) Sufficiently slow cooling reduces the birefringent effects to a negligible amount, but the rate of cooling can also be made to vary the refractive index of the glass. If a piece of glass is uniform in temperature, while being held prior to the annealing operation, the only index changes introduced into it will be the plus or minus variations due to the temperature gradient during cooling and these will probably be small, perhaps not more than a few units in the fifth decimal place. If a small piece of glass is held at, say, 550°C until the index has approached the equilibrium and is then cooled slowly from this temperature, it will have a general refractive index slightly higher than the equilibrium index for 550°C together with the

Table 5.2.1. Chance-Pilkington glass compositions (main constituents only)

	BSC517642	*MBC569561*	*DBC620603*	*SBC651586*	*SBC697562*
SiO_2	70	50	31	22	5
B_2O_3	11	6	18	23	40
BaO	1	20	49	45	—
PbO	—	3	—	—	—
Na_2O	9	4	—	—	—
K_2O	7	5	—	—	—
Al_2O_3	—	—	2	—	1
CaO	1	—	—	—	2
La_2O_3	—	—	—	11	48
ZnO	—	12	—	—	—
SrO	—	—	—	—	4

	LF579411	*DF620362*	*DEDF748278*	*BF670477*	*DEDF927210*
SiO_2	54	45	31	29	20
PbO	34	46	65	9	80
Na_2O	8	5	—	—	—
K_2O	4	4	3	—	—
BaO	—	—	—	30	—
CaO	—	—	—	9	—
ZnO	—	—	—	12	—
B_2O_3	—	—	—	10	—

	TF530512	*BORF614439*	*SBF744447*
SiO_2	44	8	10
B_2O_3	24	39	27
PbO	4	34	14
Al_2O_3	7	10	—
ZNO	—	5	1
Na_2O	6	1	—
K_2O	—	2	—
BaO	—	—	4
ZrO_2	—	—	9
La_2O_3	—	—	29
CaO	—	—	6
Sb_2O_3	14	—	—

plus or minus variations due to the gradient. If it is held at some other temperature, say 530°C, it will have an index slightly higher than the equilibrium index for 540°C together with the plus or minus variations. However, the two specimens of glass will have different refractive indices although their double-refraction pattern may be the same. If, therefore, both of these temperatures exist during the annealing operation—that is, if a temperature difference exists during the soaking period—then different parts of the finished slab, or lehr, or kiln of mouldings will have different refractive indices.

The following annealing schedules are typical for Chance-Pilkington optical glass.

Typical Annealing Schedules
BSC 510644

	Hours
(1) Heat to 550°C	17 (approx.)
(2) Soak at 550°C	12 (approx.)
(3) Reduce temperature from 550°C to 450°C at 20° per day	120
(4) Free cool from 450°C to 50°C	44
	193

HC 519604	Hours
(1) Heat to 530°	16 (approx.)
(2) Soak at 530°C	12 (approx.)
(3) Reduce temperature from 530°C to 430°C at 20° per day	120
(4) Free cool from 430°C to 50°C	23
	171

Binocular prisms, where low cost is of prime importance and minor refractive index variations unimportant, would be annealed 'on line' immediately after moulding from glass gobs flowing from the furnace. (Fig. 5.2.4.)

High-quality optical components would be separately annealed in an annealing kiln. The optical glass, in bar form, would be extruded from the furnace and broken. (Fig. 5.2.5.)

◁ **5.2.4. Binocular prism mouldings passing through the annealing lehr and receiving a sample inspection.**

▽ **5.2.5. Optical glass (MBC 569) passing from the furnace through an extrusion die on a Cane machine and being broken into convenient lengths. The bar shown is 50 mm thick.**

BRITISH OPTICAL GLASS

The author is indebted to Dr W. M. Hampton (former Director of Pilkington Optical Division and Chance Bros, Works Manager) and to Focal Press Ltd for permission to reprint extracts from the article 'British Optical Glass' published in the *Focal Encyclopedia of Photography*.

Optical glass is the raw material for all kinds of lenses and prisms for use in scientific instruments. It is an essential constituent of telescopes, microscopes, and similar optical instruments—and all types of cameras. In spite of its importance, the quantities involved are relatively small, except during a war when the demand for optical glass increases very considerably.

The essential difference between optical glass and all other forms of glass is that it must be optically homogeneous. That is to say, it must have the same optical and physical properties in all directions. In fact, first-quality optical glass is probably the most homogeneous material known, the variations throughout the mass normally being only detectable by the most elaborate and sensitive instruments.

The optics of glass

There are today many types of optical glass—largely because its applications call for various specific optical properties, in particular with regard to refractive index and dispersion.

The obvious example is a lens—essentially a curved piece of glass which redirects the light falling on it into predetermined directions. The ability of a piece of glass to change the direction (in fact, the speed) of light is specified by the refractive index, usually designated by the symbol n.

This refractive index depends on the colour of the light. The mean refractive index is usually designated by n_D and the refractive index for other wavelengths by the same symbol with varying subscripts.

With some glasses the refractive index varies more for different wavelengths of light than for others. This variation is loosely referred to as the dispersive power. More specifically, the amount of difference of refractive index for change of colour is conveniently defined by the difference in refractive index for the C and F lines of the solar spectrum, i.e., between red and blue light. The ratio of this difference to the mean refractive index, usually designated by V or the Greek letter ν, is the second quantity (in addition to the mean index) which the optical designer requires. It is important in the colour correction of lens systems, since a combination of glasses of low index but light V value and high index but low V value can yield a lens without focus differences between blue and red light.

Modern lens systems of large aperture and short focal length call for glasses of high refractive index to keep the lens curvatures within practical limits. Thus the design of the approved lenses is dependent on having new types of glass with a high refractive index and a variety of V values. As such glasses became available, designers applied them at first to known lens constructions with the object of using shallower curvatures. More recently, the basic conception of lens design has changed and it is now possible to use these glasses in more fundamental ways. Many modern lenses would be impossible without the use of some of the latest glasses to which more detailed reference is made later. In particular, the lenses needed for colour photography, where uniformity of illumination over a whole field is required, would be impossible with the older simple types of flint and crown glass.

It is therefore convenient to designate all types of optical glass in terms of n_D and V, and progress in optical design is closely linked with the development of glasses with new relationships between the two parameters.

The technology of glass

In the course of the history of the manufacture of optical glass, three processes have been used.

The 'Classical process', in general use until about 1935, involves melting the raw materials, (essentially oxides, nitrates or carbonates of the various elements), in a pot in a furnace. The pots were hand-made from various clays selected for their resistance to solution by the molten glass. The glass was melted usually at about 1300°–1400°C. The pots contained something of the order of a ton of glass, according to type, and the melting process took approximately 48 hours. At this stage the glass would be reasonably free from small bubbles, but would not be homogenous. It was at this stage that the essential feature which marks optical glass from other types was introduced, namely, a process of stirring the molten glass in order to remove these variations in composition. While the stirring process was taking place the temperature of the furnace, and the glass, was progressively reduced until the glass became sufficiently viscous for no more flow to take place, since this was likely to vitiate the homogeneity which had already been achieved. At this stage, the pot with its contents was removed from the furnace, cooled slowly and, when completely cold, broken into pieces of convenient size. These pieces were then reheated to a temperature at which the glass was sufficiently soft to be manipulated into rectangular blocks for subsequent polishing and examination. After cutting away flaws in the plates, the remaining portion was available for manufacture by the optician into lenses and prisms. The process was wasteful; usually only 25 per cent of the contents of the pot was acceptable.

With the extension of the types of optical glass using more unusual materials, and with the availability of more precise control equipment, a process was developed for making glass of optical quality in relatively small quantities. This used platinum pots and platinum stirring equipment in electrically heated furnaces. The process was not economical, however, for quantities of more than about 100 lb, and was therefore restricted to glasses which employed expensive raw materials but were wanted in relatively small quantities.

The third, and most modern, process is one of continuous manufacture, as distinct from the intermittent process, utilizing pots. Here the raw materials are fed continuously into a chamber. The glass flows from one end of this to separate compartments of the furnace. There bubbles are removed and the flowing glass is stirred continuously before being produced either in strip or block form. Through precise temperature control in the various sections of the continuous process, it is now possible to produce optical glass virtually free from bubbles and virtually homogeneous over long periods of time. However, the continuous process is only suitable for the manufacture of glasses wanted in considerable quantity.

British glass history

The position in Britain at the beginning of the nineteenth century is indicated by a statement quoted in the Chance Memorial Lecture to the Society of Chemical Industry, 1947:

> Out of every 100 lb of goblet bottoms, carafes and such-like, it is difficult to find a piece good enough to make a 3-inch objective.

In fact, at that period, the only source of glass for lenses was by selection from odd pieces of glass made primarily for other purposes. (Fig. 5.2.6.)

The fundamental invention which underlies the whole process of optical glass was made by Pierre Guinand, a Swiss woodworker who specialized in

5.2.6. A Newcastle wine glass (1750) made from lead crystal and of the type that would have been used for lens material. The base of the wine glass was suitable as a lens moulding.

clock cases and who was interested in telescopes. He built his own furnace in 1775, but only arrived at the process of stirring optical glass in 1805, after moving his factory to Benediktbeuern in Bavaria. This move was in order that he could cooperate with the famous astronomer, Fraunhofer, who was obviously interested in the production of glass for telescopes. There is no direct record of the quality of the material he produced, but it must have been a marked improvement on anything that had been available previously.

In 1827 the position in Britain was considered to be sufficiently serious—since no one had any knowledge of Guinand's process, but only of the results produced by it—for the Royal Institution to instruct Faraday to investigate methods for making glass of the required quality. Some record of these researches is kept in the archives of the Royal Institution and it is interesting to note that Faraday used a platinum crucible—one of the marked features of current manufacture. Though expensive, this material is remarkably resistant to attack by most types of molten glass. But the importance of this development was apparently not appreciated, since there is little evidence that further use was made of platinum for almost a century after Faraday's work.

The next stage in the development was due to a Frenchman, Bontemps, who had been associated with Guinand's son. He carried out further experimental work to try to improve the quality still further in his own factory between the years 1847 and 1848. However, he became involved in French politics, and in 1848 he considered it desirable to leave France and joined Chance Brothers of Smethwick.

The board minutes of that company record that Bontemps was employed in order to '... devote his exclusive services to the firm, to superintend the Coloured and Ornamental Departments, generally to advise and assist in the glass business and to carry out the manufacture of optical glass in accordance with Lucas Chance's patent of 1838'.

Lucas Chance was one of the partners of the company, and his patent, No. 7596, covered the essentials of the Guinand process, presumably as brought to the company by Bontemps. It is of interest to note that, at that date, patents were not made public, and it was not until 1857 (when all the previous patents were published) that the information contained in it ceased to be secret.

This venture was so successful that at the Great Exhibition in London, in 1851, Chance Brothers showed a flint glass disk 29 inches in diameter, $2\frac{1}{4}$ inches thick and weighing about 200 lb, of a quality which had not hitherto been seen in this country. They showed the same flint disk again in 1855 at the Paris Exhibition and also a corresponding crown disk, both of which were purchased by the French Government to make a large astronomical telescope.

Starting in 1834, the Rev. Vernon Harcourt, one of the founders of the British Association and a gifted amateur scientist, carried out investigations in the melting of various types of optical glass. Later he worked in conjunction with Sir George Stokes, President of the Royal Society, and the latter read a paper before the British Association in 1871 describing briefly their joint experimental work. Some 166 meltings were made, each probably involving several preliminary experiments, and the resultant glasses were measured by Stokes. Harcourt investigated the properties of most of the oxides which are now used in optical glass manufacture, and probably many of his still unpublished papers would repay investigation.

The next step forward came from Germany. A young physicist, Abbe, started a programme of research at Jena University to try to improve the quality of microscopes. By 1870, he had suggested the necessity for new types of glass, particularly low-dispersion flints and high-dispersion crowns, to cure chromatic aberration. Abbe published a report in 1876 pointing out that the problem was one for the glassmakers. A German glassmaker, Otto Schott, saw the report and offered to collaborate. Their stated aims were:

(*a*) to provide crown and flint glasses with strictly proportional dispersion throughout the spectrum in order to reduce chromatic aberration; and
(*b*) to obtain a higher degree of diversity between refractive index and dispersion.

This collaboration was brilliantly successful, and by 1886 the firm of Schott at Jena published a catalogue of forty-four glasses, of which nineteen were of entirely new composition. The work was not only successful scientifically, but also commercially, with the result that the manufacture of

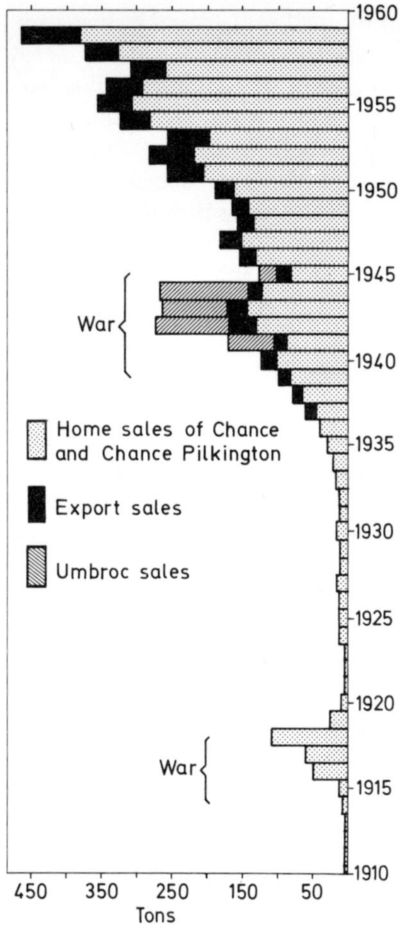

5.2.7. Sales of optical glass by weight. The histogram shows the growing production of British optical glass and its special stimulation during the war years.
A. Home sales of Chance and Chance-Pilkington.
B. Export sales.
C. Umbroc sales.
The figures relate to optical glass only, as distinct from ophthalmic glass, the total output of which is considerably greater.

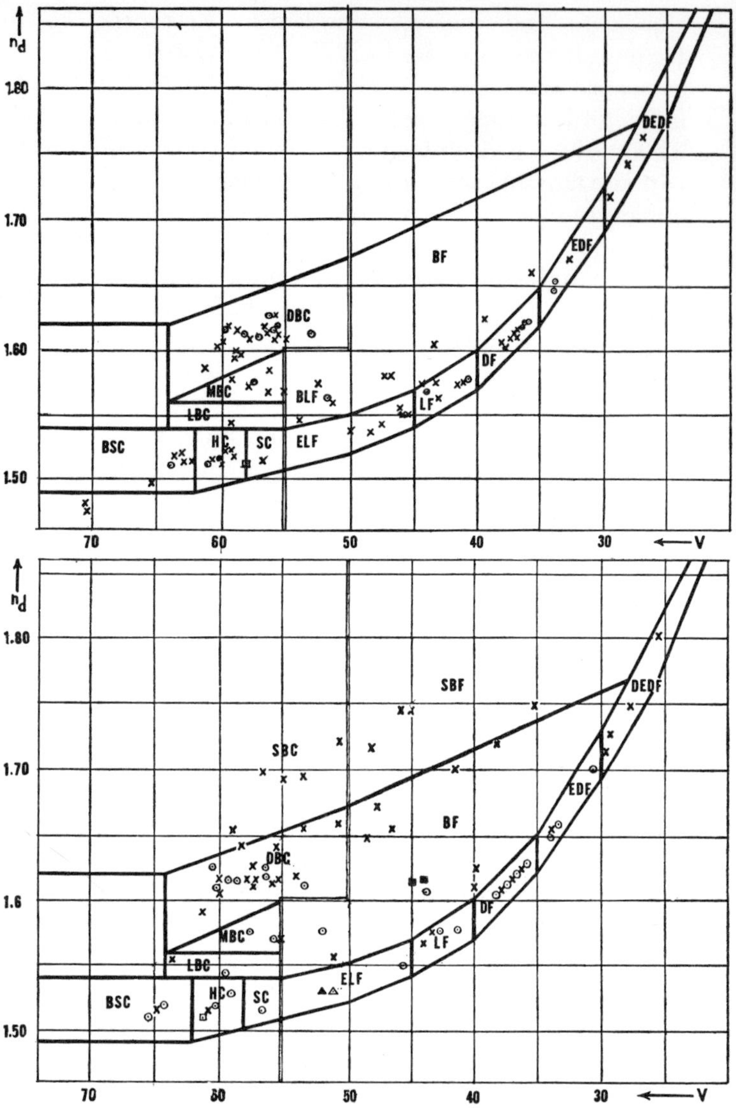

5.2.8. The growth of glass varieties. The two graphs indicate the glass types listed in the Chance 1926 catalogue (*top*) and the Chance-Pilkington 1960 catalogue. The most significant increase is in the number of new varieties among the special barium crown and flint types. In both diagrams, the glass types are located on the graph by plotting refractive index against V value. Principal types are marked with an open circle, except for zinc crown (open squares) and telescope flint (open triangles). Minor types are shown by crosses, except for telescope flint (solid triangles) and borate flint (solid squares). The actual glass types shown are: BSC, borosilicate crown; HC, hard crown; SC, soft crown; LBC light barium crown; MBC, medium barium crown; DBC, dense barium crown; ELF, extra light flint; BLF, barium light flint; LF, light flint; DF, dense flint; EDF, extra dense flint; DEDF, double extra dense flint; BF, barium flint; SBC, special barium crown; SBF, special barium flint.

optical glass became virtually a German monopoly until the 1914–18 war.

A further step forward in the development and the manufacture of varieties of optical glass of high quality occurred in 1940, when G. W. Morey in the United States published a detailed investigation of the use of various rare earth oxides, notably lanthana and thoria. These glasses (for which the process of melting in electrical furnaces in platinum pots was particularly suitable) opened up new possibilities for lens designers. The latest stage in research and development came in 1948 when the continuous process for the manufacture of optical glass, as distinct from the previous discontinuous methods, were developed in America by Corning, the Pittsburgh Plate Glass Co., and Bausch & Lomb.

New materials

A brief mention has already been made of the work of Faraday and Vernon Harcourt and of the tremendous surge forward in the variety and quality of glass available that came as a result of the collaboration of Abbe and Schott. As the records indicate, the fundamental work was carried out in England, but was not exploited as fully as it might have been.

But the researches started in 1935 had a very different result. With the support of the British Admiralty, and by the institution of committees representative of all the interests involved, intensive investigations were carried out not only into stirring but also into the supply of new materials for pot manufacture. These improved pots were the result of combining the properties of various natural clays with additional refractory crystalline materials.

The work also included the determination of the viscosity temperature curves of various glasses, which information was needed for the implementation of the results on stirring. Investigations into the production of moulded pieces of glass of satisfactory optical quality were also undertaken, since the process of reheating and shaping could upset the homogeneity of the unworked glass and special heat treatments were necessary. In order to restore the quality, researches were made on annealing (which is the process of slow cooling of optical glass in order to minimize the effect of temperature gradients across the material during the cooling process) and also on the methods of determining the durability of optical glass under all possible conditions of use.

From 1945 onwards, research was diverted towards the production of glasses with new combinations of refractive index and dispersion, using the rare earth materials which enabled, in particular, glasses of high refractive index and relatively low dispersion to be produced.

New problems arose from the use of atomic energy. It was soon found that most types of optical glass discoloured rapidly on exposure to gamma radiation and it became necessary to have new types of glass which were not subject to this darkening. A series of stabilized glasses was produced covering the whole range from crowns to flints. These have found a rapid application, both for lenses for photographing objects under conditions of high gamma energies, and also as windows to enable operators to view the 'hot cell' or chamber without harm.

Atomic energy production on a commercial scale also meant the development of modified methods needed to be of relatively immense size and of first-rate optical quality. It is now commercially practicable to produce windows of the order of 6 feet square by 1 foot thick virtually free from bubbles. These have adequate 'optical' homogeneity, but also resist darkening under exposure to radiation.

5.3 Infra-red glass

Barr and Stroud Ltd manufacture calcium-aluminate glasses which have outstanding optical and mechanical properties suitable for infra-red applications.

Glass types B.S. 37A and B.S. 39B exhibit a high transmission level to 4·0 micrometres wavelength without absorption losses. The absorption peak at 2·7–3·0 micrometres has been eliminated by removing the water during manufacture. Slightly more bubbles are formed in these glasses than in normal optical glass, owing to the processes which remove the water, but they do not usually cause any optical problems. Transmission ranges for these glasses are shown in Fig. 5.3.1.

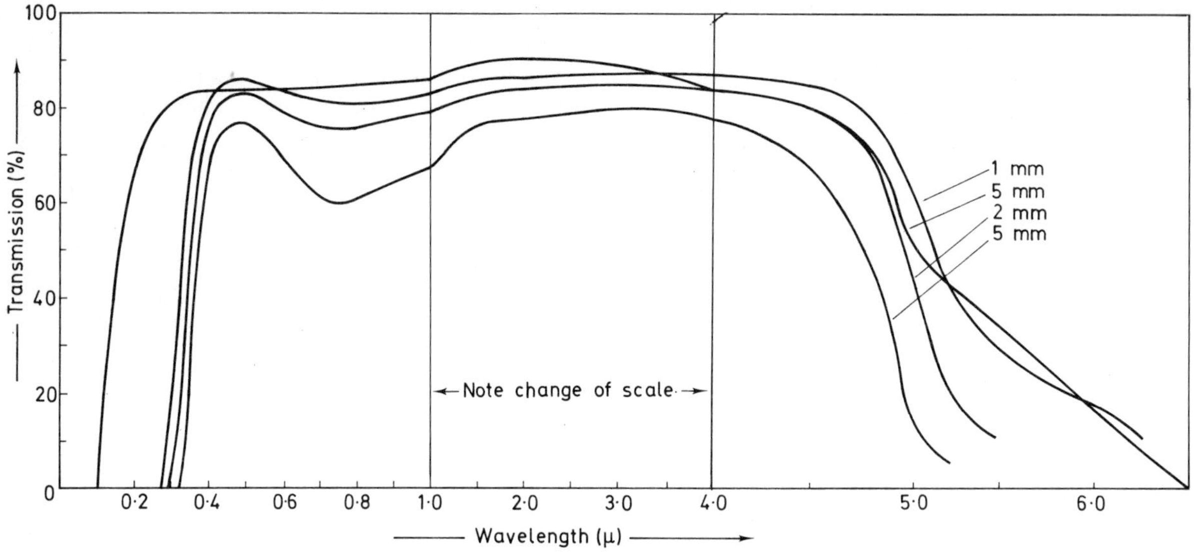

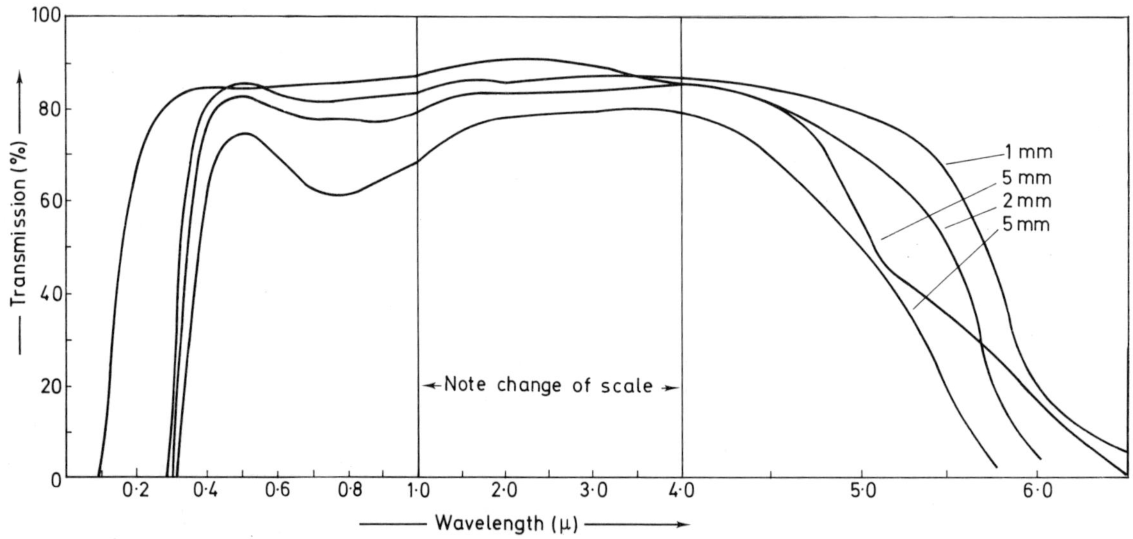

5.3.1. Calcium aluminate glasses BS 37A (*upper*) and BS 39B
The glasses have a useful transmission range from 0·3 to 5·5 µm, with complete freedom from water absorption at 3 µm. They are transparent in the visible and are mechanically strong and hard.

5.4 Rare-element optical glasses

During the period 1920 to 1933, the Eastman Kodak Company carried out research into materials which might provide glass with a very high refractive index and low dispersive power—in particular high-index crown glasses. To this end, all high atomic number cations were chosen for systematic study in silicate, borate, and phosphate glasses.

By 1933, the work had progressed to the point where silicon and phosphorus were discarded as glass-forming elements. Boric oxide had by this time proved to be the best fluxing agent. Oxides of elements, such as lanthanum and thorium found in the rare earths, and columbium, tantalum, tungsten, titanium, zirconium, and strontium, were used in major portions up to 80 per cent by weight with or without the usual barium, zinc, magnesium, and aluminium.

In 1934, samples of unusual glasses in the region of n_D of about 1·85 and a V of 43·0 were in existence with their properties measured.

The oxides of tantalum, thorium, and tungsten are soluble in the lanthanum–borate–base glass in amounts up to 35 per cent. Borate glasses are very stable, fairly hard and amenable to optical shop practices.

The first commercial production glass was delivered in June 1939. All Kodak Rare-element glasses now in production, except EK 325 and EK 497, contain thorium, of which the radioactivity may in some circumstances be a disadvantage.

The Kodak Research Laboratories have continued their study into new glasses which may be called super-flints, and refractive index values as low as 1·38 with V value 100 have been achieved. The basic materials used are silica, titanium oxide, and fluorine.

Kodak Rare-element optical glasses have been recognized as one of the most important achievements in the field of optics. These glasses, made with compounds of such elements as tantalum, zirconium, and thorium instead of the traditional silica, have higher refractive indices for given dispersions that most other optical glasses. With them, therefore, it has been possible to design lenses and other optical components with qualities never before attainable.

Each of these glasses is 'engineered' for a specific set of properties and, of necessity, is produced in relatively small batches. Many new glassmaking techniques had to be developed and many of the old techniques had to be refined. For example, instead of the familiar ceramic pots for the melts, platinum crucibles are used to avoid impurities and assure uniformity from batch to batch.

All types of Kodak Rare-element glasses are nominally clear for wavelengths from 0·400 to 2·0 micrometres. Overall visual absorption does not exceed 2 per cent per centimetre.

All types of Kodak Rare-element glasses will show a slight yellowish tinge in thick sections, indicating absorption in the blue. To specify this property, the colour index is defined in accordance with ASA Specification PH3.37-1961.

Standard Glasses

Type	*ASA Std.*
EK–110	45–0–0
EK–210	50–0–0
EK–310	40–0–0
EK–320	40–0–0
EK–325	50–0–0
EK–330	75–0–0
EK–430	80–0–0
EK–450	90–0–0

Special Order Glasses

EK–230	80–0–15
EK–497	45–0–0
EK–448	90–0–0

All Kodak Rare-element glasses, except type EK-325 (a standard glass) and type EK-497 (a special order glass), contain thorium oxide (ThO_2), as indicated:

Standard Glasses

Type	*ThO_2 by Weight*
EK–110	20%
EK–210	16%
EK–310	15%
EK–320	12%
EK–325	0%
EK–330	12%
EK–430	9%
EK–150	5%

Special Order Glasses

EK–230	20%
EK–497	0%
EK–448	28%

Because thorium oxide is radioactive, the *storage and use* of these glasses is subject to *controls and regulations at the Federal and State levels.*

The following are the catalogue values for the refractive indices of Kodak glasses in the infrared (λ = 10 140 Å) and near ultra-violet (λ = 3650 Å):

Glass	*10140* (I.R.)	*3650* (U.V.)
EK–110	1·68279	1·72742
EK–210	1·71820	1·77027
EK–230	1·73670	1·79046
EK–330	1·73778	1·79578
EK–310	1·72781	1·78643
EK–320	1·72713	1·78659
EK–325	1·72685	1·78625
EK–430	1·75821	1·82152
EK–415	1·72646	1·78862
EK–450	1·78363	1·85416
EK–448	1·85812	1·93602

Calculation of index values

Table 5.4.1 gives the index value for the sodium D line (n_D) for each glass type.

Index values for other Fraunhofer-line wavelengths can be calculated. For example, to calculate the n_C value for type EK-110 glass, proceed as follows, using values given in the table:

$$n_D = 1{\cdot}6968 \text{ (from table)}$$
$$n_F - n_D = 0{\cdot}008\ 74 \text{ (from table)}$$
$$n_F = 0{\cdot}008\ 74 + 1{\cdot}6968 = 1{\cdot}705\ 54,$$
$$n_F - n_C = 0{\cdot}012\ 41 \text{ (from table)}$$
$$n_C = 1{\cdot}705\ 54 - 0{\cdot}012\ 41 = 1{\cdot}693\ 13.$$

The value v is the partial dispersion ratio. It can be calculated from the formula:

$$v = \frac{n_D - 1}{n_F - n_C}$$

Table 5.4.1. Optical Values of KODAK Rare-Element Optical Glasses

Standard Glasses

EK Type	*Index* n_D	v	n_F-n_C	n_D-$n_{A'}$	*Partial Dispersion* n_e-n_C	n_F-n_D	n_g-n_F	n_h-n_g	*Specific gravity*
EK–110	1·6968	56·15	0·012 41	0·008 03	0·006 75	0·008 74	0·006 73	0·005 59	4·1
EK–210	1·7340	51·01	0·014 39	0·009 18	0·007 78	0·010 16	0·007 91	0·006 60	4·4
EK–310	1·7450	46·42	0·016 05	0·010 09	0·008 63	0·011 38	0·009 00	0·007 63	4·5
EK–320	1·7445	45·82	0·016 25	0·010 18	0·008 73	0·011 53	0·009 14	0·007 77	4·5
EK–325	1·7445	45·56	0·016 34	0·010 33	0·008 80	0·011 57	0·009 11	0·007 70	4·5
EK–330	1·7551	47·19	0·016 00	0·010 11	0·008 63	0·011 33	0·008 89	0·007 48	4·7
EK–430	1·7767	44·69	0·017 38	0·010 88	0·009 34	0·012 32	0·009 77	0·008 27	4·6
EK–450	1·8037	41·77	0·019 24	0·011 90	0·010 29	0·013 68	0·010 94	0·009 35	4·6

Special Order Glasses

EK Type	*Index* n_D	v	n_F-n_C	n_D-$n_{A'}$	*Partial Dispersion* n_e-n_C	n_F-n_D	n_g-n_F	n_h-n_g	*Specific gravity*
EK–230	1·7530	50·6	0·014 88	0·009 46	0·008 04	0·010 51	0·008 20	0·006 87	4·6
EK–497	1·6355	43·9	0·014 49	0·009 18	0·007 81	0·010 26	0·008 11	0·006 88	3·3
(*Short Flint*)									
EK–448	1·8804	41·1	0·021 42	0·013 26	0·011 49	0·015 23	0·012 12	0·010 29	5·9

Permission has been given by Jena[er] Glaswerk Schott and Gen, Mainz to publish the following article by Dr Carsten Eden, Mainz, on rare element optical glasses.

Advances in processing engineering of optical melting

Although new optical glasses, by using oxides of rare elements, like lanthanum, tantalum, thorium, and other materials of high refractive index have been developed since the thirties, they have only become of greater importance since the Second World War.

Hand in hand with the study of the new glasses went the technical development of new melting methods. Nearly all these glasses can no longer be made by the classical pot furnace melting process, because at their melting temperature even the highest-quality ceramic crucible materials quickly dissolve, with the exception of platinum which is practically not attacked.

Unfortunately it was not possible simply to replace the ceramic pot by the platinum crucible; a great number of technological problems had to be solved beforehand. Platinum and its alloys possess an exceptionally low heat stability at high temperatures. In view of this a great effort is required to overcome the constructional problems of large melting vessels and stirrers.

Heating by gas, in general use until then, had to be replaced by electrical heating for pot melting. Apart from the well-known methods of electric radiation heating by heat conductors or induction heating, the possibility of direct firing the glass melt through Joule's heat by platinum electrodes was investigated. It is well known that Joule's heating of glass is of special advantage if, for example, the energy supplied can be concentrated in the interior of the platinum vessel in the smallest space possible. When a 50 Hz alternating current is used, the colloidal sputtering of the electrodes causes a distinct pollution of the glass through electrode dissolution, which is too great for the high standard of quality required of optical melting. The glass should not look discoloured and no scattering effect (Tyndall effect, visible only in the dark field), which may be caused by sputtered electrode material, is allowed.

The examination of the physical-chemical processes at the transition of the alternating current from the electrode into the glass melt shows that the colloidal dissolution of platinum is to a high degree a function of the ion migration velocity in the glass melt. It was thus possible to influence the platinum sputtering by higher frequencies of the heating current. As a result of such experiments, the platinum dissolution can be shown as a function of the frequency of the various transition current densities between the electrode and the melt, for a glass with high refractive index of the heavy crown type SK 23 with a content of barium oxide of more than 25 per cent.

These measurements show that the colloidal platinum sputtering regresses strongly with increasing frequency, so that at 10 KHz, even with very high current density, no measurable platinum dissolution occurs. With the raising of the frequency of the heating current, the colloidal sputtering can not only be reduced but the forming of bubbles at the electrodes can be avoided. These can be very troublesome at the end or after completion of the fining. The investigations were made with glass SK 23, so that in glass melts of different compositions certain deviations of the measured values have to be allowed for.

One can conclude that in nearly every case, bubble-forming as well as electrode-material sputtering can be practically avoided by the use of an alternating current of 10 kHz, easily produced by a commercial converter.

The newly discovered glasses with extreme optical values are the starting point for the development of melting methods in platinum crucibles. It can soon be seen that with this process an improvement in quality can be reached for many glasses which until now had to be melted in the ceramic pot. All reactions between fire-resistant ceramic material and the glass melt which cause the formation of bubbles and striae are excluded by the use of platinum. Unfortunately, the greater expenditure on melting makes the production of glasses in the smaller platinum crucibles more expensive; therefore new manufacturing methods had to be developed.

A new process should not only improve the melting technique but, by the manufacture of lenses direct from the melting furnace, should also replace as far as possible the very expensive production method of semi-manufactured products of optical cast glass of the pot melt.

All these thoughts lead to continuous melting of

optical glass in the so-called platinum tanks. Such continuously operating melting units have naturally not the dimensions of the well-known large melting furnaces required for instance for the manufacture of window and bottle glass. The use of platinum suggests also completely new possibilities of construction methods, so that tanks for the melting of optical glass are not simply the well-known large melting furnaces on a smaller scale.

Especially important for the manufacture of optical glass in the platinum tank is the perfect homogenization through energetic stirring. The well-known task of a stirrer is to create a high local velocity gradient at as many places within the mixing vessel as possible in order to transmit shear forces. If, besides fulfilling this condition, care is taken that the whole mixture should get as uniformly as possible into the zones of maximum shear forces, then a particularly good quality free of striae can be achieved in the platinum tanks.

After the homogenization, the glass is kept further in contact only with platinum and is annealed at uniform speed to the desired processing temperature. With the help of electronically controlled regulators, the weight of the glass coming from the feeder can be exactly adjusted to within about $\pm 0{\cdot}5$ per cent with long glasses, and to within about $\pm$ 2 per cent with short glasses.

According to requirement, lenses can be pressed with the automatic machine directly at the feeder of the platinum tank. Alternatively, so-called drops of uniform weight and diameter can be produced. The automatically produced moulding can easily be recognized by its blank surface which allows checking compared to mouldings made from reheated glass.

Should larger pieces of optical glass be required, the continuous production of glass bars has proved satisfactory. The dimensions of bars cast directly at the feeder can be adjusted with a view to their future use.

To summarize, it can be said that platinum vessels do not only allow for the melting of new optical glasses for which the ceramic pot can no longer suffice, but a substantial improvement in quality can also be obtained in the case of many glasses melted in platinum. The opinion, which was still current up to a few years ago, that bubbles in an optical glass are a sign of quality, can be considered, with few exceptions, as being out of date. By the use of platinum tanks, there is the additional possibility of more rational production especially of glasses which are needed in larger quantities; furthermore the glass can be processed directly at the melting furnace by an automatic machine.

5.5 Crystal materials

NATURAL QUARTZ CRYSTALS

One of the most important operations is the selection of quartz from which it is intended to make prisms. Where quartz is pure, it is more transparent than crown or flint glass, but there is black, smoky, rose, or green quartz according to the impurity it contains. Quartz, for optical work, must be colourless and free from defects. Brazil is the most important source of supply.

Crystalline quartz belongs to the hexagonal system in which there are three equal 'horizontal' axes of symmetry at 120° to each other, and one 'vertical' axis. If polarized light is transmitted along the optical axis, the plane of polarization is rotated, the amount being 21 degrees 67 minutes per millimetre.

The conventional terminology calls a rotation 'right-handed' when it is clockwise to an observer facing the source of light, and vice versa. Two varieties of quartz are known; in one the rotation is right-handed, and in the other it is left-handed. Fig. 5.5.1 illustrates the appearance of the crystal faces in the two cases.

The most dangerous defects are those of twin crystals, for they are not always visible to the naked eye. A piece of good crystal is formed from a large number of very small crystals welded together with axes all parallel. A crystal must be examined with good polarized light equipment if one wishes to eliminate twin crystals and correctly locate the axis. (See § 11.4.)

Sawing of quartz crystal can be done with a bow saw and coarse emery or No. 80 Carborundum. A better method is with a serrated diamond saw. (Fig. 5.5.2.) After cutting, the crystal can be waxed to a plate and ground with a diamond-impregnated milling tool in the same way as a slab of glass. (Fig. 5.5.3.)

To avoid breaking the quartz by heating and cooling, the work must be conducted fairly slowly.

As quartz is harder than glass, coarser grades of emery should be used than for glass. There is very little tendency to cleave, and local hardness is only apparent when polishing a twinned surface.

In addition to prisms, natural quartz is used for windows and lenses in the air ultra-violet. The use of quartz is, however, limited by birefringence and optical rotation, and its dispersion is larger than that of fused (optical vitreous) silica.

5.5.1. ▷
Left-handed and right-handed quartz. The faces of the crystals are designated by Miller indices, for an explanation of which the reader should refer to a textbook of crystallography.

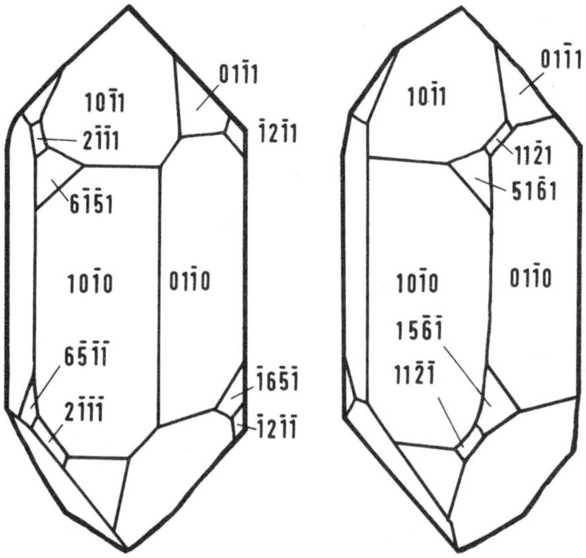

5.5.2. A diamond saw for cutting natural quartz crystals.

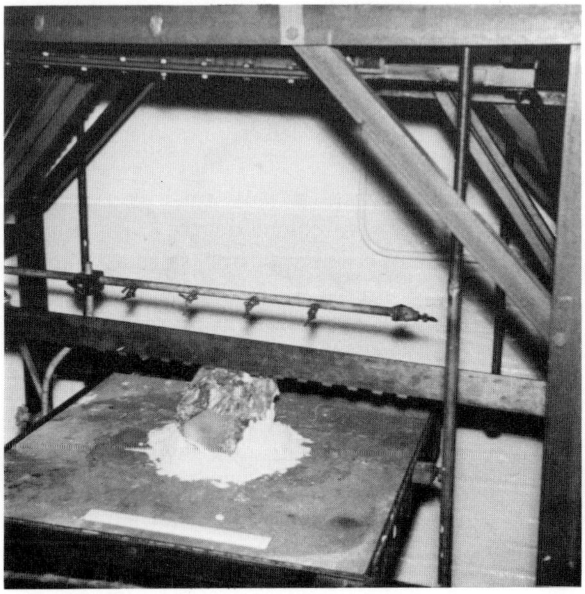

5.5.3. Grinding natural quartz crystals.

With the permission of Dr G. Hetherington, Technical Director, Thermal Syndicate Ltd, the following extracts from an article on Optical Vitreous Silica, describe the properties and manufacturing techniques of this material which is used, in many cases, to replace natural quartz as an optical material.

Introduction

In considering the properties of glasses based on silica, SiO_2, as the 'glass former' with one or more 'modifiers', it might be expected that the properties of such glasses will depend on the amount of silica present and that these properties will approach a limiting value as the silica content is increased. This effect is in fact observed in that the refractive indices decrease, thermal expansions decrease, and viscosities increase with increasing silica content (Fig. 5.5.4). Further to this, it might reasonably be expected that for an all-silica glass those values should be constant for materials made by different methods. The fact that this is not so indicates the difficulty of producing a 'true' silica glass.

A true silica glass should consist of material that is stoichiometrically SiO_2, that is with a tetrahedral structure in which each silicon atom is surrounded by four oxygen atoms and each oxygen atom is shared between two silicon atoms, and the silica tetrahedra must be arranged in a completely random three-dimensional network. It is the latter requirement which distinguishes vitreous silica from crystalline silica, in which the same tetrahedra are present but form an ordered structure (Fig. 5.5.5). It will become apparent why it is not generally possible to satisfy the above requirements of a 'perfect' silica glass exactly when the methods of manufacture are discussed.

History

The production of vitreous silica had to await the development of methods of reaching temperatures of 1700–1800°C and, although A. Marcet reported melting small crystals of quartz in an oxygen-alcohol flame in 1813, quartz was generally considered infusible until the discovery of the oxyhydrogen blow-pipe. The use of the oxyhydrogen blow-pipe was pursued by A. Gaudin, A. Gautier, C. V. Boys, A. Dufour, W. A. Shenstone, and W. C. Heraeus.

Parallel to this work was that originated by H. Moissan in 1893 on the electric fusion of quartz crystal. Moissan used an electric arc furnace, but later workers employed resistance and high-frequency induction furnaces. In 1910, H. A. Kent and H. G. Lacel invented a method of feeding powdered quartz crystal into an arc or a blow-pipe flame which allowed transparent vitreous silica to be built up in the absence of any contaminating refractory (Fig. 5.5.6).

Commercial interest in the material was now aroused and manufacturing methods were developed rapidly, different workers employing either high-temperature flames or electrical heating, until now, more than fifty years later, the industry is wide-spread.

Manufacture

Until comparatively recently, all commercial methods of producing transparent vitreous silica were based on the 'melting' of crystalline quartz. The term 'melting' must be qualified since the processes cannot be compared with processes used for melting conventional glasses. If crystalline quartz is raised to a sufficiently high temperature, the bonds which hold the tetrahedra in an ordered structure are loosened and, given sufficient time at this temperature, could form a random structure. However, as there is little difference between the energy required to loosen these bonds and that to rupture them completely to produce vaporized silica, a compromise is aimed at in practice whereby the particles of crystal are heated to just below the temperature at which silica will evaporate from their surfaces. At this temperature the viscosity of the material is still too high (see Fig. 5.5.7) to permit a sufficient degree of 'randomization' within an economic time, so that the silica glass formed contains a residue of the original crystal structure.

The above conditions apply to crystalline quartz melted electrically in a vacuum furnace (Fig. 5.5.8, for example) and one attempt to achieve a more random structure without vaporization giving bubbles in the material has resulted in the development of the method in which powdered quartz crystal is fed into an oxy-hydrogen flame and the 'melting' material built up on a

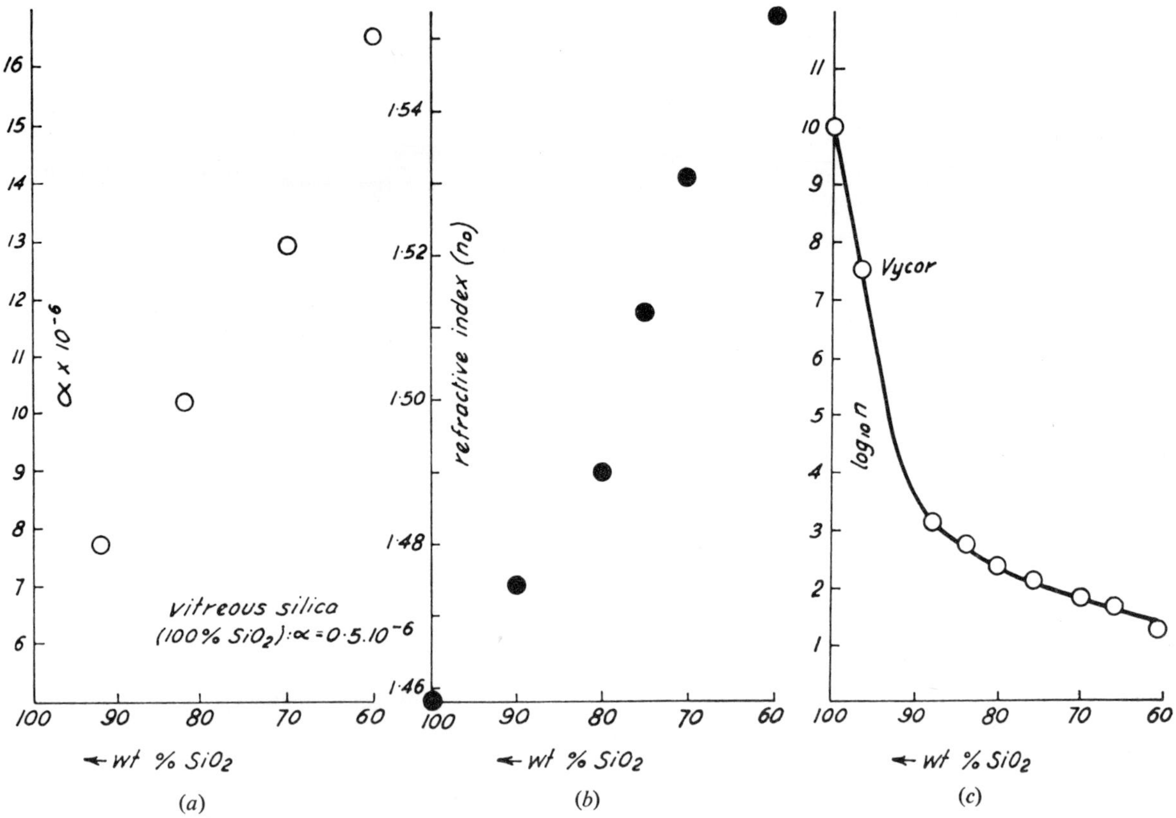

5.5.4. Variation of physical properties of glass with silica content.
(*a*) Thermal expansion of a series of sodium silicate glasses.
(*b*) Refractive index (for sodium D line) of a number of soda-lime–silica glasses.
(*c*) Viscosity (at 1400°) of a series of sodium silicate glasses (Vycor included for comparison).

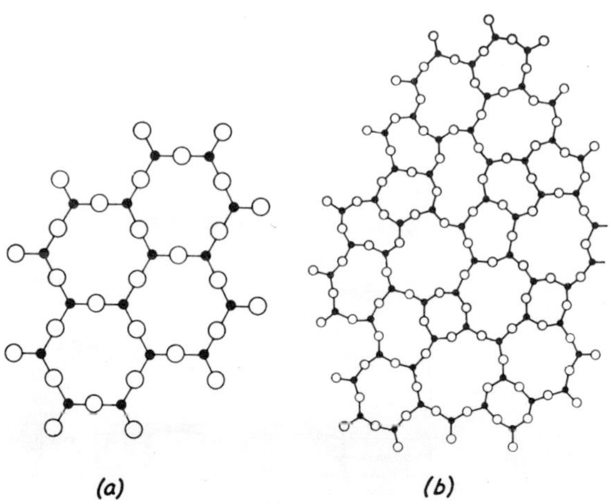

5.5.5. Two-dimensional comparison of
(*a*) ordered crystalline structure, and
(*b*) disordered structure of a glass.

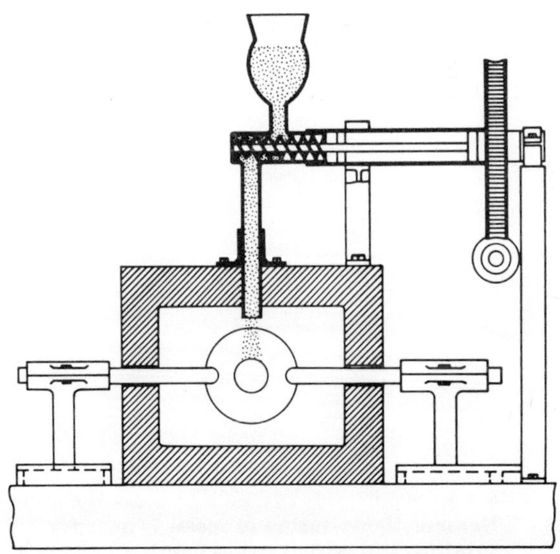

5.5.6. Kent and Lacel's process for manufacturing transparent fused quartz as rod or tubing.

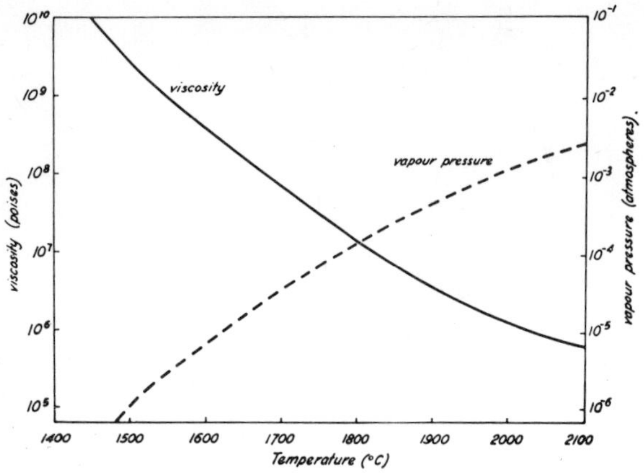

5.5.7. Variation of the viscosity and vapour pressure of vitreous silica with temperature.

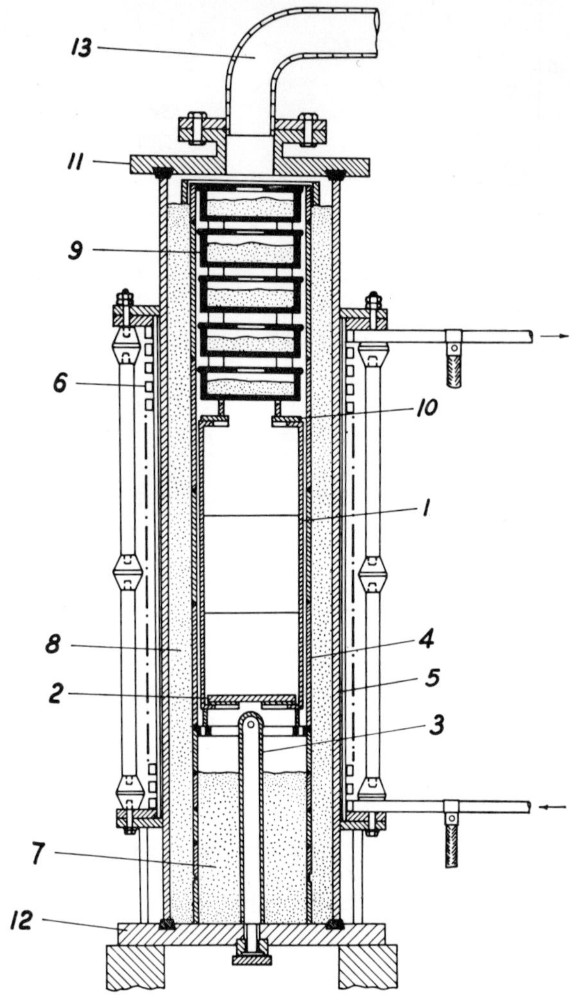

5.5.8. ▷
Quartz et Silice quartz melting furnace. 1, graphite crucible; 2, graphite plate; 3, sighting tube; 4, graphite heater tube; 5, vitreous silica vacuum jacket; 6, water-cooled induction coil; 7, silicon carbide granules; 8, silicon carbide granules; 9, trays of silicon carbide granules; 10, graphite disk; 11, water-cooled steel plate; 12, water-cooled steel plate; 13, lead to vacuum pumping system.

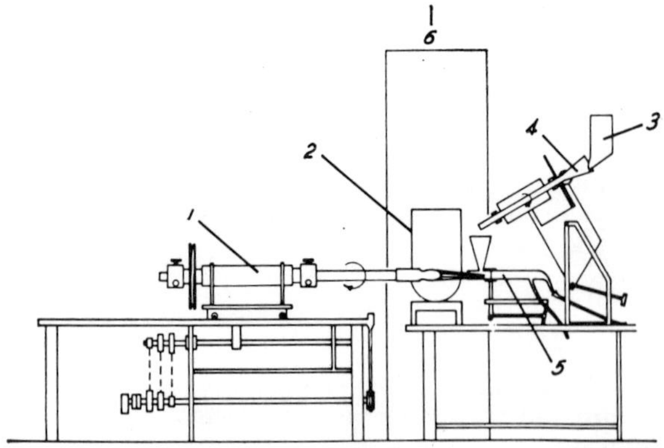

5.5.9. ▷
Heraeus flame-fusing process. 1, gear for rotating and withdrawing ingot; 2, vitreous silica muffle; 3, hopper; 4, quartz feed mechanism; 5, oxy-hydrogen burner; 6, exhaust.

comparatively cold surface (Fig. 5.5.9). Although this process does give a more random lattice than electrically fused material, it immediately invalidates the first criterion for a silica glass, since the material is no longer stoichiometrically SiO_2 but contains linkages of the type SiOH, these hydroxyls being formed by the inclusion of 'water' produced in the flame.

The above-mentioned types of vitreous silica, electrically fused (type I) and flame fused (type II), are limited in their purity by that of the original crystalline material, although, in the case of flame fusing, some impurities such as aluminium and alkali metals are reduced in quantity by partial evaporation.

This limitation has led to a third method for producing vitreous silica. This is essentially a flame fusion method, but, instead of using powdered crystal, a volatile compound of silicon, such as silicon tetrachloride or tetraethylothosilicate, is used which can be purified readily by fractional distillation, thus yielding a silica glass (type III) substantially free from metallic impurities but, unfortunately, producing at the same time an even greater concentration of hydroxyl linkages.

To maintain chemical purity and yet avoid the formation of hydroxyl linkages, a fourth type of vitreous silica has been evolved by oxidizing materials such as silicon tetrachloride in a hydrogen-free atmosphere. This process is comparatively new but shows real promise of being able to produce a true silica glass with its random structure and the complete absence of anything other than silicon and oxygen in stoichiometric proportions.

Since, however, present available grades of vitreous silica do show variations due to the above factors, the effects of these variations on the properties required in an optical material are discussed below.

Optical properties

The most important optical application for vitreous silica is the use made of its high transmission in the ultra-violet, visible, and infra-red spectral regions.

In the ultra-violet region, absorption or transmission is largely controlled by the presence or otherwise of metallic impurities whose oxidation states can be altered. Thus both crystalline and vitreous forms of silica with the same impurity content show the same order of ultra-violet transmittances.

The purest forms of both vitreous silica (e.g. Spectrosil, fused synthetic silica) and synthetic crystalline quartz are virtually free from absorption effects due to metallic impurities and this is apparent in their ultra-violet transmissions.

In the infra-red region absorptions are due to molecular effects rather than atomic, therefore metallic impurities make very little difference. However, vitreous silica containing 'water' in the form of Si-OH linkages exhibits an intense absorption band at 2·73 μm, the wavelength corresponding to the stretching vibration of this linkage, and lesser bands at 2·6 μm, 2·22 μm, 1·38 μm, 1·245 μm, and 0·97 μm, wavelengths corresponding to overtones and combinations involving the 2·73 μm fundamental and Si-O fundamental frequencies.

Transparent vitreous silica is transparent in normal thicknesses from about 0·16 μm in the ultra-violet to about 1 μm in the infra-red (Fig. 5.5.10). Thin films transmit up to about 150 μm. Vitreous silica produced from quartz crystal absorbs, however, in the ultra-violet at 0·24 μm and, as already mentioned, in the infra-red at 2·73 μm, if flame-fused. Methods have been developed to improve the transmission in these regions. The band at 0·24 μm is removed by heating the material in a non-reducing atmosphere at temperatures between 1050° and 1400°C, or by subjecting it to a high temperature electrolysis (1000 volts/cm at 1000°C). One method of improving the infra-red transmission is to replace the hydroxyl hydrogen by deuterium and so shift the absorption to longer wavelengths.

Another important optical property is the refractive index and here again it is possible to predict the type of refractive index variation from the method by which the vitreous silica has been formed. Material made by fusing crystal will have a granular structure due to refractive index changes at the surface of the original crystal grains, where 'melting' has not been quite complete. Since it is not made from solid particles, fused synthetic silica does not exhibit this 'granularity', but, since the hydroxyl content is really part of an equilibrium process, heat-treatment of the material, that is annealing, will reduce the

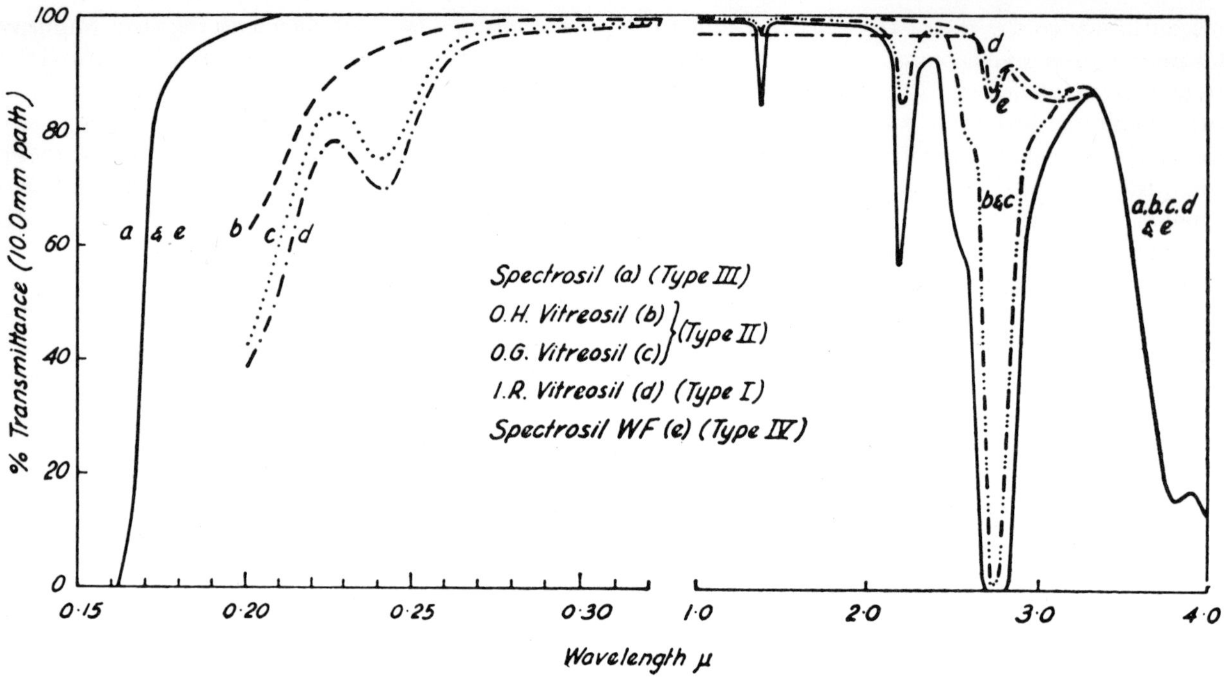

5.5.10. Ultra-violet and infra-red transmittances of vitreous silica manufactured by Thermal Syndicate Ltd.

hydroxyl content of the surface layers of the material unless precautions are taken to anneal in an atmosphere of the correct water content. Thus synthetic vitreous silica tends to have larger but more gradual refractive index changes than fused crystalline materials.

GLASS-CERAMIC MATERIALS

With the permission of Owens-Illinois Development Center, Ohio, extracts from an article by Dr G. A. Simmons on 'Development of Low Expansion Glass-Ceramic' materials describe the new group of ceramic materials which have become available since 1963.

Crystallized glasses

In the last decade, a new group of ceramic materials has become commercially available. Various generic names are used for these materials, such as 'glass-ceramics', 'crystallized-glasses', and 'melt-formed ceramics'. These materials are microcrystalline, polycrystalline ceramics capable of possessing some very useful and unique properties. The wide variety of materials in this category are all produced in essentially the same manner. Glasses of special compositions are melted and formed into the desired shapes essentially in the manner now used for conventional glasses. They are then subjected to a high-temperature heat treatment during which the glass undergoes a stage in which nuclei (incipient crystal growth sites) form throughout the glass, and a second stage in which the now nucleated glass is gradually converted into a ceramic material through the growth of crystals. Nuclei can be caused to appear very uniformly throughout the bulk of the glass and the number of nuclei occurring in a given volume can be controlled. It is possible to produce a wide variety of types of crystals and, by controlling the nucleation and crystal growth steps, to closely control the final crystal sizes. Since the ceramic is produced by crystal growth within a homogeneous glass body, the resultant body can be quite homogeneous and completely void-free—a truly superior ceramic material.

Current research, extending back over the last decade, includes effort aimed at developing a thorough understanding of the crystal species produced, nucleation and crystal growth kinetics, phase transformation phenomena, etc., as well as the development of practical techniques for controlling the processing variables encountered in the commercial production of such materials. Several thousand glass compositions which are capable of being made into crystallized-glasses have been produced and the properties of the glass-ceramics investigated. In these studies, many types of crystals have been grown—including semiconductor crystals, magnetic crystals and phosphor crystals. Through control of the basic compositions and processing, it is possible to produce a variety of properties and to alter them so that a particular desired property value can frequently be produced. Some of these materials are now being produced and marketed. It is the purpose of this article to describe the development of one of these materials possessing a very low coefficient of thermal expansion.

Extensive work has been done on producing a variety of materials having different thermal expansion coefficients. In fact, this technology is now so far advanced that a material having a given expansion coefficient can often be made.

One of the possible applications for a material having a very low thermal expansion coefficient is as a substrate for telescopic mirrors. Inquiries revealed that previous attempts to polish available opaque glass-ceramics (of the type used in the production of cooking ware) to produce a mirror for use in a telescope were unsuccessful, apparently because the polished surface could not be produced sufficiently free of small pits. Crystal diameter in such crystallized-glasses is rather large. It was believed that this pitting phenomenon might be eliminated if the crystal diameter could be greatly reduced. It was also apparent that the expansion coefficient of the desired material should be lower than that characteristic of these opaque crystallized-glasses.

There is a well-known hexagonal, trapezohedral modification of silica called β-quartz which is stable from 573° to 870°C. Pure β-quartz (SiO_2) crystals are known to possess a negative thermal expansion coefficient and display very low birefringence (optical anisotropy). Crystallized-glasses were invented containing crystals which may be considered as substituted derivatives of β-quartz in which extraneous ions (other than silicon or oxygen) exist in the β-quartz crystallographic structure. According to present theories, these modifying extraneous ions occur both as constituents of the crystal lattice structure (such as aluminium ions) and also as ions interspersed within spaces throughout this basic lattice (other carefully selected ions, including lithium). Crystals of this general type have been described in the literature as stuffed derivatives of β-quartz and, as solid solutions. The terminology 'solid solution' is preferred. It implies that variations in the concentration of modifying ions can be produced, within the 'solubility limit' for the particular ions under consideration. The choice of the word 'stuffed' appears unfortunate because it seems to imply that the extraneous ions are somehow forced into the crystals. Actually, the introduction of these ions occurs spontaneously under the right conditions because their inclusion is part of a natural transition towards a more stable (lower energy) state. It was discovered in the research work on crystallized-glasses containing such altered β-quartz crystals that the crystallized-glasses can be deliberately changed so that the expansion coefficient is reduced very close to zero and so that the change in expansion coefficient can be controlled to remain very small over a wide temperature range below about 500°C.

By 1964, samples had been produced in the research laboratory that possessed the desired very low coefficients of thermal expansion and which contained extremely small crystals. Some of these were sent to the Kitt Peak National Observatory to be finished into mirror blanks. The reaction to the finishing of these samples was quite favorable. Dr A. Keith Pierce described one of the resulting mirrors as one of the smoothest mirrors ever made. Based upon this and other studies, it was decided to proceed towards the commercial production of these materials. Mirror blanks of this material were produced in sizes up to 33 inches in diameter and many were polished into very good mirror substrates. While the reaction of the new material remained good, it also became apparent that some change would be desirable. One of the major difficulties was due to translucency, rendering it difficult to inspect the material for bubble content and stress levels through the thickness of the large mirror blanks. Therefore, in early 1966, a change in the formulation was accomplished. This new material is much more transparent, so that it can be very carefully inspected for internal defects.

Thus, a material has been created having a near-zero expansion coefficient over the temperature ranges which telescopes could be realistically expected to encounter, having adequate transparency for inspection of inclusions and defects, and possessing excellent polishing characteristics. It has been possible to do this through control of the crystal species to impart the desired physical properties and through control of the crystal size (to less than the wavelength of visible light) to impart the desired transparency and polishability.

By 1967, mirror blanks in sizes up to 42 inches in diameter were in stock and larger sizes could be produced. As with most new production facilities, there has been a steady improvement in the quality of the material with time as the production personnel have become more skilled and as improvements have been made in equipment and processes. Since the mirror blanks are formed while the material is still in the glassy state, they possess many of the defects found in glasses or fused silica. These are occasional scattered solid inclusions (stones), somewhat randomly distributed gaseous inclusions (bubbles or seeds), and occasional small optical distorting streaks or bands due to slight chemical heterogeneities (striae or cord). Since these defects do not disappear when the material is converted to the ceramic state, they appear in the finished blanks.

The two defects of most concern in reflective optics applications are seeds and striae. Although there seem to be wide differences of opinion among astronomers concerning the importance of these defects, it is generally agreed that their presence should be kept to a minimum. Mirror blanks, coded C-101, produced early in 1966 were relatively 'seedy', but distinct improvements in the seed content of the commercial blanks were made, and this improvement still continues.

A similar improvement has occurred in the stress levels present in C-101 commercial blanks produced during 1966. For this measurement, the highest stress level in the blank, almost always in a cordy streak, is located and its optical birefringence measured. This is converted into millimicrometres per centimetre by dividing the total birefringence reading by the thickness through which the measurement is made.

Properties

Four properties seem to be of definite importance for reflective optics: thermal expansion coefficient, polishability, mechanical stiffness and long-term stability. Each of these is discussed below as that property affects the use of the material as a mirror substrate.

1. Thermal expansion coefficient

There are four common questions about this property: how is it measured? how close is it to zero? how does it change with temperature? and does it vary from place to place within a blank, and if so, how much?

As the thermal expansion coefficient approaches zero, its measurement becomes increasingly difficult because the change in length with temperature (from which it is calculated) becomes smaller. The dilatometer employs a Carson-Dice electronic micrometer and compares the expansion of a material with that of fused silica. Unfortunately, the expansion coefficient of fused silica varies significantly from lot to lot and it is difficult, therefore, to get a reliable standard. As a control standard, we now use a piece of C-101 whose coefficient was determined by means of a vacuum path laser interferometric measurement, a technique which does not require an independent standard.

With the dilatometer, the change with temperature can be measured, and the curve shown in Fig. 5.5.11 is typical of that obtained on the commercial material. As explained above, it is possible to alter the position of this curve if deliberate changes are made in the manufacturing process or material. Obviously, it is the objective of the manufacturing personnel to maintain this curve in an unchanging position. However, as in most manufacturing processes, some small change from batch to batch does occur. Therefore, a measurement is made on each mirror blank produced and a certificate issued with each blank giving the measured expansion coefficient.

The existence of a small variation from blank to blank as described above does not imply the existence of a variation within a single blank. Blanks have been sectioned and many measurements made throughout their volume. To date, no blank of commercial quality has given results showing a significant variation greater than the

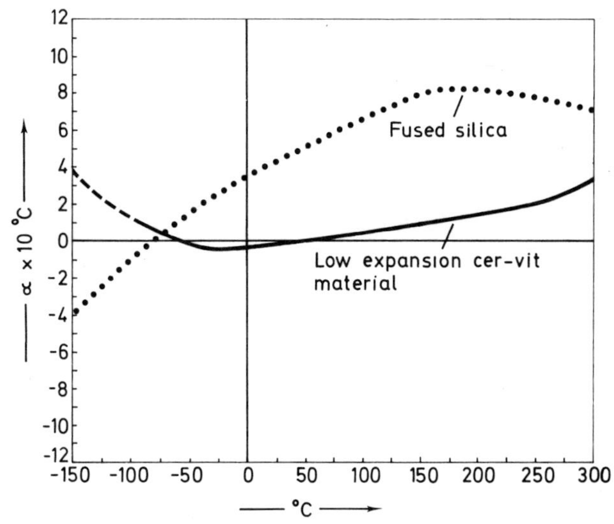

5.5.11. Thermal expansions of Cervit (C-101) and fused silica compared.

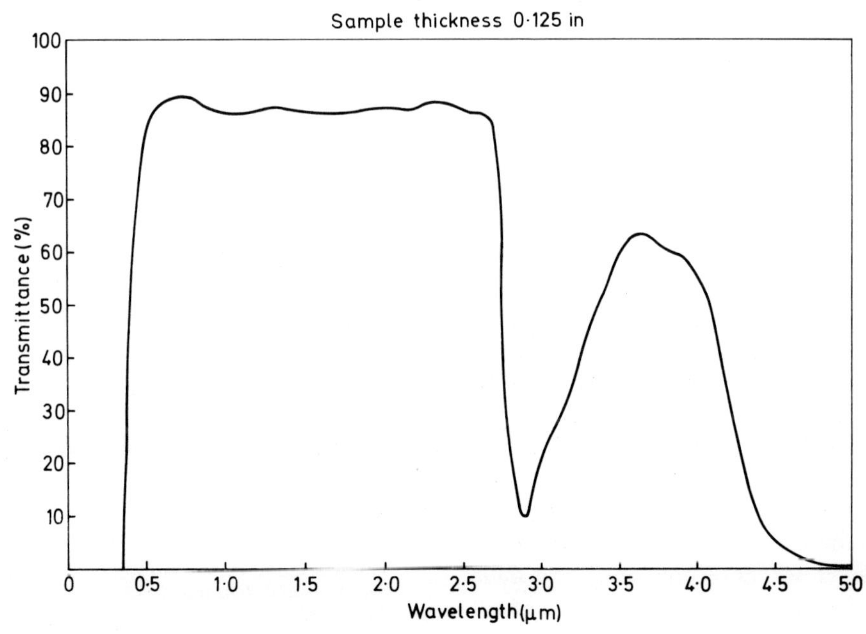

5.5.12. Transmittance curve for low-expansion Cervit (C-101).

experimental error inherent in the measurement itself. In fact, this multiple sampling process is repeated each time a new, larger size blank is first produced to insure that no new size-dependent variable has appeared in the processing. Since there has been a steady upward movement in production to larger and larger sizes with time, this examination has been repeated often. There is an even stronger reason for believing that no detectable variation occurs within the volume of a single blank. When the blanks are produced, they are at a high temperature and are then cooled to room temperature. The material loses its ability to relieve internal stresses around 1200°F. If there were any variation in the expansion coefficient from place to place in the blank, the areas with different coefficients would differ considerably in their degree of shrinkage as they cooled from 1200° to room temperature. Thus, even a small variation in the expansion coefficient would generate considerably strained areas upon cooling. Since these are not found in the blanks, appreciable variations are apparently not present. It is primarily this property, the stability of an optically figured surface under temperature change, that makes the low-expansion material of great importance in the field of reflective optics.

2. Polishability

An electron micrograph of a polished and etched surface of low-expansion material shows that crystal diameters in this highly crystalline material are exceedingly small, apparently of the order of a few tens of millimicrons. Obtaining a micrograph of this type is not an easy task. If the polished surface is examined without an etching treatment, the electron microscope does not reveal any discernible surface texture and the electron micrographs obtained are similar to those obtained on polished borosilicate glass or fused silica surfaces. Since the glass-ceramic material is quite chemically resistant, it required considerable experimentation to find a suitable etching acid in order to reveal the structure.

By now, many mirror blanks of this material have been successfully polished to yield high-quality surfaces. However, some opticians have encountered difficulties when using the exact approaches that they normally use for the polishing of fused silica. This is due to the fact that many of the traditional polishing procedures have been developed to avoid undesired side effects which normally arise from either heating or cooling of areas while fused silica and glass blanks are being polished. Since these effects do not occur with glass-ceramic mirror blanks, some alteration in the normal polishing procedures is usually needed. In addition, several of the opticians' 'tricks' which depend upon deliberate warming or cooling do not produce the same effect on these blanks. However, no significant time is wasted waiting for the blank to reach thermal equilibrium in order to test for the progress of polishing.

3. Mechanical properties

The mirror blank material C-101 differs significantly in its mechanical properties from fused silica. This is shown in Table 5.5.1 in which a comparison is given of six of the properties of both materials. It can be seen that Young's modulus (the stress required to obtain a unit deformation) is higher for the glass-ceramic material. This means that the material is stiffer and, when a large piece is edge-supported, it will sag less than an identically sized piece of fused silica under the same conditions. This greater stiffness also leads to the conclusion that a thinner piece could be used than is customarily used for fused silica mirrors to obtain the same serviceability.

4. Stability

Since this low-expansion material has only been available a short time, some question exists as to its long-term stability. There is no reason to expect change to occur with time because, being in a thermodynamically lower energy state, the materials should be more stable than glasses. The oldest optically finished glass-ceramic mirror blank of appreciable size is probably a 16-inch

Table 5.5.1 Properties of material C-101

	Material C-101	*Fused silica*
Young's modulus, psi	$13{\cdot}4 \times 10^6$	$10{\cdot}5 \times 10^6$
Rigidity modulus, psi	$5{\cdot}5 \times 10^6$	$4{\cdot}3 \times 10^6$
Bulk modulus, psi	$9{\cdot}0 \times 10^6$	$5{\cdot}3 \times 10^6$
Poisson's ratio	0·25	0·14
Density, g/cc	2·50	2·20
Hardness, Knoop (200 g loading)	540	500

diameter mirror blank finished at Kitt Peak Observatory in 1965. This blank has been periodically examined since then with knife-edge tests and no change in its optical figure has been detected. Another approach to determine stability is that of attempting to accelerate any long-term change by means of elevated temperatures. A sample of material C-100 was held at a temperature of 800°F for 8000 hours (about one year) and no change detected in any of its measurable properties, such as density and expansion coefficient. Furthermore, a measurement of the microcreep (so-called delayed elastic creep) of material C-100 and of fused silica at room temperature has been made. All glasses undergo an instant elastic deformation under mechanical load, the magnitude of which is inversely proportional to the Young's modulus of the material being tested. They then undergo a much smaller delayed deformation (microcreep) with additional time under load. This particular test was performed with a load of over 3000 p.s.i. maintained for 96 hours. It is estimated that the microcreep was complete within at least 72 hours. The measured microcreep of glass-ceramic material was less than the microcreep of fused silica under the same conditions. It thus appears that the low-expansion material is certainly as stable as, and probably more stable than, the present accepted mirror blank materials.

ZERODUR

With the permission of Jena^{er} Glaswerk Schott and Gen., Mainz, the following information is published.

Zerodur is the name given to a glass-ceramic material newly developed by Schott. Its main application is in the field of reflective optics.

Glass ceramics are generally defined as being an inorganic non-porous material containing both glass and crystalline phases. Glass ceramics are characterized by their method of production.

A glass-ceramic base glass is produced in accordance with standard melting procedures in the glass industry and is formed in the usual way by casting, pressing, rolling, or blowing. By, often very complicated, thermal treatment at temperatures generally between the points of transformation and softening of these special glasses, crystals form within the volume of the thermally-treated glass bodies, which are present in the remaining residual glass, randomly distributed. The properties of the original glass are characteristically altered by such controlled volume crystallization. Glass properties such as the steady decrease in viscosity at increased temperatures disappear, so that glass-ceramic articles may no longer be heat-formed. The properties of the material are now largely determined by the properties of the crystals separated. Compared with ordinary ceramic materials which, of course, may have some glass phase apart from crystalline phase, a glassy crystalline material produced in this way has the great advantage that it is completely non-porous. Whether the material is transparent, translucent or opaque depends on the type and size of the crystals in glass ceramics.

The main objective in producing Zerodur was to get the expansion coefficient of the material as close as possible to zero in the temperature range of its application. For telescope mirrors this application range is between 0 and 50°C. The objective is attained by adjusting a certain chemical composition which, however, may change insignificantly from charge to charge owing to the type of melting process. The fine adjustment of the expansion coefficient is obtained through modified temperature control during the ceramizing process. The values given in the tables are typical for Zerodur.

Zerodur is amber-coloured and transparent.

Optical properties

Refractive index n_d = 1·542
Abbe number v_d = 55·8

Thermal properties

Typical mean linear expansion coefficients α in the temperature range:

-30 to 70°C: $-0{\cdot}04 . 10^{-6}$ deg^{-1}
0 to 50°C: $-0{\cdot}02 . 10^{-6}$ deg^{-1}

Mechanical properties

Young's modulus	$9{\cdot}2 . 10^5$ kp/cm^2
Rigidity modulus	$3{\cdot}7 . 10^5$ kp/cm^2
Poisson's ratio	0·245
Tensile strength σ (in bending test) (Fine-ground)	$9 . 10^2$ kp/cm^2
Micro-hardness according to Vickers (50 g load)	$9{\cdot}5 . 10^4$ kp/cm^2
Density	2·52 g/cm^3

Zerodur optical elements

With the development of Zerodur, a particularly transparent glass-ceramic material was produced which completely satisfies all requirements in the field of reflective optics, owing to its almost negligible thermal expansion, its high Young's modulus, and its excellent polishing qualities.

Optical systems which incorporate mirror elements of Zerodur, because of the extremely low thermal expansion of this material, are independent of temperature changes. Therefore, optimum times of operation can be achieved and costly thermostatization is no longer necessary.

Because of the excellent temperature stability, Zerodur mirror blanks require less optical finishing work than borosilicate glasses. The high Young's modulus of Zerodur ensures outstanding dimensional stability of large mirrors, too. The transparency and the polishing qualities of Zerodur, which are in no way inferior to the polishing qualities of conventional optical glasses, are obtained by limitation of the mean size of the crystals to approximately 500 Å. A further prerequisite to transparency and good polishing qualities of Zerodur is that the chemical compositions of the crystal phase and residual glass phase in the glass-ceramic material are similar, so that there is little difference in refractive index and hardness between crystals and glass.

The remarkable expansion properties of Zerodur are due to the structural characteristics of the randomly distributed crystals of the high-quartz type structure (up to 70 per cent by weight) in these glass ceramics. SiO_2 with *h*-quartz structure, which has a slightly negative expansion characteristic, together with a number of oxide components, forms solid solutions.

All Zerodur disks or plates are subjected to optical precision annealing. Care is taken to achieve not only minimum strain, but also symmetrical strain distribution. It is ensured that all surface areas are under compressive strain.

5.6 Artificial crystal materials

Single crystals are an important addition to the conventional family of optical glasses because of:

(*a*) Good transmission in the ultra-violet and infra-red spectral regions.

(*b*) Good relation between refraction and dispersion.

(*c*) Good mechanical strength.

(*d*) High thermal stability.

Optical applications are in most cases limited to isotropic crystals. The transparency of ordinary optical glass extends very little beyond the visible region of the spectrum and therefore components designed for ultra-violet or infra-red wavelengths must be made of some other material.

Natural crystals of quartz, calcite, rocksalt (sodium chloride), sylvine (potassium chloride), and fluorite (calcium fluoride) have been used for optical components, but clear natural crystals of sufficient size are difficult to find.

Artificial crystals may be grown either from the molten material or from solution, but most optical crystals are grown from the melt. In optical qualities, these synthetic crystals compare favourably with natural crystals, as they are made from very pure materials and the rate of growth can be very carefully controlled. Artificial sodium chloride crystals show the same properties of cleavage as natural rocksalt and are less affected by a humid atmosphere.

Calcium fluoride (CaF_2) and lithium fluoride (LiF) are materials for making prisms, windows, and lenses to be used in the vacuum ultra-violet. Fig. 5.6.1 illustrates a typical prism for angular dispersion when fitted to the vacuum chamber of a spectrometer.

Some single crystal materials transmit light from the ultra-violet region (0·15 μm to 0·40 μm), through the visible region (0·40 μm to 0·70 μm) and into the infra-red region (0·70 μm to 15 μm). However, plastics and most standard glasses are somewhat restricted in their transmission properties in the upper ultra-violet range, the complete visible range and, for some speciality materials, slightly into the infra-red range.

Most of the alkali halide crystals can be grown in either Stockbarger-type furnaces (Fig. 5.6.2) or Kyropoulos-type furnaces. (Fig. 5.6.3.)

The following information on optical crystals has been provided by Mr J. Reed, Manager of the Crystal Department, Rank Precision Industries Ltd.

5.6.1. ▷
Prism for angular dispersion in the ultra-violet.

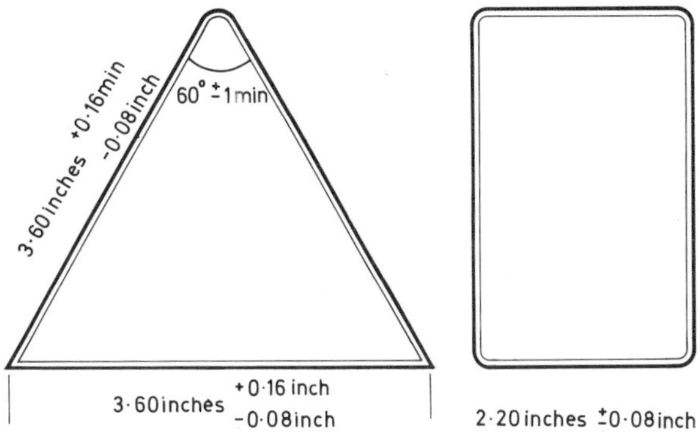

Apex of prism can be 0·10 inch radius providing the size of the prism is not less than 3·52 x 2·12 inches

Material: Lithium or calcium fluoride

Angles: To limits given

Surfaces and definition: To within two fringes by double transmission on interferometer. Faces square to base to ± 2 minutes and base to be marked

Chamfers: 0·02 inch on all edges

5.6.2. Stockbarger furnace showing the crucible in which salt is heated. The heater elements and terminal fixings are seen. A completed crystal is standing on the rim of the crucible.

5.6.3. The inside of a Kyropoulos furnace, showing the crystal formed in the pot.

The last few decades have seen a great increase in the demand for artificial crystals of many substances. Much activity has been involved in the production of new semiconductor materials, but alongside this there has been a consistently increasing requirement for optical crystals.

As one of the major European manufacturers, Rank Precision Industries has been actively engaged in improving both the availability and quality of its products in this field.

Optical crystals may be defined as being those which can be used as components in a system

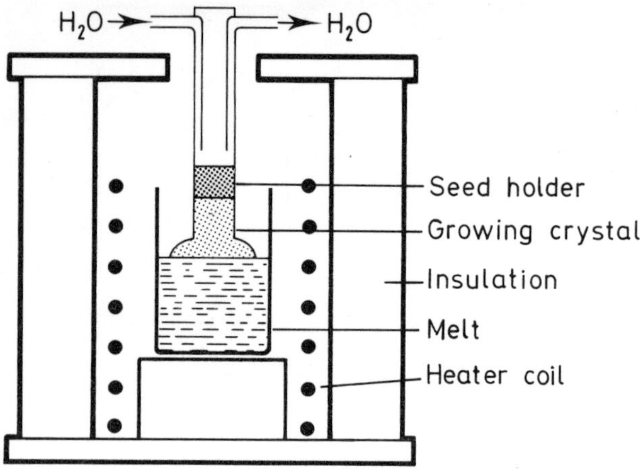

5.6.4. Kyropoulos method. The crucible containing the raw material is placed in the furnace and the heater is increased until a stable melt is obtained. A seed is introduced from above and the crystal grows on it as the heater power is reduced.

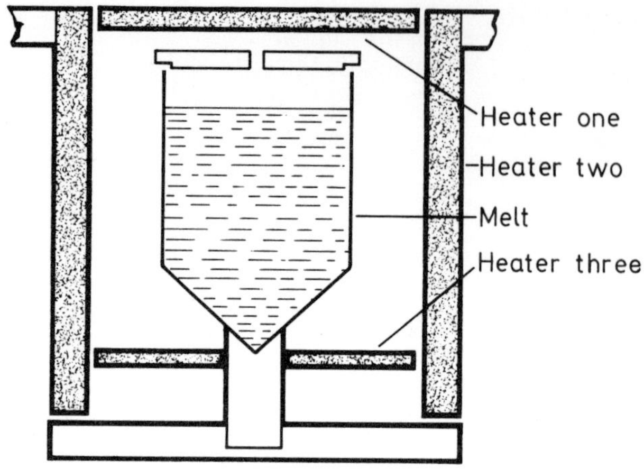

5.6.5. Stöber method. Two or three heaters may be used. The first melts the charge, the second establishes an independent temperature gradient, and the third (if used) acts as a booster, crystallization starting at the bottom of the crucible as the booster power runs down. The process is continued by controlled reduction of the power supplied to the first heater.

through which electromagnetic radiation is propagated.

In practice the region of the electromagnetic spectrum in which most 'optical' crystals are employed ranges from 0·1 μm in the ultra-violet, through the visible spectrum and out to about 4·0 μm in the infra-red. Typical of the numerous components into which such crystals are made are cell windows, prisms and multiple reflection plates.

Although used in a spectral region outside that mentioned in the preceding paragraph, crystals employed for X-ray diffraction do nevertheless satisfy the above definition and are supplied by RPI. On the other hand scintillation crystals, notably NaI (Tl), CsI (Tl), and CsI (Na), which are produced on a large scale by RPI, are energy converters rather than passive optical components and as such are not within the scope of the present book.

Table 5.6.1. General properties of optical materials

Property	*Glasses*	*Crystals*	*Plastics*
Refractive index	1·45–1·95	1·54–1·76	1·49–1·7
Infra-red transmission	Fair	Good	Very poor
Visible transmission	Very good	Very good	Very good
Ultra-violet transmission	Fair	Good	Very poor
Heat resistance	Very good	Excellent	Fair
Thermal stability	Very good	Excellent	Poor
Mechanical strength	Fair	Very good	Good
Mar resistance	Good	Very good	Poor
Weather resistance	Fair to very good	—	Poor to very good
Chemical resistance	Fair to very good	Very good	Poor to excellent
Specific gravity	Moderate	High	Low
Formability	Poor to good	Poor	Very good

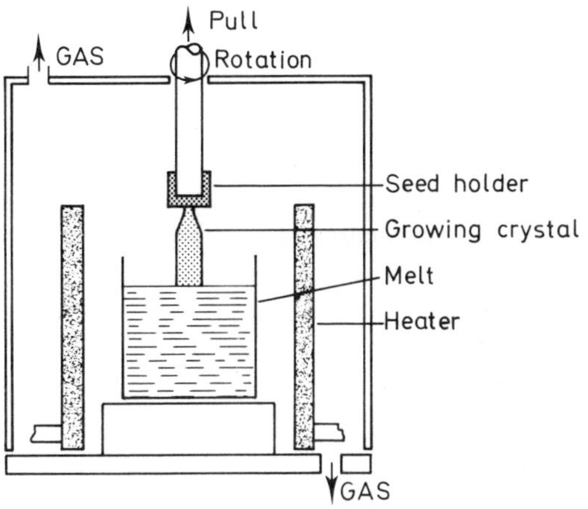

5.6.6. Czochralski method. The furnace geometry is similar to that of the Kyropoulos method, but the seed crystal is slowly withdrawn from the melt, the heater power being maintained at a constant level. Growth takes place in a stationary zone at the top of the melt as the seed is raised.

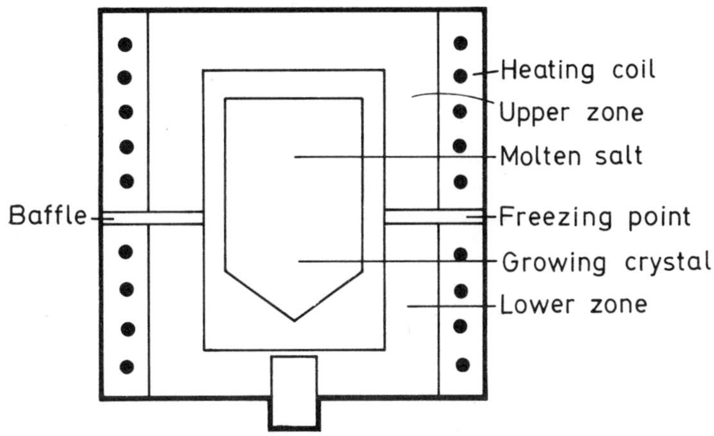

5.6.7. Bridgman–Stockbarger method. The second and more commonly used variant of the fundamental method. The crucible is held initially in the upper zone of a two-zone furnace and the power to the two zones is adjusted to give a vertical temperature gradient between them. The freezing level is in the middle of the gradient. The crystal is grown by lowering the crucible through the gradient into the lower zone.

PRODUCTION OF OPTICAL CRYSTALS

Most optical crystals are grown from the melt, the basic requirements for the process being the passing through the melt of a freezing-point isotherm. This can be achieved in three main ways: by keeping the crucible fixed and moving the temperature profile (Kyropoulos method and Stöber method), by keeping the temperature profile fixed and moving the crucible (Bridgman–Stockbarger method), and by fixing both profile and crucible and moving the growing crystal (Czochralski method).

Various methods are shown in Figs 5.6.4 to 5.6.7.

Size

The size to which crystal ingots are grown is governed by several factors including their method of growth, the use to which the resulting crystal is to be put, and the nature of the material itself, so that it is difficult to lay down a rigid specification for the maximum dimensions available.

Owing to the presence of thermal stresses in a growing ingot, the larger the diameter of the latter the poorer tends to be the structural quality of the resulting crystal. Whilst this may not be of great importance in work involving infra-red transmission, it may render the crystal useless for employment as an X-ray diffracting component.

In general, crystals in the standard range are available in diameters up to 8 inches.

By definition, the optical properties of RPI 'optical' crystals are the ones of most critical interest; however, other factors have to be taken into account when selecting a material for a particular application. These may involve such considerations as water solubility and mechanical strength.

Furnace runs

Crucibles are loaded in specially screened 'clear areas' prior to mounting in the furnace. The furnace runs themselves are rigorously controlled in order to achieve the necessary stability of growth conditions. The values of the various parameters such as temperature gradient and growth rate are chosen to give the best results for the particular material in each run and are under regular review by the development staff.

Ingot processing

The grown ingots are visually checked for clarity and monocrystallinity before processing.

Polishing is carried out under conditions of strict temperature and humidity control.

PROPERTIES OF OPTICAL CRYSTALS

Refractive index is wavelength dependent and the accompanying series of curves shows the form of the relationship for a number of optical crystals. (Fig. 5.6.8.)

A number of curves showing the relationship between dispersion and wavelength are shown in the graph. (Fig. 5.6.9.)

Some of the more important 'non-optical' properties of the crystals in both the standard and non-standard ranges are shown in Table 5.6.2.

Many of the materials listed exhibit the phenomenon known as 'cleavage' when struck with a sharpened tool suitably orientated along a cleavage direction. When a crystal cleaves, the surfaces revealed are approximately smooth and flat and can be identified with a set of planes in the crystal structure. The table lists the relevant planes in the cases of those crystals which do cleave.

The solubility in water of the listed materials is important in that it governs the conditions under which they may be handled and used. The values given represent the weight of material which will dissolve in 100 g of water at the temperature (°C) indicated. The figures quoted have been taken from *The Handbook of Chemistry and Physics*, 48th Edn, 1967–68.

Several methods exist defining the hardness of surfaces, notably those of Knoop and Mohs. None are completely satisfactory and results are occasionally contradictory. The problem is further complicated by considerable variations between different crystals of the same material due to quite small changes in impurity content. (This results in a further problem in that cleavage becomes more difficult to effect with increasing softness.)

The results quoted are thus intended only as a rough guide and are comparative rather than absolute. The standard comparison is 'general

5.6.8. Refractive indices of artificial crystals.

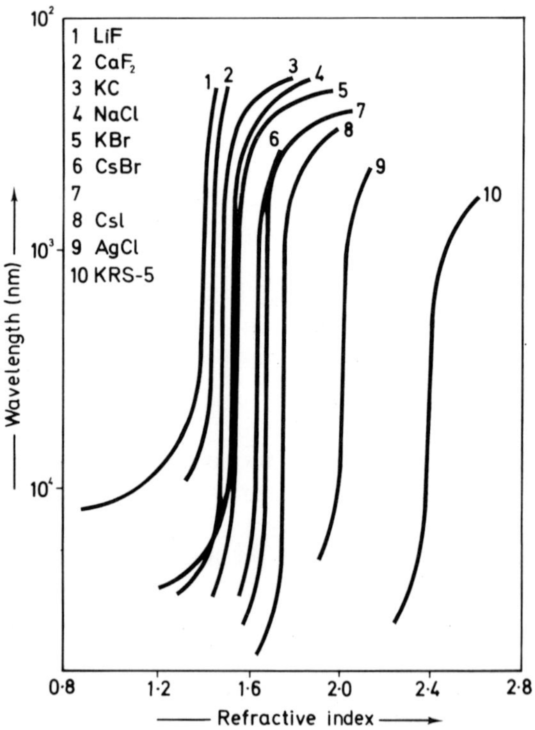

5.6.9. Dispersions of artificial crystals.

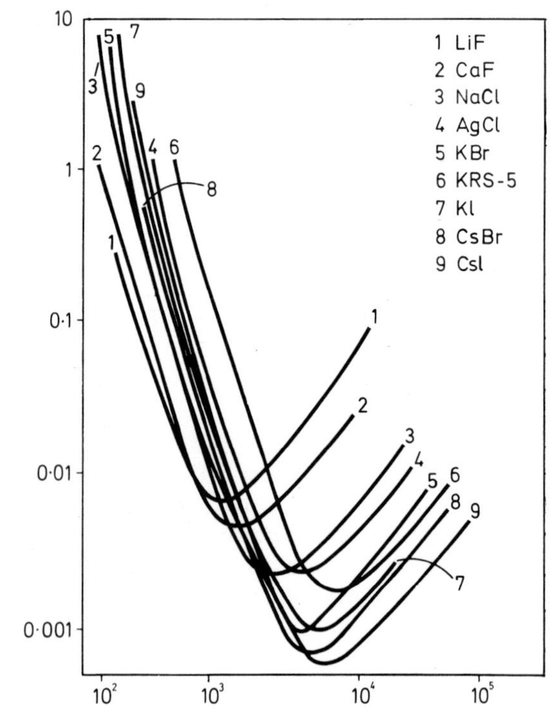

Table 5.6.2. Non-optical properties of crystals available

Material	*Structure*	*Cleavage*	*Solubility (g/100 g H_2O)*	*Melting Point °C*	*Hardness (ref: LiF)*	*Elastic moduli* C_{11}	C_{12}	C_{44}
LiF	Cubic NaCl	100	0·27 (18)	842	1·00	11·42	4·79	6·36
NaCl	Cubic NaCl	100	35·7 (0)	901	0·4	4·93	1·30	1·28
NaI	Cubic NaCl	100	184 (25)	651	0·1	3·03	0·89	0·73
KCl	Cubic NaCl	100	34·7 (20)	776	0·9	4·09	0·71	0·63
KBr	Cubic NaCl	100	53·48 (0)	730	0·07	3·45	0·49	0·51
KI	Cubic NaCl	100	127·5 (0)	725	very soft	2·75	0·45	0·37
CsBr	Cubic CsCl	None	124·3 (25)	640	0·2	3·07	0·84	0·75
CsI	Cubic CsCl	None	44 (0)	621	very soft	2·45	0·67	0·63
CaF_2	Cubic CaF_2	111	0·0016 (18)	1360	1·6	16·4	4·7	3·39
MgF_2	Tetragonal Rutile	poor	0·0076 (18)	1255	4·0	Not applicable		
KRS–5	Cubic NaCl	none	0·05 (25) *TlBr	414	0·4	3·31	1·32	0·58
AgCl	Cubic NaCl	none	0·0021 (100)	458	very soft	6·01	3·62	0·63
NaBr	Cubic NaCl	100	116 (50)	755		3·97	1·06	0·99
NaF	Cubic NaCl	100	4·22 (18)	980	0·8	9·70	2·40	2·82
KF	Cubic NaCl	100	92·3 (18)	851		6·56	1·46	1·25
RbF	Cubic NaCl	100	130·6 (18)	760		5·52	1·40	0·93
RbCl	Cubic NaCl	100	77 (0)	715	3·63	0·62	0·47	0·62
RbBr	Cubic NaCl	100	98 (5)	683		3·14	0·48	0·38
CsF	Cubic NaCl	100	367 (18)	690				
SrF_2	Cubic CaF_2	111	0·011 (0)	1400	1·5	12·35	4·31	3·13
BaF_2	Cubic CaF_2	111	0·12 (25)	1280	0·8	8·915	4·00	2·54
PbF_2	Cubic CaF_2	poor 111	0·064 (20)	822		9·34	4·40	2·10

optical' lithium fluoride, which is allocated the value unity.

The elastic stiffness constants are given in units of 10^{11} dyne cm^{-2} measured at room temperature. They relate the stress components to the components of internal strain and for cubic crystals are reduced by symmetry to three independent values, C_{11}, C_{12}, and C_{44}. There are several instances of divergences between reported results for different samples of the same material and the values given may not necessarily be the most accurate.

SUMMARY OF APPLICATIONS

Standard range lithium fluoride

A widely used optical material with useful infra-red transmission and the farthest ultra-violet transmission of all the optical crystals. It has low water solubility and a long working life. RPI make four separate grades:

(*a*) *General optical.* Standard quality for plates, infra-red windows, etc.

(*b*) *Far ultra-violet.* Specifically for transmission down to the limit of the ultra-violet range, a transmittance of 50 per cent through 2-mm thickness at 121·5 nm is guaranteed. Owing to its high purity, this material is soft and difficult to cleave.

(*c*) *X-ray.* This material is intended for use in X-ray applications and is selected for its good structural quality. Plates are usually made in a < 100 > {100} orientation but other orientations can be produced where very high reflected intensity is desired.

5.6.10. The structure of a common alkali-halide crystal (rock salt).

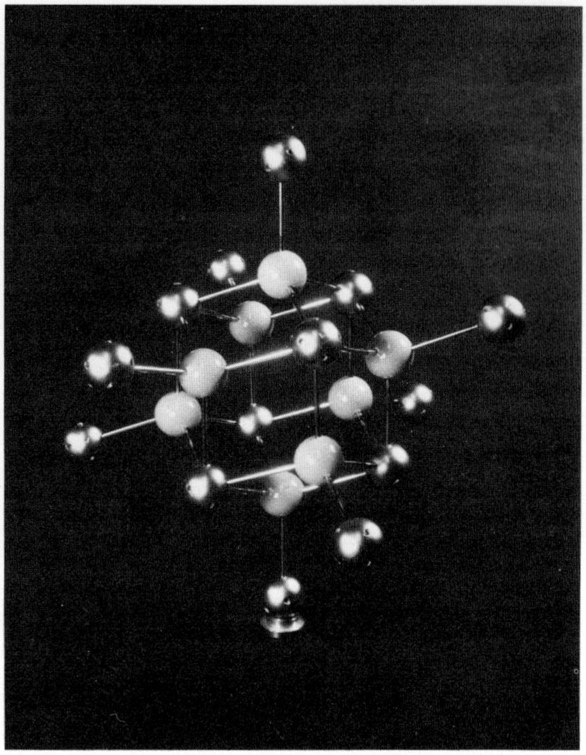

5.6.11. The structure of calcium fluoride.

(*d*) *Low dislocation density* (*LDD*). A small demand exists for crystals of very low dislocation density, mainly for research purposes, and crystals up to 1-inch diameter can be supplied.

Sodium chloride

Widely used for windows and prisms in infra-red spectroscopy, it is cheap and easy to handle. Although water soluble, it is relatively unaffected by atmospheric moisture in instruments wherein the temperature is kept only a few degrees above ambient.

Sodium iodide

In its pure form this material is sometimes used as a light guide in scintillation detectors employing thallium-doped NaI as the basic scintillator. It is also of some interest in applications for pure research, but is very hygroscopic and thus difficult to handle.

Potassium chloride

Its infra-red cut-off is beyond that of NaCl but falls short of that of KBr. Like all the alkali-halides it is used in basic research studies.

Potassium bromide

Widely used in infra-red applications where transmission beyond the NaCl cut-off is required, it suffers from surface 'fogging' by atmospheric moisture unless kept in a suitably dry environment.

Potassium iodide

Infra-red transmission extends farther even than KBr, but it is very hygroscopic and also very soft. For most practical applications KBr is preferred.

5.6.12. **A simple cubic Bravais lattice (caesium chloride structure).**

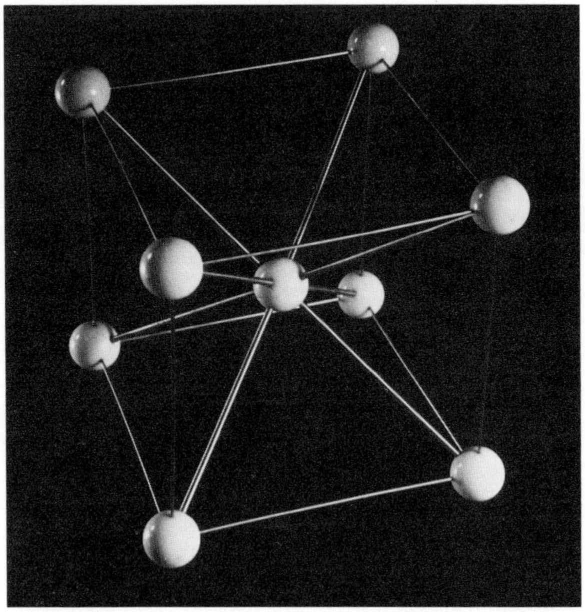

Caesium bromide

Useful as an optical component in the infra-red, it is, however, hygroscopic and finished surfaces will be attacked by atmospheric moisture. This may show up as a 'separation' along grain boundaries.

Caesium iodide

Having farthest infra-red transmission of all optical crystals, wide use is limited by the fact that it is very soft and difficult to polish. CsI components are considerably more expensive than KBr ones of similar size.

Calcium fluoride

Useful both for infra-red and ultra-violet work, especially the latter, where its transmission is bettered only by LiF and MgF_2. It is hard, takes an excellent polish and, because it is insoluble, can be left for long periods without surface deterioration.

Magnesium fluoride

Like CaF_2 it is hard and insoluble but in its molten form is less stable and therefore more difficult to grow. It exhibits birefringence (n_0 = 1·37770, n_e = 1·38950), and current applications include the manufacture of polarizing prisms for the ultra-violet.

Thallium bromo-iodide KRS–5

Transmits almost as far as CsI into the infra-red and is not water-soluble to any significant extent, making it useful as an infra-red window in 'field' applications. At the other end of the scale its ultra-violet cut-off is in fact in the visible range, so that its width of useful transmission is limited.

It has a very high refractive index (n = 2·22 at 4·0 μm) and is used in multiple reflecting 'ATR' plates.

Silver chloride

Has useful infra-red transmission applications. It is insoluble and can be used as windows in cells containing aqueous solutions. Although soft, difficult to polish and exhibiting no cleavage, it is ductile and can be rolled into sheets of the desired thickness.

5.7 Cutting, grinding and polishing crystal materials

Cutting

When cutting non-soluble crystals, such as lithium fluoride and calcium fluoride, a diamond cutter can be used and a standard cutting fluid such as Castrol Cleeredge AA 7050. (See §4.17.)

When cutting soluble crystals, such as sodium iodide, sodium chloride, potassium bromide, potassium iodide, potassium chloride, caesium iodide, caesium bromide, and KRS 5, use Castrol Honilo 430 as a cutting fluid. This fluid has a

5.7.1. Trepanning soluble crystals with a stainless steel cutter and Castrol Honilo 430 as cutting fluid.

5.7.2. Soluble crystals stored in a desiccator with phosphorus pentoxide. The worker is facing a sodium iodide crystal to thickness with emery paper.

5.7.3. Sawing soluble crystals into thin slices with Honilo 430 as cutting fluid.

5.7.4. Sawing soluble crystals with a hard-edge flexible hacksaw and Honilo 430 as cutting fluid.

5.7.5. Edging soluble crystals to correct their diameter by rolling them on water-coated rollers. Water drips from holes in the pipe over the rear roller.

5.7.6. A string saw cutting serrations in the base of a soluble crystal. Note that the string passes through a waterbath at the base of the machine.

dark red colour. Soluble crystals can be trepanned with a stainless steel cutter. (Fig. 5.7.1.) Between operations, soluble crystals must be stored in a desiccator containing phosphorus pentoxide, P_2O_5. (Fig. 5.7.2.) The phosphorus pentoxide is dangerous, because of the phosphorus content, and should be kept on a saucer lined with polythene and changed each week. When phosphorus pentoxide is used in a dry box, such as for the assembly of scintillation counters, then it should be renewed each day.

Crystals can be sawn up (Figs 5.7.3 and 5.7.4) using a bandsaw with a hard edge flexible back blade about $\frac{1}{8}$ inch (3 mm) wide 18 TPI, such as 'Raker' blade from L. S. Starrett Co. Ltd. Soluble crystals can be edged to correct diameter by rolling in a film of water. (Fig. 5.7.5.)

When cutting hygroscopic crystals, such as rocksalt, a wet cotton thread used as a saw is a good method although rather slow. (Fig. 5.7.6.) There is very little risk of splitting or damaging the crystal and, as large crystals are very expensive, this is an important virtue of the method. The thread is of 'six-cord' three-strand crochet cotton and is made into a belt by scraping the ends with a knife and then applying a waterproof household mending cement. The thread should run at 2 feet or more per second over smooth-running pulleys and bear lightly on the crystal so that the speed of actual cut depends only on the travel of the workpiece which may be an inch or so per hour for a 1-inch thickness of crystal. The thread should be barely damp when passing through the cut.

If the thread is made to pass through a water bath at the base of the machine the saturated salt solution will be removed. The thread should always be soaked in fresh water before it passes through the crystal.

It is important to handle fluorite at all times with great care to avoid temperature differences which may shatter the crystal. Fluorides can be sawn with a copper blade whilst water and Aloxite drips into the cut, but diamond impregnated blades are much quicker cutting.

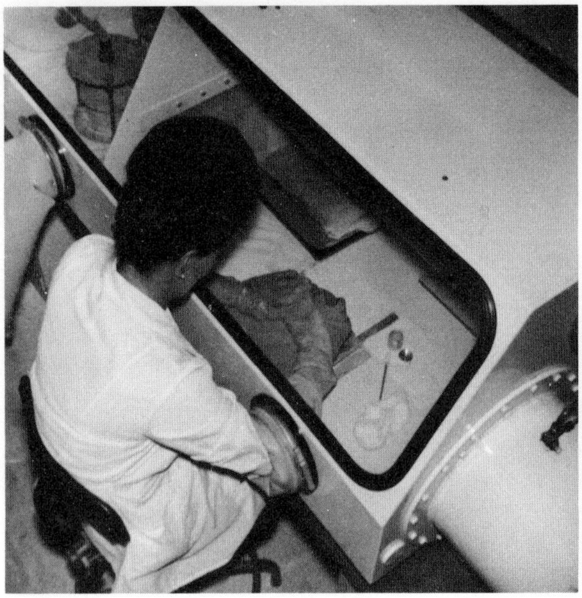

◁ **5.7.7. Polishing soluble crystals in a glove box purged with dry nitrogen and kept dry by phosphorus pentoxide.**

▽ **5.7.8. A row of glove boxes purged with nitrogen.**

Grinding and polishing

The grinding and polishing of crystals, especially the soluble crystals, is best carried out in a glove box purged with dry nitrogen. (Figs 5.7.7 and 5.7.8.) However, some of the early operations can be carried out on an open bench, but preferably in a section set aside for this specialized work. All stock and work in progress must be stored in desiccators. (Fig. 5.7.2.) Rock salt (sodium chloride) will polish by normal optical methods up to the last few strokes. The finishing strokes should be from a flat glass plate on which is stretched a piece of real silk damped with a little rouge and concentrated salt solution. (See Frontispiece.)

Lithium and calcium fluorides and sodium bromide should be polished with Linde Alumina powder (see § 4.10), of grades A and B. The correct way of using them is to mix a thick cream and to go on polishing until almost dry. Great care is necessary to avoid temperature changes which may shatter the crystals. Silver chloride presents special problems because one must work in darkness to prevent reduction of the silver chloride and its consequent darkening. Only a dull light from a tungsten lamp is permitted for purposes of inspection.

Silver chloride can be ground on iron using soap and Aloxite. The process must not be rushed or the surface will become badly charged. The polishing lap should be beeswax scored in $\frac{1}{4}$-inch (6 mm) squares. Prepare one-half saturated solution of photographic 'hypo' and apply two drops of hypo to the wax lap. Use a standard polishing stroke. Rub for a few strokes and then add two more drops of hypo. There is always a crust which is difficult to remove and afterwards the lap will probably have deposits of silver on its edges. If necessary, use Linde A powder to remove the crust. Remove the deposits by washing with hypo and scraping if necessary. If the deposit is not removed, the polishing action will be retarded and scratches result. If the piece of silver chloride being polished is thin, the solution may creep around the edges and get to the top, so if the top must be free of stain and pits, it should be covered with wax or lens varnish. After a 'wet' or rub, do not remove the work from the lap in order to look at it or stains and pits will result. Plunge the work into clear water to stop the action of the hypo and then wipe it lightly for inspection in a low light.

Potassium bromide is very apt to cleave and, as this crystal is soluble in water, mineral oil should be used when smoothing. Polish on raw pitch with rouge or putty powder. A pitch polisher coated with wax is sometimes used as the crystal is very easily scratched. Surface brilliance is obtained by rubbing finally with clean chamois leather.

Polishing lithium fluoride

With the permission of the Royal Radar Establishment, Malvern, Worcs., the following extracts are published from a paper by G. W. Fynn and W. J. A. Powell (*Journal of Physics E: Scientific Instruments 1971*, Volume 4) on the subject 'Polishing Lithium Fluoride for High Transmission in the Ultra-violet'.

It has long been recognized that wax laps give excellent surfaces on glass and some crystals (e.g. Twyman, 1943) but that the polishing process is appreciably lengthened. Moreover, they require special techniques for flattening owing to a relatively narrow-band melting range compared with that of pitch. Many mixes of wax and pitch have been used in attempts to acquire the flow-flattening properties of the latter and at the same time retain the superior surface-finishing properties of wax. Painting pitch laps with molten wax is a technique aimed at circumventing these flattening difficulties. In the techniques described here pitch laps are used to speed-up the initial removal of material, and pure wax laps are used for final polishing. This extended polishing process is, in fact, applicable to a wide range of soft optical materials.

The machine used (Fynn and Powell, 1969) is basically a Draper-type—that is, a machine in which the work is swept across a slowly rotating lap, e.g. the Multipol marketed by Metals Research Ltd. (See Chapter 18.) For this work the sweep is restricted to about 7 mm. The specimen and its associated conditioning ring are run continuously on one side of the lap (cf. the Lapmaster machine) with a typical spacing of 35 mm between work and lap centres. The sweep is used mainly because it is available on this particular type of machine, but it does have the advantage of reducing any channelling effects on laps due to excessive specimen loading. In constructing small crystal polishing machines expressly for use with conditioning rings, or jigs with stabilizing ring facings, it would be simpler to dispense with the sweep and balance the respective loads of ring and specimen carefully.

For specimens mounted on a block of less than 75 mm in diameter a 120-mm diameter lap is used and polishing is carried out within a conditioning ring of 76-mm internal diameter. The largest single specimen polished was 56-mm diameter by 10 mm thick. Extra care was needed in order to achieve flatness of $\lambda/10$ (He) and the polishing of larger specimens might well require either a relaxation in flatness specification or a larger lap. Literature on optical manufacturing processes abounds with accounts of pitch polisher preparation. Whilst most of these descriptions are directed at the hand glass-polishing field, a few are specifically for machine work (for example, Dickinson, 1968). The term pitch polisher includes tools ranging from pure pitch to blends of pitch, wax, resin and woodflour. The latter mixture seems generally better than pitch alone, since it is easier to prepare as well as giving sleek-free results on a wide range of crystals. It wears less rapidly, too, but for the same reason takes a little more working by conditioning ring techniques to wear into shape.

The mixture used is of the approximate composition (by weight): 30 parts 4 mm pitch; 5 parts woodflour; 1 part beeswax; 1 part resin. It is heated to 120°C and poured on a 120 mm stainless-steel plate which has been previously heated to 80°C. A wall of masking tape at the periphery retains the liquid mixture. When cool, the polisher is mounted in a lathe with good facing accuracy; then the surface is faced and finally recessed centrally over a diameter of 30 mm. A spiral groove is machined in it some 0·25 mm deep and at about 6 grooves per cm. It is then ready for running-in with a conditioning ring.

The wax laps used are made by pouring filtered 'Okerin 100' wax on to 120-mm diameter by 18-mm-thick cast iron or stainless-steel plates using a wall of masking tape to contain the wax. When cool, the front surface is machined on a lathe which is capable of good facing flatness.

Surface grooving can be achieved in a variety of ways ranging from a hot-nozzle machine (Brooke, 1966) to a simple 4 threads per cm chaser drawn across the wax surface and guided by a straight edge. The process which involves least contact and consequently least possibility of contamination is best. Here again we prefer to cut to 4–8 grooves per cm spiral groove in the wax immediately after facing.

Flattening can be done quite simply by 'running-in' the wax surface with a heavily loaded glass-faced conditioning ring. Alternatively, a specially prepared flattening tool can be used, which consists of a 75-mm diameter billet of stainless-steel with a spiral groove of some 4–8 grooves per cm machined in it with sharp crests. Its surface is then lapped flat while still retaining as much of the edge fineness as possible. This device—which has been idiomatically termed a 'stroller'—is run on the wax polisher for a short time till the wax surface acquires a high polish. If the lathe used for facing is known to give a small degree of either convexity or concavity then the conditioning ring running centre can be adjusted in or out respectively to reduce these errors.

The initial flatness testing can be done by using a lapped (grey) glass faced conditioning ring. The grey surface is given a polish sufficiently specular to show fringes by running it on the wax polisher using polishing alumina or cerium oxide. It will be immediately evident if it is either concave or convex, and the compensating adjustment may be made using the general rule that if the conditioning ring shows convexity its running centre must be moved out; conversely, if it is concave it must be moved into a position where the outside diameters of both ring and polisher are nearly coincident. On pitch polishers, rapid corrections can be made, but on wax, correction is slow since the polishing rate itself is slower.

The general technique for polishing LiF is to avoid, if possible, greying-off (i.e. lapping) the surfaces since the microcracks are thought to reduce the ultimate transmission of the polished specimen (Makarov and Novikov, 1967). Because LiF can be cleaved readily near to the required thickness, the final surfaces can be polished right from the cleaved faces. Since wax polishing is slow, all the irregularities of the cleaved face are polished out by some 4 hours of pitch polishing with Linde A abrasive and water (5 per cent Ha_2CO_3) as a lubricant. This leaves the surface covered with fine sleeks or scratches, which are then removed by wax polishing for about 3 hours with Linde B. Latterly, a drip-feed of water (with 5% Na_2CO_3) without an abrasive is used.

A single specimen is normally polished in a jig which is similar to that described by Bennett and Wilson (1966). A typical polishing pressure is about 120 g cm^{-2}. A reasonable balance of loading must be maintained between the specimen and the conditioning ring, since excessive load on the former will wear a track in the polisher. As a result of this, the ring and the specimen will exhibit different fringe patterns, and if these indicate an established pattern on the polisher the best remedy is to machine and reflatten it. Small errors may be corrected by running the lap for a while against the conditioning ring alone, after it has been freshly greyed. The specimen is replaced when flatness has been regained.

When a number of small disks or windows have to be polished, it is the practice to mount them spaced out on the surface of a 75-mm diameter solid work plate. Conventionally, the lowest number that can be worked is three around the centre, then one specimen at the centre surrounded by six, etc. However, these arrangements are intended for the standard poker or Draper-type polishing machines and are designed to minimize the formation of high or low zones. With conditioning rings or plate working where bias polishing is involved (Fig. 5.7.9), the specimens

5.7.9. Bias polishing.

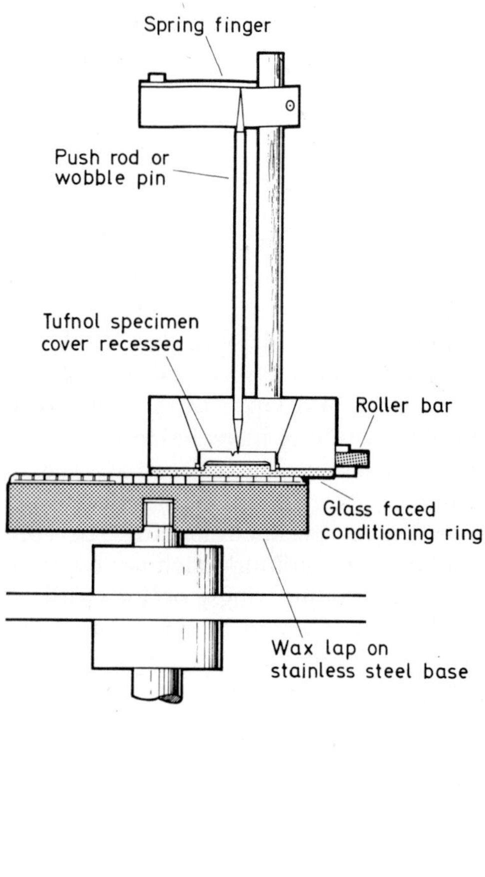

can be arranged in rings without the necessity of filling the central area: then, because the work plate is a rigid whole, no load balancing is required.

With a 20-mm LiF crystal a typical polisher speed of 120 rev min^{-1} is used, both for pitch polishing and the early stages with wax. This is reduced to about 40 rev min^{-1} for the final wax polishing stages in order to reduce the amount of heat generated by polishing and consequent reduction in time required for the specimen to attain room temperature before flatness testing. The best edge definition has been obtained as a result of re-machining the wax surface during the final polishing stage. This appears to improve the cutting characteristics of the wax polisher with an attendant reduction in edge turndown. This effect is also noticed by experienced hand optical workers who periodically scrape hard pitch polishers and thus renew their polishing qualities.

5.8 Optical plastics and moulded lenses

Optical elements fabricated from organic polymers have many advantages some of which are as follows:

(*a*) Cheapness of the manufacturing process (moulding instead of grinding and polishing).

(*b*) Relatively light weight (specific gravities below 1·5) and non-brittle character.

(*c*) Fairly good infra-red and ultra-violet transmittance in some cases.

Some disadvantages are as follows:

(*a*) Poor scratch resistance.

(*b*) Much greater thermal expansion than glass and low softening temperature.

(*c*) The grinding and polishing behaviour of non-brittle materials.

Typical materials are polymethyl methacrylate (Transpex I, Perspex, etc.) corresponding optically to crown glass, and polystyrene (Transpex II, Styron, Distrene, etc.) corresponding to flint glass.

Table 5.8.1. Some important commercially available optical plastics and their properties

Plastic	*Supplier*[a]	*Density* (*g/cm*^{3})	*Coeff. of expansion* (10^{-6}/*deg C*)	*Upper limit of stability* (°*C*)	n_D (*20°C*)	*V-value*
Allyl diglycol carbonate	(1)	1·32	90–1000	60–70	1·498	53·6
Polymethyl methacrylate	(2)	1·19	63	70–100	1·492	57·8
Polystyrene	—	1·10	80	70	1·591	30·8
Copolymer styrene-methacrylate	(3)	1·14	66	95	1·533	42·4
Copolymer methyl-styrene-methyl methacrylate	(4)	1·17	—	110–120	1·519	—
Polycarbonate	(5)	1·2	70	120–135	1·586	29·9
Polyester-styrene	—	1·22	80–150	50–120	1·54–1·57	ca. 43
Cellulose ester	—	1·30	90–100	50–60	1·47–1·50	45–50
Copolymer styrene-acrylonitrile	(6)	1·07	70	90	1·569	35·7

[a] (1) 'CR-39', Pittsburgh Plate Glass, Columbia-Southern Division; (2) 'Plexiglas', Rohm & Haas Co; 'Lucite', E.I. DuPont Co; 'Acrylite', American Cyanamid Co; (3) 'Zerlon', Dow Chemical Co; (4) 'Bavick', J. T. Baker Chemical Co; (5) 'Merlon', Mobay Chem. Co.; 'Lexan', General Electric Company; (6) 'Lustran', Monsanto Chemical; 'Tyril', Dow Chemical Co.; 'Bakelite C-11', Union Carbide Plastics Co.

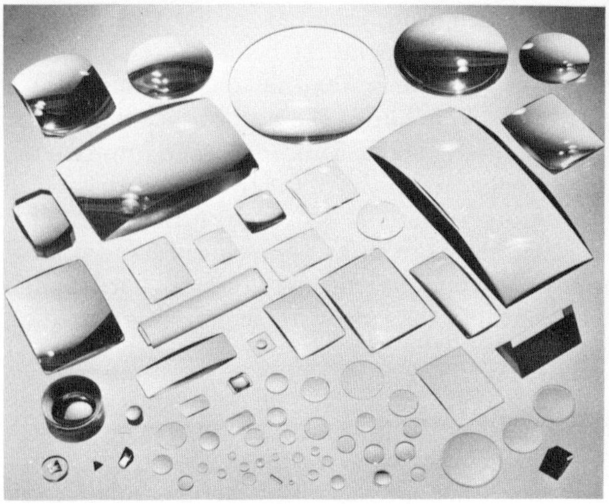

5.8.1. Aspheric magnifiers of a variety of patterns can be made from optical plastics.

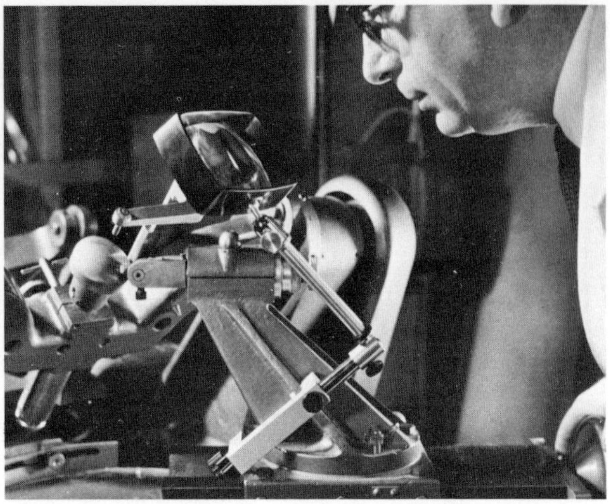

5.8.2. An aspheric magnifier in use for examination of the turning of a cylindrical tool surface.

The usual methods of manufacture of optical plastic components are:

(*a*) Moulding under heat and pressure in optically worked stainless steel moulds.

(*b*) Grinding and polishing using the normal glass working techniques.

(*c*) Casting the lens from monomer in optically worked glass moulds.

In the last case, polymerization is completed in the mould and the most important feature is the possibility of the cheap production of aspheric lenses in quantity.

Relatively cheap large-diameter lenses are also a possibility. Various devices are employed to overcome the distortion of the lens which would normally arise from shrinkage of the material during polymerization, and cyclohexyl methacrylate is preferred owing to the smaller shrinkage of this material. (Figs 5.8.1 and 5.8.2.)

When grinding and polishing plastic lenses the grinding is often done with a diamond-plated tool to avoid clogging of the diamonds. (See § 4.13.) Slow polishing speeds are used and polishers are usually made of pitch. The polishing material can be putty powder or the proprietory material Silvo. Optical plastic lenses require less time for grinding and polishing than glass ones.

Mr M. Sigmund, Managing Director of Combined Optical Industries, Slough, Bucks, has kindly provided the information in the remainder of this section.

ASPHERIC MAGNIFIER LENS

Combined Optical Industries developed the Aspheric Magnifier lens to eliminate the distortion normally experienced from all other spherical or cylindrical lens forms.

The squared pattern (Fig. 5.8.3), when viewed through a normal spherical lens will look as shown (right). This distortion is known as 'positive or pin-cushion distortion' or 'spherical aberration', and can be eliminated by an Aspheric Magnifier. Seen through such a lens, the pattern remains absolutely square right up to the edge of the frame.

FRESNEL TECHNIQUES

The original cylindrical or barrel-shaped Fresnel lenses were first used in lighthouses around 150 years ago. Several decades later, a simpler disk-shaped lens advanced railway signalling techniques where efficiency and reliability were paramount. Later a similar style of lens was extensively adopted by the manufacturers of such low- and medium-power lighting and indicating

5.8.3. ▷
(*left*) **Pattern of squares.**
(*right*) **Spherical aberration distorts the image of a pattern of squares.**

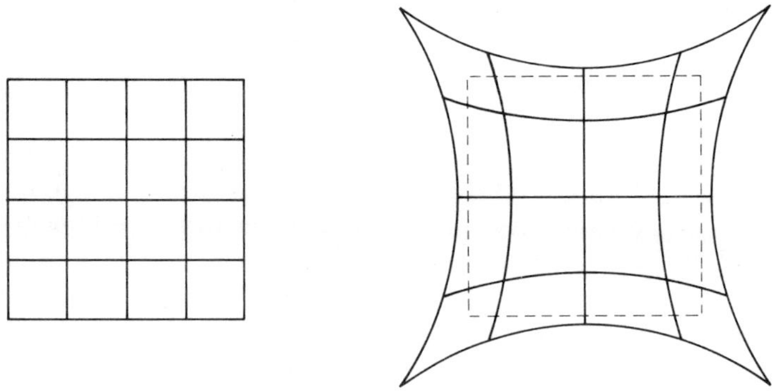

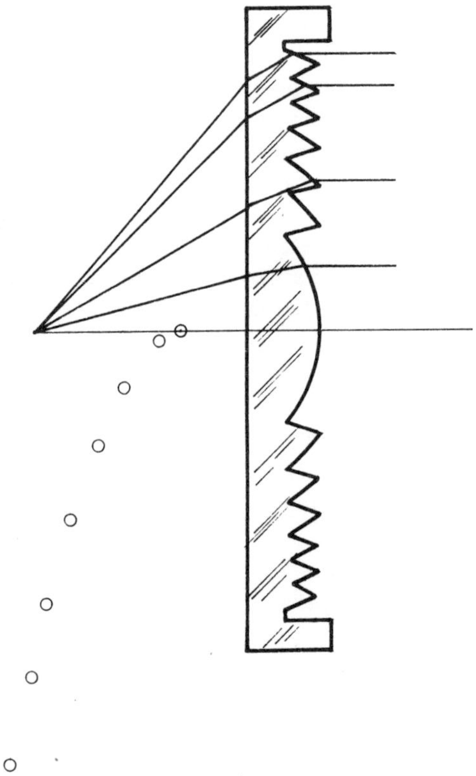

◁ **5.8.4. A Fresnel lens (J. D. Fresnel, 1788–1827) consists of a central spherical or aspherical form with circular zones surrounding the central lens, all having the same focal point and almost the same thickness. The centres of curvature of the circular zones (marked as small circles) do not lie on the optical axis. The radii of curvature increase with distance from the centre as in an aspheric lens. The lens will give parallel rays of light, but dark annular spaces are formed by the dead areas between the circular zones.**

apparatus as motor-car lamps and road-traffic signals. (Fig. 5.8.5.)

Where maximum use of available light is required, such as for searchlights, the Fresnel lens circular zones are replaced by catadioptric rings so that use is made of reflection as well as refraction. Fig 5.8.5 illustrates two refracting and three catadioptric reflecting rings which collect light from a larger angle than would be possible by refraction only.

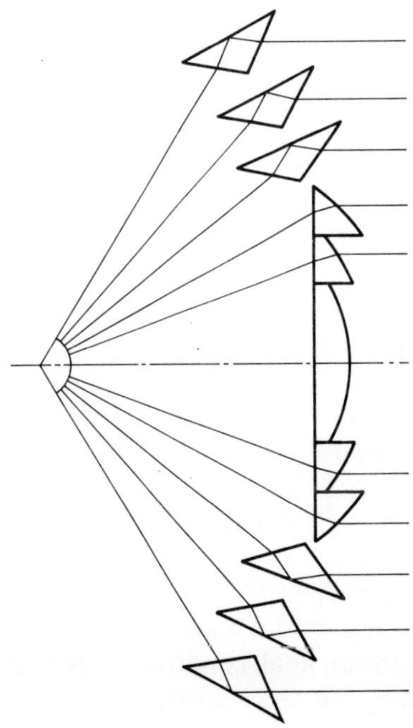

5.8.5. A Fresnel lens for maximum use of light.

Plastics Fresnel optics

Early lenses were of crude design. They were of glass, which had to be expensively hand-ground and hand-polished. In recent years, using high-quality optical plastics and precision-moulding techniques, Combined Optical Industries has been able to supply industry, the Forces and scientific establishments with a fast-expanding range of low-cost plastics Fresnel optics. These new, versatile optics are not only establishing themselves in the hitherto unchallenged domain of conventional lenses but are also penetrating new territories where glass optics are banned by prohibitive costs.

In its simplest form, a plastics Fresnel lens somewhat resembles a transparent gramophone record. There is no spindle hole, however, nor is there just one continuous groove. Instead there are many entirely separate, concentric, and usually equidistant, grooves. (Fig. 5.8.4.)

The curved volume of the spherical lens is in effect divided into numerous concentric rings which are telescoped to form the comparatively flat Fresnel geometry. The curved surface of each Fresnel ring corresponds to (but is not identical with) that part of the spherical lens surface whose function it takes over.

Materials

Fresnel optics are precision-moulded by Combined Optical Industries Ltd in two basic materials, polymethacrylate and Duroplast 01 PS. The choice depends on the rigours of the application envisaged.

Optics made of polymethacrylate are recommended for low- and medium-temperature work (except in low-pressure environments where the material is apt to emit gas). They are manufactured as readily-machineable sheets, in a variety of convenient configurations, which may be sawn, drilled or turned using standard workshop tools and machinery. Mounting and alignment are considerably easier than with unwieldy conventional optics because no elaborate clamping devices are needed.

The Duroplast 01 PS range is better suited to severe environmental conditions—it is especially recommended for applications in vacua and where operating temperatures up to 65°C must be tolerated. At greater temperatures the possibility of warping can be eliminated by heat filters or forced-air cooling.

Advantages

The most striking features of the plastics Fresnel lens are its lightness and slenderness, contrasting with the weight and bulk of the conventionally designed planoconvex glass lens; a plastics Fresnel lens may be as much as fifty times thinner and forty times lighter than a conventional glass lens. Extremely thin (0·25 mm) Fresnel lenses are available which can be temporarily deformed without risk of permanent damage. In fact, some of them can be rolled up, which simplifies packaging and reduces transportation costs.

Fresnel geometry may be easily and economically designed for almost all normal optical functions. Fresnel optics can replace mirrors, prisms, and diffractors. Much space can be saved by using multifunctional elements. For instance, two or more lens characteristics can be compounded to give a single Fresnel bifocal or multifocal.

Sensible employment of plastics Fresnel optics can reduce costs for makers of optical equipment.

Fresnel optics are less expensive than conventional optics of similar performance. Indeed, glass can be so costly in some applications that only Fresnel optics can provide an economic solution. Large-diameter condenser lens systems are an example. Fresnel optics can be miniaturized to a certain extent, so the overall cost of structural material is reduced.

Distortion-free magnification

A conventional glass lens suffers from spherical aberration—it forms an image which is an imperfectly magnified copy of the object. The image is most distorted where seen through the peripheral regions of the lens because peripheral rays are focused at points closer to the lens than the more central rays are. The complete flexibility of the manufacturing process permits aberrations to be designed out of Fresnel lenses. Virtually distortion-free magnification is possible.

Groove pitch and resolution

The imaging properties of Fresnel optics depend on the pitch of the grooves. Resolution improves as pitch decreases. Where image resolution is not a

5.8.6. A Fresnel lens of the type that would be used as a screen in an overhead lecturing projector.

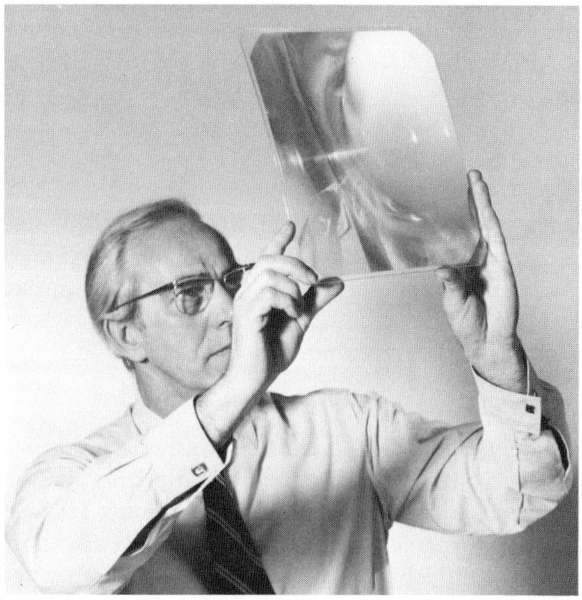

significant design factor (say in lighting fixtures, built-in exposure meters, or traffic signals) a groove pitch of 0·4 mm (about 60 grooves per inch) may be more than adequate. Where image quality is of primary importance (as in commercial magnifiers, photographic enlargers and rear-projection screens) Fresnel optics are available with pitches as small as 0·05 mm (about 500 grooves per inch). (Fig. 5.8.6.)

Fresnel beam splitters

A useful form of bifocal construction is known as an eccentric Fresnel structure because the rays are made to focus at points which do not lie on the geometric axis. In this way the twin functions of beam splitter (or beam diffractor) and focusing element are combined in a single Fresnel optic.

A noteworthy feature is that the optic is perpendicular to the incident light rays. Conventional beam splitters, which exploit the partial transmission/reflection properties of semi-silvered mirrors, must be inclined at some angle (usually about 45°) to the incident rays. Consequently much space can be saved by going over to Fresnel optics. A version is commonly used in cameras with built-in exposure meters. The chief benefit is that light representative of the whole field of view is focused on the photodetector, so the instrument responds to the average illumination.

Fresnel prisms

Miniaturized Fresnel beam-deflecting components may be used to advantage where the optical requirements are not too demanding. In large-scale instrumentation it is often desirable to monitor projected images of several indicators from a common observation point. The usual method is costly, as one prism and lens system is required in each projection path. Fresnel prisms may be inserted into the optical paths without loss of quality. But there are certain limitations—for example, it would not be wise to replace an entire lens system with Fresnel optics. Too many would cause interference, and there would be a strong likelihood of moiré fringes which would obstruct observation.

Large-diameter optics

Because of the simple manufacturing process it is possible to produce large-diameter plastics Fresnel optics (exceeding 25 mm) which can be employed where glass optics are ruled out by excessive cost. A typical and established application is the improvement of screen illumination in rear-projection systems.

The combination of Fresnel lens and ground-glass screen removes the central bright spot, increases illumination along the edges and in the corners of the screen, and sharpens the image.

5.9 Glass moulding from rod

In parallel with the development of plastic mouldings, great progress since 1950 has been made in the production of high-quality glass mouldings from fire-polished white glass rod. (Fig. 5.9.1.)

Glass rod can be reheated at one end, softened and reshaped by pressing. This has the advantage that only the minimum amount of heat is required to make the glass sufficiently hot to shape. More accurate dimensions and better surface finishes are achieved than by pot and tank pressing methods.

The surface finish on the finished moulding depends on the finish of the press tool and also on the surface of the glass rod, which must be free from draw marks, roller marks and abrasion marks from rehandling. The lower the temperature, the better the surface finish of a fire-polished blank, but the greater the force needed to shape the blank and the less the effect of the moulding annealing schedule. A good surface finish may be achieved at the expense of optical homogeneity.

5.9.1. Moulding aspheric condenser lenses from white glass rod.

Great care must be taken with the annealing of the rod, as the glass must not fracture when reheated at one end from room temperature to 1150°–1200°C in 10 minutes during the moulding process. Chance–Pilkington White glass rod (SW9) has the necessary physical qualities for flame polished mouldings suitable for:

Aspheric condenser lenses for projectors.
Inspection and viewing windows.
Spherical and aspherical mirrors.
Condenser mirrors.

The information in the remainder of this section has been provided by Mr T. J. Lawson, Managing Director of English Glass Co. Ltd, Leicester.

Englass mouldings from rod

Glass can be formed in many ways. The most familiar method is blowing, an ancient craft by which vases, drinking glasses, and bottles are formed, whilst pressed ware such as ash-trays, bowls, car lamp glasses, Pyrex ovenware, and valve bases probably come a close second.

Pressed articles are usually formed fully automatically, the glass being fed from a molten mass contained in a refractory tank. Where the quantity of mouldings is not great enough to warrant a fully automatic process, a method of hand gathering from the molten glass batch using a single tool can be employed. However, owing to the high temperature of the molten glass batch and the method of gathering the molten gob, the accuracy of the finished article is not of a high order and often chill marks occur on the surface.

The Englass rod-moulding process differs from the tank and pot method in that the raw glass material used is in the form of rods. By softening the end of the rod in a furnace and subsequently forming the soft glass between precision dies, a clean, accurate fire-polished moulding free from chill marks can be obtained. The flash, which occurs at the mould parting line, is then removed by subsequent trimming or grinding. Stresses which may occur within the glass owing to an uneven section or too rapid cooling are removed by controlled annealing.

Where it has been established that a glass moulding is required, but a moulding by the tank method does not provide the tolerance or smooth surface finish specified, precision moulding from rod may be the solution.

There are many cases where economies can be achieved by the use of glass mouldings instead of alternative products. Mould costs are low for this type of process, especially when compared with forming tools for metals or plastics, and longer service life is very often obtained, thereby saving replacement and maintenance costs.

Because of the relatively small temperature differential between mould and material when moulding from rod, high-quality surface finishes are generally achieved. Pre-conditioning of dies and raw material will result in a surface of good optical quality regardless of the complexity of contours required, thereby saving subsequent grinding and polishing operations. Typical mouldings of this type are those of aspherical and cylindrical form, condensers, biconvex and plano convex lenses.

Most glasses used for small mouldings will withstand temperatures up to 400°C. This makes coloured glass mouldings eminently suitable for use in light indicators which are necessary in an age where more automatic machinery is constantly put into use. Thus many plastics indicator caps have been replaced by glass caps whose colours will remain constant throughout the life of the component. Most glasses are chemically inert and are only attacked by hydrofluoric acid and hot phosphoric acid.

6 Production of Lenses in Quantity

6.1 Introduction

During the twenty-five years from 1945–70 very considerable advances in machinery, methods, tools, and cutting materials have been made, so that production times have been much reduced. Costs and therefore prices of cameras (including ciné cameras), have been reduced to a level such that a mass market has been reached. The micro-circuit industry has required a large number of microscopes and there has been a substantial increase in the annual demand for millions of spectacles in great variety and style.

The demand for improved methods by these large quantity producers of lenses has helped the scientific optical industry to obtain highly developed machines at prices which would have been economically impossible if limited to small-quantity production of high-grade components required only in hundreds per annum.

New techniques in diamond curve-generating, diamond smoothing, plastic polishing, continuous-flow abrasives and polishing compounds, high polishing pressures and speeds, which are now used on relatively small quantity production, were all developed for the mass market in optical components.

Some of these ideas can be traced back to W. Taylor of Taylor, Taylor & Hobson Ltd and others to F. Twyman of Adam Hilger Ltd in the years 1918–30, but it needed the mass demand and available capital to develop the machines and equipment which are now on the market and which perform each operation with the highest possible speed and efficiency. Before 1945 the conservative attitude of the optical industry delayed the large-scale introduction of new techniques.

6.2 Cutting disks from plate

Although moulded lens blanks are used in nearly all large-quantity production, the techniques of trepanning are still in common use for small quantities and where a homogeneous glass is essential for the highest class of optical instruments.

Hard Cutting

Large disks from sheet glass are usually cut with a diamond by hand (Fig. 6.2.1), and this method is also used for domestic purposes, such as when a hole is required in a window for the installation of an extract fan.

For small quantities of lenses where mouldings or trepanning cutters are not available, it is usual to cut sheet or plate into squares and then wax a few of the squares together in the form of a stack. This stack is then stuck onto the end of an edging chuck, and the normal edge-grinding procedure produces accurately ground blanks which, after removal of the pitch, are ready for curve generating. (Fig. 6.2.2.)

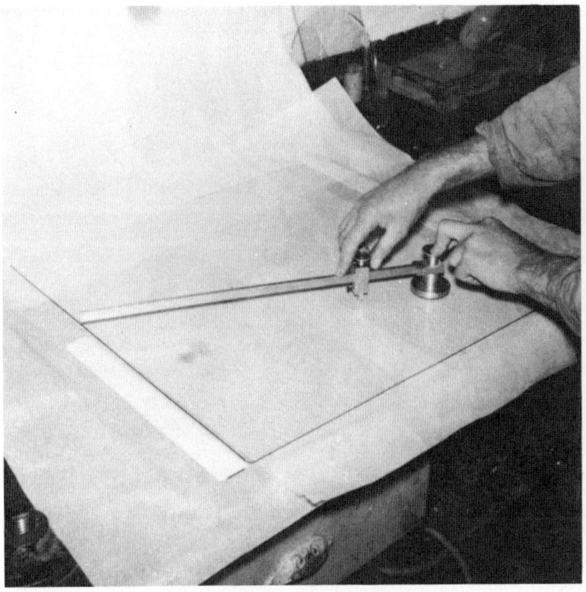

6.2.1. Cutting circles of sheet glass by hand with a diamond cutter.

6.2.2. Stack edging of lens blanks on a Taylor-Hobson edging machine—a process before curve generating.

6.2.3.▷ Typical trepanning cutters with lens blanks cut from slab glass.

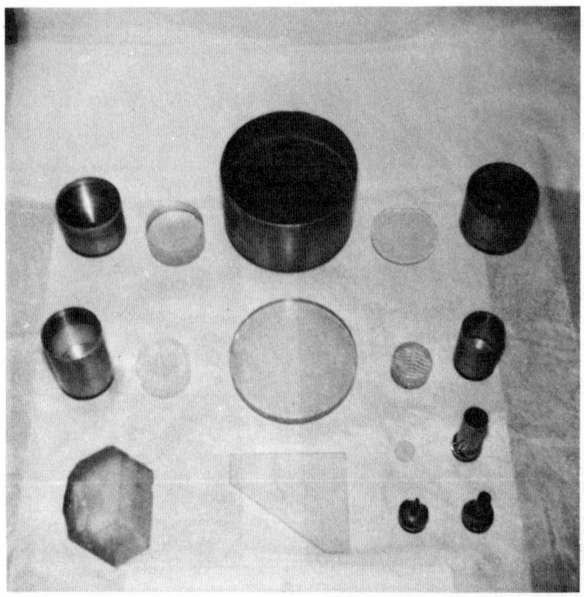

Trepanning

Glass for lenses is available as slab, and diamond trepanning cutters (usually with plated diamonds) are used to cut out the disks. (Fig. 6.2.3.)

A drilling machine with a powerful motor can be converted to provide an ample flow of lubricating liquid through the spindle. The manual adjustments to the machine are similar to those of a normal drilling machine, providing the usual stops on spindle movement are available. The glass plate to be trepanned is either clamped to the table or stuck to a sheet of plate glass; the diamond tool is fixed to the spindle and the table adjusted. A suitable lubricant for the cutter is Castrol Honilo cutting oil for quartz or Castrol Cleeredge for glass.

The trepanning tools (described in § 4.13) consist of a steel cylinder the front end of which bears a ring of impregnated diamonds thicker than the steel cylinder to provide clearance. It is essential that the power available be adequate for the diameter of the cutter and the correct speed for efficient cutting. (Fig. 6.2.4.)

6.2.4. Optibel trepanning machine cutting the centre hole of theodolite circles.

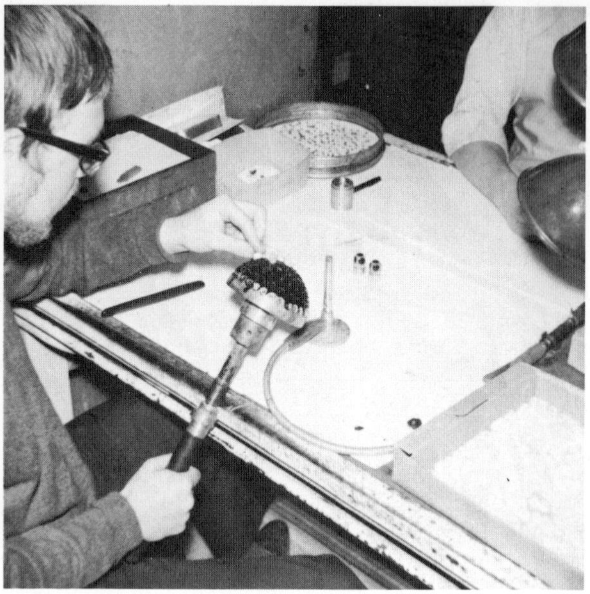

6.3.1. Lens blanks being pressed into a recessed tool before block curve generating on a Loh RF1 machine.

6.3 Moulding

Mouldings are available either direct from the glass-making machine on line for large quantity spectacle lenses, camera lenses, binocular lenses, and prisms, or alternatively as remoulds in a separate moulding furnace. (Fig. 6.3.1.)

The remould process consists of cutting and chipping slab to an accurate weight demanded by the particular moulding, heating the glass in a furnace to the correct plastic temperature, and pressing in the mould, followed by annealing in a furnace over a number of hours to restore homogeneity.

Mouldings so produced are consistent for diameter and thickness within close limits, which permits collets and fixtures for long production runs without alteration. (See Chapter 14.)

6.4 Curve generating

Before the introduction of sintered diamond tools, invented by Mr P. Neven who commenced manufacture in 1922 at his factory in Belgium, and now produced at Impregnated Diamond Products Ltd, Gloucester, the manufacture of lenses required a number of machining processes after starting from a moulding and before polishing.

At least three grinding operations were needed with increasingly fine grinding grain for rough machining. This was then followed by one or two fine-grain operations to achieve a finish suitable for starting the polishing process. Positive-contact tools and loose grinding compounds such as silicon carbide were used, which meant that during the operation the shape of the tools had to be continuously changed and corrected. The economical use of this process depended on manual skills.

These hand methods have been replaced by sintered diamond tools for milling operations. Sintered cup tools are used in precision spherical generators and the glass can be ground very much more accurately and quickly than was possible with the use of positive contact tools and loose abrasive.

The choice of bond is important for optimum service life and best surface finish on the lens. Bronze bond is in general use, but copper bond is used for the highest class of optical finish on soft glasses. Moderate wear of the tool guarantees optimum performance.

To obtain even wear of the tool, the diamond concentration is important and a concentration of 50 to 90 has been found suitable for milling lenses. Higher concentrations may increase the service life, but reduce the grinding capacity; for the blunt grains do not detach themselves sufficiently early from the bond of the tool.

Several glass types could be grouped together for the purpose of specifying optimum bond and diamond concentration. The type of coolant, the cleanliness of the coolant, and its temperature also affect the quality of the milled surface.

Modern curve-generating machines will consistently produce accurate spherical surfaces, which are then ready for fine grinding before polishing.

Curve generators are used in two ways:

(*a*) To generate the spherical surface on a blank or individual moulding.

(*b*) To generate a block filled with blanks or mouldings.

Diamond-impregnated tools are designed and constructed around the geometric principle illustrated in Fig. 6.4.1.

The P.C.D. of the milling tool is generally about three-quarters of the chordal diameter of the lens to be radiused. Where the curve of the lens is particularly steep, as for example in hyper-hemisphere work, the P.C.D. of the milling tool is critical; where the curve is shallow and in plano work, the P.C.D., if over half the chordal diameter, is relatively unimportant.

Head setting angle ϕ

From Fig. 6.4.1 it can be seen that ϕ is given by the formulae:

$$\sin\phi = \begin{cases} \dfrac{D}{2(R+r)} & \text{for convex lenses} \\[2ex] \dfrac{D}{2(R-r)} & \text{for concave lenses} \end{cases}$$

Setting dimension X

To compensate for the movement of the point of contact between the milling tool and the lens, a lateral displacement of the work piece is required. This displacement, X, may be calculated as follows (see Fig. 6.4.1):

$$X = \begin{cases} \dfrac{D}{2}\cos\phi - r\sin\phi\cos\phi & \text{for convex lenses} \\[2ex] \dfrac{D}{2}\cos\phi + r\sin\phi\cos\phi & \text{for concave lenses} \end{cases}$$

$$\therefore X = \begin{cases} \dfrac{D}{2}\cos\phi - r\sin\phi & \text{for convex lenses} \\[2ex] \dfrac{D}{2}\cos\phi + r\sin\phi & \text{for concave lenses} \end{cases}$$

since for small angles $\cos\phi \simeq 1$.

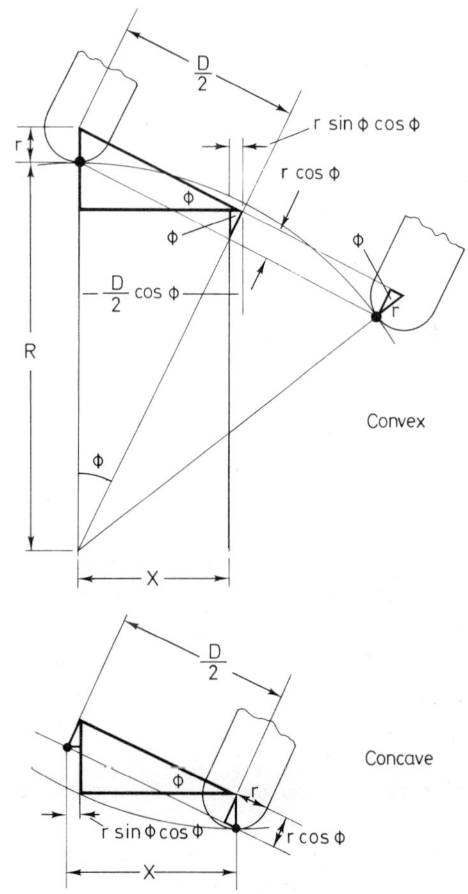

6.4.1. Geometry of diamond-impregnated tools.

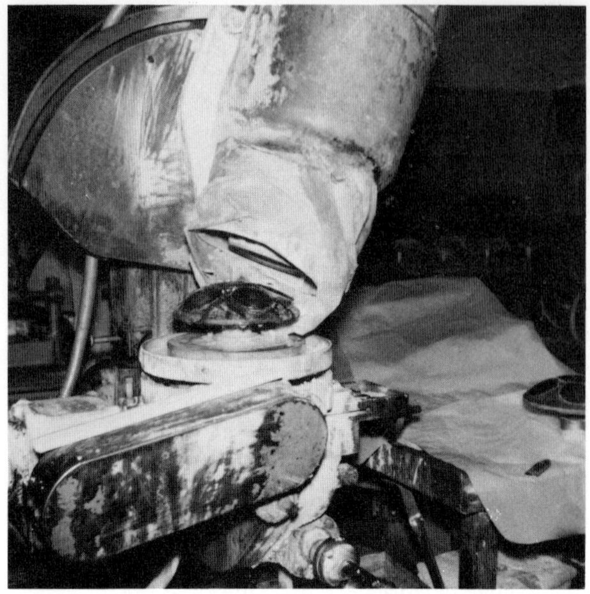

6.4.2. Curve generating a block of lenses on an Adcock & Shipley OVS curve-generating machine. The block is made of epoxy resin and the lens is the spherical side of a fire-polished aspheric condenser lens for a slide projector.

Generation of spherical surface on moulding

This is a very flexible method suitable for small batch quantities, as the full quantity can be quickly generated and then placed in stock until subsequent operations of malleting and polishing can conveniently be carried out. For best results, two spindles are used, one with a rough diamond wheel and the other with a fine diamond, so that the greatest accuracy and best surface finish is available for the fine smoothing and polishing operations. (See § 4.14.)

Some machines have an electric hydraulic cycling circuit and process timers for the rapid and accurate setting of the dwell period required before the motor is automatically shut off.

This control gives precise repetition on both curvature and thickness.

Mouldings are usually held in position on the chuck by vacuum which may be automatically cut off when the cycle is finished. Vacuum holding is essential for lenses having a diameter less than 30 mm but most larger lenses require only simple 'drop in' chucks.

Generation of spherical surface on a block of lenses

For large quantities of lenses, there can be no doubt about the economics of curve-generating blocks of lenses, whatever the subsequent smoothing and polishing processes. The problem is that the lenses, once stuck on the block as mouldings, have to stay in position until polishing is completed, and this means that a large number of blocking tools (runners) are needed. The blocking tools (with recesses for the mouldings) have to be very accurately made and are expensive. They are usually made of aluminium but can sometimes be made of steel, brass, or epoxy resin. (Fig. 6.4.2.)

Because of the large number of recessed blocking tools required for each curve there is an essential need for the subsequent operations of smoothing and polishing to be as short as possible, so that the tools can be unblocked and made available for the first operation once again.

6.5 Curve generators

The smaller range of machines available with capacity of up to say, 100-mm lens and blocks include:

Autoflow	124/158 machines
C.M.V.	60
Dama	M.F.S. 40 and F.S. 3
Loh	R.F. 1
Optibel	M.D.M.

The larger range of machines take lenses up to 550 mm in diameter and include the following:

Adcock & Shipley	0 V.S. and 2 V.S. machines
Autoflow	159 machine
Dama	F.S.K. 150 and 300
Loh	R.F. 2
Optibel	G.D.M.

Although varying between a vertical and horizontal layout the principle of operation is the same with each machine. The lens is chucked and rotated at a low speed (5–25 r.p.m.). To the centre of the lens is presented the edge of a round nosed milling tool rotated on a high-speed spindle (2000 to 16 000 r.p.m.).

Autoflow Technaphot Model 158 MK III A four-spindle curve-generator

This machine has special merit in its suitability for mass production of single lenses or small blocks. The work spindles will take a lens size from 4·75-mm to 70-mm diameter and diamond tools from 6-mm to 50-mm P.C.D. The high-speed spindle rotates at 10 000 r.p.m. and the work spindle at 20 r.p.m. The angular head has a maximum recommended angle of 45° but has a range from plano to 57°. The lens thickness can be controlled to 0·012 mm and curvature accuracy to 0·002 mm. The diamond-wheel spindle motor is 0·75 h.p. and the work spindle 0·25 h.p. The four spindles are grouped on a single casting of great strength, so that one operator can control them with a minimum of movement (Fig. 6.5.1).

The machine in its single-spindle form, Model 124 MKIIIA (Fig. 6.5.3), like the four-spindle form, has an electro-hydraulic cycling circuit with process timers (Fig. 6.5.2) for quick and precise setting of the 'dwell' period required. This control gives repetition on both curvature

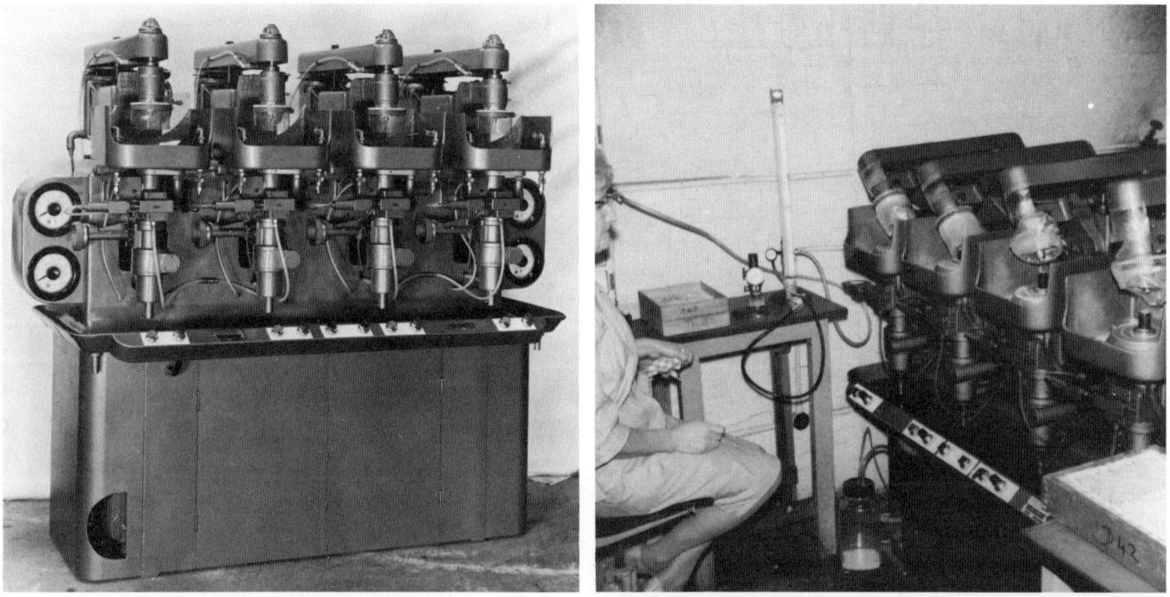

△**6.5.1.** (*left and right*). **Autoflow Technaphot Model 158 Mark III A four-spindle curve generator. Dial gauges show centre thicknesses. A Solex water gauge comparator on the table allows sphericity control.**

6.5.2.▷ **Technaphot electrohydraulic cycling circuit with process timers for 'dwell' control.**

6.5.3. ▷
Autoflow Technaphot Model 124 Mark III A single-spindle curve generator.

△6.5.4. **Two Technaphot Model 159 Mark III A curve generators set up for plano generating the surfaces of theodolite-circle blanks and reducing to thickness.**

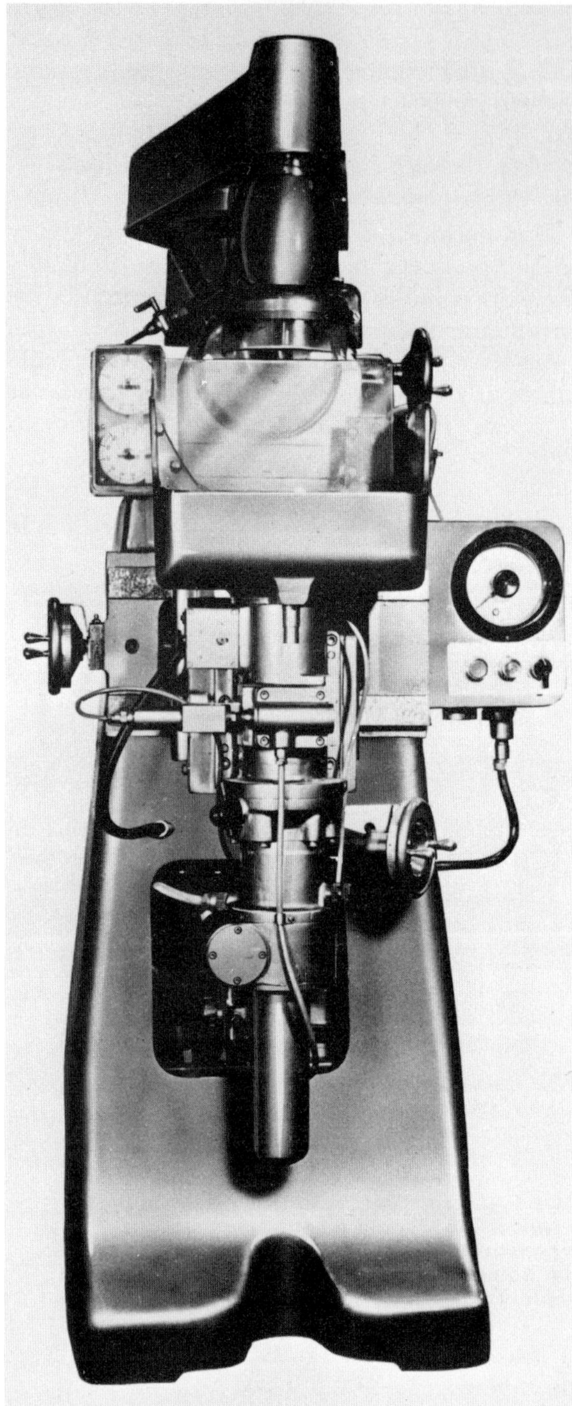

and thickness. Typically, a lens will take 45 seconds to grind, 15 seconds dwell, and 10 seconds to load and unload—a total of 1 minute per surface.

The use of clean cutting fluids is essential, and an efficient clarifier, preferably a centrifuge (see Chapter 4), must be fitted in order to achieve the best results. Vacuum equipment can be used for lenses of less than 30-mm diameter.

Autoflow Technaphot Model 159 MK III A high-capacity generator

This machine is a larger development of the Model 124/158 range and is suitable for either long runs on blocks of lenses or varied jobbing work. (Fig. 6.5.4.)

The work spindle will take blocks from 25-mm to 150-mm diameter and the maximum sag at 150-mm diameter is 50 mm. Diamond tools from 25-mm to 101-mm diameter can be accommodated.

△6.5.5. **Loh Model RF1A lens-curve generator with a magazine for automatic lens feed.**

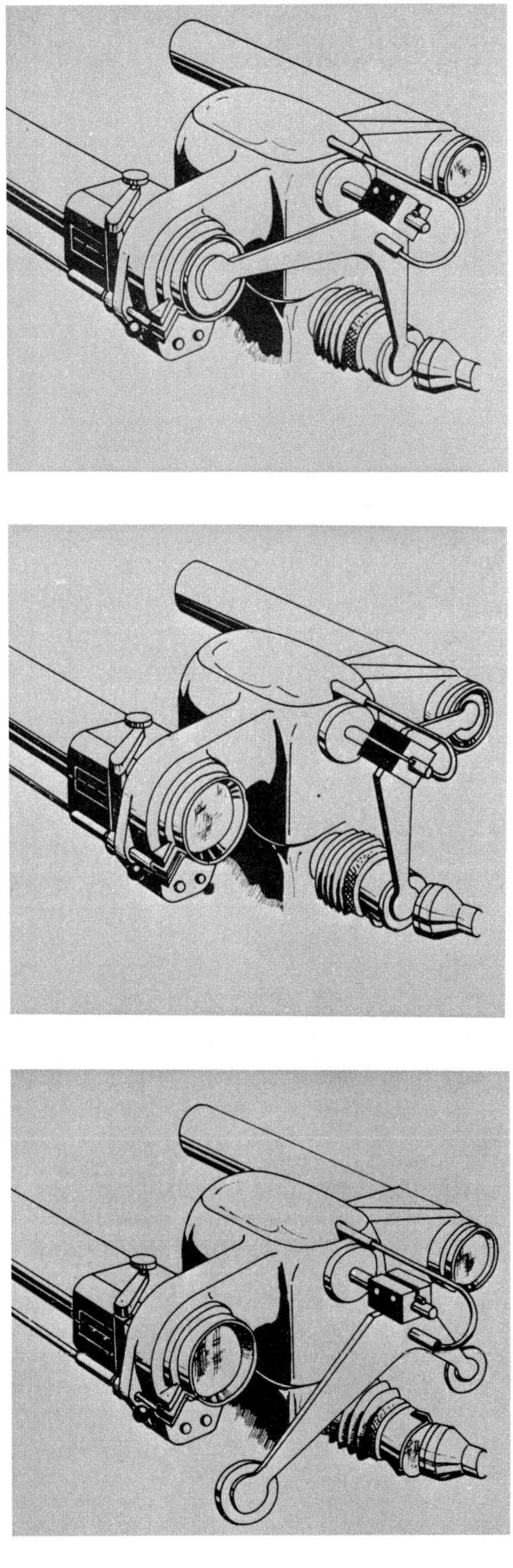

6.5.6.▷
Automatic feed of the Loh Model RF1A generator. ***Top:*** **simultaneous pick up of a blank from the loading magazine and of a generated lens from the collet chuck.** ***Centre:*** **loading of a blank into the collet chuck and discharge of a generated lens into the storage magazine.** ***Bottom:*** **The automatic generating cycle starts as soon as the vacuum transfer arm reaches its centre position.**

Low Model RF1A lens curve-generator

This machine is unique in that it has automatic lens feed from a magazine for single surface work up to 65-mm diameter. (Figs 6.5.5 and 6.5.6.)

The model RF1 has a maximum working range of 100-mm diameter for single lenses and recess blocks. The single lenses are held by collet chucks with quick release, or by vacuum. Recessed multiple lens blocks are centred by a centring flange and secured by means of a draw bar.

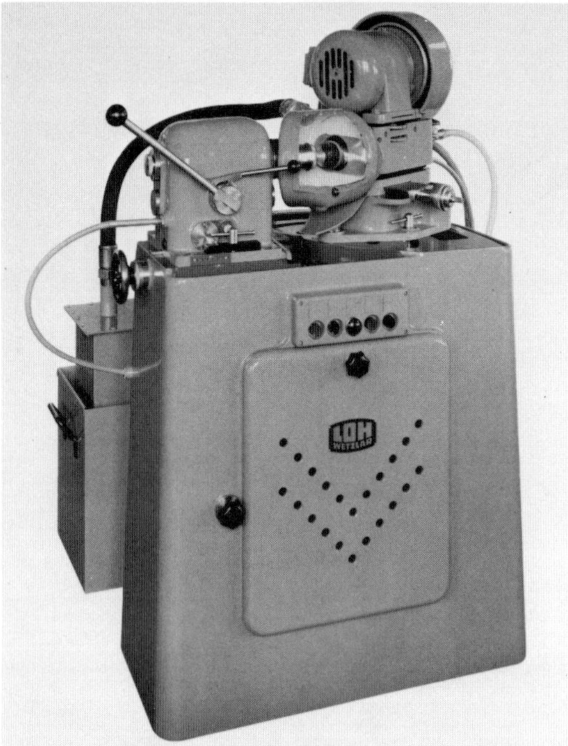

◁6.5.7. Loh Model RF1 with an attachment for generating curves beyond the hemisphere.

△6.5.8. Loh Model RF1 curve-generating of a hyper-hemisphere. Extreme accuracy of radius of curvature is possible by direct measurement across the spherical diameter.

◁6.5.9. Loh RF1 curve generator with bevel attachment.

Generation of curves over the hemisphere is a feature of the machine design. (Figs 6.5.7, 6.5.8, and 6.5.9.)

A fine-adjustment device enables exact repetition of angular settings, which reduces the set-up and down time of the machine.

The cycle time varies from 30 seconds to 4 minutes, and in the case of the automatic lens-feed model, the time is not controlled by the movements of an operator. The grinding spindle speed is approximately 12 000 r.p.m. This is a very universal machine which contributes to automation in the optical industry.

6.5.10. Strasbaugh Model 7M glass grinder and curve generator for work up to 700 mm in diameter.

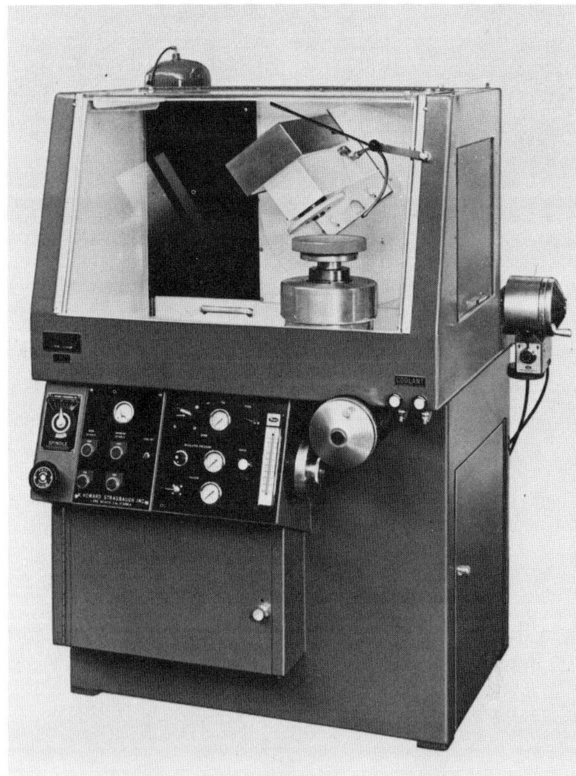

Strasbaugh Model 7 M heavy duty Universal glass grinder

The Strasbaugh Model 7 M has a work capacity up to 24-inch (700 mm) diameter and a glass thickness up to 7 inch (178 mm). (Figs 6.5.10 and 6.5.11.)

The machine will edge, bevel, generate curves, or grind flat. If required, it will also saw or grind holes. The lower spindle holding the work rotates at 150 r.p.m. and the travel is controlled by a pneumatic hydraulic system to ensure a smooth drive and so eliminate chips in the glass.

The upper grinding wheel spindle will tilt through 50°, and has a range of speeds from 1750 to 4500 r.p.m.

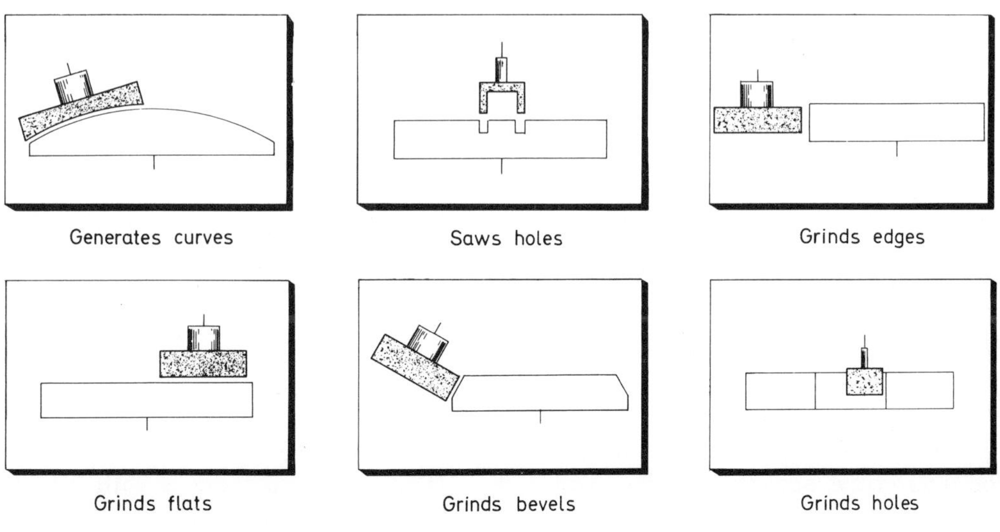

6.5.11. Operations with the Strasbaugh Model 7M grinder.

6.5.12. Loh Model RF2 lens-curve generator for spherical and flat grinding of blocks up to 400 mm in diameter.

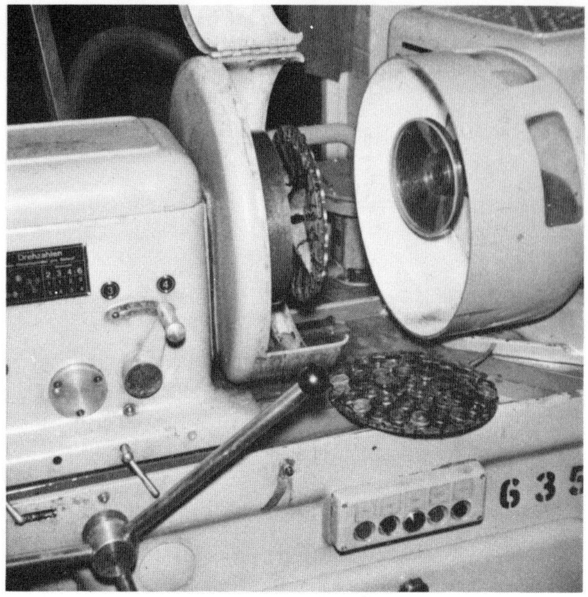

6.5.13. Loh Model RF2 curve generating a block of lenses.

Loh Model RF2 lens curve-generator for spherical and flat work

This machine has a working range up to 400-mm diameter standard or 500-mm diameter extended. It is also capable of generating hyper-hemispheres. Work pieces are held by collet chucks, on blocks, by vacuum or by a magnetic chucking device. (Fig. 6.5.12.)

The grinding spindle speed is 2000–3000 r.p.m. The lens carrier is fixed by a reception flange on the working spindle and the shape of this depends on the type of lens carrier to be used. The time for generating a block would be about 8–10 minutes, depending on the size of tool. (Fig. 6.5.13.)

Grinding of hyperhemisphere microscope O.G.s by air centrifuge and emery mill

By permission of Vickers Ltd the method of producing finish smoothed spheres 0·065 inch in diameter for microscope objectives is described. Cubes of glass of the correct refractive index, 0·08 inch in each direction, are placed in an air rolling mill. (Fig. 6.5.14.) This is a centrifuge of approximately 3-inch diameter. Around its inside surface is a band of emery cloth, except for the space at the bottom, where the air enters at a tangent. A quantity of cubes are placed in the bottom of the centrifuge, the plastic cover is lowered, and high-pressure dry air is turned on for about 1 hour. After this period the glass cubes will have become roughly spherical and are ready for coarse lapping.

The coarse-lapping tool has two concentric rings. (Fig. 6.5.15.)

The glasses are first placed in the outer lapping ring with coarse No. 320 emery until the diameter is reduced to 0·073 inch, and they are then transferred to the inner lapping ring and smoothed with No. 302 emery until the diameter is reduced to 0·069 inch. The glasses are then transferred to the smoothing tool and, with No. 303 emery, smoothed to a diameter of 0·0662 inch.

The final smoothing is with No. 303½ emery until the final spherical diameter of 0·065 inch is achieved. The smoothed sphere is then ready for mounting on a stick for polishing.

6.5.14.▷
Air rolling mill for microscope objective lens spherical grinding. The mill is mounted on the wall and the air enters tangentially at the lower left-hand side. The raised cover is shut down during grinding. The inside periphery is faced with emery cloth.

6.5.15.▷
The left-hand spindle is for coarse spherical grinding of microscope objective spheres and the right-hand spindles are for fine grinding. The instrument in the centre is a comparator gauge for measuring the diameter of the spheres.

Note. The traditional method of making microscope objective lenses is to take a piece of optical glass of suitable thickness and slit it into pieces about 1 inch (25 mm) square. These squares would then have one side ground and polished. The other side would then be ground down to a thickness about 0·002 inch in excess of the finished lens thickness. The square plates are then ready for cutting (by diamond) to a size a little larger than the finished lens. These small squares are now mounted with the polished side next to a stick, which will be used to hold the lens during grinding and polishing. The squares are now rotated by hand in a groove of the periphery of an abrasive wheel about 12–15-inch diameter, and the 'square' ground to a spherical shape. After this rough grinding operation comes a series of smoothing operations with a reduction in size each time, using 302, 303, 303½ emeries in succession, until the lens is ready for polishing with pitch and cerium oxide. For high-grade lenses, where an accuracy better than one ring is required sapphire powder is often used for polishing. Linde Alumina powders (Type A has an average size of 0·3 micrometres and Type B of 0·05 micrometres) provide smooth, scratch-free polishing. (See § 4.10.)

6.6 Malleting or pelleting lenses

Where small-quantity lens production is concerned, the usual way to support the lens, which must not move during the smoothing and polishing operations, is by a blob of fairly hard pitch. There are several methods of sticking the mallet of correct size to the lens (Fig. 6.6.3.):

(*a*) Hand method.
(*b*) Machine method.
(*c*) Pouring liquid pitch on the lens in a mould.
(*d*) Making mouldings with liquid pitch and then sticking the mouldings to the lens as a separate operation.
(*e*) Hard on blocking (negative lenses).
(*f*) Waxing into recess tools (for large-quantity production).
(*g*) Dripping pitch on to small lenses.

(*a*) *Hand method*

The well-tried hand method is mostly used for small-quantity production. The operator will require an electrically heated thermostatically controlled pitch pot, an electrically heated hot plate and preferably, a water-cooled cold plate. (Fig. 6.6.1.) The pitch will be heated to a temperature at which it will be a little stiffer than treacle, and it must be well stirred.

The lenses to be blocked are placed on a piece of white paper, which is on the electric hot plate. The lenses are warmed up. A blob of melted mallet pitch is collected on the end of a stick, which is rotated so as to prevent the pitch from falling off. The blob of pitch is then laid on a cold tool, where it is turned over to cool it. As it cools, it is formed into a flexible piece by hand, then placed on a lens and the required amount snipped off with scissors. The lens is then taken up in the hand and the mallet at the back is pressed onto the lens and worked up into a nearly hemispherical shape and placed in a box to cool.

(*b*) *Machine method*

For larger quantities, machine pelleting is more efficient. The pitch pot is under the machine and electrically heated to the correct temperature.

6.6.1. Malleting lenses by hand. A pitch pot and an electric hotplate are on the left. Lenses on the white paper covering the hot-plate are being warmed ready for the pitch to be stuck to them.

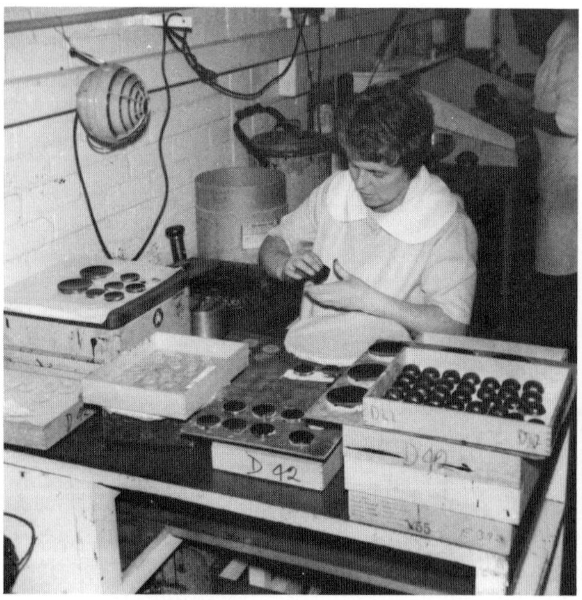

6.6.2. Malleting lenses by machine. The pitch is in a heated tank below the table surface and is forced up into the water-cooled mould that carries the lens.

◁6.6.3. **A variety of blocks of lenses, prisms, and theodolite circle blanks, illustrating the pattern of the components on the blocks during the smoothing and polishing processes.**

▽6.6.4. **Lowering the block-holder on to the malleted lenses.**

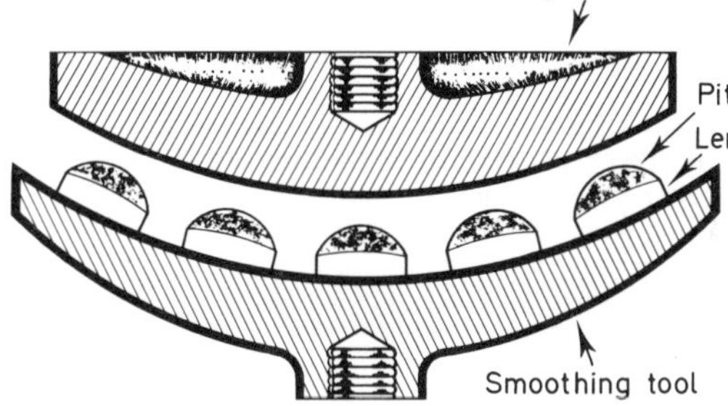

A lens is placed in a water-cooled brass mould, which is located over a nozzle through which the pitch will pass. (Fig. 6.6.2.) The mould is clamped to the table and hot pitch forced up into the mould in contact with the lens. The water-cooled mould rapidly cools the pitch and, in a few seconds, the lens with pellet can be removed ready for the next cycle. An advantage of machine pelleting is that the pellets are all the same size.

(*c*) *Pouring liquid pitch on lens in a mould*

A different version of the machine method is to have a multi-impression mould into which lenses are placed. The mould would be on a water-cooled plate under the pitch pot and the liquid pitch poured through a tap on to the various lens mould positions. In a few minutes, the plate would have cooled sufficiently for the pelleted lenses to be knocked out and the process started again. To prevent distortion of thin centre concave lenses by contraction of pitch in cooling, a small cardboard disk may be placed in the centre of the lens.

(*d*) *Making mouldings with liquid pitch*

Another version is to have a similar set up as described in (*c*) but to pour the pitch into the moulds and not on to the glass if strain or other defect is feared. As a second operation the bottom of the moulding or 'button' must be heated with a small flame and then pressed on the prewarmed lens.

After producing a mallet, pellet, or button on the lens by either of the three methods, the lenses must be mounted on the runner or block holder. The optical tool which is used for smoothing the lenses must be screwed on to a nose on the bench. (Fig. 6.6.4.) The surface of the tool must be cleaned very thoroughly and the lenses placed in the tool with the pitch mallets uppermost. The tool must be placed centrally under a blocking press—possibly an old drilling machine or arbor press.

The block-holder is heated on a gas-ring burner or electrically heated hot-plate until it is just too hot to touch. The block-holder must be placed centrally on the lenses and guided by the press slide so that the block is true to the smoothing tool. (Fig. 6.6.4.) Water can be sprayed over the pitch when the point is reached at which the mallets have been sufficiently deformed. As an

6.6.5. Radyne r.f. induction heater used to heat up an epoxy tool before sticking condenser lenses into recesses. The flow of work is from the right-hand to the left-hand side.

6.6.6. Lenses being blocked by the 'hard-on' method. The tool is warmed on a hot-plate. The wax is heated by bunsen burner and dripped on the tool. The negative lenses are then pressed in the wax. The smoothing process on a Bryant Symons machine with continuous-feed abrasive slurry is illustrated on the left-hand side.

alternative, r.f. induction heating of the block when in position under the press can be very effective. (Fig. 6.6.5.)

(*e*) *Hard-on blocking* (*negative lenses*)

Although the methods described in (*a*), (*b*), (*c*), and (*d*) apply equally to positive and negative lenses, another method for negative lenses, which gives consistent results, is the 'hard-on' method. (Fig. 6.6.6.)

In this case, the block-holder must have an accurately machined curve of a radius to accommodate the edge thickness of the lens when the smoothing operation is finished, and it is essential that the centre thickness control from curve generating be good.

The tool is heated on an electric hot-plate and the stick of wax is heated with a bunsen burner and then pressed on to the tool. The lenses are then pressed into position with the flanges hard on to the spherical tool. There are no mallets or buttons with this method but the radius on the tool must be specially machined to suit the method.

(*f*) *Waxing into recess tools*

The manufacture of recess tools is explained in Chapter 4. Whether they are made from aluminium, brass, epoxy resin, or steel, they have to be heated up before sticking in the lenses with wax. The wax may be a mixture of six parts of resin to one part beeswax, with a little black shellac for colour to aid examination of the glass.

Although gas flames and ovens are sometimes used for this process, there are advantages in the use of an r.f. induction heater, which very efficiently raises the temperature in a few seconds. (Fig. 6.6.5.) When controlled by a process timer, consistent heating is achieved and the blocks are

6.6.7. Pelleting of small lenses by the drip method before smoothing and polishing. In the background are Bryant Symons No. 9 machines fitted with the bowl method for continuous feed of abrasive or polishing compound to blocks of lenses.

ready for the wax, which has been heated for use by an electric hot-plate.

A series of operations for small lenses making good use of recessed tooling is, starting from slab, as follows:

(1) Reduce slab to thickness.

(2) Trepan holes in slab (nearly through) to suit lens diameter.

(3) Break off blanks.

(4) Wax blanks in recess tool (good side down).

(5) Curve-generate block.

(6) Smooth and polish.

(*g*) *Dripping pitch on to small lenses*

Small lenses which are polished on a stick rather than a block are best stuck in position by the use of a small Bunsen burner and stick of wax. (Fig. 6.6.7.)

Some methods for pitching small lenses consist of placing the lenses in a blocking tool and pouring the pitch on to the lenses, followed by dissolving the pitch away in the area immediately surrounding the glass.

A suitable blocking pitch for this process is six parts of hard pitch to one part of plastic vinylate. The solvent for dissolving this pitch consists of 50 per cent alcohol and 50 per cent Xylol.

6.7 Smoothing processes

Loose abrasive

When the blocking is completed, the block should pass as soon as possible to the smoothing machine to avoid movement of the pitch, as the mallets are inclined to sink unequally.

Smoothing and polishing machines vary considerably in detail of construction and capacity, but most of them have this in common: they provide for one tool to be rotated around a vertical axis while the other is moved to and fro in an approximately straight line in harmonic motion by means of a crank. (Fig. 6.7.1.)

A block should be smoothed twice in each of the grades of abrasive selected, with precautions when changing grades to avoid contamination and consequential scratching of the lenses. B.A.O. Centriforce abrasive BM302 and 303 should be satisfactory for most work. Some machines have a means of continuously feeding the abrasive slurry on to the tools and this method is usually faster and consistently good and needs less operator attention. After lapping deep-curved tools together with an appreciable thickness of abrasive, their surfaces may both be spherical, but the radii of curvature will differ by the thickness of the abrasive, so if they are cleaned and put into contact, they will touch in the middle. Although the thickness of the abrasive is only about 0·0002 inch it is quite sufficient to produce a noticeable effect. Periodically, smoothing tools must be serviced, as there is a tendency for them to wear off the true curve by degrees.

6.7.1. Traditional types of machine.

◁ (*a*) **Bryant Symons medium lens grinding and polishing machine. The operator is smoothing a block of lenses with loose abrasive.**

(*b*) **Bryant Symons heavy lens grinding and polishing machine. Note the pots of abrasive for brush application to the blocks of lenses.**

(*c*) **Taylor-Hobson polishing machine. Two of the spindles have the polisher on the spindle and the lenses on top. The other spindles have the block of lenses on the bottom.**

Diamond smoothing (fine grinding or lap milling)

When quantities are large enough, the diamond smoothing process, coupled with the use of recessed blocks of lenses, will substantially reduce smoothing and subsequent polishing times. The results on the smooth finish are very good, a semi-polished surface being produced. Diamond smoothing tools have a sintered coating in the form of pellets which have been stuck on the tool shell with Araldite. (Fig. 6.7.2.)

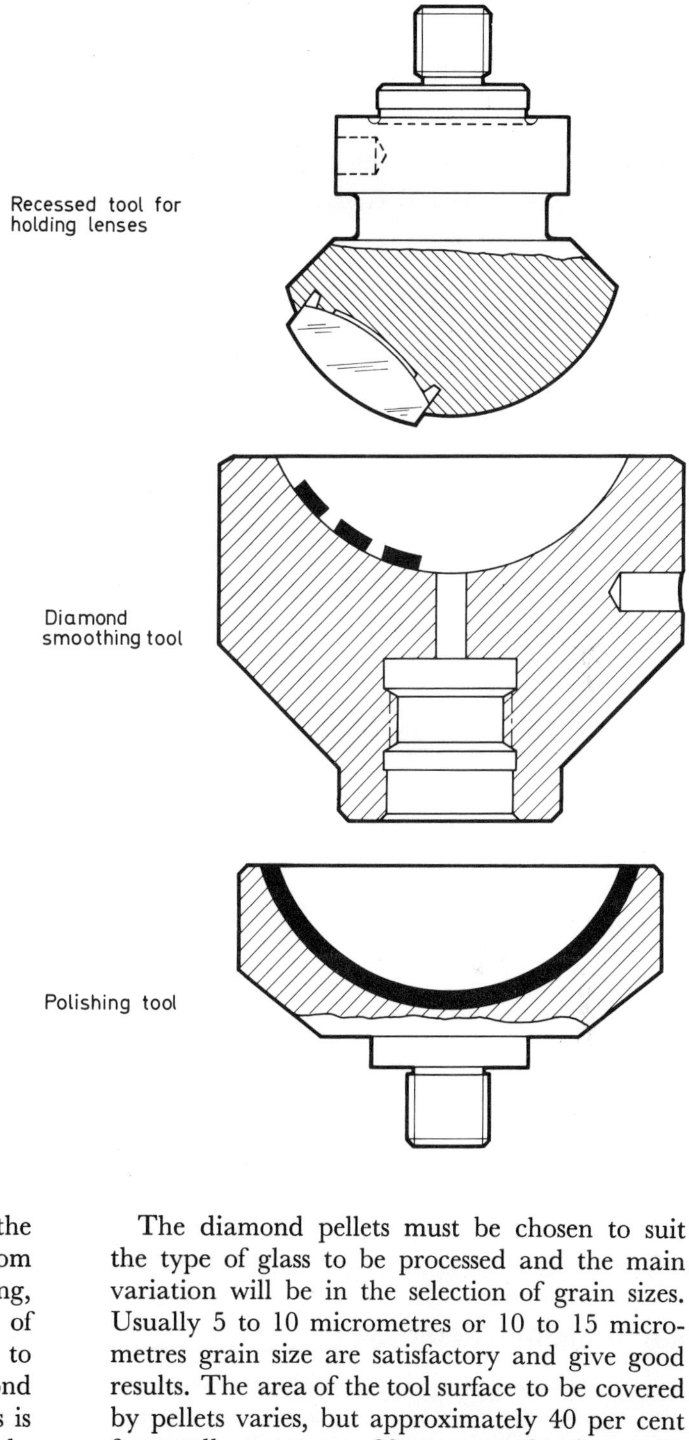

6.7.2. Recessed tool for holding lenses, diamond smoothing tool, and polishing tool.

The accuracy of the curve of the lenses in the block is very important, as any deviation from curve causes chatter during diamond grinding, resulting in chipped edges. The availability of curve generators, large and robust enough to grind recessed blocks of lenses, has made diamond smoothing a practical process; a factor in this is the long life and dimensional stability of the tools. (See § 7.9.)

Diamond smoothing gives a time saving of more than 50 per cent as compared with fine grinding with loose abrasive. (See § 4.5.)

The diamond pellets must be chosen to suit the type of glass to be processed and the main variation will be in the selection of grain sizes. Usually 5 to 10 micrometres or 10 to 15 micrometres grain size are satisfactory and give good results. The area of the tool surface to be covered by pellets varies, but approximately 40 per cent for small curvatures, 30 per cent for flat radii, and 20 per cent for flat surfaces is a good guide.

Experiments have shown the importance of clean (centrifuged) coolant of the correct specification to suit the glass being processed.

6.7.3. The Loh LP50 machine.
The machine was designed for diamond smoothing. Its rigid construction is essential to avoid faults due to vibration. (*Above*) The machine set up with the left-hand spindle for diamond smoothing and the right-hand spindle for high-speed polishing. Smoothing is effected in two minutes and polishing in three.

Uniform and geometrically accurate surfaces can be obtained regardless of manual skill, and the diamond tools enable curves accurate to within three rings to be achieved.

A very satisfactory diamond smoothing machine is the Loh LP50, which is a two-spindle machine with independent controls to each spindle. (Fig. 6.7.3.) Work pressure is pneumatically provided and can be regulated. The working range is up to plus or minus 50-mm radius. The time cycle is 10 seconds to 3 minutes and the spindle speed 2200 to 3500 r.p.m.

The Loh LP100 (Fig. 6.7.4) is a similar machine with larger capacity up to plus or minus 100-mm radius: time cycle 15 seconds to 5 minutes, and spindle speed 500 to 1500 r.p.m.

6.8 Polishing of lenses

The mechanics of the polishing process are similar to those of the grinding process but the polishing tool is lined with a specially prepared pitch or plastic, and the polishing compound is a slurry of water and cerium oxide or similar material. The polishing pitch will cold flow and take the shape of the work in a short time. The process of optical polishing of spherical surfaces depends mainly on the fact that a pair of spherical or plane surfaces are the only ones which will fit each other in all relative positions. (Fig. 6.8.1.)

As soon as a reasonable degree of polish appears, the form of the surface can be tested with a test plate under monochromatic light such as a

◁6.7.4. **The Loh LP100 diamond smoothing machine for multiple recessed lens blocks up to 100 mm in radius.**

▽6.8.1. **Lens mouldings and smoothed and polished lenses.**

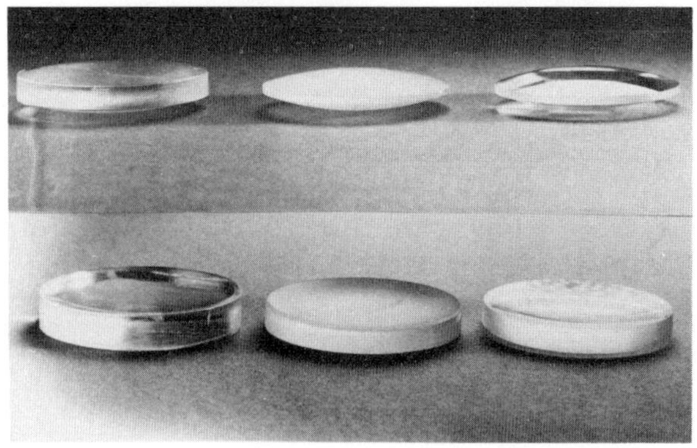

helium lamp. A set of convex lenses on the block would be tested with a concave test plate. After wiping both surfaces clean, the test-plate is slid closely on a lens so that interference colours can be seen. When large surfaces are polished, say 12-inch diameter and over, it is more convenient to use the Foucault knife-edge test described in Chapter 11.

After having tested the curvature of the lens and determined the 'figure' in formation, it is possible by machine adjustment of the stroke to control the polishing process and correct the errors. When dealing with small surfaces it may be necessary to chip away pitch from either the outside or inside surface of the polisher to lessen the wear towards the outside or centre of the block, or correct a zonal irregularity.

By careful control of these methods it is possible to produce a non-spherical polished surface. A concave mirror can be turned into a parabolic curve in this way.

Sometimes it is necessary to correct a lens for faults not due to optical aberrations such as radii of curvature, refractive indices, or separation of components. The errors may be due to irregularities of construction, lack of homogeneity of the components, or lack of true centring, and these faults are shown up on the Twyman-Green interferometer. To correct those faults, local figuring may be necessary (often carried out by local rubbing with a chamois leather pad and polishing compound) and this may be the final stage in making the finest quality of lenses or prisms.

The following machines have been proved to be reliable over a number of years' production of high-quality components.

Bryant Symons medium lens grinding and polishing machine MK II

The medium-size Bryant Symons machine has given very reliable service over many years. (Fig. 6.7.1(b).) The machine will produce, by

6.8.2. Bryant Symons lens grinding and polishing machine Mark II.

traditional polishing methods, flat work and blocks of spherical or cylindrical lenses up to 8-inch diameter.

The latest design (Fig. 6.8.2) has independent drive to each spindle and infinitely variable-speed motors with a spindle-speed range from 60 to 400 r.p.m. Automatic abrasive supply to each spindle (with thermostatic temperature control of the slurry) is available.

Bryant Symons heavy lens grinding and polishing machine

This machine has a capacity for 12-inch diameter blocks of lenses or prisms and is supplied as a four-spindle bank. It is designed for precision optical polishing and can be used for free-running blocks of high-quality flat finish.

Taylor Hobson polishing machines

These polishing machines (by traditional means) produce very high-grade optical components for professional camera lenses.

Autoflow BM4 precision scientific lens polishing machine Model AV2 (Special)

This machine will polish a shallow curve up to 15-inch (381 mm) diameter, but is equally efficient on steeper curves down to block diameters of 4-inch (100 mm) diameter. Extremely fast polishing times are possible at no expense to

6.8.3. ▷
Autoflow BM4 precision scientific polishing machine. The top left control is an independent speed control over the arm motor. The other two controls with timers are for automatic spindle speeds.

6.8.4 ▷
Autoflow BM4 precision scientific lens polishing machine. The independent motor drive to the arm and the air valve controlling pneumatic loading of the block are features of the machine.

quality. In the fully automatic version (Fig. 6.8.3) the spindle is controlled by an infinitely variable-speed motor with timer control, and, although the spindle can rotate at 800 r.p.m. maximum, the normal speed with 11-inch diameter blocks is 300 r.p.m. An independent Neco-geared motor driving the arm makes it possible by Variac control to vary the ratio of arm speed to spindle speed and so make figuring easier. (Fig. 6.8.4.)

The normal speed of the arm is 16 to 20 r.p.m. By means of an air valve, the pressure on the runner can be varied. For an 11-inch (275 mm) diameter block and polishing with Swedish pitch with pressure on the runner (excluding the weight of the tool) of about 20 lb (9 kg), output will be about three high-quality (two ring) blocks per 8-hour shift, when polishing with cerium oxide compound.

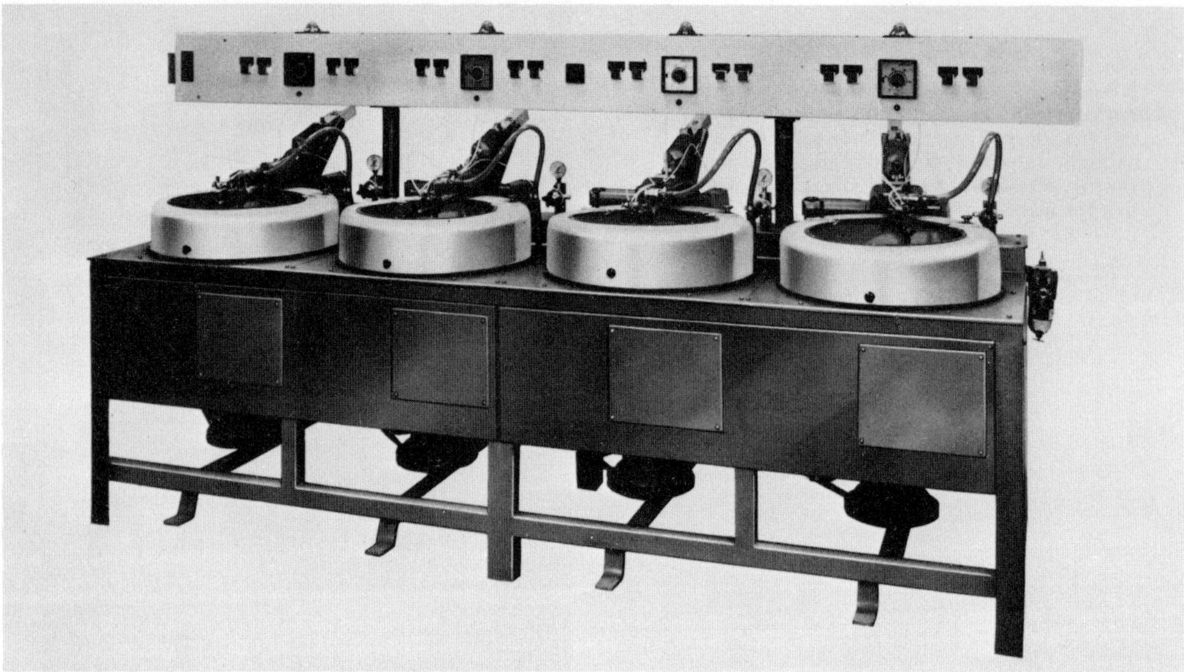

△6.8.5. **Bank of four Autoflow AV4 smoothing and polishing machines complete with automatic timers.**

A bank of four spindles (Model AV4) makes a very versatile machine suitable for emery smoothing, diamond smoothing, pitch polishing, or plastic polishing of lenses or prisms, according to the tooling and slurry used. (Fig. 6.8.5.)

Loh high-speed polishing machine PM 2

(Spindle speeds 300–650 r.p.m.)

This machine is suitable for single-surface or lens blocks with radius up to 35 mm and also over hemispherical curves. The lenses polished have accurate surfaces through constant polishing pressure, and the automatic feed of polishing medium ensures consistent results. (Fig. 6.8.6.) Up to twelve or fifteen spindles can be controlled by one operator on continuous work. Polishing times vary from 5 to 15 minutes per block.

Loh high-speed polishing machine PM 3

(Spindle speeds 85–240 r.p.m.)

With a working range up to 50-mm radius this machine will produce consistently good work.

The polishing medium is fed from a common container by means of a pump and the temperature of the liquid is controlled by a tubular immersion heater. There are many possible variations of pressure on the tool, types of polishing pitch, and spindle speed, but for high-quality

△ 6.8.6. Loh PM 2 three-spindle high-speed polishing machine.
◁ (*Opposite*) The Loh PM 2 machine is on the left and Loh LP 50 diamond-lap surfaces are on the right.

work with 5-inch (125 mm) diameter blocks, a pressure on the top of the runner (excluding the weight of the tool) of 35 lb (15·9 kg) for flints and 40 lb (18·1 kg) for crowns (with Swedish pitch and cerium oxide medium) gives good results. (Fig. 6.8.7.)

▽ 6.8.7. Two views of the Loh high-speed polishing machine PM 3.

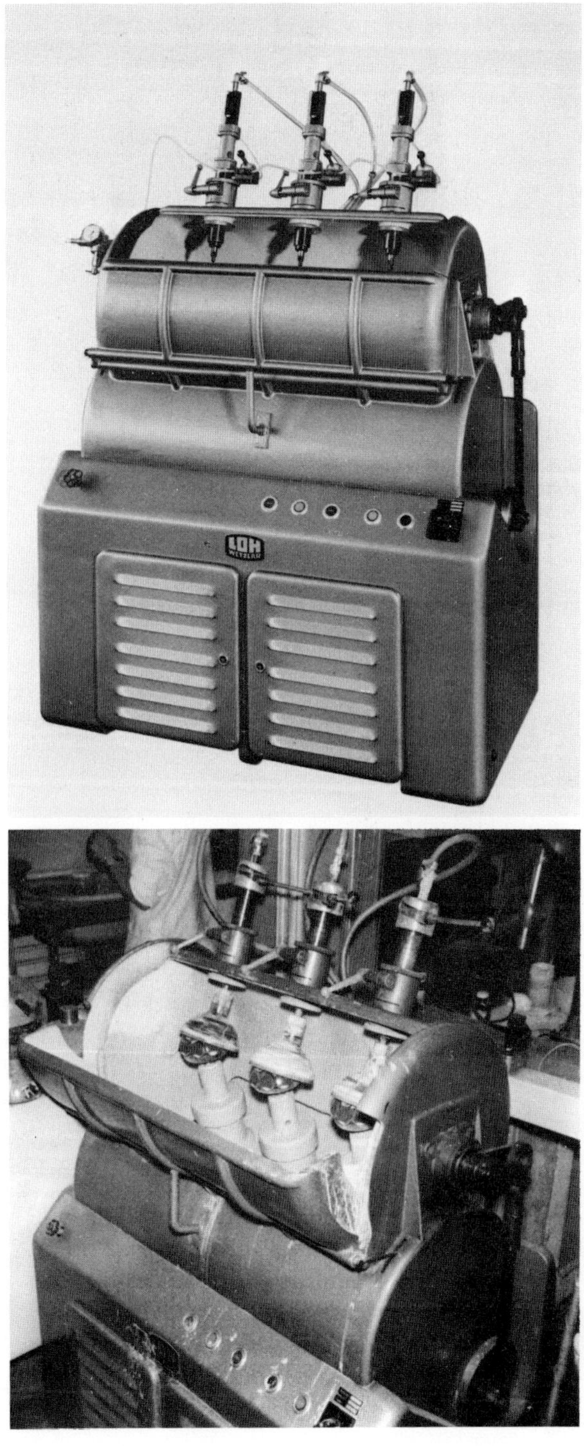

6.8.8. Loh PM 300 polishing machine for shallow curves.

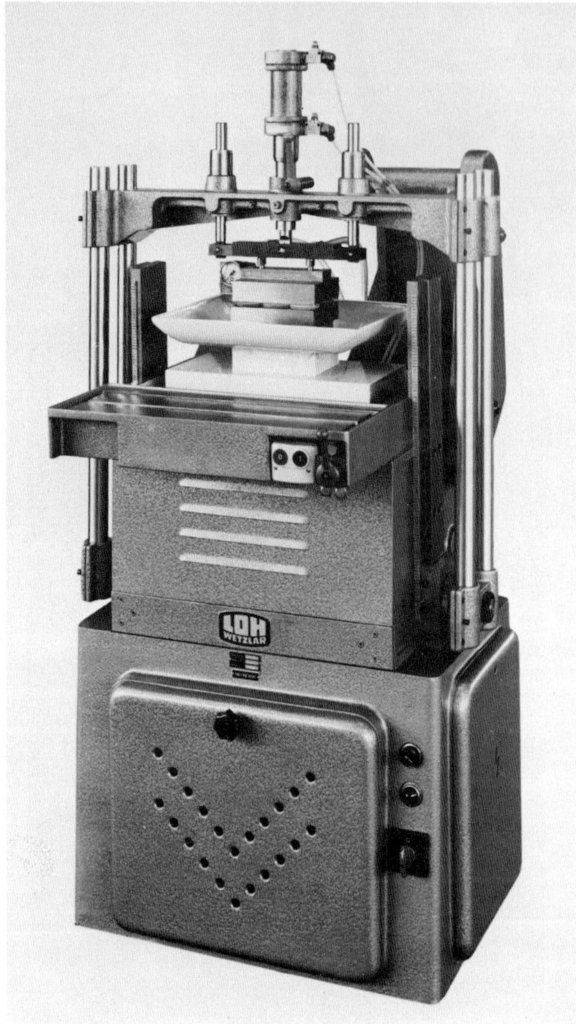

6.8.9.▷ Loh ZP 2 cylinder polishing machine.

Loh grinding and polishing machine PM 300

This machine is suitable for working shallow curves as well as plano surfaces up to 300-mm diameter. This is a good machine for high-quality work and incorporates the best features of traditional designs of polishing machines, with speed variations to suit the work, and adjustable spring loading to replace the traditional weights on the tools. Smoothing times are about 20–30 minutes per block and polishing times about 60 minutes per block. (Fig. 6.8.8.)

Loh cylinder grinding and polishing machine ZP 2

This machine meets a long-felt need for a reliable cylinder-polishing machine. Maximum working range is 500-mm radius, and maximum workpiece size 150 × 100 mm. The ZP 2 incorporates the well-tried features of PM 2 and PM 3 machines, and the constant working pressures guarantee uniform wear of tools, which is particularly important for cylindrical surfaces. (Fig. 6.8.9.)

Vickers microscope lens polishing machines for small blocks and hemispherical objectives

The machine shown in Fig. 6.8.10 is suitable for single lenses and small blocks. Lenses are tested on a spherometer, by comparison with the curve on a testplate and readings on a Mercer Clearline air-gauge unit. The Mercer Clearline

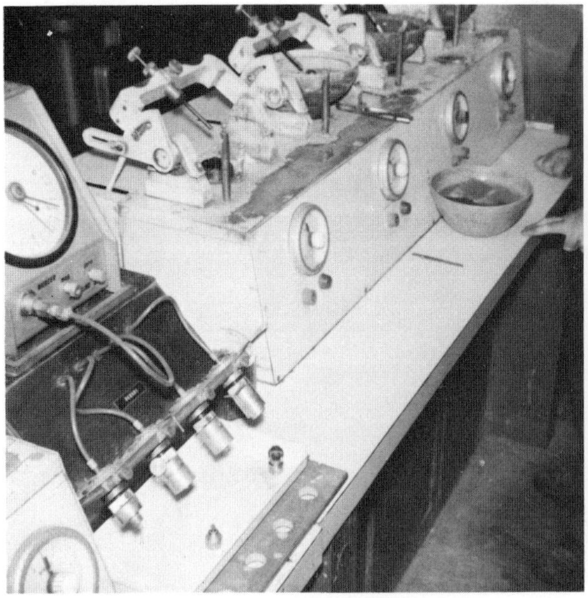

6.8.10. Polishing machine for small single lenses and blocks. Dimensional tests are by comparison with a testplate through a spherometer. The probe senses the movement caused by the spherical surface and displayed by a Clearline air gauge.

6.8.11. Polishing machine for microscope hemispherical objectives. Dimensional tests are by interspherometer and non-contact examination of fringes by means of a microscope eyepiece.

unit uses the back pressure principle of pneumatic gauging and operates from a standard compressed air supply of 4·2–8·4 kg per square cm (60–120 lb per sq. inch). This high operating pressure gives improved accuracy of measurement because of the cleaning effect on the work surface. At a magnification of 5000 each division on the scale represents 0·0001 inch with a range of ±0·001 inch, or 0·001 mm with a range of ±0·025 mm.

The machine shown in Fig. 6.8.11 is for the smallest objective hemispheres, which are tested on an interspherometer. This is an instrument which works on the principle of a shearing interferometer and makes it possible, on examination of fringes through a microscope eyepiece, to see errors in sphericity.

6.9 Removal of lenses from blocks

Polished glass surfaces are delicate and easily scratched, and so, immediately after polishing, blocks of lenses must be rinsed in water to remove the polishing compound, dried, and the surfaces either brushed with shellac or sprayed with a protective cellulose paint. Details of pitch and solvents will be found in Chapter 4.

Lenses are usually removed from mallets or pitch buttons by placing the block in a refrigerator. (Fig. 6.9.1.) After about 20 minutes at −20°C the lenses can easily be knocked off the pitch.

Lenses which are waxed in recessed tools must be removed by heating, and the blocks can be placed in an oven (Fig. 6.9.2) at 140°C for 10–15 minutes, or alternatively heated by an r.f. induction coil in a few seconds controlled by a process timer. In either of these two methods of heating, when the blocks are at the correct temperature, the lenses can be lifted out without difficulty.

If the lenses are painted with shellac, and it is desired to remove the pitch only, then they should be soaked either in Securitine non-inflammable cleaning spirit, from Securitas Co. Ltd, or G700 Solvent, from G. S. Chemicals Ltd. (Fig. 6.9.3.)

However, experience shows that it is often better

6.9.1. Blocks of lenses in a refrigerator to remove the lenses from blocks by contraction of the pitch.

6.9.2. Recessed blocks being heated in an oven before removing the lenses or, alternatively, before sticking lenses into the recesses.

6.9.3. Trays of Securitine in which lenses or prisms are placed for soaking off pitch without dissolving shellac.

6.9.4. Ultrasonic tanks (supplied by Dawe) for taking Dunkit and racks of lenses from which pitch is to be removed. The second tank is filled with Formula 861 for removing all trace of Dunkit from the lenses.

6.9.5. Cleaning lenses. (*Left*) **Racks of lenses before and after cleaning in acetone followed by isopropyl alcohol and detergent.** (*Top right*) **Racks of lenses to be cleaned are placed in an ultrasonic tank (at front of picture) containing acetone. The rack is then placed in a tank containing 50 per cent isopropyl alcohol and 50 per cent water with Quadriline detergent. The rack is then sprayed with water to rinse it.** (*Bottom right*) **Before vacuum coating, the lenses are loaded in a rack that is placed in a tank of isopropyl alcohol. The lenses are then wiped and loaded into the coating domes.**

to strip off all of the pitch and shellac (or paint), and a powerful solvent for this is Dunkit (from Penetone Co. Ltd) preferably used in an ultrasonic tank. This tank has a Dawe 25 kilocycle Soniclean Generator with high-intensity tank transducer assembly. After the Dunkit, a dip in an ultrasonic tank of Formula 861 (also from Penetone Co. Ltd) followed by a rinse in water will leave the lenses completely clean. (Fig. 6.9.4.) The lenses should then be dried, laid out in a tray, and the polished surfaces brushed again with shellac or sprayed with protective paint before malleting for second-side polishing.

An alternative cleaning process, if the lenses had been coated with cellulose paint after polishing and before deblocking, is to load the racks of lenses into an ultrasonic tank of acetone. (Fig. 6.9.5.) This is followed by an ultrasonic tank of

6.9.6. **Lenses in trays of isopropyl alcohol before being wiped and after cleaning in a vapour tank.**

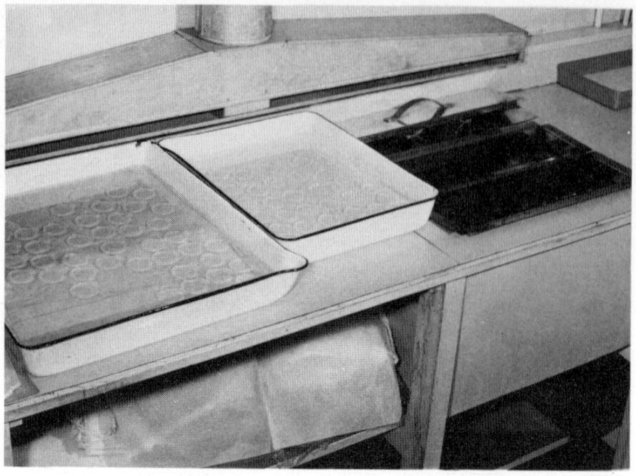

50 per cent isopropyl alcohol and 50 per cent water with Quadraline detergent (Quadraline Chemical Products Ltd). The rack should be removed and sprayed with water to rinse.

Note. Polished lenses must not be allowed to dry, or they will stain. (Fig. 6.9.6.) They must be correctly wiped or cleaned in vapour solvent such as iso-propyl alcohol.

6.10 Centring, edging and chamfering lenses

Having polished the lens on both sides to the specified radius and accuracy and cleaned both surfaces, it is now necessary to reduce the blank to the required diameter and at the same time ensure that the edge generated is co-axial with the optical axis.

There are two methods of centring lenses for normal commercial accuracy: visually by examination of the reflected images from the lens, and automatically by bell-chuck clamps.

The oldest method, and the only satisfactory method for shallow lenses, employs an accurately turned spindle and chuck. The lens is stuck on the chuck with pitch and reflections of a lamp are observed in the lens, possibly with the aid of a microscope for additional accuracy. The spindle is then rotated and, whilst the pitch is soft, the lens is moved to a position where the image of the lamp remains stationary.

Some examples of these techniques and suitable machines are as follows.

The Taylor-Hobson method

The method used by the Taylor-Hobson factory for many years requires a solid ground-steel spindle and a brass chuck very accurately turned concentric with the spindle. The method is fully described in § 6.11, as it is identical with the operation before cementing. After the lens has been accurately centred on the chuck, a very accurate grinding machine is needed for the edging if high quality is required on thick lenses which must be ground parallel.

A suitable machine is the G. Karstens Cylindrical grinding machine (Agents: Dixi and Associates). The features of this machine (Fig. 6.10.1) are:

(*a*) Hydraulic table traverse infinitely variable.

(*b*) Work rotation speeds of 30–450 r.p.m., infinitely variable.

(*c*) In-feed range (0·001–0·025 mm) 0·00004–0·001 inch, infinitely variable.

(*d*) Plunge grinding speed infinitely variable.

(*e*) A single push-button control starts the automatic grinding cycle.

Loh WG Universal edging machine

A popular machine using a precision ball-bearing spindle is the Loh WG edging machine which in addition to accepting optically centred lenses on an exchange spindle basis, can be used with bell chucks for automatic centring. (Fig. 6.10.2.)

For the operation of optical alignment and cementing the lens to the chuck, a lens adjusting bench is necessary. This may be either vertical or horizontal to suit local requirements. (Fig. 6.10.3.) The Loh WG machine will edge lenses up to 120-mm diameter.

A turning and grinding machine, Loh DSM, is

△ **6.10.1. Karstens cylindrical grinding machine for very accurate automatic edging of lenses after optical centring.**

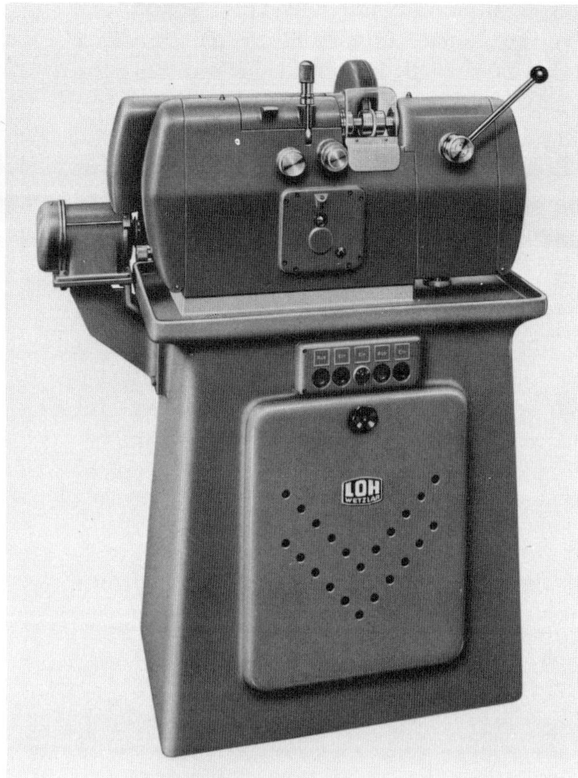

6.10.2 ▷
Loh Universal WG automatic lens edging machine.

6.10.3. Centring and edging.

△(*a*) **Centring a lens on a Loh ball-bearing spindle held in a vertical position during optical alignment.**

△(*b*) **A horizontal lens adjustment bench for optical alignment with use of a microscope and a Loh WG horizontal edging machine with a lens being edged after optical alignment.**

necessary to machine the lens chucks very accurately. For small quantities of lenses, brass chucks can be turned, but for larger numbers the chucks should be of hardened steel, ground and polished. (Fig. 6.10.4.)

The Loh DSM machine will accept spindles for the WG and LZ 25 lens-centring and lens-edge grinding machines.

Optik Dama Maschinen edging machine

This horizontal grinding machine has been adapted to take Loh edging spindles and has a precision bevel setting which is particularly useful for deep bevels, a common requirement of microscope objective lenses (Fig. 6.10.5). A typical series of operations would be:

(*a*) Stick lens blank (cylinder) on chuck.

(*b*) Bevel on machine and remove from chuck.

(*c*) Mount on polishing tool, grind, smooth, and polish both sides. Clean and inspect.

(*d*) Centre on Loh optical centring machine with aid of microscope.

(*e*) Edge true to optical axis on Optik Dama Maschinen.

(*f*) Remove lens from chuck, clean, and inspect.

Strasbaugh Model 7-H centring and edging machine

This is an extremely accurate versatile grinding machine capable of repeating to 0·0002 inch and will accept work up to 16-inch diameter. (Fig. 6.10.6.) The oscillation travel of 0–4 inch (0–101 mm) is particularly useful in permitting stacks of blanks 4 inches (101 mm) wide to be accurately edged.

The work spindle on the grinding machine can be set at angles for bevelling. A special centring machine enables the operator to optically

◁ **6.10.4. Loh DSM Centring Spindle turning and grinding machine.**

▽ **6.10.5. Precision grinding machine made by Optik Dama Maschinen adapted to take Loh spindles after a lens has been optically centred on the chuck. This machine is used to bevel microscope lenses accurately at Vickers Ltd.**

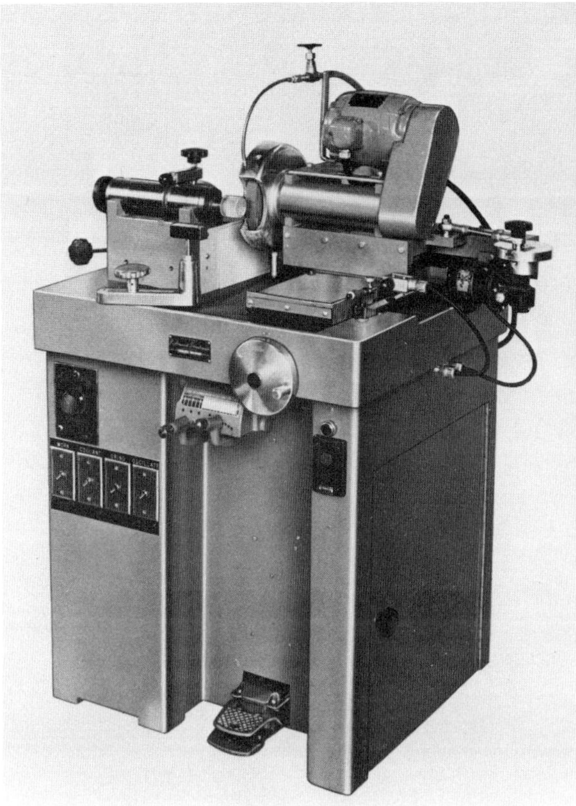

6.10.6. Strasbaugh Model 7 H centring and edging machine. The machine is very suitable for the accurate edging of thick lenses. (*Above*) The machine is shown cylindrically grinding the edges of theodolite circles. The lens centring machine is on the left and a spindle is shown in position. The centring is carried out with the spindle vertical.

mount one lens whilst the other is being ground. The spindle is held vertically, so making it easier to centre large lenses.

Strasbaugh Model 7 J centring, edging and bevelling machine

This machine will accept work up to 8-inch (203 mm) diameter and will edge and bevel or edge and face in one operation. (Fig. 6.10.7.) The work spindle is vertical instead of horizontal, and therefore does not tend to deflect or sag in the bearings. During the grinding operation, doors shut to enclose the work table completely.

Hand chamfering

This equipment consists of a vertical smoothing spindle within a sink sunk into a bench. (Fig. 6.10.8.) The lens can be held by a vacuum sucker mounted on a wooden handle. Lenses are chamfered in a negative tool, to which grade AO 302½ emery is added liberally by brush. Lenses are picked up by vacuum, roughly centred, and worked in the tool by a normal 'smoothing' stroke. The amount removed is calculated by the operator's counting the number of strokes. The lens is then thoroughly washed under a tap with cold water and passed to a storage tray containing water, which tray is passed to the cleaners when the batch is completed.

Photoelectric centring with exchange spindle

The accuracy of lens edging, with the exchange spindle method, has always depended on the accuracy of the spindle and the precision with which the lens could be centred on the chuck. Shallow lenses, thick edge lenses, and lenses with very tight centring limits have presented production problems with conventional methods and visual microscopes.

Recently, Sira Institute have developed a

△ **6.10.7** (*Left and right*) **Strasbaugh Model 7J centring, edging and bevelling machine. The foot pedal operates the work spindle at centring speed. The machine can take work up to 8 inches in diameter.**

6.10.8 ▷
Although one bevel is usually possible during edging, it is often necessary to add another bevel by hand. Bevelling is usually a hand operation. The lens is held in a rotating cast-iron hemisphere, with an emery abrasive to remove the sharp edges.

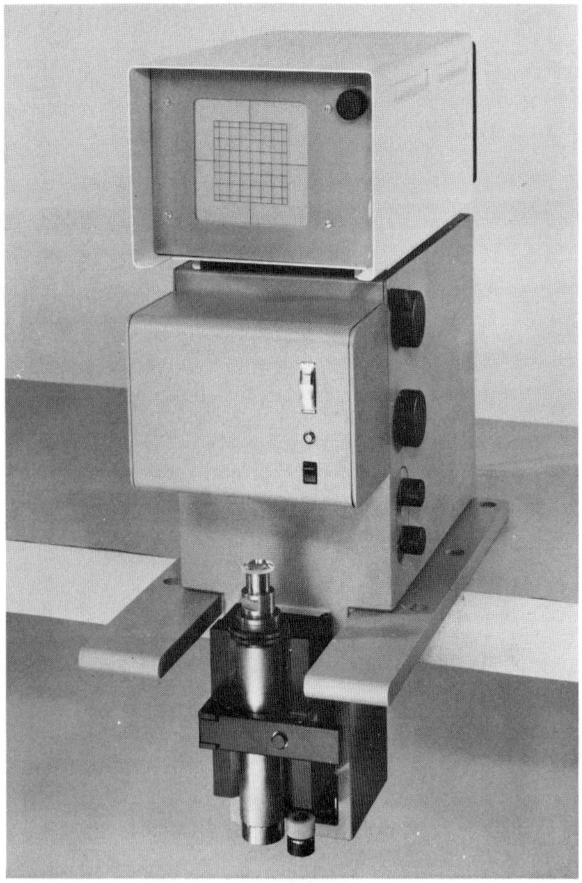

6.10.9. Photoelectric lens-centring aid for high-accuracy optical centring.

photoelectric centring aid as an alternative to the visual observation of reflections where very high accuracy is required. (Fig. 6.10.9.)

The lens is fastened with pitch or wax to a bellchuck carried on a spindle. Thus unit is put into the centring instrument. The lens is then moved with respect to the bellchuck, while remaining in contact with it, until the lens is centred. After centring, the unit is transferred to the lens-edging machine.

The only way to centre a lens held on a bellchuck is to tilt the lens about the centre of curvature of its lower surface. Centring accuracy is therefore best expressed in terms of tilt of the *upper* surface. This instrument has a constant sensitivity—in these terms—independent of the radius of curvature of the lens surface.

The centring accuracy of the *lower* surface is limited by bearing and bellchuck error, and by the seating of the lens on the bellchuck. The lower surface may be checked after the upper surface has been centred.

The instrument is simple to use. The spindle is clamped into a vee-mount on the instrument face, whereupon it will commence to rotate. The lens to be centred is fixed to the bellchuck by means of adhesive wax, warmed by the small gas flame provided. The detecting unit is then lowered until the 'focus' meter reading is a maximum. As the spindle rotates, the centring error shows up as a circular trace on the c.r.t. The motion of the spindle is temporarily arrested with one hand and the lens is moved with the other so that the spot moves towards the centre of the circle. When the spindle is released, the error trace will be reduced in diameter. A few repetitions of this process will suffice to reduce the centring error to an acceptable value.

Once the top surface has been centred, the centring of the lower surface may be checked. The detecting unit is lowered further to focus on the lower surface of the lens. This time, the size of the error trace will be comparable with that finally achieved for the top surface. If it is occasionally found to be significantly larger than this, the explanation will be either that the lens is incorrectly seated on the bellchuck, or that the bellchuck or bearing is damaged.

To facilitate centring large numbers of identical lenses, the focusing slide may be equipped with adjustable end stops. A lever mechanism would enable the upper or lower surface of the lens to be selected rapidly as required.

The oscilloscope display has a high sensitivity—for accurate centring—over its central portions, but a progressively decreasing sensitivity towards the edge, to allow for the large initial error when the lens is first attached to the bellchuck. The oscilloscope uses a long-persistence phosphor so that the error trace remains visible while the operator is adjusting the lens position. This form of display provides for the quickest and most natural operation of the instrument.

Automatic bell-chuck centring, edging and faceting

Since 1955, several centring and edging machines have come on the market (the most important manufacturers being W. Loh, Wetzlar and

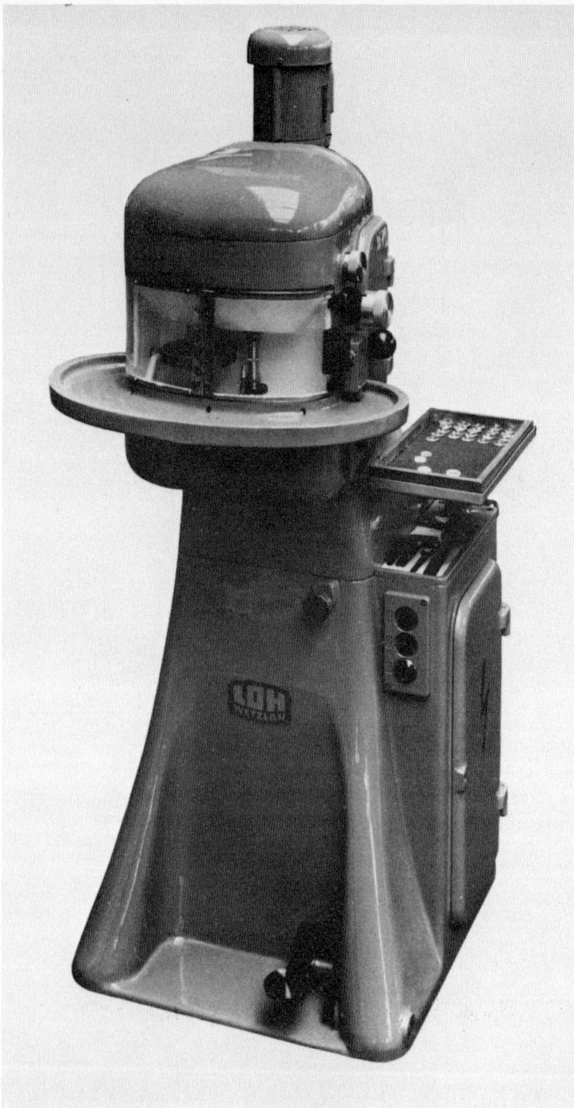

6.10.10 (*Top left*) **Loh LZ 25 automatic lens centring and edging machine for lenses up to 1 inch (25 mm) in diameter. The machine can be fitted with a magazine storing device for automatically feeding lenses.** (*Bottom left*) **Loh WG and LZ 25 lens edging machines set up for automatic centring with bell chucks.** (*Top right*) **Three Loh LZ 25 machines. The operator is wiping the lenses before passing them to inspection.** (*Bottom right*) **Loh LZ 25 A machine automatically feeding, centring, and edging lenses.**

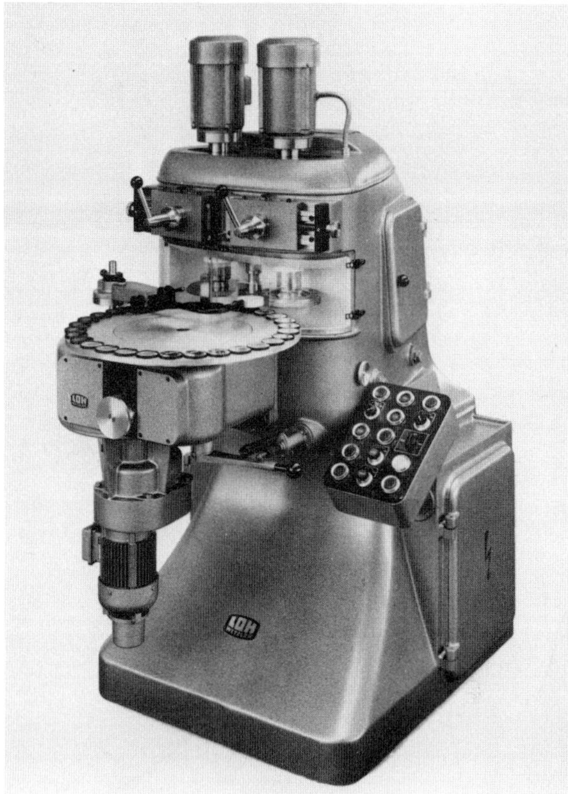

6.10.11. Loh LZ 80 A automatic lens centring and edging machine. It is similar to the LZ 25 A machine, but has a magazine capacity for diameters up to 80 mm.

W. Bothner, Brunswick) relying for accuracy of centring on bell-chuck clamps which grip the lens during the edge-grinding operation. This method is efficient only if the clamping angle on the lens is at least 17°, when a centring tolerance of 2 minutes of arc can be consistently achieved.

For production purposes (for curves and diameters within the capacity of available machines), this is a much cheaper method than visual-alignment centring.

Loh LZ 25 A and LZ 80 A machines

The LZ 25 A has a capacity of up to 1-inch (25 mm) and the LZ 80 A up to 80-mm diameter. (Figs 6.10.10 and 6.10.11.)

An automatic lens-centring machine with a magazine-type device will not only perform edge-grinding and facette-grinding automatically, but will also feed the machine with the lenses to be centred, bringing them back to the storage tray after the work cycle is completed. These storage trays are easily interchangeable and can serve simultaneously as storage and transport containers of those lenses coming from polishing to be centred. (Fig. 6.10.11.) Their capacities are as follows:

LZ 25 A

Lens diameters	*Lenses per tray*
Up to 5 mm	120
5·0 mm–11·5 mm	60
11·5 mm–18·5 mm	40
18·5 mm–25·0 mm	30

LZ 80 A

Lens diameters	*Lenses per tray*
Up to 10 mm	120
10 mm–23 mm	60
23 mm–36 mm	40
36 mm–47 mm	30
47 mm–59 mm	24
59 mm–69 mm	20
69 mm–80 mm	15

The machines consist of three main parts, the upper part with all elements needed for edge-grinding and faceting of the lenses, the storage device for automatic feeding of the lenses and the base of the machine containing controls, suction pump, cooling arrangement, etc.

For mass production of lenses, the automatic operation allows one operator to serve up to 20 machines, and the only 'hand work' is the changing

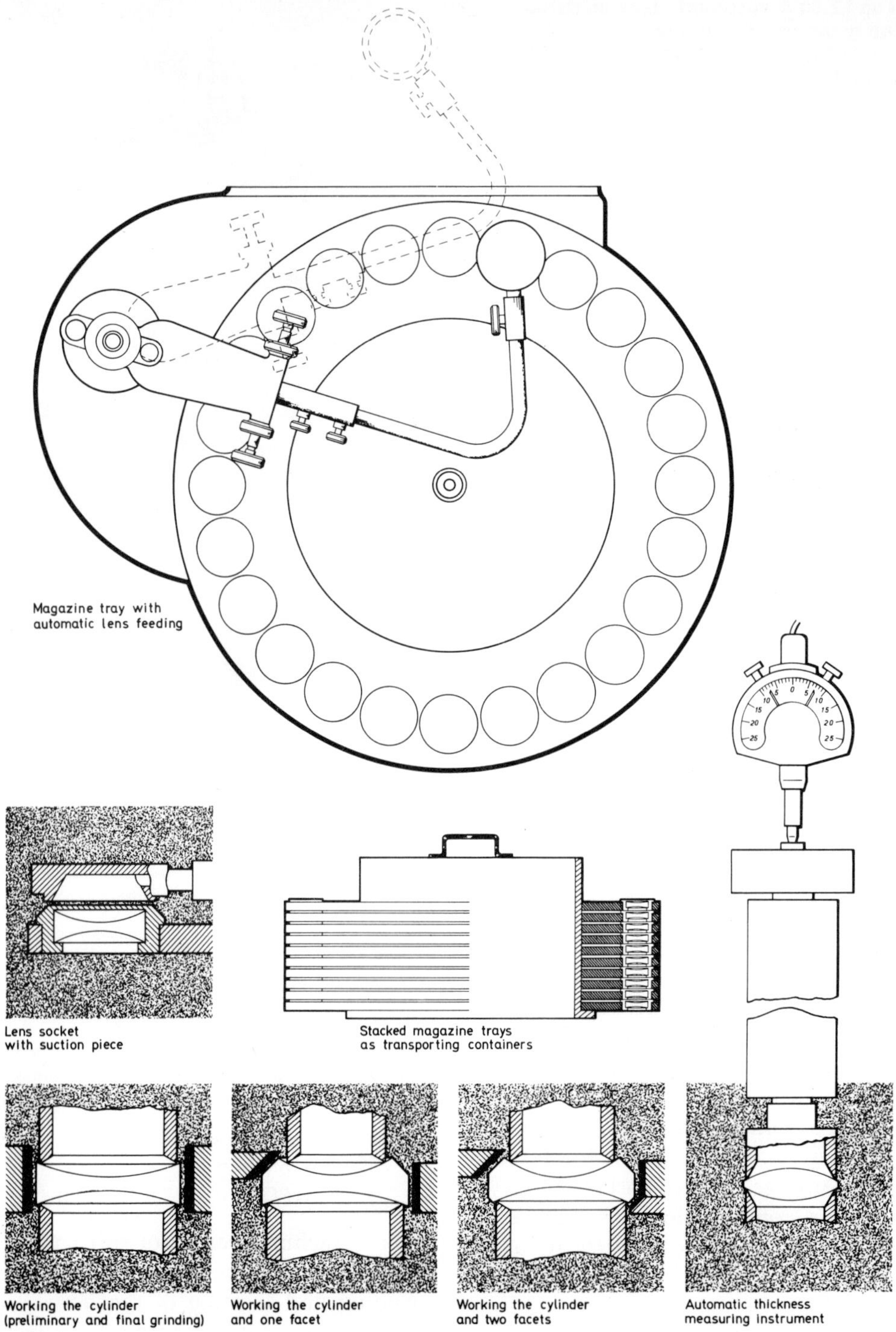

6.10.12. Automatic feeding mechanism for Loh LZ machines.

of magazines about once an hour. Alternatively, the machine becomes a spare-time job for another work section. The machine is equipped with a device to guarantee equal size of chamfers on both sides. An automatic thickness-measuring device, if fitted, will eliminate lenses out of thickness tolerance.

About 45 to 60 seconds a lens is the usual cycle time, but this depends on the size and amount of glass to be removed by grinding.

6.11 Optical cementing

If a single lens is used in an optical system designed for white light, definition is poor owing to chromatic aberration, and this can only be avoided by using a positive lens of crown with a negative lens of flint glass. The second surface of the crown is usually made the same radius as the first surface of the flint and they are cemented together to form an achromatic lens.

Most companies cement after final lens-edging, but, for the highest quality of work, it is usual to edge the lens oversize and edge again after cementing.

There are several types of modern optical cements available with different properties but the original cement was Canada Balsam (an olea-resinous exudation of the so-called Balsam Fir, *Abies balsamea*, Miller), which has as its chief constituents two separate resins and turpentine. The turpentine and other volatile constituents can be evaporated off by heating so as to obtain balsam of varying degrees of hardness.

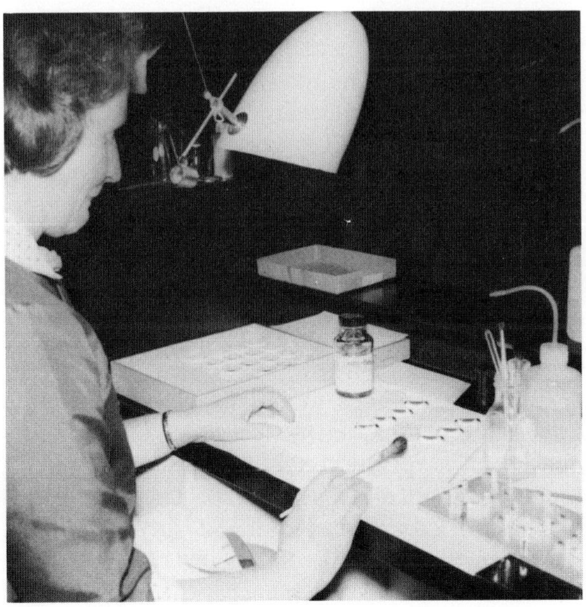

6.11.1. Cementing booth. Clean conditions for assembling cemented lenses into jigs before they are baked in an oven for polymerizing.

A synthetic substitute for Canada balsam is *n*-butyl methacrylate, manufactured under the trade name of H.T. cement and obtainable from Hopkins & Williams Ltd. Both balsam and H.T. cement require heat treatment to cure the cement.

Another important cement is Crystic 191, which is a two-part mixture especially suitable for large lenses, as it is air-drying without any heat treatment. Details of these and other cements are given in § 4.21, but the methods of application and centring are now explained.

CEMENTING OF LENSES WITH BALSAM OR H.T. CEMENT

The method of cementing after edging recommended by Mr F. Twyman is as follows (Figs. 6.11.1 and 2).:

The lens components are first paired for diameter to a tolerance fixed by the designer; this tolerance is also the amount by which the components may move relative to each other during baking.

In the case of lenses whose centring is of critical importance, it is advisable to measure the residual errors on each component and pair them again for deviation, marking the 'thick edges' which have to be opposed in assembly.

The pairs are cleaned and laid together cold, and the cement inserted. The combination is now

6.11.2. Hilger & Watts jigs for cementing object glasses with H.T. cement.

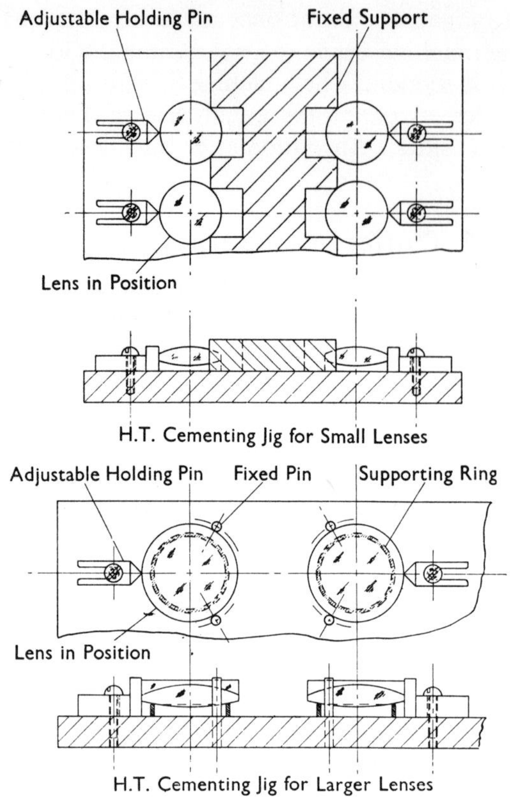

placed on a balsaming jig, which consists of a steel plate bearing a number of sets of three small locating pegs, two in each set being fixed and the third adjustable to allow for variations in diameter. (Fig. 6.11.2.) For small lenses, the two pegs are replaced by an equivalent made from sheet brass.

Lenses with a convex exterior to the flint component are laid on a thin parallel ring of metal to avoid scratching the vertex, and to ensure that the lens sits truly level.

When the jig is full, it is transferred to the oven and baked according to the time and temperature stated by the maker.

It is advisable to chamfer the contact surfaces as well as the exteriors when using H.T. cement, as a slight contraction takes place during polymerization, which is made good by the rim of cement held in the chamfer; if this is not present, there is a risk that air pockets will creep in round the edges.

H.T. cement is readily soluble in acetone, and where the cementing of a combination has not been successful, an overnight soaking in acetone is much to be preferred to heating the part to 200°C to soften the polymer.

CEMENTING OF ACCURATE LENSES WITH COLD SETTING CRYSTIC 191 CEMENT

The method, developed by Rank Taylor Hobson for edging after cementing, is reproduced by permission of Rank Precision Industries Ltd. The description forms part of a skills analysis written by Mr A. Eabry.

Bell or carousel system of truing up

For optical cementing to be effective, the optical centre of each element should coincide. Unless this is achieved, the performance of the completed lens system is impaired.

Prior to the present system, cemented doublets were trued up mechanically, by pins locating on the edges of the elements. This meant that the truth was only as good as the edge grinding and blacking.

In the current method (bell or carousel system), however, the lower element is optically trued, using the lens centring instrument to a very accurate brass chuck mounted on a steel spindle or shaft. The upper element is trued on the lower one by an identical chuck. Both chucks and elements are aligned in a jig. (Fig. 6.11.3.)

The chuck diameter required is governed by the *diameter of the polished surface* being stuck on it, in the case of the lower element, and by the *diameter*

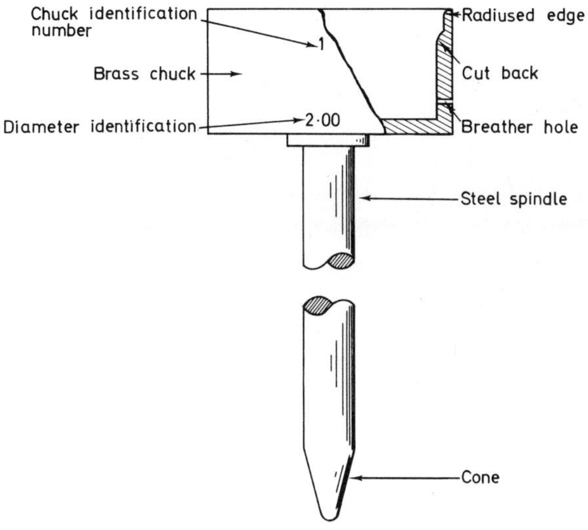

6.11.3. Edging chuck and spindle.

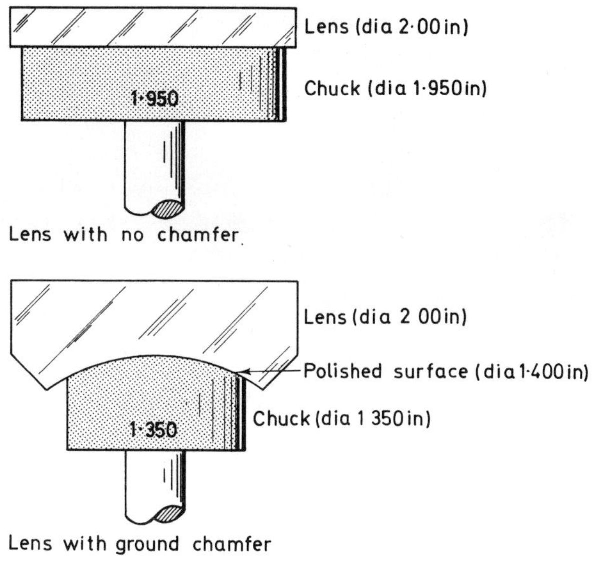

6.11.4. Chuck diameters.

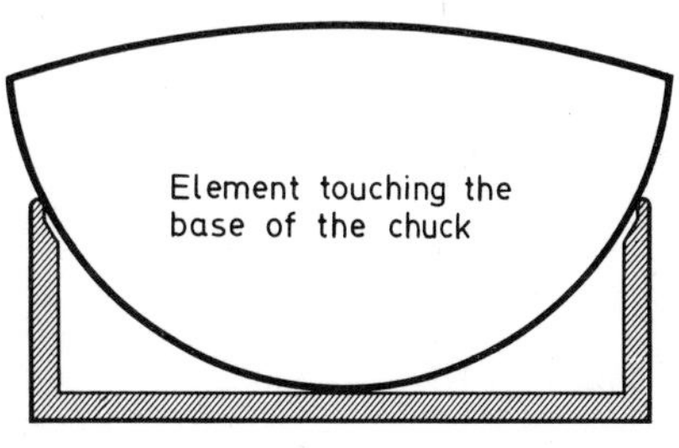

6.11.5. Chuck depth.

of the polished surface it is coming into contact with in the case of the upper element.

The diameter of the polished surface is not necessarily identical with the diameter of the lens itself. This can vary if there is a ground chamfer incorporated in the lens design. It is generally accepted that the chuck diameter is 0·050 inch less than the polished surfaces mentioned above. (Fig. 6.11.4.)

There are twelve chucks to a set, separate from the edging chucks. Four are used in the jigs on the carousel, and the remaining eight for the lower or trued-up elements. These eight are used as follows: four on the carousel, and four having elements trued up on them. This allows a continuous cycle of cementing to be maintained.

To give maximum efficiency, the chucks must be in perfect condition and turned true on a special lathe if there is any defect, such as:

(1) Bruises or damage to the radiused edge.

(2) The chuck could be misshapen as a result of a heavy blow.

(3) The chuck may not be deep enough for the curve of the element. (Fig. 6.11.5.)

(4) Dried cement on the spindle.

(*a*) **Tool forming chuck wall and cut back.**

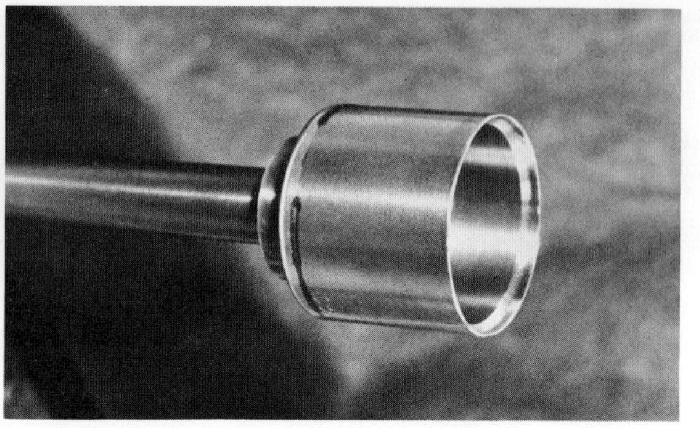

(*b*) **Edging chuck, showing radiused edge and cut back.**

6.11.6. Chuck forms.

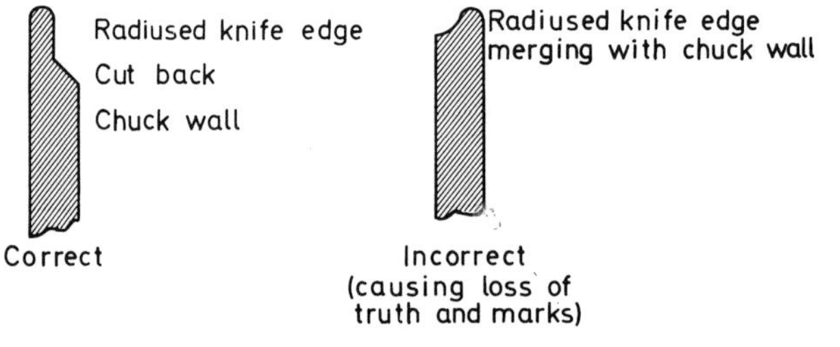

(*c*) **Correct and incorrect chuck forms.**

(5) Damage to the cone.

(6) Radiused edge merging with the chuck wall. (Fig. 6.11.6.)

Any of these defects will prevent the element from being trued-up.

A small hole in the chuck wall allows an escape route for the hot air from inside the chuck as it is heated. Should it become blocked with edging cement, there is a danger that this expanding air could lift the element from the radiused edge. A heated needle or straightened-out paper clip will clear it.

Use of the chuck trueing-up lathe

This machine is of an original design, and was developed by Rank-Taylor-Hobson. Its specific purpose is to produce to within very fine limits:

(1) The correct wall-thickness of the chuck (Fig. 6.11.7.)

(2) A radius edge to the face of the chuck which is true to the centre of an accurately ground spindle (Fig. 6.11.8) on which the lens sits. Upon the truth and condition of this radiused edge depends largely the correct truing up of the lens.

The spindle is located in a collet, and the cutter fed to the chuck face by means of the large hand-wheel rotated *anti-clockwise*. The cut is considered complete when the shoulder of the cutter just clears the face of the chuck. (Fig. 6.11.9.) Failure to stop at this point produces bruises to the face of the chuck.

The rubber-edged drive wheel is brought into

△6.11.7.▷
Chuck truing-up lathe.

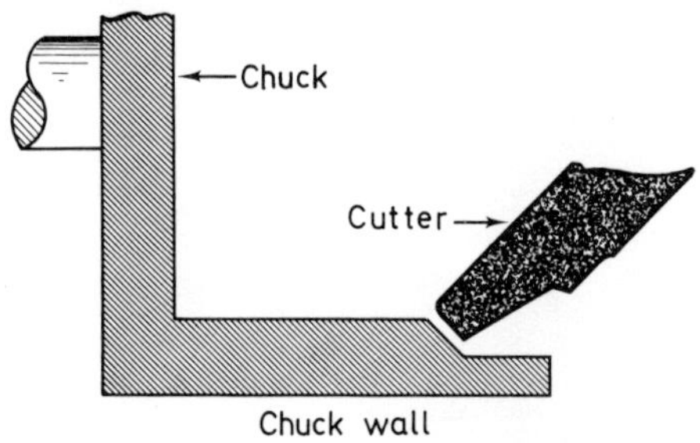

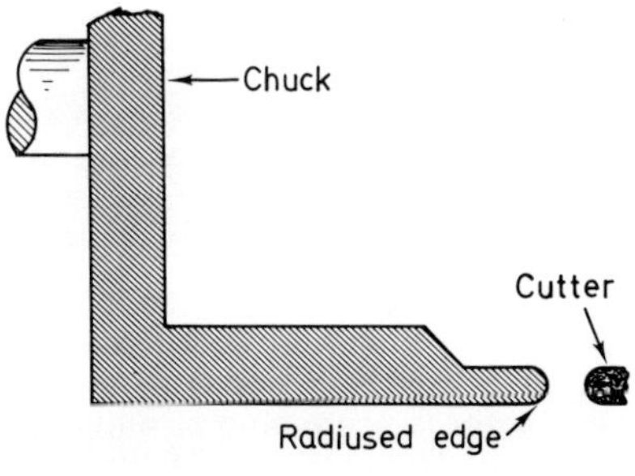

6.11.8. Radiused edge of the chuck.

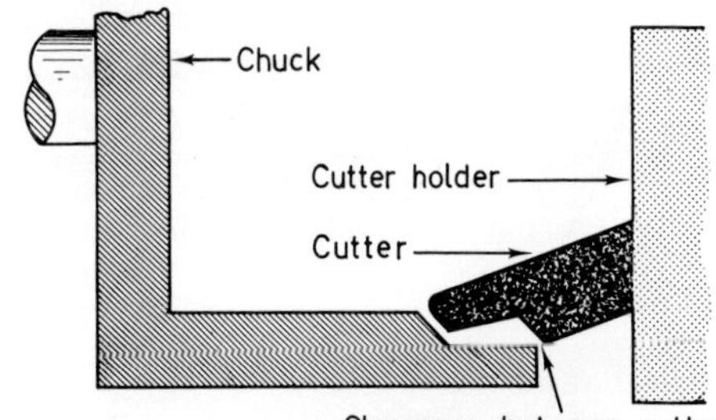

6.11.9. Turning the chuck wall.

6.11.10. Taylor-Hobson chuck truing-up lathe. (*Top*) The chuck spindle is in the lathe. The rubber-rimmed driving wheel is in contact with the knurled drive and the cutter is producing a radiused edge on the chuck. (*Centre*) The method of using a height gauge to position the knurled drive on the chuck spindle. (*Bottom*) The rubber-edged drive wheel can be seen bearing against the aluminium disk that drives the chuck spindle.

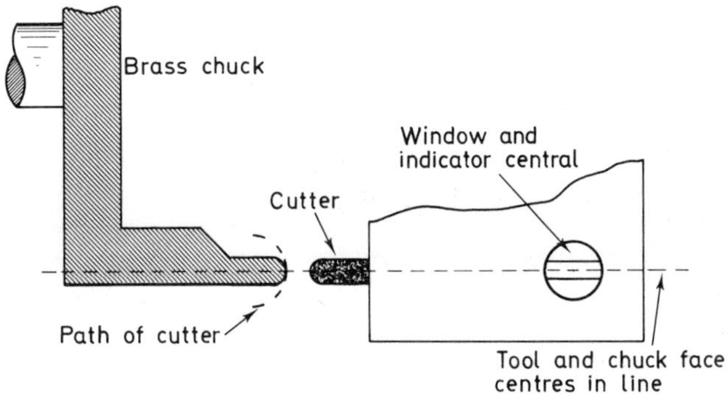

6.11.11. Centring the cutter.

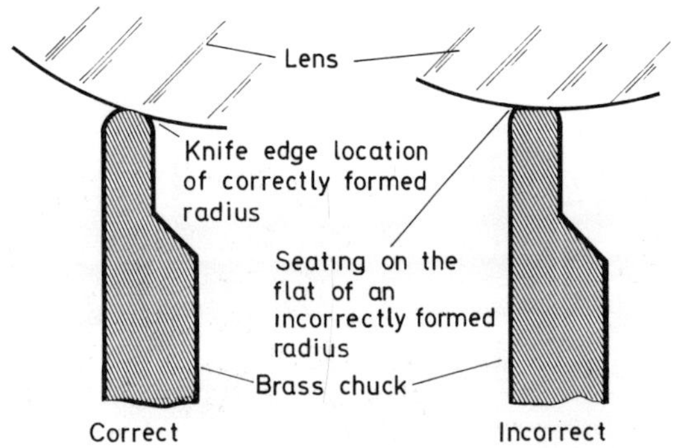

6.11.12. Correctly and incorrectly formed radius.

contact by pressing down the brass gear housing. (Fig. 6.11.10.)

In the window on the cutter housing, the white line indicator should be central. This can be achieved by operating the auxilliary motor switch found on the front of the lathe bench to the operator's right. This line when central in the window indicates the centre of the cutter is lined up with the centre of the chuck face. (Fig. 6.11.11.)

Viewing is made easier by the use of the magnifying lens positioned above the cutter tip. The lining up is important to ensure that this is the locating position between chuck and lens and is essential to successful truing-up. (Fig. 6.11.12.)

Turning may now commence, and the following procedures followed.

(1) Switch on both motors.

(2) Lock cutter head.

(3) Feed cutter in, using the fine adjustment control 0·002 inch to 0·003 inch at a time.

(4) Turn until the radius is correctly formed and is running true. The cutter moves on a predetermined radius controlled by a cam motivated by the auxiliary motor.

(5) Wind cutter out after the final cut.

(6) Check for any burns; remove with emery cloth if present.

(7) Remove spindles and chuck from lathe; inspect.

Inspection should reveal that there are no dents, or damage of any kind to the radius turned, and that it is complete and free from burns.

Use of the lens-centring instrument

The optical truing up of the lower element is achieved on the *lens-centring instrument.*

This instrument has the ability to:

(*a*) Locate and hold the chuck and spindle in the truest possible position.

(*b*) Allow observation by optical means of the truth of the lens being worked on.

For lens-truing, the spindle is placed in the vees to be found in the centre and lower half of the instrument. Mounted below the upper vee and above the lower vee are magnetic blocks which locate the spindle in the vees. (Fig. 6.11.13.)

The conical end of the spindle locates in a seating which contains an annulus and three ball bearings.

Extreme care should be exercised when loading the chuck into the instrument. Heavy impacts by the chuck, spindle, or cone of the spindle will impair the truing up process. The seating faces

△ **6.11.13.** (*a*) **Lens-centring instrument.**

▽(*b*) **The operator is rotating the lens spindle with his left hand and examining the movement of the image reflected from the lens surface through the eyepiece.**

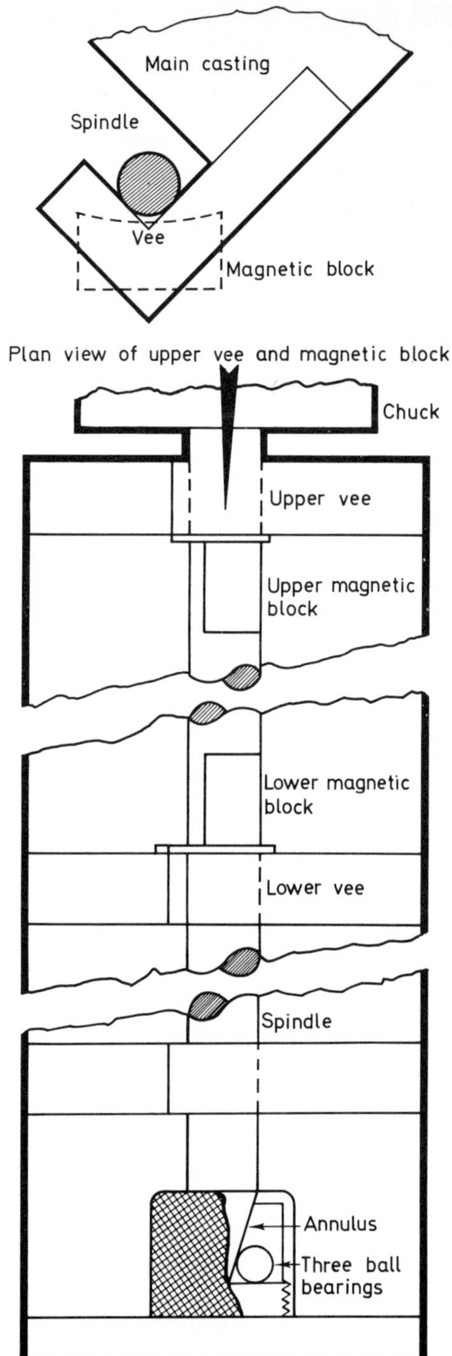

△(*c*) **The method of loading the spindle into the machine.**

(a)

6.11.14. **Applying the edging cement to the edging chuck.**

(b)

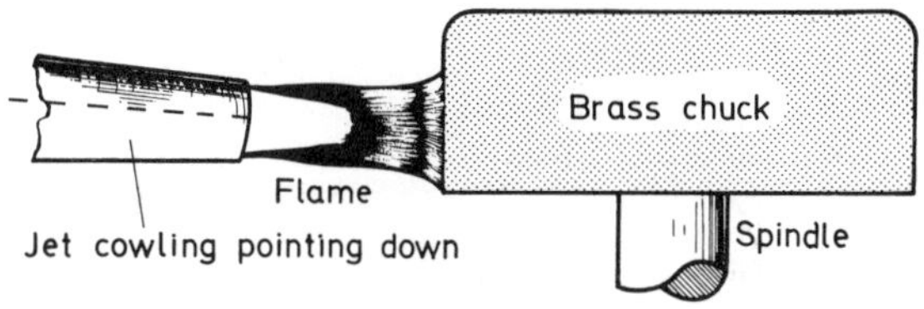

of the vees must be clean to maintain the truth.

Edging cement, used to hold the lens during truing-up, edging and cementing, is a mixture of unbaked Swedish pitch, shellac, and umber resin in the form of cellophane-wrapped sticks and containers. It has certain qualities:

(1) Low melting point.
(2) Rapid hardening.
(3) Solubility in most solvents.
(4) Great adhesive power.

This pitch has to be applied to the chuck by hand. It is heated by a gas/air jet, which must be adjusted until the jet plays on the chuck roughly half-way up its side (Fig. 6.11.14*b*).

The gas/air jet should be played on the chuck until the pitch can be applied in an even, smooth layer as the spindle is rotated. The pitch should hold its shape as it is applied and be capable of being spread with the tip of the finger, without sticking to it. For an average-size lens (1·5-inch diameter) heat may be safely applied for approximately 3–5 seconds before pitching commences.

Overheating is to be avoided because:

(1) Excessive heat may well distort the brass chuck.
(2) The pitch will run down the sides of the chuck. This is messy and time consuming.
(3) Long cool-down period afterwards results in much lens movement and subsequent untruth.
(4) Personal danger from burns.

Inspect the pitch first for dirt or grit at the end to be applied to the chuck, and if contaminated clean off by heating.

Excessive pitch application can result in pitch running down the chuck, and is again time and material consuming. Sufficient has been used when a continuous ring of pitch is seen. This is then moulded into shape with the fingers, as shown in Fig. 6.11.15(*b*). The pitch should be even and should contain no lumps. Whilst the pitch and chuck are still warm, the element to be trued should be picked up by its edge and placed, Trilac-sprayed surface down, on the pitch.

(a)

(b)

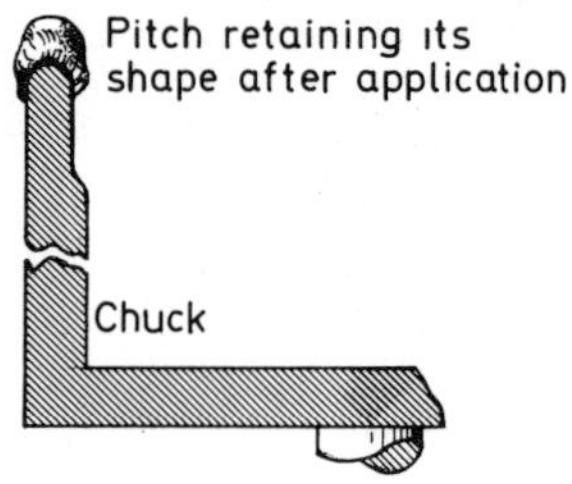

6.11.15. Placing the lens on the edging chuck after applying edging cement.

The initial placing of the element calls for care and, correctly carried out, will speed the process considerably. As can be seen in Fig. 6.11.15(*a*), the element should initially sit on a soft bed of pitch and then the element should be pressed down on the chuck until the brass radiused edge is visible through the pitch. This is observed through the upper surface of the element being trued. (Fig. 6.11.16.) Short pauses should be observed whilst this takes place, to allow the spindle to be rotated.

The spindle must be rotated anti-clockwise while the fingers of the right hand exert pressure to maintain the position of the spindle in its vees.

There are two indications that the first phase of this stage has been successfully achieved:

(1) The element is correctly located on the radiused curve. Unless the brass chuck is visible as a complete circle then the lens is not sitting squarely on the chuck and is therefore untrue. (Fig. 6.11.17.)

(2) The element is placed 'eyably' central. To check this initial truth as the spindle is rotated, observe windows or light fitting reflections in the polished upper surface of the element. In the most central positions these reflections should run steady, but if they swing wildly, reheat the chuck and gently move the lens or element until the images or reflections are steady. It is important to note that this is only a very crude method of truing up and that optical centring must follow.

Unless the element is running reasonably true, an image will not be seen in the eyepiece of the centring instrument.

To understand the truing procedure, it is necessary to have a working knowledge of the lens-centring instrument, as follows:

Light is focused through an optical system on to the lens surfaces, each surface being selected by a hand-controlled focusing system. The rays are bent and turned into a slit of light by means of a mirror and prism.

This slit is viewed through an eyepiece, wherein it is visible against an illuminated fixed graticule scale. Deviation by the slit can therefore be assessed, and the optical truth of the lens under test be checked.

The lens under test, mounted on its chuck, is held in two pairs of vees. The centre of spindle, vees, and optical system are common to fine limits.

Thus a lens running true reflects an image back along the system, which coincides with the light beam focused on it. As the lens rotates about its axis the slit of light observed in the eyepiece will appear central and stationary. (Fig. 6.11.18.)

△6.11.16. Truing up the lens on the chuck. The hand positions should be noted.

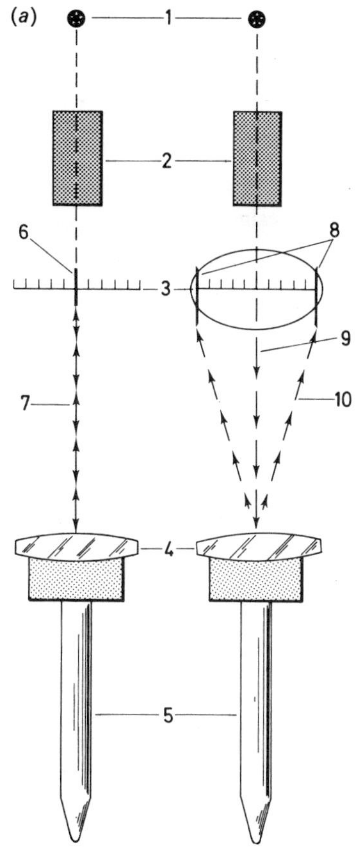

1 Light source
2 Optical system
3 Graticule seen in eyepiece
4 Lens
5 Chuck on spindle
6 Image remains steady and central as spindle revolves
7 Light focussed on top surface of lens returning along the same path
8 Image swinging from side to side as spindle rotates
9 Path of light focussed on lens surface
10 Path of light reflected from lens surface

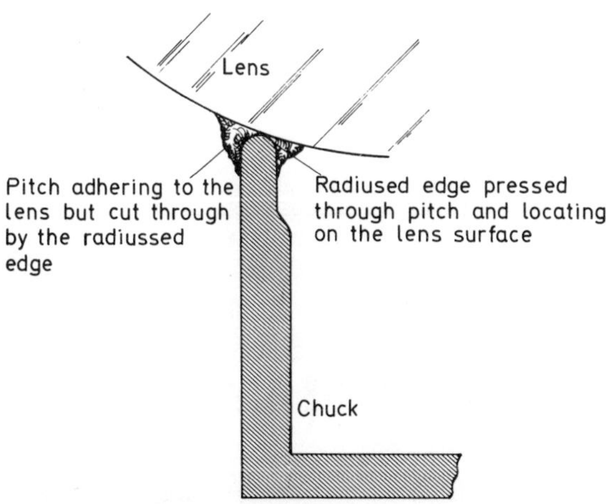

△6.11.17. A lens correctly located on the radiused curve of the chuck.

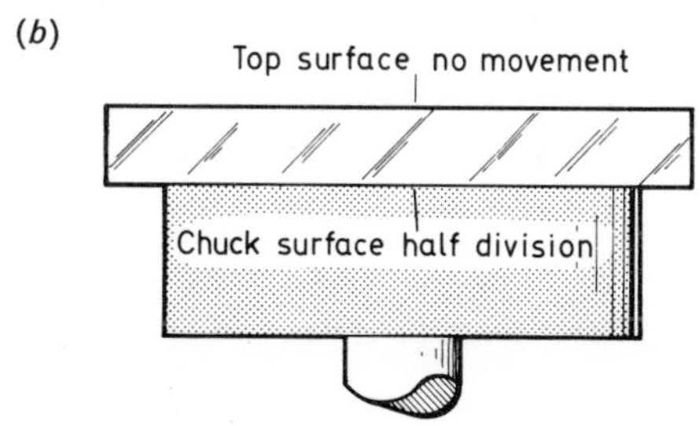

△6.11.18. Principle of the lens-centring instrument.

An untrue lens will reflect the image at an angle to the focused beam and the slit will appear offset. Rotation of the lens will result in a swinging of the slit to either side of the central line on the graticule scale. Moving the lens will bring the image central.

The tolerances required are:

Top surface : no movement during rotation.
Chuck surface: half a division during rotation.

This is the only tolerance allowed; the type of curve will affect the meaning of half a division, but as shallow curvatures indicate less sensitive lenses, the deep curves being more sensitive, the tolerance laid down covers acceptability.

The instrument is focused in the first instance on the top surface.

Focusing is carried out by the rotation of the knurled hand-wheel on the right-hand side of the instrument. By turning it clockwise, the focusing optics are raised, anti-clockwise lowered. The position of the optics may be noted by reading off the scale in the form of a 0–12-inch pointer on the focusing unit.

On most plus curves the focusing lens will be BELOW half-way on the column, whereas in the case of negative lenses they are to be focused ABOVE half-way.

As there are two surfaces, it is important to recognize which image is from what surface.

One simple way is that the coated surface, that is always the chuck image, is considerably duller than the top surface owing to the coating. Also it is sprayed with Trilac, and so in certain circumstances tends to show the dull image with distinct flares coming from it. Careful focusing will reveal the two images. (Fig. 6.11.19.)

In the case of a meniscus, if the crown curve was uppermost, i.e. opposite to the condition shown in the diagram, then the first image observed would be the chuck image.

To pick up images from a surface that has a radius of 6 inches and above, the supplementary lens has to be used. This can have the effect of altering the conditions suggested in Fig. 6.11.19.

The image would have to be searched for. One way of image identification is as follows. After the initial true-up by eye, observation and check by the eyepiece and graticule will show that one image will swing about much more than the other. This is invariably the *top surface*.

This is so because, however untrue the lens is on the chuck, the lens is locating on the radiused edge and is therefore relatively true. It is only the effect of observing this chuck image through the top surface which gives the untrue effect.

Lenses with deep minus curves, or heavy lenses, will tend to move during cooling down of the pitch. These should be checked again.

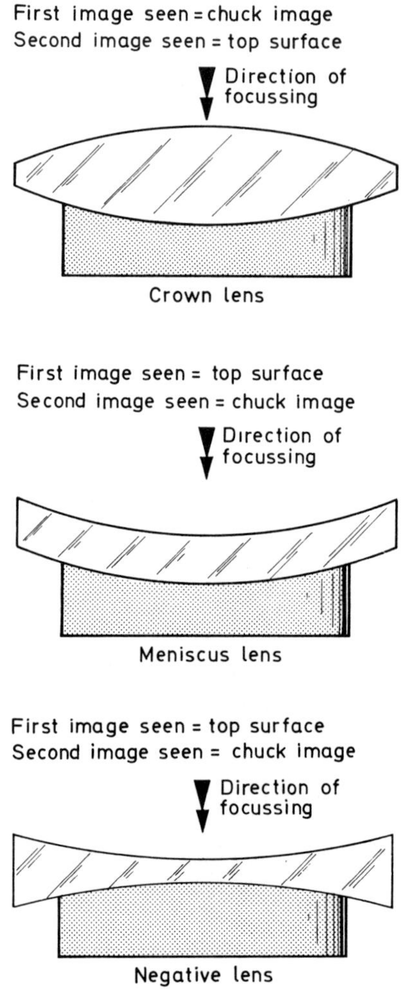

6.11.19. ▷
Effect of a supplementary lens, as explained in the text.

6.11.20. Carousel and clean cabinet. A trued-up lower element (supplied by the centring operator) is in the turntable of the cabinet. The operator is brushing the lens with an anti-static brush before applying Crystic 191 cement and placing the top element in position.

Cementing

For good acceptable results, the cementing of lenses must be carried out in a clean, draught-proof room, one that is preferably air-conditioned and segregated from the main workroom. As an extra precaution, all cleaning and cementing is carried out in clean-air-pressurized booths. Personal cleanliness is necessary, hands and clothing being required to be up to standard.

Cleaning materials used for cementing are few, *and must be absolutely clean*. They are:

(*a*) Selvyts for interface cleaning.
(*b*) Pure alcohol for interface cleaning.
(*c*) Radio-active brushes for inter-face brushing.
(*d*) Camel-hair brushes for interface brushing.
(*e*) Stockinette for rough external cleaning.

Selvyts should be changed at least once per day. Any contact with cement must be avoided; it will cure hard in the material, causing marks on the glass.

Alcohol as supplied must be used only for cleaning. Any other solvent, especially methylated spirit, distilled or industrial, has a harmful effect on the cement used.

Static-master radio-active brush must be cleaned regularly but in accordance with the instructions issued. Do not handle the bristles.

Camel-hair brushes. Clean regularly in the lab, do not handle the bristles.

Stockinette. Use each square once, discard immediately it becomes soiled.

The trued-up lower element is taken from the carousel, and placed in the turntable set in the floor of the carrier booth, its matching element is then taken from the container on top of the jig, and placed on the foam base to the right of the turntable. Both lenses are now in a very protected position owing to the way the booth functions. (Fig. 6.11.20.)

To maintain the cleanliness of the booth or cabinet interior, it must be cleaned out daily before work commences. Such objects as dusty trays should not be placed inside the cabinet. Before *raising* the protective blind the cabinet must be *switched on* to allow pressure to be built up.

All is now ready for cementing to commence, and the ordering of the Crystic 191 is done at this

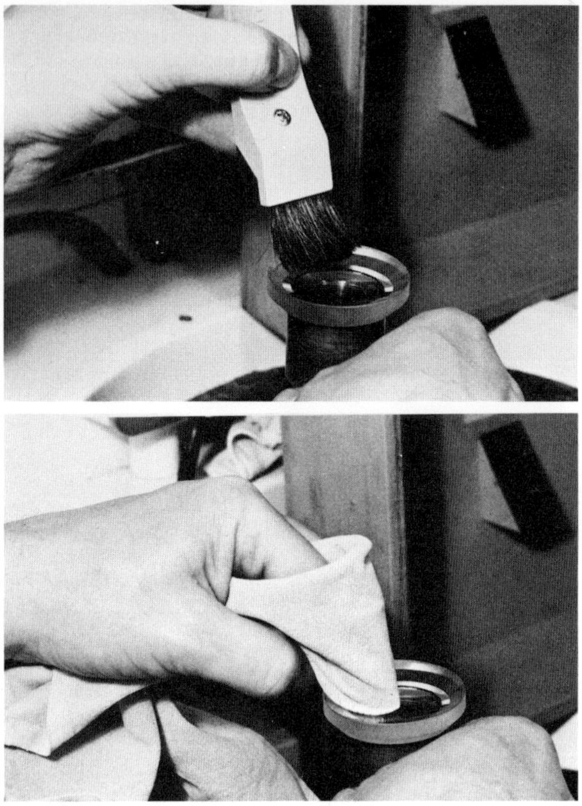

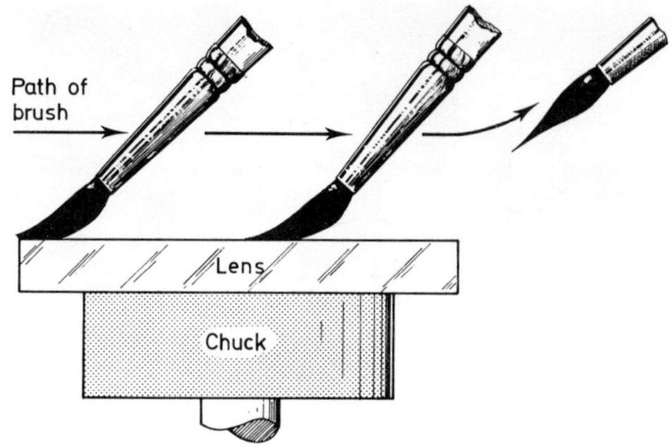

6.11.21. Wiping interfaces with anti-static brush and Selvyt cloth.

stage. Allow approximately 5 minutes for the laboratory to mix and deliver it, as the cement will usually commence to harden or gel in 15 minutes after receipt. Plan ahead to allow time for cleaning and all preparatory work.

When the small jar of cement is first received from the laboratory. it is usually full of very fine bubbles of air that is mixed into it during the blending of the two materials from which it is compounded. Use of the cement in this state will not only introduce air-bubbles into the cement layer but will make it impossible for the cement to be inspected for hairs or dust or any foreign material which may be floating in it.

The cement must be allowed to clear, then inspected for hairs against a projection light. Any hairs observed must be removed by means of a needle before cementing.

In cleaning the interfaces, the following steps should be strictly adhered to.

(i) Lightly brush interface to remove any dust or hairs. Lift brush at the end of each stroke to avoid wiping the edge of the element and chuck. This will avoid excess contamination of the brush. (Fig. 6.11.21.) Do not allow the brush body to touch the lens surface; marks are caused this way.

(ii) Pick up selvyt from cabinet base; do not drag it across the bench, picking up dust or grit which will be then wiped across the lenses.

(iii) Apply the selvyt to the alcohol bottle; sufficient solvent will be released if the valve is held down for approximately 1 second.

(iv) Wipe the lens from centre to edge, firmly but gently in circular motions until the whole surface has been wiped and the spirit evaporated.

(v) Use a dry area of the selvyt to polish the cleaned surface and remove all traces of the solvent. On no account allow the solvent to dry on a lens surface without this final polish. There is a great danger from staining if this precaution is ignored.

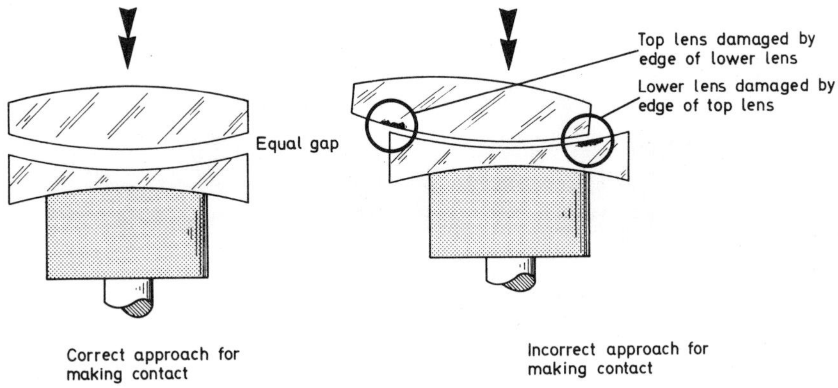

6.11.22. Contacting two elements.

Polishing of the surface should continue until all traces of smears have vanished. A good indication that any polished surface has been thoroughly cleaned is its hard, black appearance when looked down at.

If the surface has a dull grey look, this signifies there is a smear or even layer of grease and dirt still existing on the wiped surface.

(vi) Pick up the matching element by the fingers with the pads of the finger-tips holding the ground edge, and avoiding the polished surface. Visual examination will indicate the surface to clean. It is the rule that the interface is never surface coated, therefore the surface *without* the coloured hue is the side to hold uppermost and wipe. Repeat the process as indicated above from (i) to (vi).

Whilst wiping the matching element, ensure it is not held directly above the first element wiped. If this rule is not obeyed, and the second element is dropped, it will result in multiple damage. Retain hold of lens.

(vii) Both interfaces have now been thoroughly wiped. Gently brush both sides of the second element, which is still being retained in the left hand. Brush strokes should be as indicated in Fig. 6.11.21. When both surfaces have been brushed, hold in such a manner that the interface points downwards. This prevents air-borne dust from falling on the cleaned surface whilst the brushing operation is carried out on the lower element, which is still in the booth, on its chuck and wiped.

(viii) When visual examination reveals that both interfaces are completely free from hairs, place the held lens into contact with the lower element.

This is a crucial moment and every care must be taken.

(*a*) Keep a firm grip on the top element; complete control is required.

(*b*) Be certain that no hairs float between the two interfaces.

(*c*) Place the top lens gently on the lower lens; do not drop or slide on. The two lenses must go on evenly (Fig. 6.11.22) and *no noise* should be heard as the contact is achieved. The usual result when a contact noise is heard is a mark or dig in one or both interfaces due to either a violent impact between each other, or an edge of the top lens digging into the polished surface of the lower. Carry out this placing together as quickly and as safely as possible, to avoid intrusion by an air-borne contaminant.

Do not press the two elements together; allow for the top one to settle. Release both fingers from the top element simultaneously to avoid imparting a swinging movement to it. Should the fringes or colour bands not appear, do not force the top element down:

(1) There may be dirt or grit between the two elements. Pressure will grind it into the polished surfaces and cause marks.

Inability to balance colour rings due to dirt

Rings due to clinging dirt

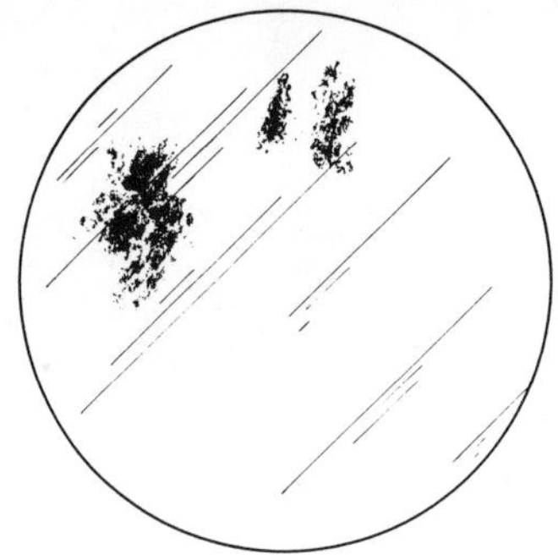

Complete failure to produce rings due to dirt

6.11.23. Dirt affects fringes.

6.11.24.▷ Moisture between elements.

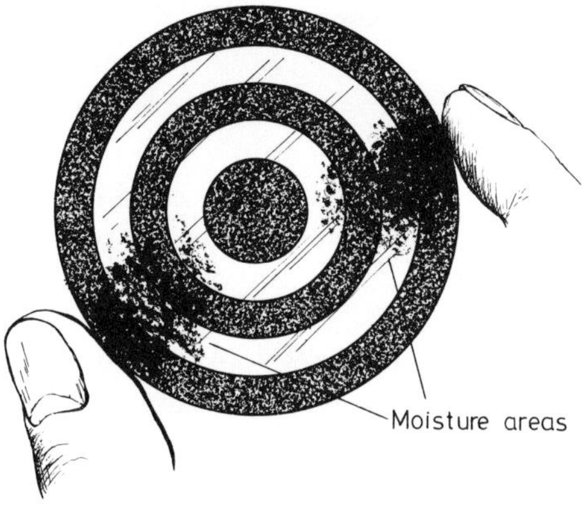

(2) The contact curves may be incorrect. Separate the two elements by gently pulling them apart; rewipe and brush. If the same fault appears, have the curves checked.

If the colour bands appear, their appearance will signify:

(1) Accuracy of the contact curves.
(2) The cleanliness of the contact.
(Fig. 6.11.23.)

The quantity of rings shown is accepted as the sum total of the error of the two elements and rarely exceeds six rings if the curves are acceptable.

The formation of the rings, whether round, oval, or distorted, indicates the shape of the polished surface.

Hair and dirt are quite easily seen and often the rings form around the piece of dirt. Also dirt can be felt, and when grit is present, the grinding of it by top-element pressure is easily felt. If a hair is observed, do not slide one lens across the other in an attempt to roll it out over the edge.

Should a dark area commence to spread across the contact face as the two elements are placed in contact, separate them immediately. This area is caused by moisture from the fingers. If allowed to spread completely across, the two elements may be stuck together and force may have to be used in separation, with the risk of damage to one or both elements. The appearance of the moisture effect is shown in Fig. 6.11.24.

This condition is known as black contact. When the elements have been parted, thoroughly clean both contact faces. As the cause is hot fingers and hands, wash them and run cold water over them before commencing wiping. Should the doublet be held fast by moisture, very gentle heating on the hot-plate will drive out the moisture, and the element can then be parted by hand.

Should a small dot appear in the centre of the lens as contact is achieved, and as the top element settles no colour rings appear, while light pressure

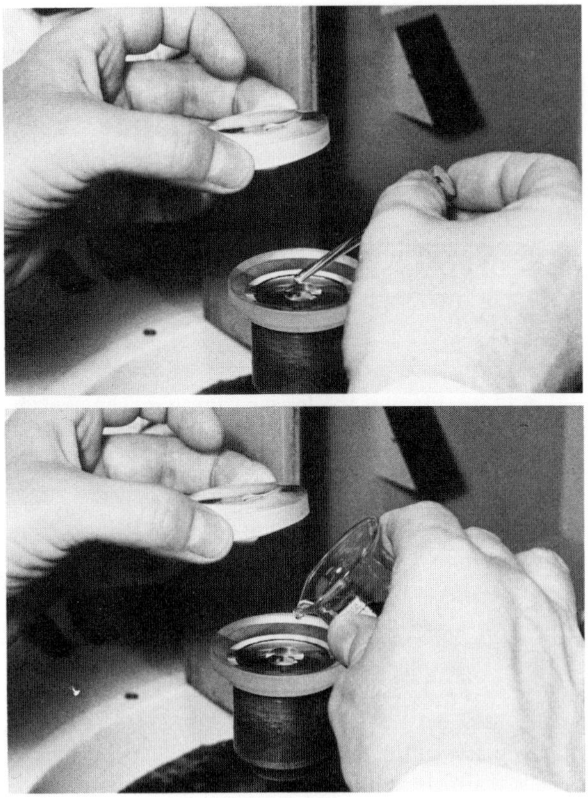

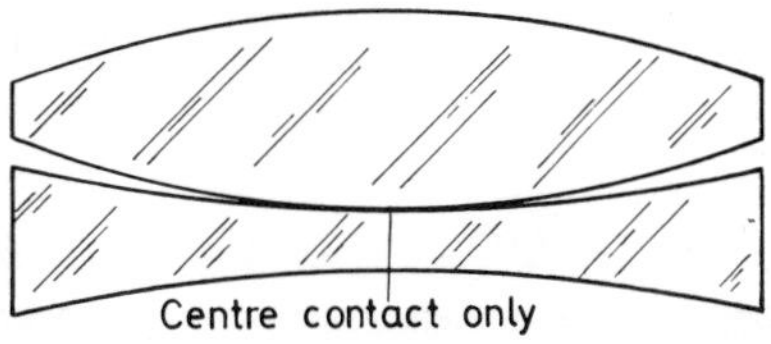

6.11.25. Applying Crystic 191 cement with a glass rod or by pouring from a beaker. The sketch explains the significance of contact at a single spot.

causes the dot to move around the centre area, separate at once. This signifies that the two contact faces are not identical and one lens is touching in the centre only, and is therefore unacceptable. (Fig. 6.11.25.) Severe edge contact due to wrong curves will produce no fringes, but in this case as the elements do not touch, no spot will be visible.

When the contact is acceptable for cleanliness and curvature, cementing proper may commence. The Crystic 191 will by this time have cleared itself and any floating hairs have been removed. The drill is as follows:

(1) Remove the top element by lifting it straight up and off. Do not scrape one lens over the other.

(2) As the lower element is exposed, the cement should be applied. The amount to use, and the method of applying, can only be given approximately, but a good guide is provided by the table below.

With the exception of prisms or truncated lenses, the cement pool should be made as near to the centre of the lower element as possible. If an excess is used, it can always be cleared off from the

Table 6.11.1. Application of Crystic 191.

Lens diameter	*Method*	*Approximate amount*
0–0·75 in.	glass rod	1 drop
0·75–1·5 in.	glass rod	2 drops
1·5–3·0 in.	pour from beaker	$\frac{1}{4}$ dia. pool
3·0–5·0 in.	pour from beaker	$\frac{1}{2}$ of mix
Prisms	pour from beaker to form an H (Fig. 6.11.26)	

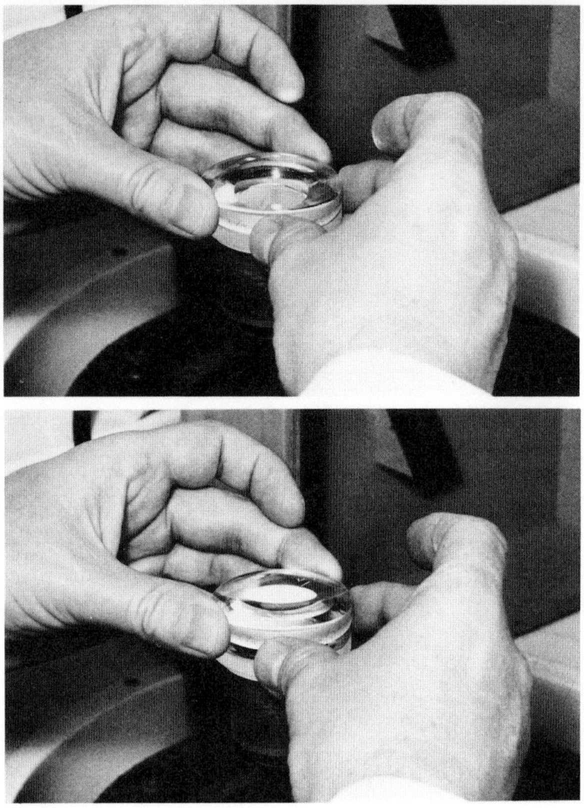

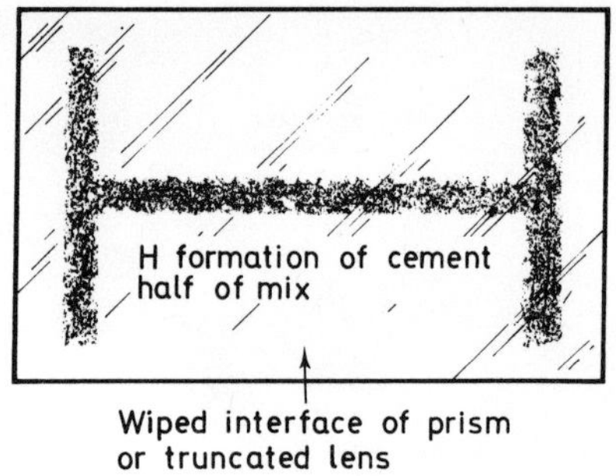

6.11.26. Placing the lenses together. The sketch shows how cement is applied to a prism in the form of an H.

edge after the pressing out. Insufficient, however, will lead to a thin film on the edge—the prime cause of edge starts and failure. The amount of cement ideally required is of the following order:

Lens diameter	*Cement thickness*
0–0·75 inch	0·0003–0·0007 inch
0·75–5·0 inch	0·001 inch

Cement layers of over 0·001 inch should be avoided. The above quantities, if pressed out correctly, will be within this tolerance.

The glass rod used should be kept, when not in use, in methylene chloride. The beaker is supplied with a glass cover, which must only be removed inside the clean-air cabinet.

Do not touch the lower element with either rod or beaker. A dirt area will be formed and there is danger of marking.

Try to avoid air bubbles in the centre of the blob of cement, as they will have to be pressed out. Replace the top element on the lower one evenly and gently; make every effort to maintain the evenness of the cement. (Fig. 6.11.26.)

If a bubble should form in the centre, it is essential that it be dispersed at once. If it is allowed to stay there, and the doublet pressed out to its finished thickness, it is very rare indeed that it can be moved.

To disperse an air bubble, apply *very light* downward pressure to the top element, and at the same time move it from side to side. The bubble will then start to move or 'swim' through the cement, until it is expelled into the atmosphere from the edge of the doublet. As the doublet is moved side to side, the bubble will become elongated; as the lens is centralized, it will again assure a round regular shape. (Fig. 6.11.27.) Only a very small sideways movement is required.

When all traces of air bubbles have been cleared the final pressing out commences. Two aims may be achieved simultaneously by the following method. (Fig. 6.11.28.)

(1) Apply vertical downward pressure to the top element, to achieve correct thickness.

(2) Rotate the lower element, to avoid a wedge of cement between the elements.

To apply the downward pressure, which will press out the excess cement, use either clean

▽ **6.11.27. Spreading cement evenly and removing bubbles.**

(*a*)

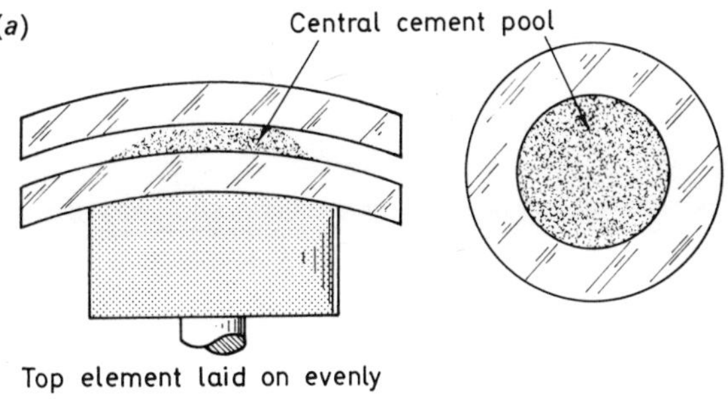

(*b*)

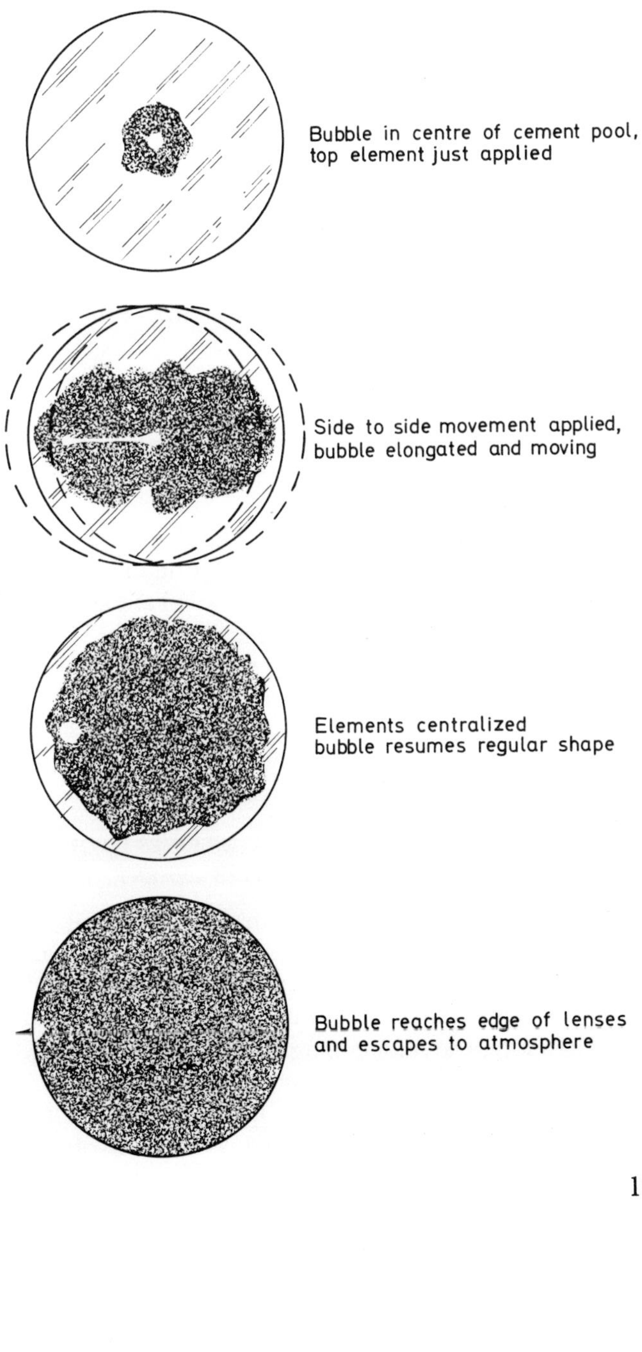

fingers or corks. There is grave danger to the top polished surface from dirt, grit or moisture being forced under pressure into the surface. Marks, digs, or stains can easily be the result.

Whether to use the fingers or a cork is decided by:

(1) Diameter of the top element.
(2) Whether the top curve is plus or minus.

A guide is as follows:

Diameter	*Curve*	*Method*
0–1 in. (0–25 mm)	Plus	Cork (small)
0–½ in. (0–12·5 mm)	Minus	Cork (small)
½–5 in. (12·5–125 mm)	Minus	Fingers (edge of lens)
1–5 in. (25–125 mm)	Plus	Cork (large)

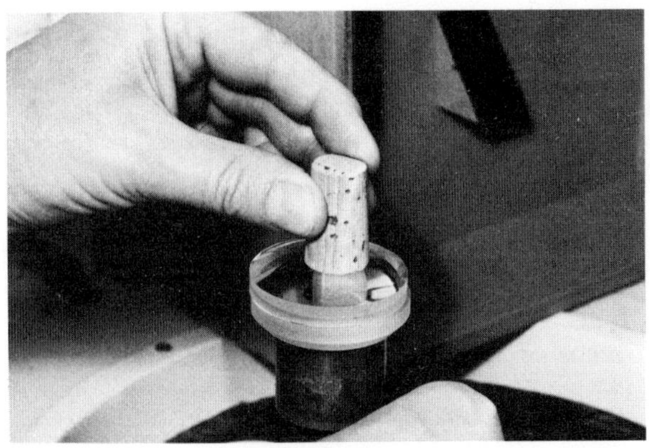

(*a*)

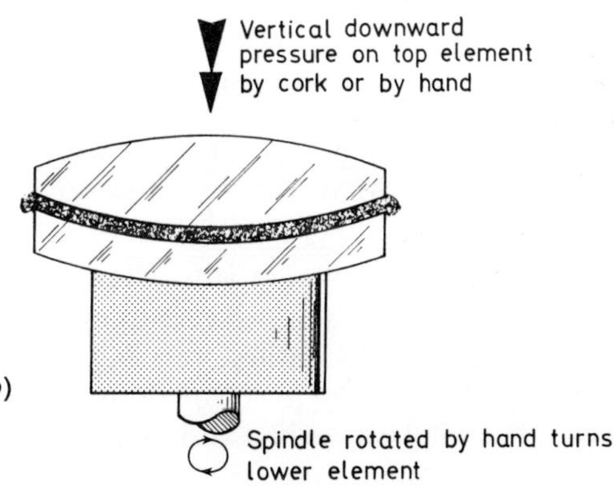

(*b*)

△ **6.11.28. Pressing out excess cement.**

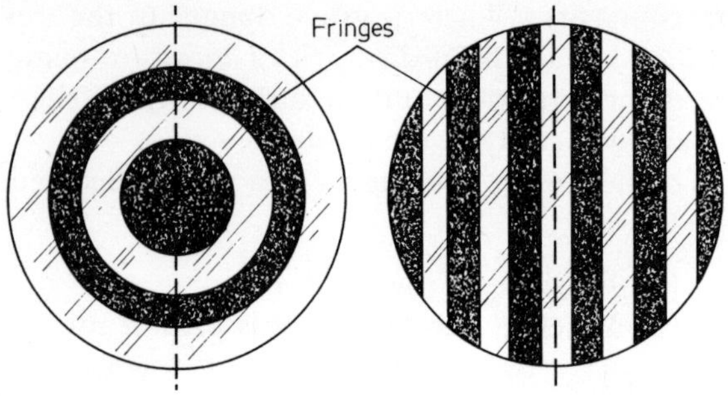

6.11.29. Correct and incorrect appearance of fringes.

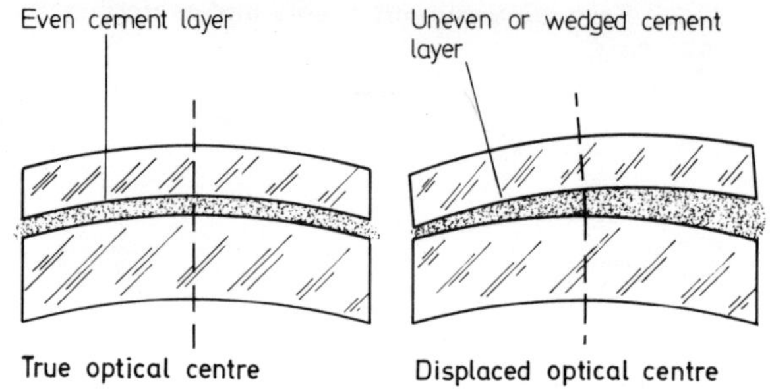

6.11.30. Correct and incorrect cementing, indicated by the correct and incorrect fringe forms of Fig. 6.12.29.

A great deal depends on the size of fingers and the dexterity of the operator in the choice of method. *Wherever possible, use the fingers on the edge of the top element, even on plus curves where the edge thickness allows.*

Rotation of the spindle will also rotate the lower element; the effect of the combined movements will eliminate a wedged cement film. A wedged condition can be observed in the fringes, and a series of concentric fringes is the target. The appearance of straight lines is to be avoided. (Fig. 6.11.29.)

Excessive wedge has the effect of placing a prism between the two elements, and Fig. 6.11.30 shows the effect on the path of light travelling through the doublet. A thick layer of cement, badly wedged, can therefore have an effect upon the truth of the doublet.

Cemented doublets are not checked for thickness until final inspection. It is up to the operator to use his or her skill to maintain the limits during cementing. When the cement is first pressed out and it reaches the edges, it is still very thick and no fringes will be visible. Continue pressing the top element and rotating the lower one until fringes appear. Observe the number and condition; at first there will be many rings visible, often twenty, and if the pressure is being applied correctly they will be central.

Further pressure will, as the excess cement escapes, reduce the number of rings very rapidly. The rings will run rapidly to the centre of the lens. When the correct thickness is reached, this movement will slow down and then cease.

Further pressure will result in a thin film, which will shrink and break as it hardens, resulting in a faulty lens. Often the number of rings now showing will be approximately six.

If, as the pressure is applied, the rings form into straight lines, indicating the formation of a wedge of cement, slightly vary the downward angle of the cork or finger. This will correct the wedge. When the rings are once again central, revert to the vertical pressure until the required thickness is achieved.

The doublet is now ready to be placed in the carousel for curing.

CEMENTING OF VERY ACCURATE LENSES WITH COLD SETTING LAMINAC 4128 CEMENT

Some lenses, such as air-survey lenses, are required to give very high resolution. The dimensional accuracy of the optical components and their subsequent mounting are very critical. Therefore, the final edging is after the cementing.

A 6-inch $f/5{\cdot}6$ air-survey lens (Fig. 13.2.10) manufactured by Wray (Optical Works) Ltd, had to be centred to within 2 seconds of arc

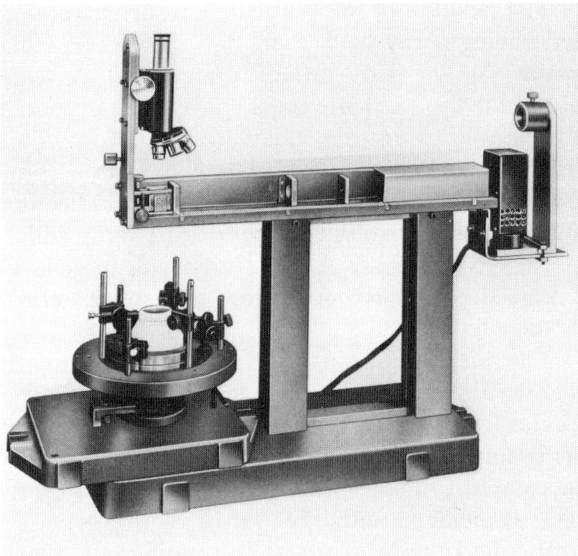

(*a*)

(*b*)

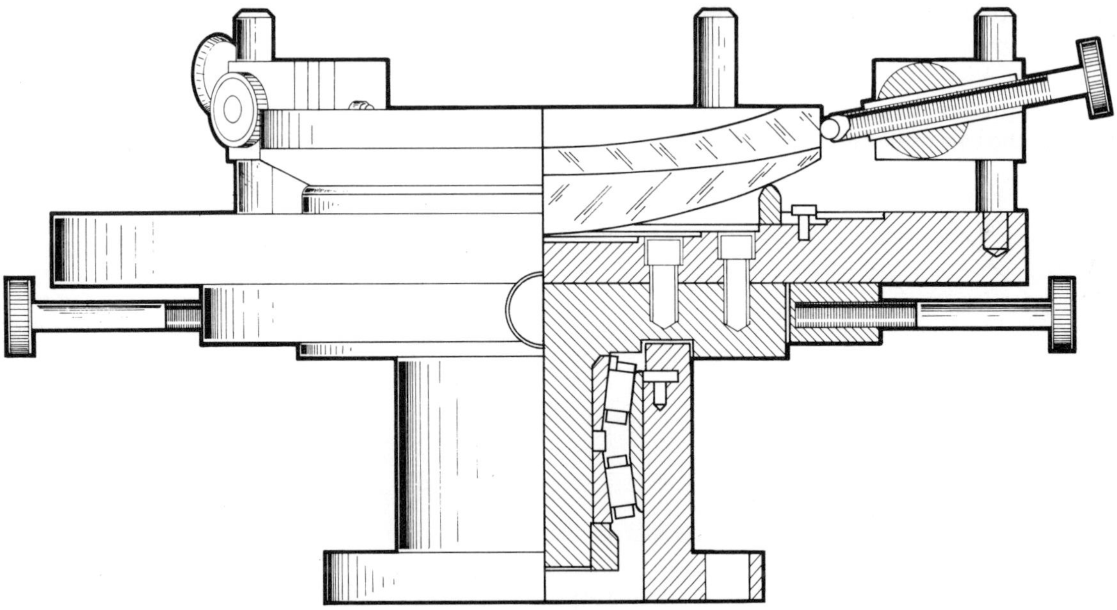

(*c*)

6.11.31. Lens centring and cementing machine. (*a*) Four cemented elements of a 6-inch *f*/5·6 air survey lens are on the turntable. (*b*) A two-turntable machine. Two lenses of different diameters are in position for cementing. (*c*) Diagram of the machine.

between the normal of any lens surface and the mechanical axis. To achieve this performance, Mr D. J. Day and Mr D. G. Monk of Wrays (with the advice of Mr H. J. Tiziani, Physics Department, Imperial College, London) developed a centring machine capable of the required specification (Fig. 6.11.31).

The machine consists of a turntable with a very accurate bearing, and an optical system for observing the decentring of each of the surfaces. The turntable and bearing are designed for easy adjustment in setting up and alignment.

An accurately centred ring is mounted on the turntable. The lens rests on this ring. When the instrument is being used for cementing, each lens component is held in position by two screw plungers (P) and one spring (S) arranged (in this case) for two levels, 1 and 2. (Fig. 6.11.32.)

The screw plungers are set at right angles to each other, and the associated springs are along the bisecting line. The optical system (Fig. 6.11.33) consists of a 250-watt high-pressure mercury-vapour lamp, a beam-splitting prism with titanium dioxide coating, and lenses A, B, and C, with space for a graticule to be placed anywhere between lenses A and C. The fixed microscope is used to view the image of the graticule after reflection from the lens under test.

The position of the graticule can be calculated by tracing a ray back from the microscope focal plane via the lens surface under test. A scale alongside the graticule track enables the graticule to be placed in the calculated position. This feature is important when there are several components to a lens, each with its own reflecting surfaces, as it enables the operator to know which surface is in focus. A filter adjacent to lens C selects the mercury green line and avoids chromatic effects.

The microscope consists of a × 10 eyepiece with a graticule 5 mm in diameter having ten concentric circles and a central cross. Two fine-control screws on the prism mount enable it to be tilted or rotated to make the image of the object graticule coincident with the centre of the eyepiece graticule. In order to set up the equipment, a lens or testplate is chosen with one radius of curvature similar to the bottom surface of the lens which is to be centred and the other surface plane. This lens is laid on the ring, curved surface down and the top surface adjusted to run true, using the screw plungers and springs. The whole turntable should be moved to bring both reflected images to the centre of the field.

Fig 6.11.31(*b*) illustrates one optical system serving two turntables for cementing lenses. While

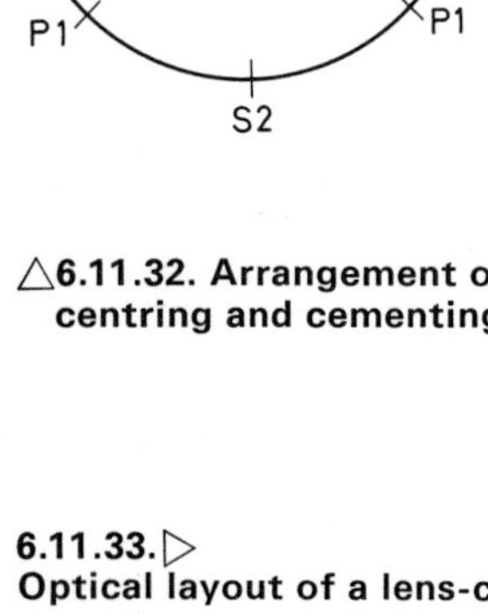

△**6.11.32. Arrangement of springs on the centring and cementing machine.**

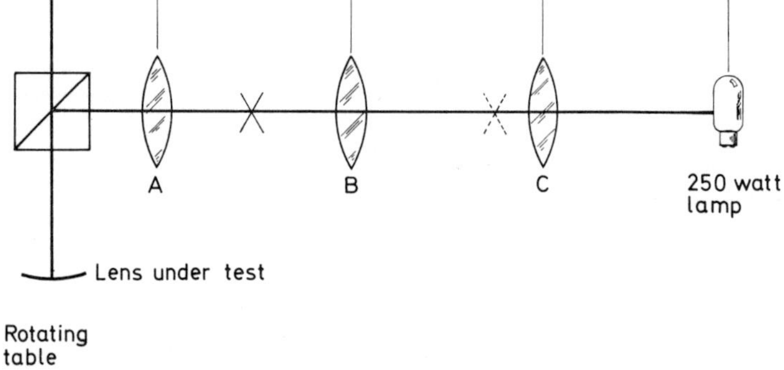

6.11.33.▷
Optical layout of a lens-centring device.

one turntable is being set up and accurately aligned, the other contains cemented lenses which are drying. When the cement has hardened sufficiently, the upper plungers are carefully removed and transferred to the next component (if any) in the same way. When all the components have been cemented, the lens must be left for at least 24 hours to dry the cement thoroughly. It can then be centred on an accurate edging chuck using a similar piece of optical alignment apparatus as for cementing, and by viewing the top and bottom surfaces.

Microscope objectives and lenses for microphotography and microprojection have to be very accurately centred and true to axis within 0·0001 inch (0·0025 mm) and they may consist of—for example—a single lens and two or three doublets (Fig. 6.11.34). A technique developed by Vickers Ltd with great success is briefly as follows:

(*a*) Cement lens elements together with Canada Balsam.

(*b*) Mount lenses in centring machine and soften the cement by heat to allow them to be accurately centred (Fig. 6.11.35.)

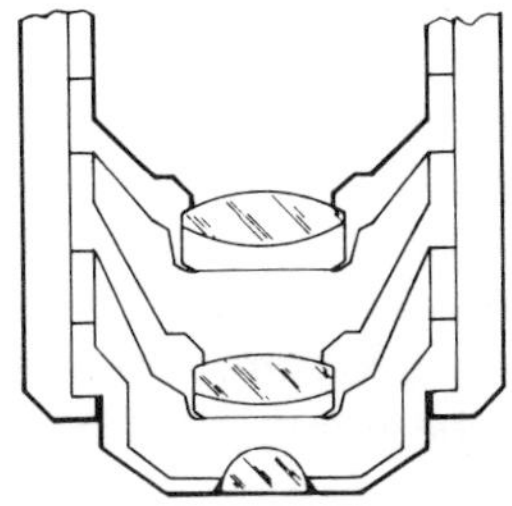

6.11.34. A typical microscope objective.

(*c*) Stick the lens doublets into an oversize cell with Araldite.

(*d*) Place the cell in the lens-cell-turning machine and accurately aligned for optical axis truth to the mechanical axis of the turntable (Fig. 6.11.36).

(*e*) Turn the mount to the finished size.

This method of mounting lenses in an oversize cell, and machining the outside of the cell whilst in the centring machine, may be worth further development in the future for optical designs where centring tolerances are critical.

6.11.35. Microscope lens-centring machine. The lens is warmed to allow the balsam to move during the centring operation. After centring, the doublet is stuck into a metal mount with Araldite.

6.11.36. Microscope mount-centring machine. The lens mount is adjusted until the lens optical axis is true with the turntable axis. The mount is then finish turned.

6.12 Edge blacking

In order to suppress unwanted internal reflections (flare) and avoid loss of contrast in photographic lenses, it is necessary to black the edge of the lens. An allowance of 0·0005 inch each side or 0·01 inch (0·025 mm) on diameter of the lens is usual and the number of coats of paint to make up this thickness varies. The usual size for an edge blacking booth is about 24 inch (700 mm) cube. In the booth is a vertical nylon chuck which can rotate by a motor controlled from a foot pedal. A pneumatic piston clamps a nylon mask on top of the lens during the blacking operation (Fig. 6.12.1.)

A Blinks–Bullows spray gun, model L 900, is suitable for spraying Postans air-drying polyurethane semi-matt black MV/Q7425 paint. This is a two-part paint, made up each day from one part of Postans Polyurethane activator D 980 with one part of Postans U.X.M.T. 2121 thinners.

6.13 Lens cell and mount machining

Hardinge lathe

This machine has been developed specifically for the machining of lens cells and is suitable for chasing accurate threads and scrolls. (Fig. 6.13.1.)

Britan turret lathe

For small lens components, cells, and barrels, this turret lathe gives a high quality finish. (Fig. 6.13.2.)

Loh automatic turning and thread-working machine

This machine has been under development since 1966 and is designed for whirling of precision threads and taking the finishing cut on locating surfaces in one operation. The method of machining and the high cutting speeds which are typical of whirling result in extremely high precision and a very good surface quality of the work. (Fig. 6.13.3.)

6.12.1. Lens edge-blacking booth. Paint is being sprayed on the edge of a lens rotating on a vertical spindle and held by a clamp. On the bench is a vacuum chuck for picking up lenses. The chuck, which contacts the glass, is made from nylon to avoid damage.

6.13.1. Hardinge lathe, automatically controlled, for accurate machining and chasing of lens cells.

6.13.2. Britan turret lathe for accurate machining of lens components.

6.13.3. Loh automatic turning and thread-whirling machine. Whirling means a cutting process with continual intermittent machining. The tool rotates at high speed and revolves eccentrically around the rotating workpiece, which itself revolves slowly and and in the same direction.

6.13.4. Whirling.
1, whirling tool; 2, mandrel for whirling (female); 3, workpiece; 4, chip; 5, rotating diameter of whirling tool nose; 6, outside diameter of thread; 7, diameter of the root of the thread; 8, mandrel for whirling (male).

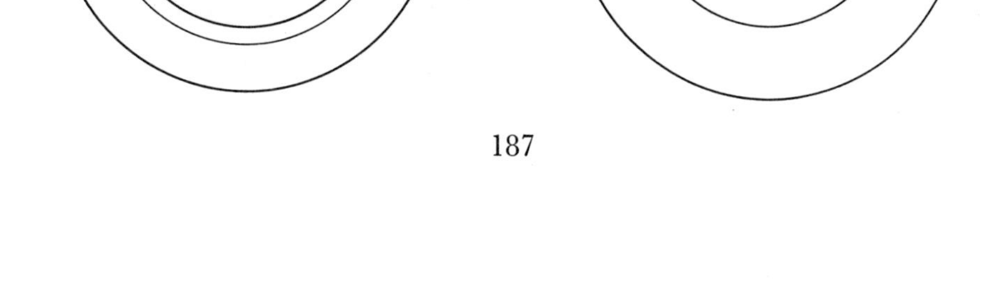

Thin wall parts or parts of plastic material can be processed without any deformation.

For thread whirling, the workpiece spindle provides for slow rotation and also for the axial feed which is effected by a mandrel rigidly mounted to the spindle and to the nut belonging to the mandrel. For the turning operation the same spindle rotates at high speed and therefore is provided with adjustable precision needle bearings.

The workpiece spindle is driven by belt for both turning and whirling, and engaged by magnetic clutch for both operations. Lubrication of all bearings of the headstock is centralized.

The workpieces are chucked by means of a quick grip device with adjustable cup spring packet which is released hydraulically by push-button control.

The spindle motor, of medium frequency, guarantees a smooth run free from vibrations, in order to satisfy the precision requirements and surface quality of the product. The frequency converter is installed outside the machine. The whirling spindle is swivel mounted and is adjustable for angle pitch of the thread to be processed. Diameter and length of the workpiece can be adjusted on a compound slide carrying spindle and whirling head. (Fig. 6.13.4.)

The compound rest of the turning head is hydraulically controlled and its motion limited by adjustable positive stops. By means of plug programmes on the crossed busbars the movement for the fully automatic turning operation can be selected. For setting up the machine, the turning head can also be moved by individual switching.

The automatic dividing attachment functions as follows. Multi-start threads are divided by means of a dividing plate rigidly connected to the screw nut. An electric lifting magnet actuates the index pin provided in a hardened steel guide which is free from any play. By moving a cover disk on the dividing plate, a number of index notches required for each type of thread are released. This arrangement makes it unnecessary to change the dividing plate for different threads. (Fig. 6.13.5.)

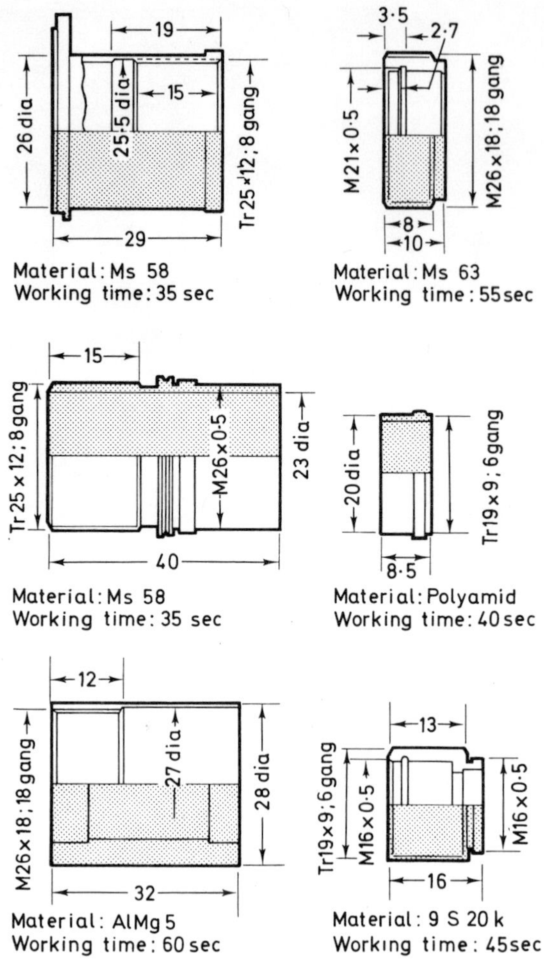

6.13.5. Whirling times for different materials.

Tarex single spindle automatic lathes

Tarex 75 and 90 automatic lathes equipped with chasing attachments and eight station turrets are very versatile machines for large volume production of lens components. A variety of special-purpose tools are available, capable of simultaneous machining operations, such as the production of screwdriver slots in lens lock-rings.

Ebosa semi-automatic turning and chasing lathes

As a second-operation machine, the Ebosa can do thread chasing (including multi-start threads), making it suitable for machining projection-lens barrels. With twin cross-slides and independent copying devices, this is a versatile machine for the production of lens-mount parts.

7 The Manufacture of Spectacle Lenses

7.1 Introduction

Although the manufacture of ophthalmic lenses is not faced with the need for exceptionally high accuracy of curvature and figure, there are a number of processes outside the normal methods of working spherical surfaces which present their own technical problems.

The processing of spherical surfaces follows the basic principles used by instrument-makers, but great emphasis is placed on high-speed operations. In addition, different techniques are used in making toric surfaces and fused and solid bifocals.

The output of ophthalmic lenses is considerable and this enables the manufacturers to develop specialized techniques which would be difficult to justify on economic grounds for small-quantity production.

7.2 Glass

The introduction of continuous melting of optical glass by Chance-Pilkington in 1957 brought a substantial change in the conventional methods of handling and preparing glass prior to grinding and polishing.

A detailed account of the manufacture of optical glass is outside the scope of this chapter, but the bare essentials are described for the sake of completeness. The choice of glass used for ophthalmic purposes was largely dictated by economic factors. Spectacle lenses are relatively cheap and this situation demands an inexpensive raw material with the most suitable physical and chemical properties. Basically, hard crown glasses meet both these requirements, and the availability of large-scale melting capacity for the production of window glass provided an obvious source of a glass suitable for ophthalmic lenses. Until 1957, the glass manufactured in England was continuously melted and rolled into sheet with a range of thicknesses suitable for remoulding. The rate of production was very high and a run would usually provide sufficient glass to last the industry for two to three years.

The quality left a good deal to be desired. It had no pretentions to being an optical glass and trouble with striae, veins, and bubbles was fairly prevalent. To make matters worse, rejectable faults could seldom be detected until the finished lenses were subjected to final inspection. Refractive index control was very good and close tolerances of expansion were also maintained. One other serious objection was an inevitable slight tint in a glass, which should be colourless or 'white', as it is called in the spectacle-lens industry. The tints usually tended towards faint pinks, greens, and blues and, although each batch was reasonably consistent in itself, there were often marked differences from batch to batch. It might be argued that small differences in the colour of a white glass could not be detected in relatively thin spectacle lenses, but the tint could be very obvious when looking at the ground edge of a lens in a rimless frame.

At one time it seemed very unlikely that a glass of optical quality would ever be available for the ophthalmic lens industry, but the substantial volume of glass used in making all forms of spectacle lenses probably provided the incentive to develop methods for continuous melting of optical glass.

The first successful process was developed by Corning in the U.S.A. and glass was produced commercially during the Second World War. Since then, the major optical glass companies have taken licences to operate the Corning process or have developed their own equipment for continuous melting of high-quality glass. The method follows the basic principles used in the continuous melting of glass for the production of sheet and other ware, but there are a number of important refinements and differences in detail.

A continuous furnace has three main zones along its length. The batch is fed into the mouth of the furnace and moves into the first zone where the preliminary fusing of the ingredients takes place. The next zone increases the temperature to complete the melting and to free the glass from bubbles, and the final zone reduces the temperature of the molten glass so that the viscosity is suitable for the final forming process. At this point the glass leaves the furnace through a suitably shaped orifice.

Melting units for commercial glassware are usually large. For the production of bottles, a furnace may contain up to 100 tons of glass while a furnace extruding sheet glass may have a capacity in excess of 1000 tons. Such furnaces are constructed with refractories, solution of which in molten glass can cause a moderately high percentage of objectionable faults, and these are totally unacceptable for optical glass.

Optical-glass melting units are relatively very small, with capacities seldom exceeding a few thousand pounds. The first zone is fired conventionally by gas or oil but the final stage of melting is achieved by passing an electric current through the glass itself. Contamination by the furnace walls is avoided by lining with high-melting-point metal alloy sheet which is virtually insoluble in molten glass. After melting is complete the glass passes through a stirring chamber which provides the degree of homogeneity required, and flows as a continuous stream which can be manipulated or formed in a number of ways to provide suitable shapes for all forms of optical components.

The production of spectacle lenses is geared entirely to the use of mouldings and the continuous process is ideal for the manufacture of large quantities of this type of product. A press unit is a fairly complex piece of equipment which consists of a large circular table carrying a number of moulds and a single press. The table is located near the glass stream, which is cut at precisely controlled intervals so that a piece of molten glass drops into a mould on the table which then indexes to the press position to form the glass to shape. At the same time the next mould is positioned ready to be filled. By the time the glass has completed a full revolution of the table, it is cold enough to be picked out automatically and transferred to a lehr for annealing. (See § 14.6.)

The pressing operation is automatic and, with the additional advantage that reheating is not necessary, it becomes a very cheap operation. However, the high costs of equipment, tooling, and setting-up make long runs absolutely essential for minimum operating costs. Prices of any particular moulding therefore vary with the quantity, and a user may have to order about 10 000 lb of any one shape of moulding to obtain the minimum price.

Every manufacturer finds that 10 per cent or more of his production involves small batches of a few thousand down even to ten mouldings, and a requirement arises for a different form of glass which can be remoulded at short notice to any shape.

One suitable form is the gob. This is a piece of unformed glass produced by allowing a molten piece to drop on the table of the press unit and form its own shape as it cools. Larger pieces may be contained by a circular mould to keep the gob to a manageable size.

While gobs are a useful source of raw material for producing moderately large quantities of a particular moulding, a fixed volume of glass does not provide very much flexibility in production unless large stocks of many different sizes can be kept. The limited versatility of gobs led to the early introduction of the new glass in strip form, which enabled the remoulders to provide any size, shape, or thickness of moulding which might be required by the industry. All glasses, white and tinted crown, and optical types used by the ophthalmic lens industry are now supplied in sheet form.

There is no doubt that the new method of glass production has provided substantial benefits for the spectacle-lens makers. The standards of glass quality are now so consistently high that it is no longer necessary to carry out routine inspection procedures to eliminate glass containing veins, striae, and seeds. The most spectacular improvements have been noticed in the tinted glasses, because it was often impossible to find minor faults in the old form of sheet owing to the high absorption of some of those glasses. It was all too common experience before 1957 to find batches of finished tinted lenses with as many as 80 per cent containing rejectable glass faults.

There is perhaps some dissatisfaction with the control of colour in tinted glasses and it does seem that it is technically impossible to produce batch after batch of a particular tinted glass without some visually detectable differences. Continuous melting has undoubtedly improved the control of colour but it must be appreciated that many factors can influence the colour such as variations in raw materials, melting conditions, the effects of remoulding and annealing, and slow changes in the finished glass due to solarization.

7.3 Remoulding

Although the great majority of mouldings used by the spectacle lens industry are supplied by the glass manufacturers, it is still necessary to remould gobs and sheet glass for fringe items in a manufacturer's range.

The present-day remoulding shop is mainly concerned with day-to-day requirements of mouldings, and the number in a batch varies from as few as ten up to a few thousand pieces. In the majority of cases, sheet glass is the starting material. It has to be cut to limits of weight so that the finished moulding is within 0·2 mm or so of the specified thickness. When the weight of a particular moulding has been determined, the sheet glass is drawn from stock and first cut up into manageable sizes so that they can be weighed. This enables the cutter to calculate the yield of pieces from each sheet. It is then marked out into square or nearly square sections and cut into parallel strips. Each strip is then cut into individual pieces which are weighed. The inevitable variations in heat-formed sheet glass results in some spread of weights and any which exceed the target are reduced by cutting off one or more corners. Any falling below the required weight are usually put aside for other requirements. As mentioned earlier, glass faults are comparatively infrequent, but it is necessary to be aware of forming faults. A fold in the edge of the sheet is one of the more common examples and unless this is trimmed off it may crack. Damage to the fire-polished surface can sometimes cause a piece of glass to split when it is put into the furnace.

When moderately large requirements arise, it is possible to use gobs, and this can lead to considerable economies if short delivery times are not essential. Gobs require no preparation prior to moulding and there is no wastage, whereas cutting from sheet glass may involve losses between 10 and 20 per cent.

As a number of high-index glasses are used for the manufacture of fused bifocals it is necessary to apply some form of identification which can be retained to as late a stage as possible. This is done by marking each type of glass with an identifying mark which will remain indelibly printed on the surface after moulding. Suitable marking inks are made by mixing various coloured oxides and applying them to the glass with rubber stamp pads.

The moulding shop is set out in a number of separate units, each consisting of a lehr with one or two moulding furnaces built on each side.

At the front of the lehr is a large rotary-hearth furnace used for moulding larger work. Two smaller furnaces are situated at the middle of the lehr and these are used mainly for producing small flint mouldings used for fused bifocals. The large furnaces are about three feet in diameter with an opening at the front. The hearth is mounted on a large casting with a central vertical spindle which projects through the bottom of the furnace and is driven by a motor. The furnace is fired by gas through air-blast burners situated round its periphery.

A massive press into which suitable dies and

plungers can be fitted is mounted slightly to one side of the mouth of the furnace, and is operated by a pedal. Another pedal operates an ejecting mechanism and is also connected to the belt striker which starts and stops the rotating hearth.

The furnace operator stacks the pieces of rough glass at the side of his furnace, and all subsequent handling is done with a pair of foils. These are pieces of $\frac{1}{4}$-inch rod flattened at one end and fitted in shielded handles. A piece of glass is picked up with the foils and laid on the hearth, the ejector pedal touched, the piece of glass carried just inside the furnace, and the hearth stopped. Another piece is placed on the hearth and the operation repeated until the first piece appears on the other side of the furnace. By this time the first piece is in a plastic condition, and is drawn by the foils down a metal chute into the metal die. The plunger is brought down by the pedal and the glass pressed to the shape of the die and plunger. The plunger is then released and the moulding ejected and slid down a chute into the lehr. Another piece of glass is then put into the vacant spot on the hearth and the cycle of operations repeated.

With the rotary-hearth furnace, a continuous cycle can be achieved by choosing a suitable number of pieces and adjusting the temperature of the furnace so that the glass is in a suitable condition for moulding after its circular journey.

When the moulding lands on the travelling mat of the lehr it is carried into the hot zone, which is thermostatically controlled to a suitable temperature for the type of glass being used. As the mat travels through the lehr the temperature gradually drops until the mouldings emerge at the end after a period of $1\frac{1}{2}$ hours at a temperature of 100°–200°C. They are then removed and packed into a box ready for examination.

The use of the two smaller furnaces down the lehr enables softer glasses to be fed into a part of the lehr which is operating at a lower temperature.

When the mouldings are received in the issue department they are sorted and put into stock ready for issuing to the various lens-working shops as required.

7.4 British Standards

The British Standard Recommendation for spectacle lenses was published in 1957, but at that time there were no recognized standards for the materials from which they were made.

It might seem that the relatively wide tolerances of the most commonly used spectacle lenses would hardly necessitate a British Standard. However, the fused form of bifocal construction places quite stringent demands on the control of refractive index. Before the introduction of continuous melting, samples of each melt of optical glass used for fused bifocal segments were tested by the user and, if approved, the complete melt was purchased. In continuous production the control of all properties is based on hourly checks and consequently batch melts are no longer available. This fundamental change of system made a British Standard highly desirable, and a committee was formed in 1958 to prepare the specification.

An important aspect of the work concerned the standardization of annealing conditions. It is a well-known fact that the rate of cooling in annealing has a significant effect on the index of a sample of glass. The indices quoted in Chance-Pilkington's catalogue refer to fine annealed glass cooled at a rate of 20°C per day. Continuous control based on a time delay of a week is obviously impossible, so it is essential to adopt a much shorter schedule for routine testing. Fortunately there is an accurate relationship between the rate of cooling and change of index, so that the value produced by one schedule can be calculated by measurements taken after a different schedule.

The formula given by Chance-Pilkington is:

$$\text{Change in index} \times 10^4 = F_A \times \log \frac{R_1}{R_2}$$

where R_1 is the first cooling rate, R_2 the second cooling rate, and F_A the annealing factor of the glass. The annealing factor of the majority of glasses varies from 4 to 13 approximately. The value can vary widely between glasses in the same type classification, so it is not safe to make any assumptions if an annealing factor for a particular glass is not quoted.

It has been found that a cooling rate of 360°C/hr produced an adequate standard of annealing for the majority of spectacle lens blanks, and this compared favourably with current practice in bifocal fusing, and so was adopted as the standard on which all measurements and controls were to be based. Calculations show that glasses with a high annealing factor can suffer an index change of 0·003 if the extremes of annealing schedules are used.

The Standard B.S. 3062 was published in 1959 and a revised one issued in 1970. This now includes standards for spectacle crown and six high-index segment glasses. The tolerances for refractive index of the most widely used ones are as follows:

	Index	*Tolerance*	υ
Spectacle crown	1·523	±·001	59
(close tolerance)		±·0003	59
EDF	1·654	±·001	33
DBF	1·654	±·0015	41

7.5 Forms of spectacle lenses

Spectacle lenses fall into two main groups—single-vision and multifocal.

The mass manufacturer has to offer a range of finished or partly finished lenses which serve the majority of day-to-day requirements. One writer estimated that the number of variables in a completed pair of spectacles could run into some thousands of millions and inevitably there are, every day, some thousands of lenses being made to special order by mass manufacturers or by the prescription manufacturers whose job it is to deal with the individual prescriptions ordered by sight-testing practitioners and dispensers.

Fortunately a very high percentage of single-vision prescriptions can be obtained from a large, but nevertheless manageable, range of finished lenses.

At this stage it would perhaps be advisable to give a brief explanation of the curvature and focal lengths used in spectacle lens work. The international standard of power is the diopter. The power of a lens in diopters is the reciprocal of the focal length in metres. Thus a lens of 500 mm focal length has a power of 2 diopters, and one of 100 mm, 10 diopters.

This is a very convenient system because the power of a lens or combination of lenses is obtained by simple summation of the power of each surface.

The radius of curvature required to produce dioptric curves is obtained by the use of the formula $r = (N-1)/F$, r being the radius of curvature, N the refractive index of the glass, and F the power in diopters.

Glass of $N = 1{\cdot}523$ being a universal standard, curvatures derived from the above formula with $N = 1{\cdot}523$ are rather loosely referred to as 'dioptric tools'.

Single-vision lenses fall into two distinct groups, meniscus and toric. Most manufacturers produce a stock range of meniscus lenses with powers between −8·00D and +7·00D in 0·25D steps, i.e. about sixty items. Toric lenses combine a cylindrical element up to 4·00D, again in 0·25D steps, with the range of spherical powers making another $60 \times 16 = 960$ items. Another 100 or so items of partly finished lenses are also required for special prescriptions. A considerable percentage of this range is also available in several tinted glasses.

Multifocal lenses are broadly of one-piece or fused construction. There are a large number of variations in each group, together with some more complicated forms combining both techniques, and attempts to estimate the number of different combinations become almost meaningless. However, by the grouping of a few basic lens forms with the required difference between the powers of the various areas (normally up to 4D), the range becomes manageable. Virtually all bifocals produced by mass manufacturers are supplied in semi-finished form to be ground and polished on the unworked surface to meet the requirements of the prescription.

7.6 Manufacture of single-vision lenses

With the exception of a very few specialized forms all single-vision lenses have one spherical surface, the other surface being spherical or toroidal. To obtain the minimum number of variables, it is fairly common practice to have a standardized range of convex surfaces or bases as they are often called. Normally eight to ten spherical bases and a similar number of toric bases are required to cover the full range of about 1000 items. Each toric base has its range of sixteen cylindrical corrections. Intermediate powers are obtained by varying the surface power of the concave sphere in 0·25D steps.

There are no special problems involved in working the concave side of toric lenses to make the handling very different from that employed in the case of meniscus lenses. Techniques for making convex spheres and torics are, of course, quite different and they are generally made in separate workshops.

7.7 Toric surfaces

Large-scale production of toric surfaces is peculiar to the ophthalmic lens industry and until the 1940s every important manufacturer developed and made his own machinery. Since then several independent firms, mainly in Britain and Europe, have designed toric machinery.

Toric surfaces are sections of the periphery of a toroidal sphere. An exaggerated sketch of the shape is shown in Fig. 7.7.1. R_1 is the radius of the base curve and R_2 the radius of the cylindrical curve. The difference in dioptric power of the two meridians gives the cylindrical power incorporated in the finished lens. Although there is no strict conformity between manufacturers, the base curves range from about 3·50D to 11·50D. These are nominal curves only and in fact are modified to compensate for the effect of the finished lens thickness on the back vertex power. The majority of the low-power lenses have base curves about 6·00D or 6·50D.

The geometry of the toroidal sphere provides a very attractive shape for the blocking process. The appropriate mouldings are stuck round the periphery of a circular metal-holder. (Fig. 7.7.2.) A hole in the holder provides the means to attach it to the grinding and polishing machines. Accurate location of the mouldings is highly important and many solutions to the problem have been devised.

The block now has to be ground to the required shape, and toric diamond generators are now used almost exclusively. The geometry of the block toric generator is illustrated in Fig. 7.7.3. The

7.7.1. Section of a toroidal sphere.

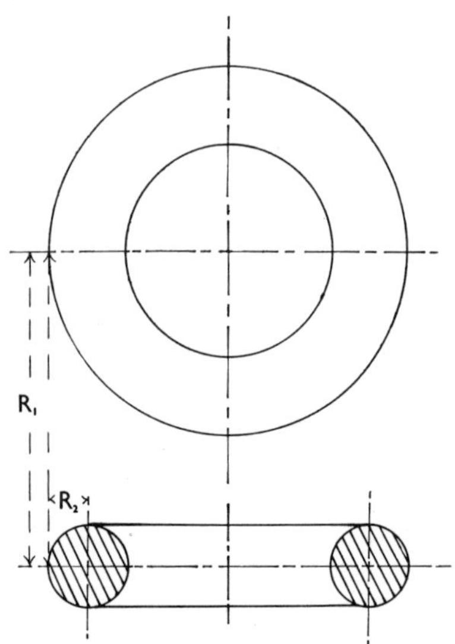

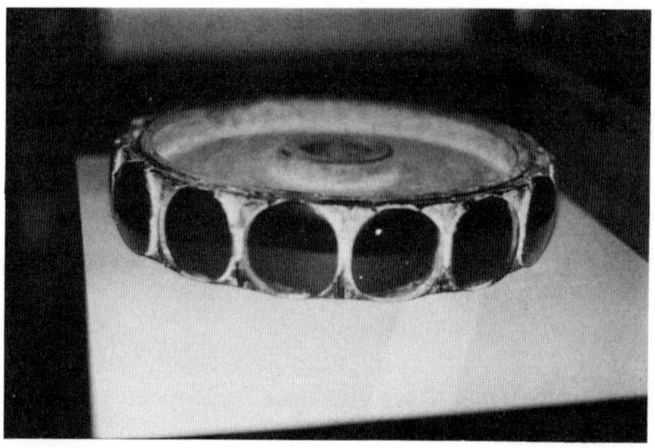

7.7.2. Block of toric lenses.

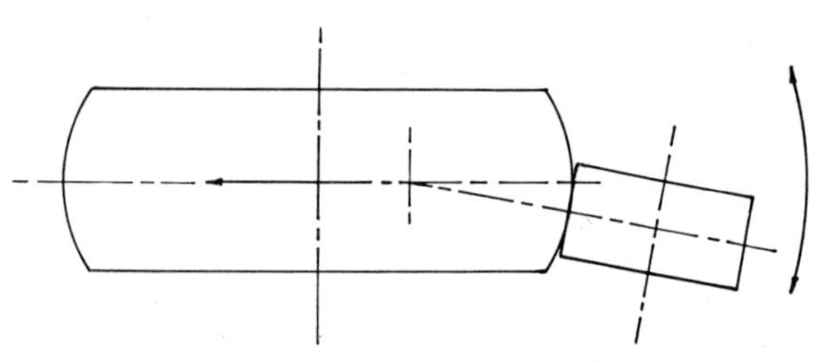

7.7.3.▷ Geometry of a toric roughing machine.

7.7.4. Autoflow toric smoothing or polishing machine.

diamond tool which may be a peripheral wheel or a cup form is swung in an arc round a pivot point. Control of the radius of the cylindrical or cross curve is obtained by a slide which can adjust the distance between the surface of the tool and the pivot point. Another slide which varies the position of the pivot in relation to the axis of the block spindle determines the diameter of the block which is twice the radius of the base curve. It will be appreciated that continued grinding gradually shortens the radius of the base curve and it is necessary to rough a toric block oversize to allow for the glass removed in smoothing—generally 0·4 to 0·5 mm. A toric generator made by Adcock and Shipley has earned a very good reputation. Smoothing and polishing are both basic lapping processes and the same machine serves for either purpose. As the surface is essentially cylindrical it is necessary to prevent the lapping tools from rotating. This is generally done by the use of a main driving pin engaging in a socket and a second pin running in a fork or slot to prevent rotation but to allow angular freedom about the plane at right angles to the driving pin.

A toric surfacing machine developed within the last few years by Autoflow Engineering is illustrated in Fig. 7.7.4. The machine is made in four-spindle units, each spindle carrying two blocks. Smoothing tools or polishers are mounted on each side of the blocks and pressure applied by air cylinders. In operation, the blocks rotate and the tools move to and fro in a straight line to provide the lapping action. A break up in the oscillation is incorporated in the driving mechanism to prevent the formation of undesirable waves on the finished surfaces. The appropriate slurries are pumped on to the blocks and returned from the catchers to the pump for recirculation. Once loaded and set running, the machine requires very little attention.

Toric lapping is something of a compromise owing to the nature of the surface. Reference to Fig. 7.7.1 shows that, while the cross curve is constant, the shape of the base curve is not. Therefore, a lapping tool which is machined to fit at the pole will lose contact when it is moved away from the centre line. A true toric can only be generated by point or line contact. It is important to avoid conditions which tend to exaggerate the loss of contact between tool and glass owing to the oscillation. The amplitude of oscillation must be kept to a reasonable minimum, and the sizes of tools and polishers should not greatly exceed the diameter of the lenses.

Cast-iron tools are most generally favoured for smoothing operations and an abrasive with a mean particle size of approximately 15 micrometres is usually satisfactory. Control of curvature presents a slightly unusual problem. The radius of the base curve depends entirely on the diameter of the block and can be influenced only to a very small extent by smoothing. The cross curve depends very much on the shape of the lapping tool and, if a trend away from the required radius is detected, correction can usually be made by local hand grinding on the face of the lapping tool.

The restriction applied to prevent the tool from rotating probably causes the greatest problems in toric polishing. The worst faults are drags from small chips on the edges of the lenses and the tendency to form lines parallel to the axis of the cylinder. These effects are minimized with well-designed machinery and the choice of a suitable polisher. A moderately soft material is generally used which helps to keep contact with the glass in spite of the peculiar nature of the surface.

After polishing, the blocks are examined and faulty lenses identified so that they can be rejected at a later stage. It is then usual to indicate the axis of the cylinder by running a narrow band of white paint round the periphery before finally applying a protective coat of transparent varnish to the finished surfaces. When this is dry the blocks are put into a refrigerator so that adhesion between the blocking compound and the glass is largely broken, thereby allowing easy removal. The partly finished lenses are then put into slotted racks to await processing of the concave surface.

7.8 Working convex spherical surfaces

Since 1950, trends in spectacle-frame designs have demanded appreciably larger lenses, and a diameter of 58 mm is now considered to be a minimum. This has decreased the number of lenses which can be made up into a convenient block, and there has been a general trend towards working spherical surfaces singly. However, there are a number of problems which make single processing of the deeper convex curves substantially more difficult than for similar concave curves. For this reason, the majority of the convex surfaces of meniscus lenses are worked in blocks of three or four lenses.

Although it is a fairly standard procedure nowadays to work toric surfaces first, there is no clear cut preference when making meniscus lenses. Much may depend on the type of equipment and organization of production which have developed in a particular company, but here we assume the convex side is being worked first.

The first operation is to diamond-generate the convex surface of the moulding, removing sufficient glass to clear up curvature errors, surface inclusions, and thickness variations. This is normally about 0·5 mm. There are no particularly close tolerances required at this stage and a relatively simple cell chuck is normally quite adequate. There are a number of excellent spherical generators available nowadays and two of these are illustrated (Figs 7.8.1 and 7.8.2; see also § 6.5.) The design of the Autoflow machine is fairly conventional and incorporates an automatic cycle which provides a rapid approach of the lens spindle to the diamond tool followed by hydraulically controlled slow feed while the glass is being ground. When the spindle has reached the thickness setting an adjustable dwell period gives a final clearing cut before the machine stops and the lens spindle retracts.

The C.M.V. machine has an unusual approach

7.8.1. Autoflow Technaphot Model 158 Mark III A four-spindle curve generator.

7.8.2.▷ C.M.V. 100 curve generator with automatic feed from a magazine tray of lens blanks

to the method of applying the feed of glass to the tool. The lens spindle is essentially fixed except for vertical adjustment to control thickness and the lap spindle is mounted on a substantial pivoted arm which is lowered and raised by a cam. While the pivoted arm is widely used on lapping machines, the application to a diamond generator solves some of the problems associated with wear and stiction in linear slides. The C.M.V. machine is also available with an automatic feeding device using an indexing tube which enables the machine to be loaded and then left unattended for some few minutes.

It is impossible to give any precise specifications for suitable diamond tools. Many factors affect the quality of surface finish, such as the bonding, concentration, type of coolant, and even the characteristics of the machine itself. Little can be gained by using the finest grits when the use of fairly simple chucks cannot guarantee a high degree of sphericity in the generated surface.

After the convex surfaces have been ground, they are blocked using cell moulds which locate from the ground surface. The lenses are first warmed on a hot plate and fitted into the cells, and a lump of hot pitch or blocking wax is placed on top of the glass. A heated metal runner is then placed on top of the pitch and located in a press to squeeze out surplus blocking compound. While the pitch is still warm the surplus is trimmed off and the block allowed to cool, until it is sufficiently firm to remove from the mould.

The surfacing machines are generally conventional, the block being mounted on the bottom spindle and the smoothing tool or polisher guided by a pivoting arm over the top of the block. The familiar lead weights of years ago are now replaced by air cylinders which can raise or lower the arm and apply any amount of pressure required in surfacing.

Abrasive and polishing slurries are now almost universally pump-fed from a sump to each individual spindle. A system may contain between 10 and 20 gallons and feed as many as thirty separate spindles. Smoothing tools are made of cast iron, and synthetic alumina of a mean particle size of about 15 micrometres is generally the most suitable abrasive. Hard felt polishers are widely used for block polishing and give fewer problems than the pitch polishers, which are essential for high-accuracy optical-element work.

The organization of all single-vision lens production depends on the working of every process to a strict time cycle, and continuous control of all conditions is essential. Pump-fed systems do enable reliable checks to be made at regular intervals, so that adjustments can be made to conditions of the slurries. The quality of smoothing slurries is best checked by measuring the glass removal obtained in the standard machine cycle. This should normally be between 0·15 and 0·25 mm. If the removal is too low, fresh abrasive must be added, and after one or two days the whole system has to be drained and recharged.

Checks on the polishing slurry are rather more elaborate. In the interests of economy control of the density is most important. A quick and reliable measurement can be carried out with a hydrometer, and it is generally accepted that a reading of 20° Baumé gives the best results. Polishing seems to be less troublesome when the slurry is warm but there is a danger if it gets hot enough to soften the blocking compound, and during hot weather it is necessary to dissipate some of the heat generated by the polishing through the means of a water-cooling coil. There has been a good deal of disagreement about the effects of pH on the polishing process. Considering the vast variety of materials and techniques used throughout the optical industry this is hardly surprising. It is often worth checking the effect of pH on a particular process because some polishing powders are sensitive. It has also been noted that some materials stay in suspension better when the slurry is held within a relatively small range of pH values.

Checks are normally carried out on an hourly basis and additions of water or polishing material are made to maintain the optimum density. Polishing releases large quantities of alkali into a system and, if control is necessary, acetic acid is added to neutralize the excess. Additional checks on the amount of glass removed in polishing may be made at much longer intervals to monitor other aspects of the process and materials.

Curve control is maintained entirely by stroke adjustments on the machines. The ophthalmic lens-maker has one advantage because the dimensions of his lenses and blocks can be standardized. Optimum tool sizes can therefore be used so that they never need be returned to correct curvature errors.

Typical cycles for smoothing and polishing are about 10 and 20 minutes respectively. After polishing the surfaces are examined and faulty lenses marked with greased pencil. The surfaces are then painted with protective varnish prior to freezing and deblocking. At this stage, faulty lenses are rejected and the good ones transferred to racks for transport to concave-side processing.

7.9 Working concave spherical surfaces

The concave-sphere department receives the partly finished meniscus and toric lenses and the first operation carried out on them is the generating of the concave surface to obtain the correct thickness and reasonable freedom from prism. Blanks with convex spheres can be held quite easily in a cell chuck with a ring location. The chucking of torics is a little more complicated and it is necessary to provide a mating surface for each toric. It is essential to cover the locations with some form of resilient material to prevent damages to the finished surface.

During the last few years there have been substantial changes in the approach to smoothing and polishing and most of the leading manufacturers now work the majority of concave surfaces singly. There are several reasons for this. As mentioned earlier, the increased size of lenses tends to reduce the advantage of blockwork. Blocks are also very heavy and it seems ludicrous when four finished lenses weighing less than 80 g in total may be held in place with a cast-iron holder weighing between 1 and 2 kg, whereas a holder for a single lens can be scaled down to 100 g or so.

From a more technical point of view the reduced size and weight of each unit enables much higher lapping speeds to be obtained. Further improvement in grinding rates can be achieved by using annular tools about 100–120 mm in diameter which makes spindle speeds of 1500 r.p.m. quite feasible.

Blocking lenses singly is a good deal simpler than making them up into blocks, and any conventional method can be used. To ease handling problems, the holders are generally made a little larger than the glass. If the glass overhangs the edge of the holder the probability of chipping or breakage is greatly increased.

The lenses can be smoothed quite satisfactorily using cast iron tools and a slurry of 15 micrometre emery. Machine cycles of about 2 minutes are typical. Owing to the relatively small size of the tools, curves do tend to change fairly quickly if there is any error in the machine setting, and regular checking and adjustment of strokes is essential.

Since about 1960, there has been a steady increase in the use of diamond tools for smoothing operations. The single lens surfacing system has done much to make this development possible.

Several machines designed for diamond smoothing are now commercially available and two of these are shown in Figs 7.9.1 and 7.9.2. Up to the present time, there appears to be very little information about the fundamental aspects of diamond grinding. Fine lapping certainly provides greater technological problems than are currently encountered in polishing.

7.9.1. Autoflow diamond smoothing machine.

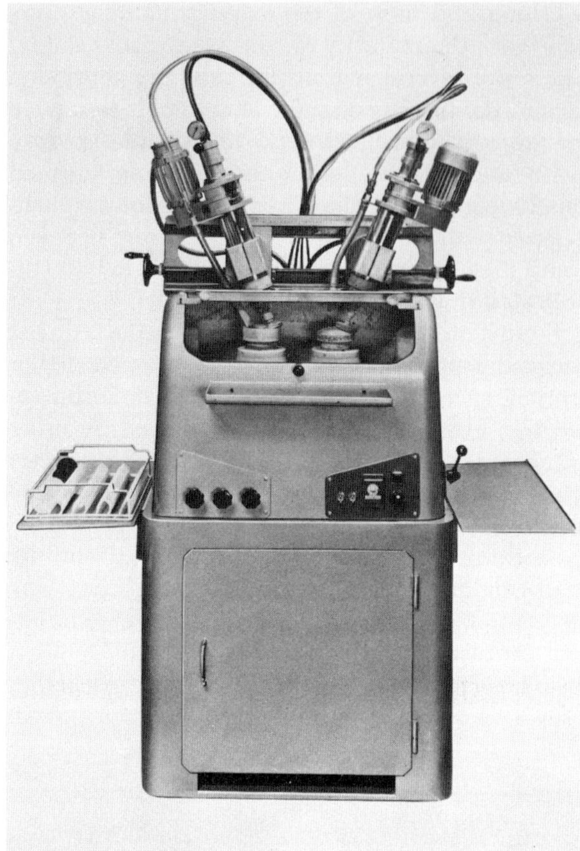

7.9.2. CMV Type ICM II diamond smoothing machine.

The two main advantages of diamond tools are very low rate of wear, which provides much more accurate control of curvature, and the elimination of the costly handling of large quantities of loose abrasive slurry.

There are very distinct differences between the type of surface produced on glass with bonded and loose abrasive. In the latter case, the surface is uniformly pitted and there appears to be a fairly well-defined optimum surface finish which can be obtained. If the size of a loose abrasive grit is reduced much below 10 micrometres the normal grinding process seems to cease and the surfaces show evidence of seizing between the glass and tool. Diamond tools, however, tend to produce a surface which consists of a myriad of fine scratches with a few isolated heavier scratches. The diamond surfacing system can be modified in several ways to obtain almost any degree of surface finish. Apart from alteration of the grit size, modification of the bonding alloy, variation of the diamond concentration, and choice of coolant can all be used to modify the surface finish. It is, in fact, possible to produce surfaces which are very near to a good optical polish subject to a penalty of a very low rate of glass removal and a relatively high consumption of diamond. A compromise has to be made between glass removal rate and surface finish and in the ophthalmic industry grades of 8–25 micrometre and 20–40 micrometres are emerging as the most popular. (See § 4.16.)

Although diamond smoothing machines are conventional in principle, they are much sturdier in construction than their predecessors because diamond tools require high operating pressures. The use of compressed air for supplying pressure is now standard practice. Spindle speeds up to 5000 r.p.m. are often provided, although good results can be obtained at much lower speeds.

Two forms of diamond lapping tools are used. One is the solid type and the other is made by sticking a number of small pellets on the surface of a metal holder with a suitable radius of curvature. Each type has its own advantages. Generally the solid tool is more suitable for highly organized, long-run production. Pellets are versatile and can be used to make up a tool of any curvature at short notice. Adjustment of diamond 'concentration' can be achieved by the number of pellets used in a given tool area. A small number of pellets per unit area is a useful means of operating a diamond smoothing system with relatively low working pressures. It is of course essential to dispose the pellets so that all parts of the glass come into contact with at least one pellet on the tool. The photograph of the Autoflow machine shows a pellet tool in use. (Fig. 7.9.1.)

Machine cycles for the diamond smoothing of ophthalmic lenses depend on several factors, including the quality of the coarse-generated surface, but are generally in the range of 10–30 seconds.

Machines for polishing single surfaces differ little in design from those used for smoothing. Spindle speeds are usually lower because very high speeds tend to throw off too much of the polishing slurry.

At the time of the publication of the final edition of *Prisms and Lens Making*, the nature of the polishing process was not really understood and the belief was still fairly widely held that glass was not removed in polishing. The tests devised shortly after the war by Sira Institute for assessing the efficiency of polishing agents quickly established that substantial quantities of glass were removed in polishing. (See Chapter 2.) The full-scale research carried out by Sira during the period 1962 to 1967 investigated many of the practical aspects of the process and it was found that the pressure applied between glass and polisher had a very powerful effect on the rate of removal. The influence of pressure had been appreciated by the ophthalmic industry and a good deal of time has been spent in searching for polishers which would stand substantially increased pressures. For some time, thick cloths and felts impregnated with waxes and resins provided the most satisfactory solution but the useful life was not considered adequate because it was seldom possible to polish more than 200 lenses before replacing the polisher. Not unexpectedly, many workers turned their attention from the natural materials to the increasing variety of synthetic substances which became available, but few seemed to have properties approaching those of the traditional materials.

During the 1960s a new material was produced by Rowland Products in the U.S.A. which represented an outstanding advance. (See § 4.5.) This is a polyurethane impregnated with zirconium oxide and is available in several thicknesses up to 4 mm and different grades of hardness. The choice of both thickness and hardness depends on the class of work being produced, but the medium-hardness material about 1·3 mm thick is generally considered most suitable for ophthalmic work. The life of the material is remarkable and under good conditions the 1·3-mm thickness can polish over 1000 lenses.

It is not resilient and will not shape itself to the curve of the smoothed surface as readily as a pitch or felt polisher. This makes it essential to machine the surface of the holder to a high standard of accuracy. As the polyurethane is not self-adhesive some other agent has to be used to stick it to the tool. The modern impact adhesives such as Evo-Stik are eminently suitable, but it is essential to press the prepared surfaces together with a mating tool using considerable pressure.

Apart from the long life of this material its nature allows much higher polishing pressures than are possible with pitch or felt. Although excellent and consistent results can be obtained with polyurethane, close attention to detail is essential. It is virtually impossible to hand-shape the polisher so that it fits the lens and therefore the smoothing must be accurate. For this reason, it is often recommended that the best results can only be obtained on diamond-smoothed surfaces when a possible drift in curve will be very slow.

Polishing cycles can be very short if the diamond-smoothed surfaces are very fine and with medium fine surfaces it takes about 90–120 seconds. Consistent production depends on regular checking of all aspects of the process as outlined in the section on blockworking. Relatively few troubles occur in polishing. The greatly improved quality of materials available nowadays has virtually eliminated the incidence of sleeks, which used to be the lens-worker's nightmare. Perhaps the main source of trouble arises from isolated scratches on the diamond-smoothed surfaces, which may be due to a minute percentage of oversize particles in the diamond powders. One such particle on the surface of the tool can scratch a large number of lenses before it is worn out or becomes detached. The grading of diamond grain has improved a lot during the last few years but it may still require a lot more development before completely scratch-free diamond tools can be made. Nevertheless, the acceptance of a very small percentage of scratches is a small price to pay for the other advantages of diamond tools. (See § 4.16.)

After polishing, the lenses are deblocked and cleaned prior to final inspection.

7.10 Fused bifocals

Fused bifocals consist of a main lens of crown glass with a small segment of flint glass fused into it. Two different powers are produced by virtue of the difference in refractive index between the two glasses.

This result is achieved in the following manner. If we take a flat piece of crown glass and grind a small depression on a radius of 32·68 mm the power of this curve in crown glass will be −16·00D (see Fig. 7.10.1). If now a piece of flint glass is worked to the same radius the resultant power will be +20·0D. By fusing the flint on the depression, as described later, and grinding and polishing both surfaces to a continuous flat surface, we get no power in the major part of the lens, but in the small circle of flint glass the power is −16·00D + 20·00D, or in other words +4·00D. If the other surface is then worked to any specified curve it gives a power in the main lens, or distance portion, and an addition to the 4·00D in the small segment, or reading portion.

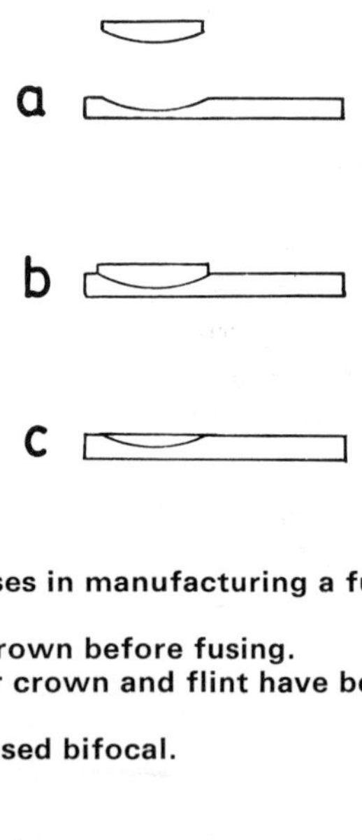

7.10.1. Processes in manufacturing a fused bifocal.
(*a*) Flint and crown before fusing.
(*b*) Blank after crown and flint have been fused together.
(*c*) Finished fused bifocal.

More conveniently a 'flint ratio' is used to calculate the depression curve required to produce a particular addition. The flint ratio is derived from the simple formula:

$$K = \frac{N_{\text{crown}} - 1}{N_{\text{flint}} - N_{\text{crown}}}$$

Thus with two glasses with indices of 1·523 and 1·654 the ratio is:

$$\frac{1{\cdot}523 - 1}{1{\cdot}654 - 1{\cdot}523} = \frac{0{\cdot}523}{0{\cdot}131} \approx 4{\cdot}0$$

If a lens is required with 2·50D addition, the surface power of the flint lens must be 4 × 2·50 = 10·00D. This power is shared between the segment side surface and the depression, so a +6·00 base lens will require a 4·00D concave depression to provide surface powers on the flint which total to 10·00D. It will also be noticed that the ratio of 4 links up very conveniently with the 0·250 intervals which are used in the ophthalmic industry and profession. It will also be noticed that the flint ratio is largely influenced by the difference in the indices of the two glasses and consequently tolerances are important. In the case shown a pairing of maximum and minimum tolerances could change the difference of 0·131 to 0·129 or 0·133—an error exceeding 1½ per cent.

Working crowns and flints

The crown mouldings are blocked up four at a time in cell moulds so that a depression is ground near the bottom of each moulding. The block is first roughed to open out the depression to 28 mm diameter, and then smoothed on a bowl machine with very fine emery until the depression is 30 mm. Inadequate smoothing causes serious trouble in the later stages.

The block is then polished and it is at this stage that all precautions have to be taken to ensure that the surface is in the right condition to stand the high temperatures used in fusing. It is essential to use a rouge which does not 'blind', and a plain hard felt polisher is used to assist in keeping the surface 'open'. This term refers to the type of surface produced by a plain felt polisher, compared with the surface polished on a hard wax or pitch polisher. In the early stages of polishing the surface appears entirely different with the two types of polisher. A lens which has apparently been completely polished on a pitch or hard wax polisher will break out into large holes if rubbed on a soft polisher, or if it is heated to a fairly high temperature.

The polishing cycle should be 50 per cent longer than the time taken to remove the last traces of grey. If all these points are observed the surface should be satisfactory.

The flint glasses D.F., $\mathcal{N}$ = 1·625, used for reading additions below 2·00D, and E.D.F., $\mathcal{N}$ = 1·654, for additions of 2·25D to 4·00D, are moulded to 28 mm diameter and one side ground flat and semi-polished. They are then made up into blocks of up to 18 and ground and polished in the same way as the crown.

The curves of the depressions vary from +7·00D on a deep meniscus moulding to −16·00D on flat mouldings. The opposite curvature is worked on the flint, but with a difference of 0·25D or 0·50D so that when placed together contact is made in the centre.

The flint glass is moulded to 29 mm round and made up into blocks of four, which are then ground, smoothed and polished in a similar way to the crowns. The normal range of depression curves runs from +8·00D to −16·00D depending on the base curve of the blank and addition required. Flints are worked to an opposite curvature so that contact is made in the centre. A difference in curvature of 0·25 or 0·50D is customary.

Fusing the flint and crown

When the flints and crowns have been polished and knocked off the runners, they are delivered to the fusing department, where they are cleaned and passed into a room which is kept as free from dust as possible by blowing filtered air into it. In this room operators put the flint and crown into contact so that there is no dust or dirt between the two surfaces. This is done by laying the crown on the bench, picking up the flint, brushing the surface of each component with a soft brush and then placing the flint carefully on the depression. Owing to the difference between the two curvatures a small interference spot is visible in the middle.

The operator now presses the top surface of the flint close to the edge with a pair of tweezers and, if all is well, the spot moves in a straight line to the edge. If, however, there is some dust between the surfaces the spot is deflected and travels in a curved path. The flint then has to be picked up and brushed again until this test, applied in a number of positions round the flint, shows that no dust is present. Two short pieces of wire are then dipped in gum and placed ½ inch apart under the flint at the bottom of the moulding. The wire has to be sufficiently thick to cause the interference spot to disappear at the top edge of the flint.

The assembled crown and flint are then placed on a carborundum slab, moulded to fit the underside curve of the crown, and placed on the travelling belt of a lehr. The hot spot of the lehr is thermostatically controlled at 640°C at which temperature the flint becomes quite soft and sinks by its own weight, gradually expelling the air from the top of the flint until the whole surface is in contact with the exception of two very small areas round the two supporting wires. As the mat travels onwards the temperature of the furnace gradually drops until the fused blanks emerge at the other end at a temperature of approximately 100°C. They are then allowed to cool and the two supporting wires pulled out.

The blanks are then checked for strain. Matching the crown and flint for expansion, as mentioned in an earlier section, is very important because any serious difference will cause the crown to flaw. Some tolerance can be allowed if the expansions are such that the flint is in compression and not in tension.

The blanks then undergo an inspection for faults in the segment. If by any chance some dirt has been left between the surfaces, a bubble is formed which may result in the rejection of the blank.

Very great care is necessary in the surfacing because the high temperature can cause the surfaces to break open, leaving a mass of tiny bubbles or scratches visible in the segment. It is rather a peculiar effect, because marks such as sleeks, which are visible before fusing, seem to flow and disappear during the process, whereas two surfaces which appear to be perfect under the most rigorous examination may be so bad after fusion that they are fit for nothing but scrap. The fault can only be minimized by paying great attention to polishing and smoothing materials and by giving ample machine cycles to both operations.

When the blanks have passed inspection they are sorted according to a code number, which is marked with a diamond on the crown, and are then put into stock ready for sale, and a certain percentage are ground and polished on the segment side to fulfil orders for the more common ranges of prescriptions. These are known as semi-finished blanks.

The code number identifies the curves moulded on the crown, the curve of the depression and the index of the flint. A chart is issued to each user so that the correct blanks can be ordered by code number to fit almost any individual prescription. The combination of prescriptions in bifocals is so enormous that they are sold in blank or semi-finished form so that the prescription houses can select and work blanks into finished lenses to suit each prescription.

Fused bifocals have the advantage in that the segment is almost invisible and they are comparatively cheap. The segments cannot be made over 22-mm diameter without introducing serious chromatic aberration.

Univis fused bifocals

The segment of the normal fused bifocal is round, but the United Kingdom Optical Co. Ltd secured the patent rights for the manufacture of fused bifocals with a variety of shaped segments.

A 32-mm depression is worked on the crown by methods the same as those used for the round segments.

The shape of the final segment is produced by a composite flint button consisting of two or more pieces of glass. Three of the stock shapes, 'B', 'D', and 'R', are illustrated in Fig. 7.10.2.

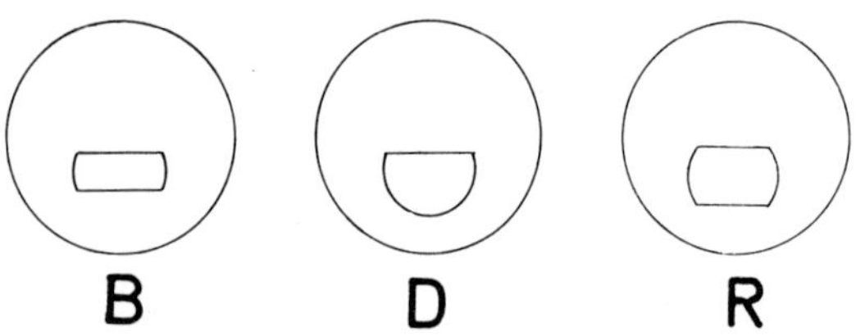

7.10.2. Three most popular Univis shapes.

The 'B' shape button consists of a narrow rectangular moulding of flint glass which is ground and smoothed on the two longer edges and two similar shaped pieces of crown glass ground and smoothed on one edge. These three pieces of glass are carefully cleaned and clamped together with the flint between the crowns and the smoothed surfaces in contact. In this condition it is passed through the lehr and fused into a solid piece of glass.

The 'R' shape is made in the same way with a wide flint, and the 'D' has one 'D' shape flint and a rectangular strip of crown.

When the button has been fused it is first ground flat and semi-polished on one side, and then ground and polished in the usual way to the required curve.

Subsequent operations are carried out as previously described, but a temperature of 680°C is required to fuse the button owing to the presence of the crown.

When the segment side is ground and polished to a continuous curve the crown component of the button becomes invisible because it has the same refractive index as the main lens and the shaped flint segment only remains visible.

It is well known that abrupt index changes, even though small, are readily visible in a piece of glass. Such a situation arises in the Univis construction where the two pieces of crown glass are fused together and any significant difference shows up as a faint ring outlining the junction of the segment side surface and the depression. Under the least favourable circumstances, the ring becomes visible when there is a difference of 0·0005 between the indices of the two separate pieces of crown. This necessitates an index tolerance of ±0·0003 for all crown glasses used for any form of fused bifocal which incorporates the crown to crown technique.

Working the segment surface

Semi-finished fused blanks are usually made with nominal standard segment diameters of 22 and 25 mm with a tolerance of ±0·5 mm. It will be appreciated that the diameter of a fused bifocal segment reduces as the segment side surface is ground away, and fairly close control is essential if narrow limits on segment diameters are required. For example, the segment diameter of a blank with a 1·00D addition will reduce from 23 to 22 mm with a glass removal of 0·04 mm.

The Univis shape presents a further problem because the location of the line in relation to the depression can affect the shape, i.e. the ratio of height to width.

In both cases the rough fused blank is ground on

a diamond generator to remove the excess flint and reduce the segment to some 3–4 mms above the required finished diameter. Blanks with round segments are then measured and graded into 0·5 mm intervals. It is then possible to relate the amount of glass to be removed so that machine cycles can be set for diamond smoothing down to finished size.

Before Univis blanks can be smoothed it is necessary to check the shape and if necessary correct it. This is done by grinding a small amount of vertical prism relative to the original rough ground surface. The amount can be quite critical and as an example an angle of about 10 minutes will alter the ratio of height to width by 0·5 mm on a blank with an addition of 1·00D. When the proportion has been corrected, the segments are measured and graded, and then smoothed to finished size as described above.

As a matter of convenience, the segment surfaces are then polished singly, so that a very critical blocking operation can be avoided. Regrettably, it is not generally feasible to polish the deeper convex surfaces on an annular polisher and it is therefore necessary to use a small cap polisher on top of the lens. This method reduces the speed of polishing substantially and cycles of 4–6 minutes are usual.

The selection of polishing materials is important because unsuitable characteristics can cause trouble due to different rates of polishing of the different glasses, which can cause a step at the junction between the crown and flint. A combination of soft felt polishers and bad technique will produce a ridge so pronounced that it can be detected with a fingernail. The choice of a hard polisher is the first essential and formulations containing pitches or hard waxes produce the best results. To a lesser degree, the polishing powder can affect the standard of finish, and high-grade cerium compounds produce marginally better results than zirconium oxides.

7.11 Solid bifocals

Solid bifocals are made by grinding and polishing two different curvatures on a single piece of crown glass, one curve being used for the distance portion (D.P.) and the other for the reading portion (R.P.).

Solid bifocals are made in toric form with standard D.P. curves of −4·00D, −6·00D, −8·00D, and −10·00D. The R.P. curve is varied to give the required addition and therefore has a longer radius of curvature than the D.P. It is thus quite impossible to use normal lens-working methods because the working tool must not be allowed to run over the dividing line between the two curvatures. Owing to these circumstances an entirely different method of working was devised by Bentzon and Emerson and patented in 1904 and 1910. This method depends entirely on the use of precision machinery to achieve the necessary accuracy and quality.

Until 1939 as many as ten different segment sizes were made as stock items, ranging from 19-mm to 50-mm diameter, but these have now been condensed to the three most useful and popular sizes, 22 mm, 38 mm, and 45 mm.

The mouldings have a D.P. curvature as mentioned above, the diameters being 73 mm for a 22-mm segment, 90 mm for a 38-mm, and 96 mm for a 45-mm segment.

Sticking on

The mouldings are first diamond-ground with a special tool which grinds the D.P. and R.P. together, and are then stuck singly, with wax, on holders which can be fitted to the machine.

The holder is an iron casting which is machined on the top to a curve to suit the various mouldings in use. Underneath, a female taper is bored in a boss, and it is by this taper that the lenses are located on the grinding, smoothing and polishing machines. It is essential that all the holders have the tapers machined to very fine limits so that there is no loss of truth when the holders are transferred from one machine to another.

Roughing

The roughing machine, in principle, is the same as the machine illustrated in Fig. 6.4.2 with a bottom rotating head and a top grinding spindle. A small precision lathe head is mounted, at the bottom, on a vertical slide which is held against an adjustable stop by two springs. The holder fits on a male taper on the lathe head. The top

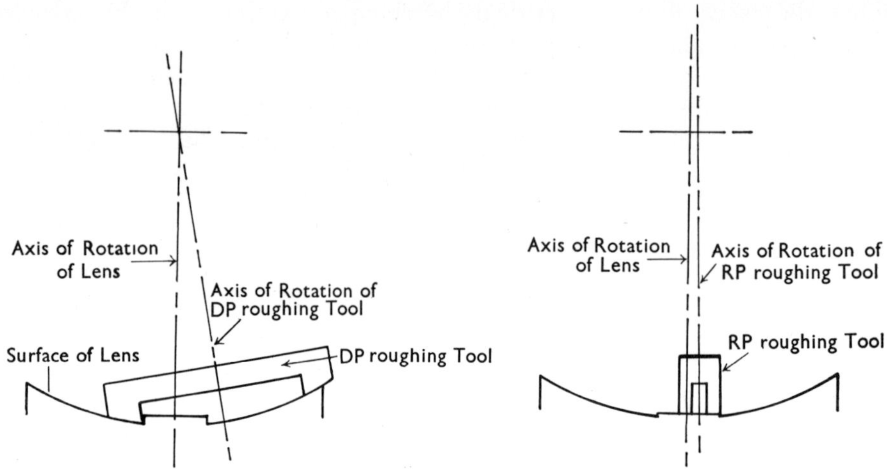

7.11.1. Principle of distance portion and reading portion roughing.

high-speed spindle can be set to any angle with the aid of a vernier and the whole of the top assembly can be moved laterally on a horizontal slide. A cast-iron tool is screwed on the top spindle, and the angle set to produce the required D.P. curve. The lateral position of the tool is then adjusted to a position to one side of the axis of the bottom head. Water and medium-grade emery is fed on the lens, and the machine started. The stop is adjusted to allow the machine to grind about 0·3 mm off the surface.

Owing to the offset of the tool, a circular area in the centre of the lens is left unground, but the outside band is ground true to the holder and correct to curvature.

The portion in the centre which becomes the R.P. is now left standing above the level of the R.P. This is ground down to the level of the D.P. by a similar machine, but this time by a tool of such a diameter that it covers the centre of the R.P. when the edge of the tool is set to the ridge or dividing line between the two surfaces. In this way a continuous spherical surface is ground on the R.P.

Fig. 7.11.1 illustrates the principle of D.P. and R.P. roughing.

If the holder is a bad fit on the bottom head, or the bottom spindle is out of truth, the R.P. will be generated on an axis different from the D.P. The effect is shown, greatly exaggerated, in Fig. 7.11.2. This is known as a lop-sided bifocal. Great attention to the condition of the taper and truth of the bottom head is necessary to reduce this defect to the minimum. Weekly inspection for truth is made and if the error exceeds 0·00015 inch the head is overhauled.

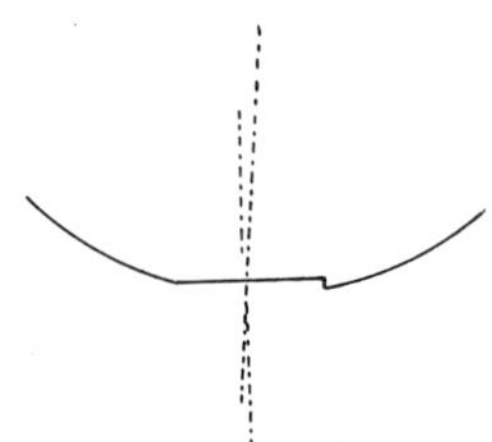

7.11.2. A lopsided solid bifocal. The diagram is exaggerated to show how the defect is caused by an untrue spindle or a badly fitted holder.

Smoothing

All subsequent operations are carried out on one type of machine. Fig. 7.11.3 shows the general principle. The bottom head is bolted on the front of a cast-iron frame and at the back there is an arm which carries the top spindle in two bearings. The arm is supported by two pivots; the top one allows the arms to be lifted up or down, and the adjusting screw moves the whole assembly backwards and forwards on the lower pivot. The top spindle has a small ball point and a collar with two driving pins. This engages with the ball

7.11.3▷
Smoothing and polishing machine for solid bifocals.

7.11.4▷
Close-up of a smoothing and polishing machine for solid bifocals, illustrating the tilt applied to the reading portion polisher.

hole and pegs on the tool as shown in the photograph. The angle of the spindle can be adjusted by loosening a simple clamping mechanism, and both the top and bottom spindles are driven by belts.

The D.P. smoothing tool is a ring tool made of brass and of the same dimensions as the roughing tools. The tool is placed on the lens and the arm lowered to engage the ball and socket and driving pins, and the position is adjusted so that the inside edge of the tool coincides with the ridge. Then the angle of the spindle is adjusted, if necessary, so that it is at right angles to the tool. The machine is started and fed with fine emery and water. Referring to Fig. 7.11.1 it will be appreciated that the roughing machine will only generate a true sphere if the axes of the two spindles are in precisely the same plane. If they are not, a toric will be produced. In production it is not possible to keep the roughing machine in sufficiently close adjustment; hence the necessity for the different type of drive, used on the smoothing and polishing machines, which allows the tool to float and to pick up its own axis in the correct plane, thereby producing a true sphere.

After D.P. smoothing the R.P. is left raised above the D.P. This is then smoothed down with a small floating tool set exactly to the ridge until it is nearly level with the D.P. again. It is sometimes necessary to adjust the curve of the R.P. This is done by setting a tilt on the top spindle, as illustrated in Fig. 7.11.4, which introduces a slight bias by the friction of the driving pins and the ball point, and increases the pressure of the tool on the centre or the edge, according to the direction of the tilt, thus shortening or lengthening the radius slightly.

The principle of the diamond-roughing tool,

which was patented by J. A. Moore, has recently been applied to an ingenious tool which enables the D.P. and R.P. to be smoothed at the same time. Two adjustable segments are fitted to the inside of a D.P. smoothing tool (Fig. 7.11.5). The R.P. curve produced by this tool is not truly spherical in practice, but only a short run with a floating tool is required to true it.

This tool has two advantages: (1) it reduces the R.P. smoothing time, (2) it ensures that the lens is entirely free from lopsidedness at the smoothing stage.

Polishing

Polishers used for solid bifocals (Fig. 7.11.5) have to be extremely hard to preserve a sharp dividing line between D.P. and R.P. There are numerous recipes for solid bifocal polishers, including hard pitch, hard waxes, shellac, wood, vulcanized fibre, and plastics. Hard waxes mixed with rouge and wood flour are generally the most satisfactory.

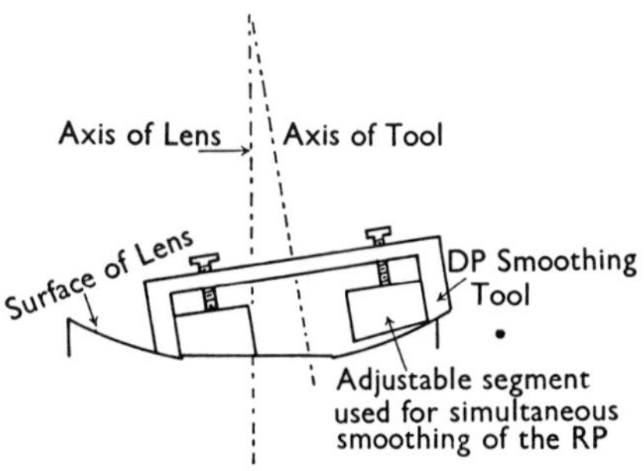

7.11.5. Principle of a combined distance portion and reading portion smoothing tool.

A piece of polishing material is stuck on a metal runner which can be fitted to the top spindle. The polisher is turned in a lathe so that it has a sharp edge and its diameter is a little greater than the width of the D.P. It is then set with its edge coinciding with the ridge, and the machine is started and fed with rouge and water until the grey has been removed. By this time, the edge of the polisher has lost its sharp edge and it leaves a narrow wavy band round the R.P. This is known as aberration or 'aber'. By re-turning the polisher and running the machine for a few minutes with the smallest trace of rouge, the band of aber is polished away leaving a clear, sharp ridge.

The R.P. is polished in a similar manner with a polisher approximately half the diameter of the R.P. When the lens is received from the polishing, there is still a perceptible ridge and the R.P. has to be polished until it is nearly level with the D.P. During this process as much as 0·001–0·002 inch of glass is polished away and, as can be imagined, it would be quite impossible to convince a solid bifocal maker that glass is not removed by polishing. When the R.P. is nearly level the polisher is returned and run for a few minutes with very low rouge concentration to remove 'aber' left by the preliminary polishing.

The bifocal surface is then complete, with the exception of a small pip which is left in the centre of the R.P. In the case of 38-mm and 45-mm segments the dimensions of the blanks are such that the pip is not included in the lens when it is fitted to the frame, but it is necessary to remove the pip in 22-mm segment blanks. This is done with a small polisher made of soft pitch moulded to the R.P. curve. It is driven by a small crank fitted to the top spindle of the polishing machine, and it is set so that the polisher does not run over the ridge. This action will cause the curvature to alter quite rapidly, but as it takes only 30–45 seconds to remove the pip, the alteration of the curve is quite negligible.

When the lens has passed the final inspection it is returned to the sticking-on department and removed from its holder and cleaned ready for a further inspection.

At all stages of manufacture the very greatest care is necessary in setting the tools and polishers, otherwise an ugly deformed ridge is produced.

After the final inspection the lenses are cut into suitable shapes for surfacing. The three types of blanks are cut from the complete circle as shown in Fig. 7.11.6.

It is claimed, quite rightly, that the manufacture of solid bifocals is one of the most delicate of all lens-working operations, and in addition to the production of the stock type of bifocals just described many other special forms of lenses can be made.

Some of these types are illustrated in Fig. 7.11.7. These are made to reduce the weight of very high-powered lenses.

The bilentic is an interesting type which is a combination of a solid lenticular and a fused

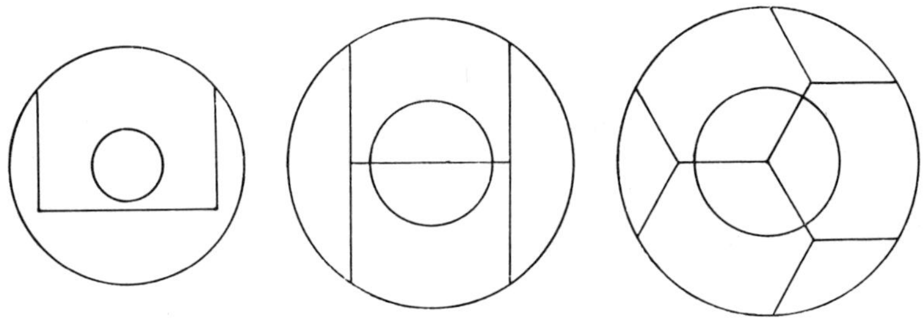

7.11.6. How solid bifocals are cut from a complete circle. A 22-mm segment yields one blank; a 38-mm, two blanks; a 45-mm, three blanks.

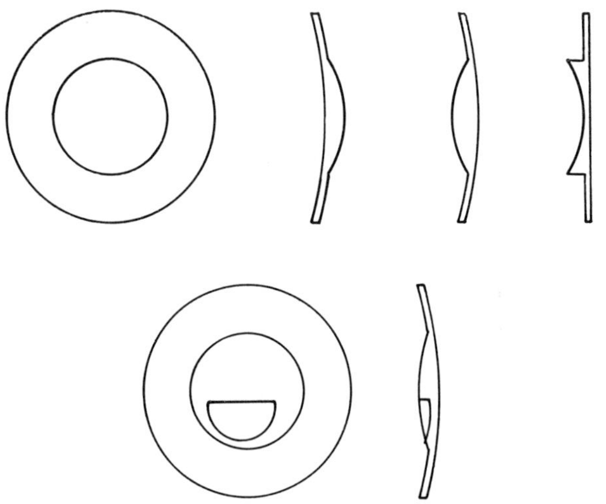

7.11.7. (*Top*) Various forms of lenticular lenses. (*Bottom*) The 'Bilentic'—a bifocal cataract lens.

Univis 'D' segment, used in cataract cases where a bifocal is necessary.

Such are the types of machines and methods of working appropriate for the mass production of stock lenses. There are a number of machines made for working single surfaces which are used for the second side of fused and solid bifocals and for making lenses which are normally outside the stock range of the mass manufacturers. This type of work, however, is not as a rule carried out by firms making ophthalmic lenses in quantity.

7.12 Testing of ophthalmic lenses

The agreed limit of error in the power of ophthalmic lenses is 0·06D. To keep the lenses within the tolerance it is necessary to work each surface of the lens to 0·03D. This limit has to be reduced still further to allow for slight variations in the index of the glass, and 0·02D can be considered a reasonable maximum.

The smaller blocks used for deeper curves are checked with accurately made profile gauges. The difficulty of producing this type of gauge with long radii makes the use of a spherometer more convenient for checking shallow curvatures. It is necessary to work to far greater accuracy than suggested above when using large blocks, because the polishing cycle would be greatly extended if a smoothed block with a slight inaccuracy were used with a correct polisher. To check within a suitable tolerance, a spherometer with two fixed legs 150 mm apart is used to compare the curve of the blocks with accurate master surfaces. The ring spherometer described in another section is not used widely in the ophthalmic industry because it is not suitable for checking cylindrical surfaces.

The base curve of a toric block can be easily and accurately checked by measuring the diameter with a slide gauge, and the curvature of the cylinder is measured with a spherometer with two fixed legs 40 mm apart.

An optical device is used for checking the R.P. curvatures of solid bifocals because, owing to the small size of the segment, a spherometer is rather clumsy and its accuracy is considerably reduced.

This instrument is in the form of an optical bench, and the surface of the lens is brought into contact with a fixed point. On the other side, mounted on a sliding carriage is a cube made up of two right-angled prisms cemented together. The hypotenuse of one prism has a semi-reflecting coating. The top of the block is fine-ground, and the rear vertical surface has a graticule on it and is illuminated by a lamp and condenser system. Light passing through the graticule is reflected by the polished surface of the glass under test and then by the semi-reflector on to the upper surface of the block. When the carriage is moved so that the graticule coincides with the centre of curvature of the surface under test an image of the graticule comes into focus on the ground surface. By fixing a scale on the slide the radius of curvature can be read very accurately.

When the lenses leave the workshops they are inspected for power, thickness and position of optical centre and surface. A very high standard of surface is required for ophthalmic lenses.

7.13 Making lenses to ophthalmic prescriptions

Lenses made to the prescription of oculists or refracting opticians incorporate elements which are spherical, cylindrical, and prismoidal. These, often in conjunction with bifocal or multifocal elements, may be required singly or in association. Furthermore, many types of glass giving a wide range of colours and depth of colour may be called for. All glass used for ophthalmic lenses is of standard refractive index (1·523), the only variant being the fused type of bifocal segments for the reading portions. Bearing in mind that the power of a lens is the algebraic sum of the effects produced by the curve of the two sides of the lens, it will be appreciated that an enormous range of combinations can be effected by a relatively large number of variants on one side in association with a relatively small range of variants on the other.

To allow for quantity and, therefore, economic production of the smaller range of variants and to reduce the stocks of lenses which would otherwise be unavoidable the common practice is for lenses to be manufactured in quantity as to one side only. The prescription manufacturer (hereafter called the P.M.) makes the second side to such curves, size, thickness and prism element in the finished form as called for by the individual prescription. Whilst the general practice is to finish the second side, the P.M. needs on a limited number of occasions to finish both sides of a lens. This is usually in cases of particularly strong powers, but double-side working is sometimes necessary when there is an element in the prescription which would make quantity production undesirable for technical or commercial reasons. The glass used for double-side working usually performed by moulding to approximate curvature is known as a blank. When finished on one side it is called a semi-finished blank. Only the fused type of bifocal is ever singly finished on both sides; all other bifocals and multi-focals have to be supplied to the P.M. with the segment side finished.

Ophthalmic lenses derive their power from the curve of the two sides, but these two sides can vary quite widely whilst maintaining any given power. For instance a power of 3 dioptres (the dioptre being the reciprocal of the focal length in metres) can be obtained with one side of the glass plano and the other a convex (symbolized as +) 3-dioptre curve, or, say, concave (symbolized as −) $1\frac{1}{4}$ dioptre on one side and $4\frac{1}{4}$ dipotres convex on the other. The normal standard for a convex lens of 3-dioptre power would be a concave curve of about 6 dioptres towards the eye and 9 dioptres on the outide. Or, in case of a 3-dioptre concave lens, a concave curve of 6 dioptres one side and a convex curve of 3 dioptres on the other. For convex lenses these examples are not strictly accurate as the centre thickness of a convex lens adds a slight degree of power which has to be allowed for.

Theoretically there is no limit to the curves used so long as the algebraic addition is maintained, but a lens with an inside curve of (say) concave 16 dioptres and an outside of convex 19 dioptres would be cumbersome and ugly in appearance. The foregoing principle applies whether the resultant power has to be convex or concave. What we have discussed is the *form* of a lens as distinct from its *power*. In a loose way lenses whose form is hardly curved, e.g. plano ⊂ 3-dioptre sphere, or 10-dioptre sphere ⊂ 2-dioptre cyl, are said to be of flat form*.

Lenses of spherical power when curved, e.g. −3 dioptre sphere ⊂ +7 dioptre sphere, giving a resultant power of +4-dioptre sphere, are known

*The symbol ⊂ means 'combined with'.

as meniscus. When a cylindrical element is present, e.g.:

$$-6\text{D Sph} \subset \begin{cases} +6\text{D Sph} \\ +1\text{D Cyl} \end{cases} = +1\text{D Cyl}$$

or

$$+6\text{D Sph} \subset \begin{cases} -3\text{D Sph} \\ -2\text{D Cyl} \end{cases} = +3\text{D Sph} \subset -2\text{D Cyl}$$

the form is known as toric. It is the function of the P.M. to decide upon the form of a lens as distinct from the power which is laid down in the prescription. This decision largely rests on the semi-finished lenses available together with a variety of technical considerations, but even at the cost of making both sides of a lens the P.M. will not depart far from the form which gives the best optical effect as distinct from focusing power. But this question of aberrations arising from lens form is beyond the scope of these notes.

There are three stages in transforming the raw glass surface to the finished surface and these are comprised in the general term 'surfacing'. They are in principle the same as previously described, which is to say roughing, smoothing (fining), and polishing, but the application of mechanical means to the production of single surfaces is substantially different in the technology applied. Historically, contra-generic tools (concave to produce a convex glass surface and vice versa) of cast iron were used for all three stages, cloth or similar medium being used to cover the tool for the polishing stage. But in modern times the roughing stage is universally achieved by means of diamond-impregnated ring tools.

The smoothing (fining) tools are of cast iron, which has been found the best material due partly to cost, ease of moulding and cutting to curve, and partly to the working surface lending itself to the use of abrasive powder. But the constant abrading of glass on the iron leads to appreciable wear of the tool and thus affects the curve accuracy. For this reason, in addition to the facility provided to cut new tools, a tool-shaping machine (Fig. 7.13.1) has become an essential piece of a P.M.'s equipment.

One tool is required for every curve or combination of curves. The spherical range is from a $\frac{1}{4}$ dioptre to 20 dioptres, mostly in $\frac{1}{4}$-dioptre steps, all both convex and concave. This amounts to say 140 tools. In the case of cylinders the range covers powers of a $\frac{1}{4}$ dioptre to 6 dioptres in $\frac{1}{4}$-dioptre steps, but all these multiplied by the spherical elements (known as the base curves), which have to be included together with the cylinder curve. The spherical elements to be covered are mostly from a $\frac{1}{4}$ dioptre to 12 dioptres each in $\frac{1}{4}$-dioptre steps, thus totalling (allowing for both convex and concave forms) some 2000 single tools.

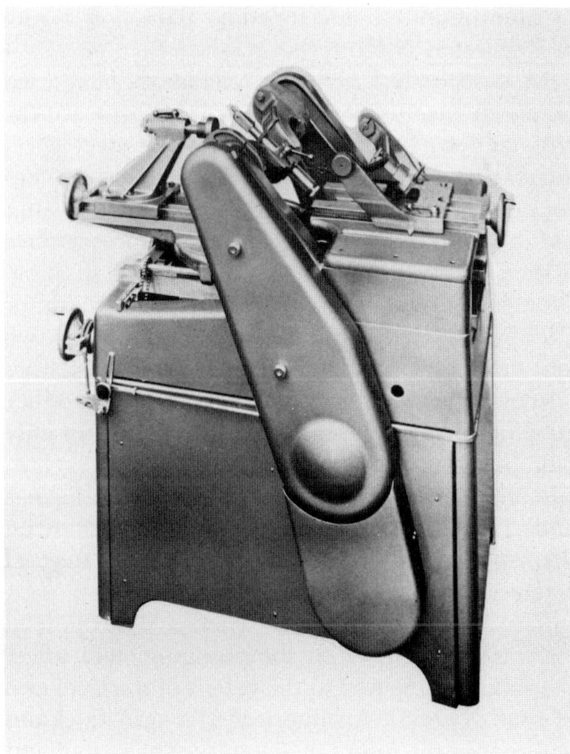

7.13.1. Autoflow toric tool-cutting machine.

7.13.2. Raphael dial gauge spherometer for measuring lens and tool surfaces.

These facts explain why diamond-faced tools to be used in the place of cast iron cannot, on account of cost, even be considered. The considerable cost in constant checking of tool powers and machining them back to accuracy provoked, during the last decade, the use of replaceable material to fit over the tool working surface so that the tool surface itself is left unworn. The first to be commercially used was fine steel mesh marketed by the American Optical Company of Southbridge, followed by aluminium foil developed by Bausch & Lomb of Rochester, N.Y.

Most recently Univis Inc. of Florida have used prepared steel sheet made to conform to the tool curves. The aluminium foil or steel sheet interfacing media are cut into shapes to encourage the abrasive powder, used with water as lubricant, to penetrate across the whole surface of the glass. Wire mesh does not need such shaping. Whereas it is not expected for wire mesh or aluminium foil to produce more than one or two lens surfaces without needing renewal, the Univis system (known as Unilap) is expected to produce 20–25 surfaces in the case of convex glass, but only half that number in the case of concave glass. The cost advantage of using so few interfacings and so much less labour in applying and removing them, apart from the necessary cleaning of the tool surfaces, is obvious.

The use of any material, whether it be of metal for smoothing or cloth for polishing, will affect the radius of the tool to the extent of the thickness of that material. An interleaf of x mm thickness will reduce the radius of a concave shape tool and increase that of a convex one by x mm. To obtain high accuracy it is therefore necessary to cut the tools initially weak or strong respectively. For a similar reason it is considered better technique to use a special set of tools cut for polishing. Nonetheless if a polishing pad of cloth or other material is placed either direct on to the tool where an interleaving is not used, or direct on top of the interleaving, only a minor inaccuracy results. The accuracy of power to be obtained is accepted as laid down by the British Standards Institute, which is to say a tolerance down to $\pm 0{\cdot}06$ dioptres.

The measurement of tool accuracy is sometimes made by metallic gauges which are placed in coincidence with the tool curve under test. Lenses, and often tools, are tested by various forms of dial gauge (Fig. 7.13.2.) Before the lens can be surfaced it has to be joined with some form of holder, which is usually a piece of mild steel somewhat smaller than the glass, the two being held together by pitch or wax.

The button used when blocking a spherical curve usually has one central depressed hole into

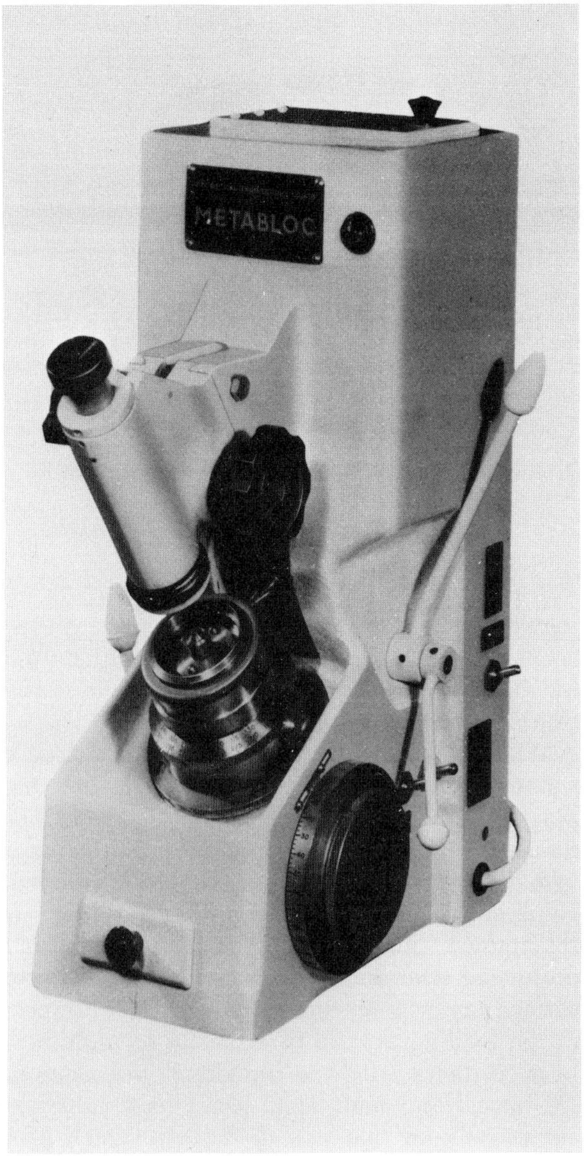

7.13.3. Metabloc prescription metal blocking machine.

which fits a pin which drives the lens across the tool. The curve to be produced being spherical, it does not matter if the lens revolves under the pin; but in the case of cylindrical curves it is essential that the glass be held so that the cylinder element during working remains parallel to the axis of the cylinder element of the tool. To ensure alignment, two or three holes are made on the top of the button, into which will fit the driving pins of the surfacing machine.

In this field of applying some intermediary between the glass and the machine itself there are two more recent and significant improvements of method. The first provides means of embracing the periphery of the button so that the pin (or pins) is not required. Apart from avoiding the nuisance of wear in the button-holes it is possible to drive the lens (whether a spherical or cylindrical curve is being processed) not from above, as in the case with pins, but on a plane sufficiently low to agree roughly with the plane of the tool surface.

The second and more revolutionary change has been the introduction of low melting point (between 136°F and 170°F) metal to supersede pitch or wax. (Fig. 7.13.3.) This is effective provided the glass contact surface is first varnished to provide a hold. The metal has three advantages. It has a narrow differential between the temperatures of the workable liquid and a solid form. It is far cleaner in use than pitch or wax. Lenses can be removed from it by small percussion without the need of freezing.

As against these factors must be considered the high initial and replacement costs since in daily use appreciable amounts of metal get wasted and continuous use brings about oxidization which calls for special treatment. The use of metal in blocking takes two forms—as an adhesive medium between the glass and the button and as a

7.13.4. American Optical Company prescription lens marking machine.

7.13.5. Autoflow prescription toric curve generating machine.

moulded button which, when the glass is varnished is its own adherent. Moulding each button having new driving-pin holes eliminates any troubles which arise from worn holes.

When blocking lenses for spherical curve surfacing there is no particular direction for the placing of the blank in relation to the button, nor indeed does it usually matter when surfacing a cylinder curve on a single-vision lens, provided, as already said, the axis once established is maintained. But when dealing with bifocal and multifocal semi-finished blanks it is necessary to surface any cylinder to a particular axis in respect of the reading portion of a bifocal or multifocal in accordance with the individual prescription. All bifocal and multi-focal lenses have spherical curves only on that side of the lens which provides the differing focal length, so any cylindrical element has of course to be provided on the alternate side. To achieve the desired location a semi-finished blank is placed in a marking device with a simple optical system. (Fig. 7.13.4.) The resulting ink marks on the glass indicate the direction of the segment and thereby of the cylinder when the blocking takes place.

Any prismatic element called for by the prescription is in effect making one edge of a lens thicker than its diametrically opposite edge. This is accomplished either by tilting the lens accordingly (as to direction and degree) at the blocking stage or by the introduction of wedge-shaped rings at the next stage to be described. Such rings produce the desired result of unlevel (wedged) surfacing. The blocked lens marked for direction is now ready for working. The process of surfacing spherical or cylindrical curves are (with one exception to be mentioned), carried out on different types of machines, although each blank goes through the same stages of roughing, smoothing (fining), and polishing.

Roughing implies grinding the glass on a curve generator by means of a ring type diamond tool to approximate curve and with a moderate smoothness of surface. The workable range is from plano to $16\frac{1}{2}$ dioptres convex and concave. The final thickness of the lens (provided the second side is being surfaced) is determined at this the roughing stage by limiting, by means of a mechanical stop, the amount the tool is allowed to grind into the glass. The second stage, smoothing (fining), is performed in the historical way of grinding the glass with abrasive powder direct on to a metal tool such as earlier described.

The machine comprises a fast-revolving vertical spindle on the top of which the base of the tool is secured. The blank with its backing of metal button is pressed downwards (usually by gravity) by a pin which fits into the button hole. The pin is part of an arm which drives, hence the blank, slowly across the face of the tool whilst abrasive powder and water lubricant is pumped on to the work. The process perfects curve accuracy and

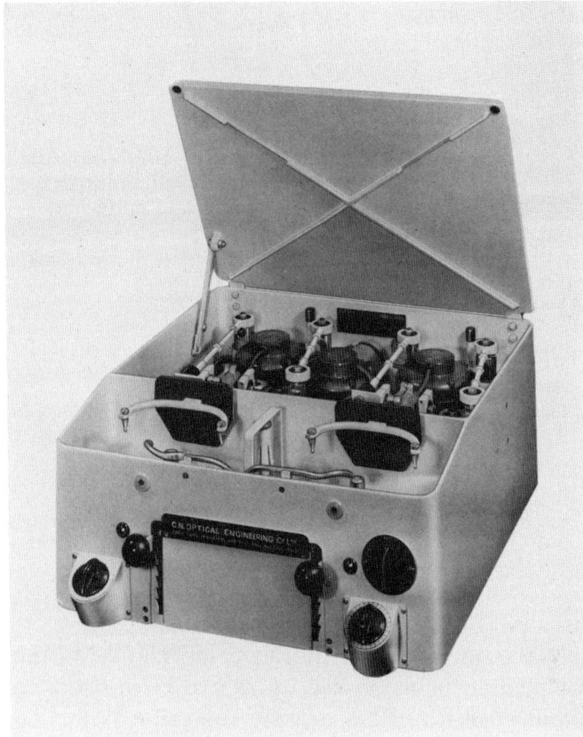

7.13.6. Optical Engineering Company CN 100 twin-spindle toric surfacing machine for smoothing and polishing lenses.

improves the surface texture of the glass, so that the polishing stage takes minimum time.

Polishing requires exactly the same machine and practice except that the tool bears cloth, or some such material, over its face and the material pumped is usually cerium or zirconium oxide. After polishing, the lens is removed from its adhering pitch, wax, or blocking metal.

Surfacing cylindrical curves follows the general procedure for each stage but the machines are somewhat different. The generator is horizontal, the diamond tool producing the spherical ingredient according to the angle that it subtends to the surface of the blank whilst the cylindrical element is introduced by moving the tool (still maintaining its angle to the blank) through a horizontal arc, the radius of which is calculated from the formula $r = (N-1)/F$. Because the length of the arc radius is restricted by practical considerations the power of the lens curve is limited in the low direction. The normal range which can be generated is up to 10-dioptre sphere combined with up to 5-dioptre cylinder, both convex and concave.

Toric generators, unlike other types of prescription lens machinery, vary greatly in sophistication and cost. Whilst no machine generates more than one lens at a time, and whilst the quality of the production from all good generators is much the same, there are variations in design which, involving cost differences up to a factor of six, give a much enhanced output per generator.

The machine illustrated allows the required scale settings to be made whilst the generating of the previous blank is taking place (Fig. 7.13.5.). The actual speed of cutting for any given quantity of glass removal is mainly influenced by the choice of composition of the diamond tool. A general rule is that the lower the concentration of impregnated diamond powder of a given size the more rapid will be the cutting, but conversely the higher the concentration the longer the life of the tool. The smoothing and polishing machines nowadays almost invariably have a tool-holder which is stationary.

The blank is impelled downwards on to the tool face by spring or compressed air pressure. The aim is to move the blank across the tool face in both meridians whilst maintaining parallelism. Two types of movement find favour (Figs 7.13.6 and 7.13.7). One is reciprocating with the greater excursion along the direction of the cylinder axis, so that the tool used must be of a rectangular shape. The other is circular with much the same excursion in either meridian. All smoothing and polishing with the so-called toric surfacers demand

7.13.7. Raphael Type 101 single-spindle surfacing machine that will smooth and polish both cylindrical and spherical surfaces.

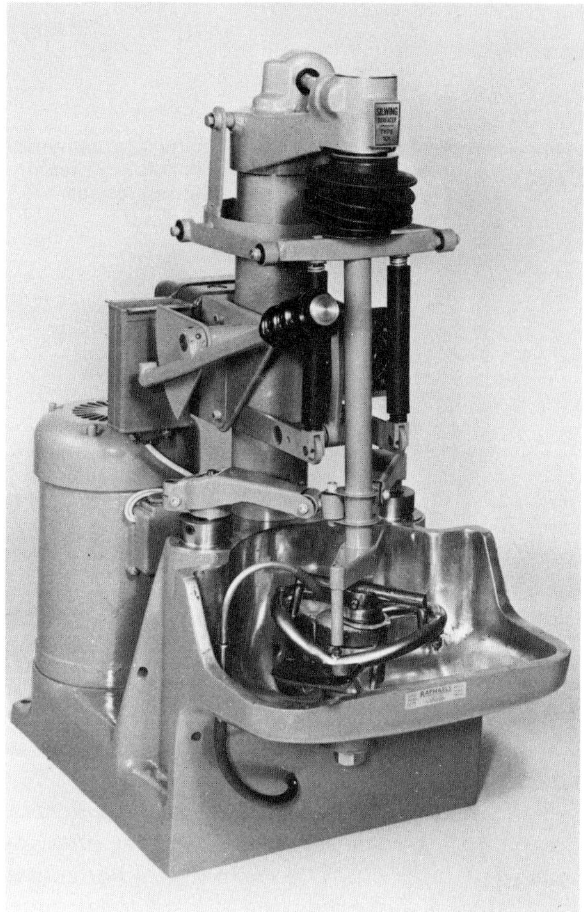

that the movement of the blank in relation to the tool should be as broken up as can be in order to avoid ever travelling exactly the same path. To this end the mechanical movement of the driving arm is made deliberately complex.

Fig. 7.13.7 illustrates a machine which operates partly by flexure and partly by positively controlled movement. Both spherical and cylindrical surfaces can be smoothed and polished (using different powders of course) on this one type, the use of which overcomes the disadvantage of separate handling and working when two different machine types are involved. The rate of spherical production is slower than obtainable with machines constructed exclusively for spherical working and the various factors as they affect any particular workshop have to be considered.

Glass safety lenses

These are of two sorts, one of which has a central sheet of plastic laminated between the two outer pieces of glass. Surfacing such lenses follows normal procedure but care has to be taken not to reduce the thickness of the central portion of the glass on either side so as to risk breaking through to the plastic sheet. Also the heat from blocking and from the friction of polishing should be kept to a minimum. Because it is of three-ply construction, the finished lens is appreciably thicker than one of homogeneous glass.

The other safety form is wholly in glass, and requires heat treatment after the surfacing and the final edging is finished. Heating to nearly melting point and then rapid cooling in a cool air blast 'case hardens' the glass. To obtain the full safety from breakage effect the lens usually has to be made somewhat thicker than normal, particularly in the case of concave lenses.

Plastic lenses

These take two forms, being made of either polymethyl methacrylate, which is thermoplastic, or allyl di-glycol carbonate (industrially known as CR.39), which needs to be cast. The refractive indices are 1·492 and 1·4905 respectively, and so the thickness for any given power and size of lens must be somewhat greater than for glass having a refractive index of 1·523. Nonetheless the weight per volume is in both cases some 50 per cent less than glass. CR.39, though later in the commercial field, is more widely accepted internationally. Both forms are more liable to scratching than glass, and the thermoplastic lenses need to receive a special hardening process to make them comparable to CR.39.

Thermoplastic lenses are produced exclusively by heat moulding in metal pressure dies of such immaculate polish that the optical properties are perfect without any further treatment. CR.39 lenses are cast between glass moulding plates and when so cast are optically perfect. But, unlike Thermoplastic, CR.39 can be surfaced by normal methods; however, more stringent conditions have to be observed than in the case of glass. This ability to surface the second sides is valuable in saving the special individual casting that would be needed for each combination of bifocal power and cylindrical axis.

The foregoing information on ophthalmic lens production is based on an end product having just a finished front and back surface. Such lens is said to be in the 'uncut' form. Before it can function the lens has to be edged and fitted to a spectacle in accordance with the requirements of the prescription.

8 Production of Prisms and Flats in Quantity

8.1 Introduction

Prisms in large quantities are usually made from mouldings, but medium quantities, where a few hundreds each year are required, are invariably made from slab glass. As the light-path length in a prism is considerably more than in a lens, it is even more important to have good homogeneity and annealing in the glass. The wastage of glass in the manufacture of prisms is very considerable (as will be explained in § 9.3), and, taking into account cutting allowances and excess materials, such as offcuts and scrap, it would be normal for the weight of a small finished prism to be less than 50 per cent of the weight of the material supplied from stores. For larger prisms the finished weight might be about 60 per cent of the weight of the original material.

In the machining of a prism, account must be taken of the angle, not only between the faces which are subsequently to be polished, but also the angles which those faces make to a common base. Where two surfaces only are to be polished, an unpolished surface is frequently relied on as a base for fixing the prisms. Where three or more polished surfaces are concerned, lack of truth may produce what is known as 'pyramidal error' which becomes apparent when, in the finishing stages, the base is ground to the prescribed limits of squareness.

When two faces are square to the base, the third (and fourth, if present) must also be so if the prism is to be free from pyramidal error. In general it can be said that for work of good quality all surfaces should be square to a common base to less than two minutes of arc.

The basic steps in making small prisms from slab are as follows:

(*a*) Slit plate or slab into squares.
(*b*) Stack a number together with wax.
(*c*) True one corner on smoothing spindle.
(*d*) Wax stack on to steel plate.
(*e*) Transfer plate with stack to grinding machine magnetic table and grind to depth.
(*f*) Remove plate and transfer stack to opposite face.
(*g*) Transfer plate with stack to grinding machine and grind to depth.
(*h*) Slit as required.
(*i*) Jig and grind angles.
(*j*) Block in plaster, smooth, and polish.
(*k*) Bloom, silver, copper, or shellac, and black.

8.2 Preparation of glass squares

Glass slab is usually cut with a diamond slitting saw into pieces not greater than 50 mm square and 75 mm long for ease of handling in subsequent operations. A slitting machine (Figs 8.2.1 and 8.2.2) designed for the purpose is used to cut the slab accurately to size.

A convenient number of the glass squares should be stuck together with pitch to form a stack and

◁8.2.1. **A glass slab being cut with a diamond slitting saw.**

8.2.3. Penta prisms stuck to a plate before being placed on a Blanchard grinder.
◁ **Other penta prisms are stuck in jigs before being ground to angle on the Blanchard grinder.**

▽**8.2.2. Autoflow slitting machine.**

then one corner trued up on a smoothing spindle. The 90° angle must be checked for accuracy with an Angle Dekkor. (See §11.19.)

The stacks of glass must now be stuck with pitch on a parallel-ground steel plate ready for the Blanchard grinder table. Alternatively, if large prisms are being made in quantities which justify jigs, then the prisms will be stuck with pitch on the jigs ready for the Blanchard grinder table. (Fig 8.2.3.)

8.3 Grinding parallel and reducing to substance

The No. 11 Blanchard surface grinder has an 11-inch diameter diamond grinding wheel and a 16-inch diameter magnetic chuck with a 15 h.p. spindle motor. This machine has revolutionized the grinding of flat glass surfaces, and in a few minutes of grinding time produces a smooth flat surface, flat to the edges, with no chipping. The steel plate loaded with prisms is transferred to the magnetic chuck on the grinder table and the machine can be set to feed automatically and give the required surface. (Fig. 8.3.1.) Steel jigs holding right-angle prisms can be set directly on the magnetic chuck and, after the first surface, turned through 90° for grinding the second side. Larger sizes of such grinding machines are capable of grinding the surface of glass slabs. (Fig. 8.3.2.)

Plano-convex lenses or windows can be easily ground on the Blanchard machine. The parts can be held in brass rings which, in turn, are stuck on the ground steel plate or placed on a piece of blotting paper, which is soaked in wax whilst on a hot-plate before being transferred to a cold-plate for setting the lenses in position. (Fig. 8.3.3.)

8.3.1. The Blanchard grinder.

(*a*) The plate of penta prisms held on the Blanchard magnetic chuck before grinding down to correct thickness.

(*b*) Jigs, in which the prisms are stuck with wax, held on the magnetic chuck ready for surface grinding.

8.3.2. Grinding the surface of a slab of glass with a diamond grinding wheel.

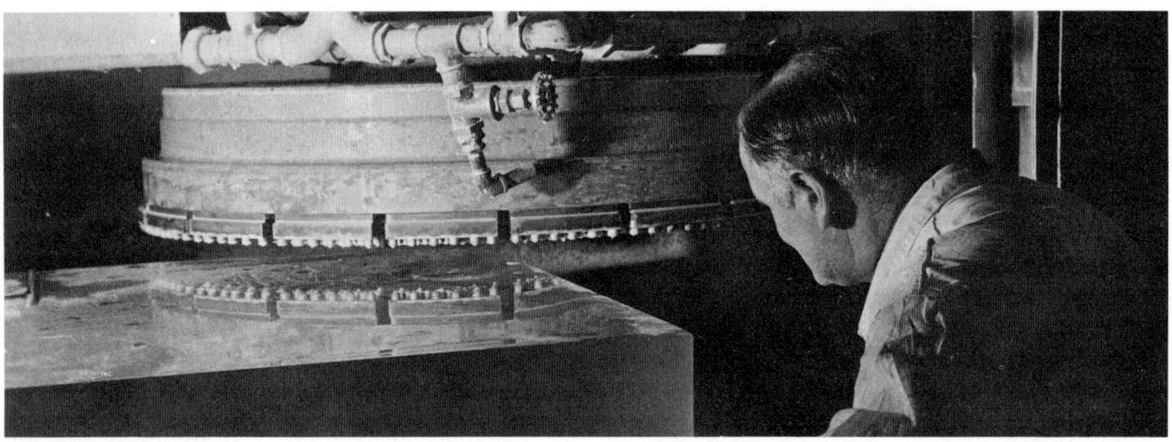

◁8.3.3. **Lenses on blotting paper on a steel tool on a hot-plate. They are awaiting waxing, after which the plate will be transferred to the cold-plate before grinding the plano side on the Blanchard No. 11 grinder.**

▽8.4.1. **Loh universal milling machine.**
(*Left*) Type UFM with horizontal milling unit for simultaneous milling of two parallel surfaces.
(*Right*) Type UFM with angle milling unit.

8.4 Large quantity production on the Loh universal milling machine

The Loh milling machine UFM with its various structural accessories can be used to work on plane, parallel opposing surfaces, angled surfaces of prisms, for milling grooves, and similar processes as well as for rounding and sawing. The maximum slide movement is 310 mm and the size of machine table 260 × 160 mm. (Fig. 8.4.1.)

The Loh milling machine UFMS 500/1000 in

8.4.2. **Loh universal milling machine, type UFMS, with angle milling unit.**

its range links up with the UFM and is as universally applicable through the use of various milling units, not only for the economic working of large prisms, glass scales, and similar work pieces, but also for large series of elements on which parallel adjacent surfaces, angle surfaces, and slots have to be worked. (Fig. 8.4.2.)

The maximum slide movement is 800 mm on UFMS 500 and 1300 mm on UFMS 1000. The size of the machine table is 900 × 290 mm on UFMS 500 and 1400 × 290 mm on UFMS 1000.

The Loh horizontal milling unit can be used not only for simultaneous milling of two parallel surfaces, but also for milling grooves, angled surfaces, and other similar work, when used in conjunction with vertical or angle milling units. It is adjustable laterally and vertically by means of two slide guides. The adjustment is made by a threaded spindle with 0·01-mm vernier scale (Fig. 8.4.3.)

An angle milling unit for milling angled surfaces consists of a drive and grinding spindle, and is adjustable for angle as well as both laterally and in height through two slide guides. The adjustment for angles is carried out by means of a high-precision measuring device, accurate to 30 seconds of arc. (Fig. 8.4.4.)

The rounding equipment as a complete unit can be easily mounted on the table of the basic machine. The rounding process is carried out in one operation. (Fig. 8.4.5.)

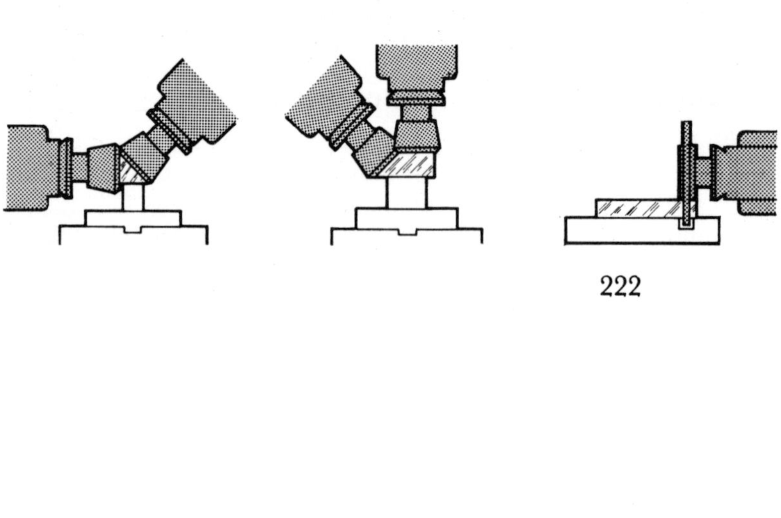

8.4.3. Operations with the Loh universal milling machine.

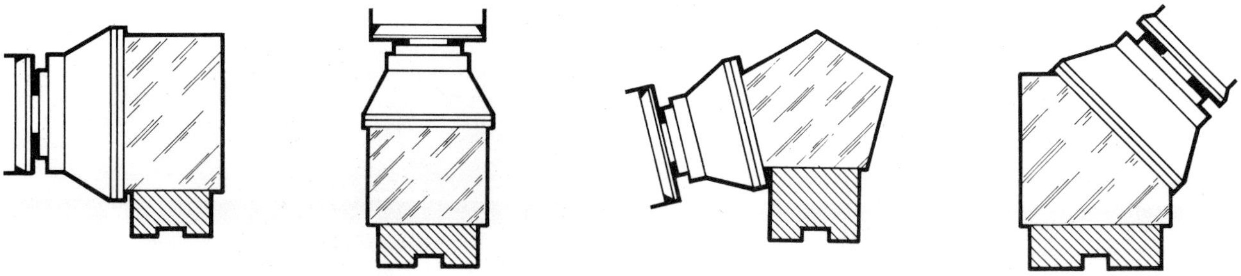

△8.4.4. Operations with the Loh universal milling machine.

8.4.5.▷
Loh universal milling machine type UFM with rounding equipment.

8.5 Blocking in plaster, smoothing and polishing

Plaster block making

Although some people will polish prisms direct on steel jigs (if the quantity required justifies the jigs) immediately after smoothing, there is a lot to be said for dismantling and cleaning the prisms after smoothing, and blocking up in plaster for polishing.

The operator responsible for plaster blocking must be given a good supply of plaster of Paris and of hydrated lime, several optical plano tools which have been carefully flattened, and several tools of the polisher holder variety into which the plaster of Paris can be poured. (Fig. 8.5.1.) The plaster of Paris must be kept dry and stored in a metal box.

One of the flat tools is screwed to a bench screw, cleaned with petrol, and then rubbed with a little clean machine oil. The trued surfaces of the prism stacks are then dusted and rubbed down into close contact with the tool surface. Any lack of care will cause errors in the finished prism. The arrangement of the surfaces should be as symmetrical as possible and equally balanced; for, if there is an excess of glass on one side of the block, the other side will grind away more quickly and the angle will alter in the grinding process. It may be necessary to add wasters in order to make up a balanced block. The polisher holder, large enough to enclose the prisms, is then laid on the tool, supported on three slips of glass about 2 mm thick to prevent the sides from touching the smoothing tool and the polisher. Melted paraffin wax or Plasticine is then placed between the tool and the ring to prevent the escape of wet plaster.

A small quantity of plaster is mixed with an excess of water to make a thin mixture which is poured in to form a layer about 2 mm thick. Alternatively, this layer can be hot wax (American method) or sawdust (German method).

Whilst this thin layer is drying out, plaster and lime are mixed together in the proportions two of plaster to one of hydrated lime in sufficient quantity to fill the polishing tool. This mixing must be thorough. The mixture is spread on a board in a heap, with a depression in the middle, into which water is poured a little at a time.

8.5.1. Plaster blocks of prisms.

(*a*) ▷
The prisms are pressed on the plano. The tin contains plaster of Paris and the lump of Plasticine is used to fill up the holes. The metal cover tool is placed over the plate of prisms and plaster poured through the four holes. The dustbins under the bench are for scrapped, used plaster and for storage of new plaster and lime.

(*b*) ▷
The block of prisms in front of the photograph is turned over to show the surfaces due to be smoothed and polished. The blocks of circles in the rear of the photograph are waxed 'hard on' to the metal block.

Mixing should proceed until the lump of material is uniformly moist and about the consistency of mortar as used by a bricklayer. The operator then puts on a pair of rubber gloves for cleanliness and takes the 'mortar' in the fingers and presses it between the prisms until the block is filled with plaster. A block so prepared should dry out in 12–24 hours—usually overnight from afternoon to following morning is satisfactory—when the block can be removed from the tool by heating the tool with a bunsen burner. The block should now be screwed to a post screw by the thread in the backing tool. The 2-mm thick layer of very wet plaster will lift off quite easily leaving the prisms projecting from the mass of set plaster. The exposed plaster must now be painted with shellac varnish or hot paraffin wax as, if unprotected, it will absorb the water used in the smoothing and polishing stages. (Fig. 8.5.2.)

Fine grinding (or smoothing) is carried out on

8.5.2. Plaster blocks, finished and inverted.

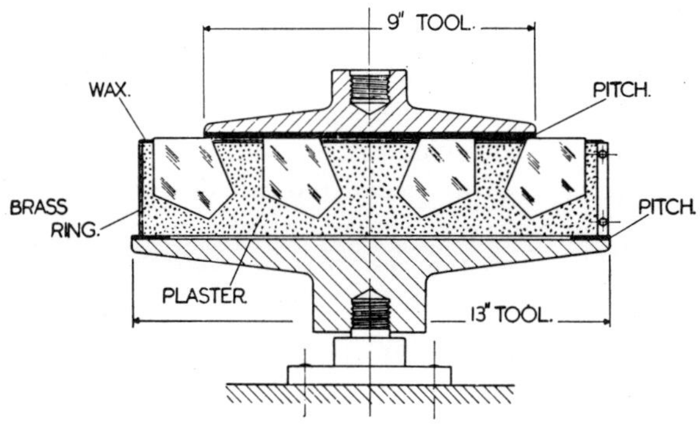

the tool on which the block was made up, and all the prisms should 'come up' during the first emery. If parts of the prisms are still coarse grey after 10 minutes of fine smoothing, then one of two things has happened. Either the prisms were not laid in good contact with the tool or the surfaces were not flat. Surfaces ready for plaster blocking should be slightly concave and never convex.

If the block 'takes' all over in a satisfactory manner the usual routine of two or three 10-minute grindings in each of the finer grades of abrasive will follow, with careful washing of the block between each stage. The block can occupy either the upper or lower position on the smoothing machine according to whether the operator wishes the tool to assume a more convex or concave figure.

Polishing plaster blocks

After washing clean, the block is screwed on the spindle of the polishing machine. A loaded polisher of the Swedish pitch and wood-flour type is warmed from the back to about 60°C, painted with cerium oxide and rubbed by hand on the block until a good, even pull is felt, free from edge binding. The machine pin is lowered and the machine switched on. It is quite feasible to feed a continuous supply of polishing compound but the traditional method of brushing is often used.

After a short period of polishing, the tool should be removed and the block cleaned for examination to see if the block is polishing evenly. If the test-plate figure is satisfactory, the polisher can be replaced and polishing continued for $\frac{1}{2}$ hour, before retesting takes place. If the block is uneven, say bright in the centre but much more grey around the edge, then the stroke should be moved more over the edge. If the central rings in the polisher have closed up they should be opened. Should the test plate figure be unsatisfactory some discretion is necessary. As a general rule, if the departure from proof plane is not serious, the grey should be removed before any attempt is made to correct the shape as, once a part of the block acquires a full polish, the rate of removal of the remaining grey is very much reduced.

If a well-polished block is found to be more convex than the limits permit, the substitution of a smaller polisher may correct the figure. However, if the testplate shows that the block is concave, then the further off-setting of the stroke with a full-sized polisher will always make the block more convex.

Removing prisms from plaster blocks

When polish and shape of figure are satisfactory, the surface of the glass must be protected by a coat of shellac or paint. The block can now be broken up by blows from a hammer on the back of the block or by a hammer and cold chisel.

Usually the prisms will fall away from the cracks in the block, and can be removed almost free from pieces of plaster. The prisms must be cleaned before blocking up for polishing the next side. Providing the initial grinding of the angles was correct to dimensions required, there should be no inaccuracies due to polishing within tolerances of 2 minutes of angle and one fringe of flatness. If a greater accuracy than this is required then the prisms must be individually finished off by hand polishing or on a free-running polishing machine where biased loading can be applied during polishing.

8.6 Correction of angle and flatness

Although skilled craftsmen are required for the finishing of very accurate prisms and flat plates (Fig. 8.6.1), an increasing volume of work is being transferred to specialized machines, such as the Hilger parallel plate polishing machine (Fig. 8.6.2) or the Lapmaster, which is an optical 'variation' on machines developed primarily for metal lapping. (Fig. 8.6.3.)

The general principle of machines for very accurate polishing is to have a slowly rotating polishing tool, on which is a corrector ring, capable of taking offset loading with the object of maintaining or correcting the flatness of the pitch polisher. The contour of the pitch plate is changed from a concave to a convex surface by moving the rings inwards or outwards and in between these conditions exists a position where flatness is maintained.

Inside the corrector ring, the prism or flat to be figured is mounted in a suitable jig. It is also possible to exert an offset load on the component being polished by means of a weight rod, as in the Hilger machine, or directly as a weight on the top of the component in the case of the Lapmaster.

It is essential to have temperature control of the atmosphere around the machine to 1°C as there is a critical balance between the hardness of the

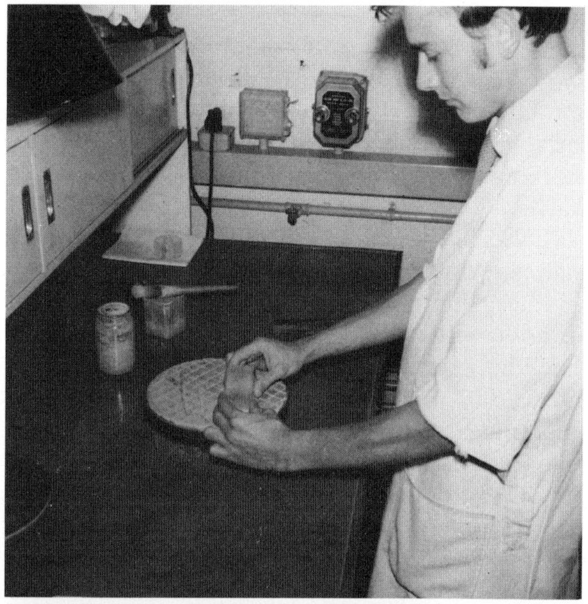

8.6.1.▷
A young craftsman figuring a prism. The polisher rotates slowly at about 1 r.p.m.

8.6.2.▷
Hilger polishing machine capable of polishing surfaces flat to 0·05 fringe. The point of the weight rod is placed in small recesses in the upper surface of the prism holder.

8.6.3. Lapmaster polishing machine for very flat components.

(*a*) ▷
A 36-inch machine.

(*b*) ▷
A 24-inch machine that will produce surfaces flat to 0·1 fringe.

pitch, heat generated by the friction of the ring, and work on the polisher. Some experience is necessary before success is achieved, as it may be found necessary to vary the weight of the corrector ring to obtain the best working conditions.

Flatness of a quarter of a fringe is readily obtained by these methods but, with perseverence, one-tenth of a fringe is practicable. Special work, such as etalons, can be matched to a higher order of accuracy. The polishing time for one-tenth of a fringe is about double that required to achieve one-quarter of a fringe.

For best working conditions the 36-inch Lapmaster should have variable speeds on the polishing table from 3 r.p.m. to 6 r.p.m., although 4 r.p.m. seems to be the optimum speed for glass polishing. A well-smoothed block of glass blanks of 9-inch diameter normally takes about 3·5 machine hours to be fully polished to graticule standard. The 36-inch Lapmaster is a suitable machine for finishing off blocks or individual prisms or flats which have been previously polished on conventional machines.

The polisher is loaded with cerium oxide only occasionally. After the initial loading, distilled water is used as a lubricant and this allows the cerium oxide to break down and give a very fine finish.

8.7.1. A precision optical polygon.

8.7 Producing optical polygons

Permission has been granted by Rank Precision Industries Ltd for details of their methods at Debden Factory, Loughton, Essex, to be published, and the photographs in this section were provided by *Machinery and Production Engineering* who published an article on 13 November 1968 describing polygon production.

Precision optical polygons (Fig. 8.7.1) provide a convenient and precise basis for angular measurements when used in conjunction with auto-collimators. Glass polygons are made in a range of sizes from 5 to 72 sides which, after polishing, are coated with a thin film of aluminium protected by a film of silicon monoxide to give high reflectivity over a long period of time. Each polygon is provided with a central hole to suit a mandrel and is enclosed in a strong cover for protection against damage.

The following description of the manufacture of polygons applies, in particular to a twelve-sided unit with 30° angles and overall dimensions of 3·2 inches across flats and 0·75 inch thick. The twelve faces are required to be flat within 0·0000025 inch (one-quarter of a fringe) and each face at 90° to the centre, within 5 seconds of arc.

The optical crown glass (H.C. 519604) is annealed in an electric furnace to ensure that no distortion will occur at a later date, during grinding or polishing, which would affect the flatness of the faces or the accuracy of the angles.

The series of operations necessary to produce the polygons is as follows:

(1) Slit slab 0·9 inch thick, 4 inch square.

(2) Drill one-inch diameter hole on Optibel drilling machine. (Fig. 8.7.2.)

(3) Face one side. Assemble six on plano and reduce to thickness by grinding on Blanchard No. 11 machine.

(4) Dismantle and clean.

(5) Mount six on spigot and edge to 3·3-inch diameter on Strasbaugh machine.

(6) Smooth both sides by hand, one at a time.

(7) Engrave.

(8) Mount stacks of six on fixtures and grind twelve facets on Adcock & Shipley O.V.S. machine.

(9) Grind first angle by hand, smooth by hand, polish by Lapmaster. Grind second angle, etc., for twelve faces.

(10) Check the stack of six polygons for angle.

(11) Chamfer and dismantle stack.

(12) Chamfer edges singly.

(13) Check each polygon for angle.

(14) Polish engraved face of polygon by Lapmaster.

(15) Clean and inspect.

(16) Vacuum aluminize and coat with silicon monoxide.

(17) Inspect.

In the illustration (Fig. 8.7.3), item A shows operation (2) completed, the blank having been drilled with a diamond cutter. As will be seen

8.7.2. ▷
Set-up on the Optibel drilling machine for producing the central hole in glass polygon blanks. A diamond-impregnated tool is used. In the photograph, one clamp is not fixed to the blank.

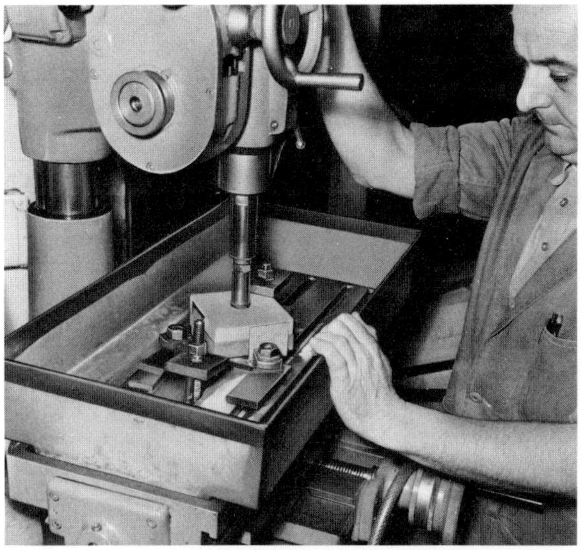

8.7.3. ▷
A glass optical polygon is made from a moulded blank, as seen at A, cut from a slab of crown glass. The blank is provided with a central hole by drilling with a diamond-impregnated tool. It is then trimmed and cylindrically ground by means of diamond-impregnated tools before the flats are worked. The optical indexing fixture, seen in the background, is used for machining the flats.

from Fig. 8.7.2, sometimes two blanks are drilled at the same setting.

The grinding of the flats is a critical operation and the fixture (Fig. 8.7.3.) will take a stack of six blanks at operation (8) and hold them rigid during the grinding of the facets. The fixture incorporates an optical-alignment device to ensure an accurate angular setting. Split bearings are provided which, when loose, enable the blanks to be rotated for indexing but can be clamped solid during the grinding operation. The mandrel can be entirely removed from the fixture for loading and unloading workpieces. An angular accuracy of within 1 minute of arc between one flat and the next is achieved by this method. Finer degrees of accuracy are achieved during polishing the stack.

After the flats have been ground the polygon is engraved to identify the angles for reference purposes. The final smoothing and polishing operations are usually a highly skilled manual operation absorbing a large number of man-hours, but mechanization of the process presents problems because the flat surfaces must be to very fine angular tolerances.

A polishing fixture has been developed suitable for use on a 36-inch Lapmaster polishing machine, which gives good results on a twelve-sided polygon, and its general principles can be applied to other sizes for operation (9).

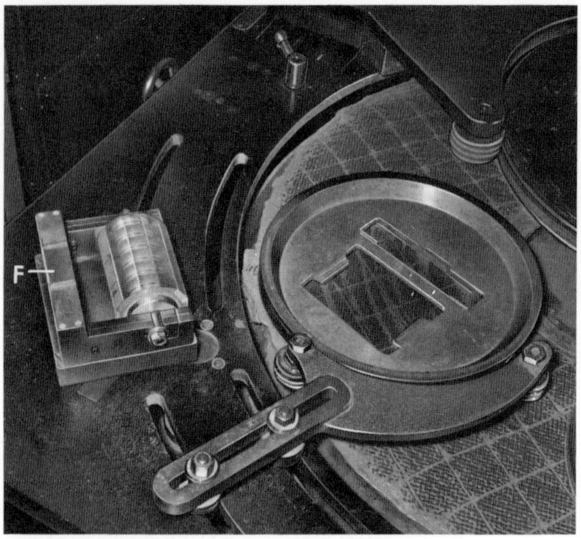

8.7.4. A view of the fixture for polishing six polygons at a time, together with the work-ring that accommodates the fixture on the Lapmaster machine. The fixture is shown inverted so that the Pyrex guide block F can be seen.

8.7.5. A Lapmaster 36 lapping machine, adapted for optical polishing, is used for finishing the polygons, which are accommodated six at a time in the special fixture seen on the plate at the left. Flatness to within 0·000 002 inch is obtained, while the angular accuracy from flat to flat is within 3 seconds of arc.

8.7.6. Concrete test table with two Hilger & Watts photoelectric autocollimators, which are used for checking the polygon angles and to facilitate the angular setting of the polygons in the polishing fixture.

A close up view of the bronze work ring (Fig. 8.7.4) and the loaded polishing fixture (illustrated in an inverted position) shows that the stack of six polygons is mounted on a mandrel with the flats accurately aligned with each other. The mandrel is then clamped in vees in the fixture with one set of polygon flats aligned parallel with the surface of a Pyrex-glass guide-block as indicated at F. This very accurately positioned Pyrex block serves to keep the fixture and polygons level during polishing.

Normally, the polishing of surfaces can be completed with not more than two intermediate checks for flatness. An element of skill is necessary, following a check on the Fizeau interferoscope, to decide if a small weight is necessary on the bridge of the fixture to adjust the pressure distribution and so correct dimensional faults.

Although polishing times vary, about 2 hours is required for each face to polish a stack of six polygons. A view of the Lapmaster machine with polygons being polished is given in Fig. 8.7.5, where the fixture, carrying six polygons, is seen in the work-ring on the left side.

The bronze work-ring in the front is loaded with four optical flats, which are being polished at the same time as the polygons. The corrector plate can be replaced by a third work-ring as simultaneous polishing of different components can be carried out if the glass specification is the same, and if there are no large differences in total surface areas of the separate items being polished.

The 36-inch polishing plate is made from Swedish pitch loaded with walnut or almond nut wood-flour and is very similar in construction to other types of optical polisher.

The work-rings have Pyrex glass pads on their under-surface where they contact the polisher. The rings are adjusted for position until the machine is polishing flat, and this is checked by a test-piece.

It is essential that the work-rings, the conditioning ring, and components are all free to rotate. To aid this freedom, the restraining rollers bearing against the rings must be only just clear of the polisher surface to reduce the tilting force on the rings. Lubrication of the polishing surface is by drip application of distilled water. Occasionally the surface of the polisher will need machining and this is done by a turning attachment.

When one set of flats has been completed, the fixture is removed from the polishing machine for indexing to the next position. The angular setting is carried out on a test table (Fig. 8.7.6) by means of the Hilger & Watts microptic photoelectric autocollimators, which are mounted at 30° to each other and will read to 0·1 second of arc. (See § 11.19.)

After polishing, cleaning, and testing, the polygons are mounted in the rotating fixture of a vacuum coating machine. (Fig. 8.7.7.) The vacuum chamber is then replaced and, during the evaporation of aluminium followed by silicon

8.7.7. Polygons are vacuum coated with a thin film of aluminium protected by a film of silicon monoxide to afford a very high degree of reflectivity of the flats. This view shows the working area of a vacuum coating unit built by Edwards High Vacuum International Ltd installed for the operation. The polygon is mounted on a work-spindle whereby it is rotated during the process. A number of polygons can be accommodated at a time. A vacuum chamber covers the working area during the coating operation.

monoxide, the polygons are rotated to ensure an even coating. When manufacture is completed, polygons are calibrated to within 0·1 second of arc.

To ensure that calibration will be accurate to within 1 second of arc it is more important that the mirror faces be flat, and square to the polygon sides, than that they closely agree with the nominal angle. Departures from square are known as pyramidal errors.

The accuracy of a polygon can be proved by first principles and, during calibration, it should be mounted on a dividing table which has a fine adjustment. (Fig. 11.19.6.) By the optical calliper method of calibration two autocollimators should be set facing the polygon at an angle of 30° (the smallest angular separation required for a twelve-sided polygon) and twelve pairs of readings taken from the polygon faces. One autocollimator 'A' may be used as a setting line, or fiducial indicator, and differences measured on a microptic autocollimator 'B'.

After the initial setting of 'A' on the zero face of the polygon, successive faces (30°, 60°, and so on) should be brought round to give exactly the same reading on 'A' autocollimator. Variations in the reading on 'B' autocollimator will correspond to variations in the 30° intervals between faces. The correct sign of the differences must be established and it is suggested that positive measurements are clockwise and negative measurements anticlockwise. The table below gives a typical series of calibration dimensions.

It should be noted that the sum of all the errors (differences from mean) must be zero and, for the same reason, the cumulative error of the 360° interval is zero as the 360° line is the same as 0° line. For the best results photoelectric autocollimators should be used from which measurements can be read to 0·1 second. An improvement in accuracy can be achieved by repeating the readings and rotating the polygon in the opposite direction. A mean of the dimensions should then be used for calibration calculations.

Table 8.7.1. Typical dimensions in the calibration of a polygon

Nominal included angle (degrees)	*Autocollimator B readings (seconds)*	*Difference from mean (seconds)*	*Cumulative error (seconds)*	*Face position (degrees)*
			0	0
0–30	0	+2	+2	30
30–60	+4	+6	+8	60
60–90	−2	0	+8	90
90–120	+1	+3	+11	120
120–150	−5	−3	+8	150
150–180	−4	−2	+6	180
180–210	−3	−1	+5	210
210–240	−4	−2	+3	240
240–270	−2	0	+3	270
270–300	−4	−2	+1	300
300–330	−4	−2	−1	330
330–360	−1	+1	0	360
0–30	+2	—	—	
		0		

8.8 Diamond smoothing of plano surfaces

Since 1956, the techniques of diamond smoothing and associated recessed tooling, or 'hard-on' blocking, have reached a high degree of development. Diamond smoothing substantially eliminates normal smoothing by cast-iron tools and emery, except for small-quantity work. (See § 6.7.)

The finish from diamond smoothing is so much better than from emery smoothing, for crown glass or green plate, that polishing time is reduced

8.8.1.▷
The two aluminium smoothing tools in the back row are of 12 inch and 9 inch diameter. They hold concentric rows of diamond pellets stuck into recesses with Araldite. In the front row are two typical blocks of glass blanks that have been diamond smoothed.

8.8.2.▷
Autoflow BM4 machine with centrifuge to clarify the cutting compound used in diamond smoothing. The diamond smoothing tool is illustrated standing up at the back of the machine. The Autoflow BM4 on the right of the photograph is used for conventional fine smoothing.

to about 40 per cent of the normal time for polishing emery-smoothed plano blocks. (Fig. 8.8.1.)

An Autoflow BM4 machine, fitted with a centrifuge to clarify the Houghton 'glassgrind' 960 coolant at a concentration of 20:1, will produce very good plano surfaces in about 3 minutes. Diamond pellets suitable for a finishing tool should have a bronze-bond with 6–12 micrometre grit at $3\frac{3}{4}$ per cent concentration (1·2 carats per cc). (Fig. 8.8.2.)

A 9-inch (228 mm) diameter aluminium plano will need about 135 10-mm diameter pellets (25 per cent of the area covered by pellets) in a pattern of equally spaced concentric rings, each pellet being Araldited into a recess. If plano smoothing by diamond is restricted to the finishing operation only (and this has some

△ **8.8.3. Loh plano diamond lapping machine PLM 200 for a maximum work diameter of 200 mm.**

8.8.4. ▷
Loh plano diamond lapping machine PLM 400 for a maximum work diameter of 400 mm.

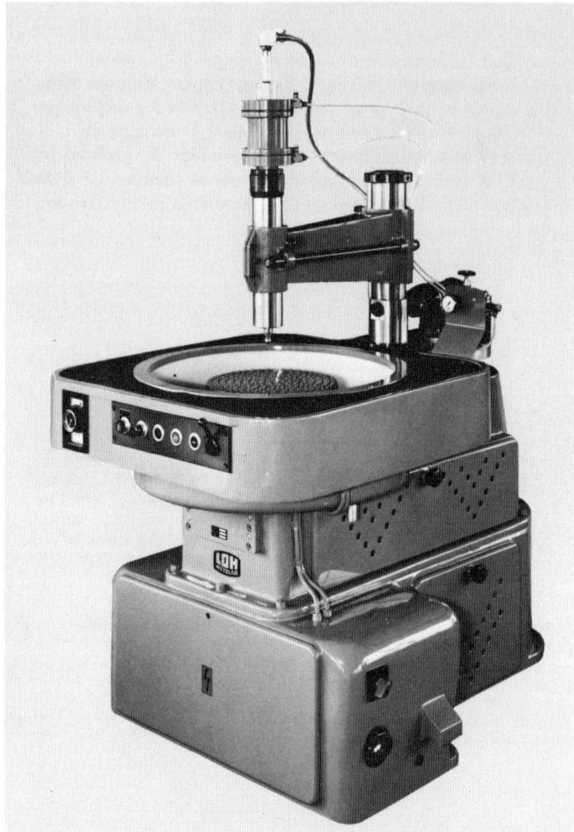

advantages on small-quantity work), then blocks of prisms or flats should be pre-smoothed for about 20 minutes with AO 302½ emery, to achieve the required flatness and surface texture.

Diamond smoothing would follow at a block speed of 530 r.p.m. for 3 minutes. This total time of 23 minutes compares with a traditional 40 minutes as an additional smoothing with AO 303 emery for 20 minutes would normally be required for the finishing operation.

A complete polish from a plastic polisher is achieved on an 11-inch block in 2 hours from a diamond-smoothed surface, as compared with 4·5 hours after smoothing with AO 303 emery.

The Loh PLM 200 and PLM 400 plano diamond-lapping machines are of robust construction and have been designed specifically for high spindle-speed, high working pressures, and good stability for absorbing out-of-balance errors such as occur on prism blocks. (Figs 8.8.3 and 8.8.4.)

Diamond pellets are used, and the grain size and concentration, as well as the amount of covering of the tool surface, are dependent on the types of glass to be worked, the type of coolant used, and on the required finish. As a guide, 25–50 per cent of the surface area should be covered with pellets. Diamond lapping of large surfaces requires high working pressures for optimum performance and on these machines the pneumatic load can be regulated to suit the work surface. Usually the tool is running on the main spindle and, in this case, the coolant supply is provided through the main spindle. For large-diameter blocks, it may be better to have the block on the main spindle and the tool running on top, in which case a separate coolant supply is provided.

The 200-mm work-diameter machine has main spindle speeds from 500–1500 r.p.m. and the 400-mm machine has spindle speeds ranging from 225–650 r.p.m.

◁8.9.1 **Autoflow precision scientific polishing machine with a plastic polisher for high-speed polishing of plano and large-radius lenses. The plastic polishers are stored in water (see metal box) to avoid drying out and hardening. Polishing compound is supplied by continuous flow.**

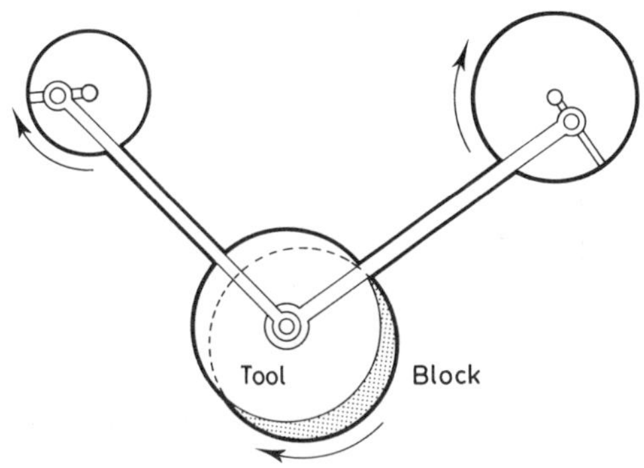

8.10.1.▷ Type of polishing machine suitable for aspheric mirrors.

8.9 Plastic polishing

Plastic polishing will give as good a surface flatness and figure as does Swedish pitch polishing, but there is a sleek problem when the highest quality of finish is required. The Autoflow precision scientific polishing machine will give good results on plastic polishing at 200 r.p.m. and 20-lb (9 kg) pressure on the top of the 9-inch plano tool. (Fig. 8.9.1.)

Under these conditions with a Rowland plastic polisher and Regipol 137 polishing powder in water as a coolant, a complete polish is achieved on an 11-inch block in 2 hours from a diamond-smoothed surface. The basic material in Rowland's plastic for optical polishing is polyurethane, foamed to give a microcellular structure suitable for carrying the polishing slurry. The polishing material contains zirconium oxide, which gives the best results, although cerium oxides and iron oxides have been tried with varying degrees of success.

After a polisher has been made with plastic material, it must never be allowed to dry out or it will crack. Tools not in use must therefore be stored under water.

8.10 Polishing machines for large mirrors

Large optical components, such as those required for astronomical telescopes, require a different type of polishing movement because of the need to generate aspheric curves. (Fig. 8.10.1.) On polishing machines of this type, the block or surface is rotated steadily on a vertical axis, while the tool is given cross-movements by one or two cranks. The tool is able to revolve freely on its own axis. By varying the eccentricities and relative speeds of the cranks, a combination

8.10.2. Strasbaugh Model 6AU 90-inch Draper polisher.
(*Top*) **The working platform in position.**
(*Centre*) **The table-tilting mechanism.**
(*Bottom*) **The table tilted into the testing position.**

of cross-movements can be arranged, which will produce circular, elliptical, figure of eight, and other movements as required. (See also Chapter 15.)

The Strasbaugh polishing machine Model 6AU (Fig. 8.10.2) has a work-table of 90-inch diameter and can be used for grinding edges and bevels or for generating curves. This Draper-type machine (which is one of a series of sizes) has a table with heavy-duty bearings, and the speed can be infinitely varied while the machine is in operation. A special powered tilting mechanism is provided to tilt the table 90° from its rotating axis for the purpose of checking the work without removing it from the table. (See § 11.18.)

The eccentric adjustment is normally made by power from the control station and can be adjusted while the machine is stopped or in motion. The speed of the eccentric can be infinitely varied while the machine is in operation.

8.11 Dry Rexine polishing

Since around 1945, the Kershaw optical factory has successfully polished prisms and mirrors with dry polishing powder impregnated in the back of Rexine cloth. (Fig. 8.11.1.) Although the polishing time is long (about 10 hours), the machines operate night and day without attention except for changing the blocks of work. The polishing tool is a circular flat plate of about 18-inch diameter to which the shiny side of the Rexine is stuck with shellac. Jeweller's rouge is loaded into the cloth back of the Rexine. The polishing machines have a 6-inch reciprocating movement for the polishing tool at 150 strokes per minute. The 18-inch diameter block of prisms revolves at about 2 r.p.m. and the polisher moves slowly from side to side across the block by a link and eccentric movement. The figure produced by the polisher is of good shape and usually two rings flatness is achieved. This is a most unusual method of polishing which has produced large quantities of flat work over many years.

The disadvantage is the long polishing time and therefore large amount of floor space needed, but in some circumstances, the low direct labour cost of running the machines would offset the higher overhead cost.

The surface finish produced by this method is satisfactory for most requirements, but does not reach the best quality obtainable from pitch polishing.

8.11.1. Kershaw dry polishing machine for flat work. The polisher is made from the reverse side of Rexine cloth loaded with rouge powder.

9 Methods Planning, Estimating and Production Control

9.1 Introduction

The outstanding feature of the optical industry is that lenses, prisms, mirrors, and graticules are required in a very wide range of sizes, shapes, quality standards, and quantities.

The large-quantity products are to be found in cameras, slide projectors, microscopes, and spectacles, whereas the small-quantity products are mostly in instruments where quality standards are usually more severe.

Different methods, tooling, and machinery are necessary for the very great quantity variations and, in any discussion on methods or estimated times, it is essential to know the quantity and quality standards to a greater extent than for equivalent mechanical components.

Large-quantity production is generally in the hands of the manufacturers of finished products, whereas, because of the specialized nature of the business, there are a large number of small companies with relatively old machinery producing good-quality components by relying on the skills of their operators and low overheads to compete with larger companies.

Optical factories which are part of large companies have to follow common administrative systems, and the methods in this chapter apply to any factory which must have knowledge of estimated costs, shop loading, departmental operating statements, and standard costing.

The method of payment of operators is very controversial because (except for large-quantity work) it is normal practice for an operator to have several jobs working at the same time on different spindles or possibly (in the case of prisms) in the same block.

Alternatively any one job may be split between several operators who may each be carrying out part of the job.

Most operations are likely to have a proportion of rejects at some time but these can usually be corrected by reworking. Polishing is the most vulnerable operation, but faults can usually be corrected if the lenses are blocked up again and repolished. With small-quantity lens-work, the next block may be in several months' time, so the storage of rejects is essential.

The number of lenses in a block varies according to the radius of curvature, and an operator will produce a varying quantity of surfaces a day, according to the type of work in progress. It is easy to be out of balance in relation to curve generation, smoothing, and polishing, and therefore not have optimum loading, and so it must be assumed that there will be idle spindles.

The manual apportioning of time by operators is therefore inevitable and this leads to errors in the booking of time to jobs and faulty job costing.

If accurate booking of time to jobs is impossible and the judgment of operators essential, then a satisfactory system of individual payment by results is not feasible and the only way to pay a bonus is by some form of group scheme, based either on the output of a section or on the total performance of the department.

If estimated floor-to-floor times are stated on job cards, then these times can be used to measure

9.1.1. Graph to illustrate the reduction in worktime that follows increasing quantities and reduced figure quality.

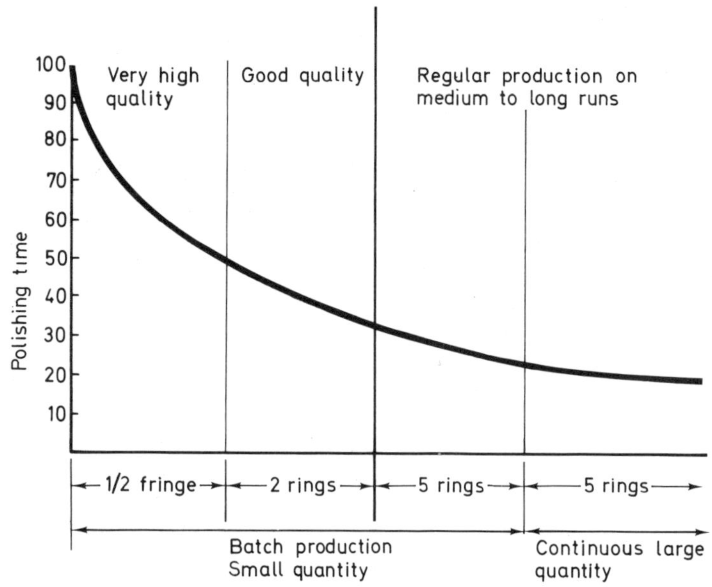

the section output and can be compared with actual times booked for the purpose of standard costing and excess costs.

However, the problem of agreeing the accuracy of floor-to-floor times will always be present with small-quantity production and made difficult by the continuously changing pattern of work.

Because of these difficulties, there is much to be said for a merit rating scheme, work study to achieve the best methods, good supervision, and no payment of a bonus based on output.

The methods described in this chapter apply to the small-quantity production of an optical department in the instrument industry. For this purpose, we shall assume batch quantities of fewer than 1000 and many fewer than 100 a year. A high quality of surface finish is required, and a figure accurate to two rings is normal. Experience shows that recess tooling, diamond smoothing followed by high spindle-speed and high-pressure polishing, together with hard pitch and continuous-feed slurry, will produce good results in reduced times. However, the best use of these techniques is obtained with large quantities and where the tolerance on figure is permitted to a maximum of five rings.

The graph (Fig. 9.1.1) indicates the way in which polishing times fall with the combination of reduced surface accuracy and large-quantity production. As the total time of the smoothing and polishing section is more than 50 per cent of the whole optical department, it can easily be seen how reduced optical costs and increased output can follow a reduction of tolerances.

9.2 Lens time standards

Large polishing machines (6 spindles)

The capacity of large six-spindle polishing machines can be taken to be four spindles for estimating purposes.

355-mm (14-inch) diameter lens block,
3 hours' spindle time.
254-mm (10-inch) diameter lens block,
2 hours' spindle time.
Smoothing time, 20 minutes.

Small polishing machines (10 spindles)

For radii up to 75 mm (3 inch). According to size of block,
15 to 80 minutes' spindle time.

Smoothing time, 3 to 10 minutes.

Note. Maintenance of polishing tools and making of blocks by operators is a time included in the above estimated times.

Smoothing and polishing

The time for blocks of lenses varies according to size of block and average times are as follows:

Block diameter	*Time in minutes per face*
Up to 1 inch	60
Up to 3 inches	80
Up to 6 inches	120
Up to 9 inches	180

When calculating the time each, the number of spindles operated has to be taken into account.

Normally one operator can run four spindles. The time each therefore in minutes is

$$\frac{\text{Time in minutes per block}}{\text{No. in block} \times 4}$$

Example 1. Smooth and polish 114 lenses on a 5-inch block, nineteen lenses per block.

$$\text{Time each} = \frac{120}{19 \times 4} = 1{\cdot}6 \text{ minutes}$$

Example 2. Smooth and polish twenty-four lenses on 2-inch blocks, three lenses per block.

$$\text{Time each} = \frac{80}{3 \times 4} = 7{\cdot}5 \text{ minutes}$$

Example 3. Smooth and polish 120 lenses on a 9-inch block, 120 lenses per block.

$$\text{Time each} = \frac{180}{120 \times 4} = 0{\cdot}4 \text{ minutes}$$

Example 4. Smooth and polish 200 prisms on a block.

$$\text{Time each} = \frac{150}{200 \times 4} = 0{\cdot}2 \text{ minutes}$$

Edging

Estimated times for lens-edging to the normal accuracy of $\pm 0{\cdot}025$ mm (0·001 inch) are as follows.

Lenses under 19 mm (0·75 inch) diameter, 1 to 2 minutes each.

Lenses over 19 mm (0·75 inch) diameter, 4·5 minutes.

Usually 2·5 mm (0·1 inch) glass on diameter should be allowed for removal during edging.

Cleaning

The estimated time for removing pitch and varnish is 0·75 minutes.

Inspection time

As percentages of the total operation time, inspection times may be expected to be as follows.

	Percentage
Prisms	5
Lenses	10
Circles	20
Graticules	100

Scrap allowances and losses

Batch quantities should be made large enough to include losses in production through various causes. The percentage additional quantities that should be allowed are as follows.

	Percentage
Lenses	10
Field lenses	25
Graticules	100

Raw glass allowance

To clean up bad glass on mouldings allow additional 2 mm on each side of moulding thickness.

Also allow 2–3 mm on diameter to provide for off-axis and inaccurate centring.

Number of lenses per block

Although the number of lenses to be included in a block is the maximum which can be accommodated without their touching one another, and can be determined by trial, it is necessary for estimating purposes to have a ready-reckoner for easily deciding the maximum quantity which can be contained in a block.

Lens blocks can be made up either with one or three lenses in the centre, and the optimum depends on the radius of curvature and lens diameter, assuming standardized sizes of tools.

The chart (Fig. 9.2.1) indicates the best pattern and quantity of lenses for a maximum tool size of 10-inch diameter and 140° maximum angle.

9.2.1. The number of lenses accommodated on a block.

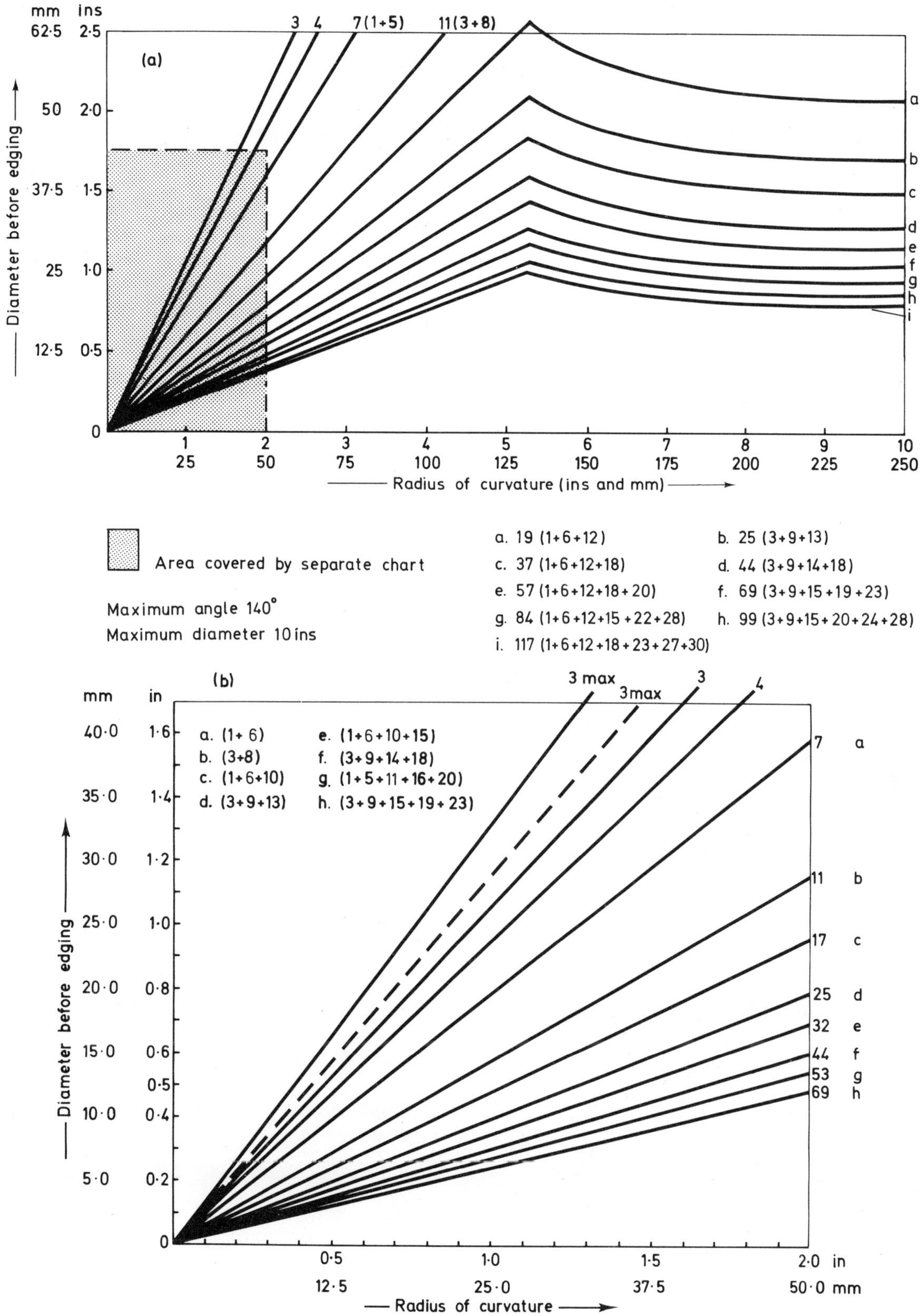

Curve generating

Times (in minutes each) for curve-generating lenses are as follows:

Size		*Load and unload*	*Grind*	*Total*
To 12·5 mm dia.	(0·5 inch)	0·4	0·6	1·0
25·0 mm	(1·0 inch)	0·4	0·8	1·2
37·5 mm	(1·5 inches)	0·4	1·0	1·4
50·0 mm	(2·0 inches)	0·5	1·3	1·8
62·5 mm	(2·5 inches)	0·5	1·6	2·1
75·0 mm	(3·0 inches)	0·5	1·9	2·4
100·0 mm	(4·0 inches)	0·5	2·4	2·9

Chamfering

Chamfering is the operation of removing the sharp corners of lenses left after grinding and polishing. The operation is performed on a rotating concave tool using a slurry of emery and water. Great care must be taken to prevent scratches on the polished surfaces due to particles of grit between the fingers and the surface. (Fig. 6.10.8)

The times taken by chamfering various sizes of lens are as follows:

Size		*Minutes per chamfer*
To 12·5 mm dia.	(0·5 inch)	1·0
25·0 mm	(1·0 inch)	1·5
50·0 mm	(2·0 inches)	1·5
75·0 mm	(3·0 inches)	2·0
100·0 mm	(4·0 inches)	2·0
125·0 mm	(5·0 inches)	3·0

9.3 Standards of practice for prisms

Right-angled prisms are usually conventional in shape, but there are a number of possible variations without departing from the basic form (Fig. 9.3.1.) Prisms are usually manufactured from a slab of plate glass. By slitting and grinding, a number of prisms can be produced from a square or rectangle of glass, first stacking a quantity of squares together. The number of prisms from any square is normally two, four, eight, or sixteen. The method of slitting is shown in Fig. 9.3.2.

For convenience of handling the side of the square should not exceed 2 inches. Thus the quantity per square and therefore the method is determined by this size. Also the length of the stack should not exceed 3 inches.

The basic steps, after the raw glass is reduced to the required thickness, are as follows:

(1) Slit plate or slab into rectangles.
(2) Stack a number together.
(3) True one corner as shown in Fig. 9.3.3.
(4) Grind opposite sides parallel.
(5) Slit as required.
(6) Jig and grind angles.
(7) Smooth and polish, bloom and silver, copper, shellac, and black.

When making very small prisms (0·3 inch or less) a different method has to be used. This is in order to provide a length of face sufficient for collimation when squaring angles.

9.3.1. Right-angled prisms.

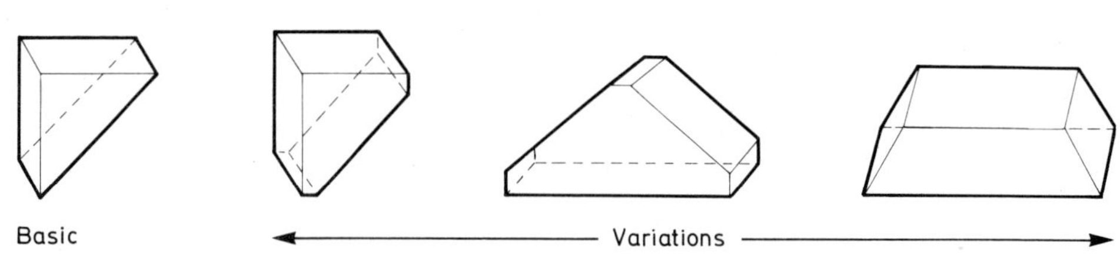

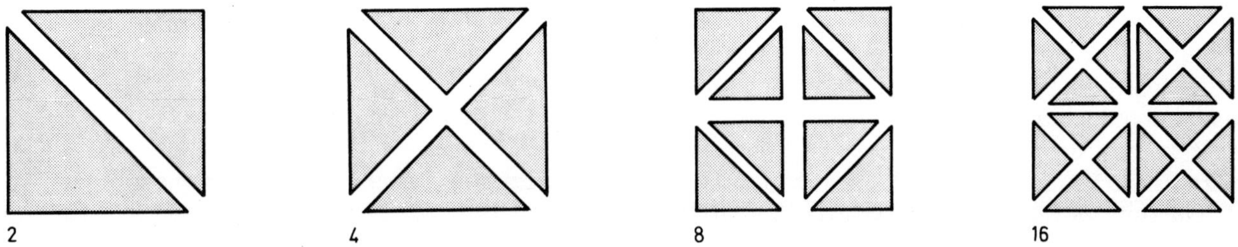

△ **9.3.2. Slitting prisms from a square slab of glass.**

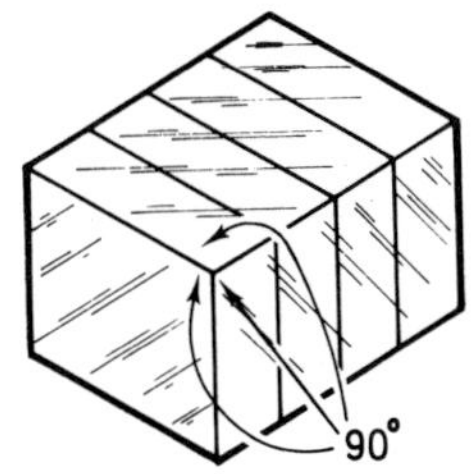

9.3.3. ▷ **Truing one corner.**

The method is as follows:

(1) Slit into squares.
(2) Grind corner 90° and grind parallel (produces the hypotenuse).
(3) Slit diagonally.
(4) Block in plaster; smooth and polish hypotenuse.
(5) Stick on waster with tissue (Fig. 9.3.4).
(6) Grind faces.
(7) Smooth and polish, etc.

It is not usual to make sixteen prisms from a square and the above method is preferred. The most common method is two or eight from a square, and this method should be used wherever possible.

When calculating the size of the raw material required and the subsequent slitting size, allowance must be made for grinding after slitting.

The standard amount of glass to leave for grinding is 0·09 inch per surface.

The polishing allowance for finishing is 0·003–0·004 inch per surface.

From this, the following expressions are derived for the side of the raw-material square, where L is the length of the 90° side of the prism.

Right-angled prism making two from each square:

$$L + 0{\cdot}56 \text{ inches}$$

Right-angled prisms making four from each square:

$$(L \times 1{\cdot}42) + 0{\cdot}76 \text{ inches}$$

Right-angled prisms making eight from each square:

$$(2 \times L) + 1{\cdot}04 \text{ inches}$$

Right-angled prisms making sixteen from each square:

$$(L \times 2{\cdot}84) + 1{\cdot}44 \text{ inches}$$

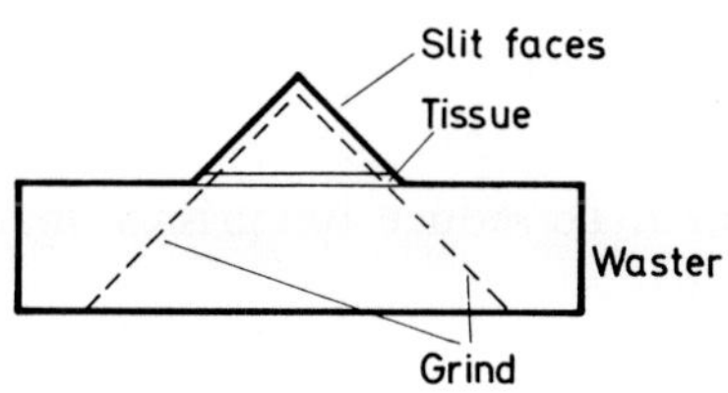

9.3.4. Sticking to a waster with tissue.

Facing one side

Facing one side is a hand process carried out in order to produce a flat face for locating on a plano tool and reducing to thickness.

It is only used when the surface of the stock plate is too rough to seat truly on the plano.

The time for the element varies according to the condition of the surface before facing.

The average time required to face by hand is as follows:

Area	*Time in minutes*
Up to 1 sq. inch (6·5 sq. cm)	0·5
1 to 3 sq. inches (19·4 sq. cm)	0·7
3 to 9 sq. inches (58·0 sq. cm)	1·5
9 to 25 sq. inches (161·3 sq. cm)	2·5
25 to 100 sq. inches (645·2 sq. cm)	3·5

Therefore the time is:

$$\frac{\text{Time in minutes (according to area)}}{N}$$

where N is the number of parts which can be made from the plate.

Example. To face one side of a plate 4 × 4 inches (100 mm × 100 mm) to produce sixteen prisms.

$$\text{Time} = \frac{2{\cdot}5}{16} = 0{\cdot}16 \text{ minute each}$$

Assembling on plano

Before grinding or reducing to thickness, one of the following processes must be carried out:

(*a*) sticking raw glass plate or squares to a plano, or

(*b*) sticking individual prisms or stacks of prisms to a plano.

A special solution of beeswax is used, or blocks of steel for holding on the magnetic table.

The average time in minutes is 0·5 per plate or piece, or stack, plus a preparation time averaging 10 minutes.

The time each is therefore:

$$\frac{\text{Preparation time} + (\text{No. of pieces loaded} \times 0{\cdot}5)}{\text{Total number of parts}}$$

Example 1. Assemble twelve stacks to plano; each stack makes 20 prisms.

$$\text{Time each} = \frac{10 + (12 \times 0{\cdot}5)}{240} = 0{\cdot}7 \text{ minute each}$$

Example 2. Assemble thirty prisms to plano.

$$\text{Time each} = \frac{10 + (30 \times 0{\cdot}5)}{30} = \frac{5}{6} = 0{\cdot}8 \text{ minute}$$

Reducing

Reducing is a grinding operation usually carried out on the Blanchard machine. The time is dependent on the amount of material to be removed. In order to calculate the time as an average, it will be assumed that an average amount of material will always be removed. The time for reducing is calculated as follows.

To reduce an area equivalent to a full plano load takes 30 minutes, i.e. a traverse area of 10 inches × 10 inches (25·4 cm × 25·4 cm) or 100 sq. inches (645·1 sq. cm).

The time is therefore calculated as follows.

Up to	25 sq. in. (161 sq. cm)	10 minutes
,, ,,	75 ,, ,, (483 ,, ,,)	20 ,,
,, ,,	100 ,, ,, (644 ,, ,,)	30 ,,

In the case of reducing raw glass to thickness:

$$\text{Time each} = \frac{\text{Time in minutes (according to area)}}{N}$$

where N is the number of parts produced from the plate.

In the case of individual parts:

$$\text{Time each} = \frac{\text{Time in minutes (according to area)}}{N_a}$$

where N_a is the number of parts in the area.

Example 1. To reduce a plate 7 inches × 9 inches (17·5 cm × 22·5 cm) which will make sixty prisms.

$$\text{Time each} = \frac{20}{60} = 0{\cdot}33 \text{ minute}$$

Example 2. To reduce 100 prisms (making a full plano load).

$$\text{Time each} = \frac{30}{100} = 0{\cdot}3 \text{ minute}$$

Slitting

The time required for slitting consists of a time for setting the slitter to the required dimension, plus the actual slitting time.

The average time for setting is 3 minutes.

The average cutting speed is 5 inches (12·5 cm) per minute, or 1 inch (2·5 cm) in 0·2 minute.

Therefore the total time in minutes for slitting complete is:

Setting + slitting

$$= \frac{3}{B} + \frac{\begin{bmatrix}\text{length of cut} + 0{\cdot}3 \text{ inch} \\ (7{\cdot}6 \text{ mm) overrun}\end{bmatrix}}{N} \times 0{\cdot}2 \text{ minute}$$

where B is the total batch quantity, and N the quantity of parts from each piece slit.

Example 1. To slit a plate into squares 1·3 inches × 1·3 inches (33 mm × 33 mm) where each square makes four prisms and the total batch is forty.

Time each

$$= \frac{3}{40} + \frac{(2{\cdot}6 + 0{\cdot}6 \text{ inches}) \times 0{\cdot}2}{4}$$

$$= 0{\cdot}075 + 0{\cdot}16 = 0{\cdot}235 \text{ or } 0{\cdot}3 \text{ minute}$$

Example 2. To slit a stack of squares (5) 2 inches (5 cm) long, where each square makes four prisms; total batch quantity is twenty.

$$\text{Time each} = \frac{3}{20} + \frac{2{\cdot}3 \times 0{\cdot}2}{20} = 0{\cdot}15 + 0{\cdot}23$$

$$= 0{\cdot}38 \text{ or } 0{\cdot}4 \text{ minute}$$

Stacking

Stacking is the operation of sticking a number of parts together to form a stack. Circular, rectangular, or square plates can be made by this method.

In the case of circular parts, a solid cylinder is formed, enabling the diameter to be ground at a faster rate than could be done singly.

Square or rectangular parts are mainly used for making prisms and, by stacking, a rectangular block is formed for grinding (Fig. 9.3.5.)

In many cases square plates may be stacked and edged to a diameter quicker and more cheaply than by disk cutting.

The adhesive used is as follows:

Prism blanks: $\frac{1}{4}$ lb flaked shellac, 2 teaspoons oil of cassia.

Edge stacking: 12 oz resin, 2 oz flaked resin, 1 oz beeswax.

The time taken is based on area and is as follows:

Size	*Time in minutes*
Up to 2·0 inches (50 mm)	1·0
Up to 3·0 inches (75 mm)	1·3
Up to 4·0 inches (100 mm)	1·5
Up to 5·0 inches (125 mm)	2·0

Example. Stack five blanks 1·6 inches × 1·6 inches (40 mm × 40 mm), each blank making four prisms.

$$\text{Time each} = \frac{1{\cdot}0}{4} = 0{\cdot}25 \text{ minute}$$

Truing the corner

Truing the corner is a hand operation carried out in order to produce two faces of a blank or a stack of blanks flat and square to each other prior to grinding. It involves the use of a truing spindle and a collimator (Angle Dekkor, § 11.19).

The average time required is 6 minutes. The time each is therefore $6/N$, where N is the number of parts in the stack.

Example 1. A stack of five squares each making four prisms.

$$\text{Time each} = \frac{6}{5 \times 4} = 0{\cdot}3 \text{ minute}$$

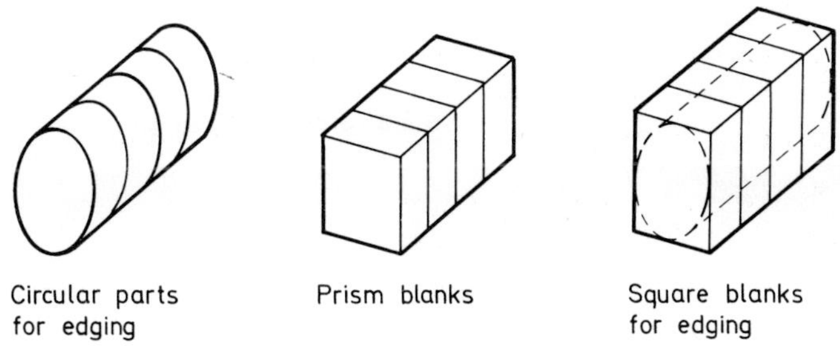

9.3.5. Stacking square plates for edging.

Example 2. A single block of glass for one pentagonal prism.

$$\text{Time each} = \frac{6}{1} = 6 \text{ minutes}$$

Grinding parallel

Grinding parallel is an operation following on from the operation 'true corner'. It produces a square or a rectangle on a stack of prism blanks.

The stack is mounted on a plano tool locating on one of the faces previously ground to a true 90° angle, and the opposite side is ground, removing approximately 0·09 inch. The operation is repeated on the remaining unground face.

The time is calculated by the area; a full plano or 100 sq. inches (645·1 sq. cm) takes 15 minutes. The time is therefore based on the following:

Size	*Time in minutes*
Up to 25 sq. inches (161·3 sq. cm)	6
Up to 75 sq. inches (483·9 sq. cm)	10
Up to 100 sq. inches (645·2 sq. cm)	15

Therefore:

$$\text{Time each} = \frac{\text{Time in minutes (according to area)}}{N}$$

where N is the number of parts produced.

Example. Ten stacks each of five blanks occupy a total area on the plano of 85 sq. inches (548·4 sq. cm). Each square blank makes four prisms.

$$\text{Time each} = \frac{15}{200} = 0{\cdot}08 \text{ minute per face}$$

Jigging on angle

'Jig on angle' is the operation of locating stacks of prism blanks or individual prisms in order to grind the facets of the prisms. The jigs are almost invariably glass blocks which are fixed to a plano surface. The blocks have an accurate angle to which the stack is fixed using jig wax, which is made up of 1 part resin and 1 part beeswax. (Fig. 9.3.6.)

The angle of the jig face determines the angle of the ground face.

The average time required to stick a single glass or a stack to the jig is 1·5 minute, including preparation. Therefore:

$$\text{Time each} = \frac{1{\cdot}5}{\text{Number of prisms produced}}$$

Example. Jig a stack of five blanks, each blank making two prisms.

$$\text{Time each} = \frac{1{\cdot}5}{10} = 0{\cdot}15 \text{ minute}$$

Jigging and grinding to height

'Jig and grind to height', or 'grind angle', is the operation of grinding after sticking on an angle jig to produce the angle of a prism. (Fig. 9.3.7.)

The time required varies with the amount of stock to be removed at an average removal rate of 0·008 inches (0·2 mm) per minute. The time is therefore calculated on the number of individual prisms loaded at any one time, as follows:

$$\text{Time each} = \frac{\text{Total stock removed}}{0{\cdot}008 \times N}$$

where N is the number of prisms in a load.

Example. To grind 0·10 inches (2·5 mm) from a load of eighty prisms.

$$\text{Time each} = \frac{0{\cdot}10}{0{\cdot}008 \times 80} = \frac{0{\cdot}10}{0{\cdot}64} = 0{\cdot}15 \text{ minute}$$

Additional allowance must be given when the

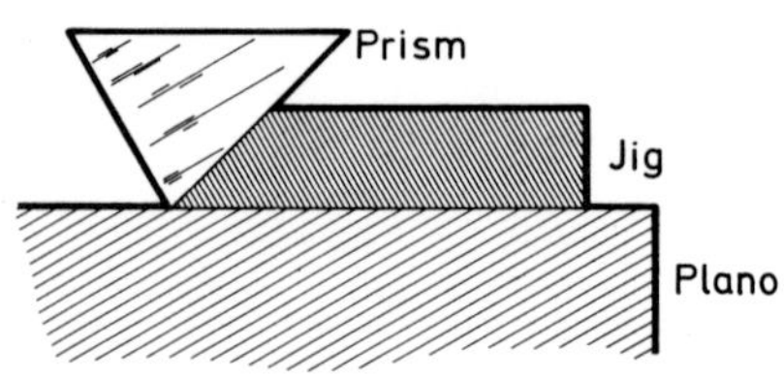

9.3.6. Jigging on angle.

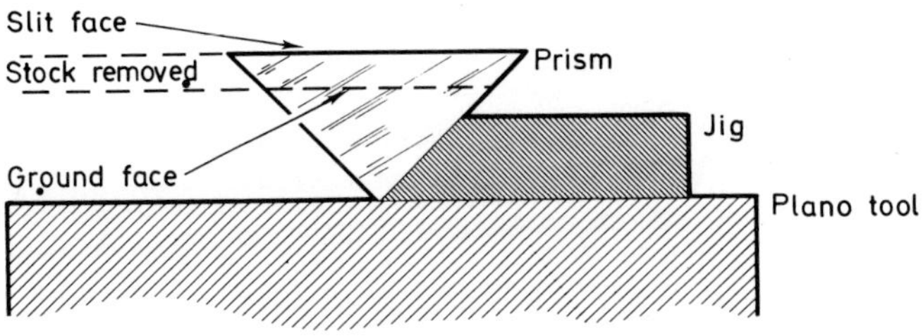

9.3.7. Grinding to height.

parts are fragile or subject to chipping, damage, etc. This can be decided only by experience.

Smoothing and buffing

'Smooth and buff' is an operation carried out on rough prisms, or on one face of a stack of prism blanks, to provide a surface suitable for collimation when squaring faces or grinding angles. On single prisms, after grinding one face on a plano tool, the whole surface is smoothed and buffed (i.e. partially polished). From this face all the remaining angles are collimated, ground, and polished.

On a stack of prism blanks it is only necessary to smooth and buff one face in each stack, and in this case the quantity of blanks smoothed and buffed is equal to the number of stacks, except where the minimum quantity of blanks to be smoothed and buffed is less than a full load, in which case a full load is done. This means that in all cases a full load or a multiple of a full load is smoothed and buffed.

Example 1. If 200 prisms are required and one blank makes four prisms, and there are five blanks to a stack, there will be ten stacks and thus only ten blanks need to be smoothed and buffed.

On very large quantities, the number of blanks to be smoothed and buffed might be more than a full load, and thus the operation would have to be carried out more than once.

The average time to smooth and buff one full load is 60 minutes. Therefore:

$$\text{Time each} = \frac{60 \times \text{number of loads}}{\text{Total number of prisms produced}}$$

In making 200 prisms, four per blank in stacks of five blanks, the minimum number of blanks to be smoothed and buffed is ten. This can be done in one load. Therefore:

$$\text{Time each} = \frac{60 \times 1}{200} = 0{\cdot}3 \text{ minute}$$

Example 2. Five hundred prisms are required, and one blank makes two prisms and is 2 inches × 2 inches (50 mm × 50 mm).

If each stack contains five blanks, then fifty stacks are required and fifty blanks need to be smoothed and buffed.

It would only be possible to load ten 2-inch × 2-inch (50-mm × 50-mm) blanks on one plano tool, and thus five loads require to be smoothed and buffed.

The smooth and buff time would be multiplied by five before dividing by 500.

In making 500 prisms, two per blank in stacks of five blanks, the minimum number of blanks to be smoothed and buffed is fifty.

Maximum number of blanks per load = 10

Number of loads = 5

$$\text{Time each} = \frac{60 \times 5}{500} = \frac{300}{500} = 0{\cdot}6 \text{ minute}$$

Example 3. To make sixty prisms, four per blank in stacks of five.

Minimum number of blanks to be smoothed and buffed is three.

This is less than a full load, so a full load would be done.

$$\text{Time each} = \frac{60 \times 1}{60} = 1{\cdot}0 \text{ minute}$$

Blocking in plaster

'Block in plaster' is the operation of placing and fixing a quantity of glasses on a tool in order to smooth and polish the faces. Prisms are located on

9.3.8. Filling the block with wasters.

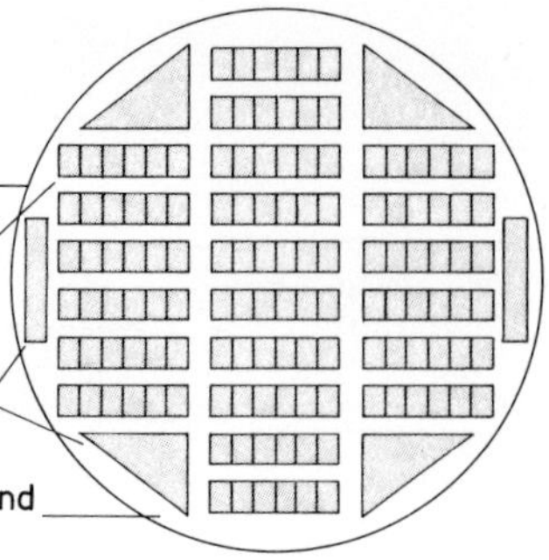

a flat tool, resting on the face to be polished. A retaining wall is placed over the glass and a film of wax is poured in to cover the face of the flat tool. A mixture of plaster is made up and poured in to fill the remaining space. When the plaster has set the holder is inverted and the wax is smoothed down to leave the prism faces standing proud for polishing. (See § 8.5).

In order to provide a good area for polishing, wasters are used to fill any areas not occupied by the prisms. These are pieces of scrap glass or prisms which are kept for this purpose. (Fig. 9.3.8.) Where the quantity of prisms is not sufficient to fill a complete block it is possible to put more than one type of prism on the same tool. This should not occur if production control load batches which are multiples of block quantities.

The time taken to block in plaster is calculated from the time for mixing, pouring, and smoothing the wax down, plus the time for locating each prism or stack of prisms, plus an additional allowance for wasters. The time for preparing the plaster and wax and pouring is 15 minutes on average. The time for locating each part or stack of parts is 1 minute. Thus the time each for single prisms is:

$$\left(\frac{\text{Prepare plaster}}{\text{No. of parts}}+1\right)+25 \text{ per cent for wasters}$$

The time each for stacked prisms is:

$$\frac{\text{Prepare plaster}+\text{No. of stacks}}{\text{No. of parts}}+10 \text{ per cent}$$

Example 1. Block up eighty prisms in twenty stacks of four per stack.

$$\text{Time each} = \frac{15+20}{80}+10 \text{ per cent}$$

$$= \frac{35}{80}+10 \text{ per cent} = 0{\cdot}45 \text{ minute}$$

Example 2. Block up 120 prisms in twenty stacks of six per stack.

$$\text{Time each} = \frac{15+20}{120}+10 \text{ per cent}$$

$$= \frac{35}{120}+10 \text{ per cent} = 0{\cdot}33 \text{ minute}$$

Example 3. Block up twenty-four prisms in stacks of two per stack.

$$\text{Time each} = \frac{15+12}{24}+10 \text{ per cent}$$

$$= \frac{27}{24}+10 \text{ per cent} = 1{\cdot}2 \text{ minute}$$

Example 4. Block up twenty prisms individually.

$$\text{Time each} = \left(\frac{15}{20}+1\right)+25 \text{ per cent}$$

$$= 1{\cdot}75+25 \text{ per cent} = 2{\cdot}2 \text{ minutes}$$

An average time for loading purposes can be taken from the table below.

Quantity per block	*Time in minutes* *Stacked prisms*	*Single prisms*
Up to 10	0·8	3·5
Up to 20	0·8	2·8
Up to 30	0·8	2·3
Up to 40	0·8	1·8
Up to 50	0·8	1·3
Up to 100	0·4	—
Up to 200	0·3	—
Up to 300	0·2	—

Smoothing and polishing

Smoothing and polishing times are extremely variable according to the quality of finish and tolerances required on the finished lens or prism. A larger variation occurs on lens polishing owing to greater problems of 'figuring' or producing the required curve on the glass. It is therefore only possible to base times on averages.

Smoothing and polishing a block or prisms or any flat parts on conventional machines (i.e. other than hand work or Lapmaster) takes an average of $2\frac{1}{2}$ hours or 150 minutes. This time includes the normal preparation of tools, etc.

Hand chamfering

'Hand chamfer' is the operation of removing the sharp corners of the prisms or stacks of prisms after polishing. The operation is performed on a flat, rotating plate using a slurry of emery and water. Great care must be taken to prevent scratches on the polished surface due to grit on the fingers. On stacked prisms, chamfering is done on all corners of the stack, which is then dismantled and the remaining corners chamfered separately.

When calculating the (total) time each, each chamfering operation is taken separately and the dismantling time is added. The time each is calculated as follows:

Length of chamfer	*Time in minutes per chamfer*
Up to 1·0 inch	0·2
Up to 2·0 inches	0·4
Up to 3·0 inches	0·5
Up to 4·0 inches	0·6
Up to 5·0 inches	0·8

Time each for dismantling is 0·7 minute average. The total time each is therefore:

$$\frac{\text{Time per chamfer} \times \text{No. of chamfers}}{\text{Number of prisms in stack}} + 0{\cdot}7 + \text{time per separate chamfer}$$

Example 1. Chamfer, dismantle, and chamfer right-angle prisms 0·5 inch, five per stack $2\frac{1}{2}$ inches long.

$$\text{Time each} = \frac{3 \times 0{\cdot}5}{5} + 0{\cdot}7 + 5 \times 0{\cdot}2 = 0{\cdot}3 \text{ minute}$$

Example 2. Chamfer 1·25-inch pentagonal prisms singly.

Number of corners to be chamfered is 15.

$$\text{Time each} = 15 \times 0{\cdot}4 = 6{\cdot}0 \text{ minutes}$$

Estimate and load sheet

Table 9.3.1 shows how an estimate and load sheet is drawn up, for the prism of Fig. 9.3.9.

9.3.9. Working drawing for a prism.

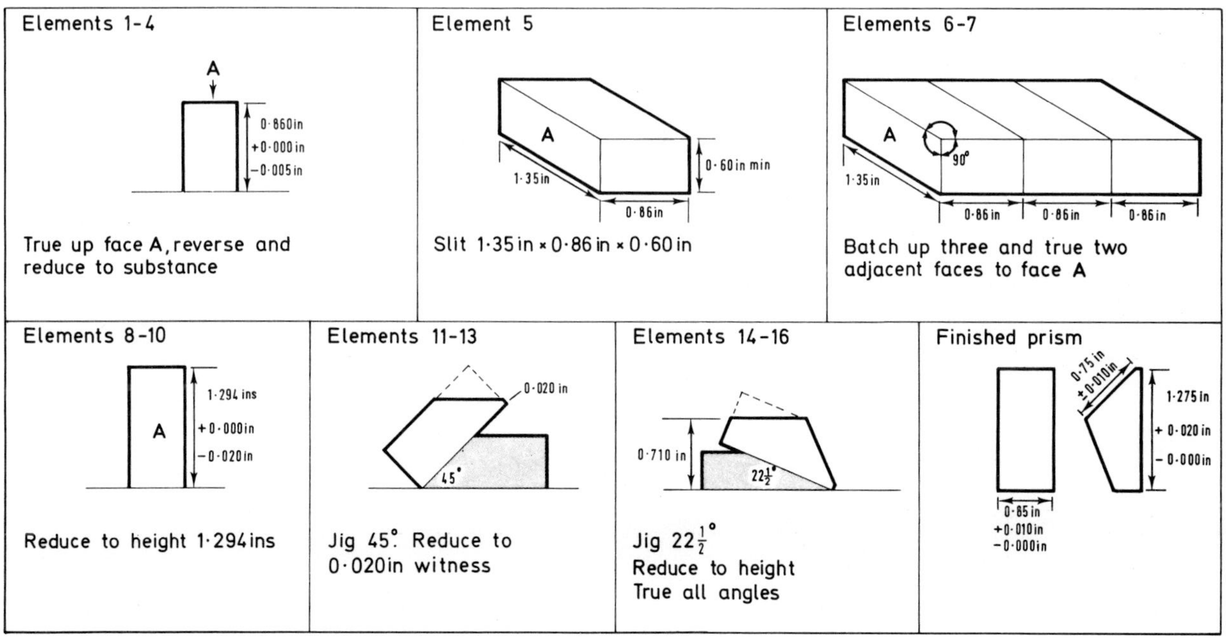

Table 9.3.1. Estimate and load sheet for the prism shown in Fig. 9.3.9

	Estimate and load sheet						Description of prism				Part no. 1234	
	Planned by	Date			Batch		Material		Specn.		Cost	
					30		1·45″ × 0·7″ × 1·0″		HC 519604		20 P	
	Process element	Block quantity	Machine type	General labour	Grinder	Slitter	Polisher	Truing	Coating			
1	Face one side		1		1·0							
2	Assemble on plano	30	0	0·5								
3	Reduce 0·855″		1		1·5							
4	Dismantle and clean		0	1·0								
5	Slit to 1·35″ × 0·6″		1			4·0						
6	Batch up (3) × 10	30	0	0·5								
7	True up 2 adj. faces		1		2·5							
8	Assemble on plano		0	0·5								
9	Reduce to height		1		2·5							
10	Dismantle and clean		0	1·0								
11	Jig 45°	30	0	0·5								
12	Reduce to 0·020″		1		3·0							
13	Dismantle and clean		0	1·0								
14	Jig 22½°	30	0	0·5								
15	Reduce to 0·710″		1		3·0							
16	Dismantle and clean		0	1·0								
17	True angles		1		4·0							
18	Collimate S1		3	3·0								
19	Block in plaster	30	0	2·0								
20	Smooth and polish		3				4·0					
21	Dismantle and clean		0	1·0								
22	Collimate S2		3	3·0								
23	Block in plaster	30	0	2·0								
24	Smooth and polish		3				4·0					
25	Dismantle and clean		0	1·0								
26	Collimate S3		3	3·0								
27	Block in plaster		0	2·0								
28	Smooth and polish		3				4·0					
29	Dismantle and clean		0	1·0								
30	Chamfer		3					1·0				
31	Dismantle		0	1·0								
32	Chamfer		3					2·0				
33	Clean and inspect		6	—								
34	Coat and inspect		6						10·0			
35	Silver and inspect		9	11·0								
	Total minutes 83·0			36·5	17·5	4·0	12·0	3·0	10·0			

9.4 Detail estimates and machine loading

In order to establish the total production time and machine capacity which would be committed to the manufacture of a component, all the process elements must be listed and estimated.

From the example (Table 9.3.1 and the working drawing in Fig. 9.3.9) of a simple prism, it will be seen that, including a contingency allowance, about $1\frac{1}{2}$ hours' labour time is needed for each prism.

However, the elemental labour time is not necessarily a true indication of machine load, because in some cases the machine will be working automatically without the attention of the operator and, in these cases, the machine commitment in terms of capacity is greater than the elemental labour time.

A guide to the factors which should be applied to the various machine element totals is as follows:

Grinder	1·5 × labour
Curve generator	2·0 × labour
Polishing spindles	3·0 × labour

This ratio of machine time to labour time will differ from one company to another and will possibly change with the mixture and size of batch quantities.

Labour load can be calculated by multiplication of programme quantities with the elemental labour totals plus a contingency allowance.

Machine load can be calculated as for labour load but, where applicable, the machine factor must be applied where more machine hours will be used than labour hours to avoid under-estimation of machine capacity required.

9.5 Production control and section loading

There are more than thirty distinct processes in an optical department. From the example in § 9.3, Table 9.3.1, of a simple prism, it will be noted there are thirty-five process elements. It is therefore essential for production control and section loading that a short list of not more than ten sections be formed where processes can be combined together and groups of operators be formed who can book their time (if necessary) on the same job. Such a grouping may be as follows:

Section 0 Cleaning.
1 Grinding, drilling, curve-generating.
2 Edging.
3 Flat polishing.
4 Spherical polishing of O.G.s.
5 Small-lens polishing.
6 Inspection.
7 Precision hand-polishing.
8 Cementing and blacking.
9 Silvering.

Operations such as coating and mounting may be carried out in separate departments. If this proposal is followed, the thirty-five process elements in the example of the prism are reduced to six operations, each in a different section, and this makes feasible the issue and control of factory forms.

The six operations in making the prism would be as follows:

		Time in minutes
1	Reduce, slit, true, grind and true angles (elements 1–17)	28
2	Collimate, smooth and polish, check angles (elements 18–29)	30
3	Chamfer, separate, chamfer (elements 30–32)	4
4	Clean and inspect (element 33)	0
5	Coat and inspect (element 34)	10
6	Silver and inspect (element 35)	11
		83

Operation 4, 'clean and inspect', is treated as an overhead with no specific time.

Note. The estimate and load sheet showing process elements and the working drawing should be available for the use of operators.

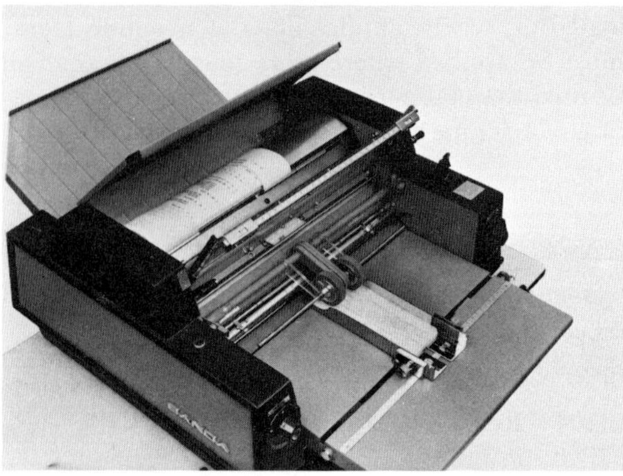

◁9.5.1. *(Top)* **Linamatic spirit duplicating machine for factory form production.**
(Bottom) **The master copy is shown clamped to the drum.**

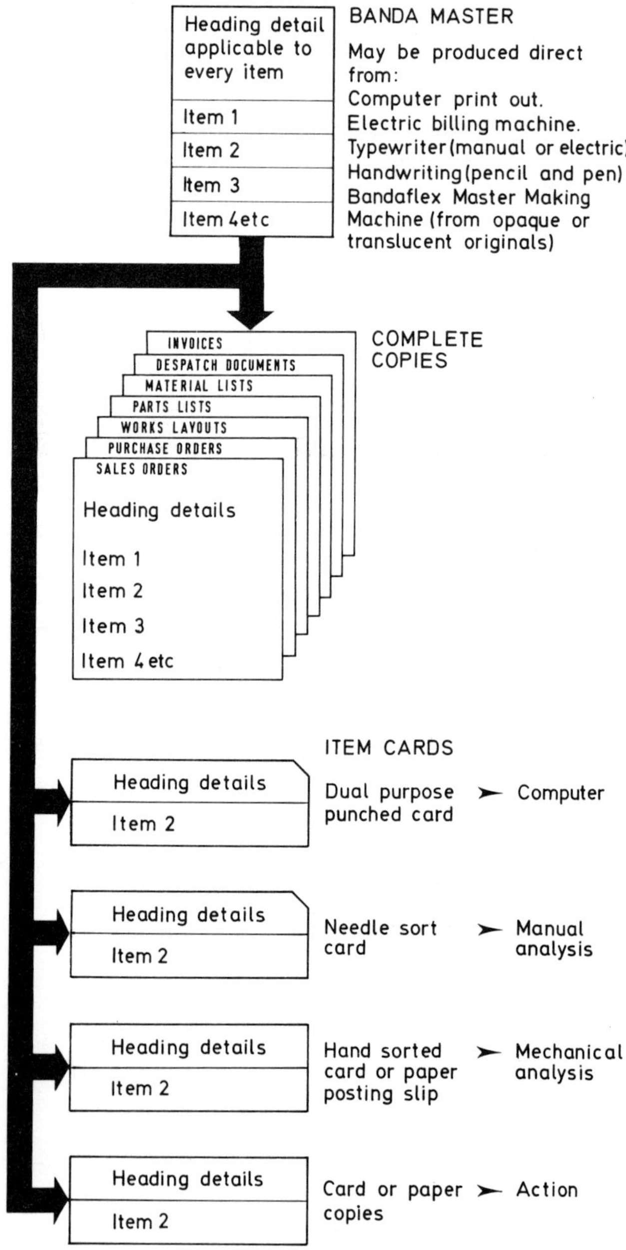

The method▷
The Banda Linamatic 1700 is a multi-purpose spirit duplicating machine incorporating the functions of a standard reproducing machine with developments in selective data reproduction. It will reproduce automatically individual lines or groups of lines or items from a master list. At the touch of a button, it spaces from one selection to the next and has finger-tip adjustment to vary the area selected. It offers a simple means of reproducing on office or factory routine forms selected data from one original without risk of errors in transcription.

The production of factory forms

Although many small optical factories use manually prepared job cards, most large factories use a spirit duplicator for factory form production, and the Block & Anderson Model Linamatic is typical of the type of machine used for this purpose. (Fig. 9.5.1.) In principle, for the production of forms a typed constant master with all the fixed information is used in combination with a variable master which carries all the information specific to the job, such as batch numbers, quantity, production order number, week due, etc.

The combined information from the constant and variable masters is used to print the labels, job cards, requisitions, route cards, and cost sheets, which are needed for the progress, control, and costing of the batch.

The forms illustrated in Fig. 9.5.2 were designed specifically for an optical department, but are compatible with the system required in any other department of a factory.

Form 1, Material Requisition

This contains the variable information in the top two lines, followed by the constant material specification, including weight per part, standard material cost and code number. Production Control must decide number off required, and enter manually the material details in the section marked 'Quantity Required'. The material requisition will be sent by Production Control to Stores as authority to batch up and issue material to the operator, who presents his Operator's Progress Record Form 4. The operator will sign the requisition as a receipt. The quantity issued will nearly always differ a little from that authorized, and this must be recorded so that stock records can be adjusted. The material requisition must then be sent to the Cost Department for calculation of excess material costs.

Form 2, Identification Label

This form contains the variable information in the top two lines and then a constant record of planned movement of work between sections of the department. A manual record of quantity passed per operation is necessary. The Identification Label always stays with the work. Additional labels must be prepared manually if a batch has to be split.

Form 3, Tool Requisition

For each operation where tools are needed, a Tool Requisition is required as authority for the Storekeeper to issue the tools and for him to have a record of which operator has borrowed the tools. The planner should ensure that on the constant master all tools are listed so that the Storekeeper can easily find the tools for the job.

Form 4, Operator's Progress Record

For every operation the Operator's Progress Record is his authority to do the job and to book time on a Time Allocation Card, Form 6. On the reverse side, the Inspector or Supervisor can record the quantity passed or rejected. This becomes a valuable record for the operator in case of queries at a later date.

When the work has been passed by the Inspector and Supervisor, the Operator's Progress Record must be sent to the Time Office so that quantity passed or rejected can be transferred by the time clerk to Form 5, Job Time Card.

Form 5, Job Time Card

This is a very important card used for calculation of a bonus (if a bonus system is being used), allocation of time to jobs, and costing. This card is retained in the Time Office during Work in Progress and is posted from the details of time on the Operator's Time Allocation card, and from the Operator's Progress Record. After the operation is completed, the Job Time Card must be passed to the Cost Department for calculation of excess labour costs and scrap.

Form 6, Operator's Time Allocation Card

As explained in § 9.1 of the present chapter, because of the type of work, it is essential that the system allow more than one operator to book on a job, or any job to be split between several operators. An Operator's Time Allocation Card is given to each operator each week and he is responsible for a fair apportionment of his time to each job on which he is working. If bonus earnings are not involved with this calculation, then reasonable accuracy is possible. However, the nature of the work is such that there will be substantial variations from one batch to another and it is important that these be recorded so that excess labour costs can be calculated.

The total time each day on the Card must equal the time on the operator's 'Gate Card' with provision for 'overhead' bookings, such as cleaning up the shop, where these are appropriate. After reconciliation, the Operator's Time Allocation Card can be used by the Accounts Department for payment of wages and for writing off non-productive work to the appropriate accounts.

Form 7, Component Route Card

This contains a complete record of the information on both Constant and Variable Masters and should be manually posted by the time clerk so that Supervision can tell at a glance the position of every job in the shop and be aware of rejections.

Forms 8, 9 and 10

These forms illustrate how two operators can be booked on to one job—possibly one man collimating and checking angles while the other is blocking-up, smoothing, and polishing.

Form 11, Inspection Report Ticket

In view of the large number of process elements it would be quite impracticable to have inspectors to check each stage (except in the case of large-quantity production), so the responsible Supervisor signs for the majority of operations. However, the planner will decide which operations must be passed by Inspectors, and in these cases an Inspection Report Ticket will be printed and on its reverse side the type of defect will be noted. These tickets must be passed to the Time Office for the recording of Job Time Cards and Component Route Cards and passed to the Accounts Department for the costing of scrap.

Form 12, Job Time Card (Final)

This card is identical with Forms 5 and 9, except that it is for the last operation. It is necessary for the Cost Department to know that it is the last operation on the job so that the job can be closed. For this reason it is overprinted with the word FINAL.

Form 13, Component Cost Sheet

This is identical with Form 7, Component Route Card, except that it provides for an extension on the form for operation standard costing. This is the basic information necessary for the costing of Scrap on the reverse side of the Inspection Report Ticket.

9.5.2. Forms designed for an optical department (above and on pages 255 to 260).

1

BIN No. 14		ALLOCATED		MATERIAL REQUISITION			STORES STOCK RECORDS COSTS
WEEK DUE TO START 18	FINISH 28	INSTRUMENT No. 184297	SUB ASSEMBLY No.	JOB CARD No. 1440	SPLIT 2	BATCH No. 3	PRODUCTION ORDER No. X4120 4
No. OFF 90 1	PART NAME Prism		DRG. ISSUE		FINISH		PART OR SYMBOL No. 1234 5
PART DESCRIPTION OR MATERIAL SPECIFICATION 1.45" x 1.0" x 0.7" HC.519604			SCPM OR X M	QTY. PER PART 42	UNIT OF MEASURE grams		PART OR CODE No. £0.08 AB 106 7
			QUANTITY REQUIRED 3780	QUANTITY ISSUED 3795			UNIT OF ISSUE GRAMS 6
ISSUED BY	DATE	RECORDED BY	DATE	RECEIVED BY J. ADAMS	DATE 19		ENTERED STOCK RECORDS

FORM S 12 MK. II

2

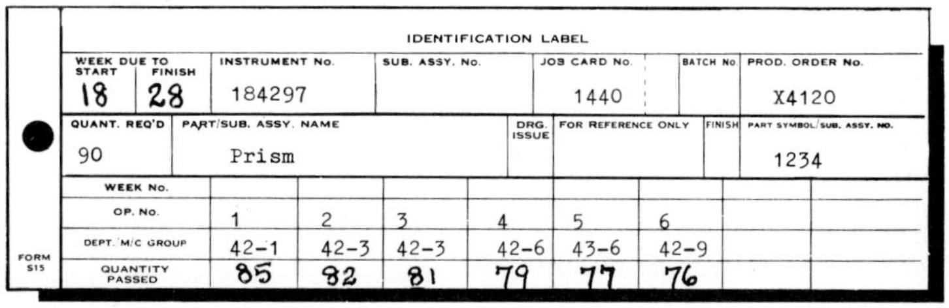

IDENTIFICATION LABEL

WEEK DUE TO START	FINISH	INSTRUMENT No.	SUB. ASSY. No.	JOB CARD No.	BATCH No.	PROD. ORDER No.
18	28	184297		1440		X4120

QUANT. REQ'D	PART/SUB. ASSY. NAME	DRG. ISSUE	FOR REFERENCE ONLY	FINISH	PART SYMBOL/SUB. ASSY. No.
90	Prism				1234

WEEK No.								
OP. No.	1	2	3	4	5	6		
DEPT. M/C GROUP	42-1	42-3	42-3	42-6	43-6	42-9		
QUANTITY PASSED	85	82	81	79	77	76		

FORM S15

3

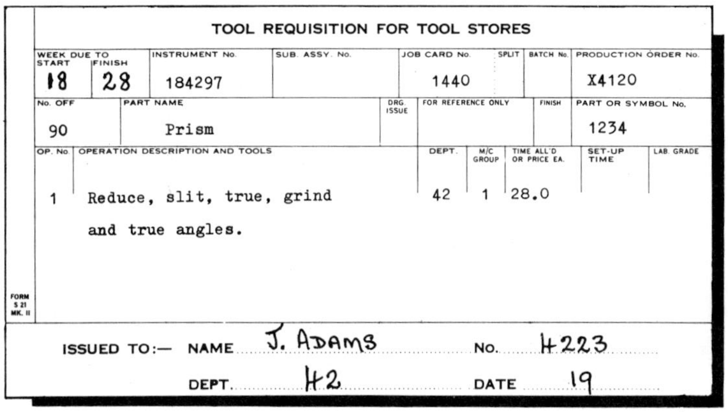

TOOL REQUISITION FOR TOOL STORES

WEEK DUE TO START	FINISH	INSTRUMENT No.	SUB. ASSY. No.	JOB CARD No.	SPLIT	BATCH No.	PRODUCTION ORDER No.
18	28	184297		1440			X4120

No. OFF	PART NAME	DRG. ISSUE	FOR REFERENCE ONLY	FINISH	PART OR SYMBOL No.
90	Prism				1234

OP. No.	OPERATION DESCRIPTION AND TOOLS	DEPT.	M/C GROUP	TIME ALL'D OR PRICE EA.	SET-UP TIME	LAB. GRADE
1	Reduce, slit, true, grind and true angles.	42	1	28.0		

FORM S 21 MK. II

ISSUED TO:— NAME J. ADAMS No. H223

DEPT. 42 DATE 19

NAME J. ADAMS NO. H223

Date	To Inspn.	Pass	Rej.	Total	Date	To Inspn.	Pass	Rej.	Total
20		85	5	90					

4

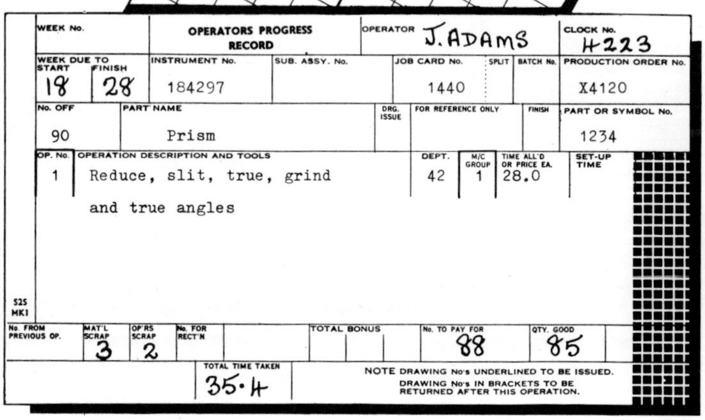

WEEK No.	OPERATORS PROGRESS RECORD	OPERATOR	J. ADAMS	CLOCK No.	H223

WEEK DUE TO START	FINISH	INSTRUMENT No.	SUB. ASSY. No.	JOB CARD No.	SPLIT	BATCH No.	PRODUCTION ORDER No.
18	28	184297		1440			X4120

No. OFF	PART NAME	DRG. ISSUE	FOR REFERENCE ONLY	FINISH	PART OR SYMBOL No.
90	Prism				1234

OP. No.	OPERATION DESCRIPTION AND TOOLS	DEPT.	M/C GROUP	TIME ALL'D OR PRICE EA.	SET-UP TIME
1	Reduce, slit, true, grind and true angles	42	1	28.0	

S25 MKI

No. FROM PREVIOUS OP.	MAT'L SCRAP	OP'RS SCRAP	No. FOR RECT'N		TOTAL BONUS	No. TO PAY FOR	QTY. GOOD
	3	2				88	85

TOTAL TIME TAKEN 35·4

NOTE DRAWING No's UNDERLINED TO BE ISSUED. DRAWING No's IN BRACKETS TO BE RETURNED AFTER THIS OPERATION.

5

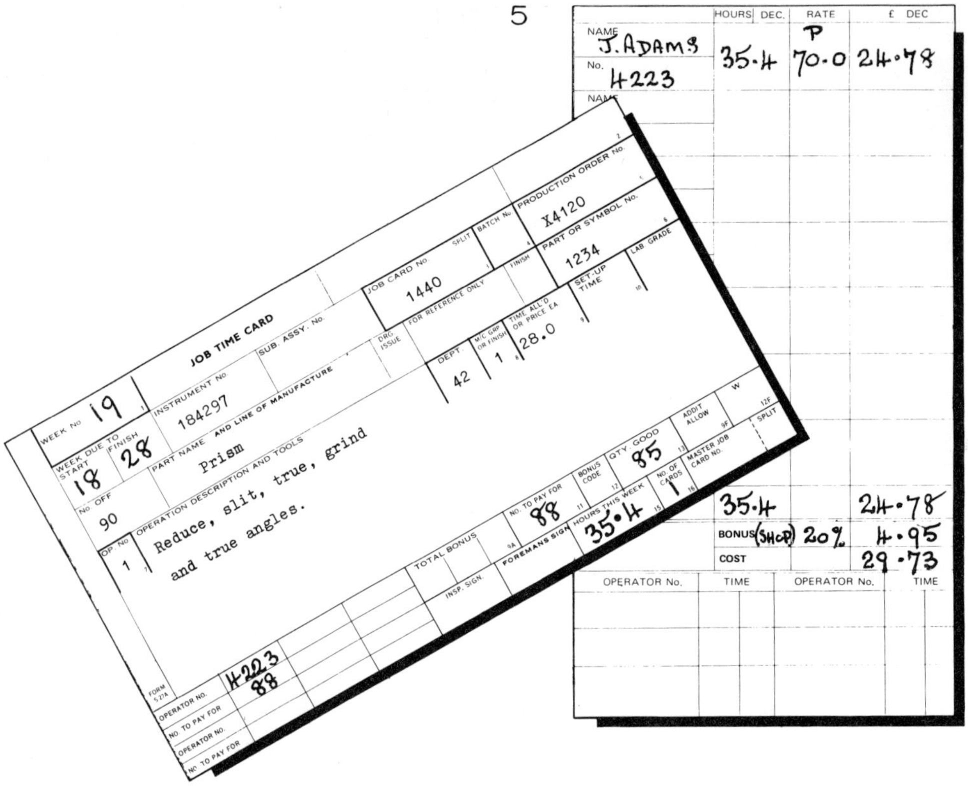

OPERATORS TIME ALLOCATION CARD		NAME J. ADAMS		CLOCK No. 4223				WEEK 19			
JOB CARD No.	Batch	PRODUCTION ORDER No.	PART No.	Opn	HOURS SAT.	MON.	TUES.	WED.	THURS.	FRI.	TOTAL
1440		X4120	1234	1		8	5	8	8	6.4	35.4
			OTHER WORK				3			1.6	4.6
											40.0
TOTAL HOURS											

6

7

COMPONENT ROUTE CARD

LOCATION								

WEEK DUE TO START	FINISH	INSTRUMENT No.	SUB ASSY No	JOB CARD No.	SPLIT	BATCH No.	PROD. ORDER No.
18	28	184297		1440			X4120

NO. OFF	PART NAME	DRG ISSUE		FINISH	PART OR SYMBOL No.
90	Prism				1234

PLANNED BY	PART NAME	CYCLE TIME	STD. BATCH QTY	PART OR SYMBOL No. (CHECK)
	Prism	10	30(x3)	1234

OPERATION No	1	2	3	4	5	6		
DEPT. / M/C GROUP	42-1	42-3	42-3	42-6	43-6	42-9		

MATERIAL SPECIFICATION	QUANTITY	UNIT	MAT'L CODE
1.45" x 1.0" x 0.7"	42	grams	£0.08
HC.519604			AB 106

QTY. GOOD	OP'TOR'S No.	OP No	OPERATION DESCRIPTION AND TOOLS	DEPT.	M/C GROUP	TIME ALL'D EACH	SET-UP TIME	LAB GRADE
85	4223	1	Reduce, slit, true, grind and true angles.	42	1	28.0		
82	4231 4274	2	Collimate, smooth and polish Check angles.	42	3	30.0		
81	4260	3	Chamfer, separate and chamfer.	42	3	4.0		
79	4271	4	Clean and inspect	42	6	0.0		
77	4340	5	Coat and inspect.	43	6	10.0		
76	4220	6	Silver and inspect.	42	9	11.0		
						83.0 minutes each		

FORM S 5 MK II	DATE PLANNED	ISSUES						

8

OPERATORS PROGRESS RECORD

WEEK No.	OPERATOR	CLOCK No.
21	T. JONES	4231

WEEK DUE TO START	FINISH	INSTRUMENT No.	SUB. ASSY. No.	JOB CARD No.	SPLIT	BATCH No.	PRODUCTION ORDER No.
18	28	184297		1440			X4120

No. OFF	PART NAME	DRG. ISSUE	FOR REFERENCE ONLY	FINISH	PART OR SYMBOL No.
~~90~~ 85	Prism				1234

OP. No.	OPERATION DESCRIPTION AND TOOLS	DEPT.	M/C GROUP	TIME ALL'D OR PRICE EA.	SET-UP TIME
2	Collimate, smooth and polish Check angles.	42	3	30.0	

S25 MKI

No. FROM PREVIOUS OP.	MAT'L SCRAP	OP'RS SCRAP	No. FOR RECT'N	TOTAL BONUS	No. TO PAY FOR	QTY. GOOD
	3				85	82

TOTAL TIME TAKEN

NOTE DRAWING No's UNDERLINED TO BE ISSUED. DRAWING No's IN BRACKETS TO BE RETURNED AFTER THIS OPERATION.

NAME	T. JONES
Date	21
To Inspn.	85
Pass	82
Rej.	3
Total	85
NO.	4231
Date	
To Inspn.	
Pass	
Rej.	
Total	

9

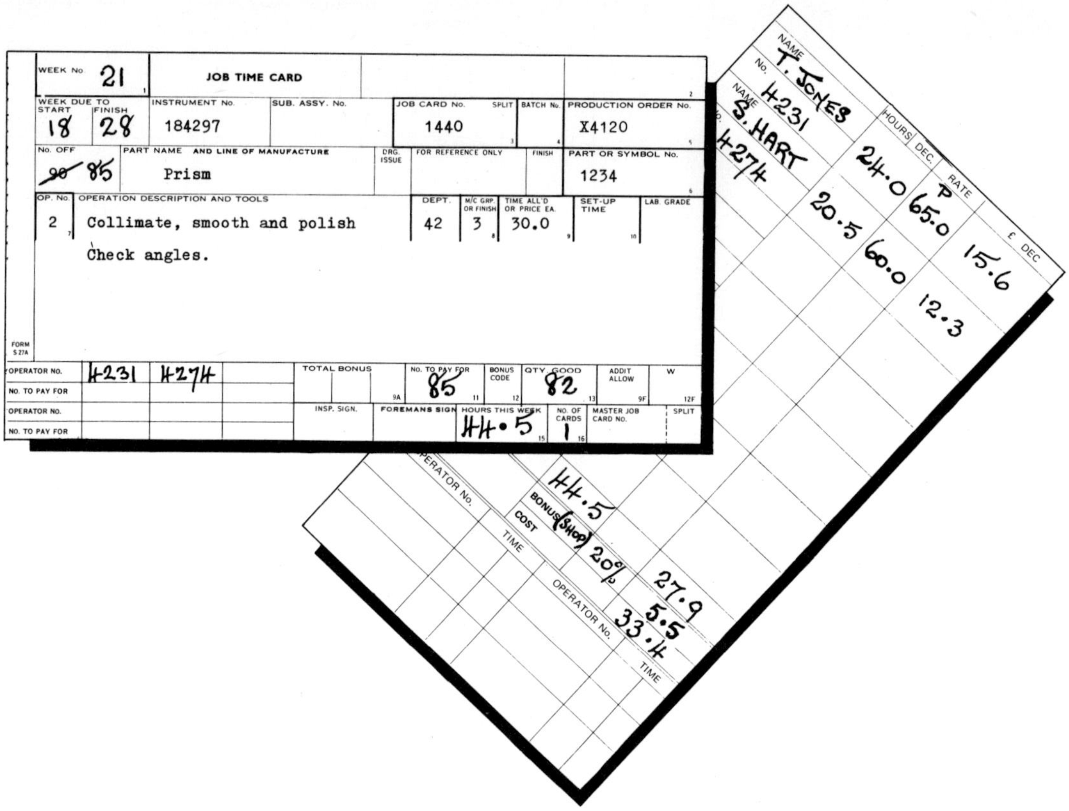

JOB TIME CARD

WEEK No.	21
WEEK DUE TO START	18
FINISH	28
INSTRUMENT No.	184297
SUB. ASSY. No.	
JOB CARD No.	1440
SPLIT	
BATCH No.	
PRODUCTION ORDER No.	X4120
No. OFF	~~90~~ 85
PART NAME AND LINE OF MANUFACTURE	Prism
DRG. ISSUE	
FOR REFERENCE ONLY	
FINISH	
PART OR SYMBOL No.	1234
OP. No.	2
OPERATION DESCRIPTION AND TOOLS	Collimate, smooth and polish. Check angles.
DEPT.	42
M/C GRP OR FINISH	3
TIME ALL'D OR PRICE EA.	30.0
SET-UP TIME	
LAB. GRADE	
FORM	S 27A
OPERATOR No.	4231, 4274
NO. TO PAY FOR	
TOTAL BONUS	
No. TO PAY FOR	85
BONUS CODE	
QTY. GOOD	82
ADDIT ALLOW	
W	
INSP. SIGN.	
FOREMANS SIGN	
HOURS THIS WEEK	44.5
NO. OF CARDS	1
MASTER JOB CARD No.	
SPLIT	

Card beneath:

NAME	No.	HOURS DEC.	RATE P	£ DEC
T. JONES	4231	24.0	65.0	15.6
S. HART	4274	20.5	60.0	12.3

OPERATOR No. · TIME · OPERATOR No. · TIME

	44.5
BONUS (SHOP)	20%
COST	27.9
	5.5
	33.4

10

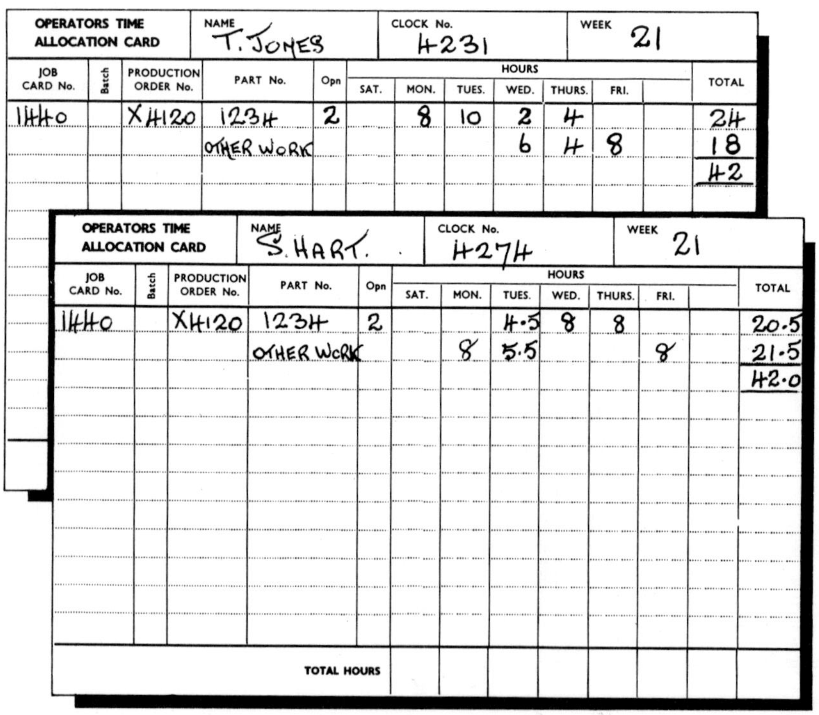

OPERATORS TIME ALLOCATION CARD

NAME: T. JONES — CLOCK No.: 4231 — WEEK: 21

JOB CARD No.	Batch	PRODUCTION ORDER No.	PART No.	Opn	SAT.	MON.	TUES.	WED.	THURS.	FRI.		TOTAL
1440		X4120	1234	2		8	10	2	4			24
			OTHER WORK					6	4	8		18
												42

OPERATORS TIME ALLOCATION CARD

NAME: S. HART. — CLOCK No.: 4274 — WEEK: 21

JOB CARD No.	Batch	PRODUCTION ORDER No.	PART No.	Opn	SAT.	MON.	TUES.	WED.	THURS.	FRI.		TOTAL
1440		X4120	1234	2			4.5	8	8			20.5
			OTHER WORK			8	5.5			8		21.5
												42.0
			TOTAL HOURS									

11

DATE 24			PART No. 1234			REF No.
DESCRIPTION PRISM						No. REC'D 81
REC'D FROM						No. ACCEPTED 79
No. REJECTED MATL. SCRAP	No. REJECTED OP'RS. SCRAP	FAULT	RECTIFY	SCRAP		
	2	BAD MAT.				COST OF SCRAP
		DIMENSIONS		1		EACH / TOTAL
		DEFINITION				MATL 8.0 / 16.0
		SURFACE MARKS				LAB 77.5 / 155.0
		GREY		1		OHD 200% / 310.0
		CHIP				TOTAL / 481.0
		COATING				
		SILVERING				
		CEMENTING				
		TRANS TO	INSPECTION TIME			INSPECTOR

INSPECTION REPORT TICKET

WEEK No. 24		INSPECTION REPORT TICKET		OPERATOR J. Kitt		CLOCK No. H271
WEEK DUE TO START 18	FINISH 28	INSTRUMENT No. 184297	SUB. ASSY. No.	JOB CARD No. 1440	SPLIT / BATCH No.	PRODUCTION ORDER No. X4120
No. OFF ~~90~~ 81		PART NAME Prism	DRG. ISSUE	FOR REFERENCE ONLY	FINISH	PART OR SYMBOL No. 1234
OP. No. 4		OPERATION DESCRIPTION AND TOOLS Clean and inspect	DEPT. 42	M/C GROUP 6	TIME ALL'D OR PRICE EA. 0.0	SET-UP TIME / LAB. GRADE
No. FROM PREVIOUS OP. 81	MAT'L SCRAP	OP'RS SCRAP 2	No. FOR RECT'N	No. TO PAY FOR —	QTY GOOD 79	SENT TO

FORM S 19A MK. II

INSP. SIGN

INSPECTOR'S REPORT TO BE MADE ON THE REVERSE OF THIS TICKET

12

JOB TIME CARD

WEEK No. 27		JOB TIME CARD				
WEEK DUE TO START 18	FINISH 28	INSTRUMENT No. 184297	SUB. ASSY. No.	JOB CARD No. 1440	SPLIT / BATCH No.	PRODUCTION ORDER No. X4120
No. OFF ~~90~~ 77		PART NAME AND LINE OF MANUFACTURE Prism	DRG. ISSUE	FOR REFERENCE ONLY	FINISH	PART OR SYMBOL No. 1234
OP. No. 6		OPERATION DESCRIPTION AND TOOLS Silver and inspect	DEPT. 42	M/C GRP OR FINISH 9	TIME ALL'D OR PRICE EA 11.0	SET-UP TIME / LAB. GRADE

FINAL

FORM S 27A

OPERATOR No. H220	TOTAL BONUS	NO. TO PAY FOR 76	BONUS CODE	QTY. GOOD 76	ADDIT ALLOW	W	
NO. TO PAY FOR							
OPERATOR No.	INSP. SIGN.	FOREMANS SIGN	HOURS THIS WEEK	NO. OF CARDS	MASTER JOB CARD NO.	SPLIT	
NO. TO PAY FOR							

13

COMPONENT COST SHEET

SEND TO COST OFFICE

REMARKS

WEEK DUE TO START	FINISH	INSTRUMENT No.	SUB-ASSY No	JOB CARD No.	SPLIT	BATCH No.	PROD. ORDER No.
18	28	184297		1440			X4120

NO. OFF	PART NAME	DRG ISSUE	FINISH	PART OR SYMBOL No.
90	Prism			1234

	CYCLE	STD. BATCH QTY	PART OR SYMBOL No. (CHECK)
Prism	10	30(x3)	1234

DATE REC'D

OPERATION No	1	2	3	4	5	6
DEPT./M/C GROUP	42-1	42-3	42-3	42-6	43-6	42-9

MATERIAL SPECIFICATION	SCPM OR X	QUANTITY	UNIT	MAT'L CODE	STANDARD PRICE	STD. UNIT MATERIAL COST	
1.45" x 1.0" x 0.7"		42	grams	£0.08		8.0	PENCE
HC.519604				AB 106			

NOTE
A NEW CARD TO BE SENT TO COST OFFICE FOR EACH MODIFICATION

OP No	OPERATION DESCRIPTION AND TOOLS	DEPT.	M/C GROUP	TIME ALL'D EACH	SET-UP TIME	LAB GRADE	STD. RATE	STD. UNIT LABOUR COST	CUM. UNIT LABOUR COST
1	Reduce, slit, true, grind and true angles.	42	1	28.0			75·0	35·0	35·0
2	Collimate, smooth and polish Check angles.	42	3	30.0				37·5	72·5
3	Chamfer, separate and chamfer.	42	3	4.0				5·0	77·5
4	Clean and inspect	42	6	0.0				—	77·5
5	Coat and inspect	43	6	10.0				12·5	90·0
6	Silver and inspect.	42	9	11.0				13·75	103·75
				83.0 minutes each					103·75

FORM 54 MK II

IBM CHECK TOTAL

DATE PLANNED

ISSUES

9.6 Scrap and reworks

There are three requirements for dealing with scrap and reworks:

(1) Recording.
(2) Costing.
(3) Physical control and handling.

Scrap is simple to deal with, as all that is required is an Inspection Report Ticket which notes:

(*a*) Quantity scrapped.
(*b*) Reason.
(*c*) Operation at which scrap occurs.
(*d*) Cost of scrap.

Reworks require a ticket as above, but attention must be paid to the following:

(*a*) Stock awaiting re-processing.
(*b*) Job tickets for re-processing.
(*c*) Cost of re-processing.

In the case of reworks the following procedure is proposed.

(1) All rejects are written off at cost of scrap.
(2) Reworks are stored in W.I.P. Stores.
(3) A weekly list of available reworks is sent to the Foreman.
(4) Foreman decides whether to treat reworks as a separate batch or to mix with the next new batch of the same part.
(5) In either case, a new book of tickets to

9.7.1. Optical section loading by cards on a pegboard. The time scale is horizontal. Capacity groups are vertical.

be made out manually, starting only at the process required.

(6) If the reworks are dealt with as a separate batch, the job tickets are treated as for a normal batch.

(7) If the reworks are mixed with a new batch the operators should have two tickets for each operation, one for the new batch and one for the reworks.

(8) The operators apportion the time taken between the two tickets.

(9) Any subsequent scrap is deducted from the main batch to avoid confusion.

(10) When the job tickets are finally costed they will show less time taken than standard (having missed all the operations prior to the one on which the parts were first rejected).

(11) The difference between the actual rework time taken and the total standard time should be credited to the excess labour and material account by the Cost Department.

9.7 Section loading

As explained in § 9.4 above, the static forward load is calculated from the production programme and the sum of the process elements of each product to establish the relationship between resources required and resources available.

When the Factory Forms have been printed, a dynamic load control is necessary to ensure that jobs pass through the department in the correct order to ensure optimum balance and, for this purpose, a Loading Board (Fig. 9.7.1) and Loading Tickets are required. A Loading Ticket (a piece of cardboard 45-mm square with a hole in it) is made out for each operation with details of the job and its Capacity Section. The Loading Board is a peg board with a time scale across the top and capacity groups in the vertical direction. By loading the board with the tickets, it is possible to see at a glance the position of jobs in relationship to the capacity and so forecast rate of progress and delivery dates. The tickets on the Loading Board should be moved daily from details on a Progress Record Sheet, prepared by the Work Issue and Transit Store, as jobs move from one operation stage to another.

9.8 Optical factory layout

Efficient production control resulting in minimum costs can only be achieved by good methods, good layout, and efficient working conditions. A good balance of work load and the elimination of unnecessary delays or movements of work in progress is essential.

Fig. 9.8.1 illustrates an ideal optical shop layout for high quality lenses and prisms, and the main features are as follows.

(*a*) Supervision, work issue, material and tool storage are in a central position.

(*b*) Inspection, cementing, blacking, and coating are all in clean rooms. A small store for coating jigs and space for cleaning vacuum

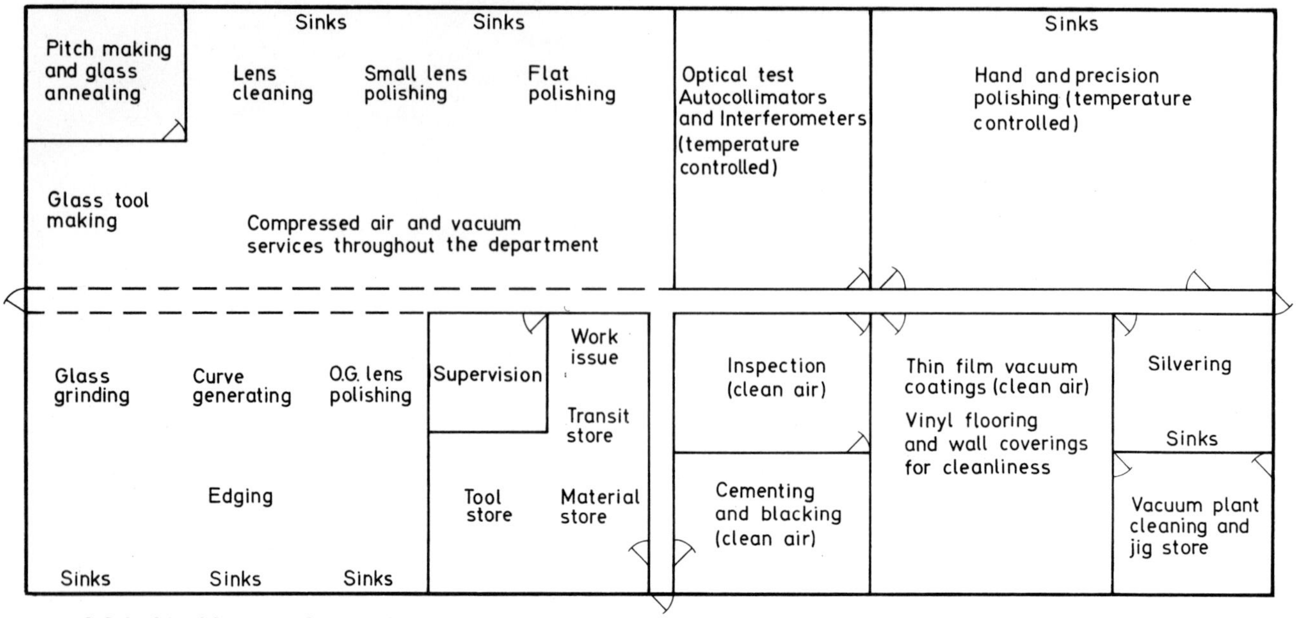

9.8.1. Ideal layout of an optical department.

plant is next to the coating room with access to the outside.

(*c*) Silvering is next to an outside wall for extraction of fumes.

(*d*) Pitch making and glass annealing are next to outside walls for extraction of heat and fumes.

(*e*) Hand working and optical test sections are in temperature-controlled areas.

(*f*) The flat polishing machines are near to the air-conditioned hand-working section as sometimes final figuring is necessary after initial polishing by normal machinery.

(*g*) An adequate number of sinks, with hot and cold water supply, in order to wash blocks of lenses, prisms, and tools. Special provision is necessary to remove abrasives and polishing compound from the drains.

9.9 Standard costing

Although traditionally optical work has usually been job costed, because of the variations in time which are impossible to foresee, it is considered more reliable to set standard times (and therefore costs) based on elemental times and to provide for adequate excess costs in estimates.

Excess labour and material costs are not normally included in overheads but are written off to cost of sales.

It is very important that the correct material cost for the glass specification is used for setting the standard as variations in cost vary 10:1 on volume according to the specification (variations of glass density vary as much as 3:1).

Material

Cutting allowances will normally be included in standard quantities of material per component. Excess materials will be due to loss in stores due to offcuts, scrap, price variances and quantity variances, provision of wasters, etc.

Labour

Standard times should include personal allowances and contingencies, but scrap and reworks would be classed as excess labour.

Note. Material prices can vary 50 per cent according to the quantity and form in which the material is purchased. If it is necessary to purchase from abroad, then the price will include duty. Excess material costs may therefore be as much as 20 per cent.

Labour variations will also be far greater than are likely to be expected and at least 20 per cent excess labour will probably be normal, but only experience based on the times set as standard and working the system will give the correct excess labour cost.

Investment in stock and work in progress

Because of the relatively small quantities of precision optics required, and the essential production feature that as many components as possible are assembled on a block, the number of batches per year of a given component is small. Also because of the need to have a consistently accurate refractive index of the glass it may be necessary to stock substantial quantities of raw material so that a melt of glass lasts as long as possible.

If batch sizes are relatively high this means that the rate of turnover of both raw glass and work in progress is relatively slow by comparison with the manufacture of metal components. This slower rate of turnover of money invested in stock and work in progress must be taken into account when fixing selling prices.

10 Non-spherical Surfaces

10.1 Introduction

All optical surfaces may be divided into two main groups:

(1) Surfaces of revolution having an axis of symmetry.
(2) Surfaces having two planes of symmetry.

Generally aspheric surfaces used in optical systems are in the first group, which includes paraboloids, ellipsoids, and hyperboloids. The second group includes toric and cylindrical surfaces. (See Chapter 7.) René Descartes was probably the first man to consider what kind of surfaces would give freedom from aberration. In his discourse *La Dioptrique*, published in Paris in 1638, he gives the geometrical construction of a lens free from spherical aberration.

After giving the proof of the property of the ellipse of focusing parallel rays Descartes continues:

> Because every ray which is directed towards the centre of a sphere suffers no refraction, if with centre *I* (Fig. 10.1.1), one draws a circle *BQB* of any desired radius, then the lines *DB* and *QB* turning about the axis *DQ* will describe the shape which a lens should have to focus, in the air, at the point *I*, all the rays which shall have been parallel to the axis before falling on the lens.

For more than two hundred years, nobody actually made lenses of the type he described. Newton observed that the chief fault in lenses was chromatic aberration, and to avoid this problem gave his attention to the reflecting telescope. Among those which he made was an instrument of 6-inch focal length which he presented to the Royal Society in 1671. The formation of a parabolic surface was first achieved by Short in 1732, who made reflecting telescopes with parabolic and elliptical figures by retouching or figuring. These were called 'Gregorian' telescopes after James Gregory, who, in 1663, made it clear that conic sections would correct spherical aberrations.

In the period 1899 to 1926, Carl Zeiss was active in inventing aspheric systems and methods of producing non-spherical curves suitable for binocular eyepieces, but the accuracy achieved was not sufficiently good for camera lenses. Machines could give approximately the required shape but the interferometer provided the means of finally correcting them by hand, or by a machine simulating hand polishing.

Newton, Cassegrain, and Gregory used aspheric mirror surfaces in their telescopes, but it is only since 1960 that machines have been available for grinding camera lenses with aspheric surfaces of

10.1.1. Aplanatic lens with one elliptical surface.

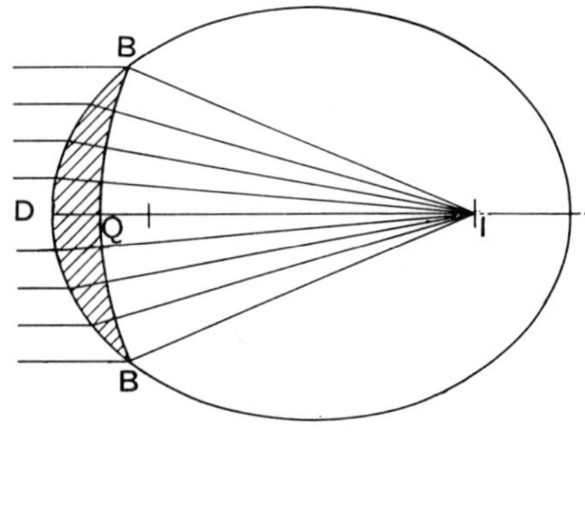

such accuracy that subsequent polishing did not have to include any retouching.

With the development of electronic computing machines, not only is it possible to make the large number of calculations for complex optical systems, but also to feed back correcting information to an aspheric curve generator controlled by a punched tape.

Aspheric surfaces may be produced by one of the following methods:

(1) The shaping of glass surfaces by removing surplus material by turning, grinding or polishing. (See §§ 10.3 and 10.8.)

(2) The shaping of the surfaces by pressing or moulding from plastics. (See § 5.8.)

(3) The change in the shape of the original glass surfaces by heat and pressure. (See § 5.9.)

(4) Modifying the original glass shape by applying an additional layer of material by vacuum deposition or by replication with a plastic coating. (See § 10.5.)

(5) Eroding the original glass shape by removing a layer of glass by an ion beam. (See § 10.7.)

All these methods of shaping the surfaces have their own advantages and disadvantages, and are used in the manufacture of optical components.

Aspheric surfaces enable a designer to obtain optical advantages such as:

(*a*) Very wide-angle objectives with complete elimination of distortion.

(*b*) Wide-angle eyepieces with flat fields.

(*c*) Reduced number of components in existing designs, but with equally good or improved performance.

(*d*) Reduced bulk of existing designs.

(*e*) High aperture condensers as moulded glass aspherics in slide projectors.

(*f*) Large magnifiers for industrial use as moulded plastic aspherics, also stepped lenses (such as those invented by Fresnel in 1822).

Figs 10.1.2 and 10.1.3 show aspheric elements having the form of surfaces of revolution with one axis of symmetry.

Mr M. Sigmund, Managing Director of Combined Optical Industries Ltd, Slough, Bucks, has kindly provided the information in the remainder of this introduction.

SOME SIMPLE PROPERTIES OF ASPHERICS

Surface shapes in optical designs may be spherical, plane, cylindrical, toroidal, conicoidal, or some more general aspheric. The vast majority of surfaces employed in optical instrumentation are rotationally symmetric, and many are spheres (a plane being a sphere of infinite radius). The simplest aspheric shapes are the ellipsoids, paraboloids, and hyperboloids of revolution, but complex aspherics depart from these basic shapes by figuring terms not usually involving more than the tenth power in the surface aperture. The description 'aspheric' is normally reserved for smooth, rotationally symmetrical surfaces.

Figuring a sphere within an optical system will leave its primary chromatic correction and field curvature unchanged, while altering its spherical aberration, coma, astigmatism, and distortion properties. When the aspheric is at the aperture stop, only the spherical aberration is affected, and this property can be easily employed to produce an immediate improvement in the axial correction of the system. Increasing the stop to aspheric separation increases the effect of the figuring on the extra-axial aberrations.

From a design point of veiw, the final surface of the system is often the simplest to aspherize, and various methods can be adopted for calculating the figure of the last surface to improve image quality.

THE RANGE OF APPLICATIONS OF ASPHERICS

Many types of optical systems can have their performance enhanced or their number of elements reduced by the employment of one or more aspheric surfaces. Some of the broad fields of application are mentioned below, and these are supported by specific examples in subsequent paragraphs.

The commonest use of aspherics lies in the figuring of a single surface within the system to correct a particularly recalcitrant aberration. These cases include:

(*a*) Spherical aberration in condenser systems (for projectors and microscopes), spherical mirror systems, and singlet plastic camera lenses.

(*b*) Astigmatism in wide-angle eyepieces and large field magnifiers.

(*c*) Distortion in aerial survey mapping lenses.

(*d*) Spherical aberration of the pupil in certain visual systems (this appears as shadows in the field of view).

However, with the modern state of sophistication in optical design, production, and testing techniques, systems including several aspherics are becoming increasingly numerous. Each figured surface contributes to the control of several aberration types, and the extra degrees of freedom introduced can result in high-performance systems, perhaps employing fewer elements than their predecessors.

While optical glass remains unchallenged in terms of its range of indices and dispersions, its high scratch-resistance, and its low thermal expansion, there are many aspheric applications in which plastic offers advantages over glass. When fairly large production runs are involved, the cost of the master for the moulding is written off over a large number of elements, resulting in very large cost savings over the hand-figured and tested glass counterpart. The plastic component also offers a great saving in weight, and is effectively shatterproof. In some systems these are important considerations.

The glass–plastic hybrid is a useful and expanding category of aspheric optics. These systems use spherical glass elements and aspheric plastic elements to combine the separate advantages of the two materials. Thus a single glass component may be located at a position of heavy wear and tear, while the addition of aspheric plastic components contributes to the production of a high-performance, robust, and lightweight device. Also, in a mass-production design employing only a single aspheric, cost grounds alone may necessitate the moulding of this element in plastic.

SOME PARTICULAR APPLICATIONS

Just a few examples of specific aspheric systems are given here to illustrate some of the points of the previous sections.

The Schmidt camera

In the early 1930s, Schmidt demonstrated a wide-aperture, wide-field reflecting telescope using a spherical mirror (expensive paraboloidal mirrors had been used on all previous instruments). The coma of the mirror is removed by locating the stop at the centre of curvature, and the large amount of spherical aberration is corrected by an afocal aspheric plate, also located at the centre of curvature. The resulting system has a usable field diameter hundreds of times greater than a parabolic mirror of the same relative aperture.

The Schmidt principle can be employed in a variety of applications, the ultra-violet spectrograph being of particular importance.

[The Schmidt camera is discussed in § 10.2.]

The projection condenser

The function of a projection condenser is to get as much light as possible from the source, through the film or slide, and into the entrance pupil of the projection lens. Large amounts of aberration can be tolerated, but if the spherical aberration is too great, a final image of non-uniform illumination results on the screen. The condenser aperture must be made large to collect as much light flux as possible, but primary spherical aberration increases as the fourth power of the aperture. If a bright, evenly illuminated image is to be obtained, it follows that some effort must be made to reduce the spherical aberration of the condenser.

Replacing a 'best-shape' singlet by a pair of half-power elements reduces the spherical aberration by a factor of four, and this system is widely used in photographic enlargers. However, a correctly aspherized singlet condenser has a superior performance and is much less bulky. This application is ideally suited to plastic moulding.

The wide-angle eyepiece

This is a good example of the powerful design possibilities of a single aspheric element. In many visual systems, but particularly in binocular applications, a system designed with a wide-angle eyepiece offers the advantages of greater information presentation, more efficient target acquisition, reduced 'instrument fatigue', and a better impression of depth.

The major difficulties of eyepiece design stem from the requirement that the aperture stop (the pupil of the observer's eye, in normal conditions) has to be a significant distance outside the system, and this results in a field of view which is usually limited by the astigmatism of the eyepiece. In

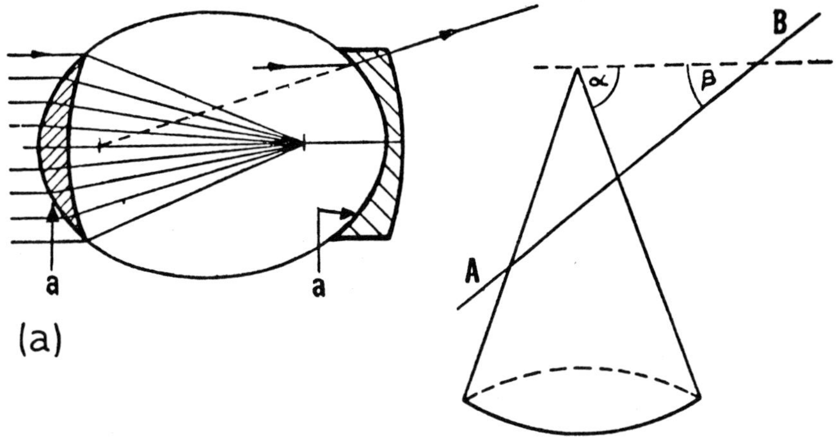

(a)

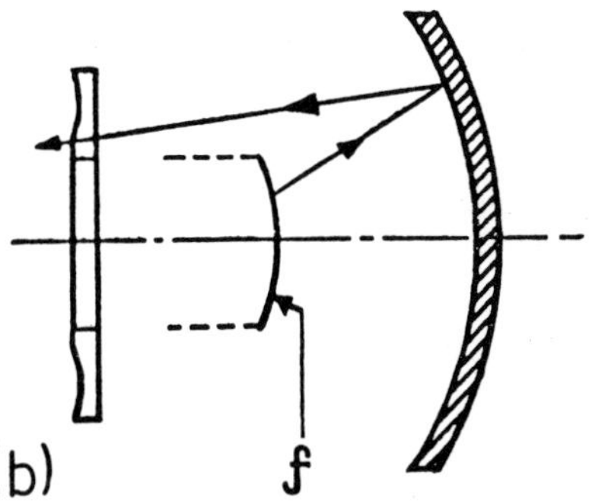

(b)

10.1.2.
Non-spherical surfaces.

(*a*) Simple non-spherical surfaces are generated by revolution of conic sections, i.e. cuts (AB) across a cone. If the angle $\beta=0$, the conic section is a circle. When β is less than α, the conic section is an ellipse. When $\beta=\alpha$, the section is a parabola. When β is greater than α, the section is a hyperbola. The simplest aplanatic lenses with non-spherical surfaces have one surface each that is elliptical (a).

(*b*) The classical Schmidt system uses a spherical mirror to project the image on a curved focal plane (f). It has an aspheric corrector plate (the surface here is exaggerated).

(*c*) The dioptric Schmidt *f*/0·7 lens has a front component with an aspheric surface (a), shown exaggerated, and produces its image in a curved focal plane (f).

(*d*) In the 6-inch *f*/5·6 triple aspheric triplet, each element has one spherical curve and one aspheric (a). Yoshido's five-element triplet (right) achieves an aperture of *f*/0·519 with comparatively few elements, one having an aspheric surface (a). The dotted line indicates the spherical curve. The focal plane (f) almost touches the strongly curved rear lens.

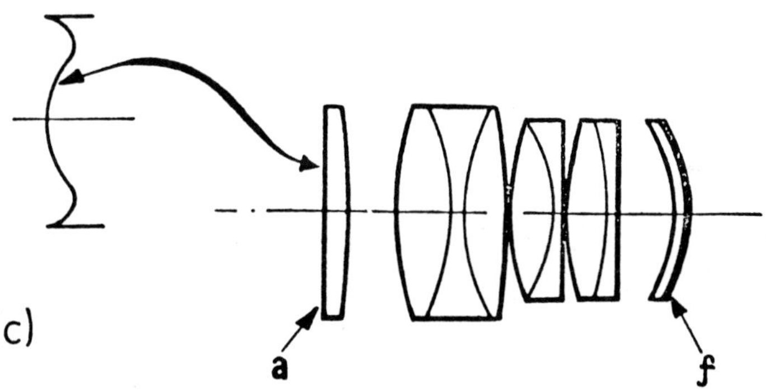

(c)

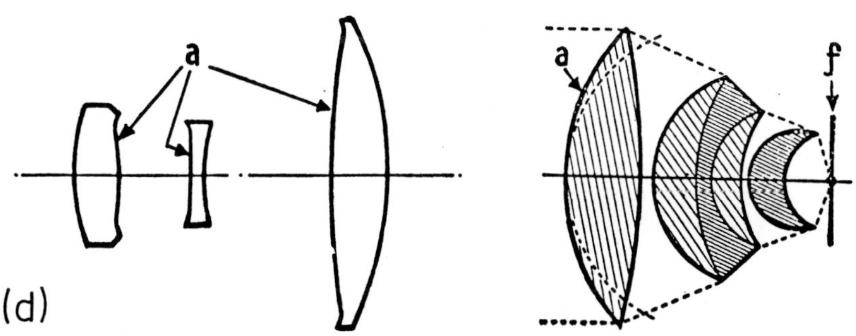

(d)

10.1.3.
More advanced aspheric systems.

(*a*) The 12-mm Elgeet Golden Navitar f/1·2 uses an aspheric surface (a) as a corrector plate taking the place of the last component. With spherical components only, extra elements would be required.

(*b*) The Rank Taylor Hobson 25-mm Cooke Speed Pancro Series III for 35-mm cinematography further illustrates the saving possible with aspheric elements. The original design in the upper diagram had all spherical surfaces and nine elements to yield an aperture of *f*/1·8. An inverted telephoto construction is used to provide the necessary back focus for use on a reflex camera type; the lens is therefore rather long and bulky and leads to obstruction of the view of other lenses on the camera lens turret. The later design in the lower diagram, on the same scale, uses only seven elements and has one aspheric surface (a). The aperture is *f*/2, but the use of fewer elements leaves the effective transmission (*T*2·2) unchanged. The lens is shorter and less bulky.

(*c*) In the inverted telephoto system of the 5·8-mm Kinoptic for 16-mm cameras, the front component has a parabolic surface (a). It demonstrates the intelligent use of an aspheric surface to control the higher order effects of astigmatism and distortion, which are usually the biggest problem in an inverted telephoto lens covering such a large angle as 103 degrees.

(*d*) Aspheric lenses also help in zoom lens design. This is the Rank Taylor Hobson 4-inch to 40-inch Varotal III *f*/4 for television cameras. There is one aspheric surface at (a) and the lens has two zoom ranges; the normal range from 4 to 20 inches at *f*/4 and a further range of 8 to 40 inches at *f*/8, obtained by moving the negative member in the rear to the dotted position.

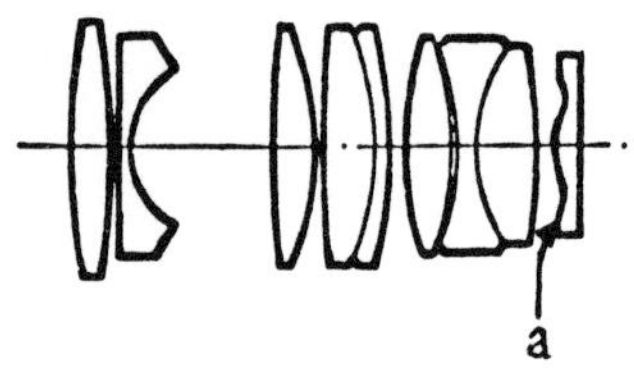

(a)

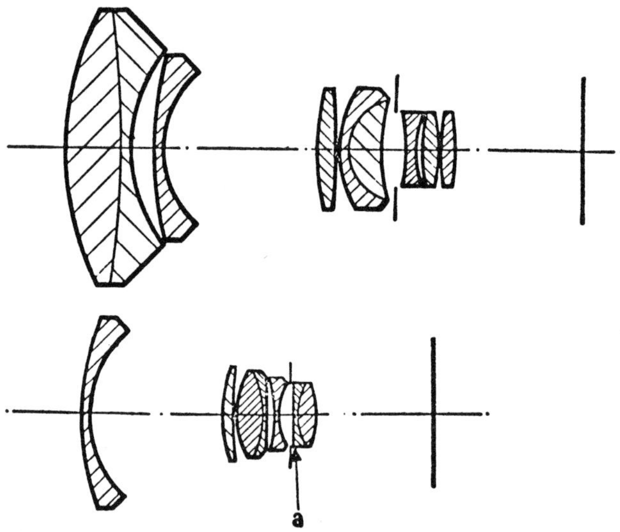

(b)

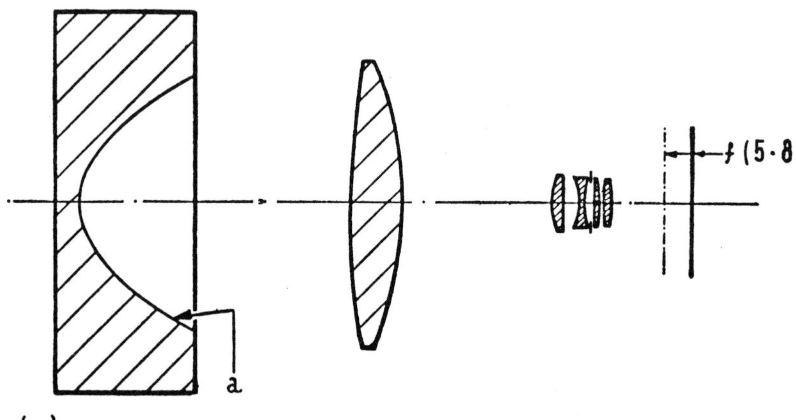

(c)

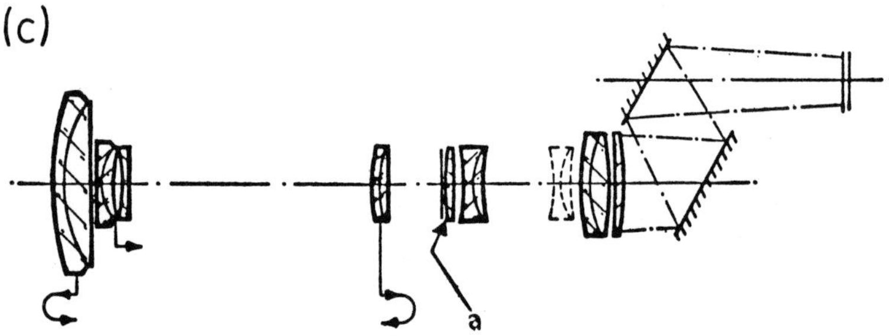

(d)

conventional binocular eyepieces the field is 40°–50°, but high-quality wide-angle devices must provide fields of up to 90° and permit an eye-roll of $\pm 30°$ or $\pm 40°$ (to look at the edge of the field) without objectionably affecting the image illumination.

This requirement for excellent field correction and a large exit pupil can only be met by incorporating an aspheric (typically the final element) into a good design type.

The lightweight eyepiece

In many eyepiece applications the specification of field coverage may not be out of the ordinary, but high performance and extreme lightness may be vital, such as in military sights, certain instrumental viewing systems, and hand-held optics. In this case, and especially when low-cost mass production is specified, the use of a glass/plastic hybrid is indicated, with a spherical-surface glass eye-lens and aspheric plastic components.

10.2 The Schmidt camera in astronomy

The following account of the problem of spherical aberration and its solution in the Schmidt camera is extracted from F. Twyman's *Prism and Lens Making*.

It is convenient first to consider one or two basic properties of ordinary spherical concave mirrors and of paraboloids. When a parallel beam of light is reflected from a spherical concave mirror, as shown in Fig. 10.2.1, the rays near the axis *CA* are brought to a focus *F* at a distance from the mirror surface equal to half its radius of curvature. However, the rays reflected from the outer parts of the mirror aperture meet the axis *CA* slightly nearer to the pole *A* of the mirror surface than those reflected from the parts near the axis. The rays of light arriving from a star may be regarded as parallel when they reach our telescopes; and a spherical mirror held with its axis pointed directly towards a star will form an image which is not quite sharp, but suffers from the defect known as spherical aberration.

Ordinary reflecting telescopes therefore do not use spherical mirrors, but paraboloidal ones, to receive the parallel rays. When a corrected paraboloidal mirror is pointed directly towards a star, the rays are all brought sharply to a single focus. (See Fig. 10.2.2.) The outer parts of the paraboloid are in fact slightly bent back or 'turned down' relative to those of the nearest sphere, and this causes a lengthening of focus in the rays reflected from the outer zones of the mirror surface, while the focus of the inner zones is slightly shortened. (Figs 10.2.2 (a) and (b).) The amount of the discrepancy between the paraboloid and the sphere is very small; for a mirror of diameter 12 inches and focal length 60 inches its maximum only amounts to $2{\cdot}3 \times 10^{-5}$ inch. Nevertheless it

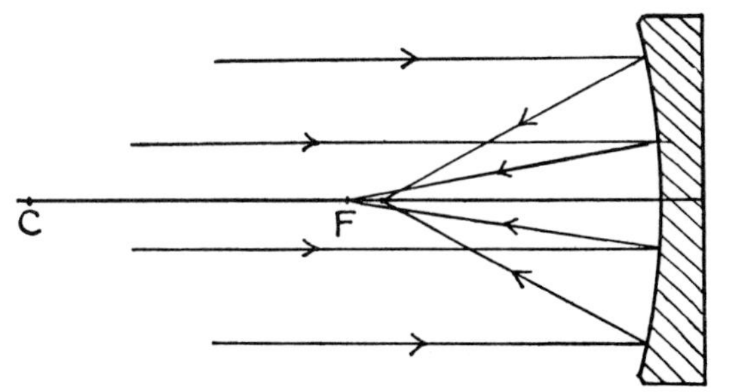

10.2.1.▷ Image formation by a spherical mirror.

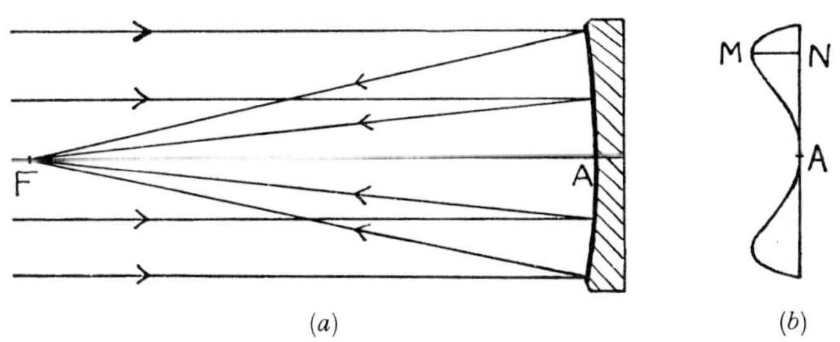

◁10.2.2. The paraboloidal mirror. (*a*) Image formation. (*b*) The difference between the paraboloid and the sphere.

makes a big difference to the performance of a 12-inch $f/5$ Newtonian telescope (Fig. 10.2.3), which consists of a paraboloidal primary mirror M_1 together with a diagonal flat mirror D to bring the image into an accessible position F' at the side of the telescope tube. The operation of converting a spherical mirror into a paraboloid is known to telescope-makers as 'parabolization' or 'figuring'. The greatest gap between a paraboloid and that of the sphere which touches it at the centre and meets it at the edge is given to a good approximation by the expression

$$\frac{f}{4096F^4}$$

where f is the focal length of the mirror, d its diameter, and F the focal ratio f/d.

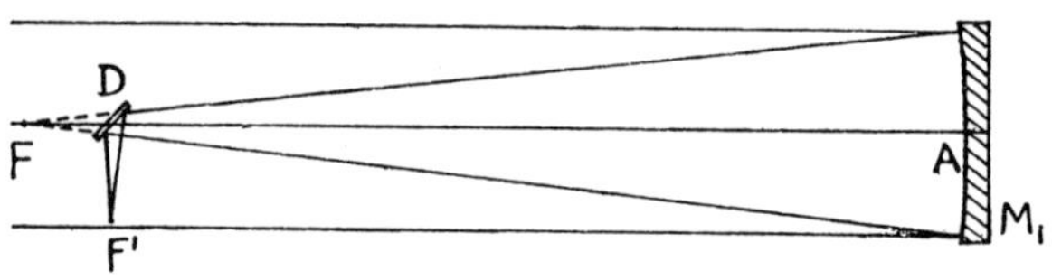

10.2.3. Newtonian telescope.

If we try to photograph a field of stars by placing a photographic plate at the prime focus F of a Newtonian telescope mirror (Fig. 10.2.3), the images close to the centre of the field are very good, but those farther out show an asymmetrical, comet-shaped form; the optical system suffers from 'off-axis coma'. This coma not only upsets measurements of position, but also results in many faint star-images being lost altogether, through the intensity of their diffuse image-patches falling below the threshold sensitivity of the photographic emulsion.

The Schmidt camera

A new era in astronomical camera design began in 1931, when Bernhardt Schmidt introduced the new type of optical system which bears his name. Schmidt was a remarkable personality. Born in 1879 on Nargen Island in Esthonia, he attempted as a boy to make a concave mirror by grinding together the flattened lower parts of bottles with sea-sand from the beaches of Nargen. Later, he studied engineering in Götenburg and Mittweida and, about 1900, began to make telescope mirrors for amateur astronomers. His mirrors were soon recognized as outstandingly good ones and in 1905 he created a sensation with an $f/2{\cdot}26$ paraboloid of 40 cm aperture, made for the Potsdam Astrophysical Observatory, which in the perfection of its figure far surpassed other astronomical mirrors of that time. Schmidt carried out all his work with the simplest means, and did his polishing and figuring by hand—with his left hand, indeed, for he had lost his right arm in an accident in early youth. He was not only a supreme master of the difficult art of mirror-making, he was also an enthusiastic astronomical observer. During the last part of his life, he was a voluntary worker at the Bergedorf Observatory near Hamburg, and it was there that he invented, constructed and put into use his new camera.

The novel idea embodied in the Schmidt camera may be explained as follows. As we have seen, the image formed by a spherical mirror of a star situated at a great distance on its axis is imperfect, because the rays from the outer zones of the mirror come to a shorter focus than those from the inner zones; the image suffers from 'spherical aberration'. (Fig. 10.2.1.) This aberration can be cured by subjecting the mirror surface to the deformation shown in Fig. 10.2.2. The effect of the deformation is to correct the spherical aberration by slightly lengthening the focus of the rays reflected from the outer zones of the mirror, while it slightly shortens the focus of rays reflected from the zones near the centre. Now there are other ways of producing this effect besides figuring the mirror. We could, for example, place a plane-parallel plate of glass (C, Fig. 10.2.4) in the incoming beam and 'deform' one of its surfaces by optical figuring so that the rays passing through the outer part of the plate were made slightly divergent, while those passing through the inner parts became slightly convergent. To secure this, the surface profile of the plate would have to be made convex near its centre and concave near its edge. Since it operates by refraction instead of by reflexion, the geometrical deformation of the plate surface would need to be about four times the parabolization-deformation of the mirror surface; for an $f/5$ mirror of 12 inches aperture it would amount to about $1{\cdot}5 \times 10^{-4}$ inch.

10.2.4. Schmidt camera.

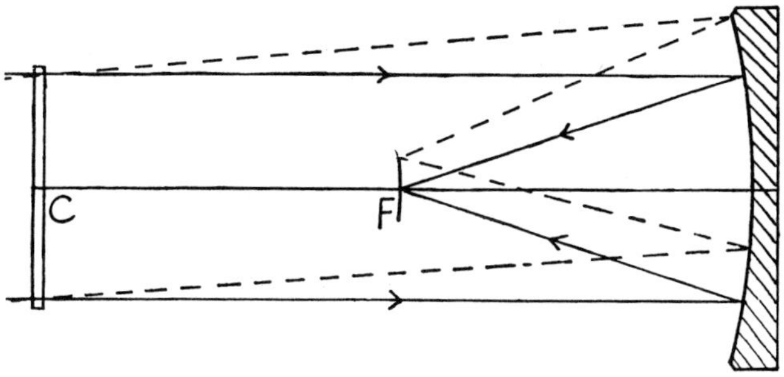

So far, it is not apparent that anything has been gained by carrying out the figuring process on the plate instead of on the mirror. In both cases, the effect is to bring sharply to a focus the rays of a parallel beam of light entering the system parallel to its axis. But when we consider, besides this on-axis pencil, the parallel pencils of rays which, entering the system at various angles with its axis, form images in the outer parts of the field, the advantages of the aspheric corrector plate become clear. For when the plate is placed at the centre of curvature *C* of the mirror (as in Fig. 10.2.4) then an off-axis pencil will be focused very nearly as sharply as the on-axis pencil, i.e. very nearly perfectly. In fact if we were to turn the plate square on to any particular off-axis pencil, this pencil would be focused perfectly. The defects in the off-axis images are therefore due to the fact that the plate does not lie squarely across the off-axis pencils; and since the plate is very nearly plane-parallel, the effect on the images of this lack of squareness is very small.

Thus the system consisting of a spherical mirror and a figured corrector plate located at its centre of curvature will give good images over a wide field. This system is the Schmidt camera; its off-axis aberrations are small even at very high apertures and are of a symmetrical character.

Modified Schmidt cameras

In some of the large Schmidt cameras now in use in the U.S.A. thin glass photographic plates replace the cut film; these can be bent to the required curve without breaking and spring flat again when removed from the camera. Nevertheless the curved field is a serious inconvenience, and it was not long before modified designs were suggested in which a flat image field was obtained at the cost, however, of some loss of image quality. The simplest of these modifications consists in the addition to the system of a plano-convex lens of suitable curvature placed immediately in front of the photographic film, which compensates the curvature of the system. (Fig. 10.2.5.)

Solid Schmidts

Still faster cameras for spectrographic work are provided by the solid Schmidt system, shown in Fig. 10.2.6, in which the space between corrector-plate and mirror is filled with a solid block of glass. The solid Schmidt, constructed in glass of refractive index *n*, is n^2 times faster than the

10.2.5. Field-flattened Schmidt camera.

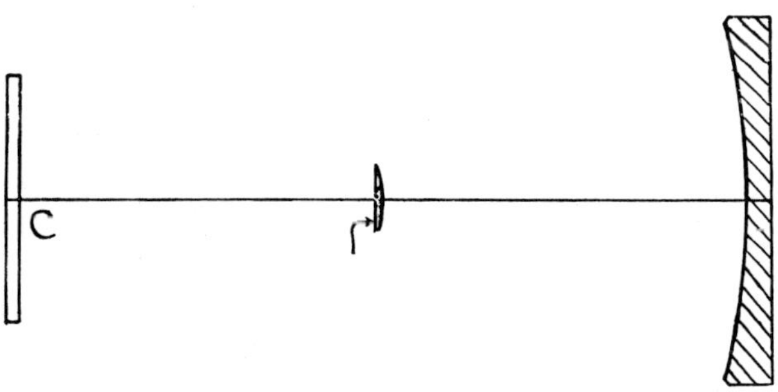

10.2.6. Solid Schmidt camera.

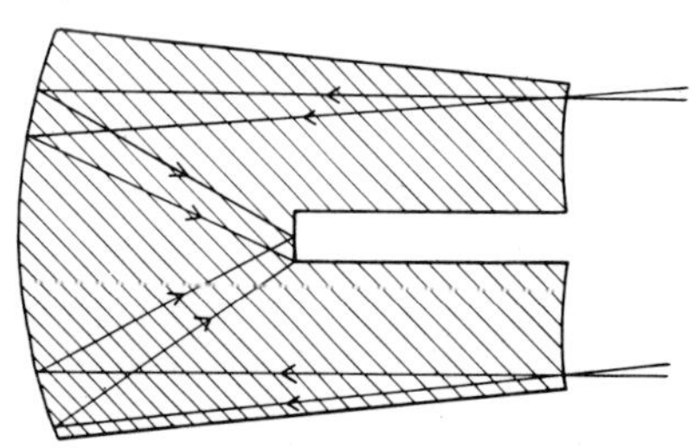

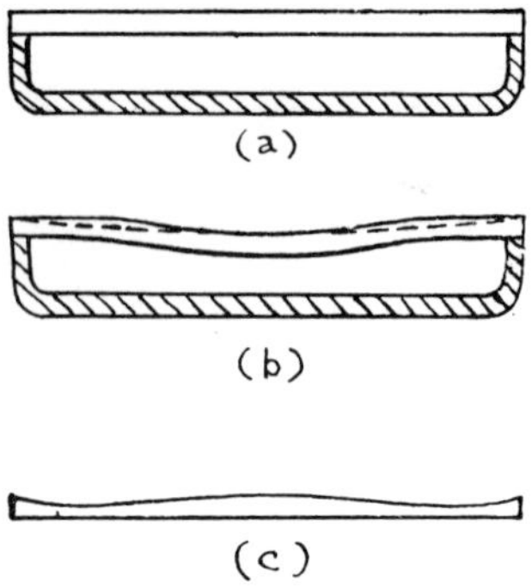

◁10.2.7. **Schmidt's method for making corrector plates.**

10.2.8. ▷ **Achromatized Schmidt plate.**

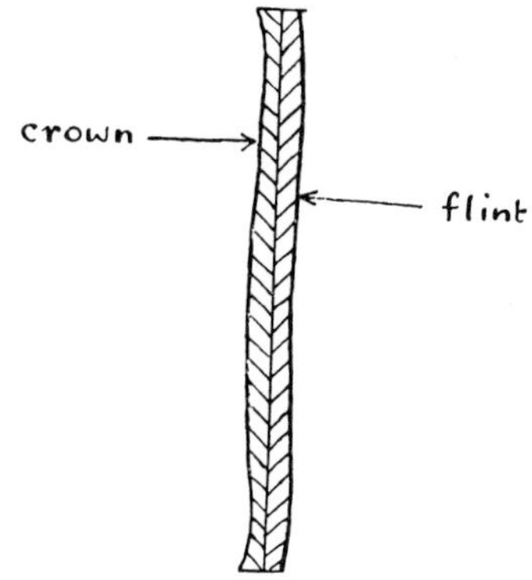

ordinary Schmidt of similar dimensions, and effective speeds of $f/0{\cdot}35$ are practicable in small sizes. Its astronomical usefulness is restricted to spectroscopy, since the initial refraction into the glass is accompanied by dispersion, resulting in a chromatic difference of magnification. The glass is cut away behind the field surface to allow access to the photographic film, which is oiled on to the spherical field surface.

Corrector plate techniques

Schmidt made his first corrector plate in a simple and ingenious way (Hodges, Paul C., 1948; *Amer. Jour. of Roentgenology*, **59**, 122). He cemented a thin plane-parallel glass plate to a metal drum, as shown in Fig. 10.2.7, so that its border was supported by the flat rim of the drum, and pumped out some of the air in the drum. This caused the plate to sag by a few hundredths of an inch, and the distorted plate was then ground and polished to a near-flat spherical curve in the ordinary manner; the dotted line in Fig. 10.2.7(b) shows the new plate surface. Finally, air was readmitted to the drum and the plate removed; as the strain in the plate was removed, its upper surface took on the shape shown in Fig. 10.2.7(c). Schmidt's knowledge of elasticity theory enabled him to foresee that the final form of this surface would be a sufficiently good approximation to that needed for the corrector plate of his camera.

Schmidt's method has rather fallen into disuse, and a more modern technique for making corrector plates of moderate size may be of interest. A plane-parallel glass plate is mounted in a rotating holder and ground with successively finer grades of carborundum by means of a flexible lap, which is traversed backwards and forwards with a rather short stroke as the plate rotates. The working surface of the lap is built up of lead facets, cemented with gold size to a disk of sponge rubber about $\frac{3}{8}$-inch thick. The facets do not cover the whole of the rubber disk but are arranged in petal-shaped areas (see Fig. 10.2.9) so that the effective grinding area which works on each zone of the plate is proportional to the depth of glass to be removed from that alone.

Curvature of field in Schmidt systems

As has been mentioned elsewhere, the curvature of field in Schmidt systems can be corrected by the addition of an extra lens, but Messrs Taylor, Taylor and Hobson prefer not to use such a lens in the systems they have built for television, since it increases the aberrations; moreover the

10.2.9. Flexible lap for grinding corrector plates.

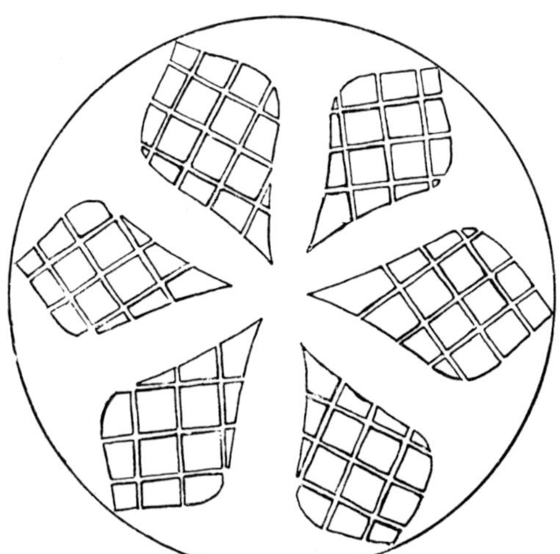

end of the cathode ray tube can readily be given a curvature appropriate for the production of a flat field in the projected image. They test these systems on a curved field in the following way. The image of a light source falls on to a concave surface of the appropriate radius of curvature on which are placed a number of small steel balls stuck to the surface with black paint. This is used with the complete telescope system and with it an image of the small images of the light source formed within the steel balls is projected on to a flat surface. This enables them to see the definition of the outermost parts of the field; the distance between the Schmidt plate and the remaining part of the system is adjusted until the best definition is obtained all over the field.

10.3 Generating machines for Schmidt plates

Before 1950, a number of companies, including Adam Hilger Ltd, Sir Howard Grubb Parsons & Company Ltd, and Taylor, Taylor and Hobson Ltd, developed and manufactured machines for grinding Schmidt plates. (Figs 10.3.1, 10.3.2, 10.3.3, and 10.3.4.)

10.3.1. Schmidt plate aspheric grinding machine at Sir Howard Grubb Parsons & Company Ltd. Note the cam that controls the movement of the cutter.

10.3.2. Rank Taylor Hobson No. 1 engine, cam driven and mostly used for Schmidt plates.

10.3.3. Rank Taylor Hobson No. 2 aspheric machine on the left, with the controller on right centre.

10.3.4. Rank Taylor Hobson No. 2 aspheric machine controller. A plan view showing the photographic track on a glass plate. The method replaced the metal cam, but has itself now been replaced by punched paper tape.

At that time, Schmidt plates were needed for projection television receivers as well as for spectrograph cameras. Fig. 10.3.5 illustrates the principles of the Taylor Hobson generating machine which is described in British Patent No. 947174 and Patent Nos 947175, 936951, and 936952.

Fig. 10.3.2 is a photograph of the Schmidt-plate generating machine which is capable of generating curves on flat or nearly flat surfaces up to 14-inch diameter, with an asphericity of 3 mm. Flexible laps for the polishing of Schmidt plates are described in § 10.2.

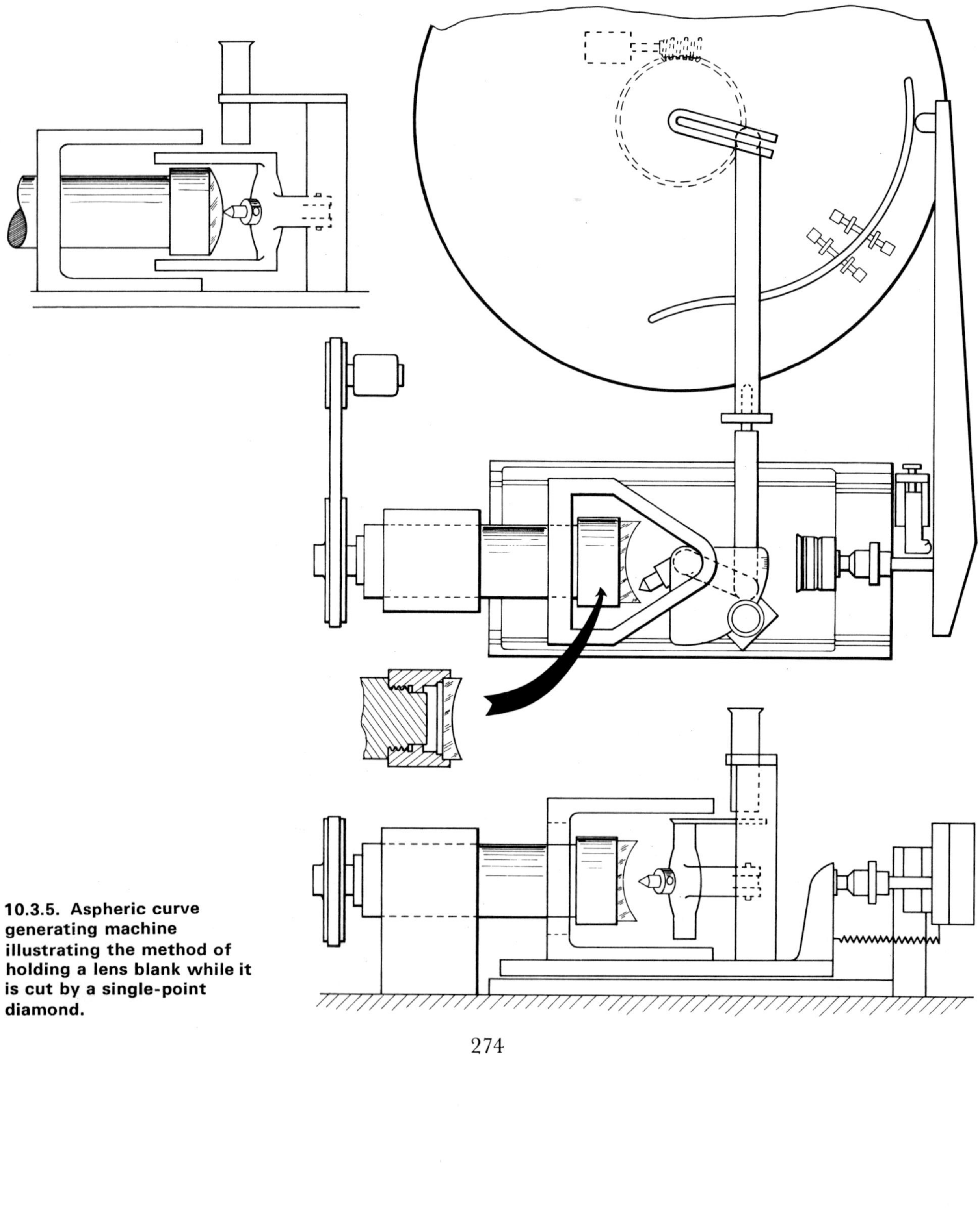

10.3.5. Aspheric curve generating machine illustrating the method of holding a lens blank while it is cut by a single-point diamond.

10.4 Aspheric mirrors

(See also Chapter 15.)

Reflecting surfaces generated by conic sections, such as paraboloids and ellipsoids, share a valuable optical property, which is that a point object located at one focus is imaged at the other focus without spherical aberration. For this reason, mirrors used in spectrographs are often specified as off-axis paraboloids or ellipsoids and, for ultra-violet instruments, a high order of accuracy is required. The tolerances for infra-red instruments are not so critical.

Aspheric surfaces are much more difficult to make than spherical surfaces and the cost can be ten times greater, so considerable care must be taken before specifying an aspheric instead of a spherical component. It may be much cheaper to have several spherical components instead of one aspheric, unless the size of the finished assembly becomes too large or too heavy.

The manufacture of off-axis paraboloids and ellipsoids follows the usual practice for astronomical objectives with the difference that the components required form only part of the mirror surface (Fig. 10.4.1).

After determining the block diameter and thickness, the components (mirror blanks) are trepanned out and then stuck back in the block again with pitch. The four types of mirror illustrated have typical dimensions as follows:

A. High-quality off-axis paraboloid (4 per block)
Block diameter: 14·5 inches (355 mm).
Mirror centre: 4·42 inches (111 mm) away from pole.
Radius R: 47·6 inches (1219 mm).
Angle A: 10° 37′.
Mirror diameter: 5·25 inches (132 mm).
Accuracy: Width of circle of confusion at best focus over any 4-inch-wide strip, 0·0008 inch.

B. High-quality off-axis paraboloid (3 per block)
Block diameter: 16·0 inches (406 mm).
Mirror centre: 4·42 inches (111 mm) away from pole.
Radius R: 47·6 inches (1219 mm).
Angle A: 10° 37′.
Mirror diameter: 7·0 inches (177 mm).
Accuracy: Width of circle of confusion at focus over any 4-inch-wide strip, 0·0004 inch.

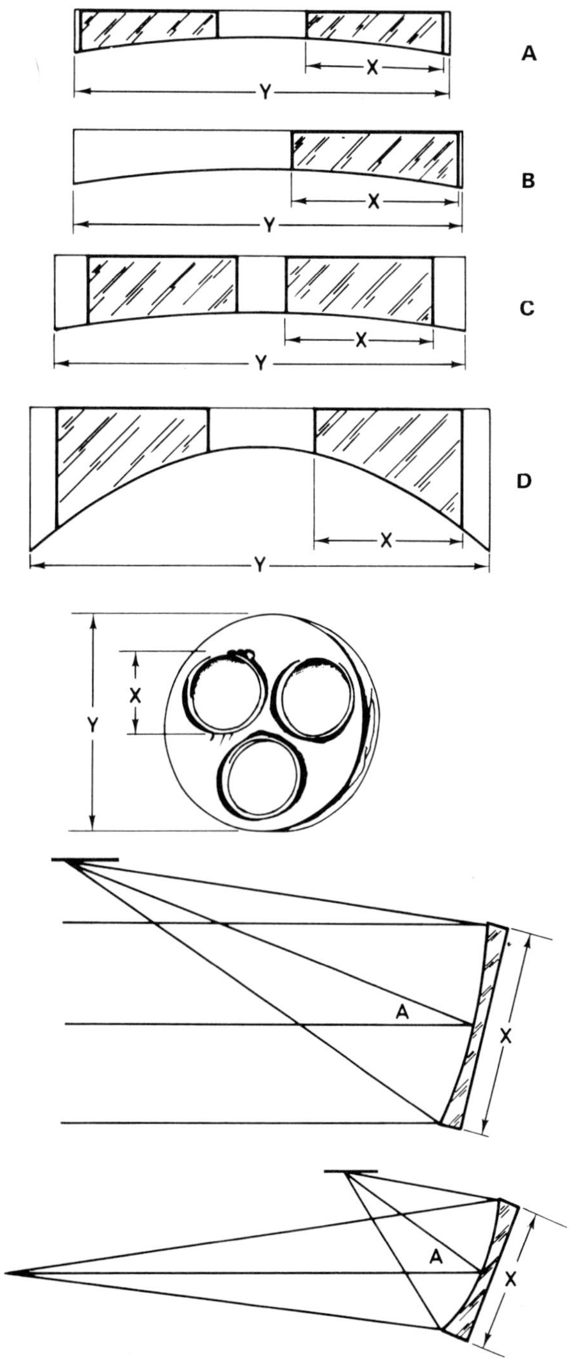

10.4.1. Off-axis paraboloid and ellipsoid mirrors, illustrating how they are ground, polished, and tested in a glass block of the same curvature. The mirror blanks are held in the block by pitch.

10.4.2. **Grubb Parsons polishing machine for aspheric mirrors. Note the polishing tool constructed from small squares of pitch, which can be removed to vary the figure required on the mirror. See also Fig. 15.2.11.**

C. Medium-quality off-axis paraboloid (2 per block)
Block diameter: 5·5 inches (150 mm).
Mirror centre: 1·3 inches (33 mm) away from pole.
Radius R: 30·0 inches (762 mm).
Angle A: 10°.

D. Medium-quality off-axis ellipsoid (2 per block)
Mirror diameter: 2·0 inches (50 mm).
Block diameter: 6·2 inches (157 mm).
Mirror centre: 1·76 inches (44 mm).
Radius R: 4·0 inches (100 mm).
Angle A: 35°.
Mirror diameter: 2·0 inches (50 mm).

The first stage is to grind the block spherical, with tools the same size as the mirror to approximately the curvature required. The grinding tool for large mirrors is often made of ceramic tiles stuck on with pitch. These are easily removed for correction of sphericity or for introducing asphericity into the surface being ground.

The second stage is to shorten the radius of curvature of the inner zones of the surface and this is accomplished by the use of full-size polishing tools. The polisher consists of a large number of squares of pitch stuck on the tool, and these are easily removed so that it is possible to give a large excess of polishing area in the central part of the tool. Alternatively, small polishing or figuring tools can be used to work the central parts of the mirror to a greater extent than the zones towards the outside edge. The form of the polishing tool can be as illustrated in Fig. 10.2.9 but with the polishing area altered to suit the particular requirement of the aspheric curve. The length of stroke and amount of side throw are also important factors in controlling the figure of the mirror. (Figs 10.4.2, 10.4.3, and 10.4.4; also 15.12.2.)

The block will be ground, polished, figured, and tested as if it were one piece of glass. High-quality off-axis aspheric mirrors must be of a standard equal to the best astronomical quality and generally to at least 0·25 fringe accuracy. The last stages must be checked using Foucault's test and figured by skilled craftsmen. It is very difficult to test the mirror segments after they have been removed from the block. If they are found to be inaccurate when fitted to an instrument, the mirrors will probably be scrapped as rectification is almost impossible.

10.5 Replication of concave aspheric mirrors

Providing good masters can be made, it is feasible to produce plastic copies almost as accurate as the original and certainly much cheaper than mirrors aspherized in glass by normal methods. The major problem is in making an accurate master of the opposite sense to the copies required for instruments (convex masters for concave copies), and even more difficult is the problem of testing the master and making corrections.

However, assuming accurate masters have been

△10.4.3. **Strasbaugh Model 6AC Draper-motion polishing machine for aspheric mirrors up to 60 inches (1520 mm) in diameter.**

▽10.4.4. **Strasbaugh Model 6AC Draper-motion polishing machine with its worktable tilted at 90 degrees from its rotating axis for checking work without removing it from the table.**

made, the following replication processes, which give a plastic aspheric surface to a spherical glass blank, are satisfactory. One method is suitable for concave infra-red condenser mirrors and the other for ultra-violet mirrors.

Infra-red aspheric mirrors

Infra-red condenser mirrors of concave elipsoid shape can be made by fitting the mirror blank (with a spacer under it) in a jig, and then pouring a layer of Araldite and aluminium powder on the blank and placing the master above it. Allow the Araldite and powder to set in an oven at 60°C for 16 hours. Separate the master from the copy using a vice and physical force. Change the spacer to another slightly thinner. Pour Araldite only on the blank and place the master above it. Allow the Araldite to set in an oven at 60°C for 16 hours.

Separate the master from the copy. Change the spacer to the final separation and once again pour Araldite on the blank and place the master above it. Allow the Araldite to dry for three days at room temperature. Separate the master from the copy. Place the copy in a vacuum plant and coat it with aluminium followed by a protective layer of silica. The silica has to be evaporated from a fine silica rod, previously aluminized to provide electric contact. The silica rod (1-mm to 2-mm diameter from Thermal Syndicate Ltd) should be surrounded by a spiral tungsten filament.

Ultra-violet aspheric mirrors

Ultraviolet aspheric surfaces can be replicated providing the deviation from a spherical form does not exceed 0·001 inch (0·025 mm).

In this process, a master optical surface is first vacuum-coated with a number of layers of material, one of which has a low mechanical strength. A smoothed backing blank is cemented to the top coated layer and, after the cement has set, the master surface and the backing blank are separated by introducing stress; parting takes place along the layer of low mechanical strength. The master should be chemically cleaned, mounted in a vacuum chamber, and discharge-cleaned. Coat it with aluminium under high vacuum, until just opaque. Break the vacuum and prepare for coating Type 704 silicone oil, aluminium, and silicon monoxide. Load the master into a jig. Evacuate the chamber to 5×10^{-5} mm of mercury. Evaporate the silicone oil. If specified, evaporate silicon monoxide as a protective layer. Evaporate the aluminium. Evaporate the silicon monoxide again for 10 minutes to give a protective backing layer between the aluminium and the resin of about 1 micrometre thickness. Break the vacuum. The plate is now ready for cementing on the backing blank and this operation should be completed within 1 hour.

The resin to be used, in parts by weight, is:

Marco Resin SB18c	100
Catalyst paste H	4
Accelerator E	4

Mix, filter and use immediately.

The radius of the blank must closely match that of the master. The blank should be smoothed with No. 303 emery and lightly buffed, so that the surface can be checked with a test plate.

Place the concave spherical blank on a table, level it with bubble gauge, and pour resin on the centre of the surface. Gently lower the convex master on to the concave blank and allow to gel for 30 minutes. Remove surplus resin from the join and leave in a temperature of 20°C for 48 hours.

The master can be separated from the copy by the thermal shock of contact with dry ice or by mechanical stress, using robustly made jigs and a hydraulic press.

10.6 Aspherizing mirrors with vacuum-coated films

In 1935, Strong and Gaviola described a method of figuring mirrors by the controlled deposition of aluminium in a vacuum. A variety of screens for use in the vacuum chamber were designed and, in turn, rotated by an electromagnet situated outside the vacuum chamber with the object of controlling the deposit on the mirror. (Fig. 10.6.1.)

The problem at that time was how to be able to deposit sufficient aluminium in one loading. The thickness of deposit required was 2·6 wavelengths and this could not be obtained in a single operation. Several reloadings were needed. The deposits therefore consisted of aluminium separated by aluminium oxide formed spontaneously when the vacuum was destroyed to reload the evaporation coils. Since 1936, other research workers have made attempts to overcome the problem of measurement of the deposit during the process, by the use of Foucault knife-edge monitoring, but the problem of depositing a sufficient mass of evaporated material remains unsolved. If efficient monitoring of the thickness coupled with continuous evaporation of material could be achieved, then an automatic servo-controlled plant might give the answer to the problem of providing low-cost aspheric mirrors.

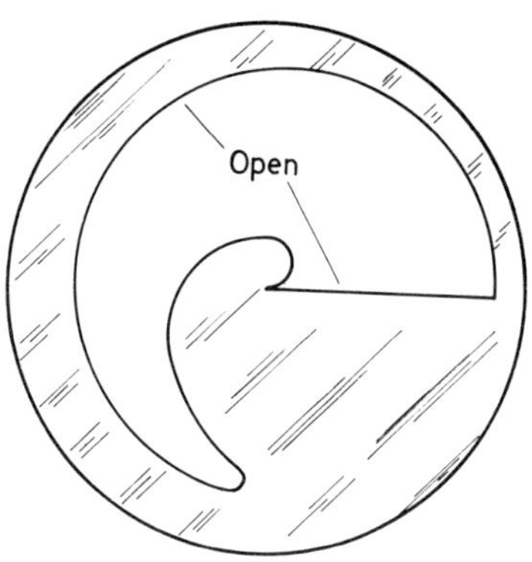

10.6.1. Screen for aspherization of a surface by vacuum deposition.

10.7 Aspherizing by removal of glass with ion beam

Because of the increased demand for low-cost aspheric optical components, new methods of generating accurate surfaces have been investigated. Since 1965, research in America and Great Britain, associated with atomic energy experiments and ion implantation and semiconductors for the microcircuit industry, has led to the development of powerful ion accelerators.

The availability of these machines has led to further research into the possibility of controlling ion beams in such a way that they may be used to chip glass away atom by atom.

There are three possible fields of application for ion beam polishing.

(*a*) Controlled polishing of aspheric surfaces.
(*b*) Controlled correction of very flat surfaces.
(*c*) High-surface-finish polishing of components such as laser mirrors.

In order that ionic polishing be feasible it is necessary to measure continuously the departure of the surface from the required shape. The measured errors must then be converted into signals which can be applied to the controls of the ion stream.

It seems inevitable that the error measurements be made interferometrically, and it is this problem of measurement and control that will require large development expenditure before ion-beam polishing becomes a feasible production method. To ensure that dimensions and accuracy requirements are understood, Fig. 10.7.1 illustrates the design of the optics for an ultra-violet $f/1{\cdot}5$ Schmidt camera where an aspheric surface is used to control all the spherical aberration of a high-aperture optical system.

10.7.1. Schmidt camera, with equation for figured plate.

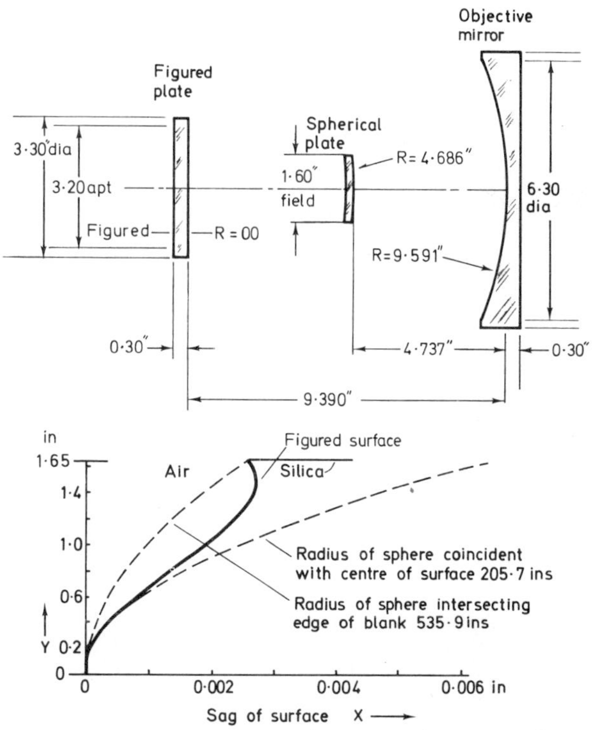

The specification calls for resolution in excess of 200 lines per mm over a 20-degree field in the wavelength region 2000 angstroms to 2800 angstroms. The back of the Schmidt plate is flat, but the front surface has a depth of figure of 0·00106 inch (0·026 mm) and the aspheric surface varies from forty-five bands to 144 bands of mercury light from the 535·9 inch (13 500 mm) radius of a sphere intersecting the edge of the blank.

At present this 3·3-inch (84 mm) diameter corrector plate is aspherized by grinding and polishing in about 40 hours by a skilled craftsman. If it were possible to have an automatic controllable method, such as ion-beam polishing, this would be an advantage and would save experienced labour, which is in short supply.

Mr L. H. Narodny, who was Chief Scientist of the Space Division, Kollsman Instrument Corporation, in 1967, has been engaged in theoretical and experimental investigations on the controlled erosion of various dielectric crystalline and metallic substances with argon ions (positive charged atoms) using low accelerating potentials up to 50 kilovolts.

Fig. 10.7.2 illustrates the interference pattern he achieved on an *f*/6, 10-cm diameter Pyrex paraboloid at its centre of curvature. Erosion was performed on a long radius (120 cm) spheroid blank using mechanical scanning techniques. Mr Narodny states:

> The depth of material to be eroded at any radial zone was calculated from the best fit between the spheroid and a reference paraboloid tangent to the spheroid at the central axis and intersecting it at the edge of the aperture. This case represents the least amount of material to be removed and therefore affords the greatest economy of time in the use of the ion-beam accelerator facility. The ion-beam dwell time as a function of mirror radius was programmed to yield the calculated depth of erosion. Up to four or five wavelengths of material thickness can be precisely removed in this way.
>
> The mirror blank was exposed to a focused argon ion beam approximately 3 mm in diameter which traced out a spiral path on the substrate surface. The exposure time was 11 hours at 30 kV accelerating potential and 100 microamperes beam current. The exposure time could have been reduced to approximately 1·5 hours by increasing the beam current, although no attempt was made to do so in this project. A small central zone approximately 1 cm in diameter was left uneroded to serve as a reference area for metrological purposes.

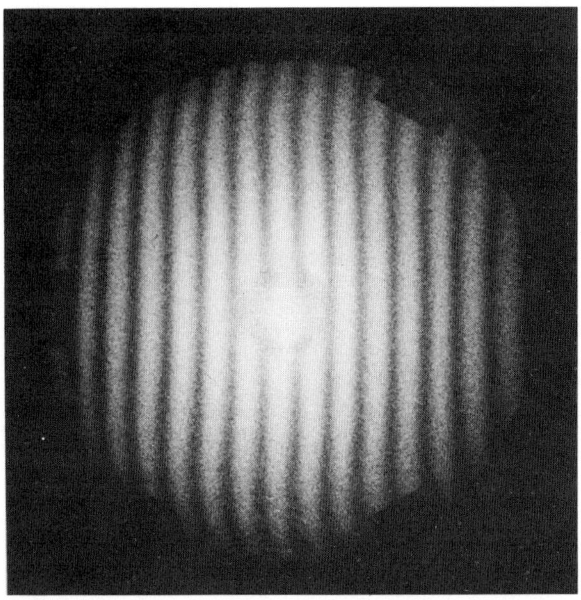

◁ **10.7.2. Scatter-plate interferogram of ion-beam-figured paraboloid.**

▽ **10.7.3. Ion accelerator and work chamber.**

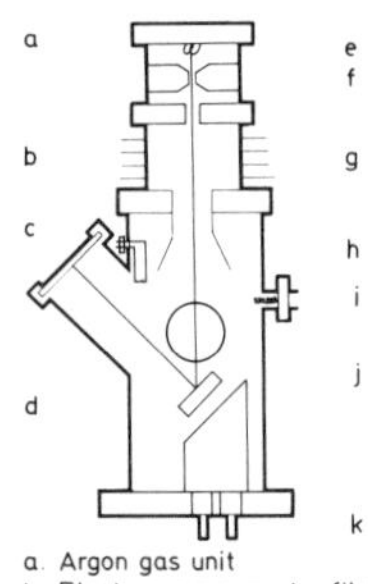

a. Argon gas unit
b. Electron compensator filament
c. Viewing port
d. Workpiece
e. Ionization filament
f. Focussing structure
g. Accelerating dynodes
h. Single axis deflection plates
i. Ionization gauge
j. Stainless steel work chamber

A computer programme was devised to obtain a contour plot of figure deviations at selected *X–Y* coordinates on the mirror using data obtained from the scatter plate interferograms of the figured surface. The results of the computations showed that an r.m.s. value (at 5500 Å) of 0·046 wavelengths figure deviation from a best fit paraboloid has been obtained.

Some interesting details of this process are that the rate of glass removal is quite uniform and depends on the material. Heating of the surface is negligible. The ion spot size on the specimen is usually between 0·5 and 1·0-mm diameter, the specimen is rotated at 10–40 r.p.m. and a spiral 'cut' taken with substantial overlap. The ion beam can be accurately deflected or scanned. An electron gun is used to prevent the blank from becoming charged. The vacuum in the specimen chamber is kept between 10^{-6} and 10^{-7} torr.

It is visualized that, when 'machining' an aspheric surface, an initial high removal rate controlled only by a fixed time programme would be followed by a slow removal rate under interferometric control.

By this method it should be possible to do in 2 hours a surface which might take two days by conventional means. The specimen could be rotated, moved on an accurate slide, or left stationary.

Fig. 10.7.3 illustrates an experimental ion accelerator used in glass polishing experiments by Mr Narodny. Further details can be obtained from the book *Ion Beam Technology*, by Leo H. Narodny (Gordon and Breach, 1971).

A machine recently developed by Edwards High Vacuum International consists of an ion accelerator, an isotope separator, and a target chamber. Argon ions produced in an R.F. source can be accelerated up to an energy of 30–35 keV and the '30-degree' mass separator magnet ensures that the ion beam at the target has a well-defined energy and is relatively free from impurities. A maximum beam current of 100 microamperes of argon ions can be produced and it is possible to focus the beam down to 1-mm diameter spot on the target. Scanning of the beam over the target surface is accomplished by means of two sets of electrostatic plates. (Fig. 10.7.4.)

10.7.4. 30-keV ion beam accelerator suitable for the controlled erosion of dielectric surfaces and capable of the aspheric polishing of glass.

10.8 Aspheric lens curve-generating machines

Before 1950, Taylor Taylor & Hobson Ltd and the Bell & Howell Company, Chicago, realized that the future development of high-quality 35-mm and 16-mm ciné camera and television zoom lenses would demand aspheric components. A design and development programme was introduced in both companies to manufacture a high-quality aspherizing machine which would produce lens blanks smooth and accurate enough to proceed direct to polishing.

Both machines were very expensive to develop but, by quite different engineering principles, achieved their objective and were in production by 1960. A brief comparison of some features may be of interest:

	Taylor Hobson	*Bell & Howell*
Lens capacity (diameter)	7 inch	6 inch
Spindle speed	1400 r.p.m.	800 r.p.m.
Cutting method	single point turning by diamond (10 micron radius)	diamond burr grinder rotating at 33 000 r.p.m.
Geometry	polar coordinate	Cartesian coordinate

THE TAYLOR HOBSON ASPHERIC MACHINE

By kind permission of Rank Precision Industries Ltd the following description and photographs (Figs 10.8.1 to 10.8.6) of the Taylor Hobson aspheric machine are published.

The conventional method of producing individual aspheric lenses or Schmidt corrector plates is to form the grey non-spherical surface either by patient hand-lapping, using flexible laps with graduated facets, or by means of cam-guided grinding machines. The grey lens or corrector plate thus produced then has to be polished and figured, again using flexible laps consisting generally of pitch, or pitch and felt facing on sorbo rubber. This is an uneconomical process commercially and involves careful optical testing and repeated refiguring of the aspheric surface. To produce aspherics in commercial quantities and in a manufacturing time which compares favourably with the time involved in producing a normal spherical lens, costly machinery of extreme accuracy is required.

The method at present used in Rank Taylor Hobson for producing aspheric elements for use in 35-mm ciné camera and T.V. zoom objectives is as follows.

The generating machines have spindles of extreme accuracy with optically polished thrust and radial bearings. In order to avoid all bias, the spindles are driven (without direct mechanical contact) through magnetic drive units. The aspheric blanks are accurately centred in chucks and whilst the spindles rotate at fast speeds the cutting tools (single-point diamonds) traverse the lens surfaces in such a manner that the required aspheric profile is produced to within an accuracy of $\pm 0{\cdot}00003$ inch. In order to work to this degree of accuracy, very careful design is required of all slideways and pivots (which again are optically

10.8.1. Rank Taylor Hobson No. 3 aspheric generating machine illustrating (*a*) punched tape controls on the left side of the machine and (*b*) slideways on which diamond tool and spindle head move for setting to suit positive or negative curves on lenses.

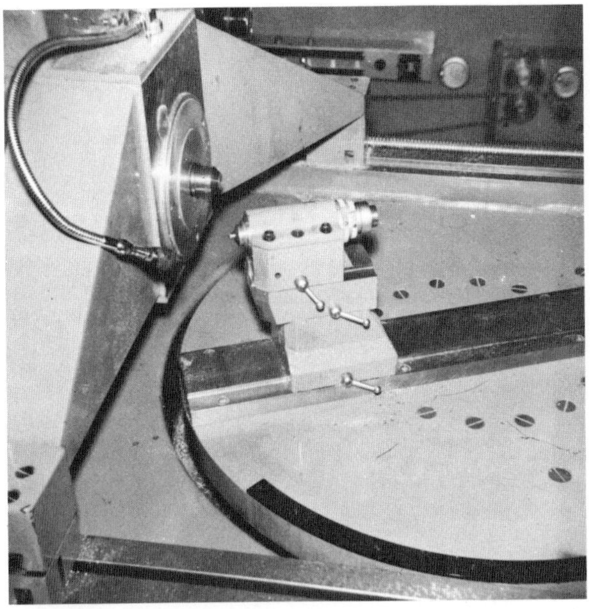

10.8.2. Rank Taylor Hobson No. 3 aspheric generating machine illustrating (*a*) chuck holding lens to be generated and (*b*) diamond tool holder.

10.8.3. Chuck, tool holder, and Talyrond probe on the No. 3 aspheric machine.

10.8.4.▷ Control tape and diagram of aspheric surface.

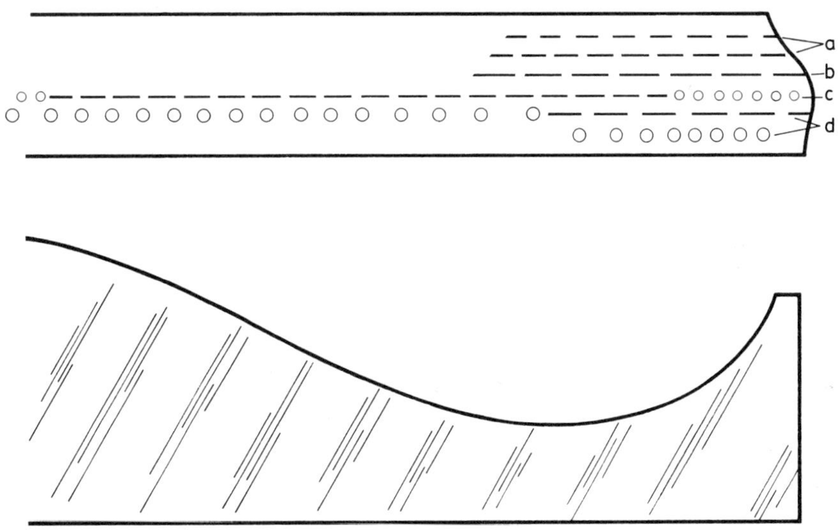

a. Two tracks used for polishing correction
b. One track used for turntable stop control
c. Sprocket holes
d. Two tracks used for aspheric data

polished) on the generating machines, and also a constant temperature must be maintained in the shop in which the machines are used.

The aspheric generators are computer controlled and data from the computer, in the form of punched tape, is fed into the machines as either rectangular or polar coordinates. As many as 400 sets of coordinates, each to five places of decimals, may be used for a lens no more than 1 inch in diameter. This information is, by means

10.8.5. Punched-tape control on the No. 3 aspheric generator.

of electronic controls, transmitted to the moving elements in the machine in order to superimpose the asphericity of the lens on to either a basic circular or a straight-line movement of the diamond tool-holder. The asphericity (in the case of a photographic lens element) is defined as the departure from the best fitting sphere, or (in the case of a Schmidt plate) as the departure from plano. Aspheric lenses may be made up to 7 inches in diameter, and Schmidt type corrector plates up to 11 inches in diameter.

The grey diamond-turned surface produced by the aspheric generator is of sufficiently fine texture to permit direct polishing without any intermediate smoothing. The polishing is not a figuring process, and no inspection of the aspheric form is made during polishing. The aspheric lenses may be polished singly or in blocks, but, before polishing, the lenses are accurately mounted in glass surrounds to avoid edge effects. The polisher consists of a flexible diaphragm which is maintained in contact with the aspheric surface or surfaces by means of pneumatic pressure. Specially designed polishing machines are used which ensure the minimum change of figure during the process. (Fig. 10.9.4.)

The procedure is to accept some change of figure but to control the polishing process with sufficient accuracy to ensure that this change is always constant from one lens, or block of lenses, to the next.

When the first lens of a production batch of aspherics has been polished it is again placed (accurately centred) on the spindle of the aspheric

10.8.6. Electrical controls and punched tape for the No. 3 Rank Taylor Hobson aspheric generating machine.

generator. An electrical transducer is mounted in place of the diamond turning tool with the stylus of the transducer in contact with the polished glass surface. The stylus is then traversed across the lens surface and the error from the specified aspheric profile is graphed on a recorder.

This error is now applied as a correction to the data on the original generating tape in such a manner that excess glass is left on the lens (for example, the centre) where the polishing process has removed too much. All future aspherics of this design are now generated using the corrected tape and, providing the polishing process is controlled so that it remains constant in cycle time, stroke, pressure, and slurry feed, the aspheric lenses will remain constant in figure and will now be of the correct aspheric profile.

The process for the quantity production of aspheric lenses has been under development by Rank Taylor Hobson since 1945 and has evolved from the early use of flexible steel strips to produce the generating cams, followed by the later application of photographic cams used in conjunction with photoelectric followers. The present method of feeding tape direct from the computer into the generator is of recent development and has resulted in a considerable shortening of set-up time, together with a broadening of the range of aspherics which can be produced.

THE BELL & HOWELL ASPHERIC MACHINE

By kind permission of the Bell & Howell Company this survey by Dr A. Cox and M. F. Royston of the Bell & Howell aspheric machine is published.

This account of work on the manufacture of precision aspheric surfaces is essentially an interim report describing the success that has been achieved in critical manufacturing processes. The basic concept of the process which is being developed is that the shape of the surface must be accurately generated in a machining operation with such a surface texture that shape is not lost in subsequent polishing operations. Means have been devised to position a tool carriage within 1 or 2 micro-inches using a so-called 'electrostatic nut' as a measuring device. A system of control using punched papertape as the original input has been devised which permits the movement of the tool carriage to be programmed so that an initial attempt may be made to generate the aspheric contour and so that corrections may be made if the contour produced on the machine does not accurately match the contour desired. Two basic types of glass machining have been devised. The first involves the turning of glass with a single-point diamond in a tool of special construction. The second depends on the use of a high-speed diamond burr. With each type of machining a surface finish is produced which permits polishing in a short time without significant change in surface contour.

The aim of this programme was to develop a method of manufacturing precision aspheric surfaces, in reasonable quantities and at a reasonable cost, without recourse to the skilled craftsmanship that was previously required to produce them. The precision required can be related to the end-use of the surfaces: they are to be capable of being used in military equipment in such a way as to provide higher standards of performance than can be realized under the same conditions with spherical surfaces only. Included in this category are such items as high-aperture and wide-angle aerial reconnaissance lenses. The manufacturing time, and the associated cost, are to be considered in the proper context. There was no intention of attempting to develop a process that would lead to the production times that are common in commercial work. What was sought was a process that would be faster, by a whole order of magnitude, than the currently available techniques.

Basic principles

Two basic principles were adopted at the beginning of the development programme:

(1) The precise contour of the surface would be generated in a machine tool, the surface texture being such that the polishing time would be held to a minimum, and the chances of losing the figure during polishing would be minimized.

(2) The machine tool should be capable of adjustment, so that if the contour generated in actual operation of the machine did not conform exactly to that desired, it should be a simple matter to make corrections.

With regard to the first basic principle, we drew heavily on our experience at Bell and Howell in the manufacture of optical elements for ciné cameras and projectors. The rapid expansion of the demand for these elements, and the very competitive market, had forced upon us a re-examination of the basic manufacturing processes. There was not the supply of skilled manpower available, nor could the cost be accepted, for a classical manufacturing process, in which the final lens figure was generated during the polishing operation. In place of this a process was developed in which the figure was precisely controlled prior to the polishing operation. The latter then became more of a burnishing operation, and stock removal for establishment of the final figure was eliminated. At the same time polishing cycles were drastically reduced. We believed that this line of approach could, and should, be carried over into an aspheric programme.

With regard to the second basic principle, we believed that in working down to an accuracy of a few micro-inches it would not be possible to predict with absolute certainty what would be the behaviour of a machine tool. Much of the value of the programme would be lost if adjustment of the machine were laborious and time consuming, and it was therefore held to be essential to incorporate some easy adjustment. In passing it may be noted that at an early time it had been decided not to limit the machine to surfaces such as conicoids of revolution: they do not provide enough freedom for the optical designer.

With these considerations in mind a survey was made of the machines that were available, in both the optical and metal-working industries, and after discussions with machine tool manufacturers the general character of the machine was decided upon.

General features of the design

Essentially the machine proposed comprises two carriages, each riding on its own ways, that are fixed on the bed of the machine, as shown in Fig. 10.8.7, namely a work carriage and a tool carriage. For the sake of brevity we shall call the direction of movement of the work carriage, on the cross slide, the X-direction, while the movement of the tool carriage is in the Y-direction, at right angles to the X-direction. The work carriage,

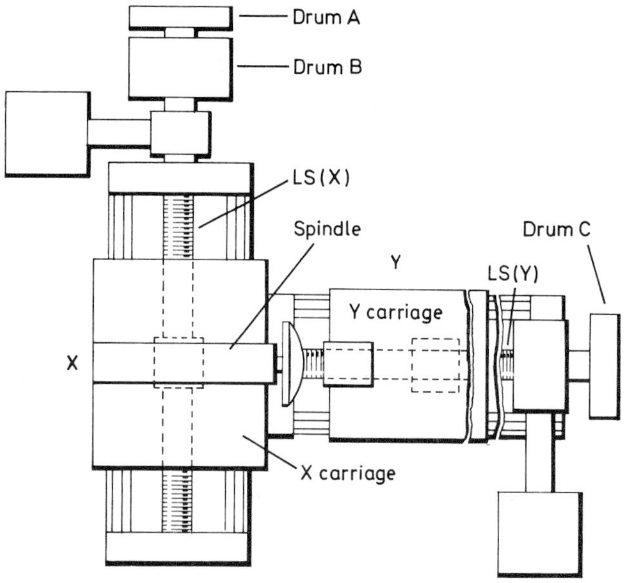

10.8.7. Basic machine schematic.

or X-carriage, carries a spindle upon which is mounted the glass to be machined. The tool carriage, or Y-carriage, carries either a tool, with which the glass machining may be performed, or a probe with which the surface contour may be surveyed. The movements of the carriages are effected through precise leadscrews, $LS(X)$ and $LS(Y)$. The X leadscrew is driven by a variable speed motor, the speed changes of which may be programmed, if it is so desired. The Y leadscrew, on the other hand, is driven (through appropriate gearing) by a stepping motor. This type of stepping motor is characterized by the fact that its rotor moves through a discrete rotation when a command impulse is fed into it from an appropriate circuit. The sequence of commands to the stepping motor is derived from an input tape. Errors in the contour which is generated on the first set-up of the machine are corrected by changing the commands on the input tape.

This form of machine was chosen because the problems which arise in translating it from a statement of operating principles to a practical reality, with the degree of precision that is required in the end product, appear to be soluble with the resources of present day technology. For the remainder of this discussion we will be concerned with the details of construction and with the techniques which make it possible to realize a practical machine of this type.

Tape control of Y-motor

Before discussing problem areas, however, it is worthwhile considering in more detail the way in which the *Y*-motor commands are related to the tape input instructions, and how this eliminates some potential problem areas.

On the *X*-carriage leadscrew there are mounted two magnetic drums, concentric with the axis of the leadscrew. Call these the *A* and *B* drums. A similar drum, the *C* drum, is mounted on the *Y*-carriage leadscrew, as shown in Fig. 10.8.7. The *A* and *C* drums share the common feature that the read heads associated with them have no relative movements along the axes of the drums, and the latter degenerate almost to disks. On the other hand the *B* read head moves relative to the *B* drum, as the latter rotates with the *B* lead screw, and traces out a spiral path on the drum surface. Read-out pulses from the *C* and *A* reading heads correspond to approximately $2{\cdot}5 \times 10^{-6}$ and 1×10^{-6} inches of travel of the *Y* and *X* carriages respectively. (In order to achieve this result, with practical packing densities of magnetic pulses on the *C* drum, it is necessary to use two reading heads in quadrature on each track and to generate read-out pulses from both the leading and trailing edges of the pulses on the drum.) A computer, used in the calculation of optical systems incorporating aspheric surfaces, is programmed to produce an output tape which records how many *A* pulses, or incremental movements of the *X*-carriage, must be recorded before the surface contour calls for a *C* pulse, or one increment of movement of the *Y*-carriage. This tape is fed into a translating device, and the numbers recorded on it are stored in counters within the translator. The drive motor for the *X*-carriage is then started up, and the pulses obtained from the *A* reading head are fed into the translator so that they reduce the number stored in the translator counter. When the contents of this counter reach zero a pulse is emitted that is recorded on the *B* drum. Subsequent operation of the machine is controlled by these *B* pulses, as detailed below, and with this mode of operation the machine is made self-contained and is divorced from the computer. The translator costs only a fraction of the simplest computer cost. Once the *B* pulses are recorded the *A* pulses serve no further purpose.

In operation of the machine, an impulse generator delivers pulses to the *Y*-carriage motor through a gating system. This gate is opened when a *B* pulse is fed into the system. It is closed when the rotation of the *Y*-carriage leadscrew, and of the *C* drum, sends a counteracting pulse into the system. The fact that the machine operates with information fed back from the *Y*-carriage leadscrew directly means that backlash in gearing, and other types of compliance in the drive, is eliminated. To secure smooth operation the *Y*-stepping motor is so designed that four pulses from the pulse generator are needed to move the carriage through one increment, i.e. $2{\cdot}5 \times 10^{-6}$ inches. The frequency of the pulse generator may be varied to suit working conditions.

The electrostatic nut

In the machine as originally conceived, it was proposed to determine the carriage positions, particularly that of the *Y*-carriage, by cube-corner interferometers. Later developments caused a change in this thinking. One such development was the realization of the fact that an adequate accuracy of positioning of the *X*-carriage could be achieved by using commercially available precision leadscrews. The slope of the surface is not expected to exceed 0·25, and this means that the movement along the *X*-axis need not be controlled with the precision needed for the *Y*-axis movement. The second, and most significant development, was the invention by Gerhard Lessman, of the Bell and Howell Company, of the so-called 'electrostatic nut'. The invention was quite independent of the work now described, but found an immediate application in the development of the aspheric machine.

The principle of the electrostatic nut is shown in Fig. 10.8.8. There is a clearance of a few thousandths of an inch between the leadscrew and the electrostatic nut, the latter being supported by the carriage whose position is to be determined. The nut itself is made of an insulating material, but the flanks of the threads are coated with a conducting layer. This layer is broken at the roots and crests of the threads, so that a double spiral is formed, each branch of which is insulated from the other. The two branches of the spiral together with the leadscrew form two arms of a capacitance bridge. The bridge is balanced when the threads of the leadscrew lie midway between the conducting flanks of the electrostatic nut. Any deviation from this position causes an imbalance of the bridge. The electrostatic nut extends over perhaps twenty threads of the leadscrew, and averages out the random errors of the latter. It constitutes a rugged device, comparatively insensitive to vibration, to eccentric location of the leadscrew axis relative to the nut axis, and to drunkenness of the leadscrew, while at the same time a sensitivity to axial displacement of 1×10^{-6} inches is readily achieved.

Early electrostatic nuts were made of glass or Pyrex, with a conducting film evaporated on the threads. This film was subsequently scraped away at the roots and crests of the threads, to form the double spiral. Later nuts were made by an improved technique, in which the nut itself is formed by epoxy compound that is allowed to set in the cavity between two brass members. The excess brass is turned away and there remains an epoxy nut carrying two thin brass spirals on its flanks. This procedure is made possible because of the way in which the brass adheres to the epoxy.

The use of the electrostatic nut provides a means of eliminating one problem that might otherwise have arisen, namely the fact that the

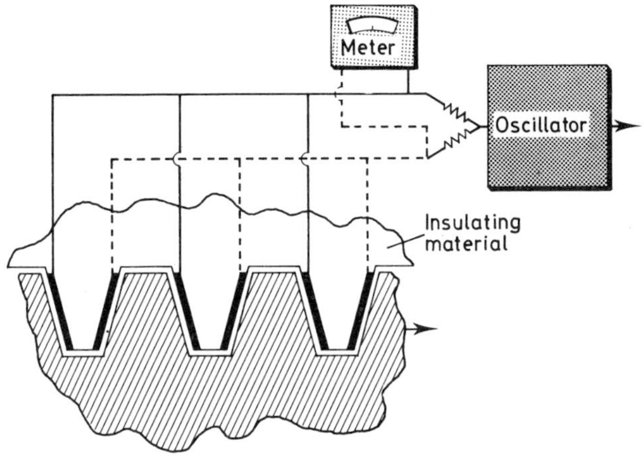

10.8.8. Principle of the electrostatic nut.

advance of the Y-carriage with rotation of the leadscrew might not be sufficiently uniform. The method of employing the electrostatic nut to secure this result is shown in Fig. 10.8.9. The Y-carriage is divided into two parts Y_1 and Y_2, while the leadscrew is separated into two co-axial parts $LS(1)$ and $LS(2)$. The two parts of the leadscrew are joined by a bellows coupling, having compliance for relative axial movement but without compliance for relative rotation of the two parts $LS(1)$ and $LS(2)$. Y_1 carries the electrostatic nut, while Y_2 carries the power nut which engages with $LS(2)$ and provides the means by

10.8.9. Leadscrew-error-correcting loop.

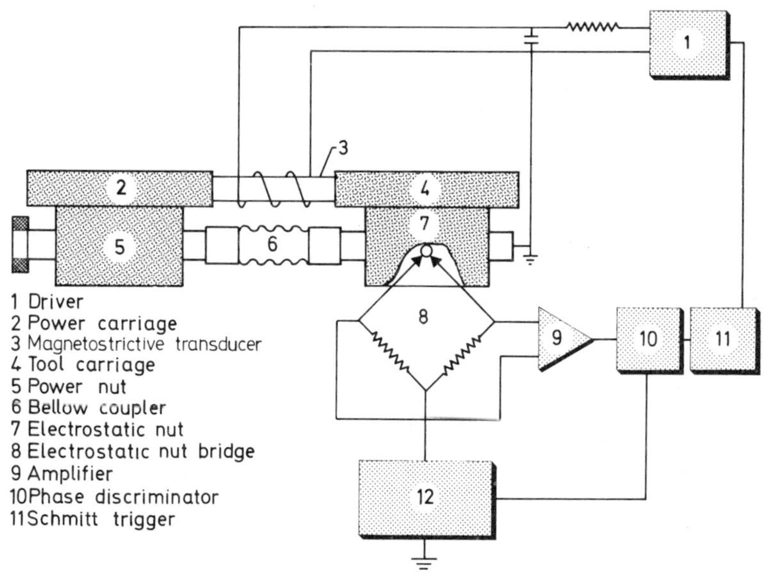

which the carriage is moved. Y_1 and Y_2 are tied together by a magnetostrictive unit. If the movement of Y_1 is not strictly proportional to the rotation of the leadscrew, because the movement of Y_2 is not uniform, then the bridge of the electrostatic nut becomes unbalanced and current is fed to the windings around the magnetostrictive element. The latter varies the relative separation of Y_1 and Y_2 until the former takes up its proper position relative to the leadscrew. The whole system of sensing device and power input to correct the error that is sensed constitutes, of course, a servo loop. It is not possible to design this servo system entirely on paper, because of the presence of too many unknown parameters, but it will be shown below that by the choice of proper circuit constants and proper amplifier gain in the servo loop, the Y_1 carriage can be correctly positioned within less than 1×10^{-6} inches. To minimize errors, the thrust bearing for the leadscrew *LS* (1) is mounted directly below the mean plane of the surface to be machined. The general layout is shown in the system schematic on p. 293.

The carriage ways

The next area to be discussed, in which problems may arise, is concerned with the type of ways on which the carriages ride. The prime requirements for these carriage ways are that the movements of the carriages on them must be accurately repeatable, and that the carriages may be set in motion with the least amount of initial sticking. A low frictional drag on the carriage movement, once it has been initiated, is desirable but not essential. It can be taken care of by a suitable choice of amplifier constants and amplifier gain in the servo loop embodying the magnetostrictive unit. Stickiness and inconsistency, on the other hand, cannot be handled in such a simple way. Considerable time was spent in discussion of the type of ways to be adopted and the final decision was made largely on the basis of the experience of the Moore Special Tool Co. and on an evaluation of their manufacturing techniques in the field of high-precision equipment.

In the final choice the ways on which the *X*-carriage rides were taken to be mating pairs of *V*-ways, constituting an extremely precise friction slide system. This provides a very stiff system, particularly with regard to any casual movement at right angles to its stipulated direction of movement. Break-away forces are of less importance for the *X*-carriage since this is in constant movement during the machining of the glass. The conditions under which the *Y*-carriage operates are very different, particularly with regard to break-away behaviour, since this carriage moves with a stop-and-start regime. Accordingly, the *Y*-carriage ways were stipulated to be *V*-ways with extremely precise rollers interposed between the mating surfaces. There are stepping motors available with an adequate torque at speeds greatly in excess of 2000 steps per second, so that the speed of movement of the *Y*-carriage is nowhere near a limiting value. The rate of movement of the *X*-carriage presents no problem, since this carriage is in continuous motion and is driven by a 400-cycle motor with more than adequate torque output. In judging the adequacy of the torque output of these motors, the frictional forces needed to move the *X* and *Y* slides of the test machine, as obtained by measurement, were combined with very conservative figures for the gearing efficiency, obtained by measurement on systems closely approximating the final gear systems.

Glass machining techniques

Other areas in which problems may develop are those concerned with the work spindle and with the machining of glass. These represent an interlocking set of problems that are best discussed by starting with a consideration of glass machining techniques.

The basic aim of the glass machining methods proposed was to develop a surface with a finer texture than that produced by the finest grade of emery in normal use ($303\frac{1}{2}$), and with a subsurface structure that would permit rapid polishing.

The first method that was tried was single-point diamond turning with a tool based upon one shown to one of the authors (A. C.) by Dr Leonard Sayce of the N.P.L. in England. The notion of diamond turning of glass is not new and has been tried with varying degrees of success by a number of workers. In general, however, the surface texture is very coarse and not at all susceptible to rapid polishing. With the use of the Sayce tool, however, some striking results were obtained. The Sayce tool essentially is a three-pronged unit: on the centre prong is mounted the cutting diamond, while the flanking prongs are faced with neoprene rubber. When the diamond reaches its cutting depth the neoprene facing pads are under compression, and their internal friction dampens out high-frequency vibrations. Experiments were carried out, using Sayce tools with detailed design variations, on a rebuilt Hardinge lathe. Two types of glass were used, borosilicate crown and dense barium crown. The Hardinge lathe, while it is an excellent machine of its type, does not pretend to achieve the accuracy of the machine tool for the manufacture of aspherics, and therefore the only aspect of the work given serious consideration was the quality of the surface finish. Surface contour was not considered.

The polishing time with a flexible lap for a 4-inch diameter disk ground with $303\frac{1}{2}$ emery was about 75 minutes; the polishing time for the same glass machined with a single-point Sayce tool was 20 minutes.

The single-point diamond, used in this manner, has some disadvantages. In the first place, even with X-ray orientation of the diamond, the tool tends to have a short life, and small chips which may break out of the tool tend to create deep cuts or scratches. In the second place the r.p.m. of the work spindle determines the maximum cross-feed rate of the diamond tool, and thus the time it takes to machine a given diameter of lens. At 2000 r.p.m. it takes 2 hours to machine a 4-inch diameter lens. And finally, a heavy damping pressure on the Sayce tool is needed to secure an adequate damping effect. This last factor was regarded as being of particular importance in view of the problems of designing a satisfactory work spindle.

Thus, in spite of the initial success that attended the use of a single-point diamond, it was necessary to seek further improvements. The first attempt along these lines was to give the Sayce tool a vibration of ultrasonic frequencies, in a direction at right angles to the tool axis. This approach did not produce any improvement in surface finish but it did substantially reduce the damping pressure needed. Work in this direction, however, was suspended when more promising results were obtained with the diamond burrs, in the work described immediately below.

The diamond burrs finally used comprised a

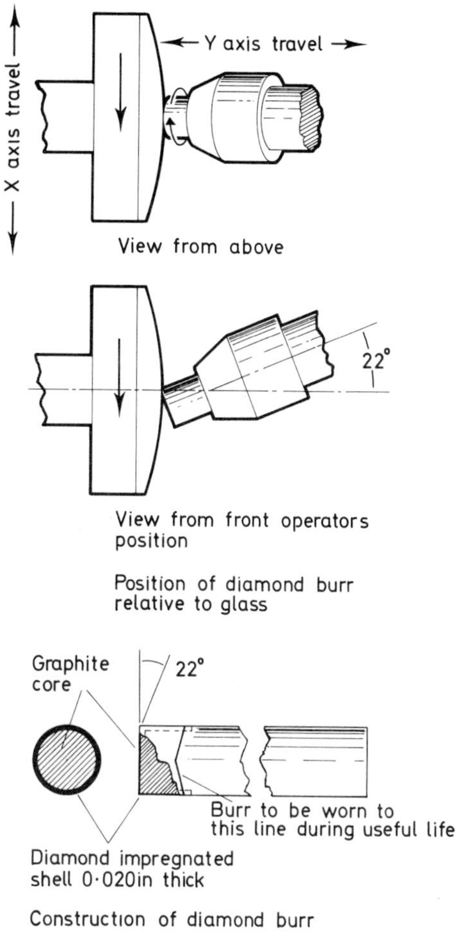

10.8.10. Diamond burrs.

thin shell of diamond-impregnated bonding material and a graphite core, as shown in Fig. 10.8.10. A thin shell was used so that wear conditions during use could become stabilized. More work is needed on the exact dimensions of the diamond burr, including wall thickness and bonding material, to establish the wear conditions that will prevail in use, but this will have to wait until the final machine is ready. The Hardinge lathe used for texture experiments did not have enough accuracy for this work. The wear problem is not expected to be too severe since it is proposed to take a final dressing cut of about 0·001 inch. The thin walls that will be used in the burr do not provide enough rigidity on their own. For this reason a graphite core is used to make a rigid tool. Graphite has an advantage in that, as the bond of the burr wears, the core material does not abrade the glass surface and readily wears away with the bond. Various materials for the bonding material have been tried, as well as possible substitutes for diamond, but the best results were obtained with the 1-M bond of Super-Cut Inc. of Chicago and 5000 grit diamond dust.

The orientation of the burr relative to the glass surface is shown in Fig. 10.8.10. With the $\frac{3}{8}$-inch diameter burr driven by a standard 1 h.p. air turbine motor made by the Moore Special Tool Company the surface roughness obtained on a 4-inch diameter disk of glass is shown on the bottom line of Fig. 10.8.11. The improvement even over the texture obtained with a single-point diamond is evident. This was reflected in the polishing time which decreased from 20 minutes to 12 minutes. The diamond burr possesses a number of advantages relative to the single-point Sayce tool. One of the most important is that the pressure of the tool against the glass is greatly reduced. Another is that the machining speed depends more on the rotational speed of the burr than on the speed of the work spindle and the latter can be greatly reduced. Thus with the burr rotating at 33 000 r.p.m. the work spindle needs to rotate at only 800 r.p.m. for a cross-feed of one inch in 15 minutes. The machining time for a 4-inch diameter lens is 30 minutes, but we have hopes of reducing this. The combination of low thrust on the work and comparatively low rotational speeds greatly simplifies the problem of spindle design, and is conducive to the production of an accurate spindle, with long life and virtually free from vibration. The life of the burr is expected to exceed by many times the life of a single point diamond.

At present the cross-feed speed is taken as 1 inch in 15 minutes, or about 0·0011 inch per second. The tentative limitation placed on the maximum slope of the aspheric surface is 1:4. This gives a maximum rate of advance along the *Y*-axis of 3×10^{-4} inches per second. Since each incremental step along the *Y*-axis is close to $2{\cdot}5 \times 10^{-6}$ inches, corresponding to one *C*-pulse, the maximum rate of occurrence of *C*-pulses is 120 per

10.8.11.
Surface finish.

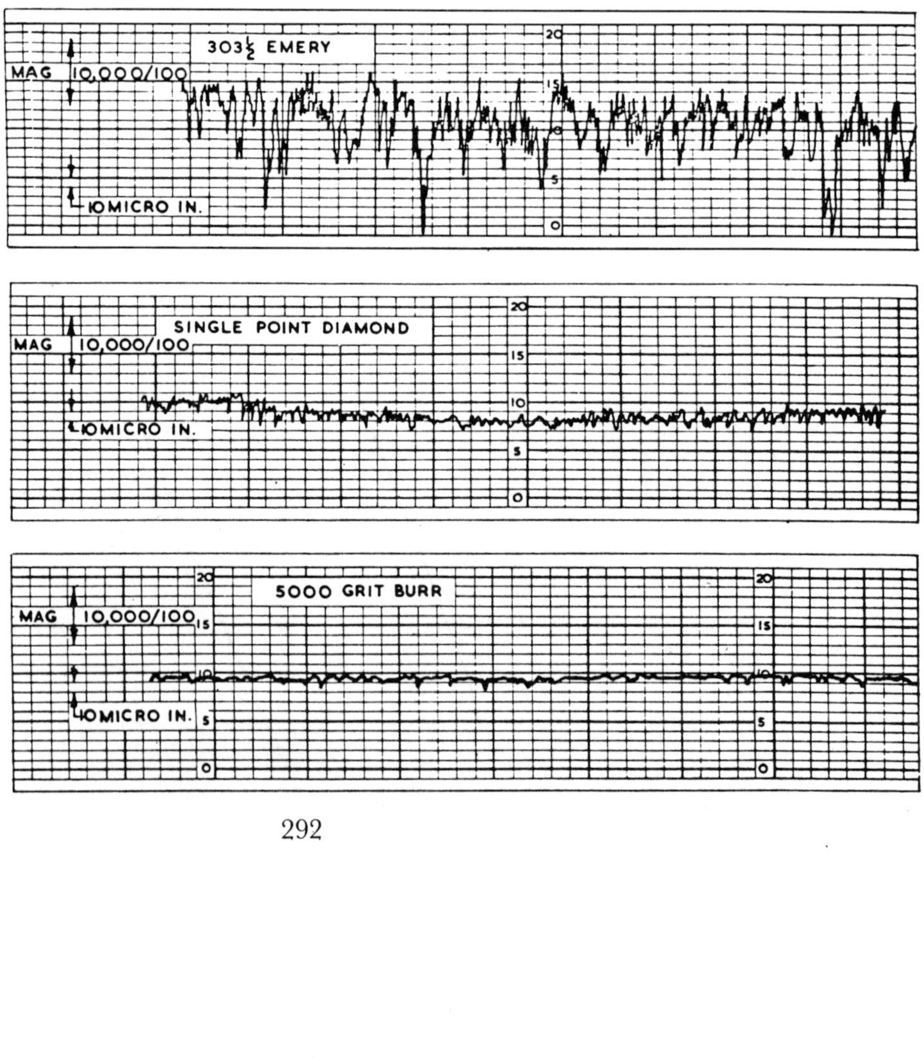

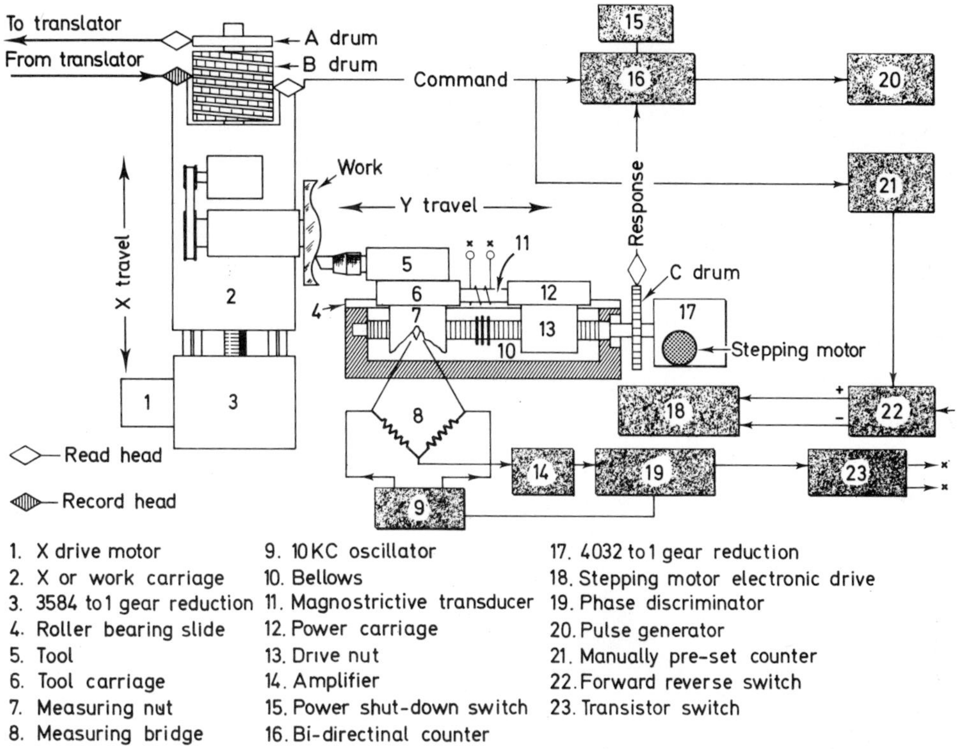

10.8.12.
System schematic.

second. There is thus a very considerable margin to spare, since the machine operation has been shown to be satisfactory from 1 pulse per second up to 2000 pulses per second.

The work spindle

The design of the work spindle has always been regarded as one of the critical elements of the machine design, both from the point of view of its effects on the surface contour, as well as from the contributions that it may make to surface texture. The work with the Hardinge lathe and the diamond burrs has considerably relieved the concern that was felt concerning surface texture. At the same time this work has more closely defined the conditions of use that the spindle will encounter, namely low thrust pressure and comparatively low rotational speed.

In the design of the spindle, three factors have to be taken into account, namely: eccentricity of rotation, movement against the thrust bearing, and method of holding the glass. Between the shaft of the rotor, and the sleeve of the stator, both of which can be machined by the Moore Special Tool Company to extreme degrees of straightness and roundness, there are trapped a large number of rollers that are cylindrical within 5×10^{-6} inches. There is a small interference fit of the rollers in the space provided. It is the ability to produce precise parts that makes it possible to design a spindle such that the rollers can be uniformly loaded by the employment of mild interference fits. This results in an essentially frictionless bearing and eliminates any possibility of radial lost motion.

This thrust bearing consists of a single hardened hemisphere ground on the end of the spindle and pulled against a flat plate. To obtain this traction, a permanent magnet system is used. This system of creating the holding force on the spindle eliminates problems that would be encountered with springs and the devices needed to transmit the spring pressure to the spindle.

In order to hold the glass, the front end of the spindle constitutes a magnetic chuck of novel construction. An initial holding force is created by a small momentary current through the energizing coils, while a second, or clamping force, is provided by a larger momentary current that saturates the Alnico. The glass to be machined is

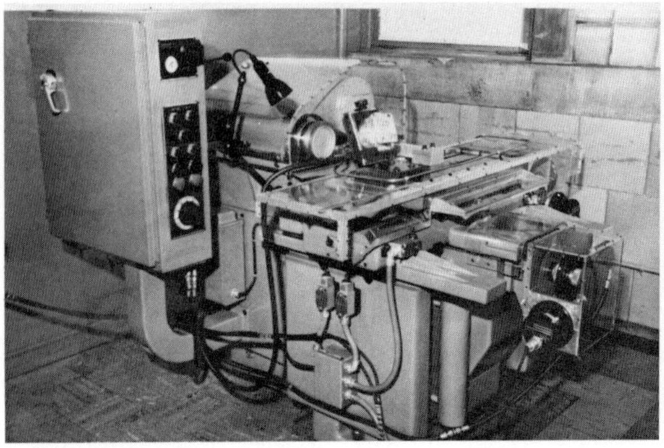

△ **10.8.13. Bell & Howell aspheric generator.**

△ **10.8.14. Bell & Howell aspheric generator from the front.**

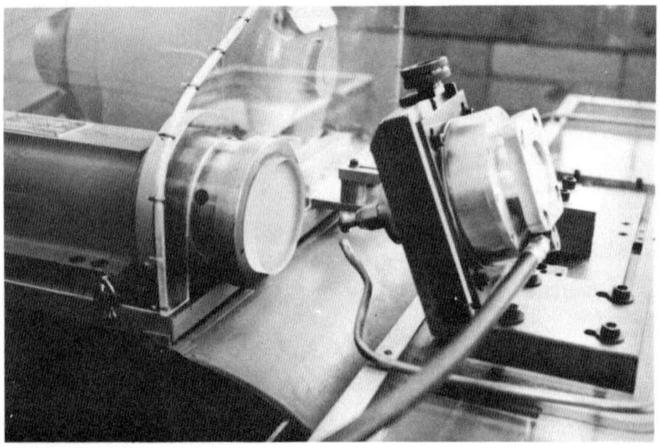

◁ **10.8.15. Bell & Howell aspheric generator, showing the work spindle and the high-speed diamond burr.**

attached to a steel plate by any of the classical methods of the optical industry. This plate is then mounted on the magnetic chuck with only the first or holding force in action.

It is moved about until the glass is centred, as shown by a sensitive indicator bearing on the edge of the glass, and the clamping force is then applied while the glass is machined. When the machining process is complete the clamping force is relaxed by a neutralizing current and the plate carrying the glass is released and may be slid off the magnetic chuck. The spindle itself is driven by a 400-cycle two-phase servo motor, with a locked torque of 2·7 inch ounces, through a flexible belt to minimize the effects of any vibrations.

Contour error correction

As was pointed out earlier, it is considered practically impossible to predict with certainty what the machine behaviour will be when dimensions of only a few micro-inches are involved. To overcome this difficulty, a tape is prepared which gives the tool carriage movement if it carries a probe that contacts the aspheric surface, and this tape is fed into the machine in the usual way, via the translator. The diamond burr is then replaced by a Talysurf probe and the machine set in operation. Contour errors are recorded on the Talysurf tape. These errors are fed into the computer that is used to prepare the control tape for the diamond burr movement and a corrected tape is prepared. This process may be repeated, if necessary.

The system schematic

The components described in the above pages have been brought together in the design for a finished machine tool. A schematic diagram of this is given in Fig. 10.8.12. A general idea of the appearance of the machine may be obtained from Figs. 10.8.13 to 10.8.15.

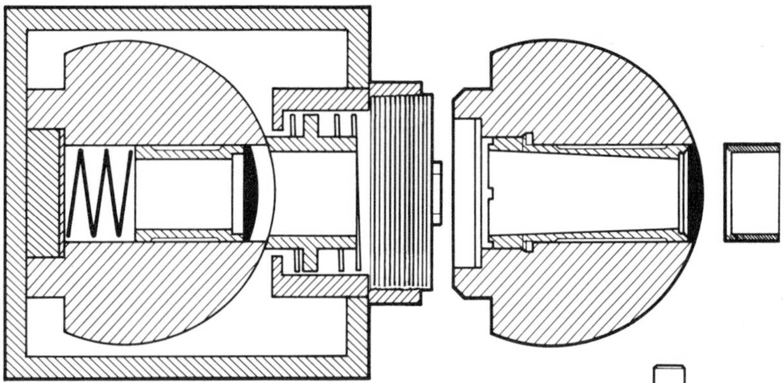

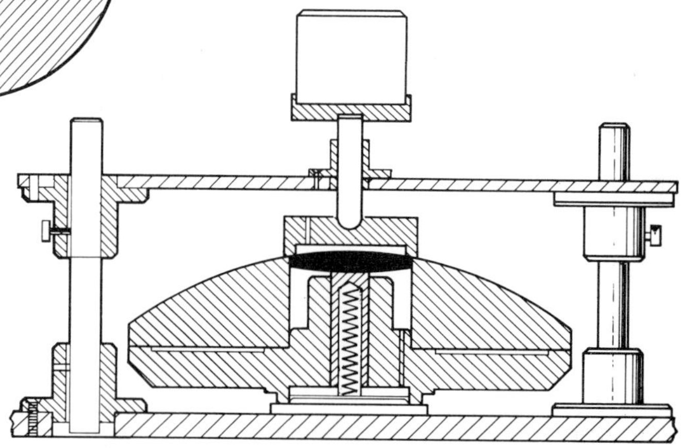

10.9.1. Workholders for aspheric lenses. The lenses are held in position by pitch and set in the correct position by the fixture.

10.9.2. ▷
Block holding three aspheric lenses. The lenses are held by pitch and set in the correct position by the fixture.

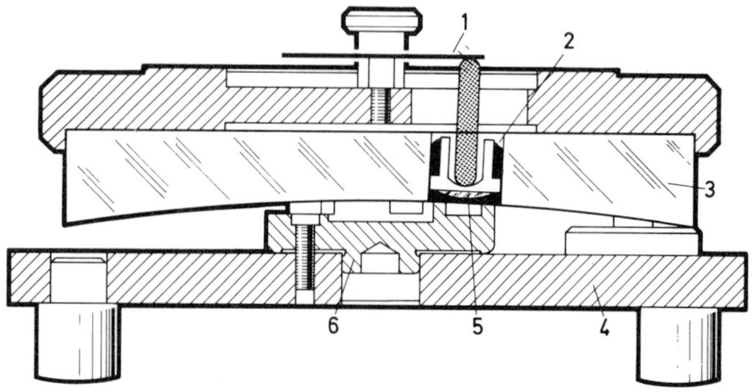

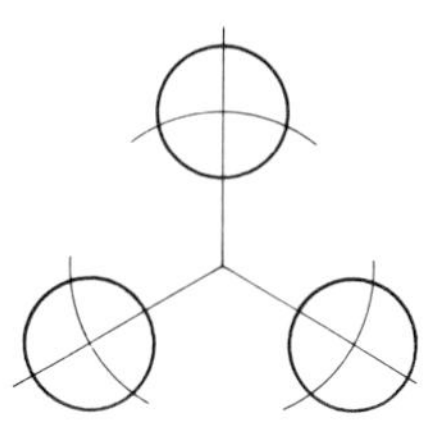

10.9 Aspheric lens polishing machines

After generating the aspheric curve on the lens, it is most important that the lens be correctly held during the subsequent polishing operation or the figure will be spoiled. Rank Taylor Hobson developed a method of holding the lens blank which, in principle, is similar to that described in § 10.4 for aspheric mirrors.

Figs 10.9.1 and 10.9.2 illustrate the method of locating the aspheric lens in recesses in a glass block which has a spherical radius of curvature,

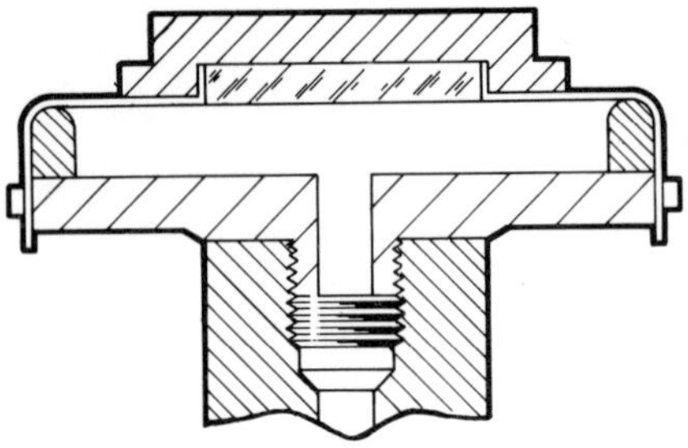

substantially an extension of the surface to be polished (Patent No. 947732).

After location, the jig is warmed to a suitable temperature for pitch or wax to be applied so that the aspheric lens is stuck to the glass block. After cooling, the jig can be removed, and the block surface cleaned prior to polishing.

The object of the special fixtures for locating the lenses and the flexible polishers is to ensure that each lens is polished evenly from its periphery to its centre.

The polishing device (Patent No. 947176) consists of a nylon diaphragm covered with a thin flexible layer of pitch which is kept in contact with the block of aspheric surfaces by means of air pressure. This ensures an even pressure over the whole of the surface to be polished. (Fig. 10.9.3.)

In order to ensure that the polishing of the lenses is as even as possible an improved polishing machine was developed (Patent No. 947177) which provides for a rocking movement suitable for this particular type of block and polisher. (Fig. 10.9.4.) The usual types of polishing compounds can be used with this machine and degradation of the aspheric curve during the polishing process does not exceed about 2 fringes of mercury green light. Surface-shape loss due to polishing is corrected in anticipation during diamond turning.

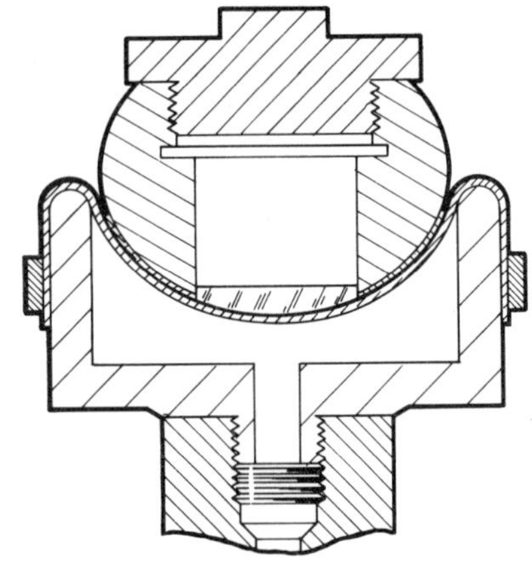

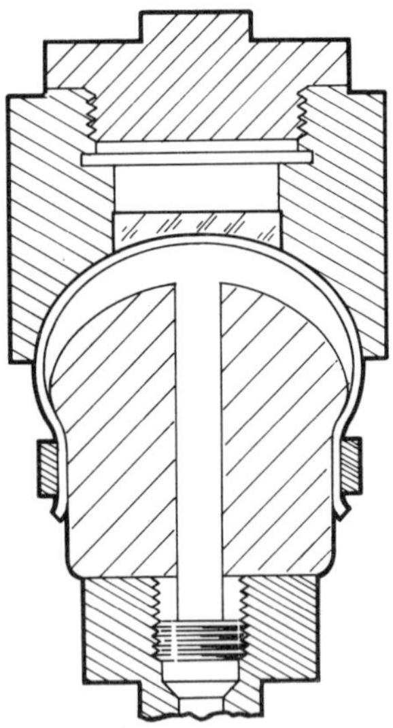

10.9.3. The hydraulic polishing tool for aspheric lenses.

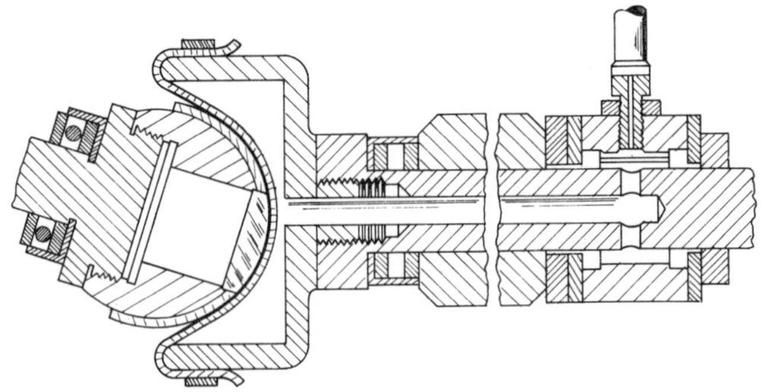

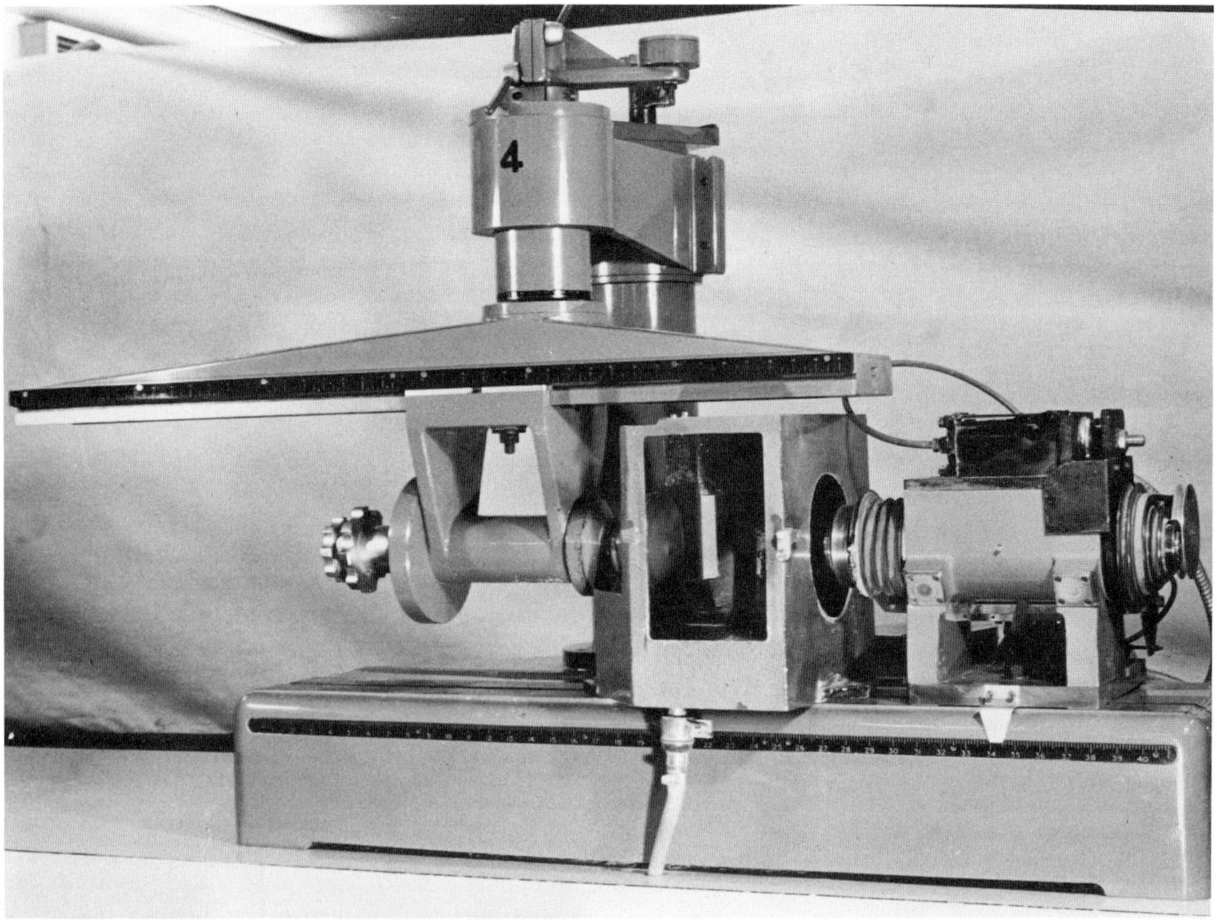

10.9.4. Two views of the Rank Taylor Hobson aspheric polishing machine. The lens block holding the aspherics is on the left and the flexible polishers are on the right.

11 Testing Optical Components

11.1 Introduction

For many years those closely involved in the manufacture of optical components have appreciated the need for a standardized code of practice in the dimensioning of drawings.

In Germany, DIN 3140 (October 1958) was published after fifteen years of research and investigation, followed by the United States MIL-STD-34 (November 1960) and the British BS 4301 (1968).

All of these drawing standards have the object of:

(*a*) Reducing the product rejection at an advanced stage of construction.

(*b*) Elimination of unnecessarily close tolerances which result from an insufficient knowledge of constructional parameter relationship to image quality.

(*c*) Elimination of subjective methods of inspection.

(*d*) Reduction of contingency factors if the specification is precisely defined.

The British Standard BS 4301 follows DIN 3140 and the system is essentially 'symbolic' in nature. The following definitions are quoted from BS 4301.

(1) *Element.* An optical unit which cannot be subdivided into other optical units.

(2) *Component.* An optical unit composed of one or more elements.

(3) *Field stop.* The aperture on any of its images limiting the extent of the object viewed.

(4) *Pupil.* The aperture on any of its images limiting the bundle of rays entering the system from an object point at the centre of the field.

(5) *Magnification.*

(*a*) *Linear magnification.* The ratio of a length on an image plane to the corresponding length on the object plane.

(*b*) *Angular magnification.* The ratio of the angles subtended by the image and the object, both as seen from the centre of the exit pupil.

(6) *True field of view.*

(*a*) *Linear.* The size of the field accepted at a given distance from some fixed point in the system such as the centre of the entrance pupil.

(*b*) *Angular.* The angular subtense of the field accepted as seen from some fixed point in the system such as the centre of the entrance pupil.

(7) *Optical axis* (*of a system*). The ray path passing through centres of pupils and field stops.

(8) *Mechanical axis* (*of an unmounted lens*). The axis of the edge surface of a lens by which it is to be mounted.

(9) *Equivalent focal length* (*of a system*). The ratio of the size of an image of a small distant object near the axis to the angular extent of that object in radians.

(10) *Test region.* An optically effective surface or space, or both, subject to testing to ascertain errors and flaws. (See Fig. 11.1.1.)

(11) *Test field.* That portion of a test region which is effective in the system at any one time.

(12) *Optical inhomogeneity.* Lack of uniformity in optical properties throughout a material.

(13) *Birefringence.* The variation of the refractive properties of an optical material with the direction of propagation and state of polarization of light within the material.

(14) *Optical path-length.* Two distances are said

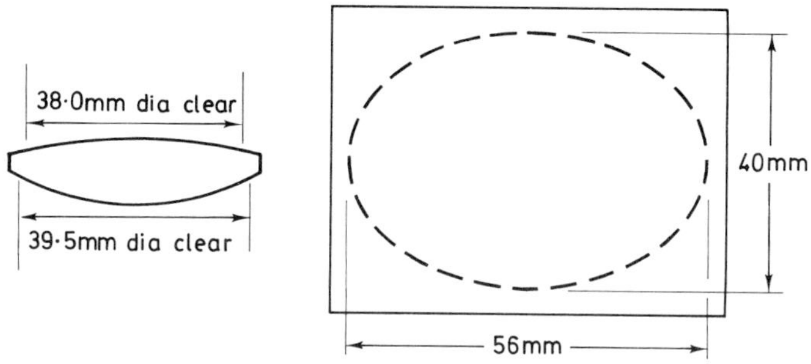

11.1.1. Examples of test regions.

to be optically equivalent when light takes the same time to pass through them. The optical length of a path is defined as the length of its equivalent *in vacuo*.

(15) *Optical parallelism.* Two nominally plane surfaces are optically parallel when, for all light rays of a given wavelength and direction, the optical path-length between the surfaces is constant at all parts of the aperture.

(16) *Centring error of a lens.* The condition which exists due to faulty manufacture when centres of curvature do not lie on the mechanical axis or when a plane surface is not normal to the mechanical axis.

(17) *Veiling glare.* Unwanted light within the system, tending to diminish the contrast of the image formed by the main beam.

(18) *Error of form.* The extent to which an optical surface departs from its specified geometrical form.

(19) *Total error of form* (of a component or system). The combined effect of form error at all the optically effective surfaces and of optical inhomogeneities of the material.

(20) *Grey surface.* A surface containing the characteristic irregularities resulting from the grinding process, which, on glass, give a grey appearance.

(21) *Coarse polished.* A surface which may contain traces of the previous machining (or other process).

(22) *Fine polished.* A surface from which all traces of earlier machining (or other processes) have been removed.

(23) *Surface texture.* Those irregularities with regular or irregular spacing which tend to form a pattern or texture on the surface. This texture may contain both roughness and errors of form.

(24) *Surface roughness.* The irregularities in the surface texture which result from the inherent action of the production process.

(25) *CLA (centre-line-average height) values.* Numerical assessments of the average height of the irregularities constituting surface texture or roughness.

In addition to the definitions stated, there are other surface defects, which may be described as follows:

(*a*) Scratches occurring in the roughing or truing process not removed by the smoothing.

(*b*) Deep grey left from the early grades of emery; this can be distinguished by being coarser than (*c*).

(*c*) Uniform fine grey due to insufficient polishing.

(*d*) Sleeks. These are strong at one end and trail away to vanishing point at the other end. They are caused by the polishing operation—a fault either in the pitch or the slurry.

(*e*) Polishing marks, caused by poor contact between the polisher and the work.

(*f*) Marks made by using the test-plate, which always result from insufficient care in cleaning and dusting the test-plate or glass to be tested.

(*g*) Marks caused whilst cleaning with a rag.

(*h*) Marks made during blocking the glass-work.

Table 11.1.1 gives the code used in referring to typical defects, and Fig. 11.1.2 shows some examples of its use. When a given defect is permitted, the tolerance code is followed by a dash; e.g. 0/–. Fig. 11.1.3 shows examples of the form error coded as 3/0·5.

Table 11.1.1 Tolerance code

Code	*Defect*	*Explanation*
0/	Birefringence	Variation of refractive properties when light is polarized; graded by maximum difference in optical path length, in *nanometres* per centimetre of glass path for light polarized in perpendicular directions.
1/	Inclusions	Bubbles, seeds, etc.; graded by maximum *number* permitted × maximum dimension in *millimetres*.
2/	Homogeneity	Lack of uniformity in optical properties, veins, etc.; graded so that no defect is detected when examined against (i) point source of light (ii) light/dark boundary (iii) light background
3/	Form error*	Extent from which surface departs from its geometrical form; graded by *number* representing, in fringes, the maximum radial separation of concentric spheres between which the surface may be contained.
4/	Centring error	Condition existing in a lens when the optical surfaces are not true to the axis defined by the ground edge; graded by the maximum deviation, in *minutes*, of a ray incident along the axis of the ground edge.
5/	Surface quality	Surface defects, scratches, pits, etc.; graded by *number* permitted × illumination reference *number* × viewing *magnification*. (Special equipment is available with controlled illumination conditions.)

* *Total form error* of a component is the combined effect of the form error of all optical surfaces and homogeneity of the material. It allows several errors to compensate and may be specified by performance on an interferometer or other test equipment.

11.1.2. Examples of the tolerance code as used on a working drawing of a lens.

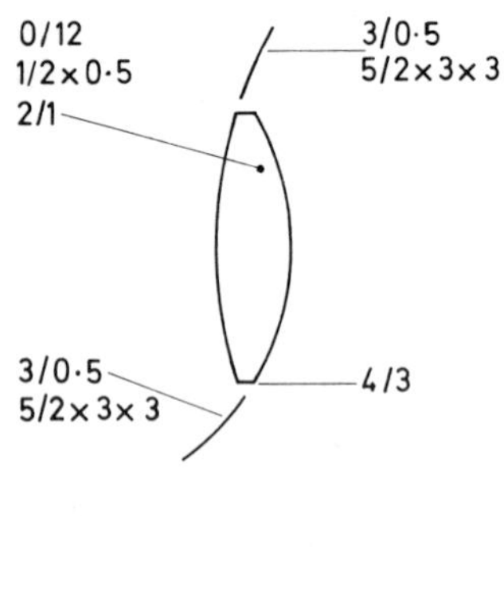

11.1.3. Two examples of the form error 3/0·5.

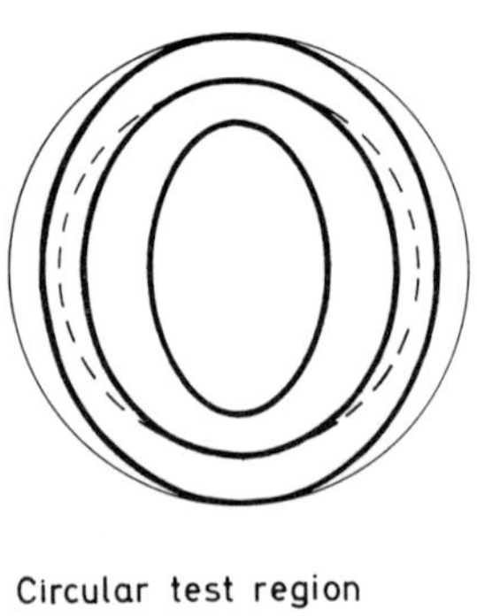
Circular test region

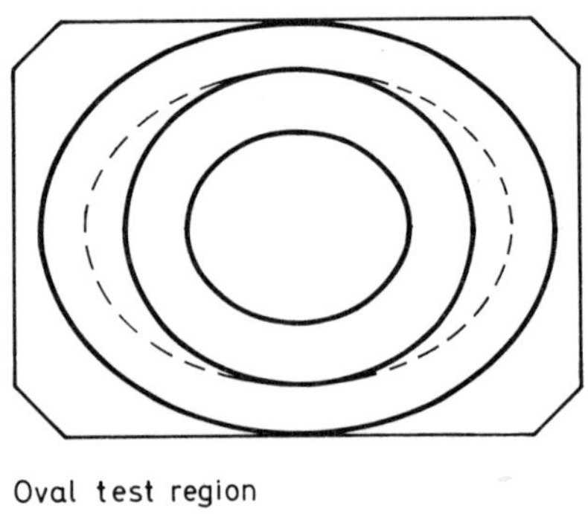
Oval test region

Table 11.1.2 Symbols for surface coatings and textures

Symbol	Description	Symbol	Description
	Rough surface—non-transparent, moulded, cast, etc.		Fine polished surface
	Bright surface—transparent, fire polished, drawn, cut, etc.		Rear surface metallized, e.g. silvered
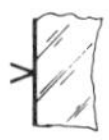	Milled or rough ground surface (63–250 μin. C.L.A.) (1·57 μm–6·25 μm)		Front surface metallized, e.g. aluminized
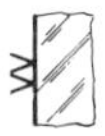	Medium ground surface (20–63 μin. C.L.A.) (0·5 μm–1·57 μm)		Partially reflecting coating, e.g. titanium dioxide
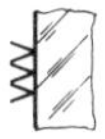	Fine ground surface (less than 20 μin. C.L.A.) (0·5 μm)		Anti-reflecting coating
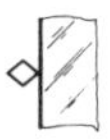	Rough polished surface, e.g. buffed		

The symbols used for surface textures and coatings are given in Table 11.1.2.

Typical manufacturing tolerances

(*a*) *Good quality 16 mm zoom cinematograph projection lenses* (designed specifically for large-quantity production with curves suitable for bell chuck centring—even at the expense of additional elements). Spherical surfaces within 5 rings centred within 3 minutes of arc.

(*b*) *Orthoscopic eyepiece.* Spherical surfaces within 10 rings centred within 3 minutes of arc.

(*c*) *Focusing telescope.* Spherical surfaces within 2 rings centred within 2 minutes of arc.

(*d*) *Collimating lens for high-grade instrument.* Spherical surfaces within 1 fringe, centring error less than 15 seconds of arc.

(*e*) *Aerial survey camera lens.* Spherical surfaces within 1 fringe. Centring error less than 15 seconds of arc.

(*f*) *Refractor block.* Flat surfaces within $\frac{1}{2}$ fringe. Surfaces parallel to within 2 seconds of arc. Optical flats to $\frac{1}{2}$ fringe and parallel to within 5 seconds of arc. (Fig. 11.1.3.)

(*g*) *100-mm diameter theodolite circle.* Flat surfaces within 5 fringes. No blemish to be visible under × 30 microscope with light-field illumination.

(*h*) *Pentagonal prism* (*optical square*). Flat surfaces within $\frac{1}{4}$ fringe. Right-angle deviation within 1 second of arc. (Fig. 11.1.4.)

(*i*) *Polygon.* Polished flat surfaces within $\frac{1}{4}$ fringe. Faces equispaced within 30 seconds of arc. After manufacture polygons are calibrated to 0·1 seconds of arc.

Table 11.1.3 converts angles into gradients. With the widespread use of powerful computers for calculating the dimensions of optical components, more realistic tolerances to give maximum manufacturing latitude can be expected.

However, there is much investigation still to be done, particularly in the area of tolerances on decentring, inhomogeneity and surface polish defects. Surface defects can be specified as grey, orange peel, digs, scratches, sleeks, stains, coating defects, and surface contaminants.

Table 11.1.3 Angles expressed as gradients

Angle	*Inch per inch*	*Inch per foot*	*mm per metre*
1 sec	0·000005	0·00006	0·005
2 sec	0·00001	0·00012	0·01
3 sec	0·000014	0·00017	0·015
4 sec	0·00002	0·00023	0·02
5 sec	0·000025	0·0003	0·025
6 sec	0·000029	0·00034	0·03
7 sec	0·000035	0·0004	0·035
8 sec	0·00004	0·00046	0·04
9 sec	0·000044	0·00053	0·045
10 sec	0·000048	0·0006	0·05
20 sec	0·0001	0·0012	0·10
30 sec	0·00014	0·0017	0·15
40 sec	0·00019	0·0023	0·20
50 sec	0·00024	0·0029	0·25
1 min	0·0003	0·0035	0·30
2 min	0·0006	0·007	0·60
3 min	0·0009	0·010	0·90
4 min	0·0012	0·014	1·20
5 min	0·0014	0·017	1·50
6 min	0·0018	0·020	1·80
7 min	0·0021	0·024	2·10
8 min	0·0024	0·028	2·40
9 min	0·0026	0·031	2·70
10 min	0·0029	0·035	3·0
30 min	0·0086	0·105	9·0
1 deg	0·0175	0·209	17·5

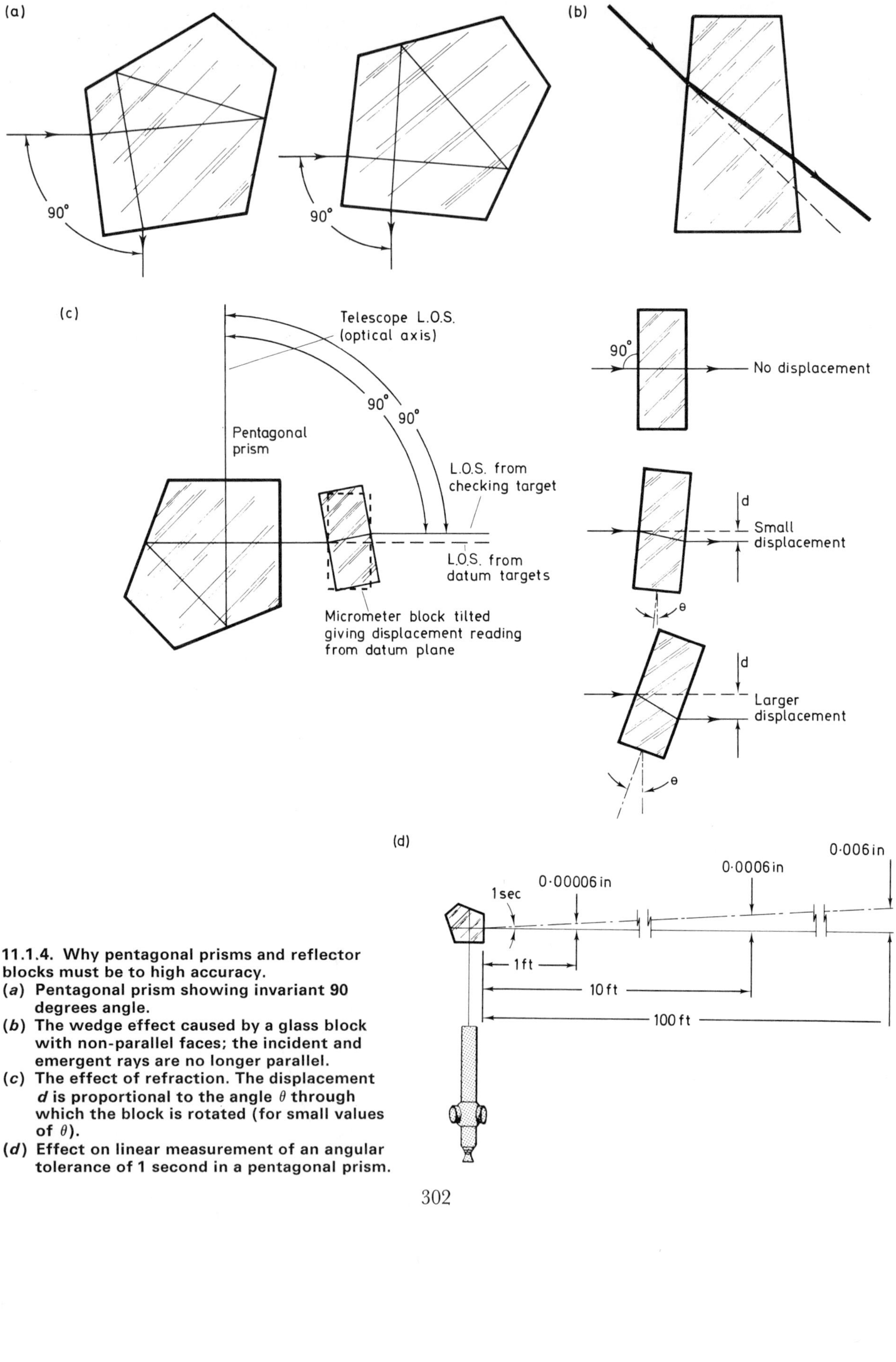

11.1.4. Why pentagonal prisms and reflector blocks must be to high accuracy.
(*a*) Pentagonal prism showing invariant 90 degrees angle.
(*b*) The wedge effect caused by a glass block with non-parallel faces; the incident and emergent rays are no longer parallel.
(*c*) The effect of refraction. The displacement *d* is proportional to the angle θ through which the block is rotated (for small values of θ).
(*d*) Effect on linear measurement of an angular tolerance of 1 second in a pentagonal prism.

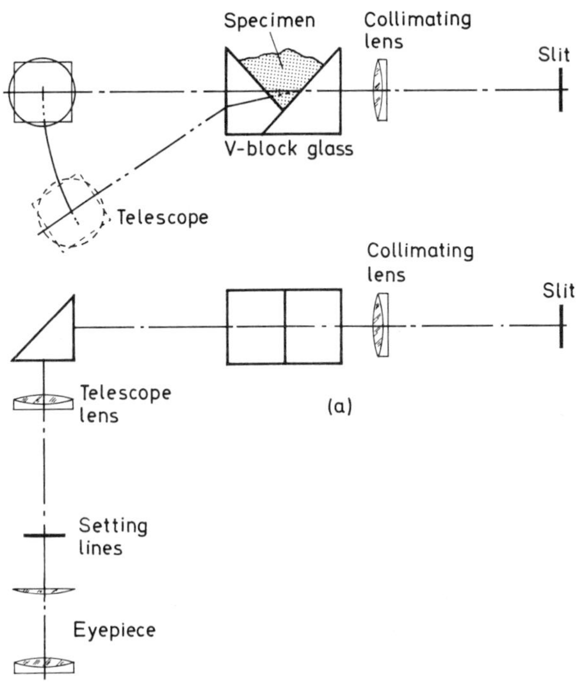

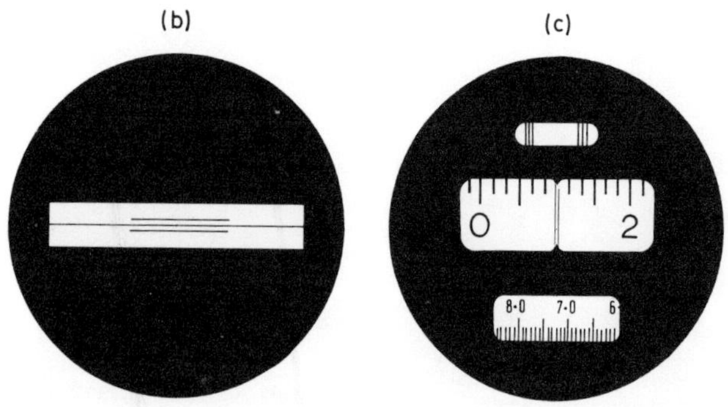

11.2.1. The Hilger–Chance refractometer.
(*a*) Elevation and plan of optical system.
(*b*) Field of view in telescope eyepiece.
(*c*) Field of view in reading eyepiece.

11.2 Testing the refractive index of optical glass

The Hilger–Chance refractometer

Although modern optical glasses are so good and true to specification, it is still necessary to be able to test for refractive index where this is a critical element in very-high-quality lenses.

The Hilger–Chance Refractometer was developed in the laboratories of Messrs Chance Brothers Ltd and made by Hilger and Watts Ltd for the accurate determination of refractive index and dispersion of optical glass and liquids.

Readings of the refractive index are obtained in less than a minute and to an accuracy of ± 2 in the fifth decimal place, while the specimen only requires to be worked approximately to angle and flatness of surface.

The optical parts are illustrated in Fig. 11.2.1 and the instrument in Fig. 11.2.2.

The refractive index is measured in terms of the angle through which a ray of light is deviated in passing through a prism block—the vee-block—and the specimen. The index is then obtained by reference to tables supplied with the instrument.

The vee-block consists of two glass prisms—one a complete 45° prism, the other a 45° prism with one end truncated—which have been worked to a high degree of accuracy in angle and surface flatness. They are joined by heat treatment to form (in effect) a single block with a V-shaped niche in the top, the sides of this niche being at right angles to each other.

The specimen is roughly prepared as a right-angle prism and then placed in this niche. Small surface irregularities in the specimen are compensated for by using a contact fluid between it and the vee-block. (Fig. 11.2.2.)

The contact fluid differs according to the refractive index of the glass under test as follows:

Contact fluid	*Refractive index of glass*
Monobromonaphthalene	1·648
Diiodomethane	1·740
Paraffin	1·480

Any two fluids are to be mixed in linear ratio for intermediate refractive indices. For very high index, sulphur should be mixed in diiodomethane until the solution is saturated. (See §11.4.)

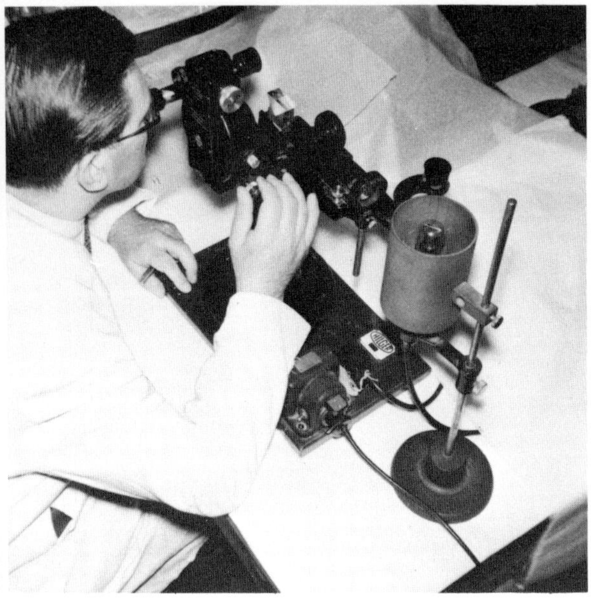

11.2.2. The Hilger–Chance refractometer. In the close-up, the operator is placing a drop of contact fluid between the prism block and the specimen.

The accurately finished vee-block does away with the need for special correction in calculating the refractive index, whilst preparation time is cut down as only rough preparation is necessary.

The refractive index calculated is that of the mass of the specimen, not that of the surface region only. Several vee-blocks, each with a different refractive index range, are available and can be rapidly interchanged.

The standard vee-block supplied with the instrument is suitable for the majority of samples; it is easily replaced by one of two other blocks for samples with higher or lower refractive indices. A vee-block with fused side plates which make it into a small trough is available for liquids. This block has a metal jacket within which circulating water maintains an even temperature for the specimen. The accuracy and range of these four blocks are given below.

The normal light source is a sodium or mercury lamp, although other discharge lamps are available. A prism can be swung into the optical path

Vee-block	*Catalogue number*	*Range*	*Accuracy*
I (Standard)	M 443	1·40 to 1·85	Accurate over the entire range to 1 or 2 in the 5th decimal place; most accurate in the centre of the range (applies to I, II and III)
II (High)	M 415	1·5 to 1·95	
III (Low)	M 437	1·3 to 1·73	
For Liquids	M 411	1·3 to 1·70	Better than 1 in the 4th decimal place

to introduce a second light source if necessary. A range of seven colour filters is available to be fitted to reduce the intensity of neighbouring spectral lines.

From the light source the beam of light passes to the condensing lens and thence to a tube containing, at the near end, the slit, and, at the other, the collimating lens. The slit is horizontal with a fixed width. From the collimating lens the light passes through the vee-block and specimen. The refracted beam is deflected by a prism into the telescope, where it is seen through an eyepiece. This telescope moves round the circumference of the divided measuring circle and is parallel to its axis.

All these components are supported on a firm metal plate, which is mounted on a strong pillar. The whole instrument stands on a polished wooden base, which also carries a small transformer to provide current for the scale illumination.

Checking sample angles before R.I. measurements

Permission has been granted by Pilkington Bros Ltd to publish the following paper, 'A Comparative Method of Checking Angles on Glass Samples', by Mr D. W. Harper of Pilkington Bros Ltd, Research & Development Laboratories.

The method to be described was developed specifically for checking the angles on glass samples, the refractive indices of which were to be measured on the Hilger–Chance refractometer (Hughes, 1941). For ease and speed of sample preparation the surfaces on the sample are fine-ground. The angle, which should nominally be a right angle, must, for many routine measurements, be accurate to ±30 seconds of arc. It is not possible to check to this accuracy using a precision square, the limit being about 1 minute of arc for an angle with faces 1 inch long. Furthermore, an improved refractometer is under development and, for measurements of the highest accuracy, it will be necessary to provide samples with an angle of 90° ±10 seconds.

The various methods of checking polished angles (see Twyman, 1952) were considered unsatisfactory for this purpose. These methods are meant for samples with well-prepared (i.e. flat) polished faces. The accurate location of one of the faces (e.g. on an optical flat) is then much easier to achieve than it would be with a fine-ground face with no guarantee of its flatness. The method developed makes no assumptions about the position of either of the faces of the angle.

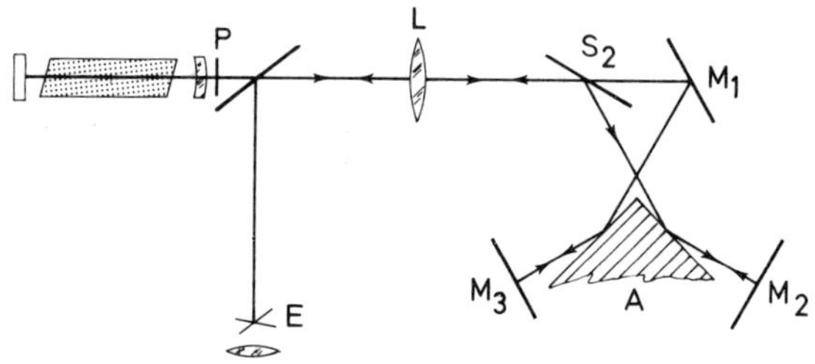

11.2.3. Checking the angle of a glass sample.

The principle of the system is shown in Fig. 11.2.3. A laser beam (6328 Å) from a Ferranti d.c. helium-neon hemispherical cavity gas-laser (giving about 0·15 mW) is collimated by the lens L. Part of the beam is reflected at the semireflecting mirror S_2 and the rest at the fully reflecting mirror M_1. These mirrors are arranged so that the reflected beams are again reflected off the fine-ground surfaces of the angle A at close to grazing incidence. Under these conditions a specular reflection is obtained (Twyman 1952, Voishvillo 1967). (The optimum angle for obtaining a good reflection varies with the type of glass and the grinding powder used but, for the conditions described here, is between 10° and 16° from grazing incidence. The apparatus is set up with the angle at about 12°–13°. This gives satisfactory results with most glasses.) The beams are then returned approximately along their own paths by mirrors M_2 and M_3.

The returned beams are reflected by the semireflecting mirror S_1 into the filar micrometer eyepiece E.

The gas-laser beam was used because with the small diameter of the beam (5 mm) and the long focal length of the collimator lens (100 cm), together with the loss in intensity on reflection from the fine-ground surface, measurements were easily made with the high intensity of the laser beam but would be extremely difficult with a conventional source.

The overall intensity of the beam (which, being a laser beam, is polarized) is adjustable by rotating polaroid P between the laser and mirror S_1. If the beams are of unequal intensity after division at S_2, measurement is often more difficult owing to the different intensities of the spots observed in the eyepiece. However, a suitable neutral density filter or polaroid placed between mirrors S_2 and M_1, or S_2 and M_2, can be used to equalize the intensities.

The rotatable polaroid between the laser and mirror S_1 is used as a safety device. When fine-ground surfaces are investigated the polarizer can be oriented parallel to the plane of polarization of the laser light, since the intensity of the spots in the eyepiece is such that they are just visible. However, when a polished sample is under investigation it is necessary to reduce the intensity of the laser beam by rotation of the polaroid. (For laser safety recommendations see *Laser Systems—Code of Practice*, issued by the Ministry of Aviation.) The procedure employed is to rotate the polaroid from the extinction position to the point where the spots are just visible.

The system is initially set up by means of a polished glass sample, the angle of which has been measured to 1 second of arc on a goniometer. The two spots observed in the eyepiece are set on the centre vertical crosswire, one (the 'reference' spot) just above the horizontal crosswire and the other just below. The sample, whose angle is to be checked, is placed on a table which has provision for locating the sample. The position of the sample is not critical but it was considered desirable in this case to be able to locate it with reasonable accuracy (i.e. within 0·05 cm) as this then relaxes the tolerances on the flatness of the two mirrors M_2 and M_3. Levelling screws on the table enable the two spots to be brought into their correct positions vertically, i.e. just above and just below the horizontal crosswire. (This adjustment ensures that the centre lines of the two beams after reflection from the faces of the sample are travelling in approximately the same angular direction as when the reference sample was in position.) The sample table can then be rotated roughly about an axis coincident with the apex of the angle. This enables the reference spot to be made coincident with the vertical crosswire. The direction and distance of the other spot (measured with the movable cursor in the filar micrometer eyepiece) gives the positive or negative deviation from a right angle. The sensitivity in the eyepiece corresponds to 50 seconds of arc per mm (the angle measured is 4α, where α is the deviation from the right angle). In principle 2 seconds of arc should be detectable but the fine-ground surfaces degrade the spots in the eyepiece somewhat and about 5 seconds of arc would seem to be the practical limit. It has been found in practice fairly easy to detect a difference in angle of 5 seconds of arc.

At first sight it would seem that the mirrors in the system should be of fairly high optical quality, otherwise false readings could be obtained due to angular deviations caused by the mirror surfaces. This is not the case here since the system is being used as a 'null system' and very small areas of the mirrors are used. The reasonably accurate location of the sample means that practically the same areas on mirrors M_2 and M_3 are being used for all samples. If no provision had been made for sample location the flatness gradient tolerance on the mirrors would be approximately $\lambda/100$ per mm.

The method has been found extremely easy and convenient to use in practice. The angles can be checked to about 2 minutes of arc by use of a precision square. The angle can then be corrected to better than 10 seconds of arc in about 10–15 minutes. The fine grinding is easily carried out on a piece of $\frac{1}{4}$ inch plate glass, usually with grade B.S. 303 grinding powder and water. If the fine-ground surface is (for example) curved, the spot spreads in the eyepiece and this can then be corrected for in the grinding process.

The method can, of course, be applied to angles other than right angles and can obviously be used on polished surfaces. In the latter case it is not necessary to reflect off the surfaces at near grazing incidence and a sensitivity of 2 seconds of arc can be achieved. (This could be improved upon considerably if a precision autocollimator was used instead of the laser and eyepiece.)

There are many possible alternative layouts of the system. The system could also be made interferometric, although the range of usefulness would then be reduced and it would not be as convenient to use in practice. The two beams of the interferometer would be the beams reflected off the two faces of the sample. Any departures of these two surfaces from perfect flatness would confuse any fringe pattern observed.

11.3 Testing for strain in optical components

Strain in glass is a condition where molecules are separated by a greater distance in one direction than in the directions at right angles to it, and this causes the electromagnetic waves, of which luminous radiation consists, to separate into two waves travelling through the glass at two different speeds, resulting in double refraction similar to that seen in Iceland Spar.

The degree of double refraction varies with the strain from point to point and may be detected by a simple form of polariscope, known as a strain viewer, illustrated in Fig. 11.3.1. Glass which has not been sufficiently well annealed after the chilling which takes place in moulding shows double refraction, but the effect can be quite marked in a strain viewer without, in itself, causing optical parts made from the glass to be defective.

The form of strain viewer normally used embodies a half-wave plate (first described by Brewster in 1830) which shows up regions of stress in vivid colour contrast. Well-annealed specimens have no effect on the colour of the magenta background while regions of stress become a light blue or red according to the direction of stress.

11.4 Testing of quartz crystals

For some instruments, such as quartz spectrographs, there is no material as satisfactory as natural quartz for making the prisms, also there is no other suitable material for polarizing microscope wedges. Natural quartz is extremely expensive and the labour of cutting it and getting it 'to axis' is a considerable part of the whole labour in making the prism or wedge.

Examination of the crystal for internal flaws and approximate alignment for initial cutting is carried out in a glass tank filled with methylsalicylate (oil of wintergreen) or alpha bromonaphthalene ($\mathcal{N}$ = 1·66). These liquids have a refractive index close enough to quartz to enable even small defects in the interior of the crystal to be located and avoided.

High-pressure mercury light is used for this examination of the crystal. Fig. 11.4.1 illustrates a box of natural crystals next to a glass examination tank. After cutting the initial two parallel

11.3.1. A strain viewer detects strain in finished components.

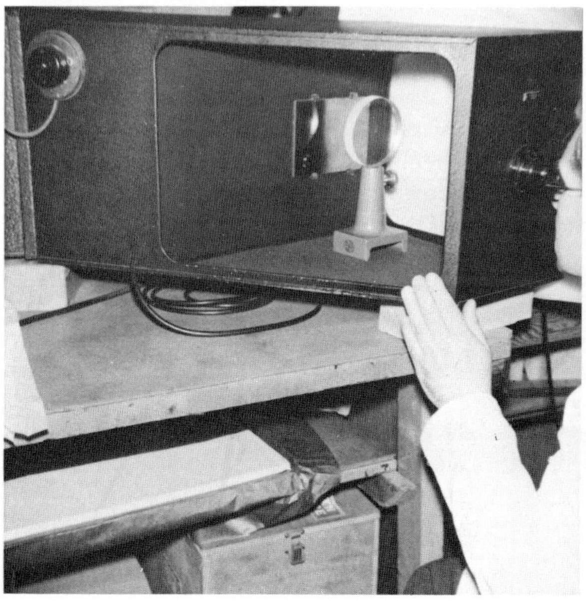

11.4.1. Quartz crystals and a tank for examining the crystals to find their optical axes.

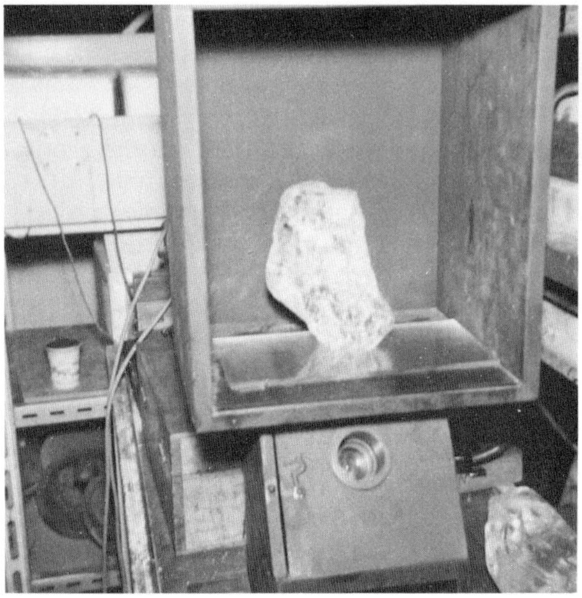

11.4.2. A polariscope for finding the optical axes of quartz crystals. Sodium light is projected down from the top of the box.

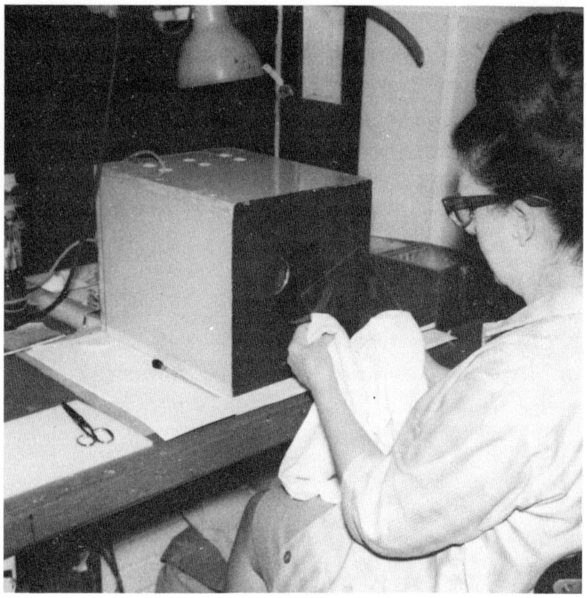

11.5.1. Dark-background inspection of glass surfaces under strong illumination.

faces on the crystal, it is examined again under sodium light. (Fig. 11.4.2.)

Immersion oil between the lower crystal face and the transparent table makes it possible for fringes to be seen in the eyepiece. The two parallel surfaces on the crystal must be lapped by hand until the fringes are concentric and line up with the circles on the graticule in the eyepiece. By the use of this instrument the optical axis can be made true to within 0·5°.

11.5 Testing of glass surfaces

(a) Dark-background surface inspection

The examination of glass surfaces under a high-power light against a dark background (Fig. 11.5.1) is one of the best ways of finding marks of any kind. The examiner has to decide, with the help of the specification, whether the marks he sees are acceptable or not. Experience plays an important part in this judgment of what is acceptable, but the maintenance of standards for optical surface quality has always been a great problem.

There have been many attempts to develop quantitative methods, but no common agreement exists at international levels.

(b) Surface-quality inspection standards

B.S. 4301:1968 states that surface defects fall into three categories.

(1) Those which are objectionable solely on aesthetic grounds.

(2) Those which are situated in the immediate neighbourhood of a focal plane.

(3) Those which contribute objectionable amounts of veiling glare.

These defects can be assessed in terms of various criteria; this standard uses their visibility under certain conditions of illumination and magnification. Visibility can be determined by comparison with the appearance of scratches having known dimensions and profile.

These master scratches are of square profile, constant width (15 micrometres) and in four standardized depths (0·09, 0·175, 0·35, and 0·70 micrometres).

The test equipment (Figs 11.5.2 and 11.5.3) illuminates the specimen obliquely so that the

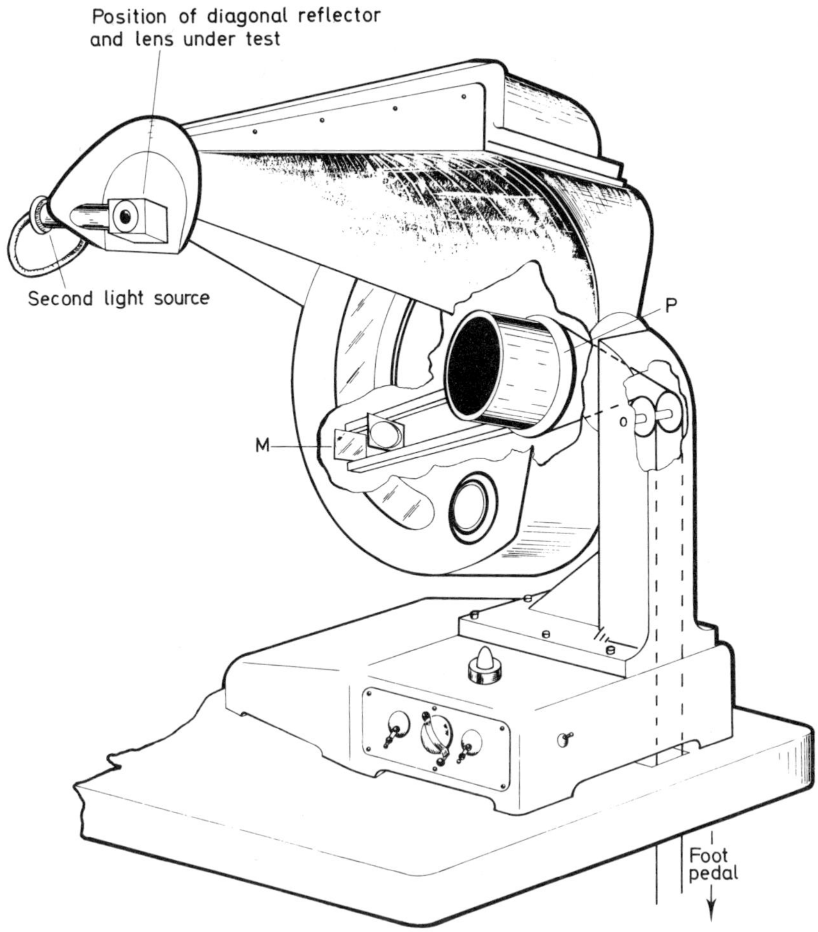

◁ **11.5.2. Surface-quality inspection instrument.**

11.5.3. ▷
Lens mark instrument for surface-quality inspection.

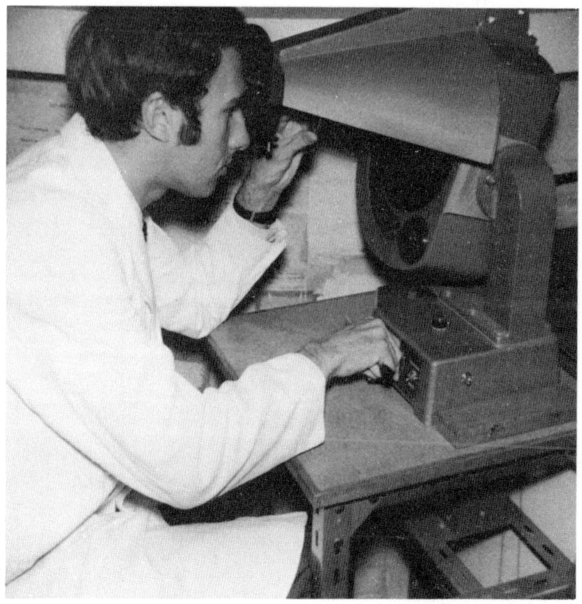

defect appears bright against a dark background, and its contrast is reduced by introducing an overall veiling glare of controlled amount.

The intensity of the oblique illumination can be decreased automatically as the veiling glare is increased until the appropriate standard scratch becomes only just visible.

The component to be inspected is held in the hand with the back of the hand held against a supporting bracket close to the eyepiece. Instead of moving the component to search for marks, the direction of the incident light is varied by operating a foot pedal. When the pedal is moved, a cable passing round a vee-pulley *P* swings a mirror *M* in an arcuate path in such a manner that light from the central lamp reaches the component from constantly changing directions. The light is reflected from the mirror *M* through the glass-covered semi-circular opening in the casing. The brighter this light, the more readily visible are marks on the part under test.

A second light source is contained in the tube shown. This illuminates a diagonal piece of glass between the eyepiece and the part under test, producing an illuminated field of view. This is necessary in order that inspection may be independent of variations in room lighting. The more intense the light from this second source, the less easily visible are marks on the surface under inspection.

A double-bank four-way rotary switch on the front panel of the instrument varies both these sources of illumination simultaneously, the one being brightened whilst the other is dimmed. There are therefore four combinations of illumination under which components may be examined, and these correspond to four previously chosen grades of surface quality. In order to ensure that these grades remain constant a set of master standards is provided.

The instrument is adjusted so that the standard lines are just visible when the four way-switch is set to correspond with the particular grade of standard under inspection, the lines becoming invisible on switching to the next grade or lower standard of inspection.

Lenses may also be inspected by reflected light, using a third lamphouse for the detection of stain on the glass surface.

(c) High-quality surface inspection

Graticules, theodolite circles, and scales which are going to be viewed in an instrument under a microscope must be tested under conditions more critical than the normal operating conditions. For these products the tests described in (*a*) and (*b*) will not be sufficiently severe. High-quality surface-finish examination must be against a dark background and the whole of the required high quality area viewed by a microscope.

To avoid contamination after test, the components may be stored in distilled water or dipped in 'Sealac' to protect the surfaces.

11.6 Newton's rings and the testplate

If a shallow convex glass surface is laid on a flat one, a system of rings is seen around the point of contact. These are called 'Newton's rings' or 'fringes'. When two flat surfaces are put together, the one being very slightly inclined to the other, the colours are not arranged in rings but in more or less parallel lines or curves and they are then called 'Newton's bands'.

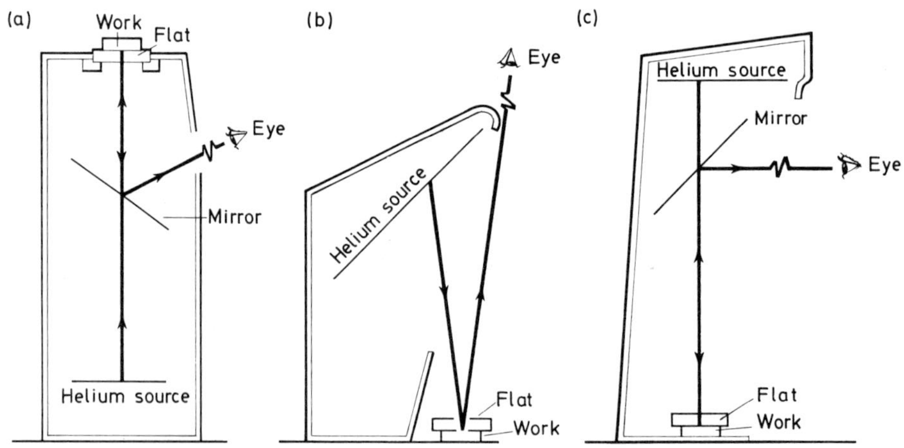

11.6.1. Monochromatic lights used in testing work by means of Newton's rings.
(*a*) Reflex viewer for production.
(*b*) General purpose light.
(*c*) 'L' light for laboratory use.

It was Sir Isaac Newton (*Opticks*, Book II) who first found a relation between these colours and the corresponding thicknesses.

The thickness of the film of air at any point of such a system of rings or bands can be determined by counting the number of rings from the point of contact and for practical purposes the thickness change from one band to another is half the wavelength of light; e.g. 0·28 micrometre (0·000 011 inch) for helium light in the apparatus shown in Fig. 11.6.1. Figs 11.6.2 and 11.6.3 show how curvature is calculated from Newton's bands.

11.6.2. Newton's bands.
Helium light has a half-wavelength of 11·6 micro-inches. Band curvature can be estimated to a tenth of a band interval, or to a millionth of an inch (0·028 μm).

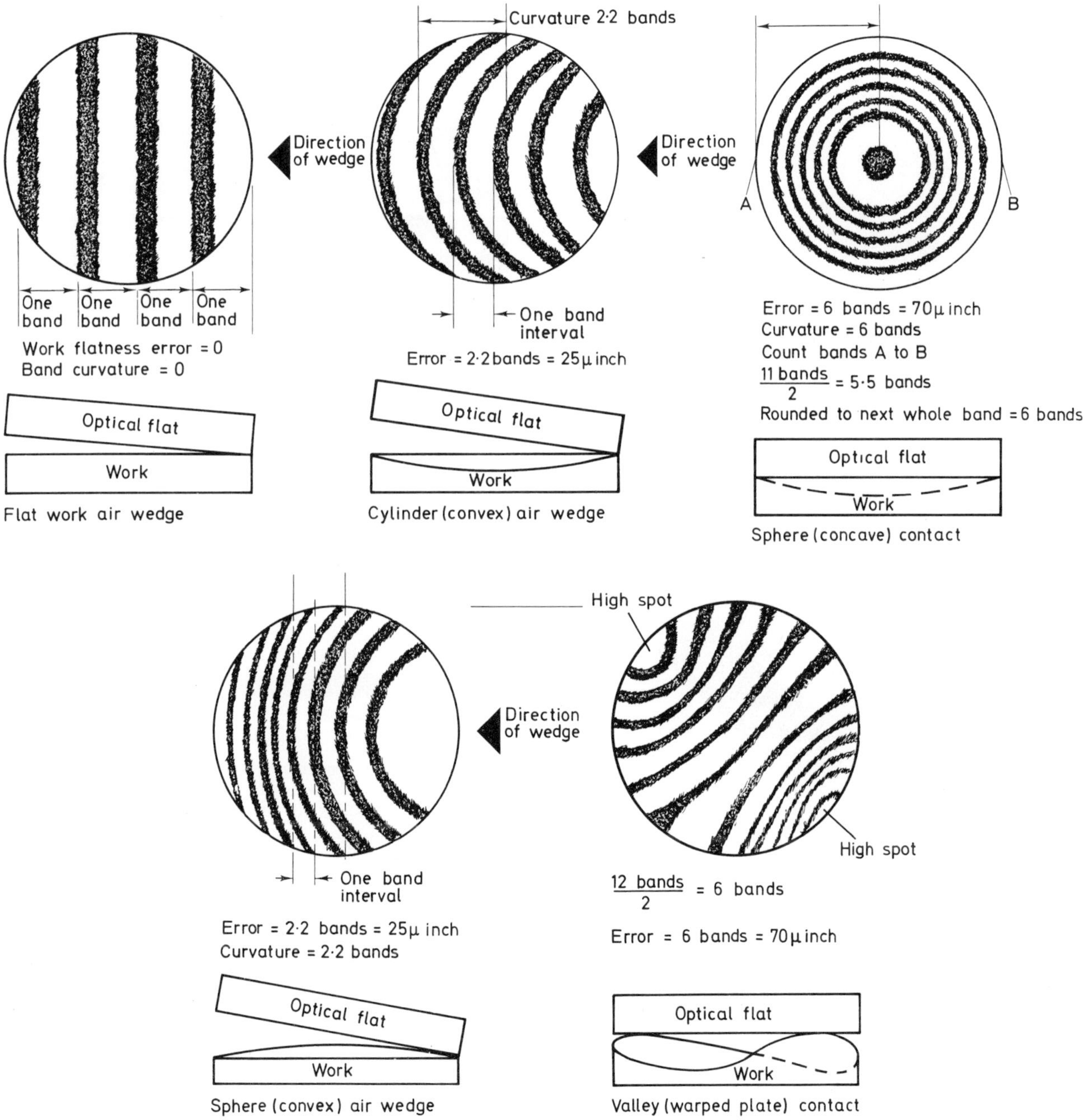

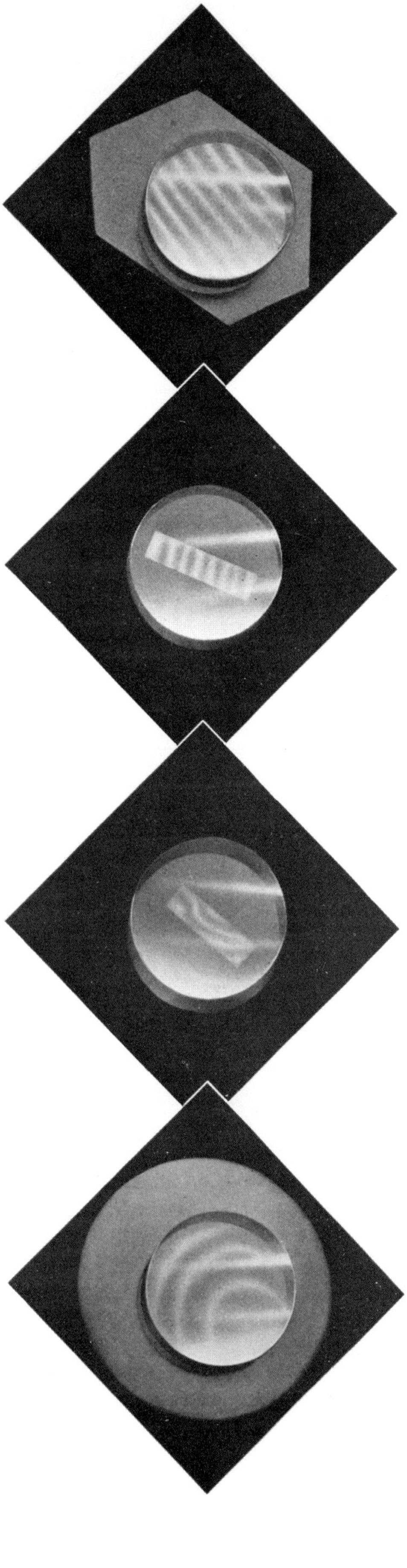

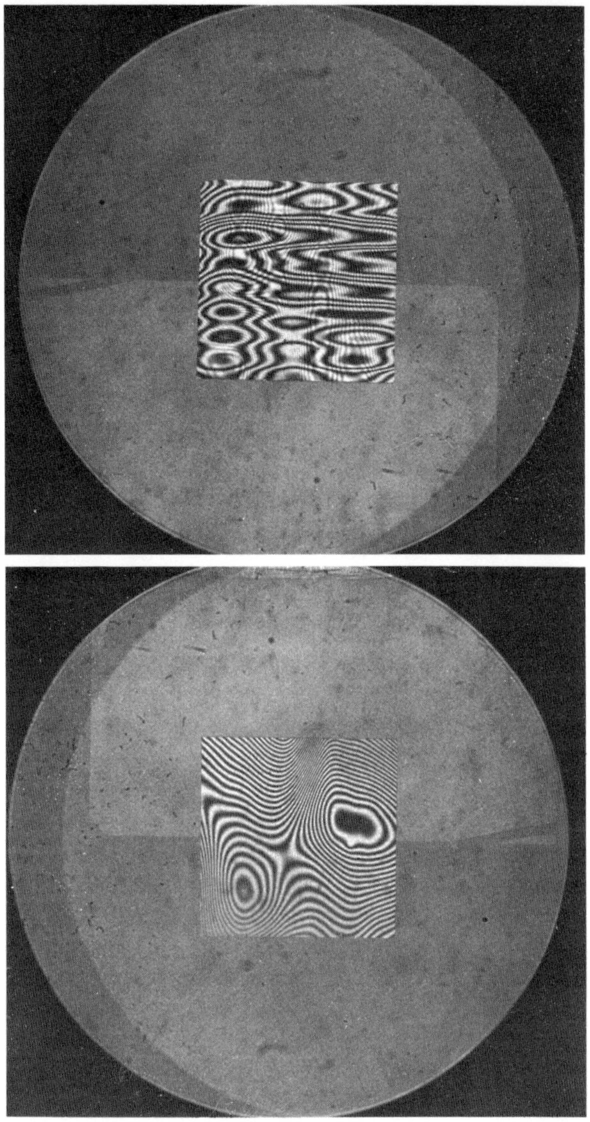

11.6.3. (*Left*) Newton's bands showing, from top to bottom, a slight concavity or convexity of about 0·000 005 inch, practical straightness indicating surface flatness to within 0·000 002 inch, surface variations up to 0·000 02 inch, and convexity of 0·000 03 inch.

(*Above*) The photographs illustrate lack of flatness on good-quality photographic plates. Except for such very-high-resolution requirements as microcircuit photocopying processes, the irregularity of figure is unimportant because of the thickness of the emulsion.

Fig. 11.6.4 shows the convex surface of a lens resting on a plane surface. At the point of contact the difference in the optical paths reflected from the upper and lower surfaces is zero. The phase change on reflection from the lower surface causes the beams to rejoin exactly out of phase, resulting in complete cancellation and the appearance of the central 'Newton's Black Spot'.

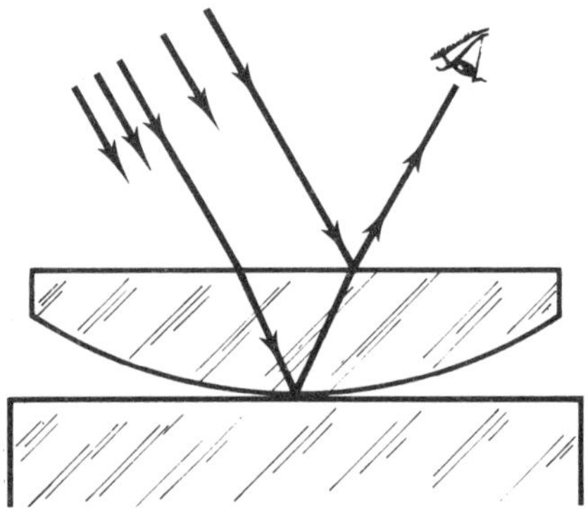

11.6.4. Formation of Newton's black spot.

Some distance from the centre, the surfaces will be separated by exactly one-quarter wave-length, and this path difference of one-half wave-length plus the phase change results in reinforcement producing a bright ring. A little further from the centre, the separation is one-half wave-length, resulting in a dark ring, and so on.

If instead of applying a shallow curved surface to a flat one, we apply a truly spherical surface to one of slightly different curvature, the width of separation of the rings indicates how nearly the second surface matches the first one, while any departure of the rings from circularity indicates a departure from sphericity of the second surface.

Accurate plates, whether flat or curved, used for testing surfaces which should fit them are known as testplates or proof-plates. Before contacting a testplate with a lens to be tested it is important that both surfaces be clean and free from grease. A camel-hair brush can be used to remove specks of dust. A piece of tissue paper placed between the two surfaces, and then removed whilst maintaining a slight pressure on the upper plate, will also remove the dust.

Although fringes can be seen in white light a far better result is achieved by the use of a low-pressure mercury vapour lamp light or helium-filled tube source. The mercury light is nearly monochromatic green, and helium gives a yellowish orange.

As the spacing of the bands is the standard of measurement, it does not matter whether the bands are closely or widely spaced, but the test is, of course, much more sensitive with widely spaced bands. Complete accounts of the phenomenon of Newton's rings are given in many optical text-books but it is sufficient to say that there will be a black spot in the middle surrounded by coloured rings for white light, while with monochromatic light the rings will be dark in colour, sharper and easier to count.

11.6.5. Testing curves on a block of lenses under a helium lamp.

11.7 Fizeau's fringes and the interferoscope

Fizeau's fringes are caused by the interference of reflections between two surfaces with an air space in between. The way in which Fizeau described his system is as follows (Fizeau, 1862):

> A convex lens of very long focus was placed on a glass plane so as to give—with white light—large coloured rings as in Newton's well-known experiment; but here (viz. in Fizeau's experiment) the lens was fixed in a metallic mount movable by a micrometer screw perpendicular to the plane of the lens and of the stationary glass plane. By rotating the screw-head one can vary the distance between plane and lens. The observer, looking in a direction parallel to the lens and the system being illuminated by the light from a Brewster lamp (viz. a sodium burner) reflected from the hypotenuse of a small prism placed near the eye, perceives between the two glasses large rings of the greatest beauty; rings which result, as one knows, from the interference produced between the rays reflected towards the eye by the two neighbouring surfaces.
>
> It is easy to see that to render these appearances visible over the whole surface of the lens at once one should add near it a convex lens of focus equal to the distance from the lens to the eye; one then sees the phenomenon in its entirety, the actual surface of the lens being covered to the margin by rings of the greatest sharpness.

The method of testing flat surfaces described by Fizeau has the great advantage of non-contact between the reference flat and the surface being tested and therefore the tests can be completed without risk of damage. The further advantage is the ease with which large flats, up to about 12-inch diameter, can be tested to a high order of

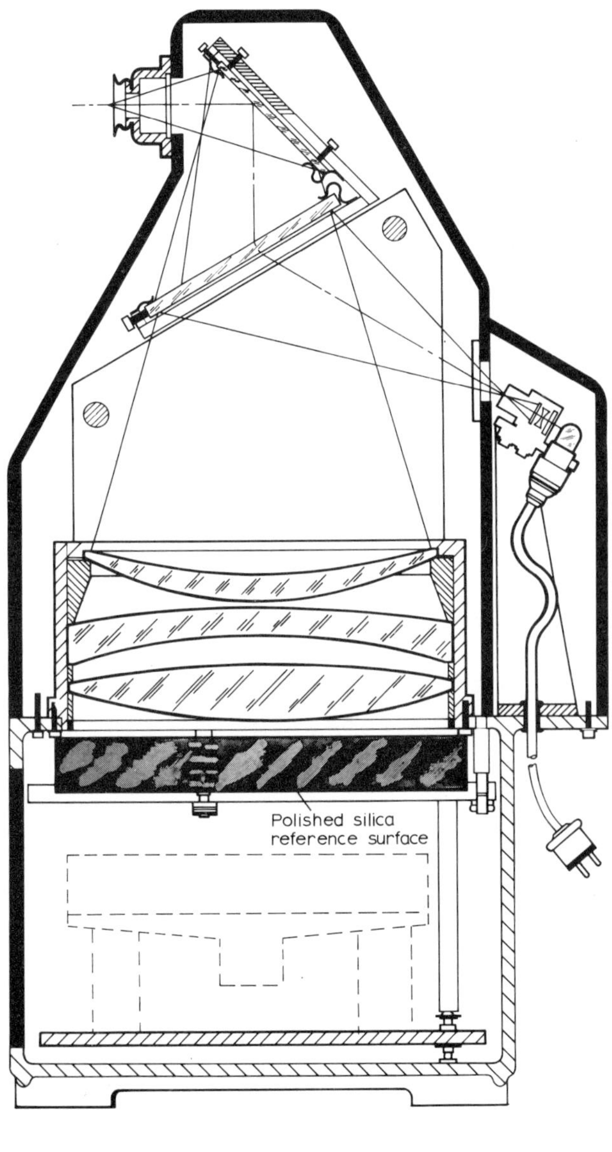

11.7.1. A 12-inch-diameter interferoscope. The test-object stands on a suitable support on the plate at the bottom of the instrument below the reference surface.

11.7.2. Twelve-inch interferoscopes for testing flats to 0·1 fringe.

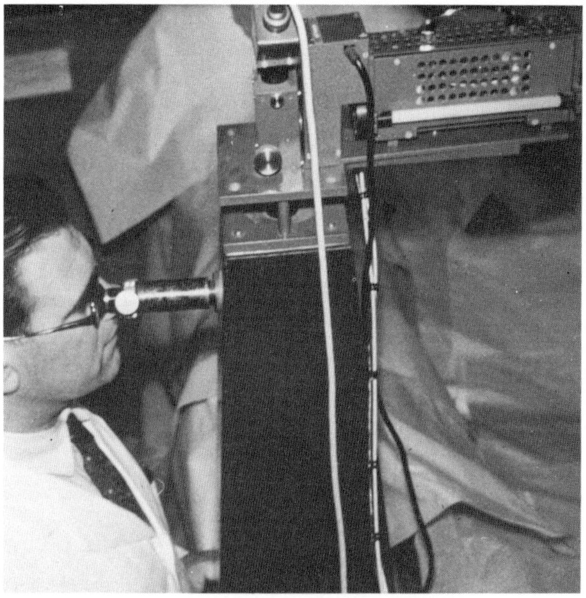

11.7.3. A 7-inch interferoscope with a telescope for testing flats to 0·02 fringe by helium-neon laser illumination (6328 angstroms).
(Below) A Duo-illuminant for interferoscopes.

accuracy, which would be difficult if not impossible by any other method.

Fig. 11.7.1 illustrates the design of a 12-inch aperture three-component collimating system with silica referencc flat, intended for the routine inspection of blocks of prisms and flats as well as large components such as interferometer mirrors polished to an accuracy of one tenth of a fringe. Fig. 11.7.2. shows a pair of 12-inch interferoscopes in a temperature-controlled room being used for routine testing.

The light source for interferoscopes is usually a mercury Sieray lamp (5461 Å) which is adequate where the separation between the component being tested and the reference surface is only a few millimetres. However, some components, such as etalon spacing consisting of a hollow cylinder of fused silica, require a separation of over 12 mm and it is essential to use a light source with a longer wavelength, such as a helium neon laser of 6328 Å wavelength.

Fig. 11.7.3 illustrates a lamp-house which contains a laser as well as a mercury lamp. It consists of five parts:

(1) A single-frequency helium–neon laser of not more than 1 milliwatt power.

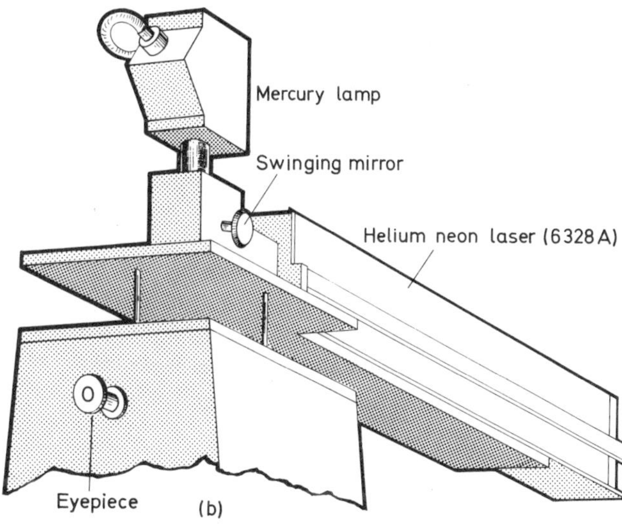

(2) Anti-clutter unit.
(3) Sieray mercury lamp.
(4) Fully aluminized swinging mirror.
(5) Common condenser.

By swinging the mirror, either 5461 Å or 6328 Å radiation can be directed into the interferoscope.

Some optical surfaces such as laser mirrors and

Fabry–Perot interferometer etalons must be very accurate flats and, to keep the fringes sharp and truly circular, it is necessary for the reflecting surfaces to be flat to a small fraction of a wavelength.

Etalons (with apertures up to 100-mm diameter) are usually slightly wedge-shaped, so that the outer unsilvered faces of the plates are not exactly parallel to the inner silvered faces. The wedge angle varies between 1 and 10 minutes of arc, which is enough to reduce the effect of the interference patterns formed by reflections at the unsilvered faces without materially affecting the observed fringe system.

For an accuracy closer than one-tenth of a fringe, the following conditions are required:

(*a*) The surface of the polished component must have a silver coating of approximately 5 per cent transmission (removed by nitric acid before repolishing).

(*b*) The interferoscope must have a telescope eyepiece with cross-wires in order to be able to measure fringe spacing and sag.

(*c*) The test must be carried out in a temperature-controlled room after the component has had time to settle down and 'soak' in the conditions under which it will be tested.

(*d*) The interferoscope must have anti-vibration mountings so that fringes are still.

Flat supports

Measurements of high accuracy can be grossly distorted because of deflection of the flat caused by gravitational influence and as no obstruction in the field of view can be tolerated. A flat for a 12-inch diameter interferoscope weighs 16 lb and the weight causes a distortion on the surface of about 0·5 fringe. The upper flat must be supported at the edge and three points or small pads equally spaced round the periphery are best employed (Dew, 1966). (Fig. 11.7.4.) However, this method does tend to produce a non-uniform, if consistent, sag and a method frequently adopted is to support around the whole of the periphery with soft rubber or polyurethane foam. The inconsistency of this method is often not as serious as the consistent sag of a three-point support.

11.8 The measurement of optical flatness

In order to achieve accurate measurement it is, of course, essential to have good optical design and manufacture of interferoscope components, which are essentially the collimating system and the reference flat. The problem is simplified because the lenses are used with monochromatic light, but the number of components necessary for the elimination of spherical aberration and coma from collimators depends on the refractive index of the glass. Usually two- or three-component systems polished spherical to one band are satisfactory for high accuracy testing.

Reference flat testing

The reference flat manufacture is quite another problem, as it should be more accurate than the flats it has to test and, more important, the errors in the reference flat must be known.

In order to calibrate the reference flat, it must be compared with a surface of unimpeachable planeness, such as a liquid surface.

Some reservations concerning the quality of flatness achieved by a liquid surface are based on the effects of the curvature of the Earth, capillary curvature, and vibrations or other external disturbances.

For all practical purposes, the first effect is negligible, as the curvature on a liquid surface of 10-inch diameter is only 12 angstroms.

The effect of capillary curvature vanishes approximately 2·5 cm from the boundary and causes no difficulty if the diameter of the liquid reference surface is about 7·5 cm larger than the surface to be tested (Bünnagel, 1956).

External disturbances are cut out by using a suitably viscous liquid as the reference surface shielded from air draughts. Both liquid paraffin (B.P.) and silicone oil (DC 705) about 3 mm thick have been found satisfactory. Temperature of the air must be controlled to 0·1°C and the flats left undisturbed for 24 hours before being tested.

Fig. 11.8.1 illustrates the optical arrangement of a Fizeau interferometer using a liquid plane surface as a master reference. By the use of a camera, a photographic record or interferogram can be obtained of the fringe pattern.

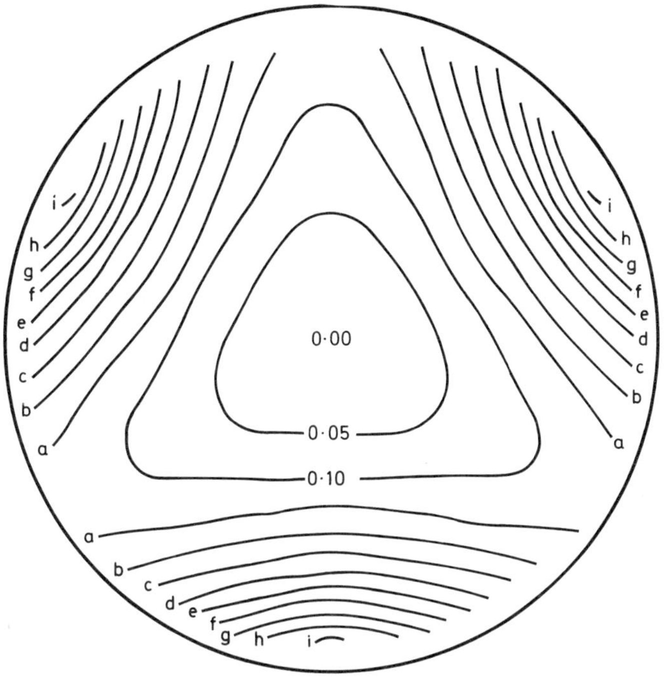

a. 0·15 b. 0·20 c. 0·25
d. 0·30 e. 0·35 f. 0·40
g. 0·45 h. 0·50 i. 0·55

◁ **11.7.4. The sag of a 30-cm flat, 3·8 cm thick, supported at three edge points (one-twentieth fringe contours).**

▽ **11.8.1.** ▷
The 45-cm N.P.L. Fizeau interferometer with liquid reference. In the photographs, the instrument is shown in raised position and in working position.

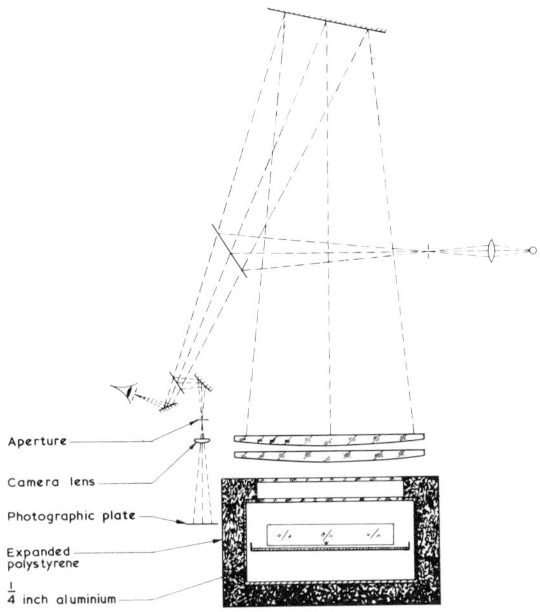

NATIONAL PHYSICAL LABORATORY

Teddington, Middlesex, England

REPORT

ON THE EXAMINATION OF ONE OPTICAL FLAT

for: Hilger & Watts Ltd.,
98, St. Pancras Way,
Camden Road,
London, N.W.1.

Identification

The flat is 13.04 in (33.1 cm) in diameter and its thickness varies from 1.53 in (3.89 cm) to 1.58 in (4.01 cm). It is marked with the manufacturer's number WO 3431/1 and the diameters A-B and C-D are defined by marks on the periphery.

Flatness

The flat was supported horizontally, with the worked surface uppermost, on three small pads close to the edge, one at A and the other two at 120° separations from this point. The surface was compared interferometrically with a master flat of known figure using mercury green light of wavelength 0.5461μ (1 fringe = 0.273μ or 1.07 x 10^{-5} in).

The errors of flatness are given in the following table in hundredths of a fringe for points spaced at 2 cm intervals from the centre. A negative value indicates that the height of the surface at the point concerned is less than that at the centre. It is estimated that the figures quoted are accurate to within one fiftieth of a fringe except at the marginal points where the accuracy may be slightly less.

A

	Cms	14	12	10	8	6	4	2	0	2	4	6	8	10	12	14	
	14							$\overline{04}$	$\overline{04}$	$\overline{07}$							
	12					$\overline{12}$	$\overline{03}$	02	03	00	$\overline{09}$	$\overline{20}$	$\overline{36}$				
	10			$\overline{30}$	$\overline{15}$	$\overline{05}$	01	03	03	01	$\overline{04}$	$\overline{13}$	$\overline{23}$	$\overline{39}$			
	8		$\overline{39}$	$\overline{21}$	$\overline{09}$	$\overline{02}$	01	03	02	01	$\overline{04}$	$\overline{07}$	$\overline{17}$	$\overline{28}$	$\overline{46}$		
	6		$\overline{27}$	$\overline{14}$	$\overline{06}$	$\overline{02}$	00	00	01	00	$\overline{02}$	$\overline{07}$	$\overline{12}$	$\overline{21}$	$\overline{37}$		
	4	$\overline{40}$	$\overline{21}$	$\overline{10}$	$\overline{05}$	$\overline{03}$	$\overline{01}$	00	00	00	$\overline{02}$	$\overline{04}$	$\overline{09}$	$\overline{14}$	$\overline{26}$		
	2	$\overline{32}$	$\overline{16}$	$\overline{06}$	$\overline{02}$	$\overline{01}$	$\overline{01}$	00	00	00	00	$\overline{01}$	$\overline{04}$	$\overline{10}$	$\overline{16}$	$\overline{35}$	
D	0	$\overline{26}$	$\overline{11}$	$\overline{03}$	$\overline{01}$	00	00	00	00	01	02	02	01	$\overline{03}$	$\overline{11}$	$\overline{26}$	C
	2	$\overline{23}$	$\overline{08}$	$\overline{01}$	01	01	01	01	01	02	04	05	04	02	$\overline{04}$	$\overline{19}$	
	4	$\overline{24}$	$\overline{07}$	01	02	02	01	01	02	03	05	06	07	06	00	$\overline{14}$	
	6		$\overline{07}$	00	01	01	00	00	02	03	05	07	08	07	00		
	8		$\overline{18}$	$\overline{08}$	$\overline{03}$	$\overline{02}$	$\overline{02}$	$\overline{01}$	$\overline{01}$	01	02	05	06	04	$\overline{05}$		
	10			$\overline{21}$	$\overline{13}$	$\overline{10}$	$\overline{08}$	$\overline{08}$	$\overline{07}$	$\overline{05}$	$\overline{04}$	$\overline{03}$	$\overline{04}$	$\overline{02}$			
	12				$\overline{33}$	$\overline{25}$	$\overline{21}$	$\overline{19}$	$\overline{17}$	$\overline{15}$	$\overline{15}$	$\overline{17}$	$\overline{20}$				
	14						$\overline{43}$	$\overline{40}$	$\overline{35}$	$\overline{36}$	$\overline{36}$						

B

DATE 12th July 1966

REFERENCE E.2551

for Director

J. DYSON,
Superintendent, Light Division

11.8.2. N.P.L. report on the flatness of a silica reference flat.

After evaluating the flatness of the reference compared with the surface of a liquid, and having recorded the fringe pattern photographically, it is then desirable to transpose the information into a contour map for easy understanding or where the accuracy obtainable by visual examination is inadequate.

A photoelectric method (Dew, 1964) is currently being used to establish a precise standard of optical flatness which is capable of measuring fringe displacements of less than 0·01 fringe. Fig. 11.8.2 is a report by the National Physical Laboratory on a 13-inch diameter silica reference flat used for calibrating other flats to be used as interferometer references.

INTERFEROMETRY

The rest of this section consists of a report by Mr G. D. Dew on 'Optical Flatness Measurement', published by kind permission of the National Physical Laboratory.

Photographic recording

Interferograms which are required for purely visual appraisal can be adequately recorded on film-based material, but if the intention is to make precise measurements by the method outlined, then photographic plates are preferred by reason of their higher dimensional stability. A medium-to-fast orthochromatic emulsion is the most suitable and has the advantage that it is insensitive to any red emission from the lamp which is transmitted by the Wratten 77A filter. Exposure times depend on so many factors that it is impossible to be specific: the NPL instruments require exposures of 10–30 seconds on a Kodak 0·800 plate.

Before the instrument is used for measuring purposes, it should be verified that the camera system is free from distortion. The conventional method, of photographing a scale in the work-plane of the interferometer, lacks the necessary precision, and a more sensitive technique can be employed which will indicate directly the errors which result when the interferogram is evaluated. A pair of good-quality surfaces is set up and photographs are taken of two interference patterns in which the direction of the air-wedge is reversed but the fringe frequency is approximately the same. Because of the reversed wedge, the errors due to distortion in these two photographs will be of equal magnitude but opposite sign. If therefore they are evaluated along a common diameter, the difference between the two results will, when the effects of the wedges have been removed arithmetically, equal twice the error in measurement caused by distortion in the optical system. In a well-corrected system this can be reduced to $\pm 0{\cdot}005$ fringes.

Evaluation of the interferogram

Gross errors of flatness (of several fringes magnitude) are most conveniently assessed by adjusting the surfaces to give as symmetrical a fringe pattern as possible. The resulting interferogram can then be regarded as a map in which the contours are at intervals of half a wavelength. Convexity and concavity can be distinguished by making small changes in the separation of the flats and observing the effect of the fringe pattern. For instance, if the air-gap is reduced by gently pressing downwards on the upper (reference) flat and the fringes move outward from centre to edge, then the test flat is convex; conversely if the fringes move towards the centre, the test flat is concave.

The majority of flatness measurements are concerned with errors of less than one fringe, and it is then more convenient to employ a 'wedge' interferogram obtained by inclining the surfaces at a slight angle to each other. Two perfectly plane surfaces will, under these conditions, give a set of straight, parallel and equidistant fringes, and any departure from this ideal pattern will represent an error in one or other of the flats. A wedge interferogram can be evaluated in one of two ways:

(*a*) by measuring the departure from straightness of a fringe in terms of the fringe separation, or

(*b*) by measuring the relative position of the fringes along an ordinate at right angles to them.

Visual estimates

Visual estimates are best made using the first method with the assistance of a nylon thread stretched across the field of the interferometer. The sign of the error can in this case be inferred from a knowledge of the wedge's direction:

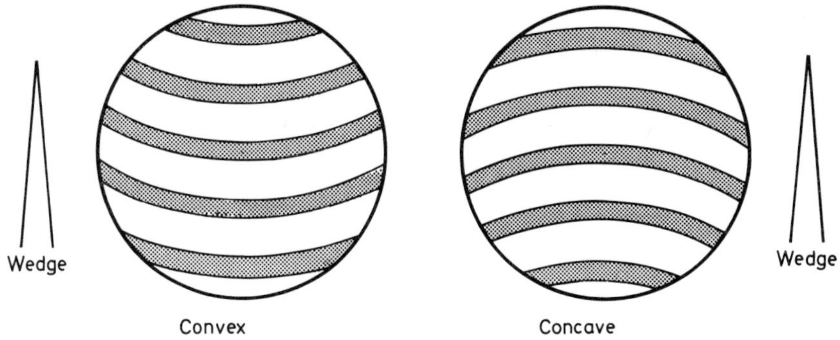

11.8.3. Inferring the sign of the test surface from the wedge direction.

examples are given in Fig. 11.8.3. It should be remembered that the pattern represents the sum of the errors of the test and reference flats, and the errors of the latter should therefore be deducted (taking account of sign) from the estimate made.

An accuracy of a fifth of a fringe is in most cases the best that can be achieved by direct visual assessment; slightly more refined measurements can be made with a rule on an enlarged photograph and, by this means, an accuracy of a twentieth of a fringe can be obtained. In either case a severe limitation is imposed by the fact that two-beam fringes are not sharply defined but have a $\cos^2$ intensity distribution. Considerable sharpening of the fringes results from the use of the multiple-beam technique due to Tolansky, but, for present purposes, there are two disadvantages:

(*a*) the separation of the surfaces must be reduced to a few wavelengths, and

(*b*) both surfaces must be silvered by vacuum deposition, the lower surface to give maximum reflectivity and the upper, a transmission of a few per cent.

It must therefore be stipulated that neither the deposition of the coatings nor their subsequent removal affects the contours of either surface.

Photoelectric assessment

For the testing of uncoated surfaces, the use of two-beam fringes is therefore unavoidable and, if a precision of better than a tenth of a fringe is required, some photoelectric means of assessment must be employed. Such a technique is in use at the NPL. It employs the second method of evaluating a wedge interferogram referred to earlier in this section. A series of ordinates at convenient unit separations is constructed perpendicular to the fringes (Fig. 11.8.4; $x = 1$ to 13) and the distance from an arbitrary origin at which each fringe intercepts these ordinates is measured. The fringes are numbered sequentially (giving due regard to the direction of the wedge) and the fringe value corresponding to each of a series of points at unit separation along each ordinate is determined by a process of interpolation. This establishes the magnitude of the air-gap (less an arbitrary constant) at a network of points uniformly distributed over the surface. Subsequently corrections are applied for the errors of the reference flat (which will have been previously determined), and the effect of the wedge is removed arithmetically. The results then represent departures of the test surface from an appropriate mean plane. The selection of this plane is of course arbitrary; it is generally chosen to pass through and be approximately tangential to the centre of the test surface, but surfaces which exhibit marked asymmetry may require special treatment. The results so obtained can be conveniently presented in the form of a map in which the contours are at intervals of a tenth or a twentieth of a fringe.

The measurements on the photographic plate are made using the equipment illustrated diagrammatically in Fig. 11.8.5. The negative is mounted on a coordinate stage so that the fringes (normal to the plane of the paper in the diagram) are approximately parallel to one of the movements. A magnified image of the interferogram is then projected on a screen in which are cut two rectangular apertures whose centres are separated by a distance equivalent to approximately half the fringe spacing. Behind each aperture is a diffuser and a photoelectric detector. In operation,

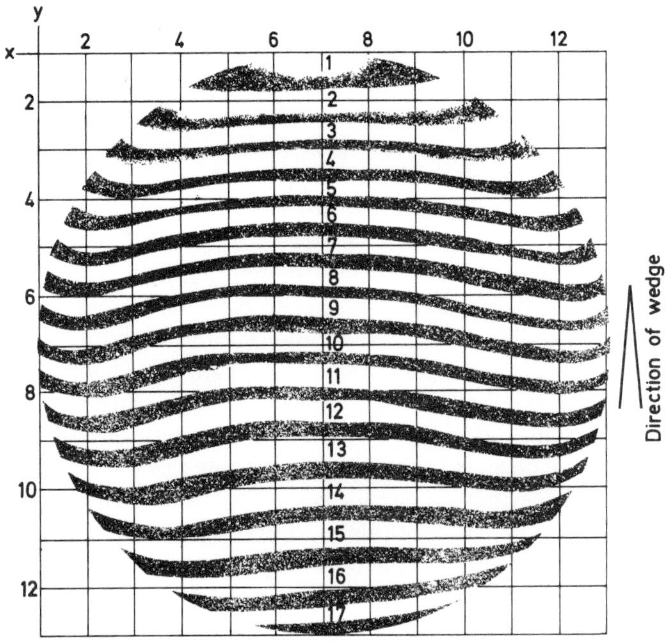

◁ **11.8.4. Specimen interferogram illustrating the method of evaluation.**

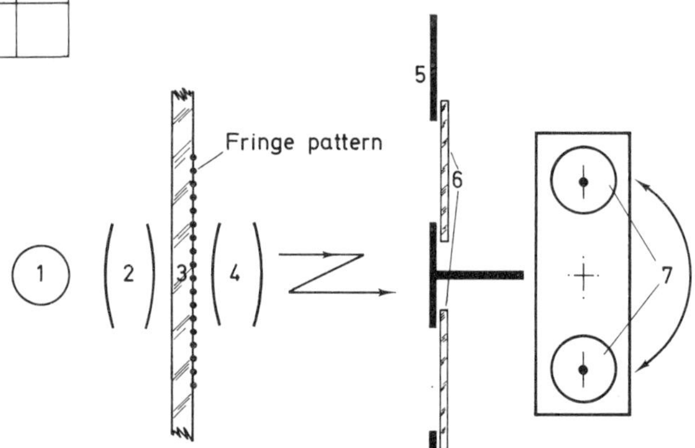

1. Lamp
2. Condenser
3. Negative
4. Projection
5. Image plane
6. Diffusers
7. Detectors

11.8.5. ▷
Equipment for photoelectric evaluation of the interferogram.

the position of the plate is adjusted by means of the appropriate micrometer (i.e. in a direction perpendicular to the fringes) so that each fringe along the ordinate under examination is in turn brought on to the optical axis of the system. Each setting corresponds to a condition under which equal responses are obtained from the two photocells and is identified by a nul-detection device. The detectors employed are Ferranti M.S.I. silicon diodes and they are connected in parallel to a 450-ohm 'spot' galvanometer with their outputs in opposition. It is essential that the sensitivities of the two halves of the system should be precisely equated. The photocells as purchased are, however, only approximately matched, and they are therefore mounted on a pivoted beam, the rotation of which causes the distance of the photocells from their respective diffusers to vary in inverse ratio. Before a set of measurements is made, this adjustment is used to obtain a balance with a piece of clear negative glass in the field of view. An accuracy of 1/100 fringe is readily obtainable with this equipment, and only the effects of 'plate-noise' and random blemishes on the negative prevent an accuracy five times better than this from being realized.

Testing aluminized surfaces

All that has been said hitherto refers to the interferometry of uncoated glass or fused silica surfaces with reflectances of 0·04–5. There are, however, many optical and engineering applications in which the testing of polished steel and aluminium-coated components is important. If such surfaces are examined on a Fizeau interferometer equipped with an uncoated reference surface, it will be found that the fringe contrast decreases as the reflectance of the test-surface increases; with fully coated aluminium or silver specimens the fringes are barely visible to the naked eye. There are two reasons for this: firstly the fringe contrast is a function of the relative amplitudes of the two interfering beams and becomes a maximum when they are equal: secondly and more significantly, multiple reflections between the two surfaces produce secondary interference patterns which are of contrary phase and therefore tend to reduce the contrast of the primary pattern.

If the two surfaces are uncoated, the amplitudes are closely matched and the multiple reflections are of negligible intensity: good contrast fringes are therefore obtained. If the test-surface is of polished steel (reflectance 0·4), the contrast is reduced, since the amplitudes are no longer matched and the intensity of the multiple reflections has increased. The visibility of the fringes is however tolerable under these conditions, although a significant improvement can be effected by increasing the reflectance of the reference surface, e.g. by applying a coating of bismuth oxide or titanium dioxide.

If the reflectance of the test surface is as high as 0·92, as will be the case with fully aluminized work, the first multiple reflection is comparable in intensity with the primary reflection from the reference flat. Under these conditions, fringes of very low contrast are obtained even with a bismuth oxide coated reference surface. To overcome this difficulty, a double-layer coating has been specially designed; it consists of an absorbing layer of 70 Å of bismuth with an anti-reflection coating of 250 Å of bismuth oxide. The photometric characteristics of such a film for a wavelength of 0·5461 μ are: reflectance at the glass/film interface, 0·10; reflectance at the air/film interface, 0·06; transmittance, 0·33. Used in conjunction with an aluminized test surface, the amplitudes of the two primary interfering beams are closely matched, whereas those of the multiple reflections are small by comparison. Symmetrical fringes of exceptionally high contrast are obtained.

The coated reference flat must of course itself be calibrated by comparison with an uncoated standard flat or a liquid surface and, in order to avoid corrections for sag, it should for this purpose be mounted in the same attitude as it will assume when used to test aluminized surfaces. Under these conditions, i.e. mounted face downwards above a surface of 0·04–5 reflectance, multiple reflections are of negligible intensity, although the primary interfering beams are not now matched. The contrast of the fringes is therefore reduced, but it is still comparable with that obtained between two uncoated surfaces; measurements to an accuracy of 1/100 fringe can be made without difficulty.

There is a second and more cogent reason for not reversing the coated flat for calibration purposes in spite of the fact that, by this means, fringes of higher contrast are obtained. Phase changes at the air/film interface are very dependent on the thickness of the film. For this reason, the small thickness variations which inevitably result from the sputtering process can introduce errors of several tenths of a fringe if opposite sides of the film are used in calibration and testing.

11.9 The measurement of testplates

The instrument most generally used to measure radii of curvature is the spherometer, and this is essential for the initial manufacture of test plates. Once test plates have been made, and the radii have been measured, then they can easily be copied directly by contact and the examination of Newton's rings. Initial measurement demands a precision spherometer. (Fig. 11.9.1.)

11.9.1. Watts precision spherometer for measuring the radii of testplates.

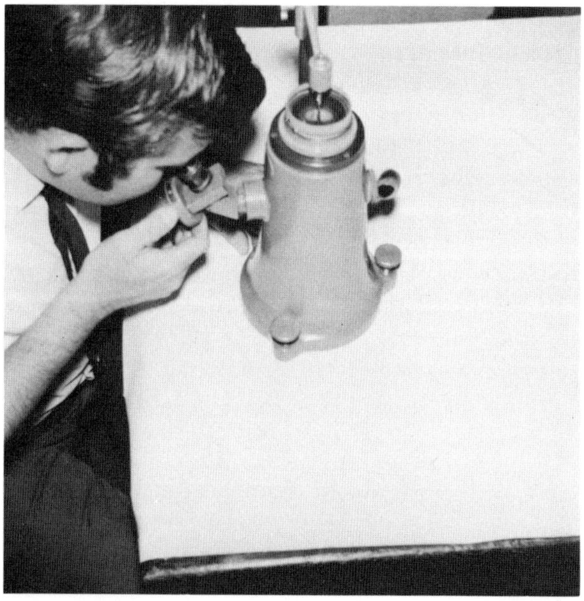

The Watts precision spherometer

This instrument is a variation of the Abbe type (see Glazebrook's *Dictionary of Applied Physics*, Vol. IV, p. 787). A cast Meehanite body carries a suitable support for the curve to be tested and a central ground steel plunger is kept in contact with the curve by a counter-balancing weight with cords and pulleys. A glass scale is rigidly mounted in the plunger and is viewed by a reading microscope mounted in the body of the instrument.

The position of the plunger is recorded for the test curve, and similarly when replaced by a perfectly flat surface. The displacement of the plunger due to the curve may be thus determined and, knowing the constants of the support, the radius of the curve may be calculated.

Opaque specimens such as grinding laps or ground-glass surfaces, may be tested as well as those that have been optically polished. If they are light in weight, a clamp is fitted which holds them securely in place under about one pound pressure exerted by a spring plunger.

Large specimens can be tested if this clamp is removed, but it is recommended that their weight be counterbalanced so as to reduce excessive load upon the instrument.

Constant pressure of the plunger upon the test surface is ensured by mounting it between the ball races. These provide a perfectly free motion with no serious degree of lateral shake. A milled head at the side of the body enables the plunger to be lowered against the upward force due to its counterweight but does not operate in the reverse direction and cannot affect the pressure.

Levelling screws and a spirit level are fitted in order that the plunger and counterweight may be set vertical.

High reading accuracy is achieved by the microscope and glass scale. Instruments can be supplied to read in inches or millimetres, the micrometer microscope enabling sub-division to 0·00005 inch or 0·001 millimetre. With suitable care, accurate readings may be attained by estimation to one-fifth of these values.

Permanence of accuracy is guaranteed owing to the stable nature of a linear glass scale, as opposed to the wear that may affect a micrometer screw. This permits the satisfactory use of a calibration chart if extreme precision is desired.

Permanence of the support constants is achieved by a kinematic design. Each curve support is made of hardened steel and truly located in a hardened steel insert in the spherometer body by a ground parallel fitting. Three steel balls rest in an annular groove ground into the support, and clips maintain their position but permit them to turn freely.

Clearance of the support in the testing of deep

convex and concave curves is ensured by the design and a support will accept any radius of curvature greater than three quarters of the annular groove diameter.

The support constant is the same for convex and concave curves, thereby reducing the risk of error due to using a wrong value. The largest possible annular groove diameter should be used for the test curve in order to achieve maximum accuracy.

The formula for calculating radius of curvature (r) is as follows:

$$r = \frac{A}{h} + \frac{h}{2} - a \quad \text{(convex surfaces)}$$

$$r = \frac{A}{h} + \frac{h}{2} + a \quad \text{(concave surfaces)}$$

where h is the measured displacement of the plunger, a the radius of the steel balls, and A the constant of the support (it represents half the square of the radius of the annular groove).

In an instrument of high precision it is preferable that the value of A be obtained indirectly by calibration.

Calibrating of the curve supports is carried out before they leave the factory by using standard curves. Should future calibration be desired, the following procedure is suggested.

Plunger displacements are determined using known curves and A is calculated from the equation

$$A = h\left(r \pm a - \frac{h}{2}\right)$$

As standard curves, large-diameter steel balls of high quality are available whose diameters can be very accurately determined. The larger balls may weigh several pounds and a counterbalancing system must be employed. The standard curve should be suspended in a cradle from a spring-balance vertically above the spherometer and gradually lowered on to the instrument until the balance records a pound less than the weight of the ball. This ensures a constant load of weight.

Final accuracy of the instrument depends upon the accuracy of determining both A and h. Laboratory tests have proved that A will be accurate to within 0·01 per cent

Although plunger displacements can be accurately measured to 0·00001 inch, a small value of h (such as would be obtained from a large radius) must, of necessity, impose certain limits on the percentage accuracy.

For example, when testing a convex curve on a support with annular groove diameter of 1·25 inches and ball radius 0·125 inch, a plunger displacement of 0·01000 inch is measured. By calculation, the resulting radius of curvature is then found to be 19·41 inches. Since the plunger displacement can only be accurate to 1 in 1000, this must also be the accuracy of determination of the given radius.

In the general case, the following formula may be applied:

$$\text{Percentage accuracy} = \frac{0{\cdot}001}{r - \sqrt{(r^2 - 2A)}}$$

11.10 The measurement of curvature on large lenses and mirrors

The tripod form of spherometer consists of a triangular metal frame with three pointed legs fixed at the corners of an equilateral triangle. Equidistant from these legs is a pointed micrometer screw with precision head and screw.

This type of spherometer is used to measure large optical components (Fig. 11.10.1), but has two main sources of error. These are the difficulty of determining the effective radius of the circle passing through the points of contact of the three legs, and the exact location of the point of contact between the central screw and the test surface. (See § 15.4.)

A far more accurate method of checking large, shallow lenses and mirrors is by using an autocollimator, a penta prism or optical square, and self-centring rails. (Figs 11.10.2 and 11.10.3.)

11.10.1. ▷
Spherometer for checking the curve in an astronomical mirror being made from a Cervit blank.

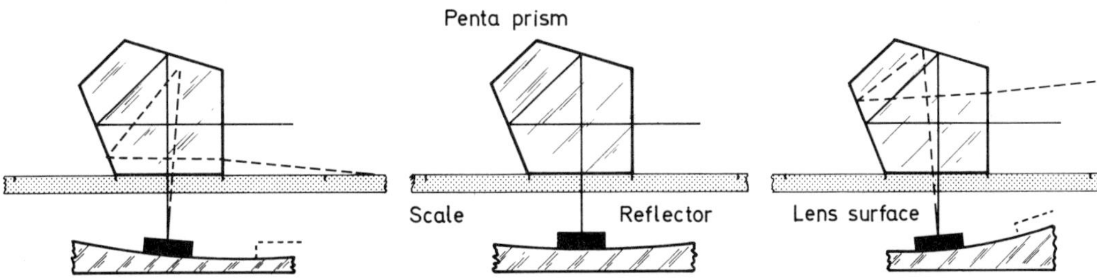

11.10.2. Using a penta prism to obtain the profile of a surface. The prism moves in steps along the lapped rails. The angular deflectors of the return beam is read from either a visual-setting or a photoelectric-setting autocollimator.

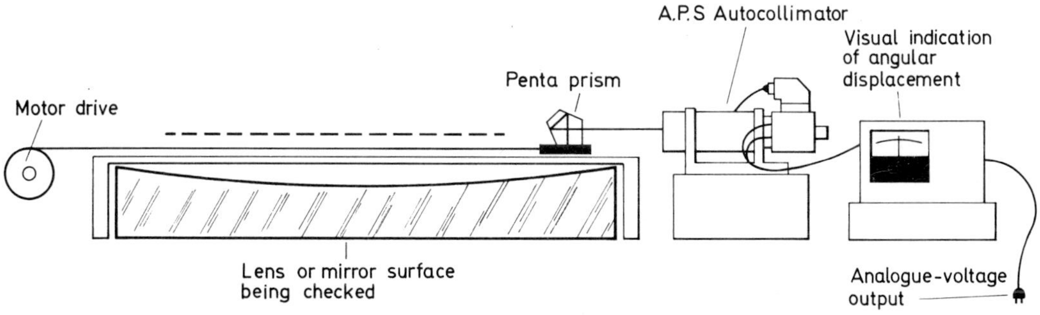

11.10.3. Continuous recording of a profile. Continuous movement of the prism is given by the geared motor-drive coupled to the carriage. The autocollimator detects angular displacements of the return beam; the analogue voltage output may be integrated and connected to a chart recorder.

11.11 Comparator gauges for spherical surfaces

When turning the spherical surfaces on glass working tools or generating the spherical surface on lenses by diamond laps, it is necessary to have an accurate spherometer which will not be damaged by contact with grey surfaces.

The air-operated gauge made by Solex (Gauges) Ltd, 235 Marylebone Road, London, NW1 (Fig. 11.11.1) consists of a comparator unit placed in the centre of a rim or listel (C). The stem of the comparator unit, having a flat tungsten-carbide contact (D), rests on the surface of the part to be measured. Using the rim contact as a basis of reference, variations of curvature will lift the stem of the comparator unit to a greater or less extent. The consequent variations of air flow through the valve inside the unit cause the water level in the column indicator to record variations in the distance (*h*) from that of the master reference.

The Solex spherometer consists of a comparator unit, three standard listels (C) of 10 mm, 20 mm, and 40 mm diameter, and a column indicator (Fig. 11.11.2) fitted with one of two suitable control jets and scales graduated in 0·0001-inch divisions.

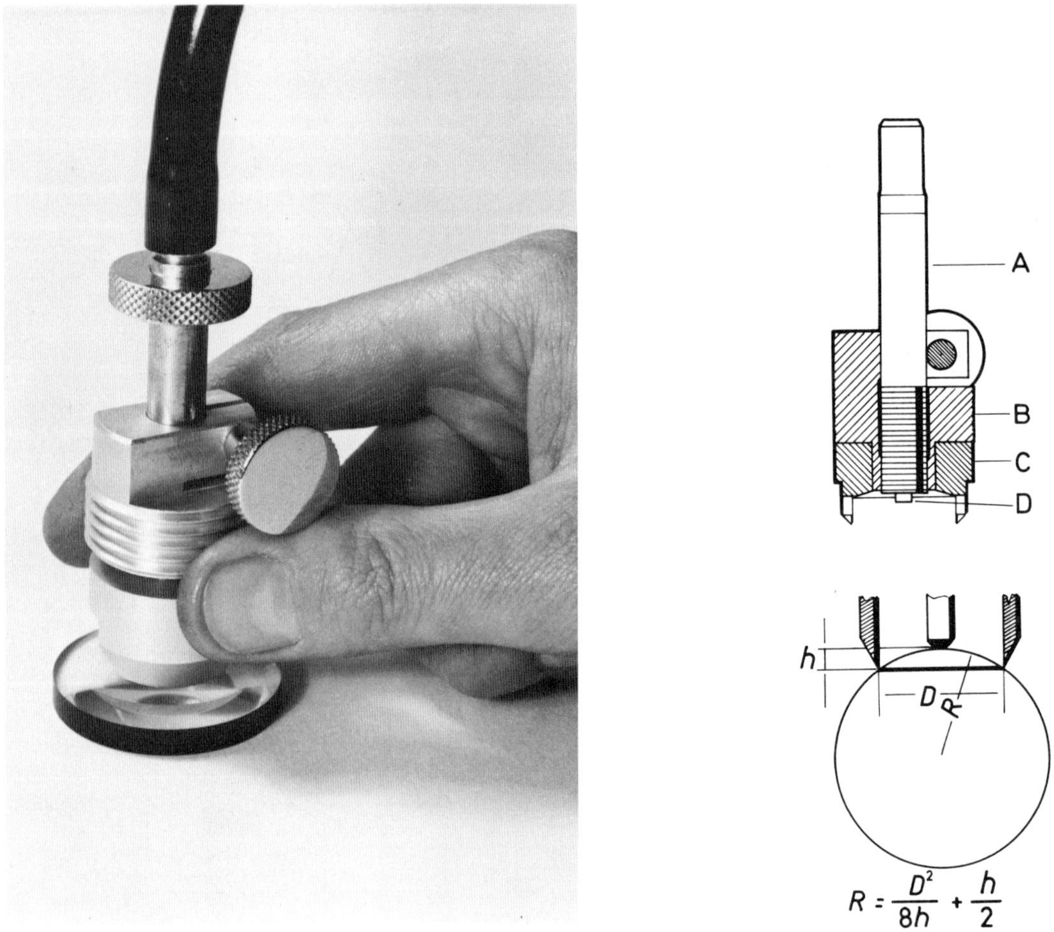

11.11.1. Comparator gauge in contact with a spherical glass testplate. The diagrams show the construction and the geometrical principle of the instrument.

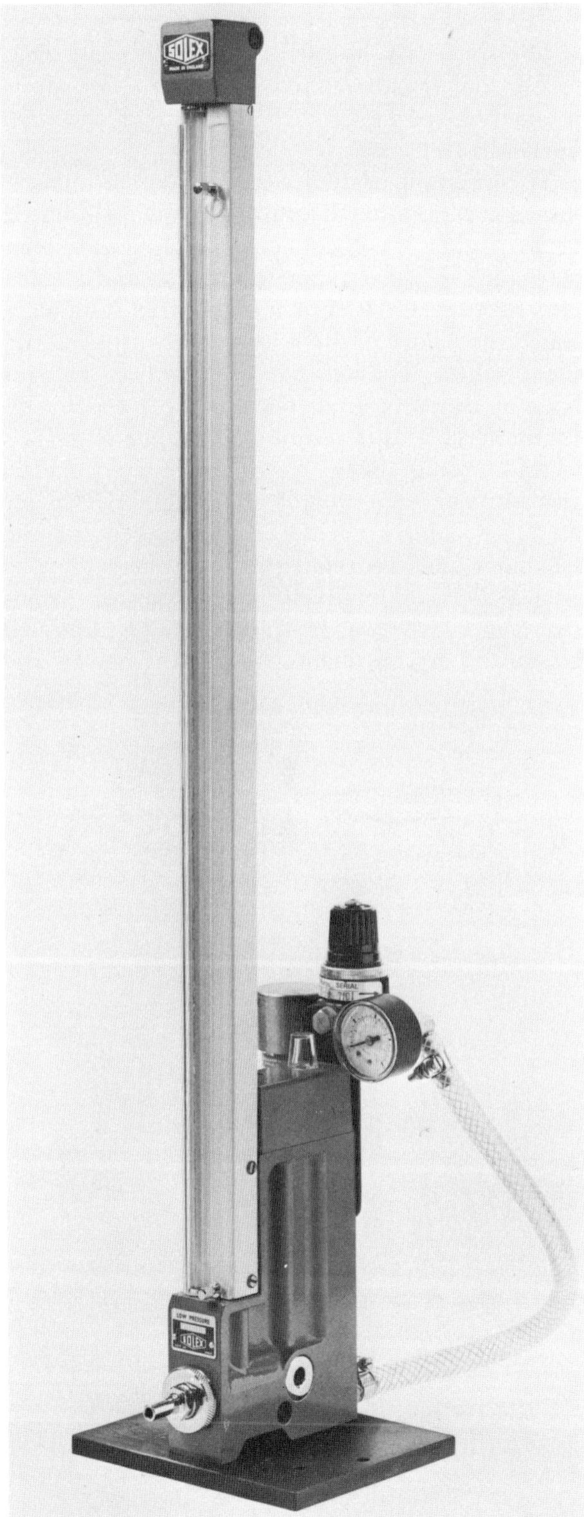

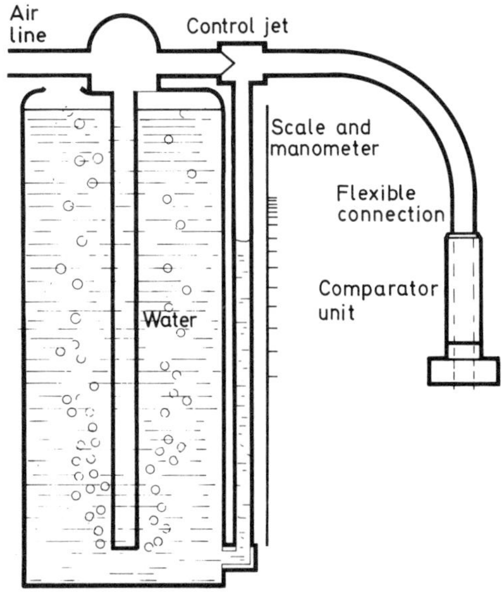

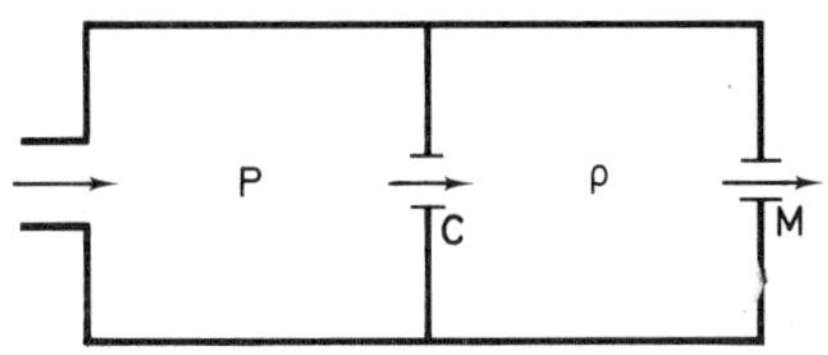

11.11.2. The Solex column indicator is graduated in 0·0001-inch divisions. It consists of a vertical cylinder containing water. A restricting control jet leads to a glass manometer tube and to the comparator unit. A calibrated scale is set beside the glass tube. The pressure *p* in the manometer tube is regulated by the relative rates of escape of air through the control jet C and the jet in the comparator unit M. If the jet in the comparator unit is completely closed, the manometer tube water level is depressed to the bottom of the tube. Pressure *P* in the airline should be about 20 lb/sq. inch (1·40 kg/sq. cm).

11.12 The Hilger & Watts sphericity interferometer

The Hilger & Watts sphericity interferometer is for examining steel and glass spheres to determine departure from sphericity. The instrument was developed from a design by the National Physical Laboratory, and it is capable of examining spheres from $\frac{1}{8}$-inch (3·2 mm) diameter up to 1-inch (25·4 mm) diameter. (Fig. 11.12.1).

Compared with the more usual testplate optical-interference method for testing sphericity, the instrument:

(1) overcomes the need for individual comparison test-plates of known curvature,

(2) enables the maximum area of a sphere to be examined at one time,

(3) avoids the risk of damaging either the reference surface, or the sphere under test.

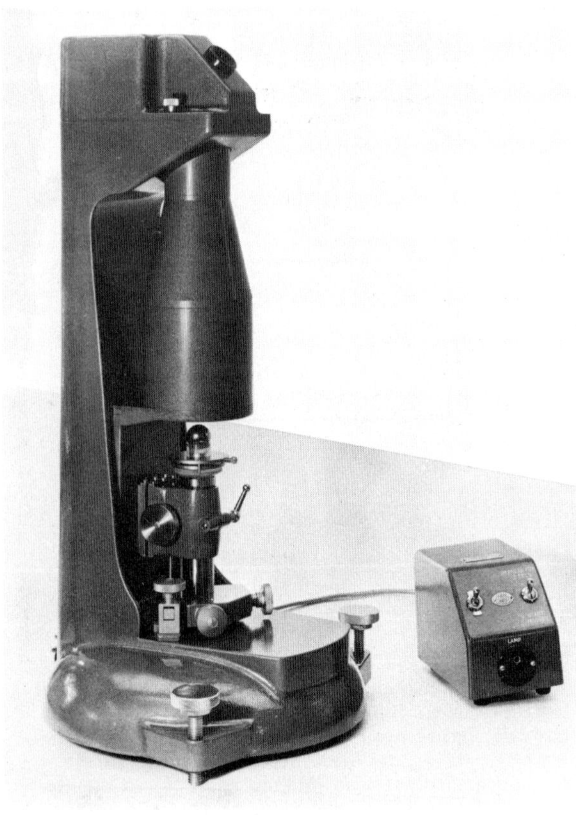

The sphericity interferometer is so convenient to use that it can readily be applied to routine production-testing of such products as precision steel balls and lenses.

The principle of the sphericity interferometer is shown in the optical diagram. The sphere is placed beneath the reference hemisphere, and monochromatic light from a mercury lamp and a green filter is converged by a multiple-lens system towards the centre of the reference hemisphere and the specimen. The surfaces of both reflect the light back along its original path, causing the beams to interfere. The resultant interference pattern is viewed through the eyepiece of the interferometer, and is analogous to that of a Fizeau interferometer used for comparing flat surfaces. Because the specimen and the reference hemisphere may be separated by a relatively large distance, the instrument allows a range of spheres to be examined by means of the one reference surface. It also

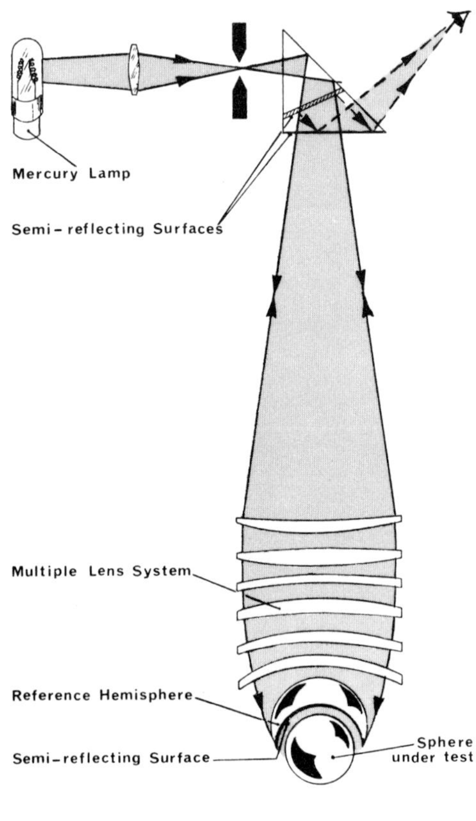

11.12.1. View of the Hilger and Watts sphericity interferometer showing mains-operated transformer for mercury light. (Right) Optical diagram.

11.12.2. Typical interferograms.

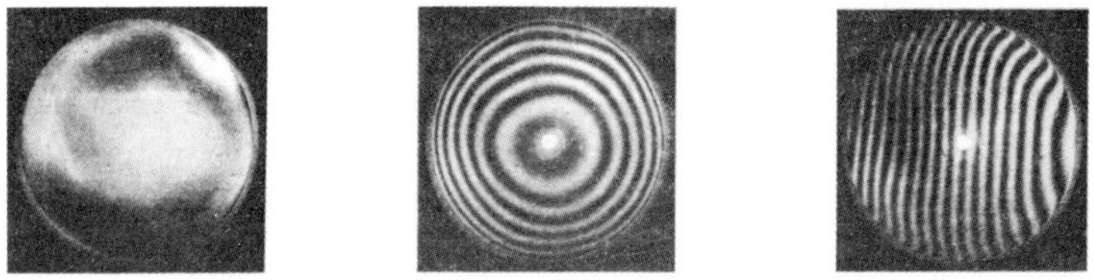

A B C

includes a standard reference hemisphere suitable for examining spheres of $\frac{1}{8}$-inch to 1-inch (3·2 mm to 25·4 mm) diameter.

From the interference patterns seen through the eyepiece, departures from sphericity are determined. Three typical interferograms produced by a pitch-polished steel ball are illustrated. (Fig. 11.12.2.) If the sphere being examined is perfectly round, and is correctly positioned to coincide with the centre of curvature of the reference hemisphere in all planes, the interference pattern is 'blacked-out' or confined to one interference fringe (A).

Separating the two surfaces axially produces a pattern of rings the shape of which indicates departures from sphericity (B).

Separating the two surfaces laterally produces light and dark cross-bands which, depending on departure from straightness, indicate errors of curvature of the test surface (C).

The actual distance separating the fringes is equal to half the wavelength of the light source—which for mercury-green light is 10·8 micro-inches, or 0·27 micrometre. By means of this natural unit of measurement, irregularities of the sphere are determined. For example, ovality of rings to the extent of a normal spacing indicates an error in sphericity of approximately 10 micro-inches. Departure of the cross-bands from straightness to the extent of a spacing, indicates an error in curvature of approximately 10 micro-inches. Since variations of one-fifth of a spacing can be estimated, measurements representing 2 micro-inches, or 0·05 micrometre are possible.

11.13 Precision thickness measuring machine

Some optical components have to be produced to a very close thickness tolerance.

Fabry-Perot etalons are used in the hyperfine study of spectral lines. They consist of two plates separated by a distance piece made of a hollow cylinder of fused silica. These distance pieces are supplied in varying lengths and their dimensions must be closely controlled.

A suitable instrument for this is the microptic vertical measuring machine which is capable of an accuracy of 0·002 mm (0·000 01 inch).

Fig. 11.13.1 illustrates the instrument being used for precision thickness measurement on an optical component.

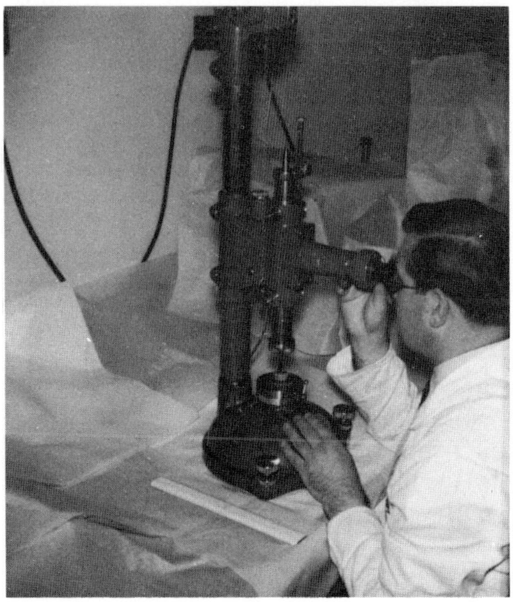

11.13.1. Microptic vertical measuring machine for precision inspection of components.

11.14 Testing for parallelism of glass plates

Some optical components such as Twyman–Green interferometer beam-splitters and compensators, have to be flat to 0·1 fringe and, in addition, must be parallel to within 5 seconds of arc. The method of testing by interference (reflection) from two nearly parallel surfaces separated by air or by glass will give the measurements listed in the table.

The relationship between the number of bands and the deviation from the parallel is as follows:

One band by mercury green light (air between the surfaces) indicates 0·000 0107 inch out of parallel.

One band by mercury green light (glass, refractive index 1·52, between the surfaces) indicates 0·000 0071 inch out of parallel.

Angle between surfaces (A)	*No. of dark bands per inch by reflection (air between surfaces)* $=2t/\lambda$	*No. of dark bands per inch by reflection (glass between surfaces)* $=2nt/\lambda$	*Deviation (produced by back-surface reflection)* $=2nA$	*Deviation (single transmission)* $=(n-1)A$
1 sec	0·45	0·7	3·0 sec	0·5 sec
5 sec	2·2	3·4	15·2 sec	2·6 sec
10 sec	4·5	6·8	30·4 sec	5·2 sec
15 sec	6·7	10·2	45·6 sec	7·8 sec
20 sec	8·9	13·6	60·8 sec	10·4 sec
25 sec	11·2	17·0	76·0 sec	13·0 sec
30 sec	13·4	20·4	—	15·6 sec
35 sec	15·6	23·8	—	18·2 sec
40 sec	17·9	27·2	—	20·8 sec
45 sec	20·1	30·5	—	23·4 sec
50 sec	22·3	33·9	—	26·0 sec
55 sec	24·6	37·3	—	28·6 sec
60 sec	26·8	40·7	—	31·2 sec

Refractive index, $n=1·52$.

11.15 Gauge interferometer for demonstrating interferometry

The Hilger & Watts Gauge Interferometer TN 100 (Fig. 11.15.1) was designed to demonstrate the phenomenon of optical interference and how interference fringes produced by different wavelengths of light are used as very finely divided scales having different graduation intervals.

The instrument is a 'Fizeau' interferometer incorporating an inclinable dispersion prism for splitting light from the cadmium lamp into red, green, blue, and violet radiations. The optical components are of the highest quality and workmanship, and are enclosed within the tubular body of the instrument. A fiducial setting line is visible through the eyepiece. The light source and condenser lens are mounted on an arm secured to the instrument stand. The arm may be located to the left of the interferometer, as illustrated, or at the back of the stand, depending on the dictates of available space. For the latter position, the monochromator unit is rotated through 90 degrees.

The cast-iron base has a levelling platform on which is mounted either the micro-mover or a work table having a 2-inch diameter lapped-steel platten.

Light from the cadmium lamp is focused by the condenser lens (a) on slit (b), and is deflected by mirror (c) to the dispersion prism (d). The prism separates the light into its four constituent radiations (red, green, blue, and violet) which are deflected downwards at different angles by the mirror surface on the end-face of the prism.

Adjustment of the inclination of the prism, by means of an external lever, permits any of the four radiations to be positioned normal to the perpendicular optical path. Light of the selected radiation is focused by the lens (e) on to slit (f), and passes through the semi-reflecting beam splitter (g) and the collimating lens (h). Before reaching the lapped surface of the work-table (k) the beam of light passes through the optical reference flat (j). Interference of the beam is

11.15.1. ▷
The optical system of the Hilger & Watts gauge interferometer.

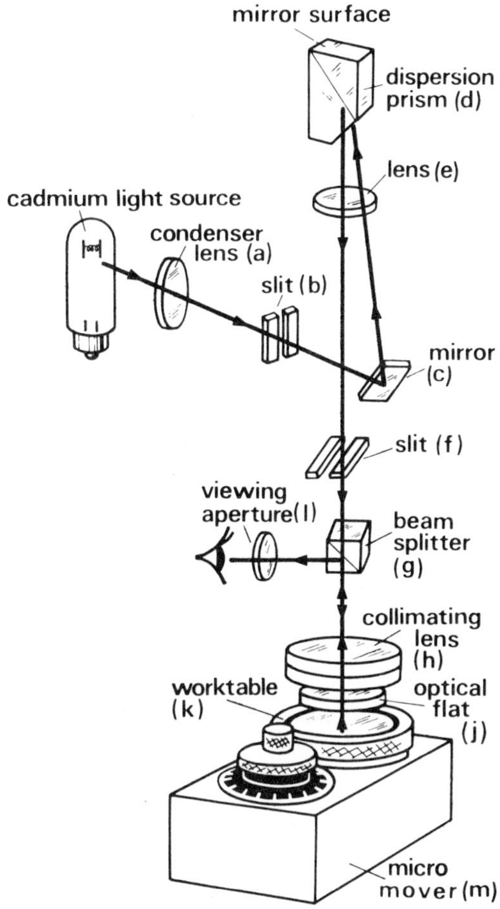

▽ 11.15.2. The micro-mover of the gauge interferometer.

produced by light reflected by the surface of the work-table and the underside of the optical flat; the interfered beam is reflected back into the optical system and is deflected by the beam splitter to the viewing aperture (l) through which interference fringes are observed.

A simple 1:1 extension eyepiece is available as an optional accessory, to permit observations through the viewing aperture at a distance which is desirable for minimizing the effects of body heat when making measurements to the highest accuracy.

The micro-mover (m) consists of a rotatable micrometer screw which, through a lever system, imparts vertical motion to the worktable at a reduction of 140:1. Measurements of the travel of the table are given direct to 1 μinch (0·000 001 inch) on a circular scale graduated from 0 to 150 μinches. The scale is figured every 10 μinches. A metric scale is also available, graduated every 0·0001 mm and figured every 0·0005 mm.

The scope of the interferometer may be demonstrated as follows (Fig. 11.15.3):

(1) Observe the platten of the micro-mover, through the viewing aperture, and adjust the interferometer so that interference fringes are seen. Select each of the four radiations, in turn. In this way, students can prove that the spacing of fringes varies with different radiations.

(2) Slowly raise and lower the platten of the micro-mover, so that the pattern of interference fringes moves laterally across the field of view.

Students can determine the wavelength of a radiation by noting the difference in micrometer readings for the passage of 10 fringes past the fiducial line visible through the viewing aperture.

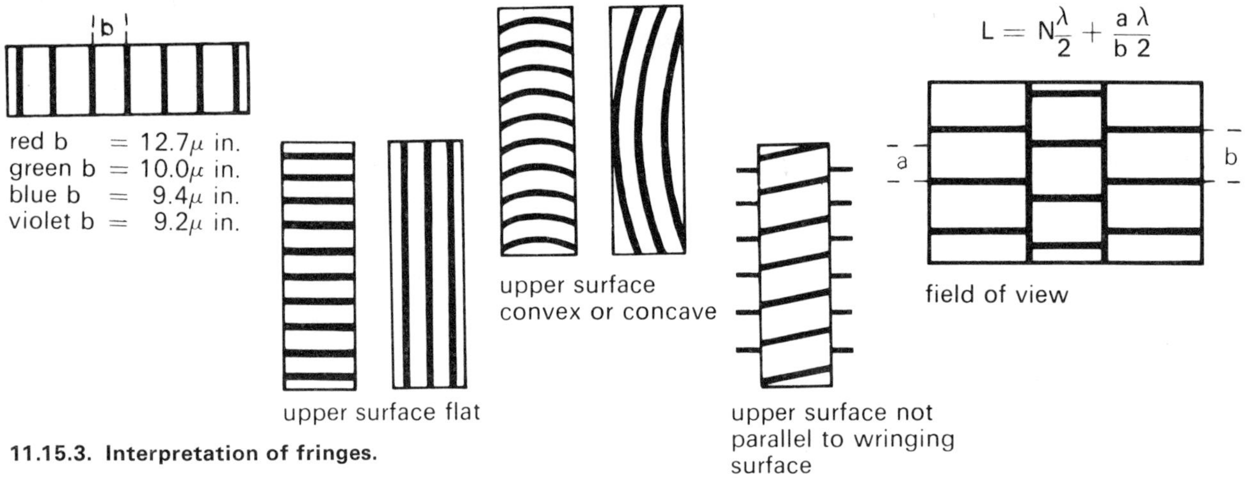

11.15.3. Interpretation of fringes.

(3) Wring a slip gauge to the worktable and demonstrate that the interference fringes are perfectly straight if the surface of the gauge is truly flat; that they are curved if the surface is concave or convex; and that they are inclined if the surface is not parallel to the wringing surface.

Students can determine the errors in flatness or in parallelism, by estimating the departure of fringes from straightness, or from the horizontal, expressed as a fraction of the fringe spacing.

(4) Wring a slip gauge of approximately known length, on the surface of the worktable. Explain how the two patterns of interference fringes seen through the aperture are produced by the upper surface of the gauge, and by the lapped surface of the worktable. Demonstrate how the misalignment of the two patterns, relative to each other, depends on the wavelength of the radiation used.

Students can measure the precise length of the gauge by estimating the amount of displacement between fringes from each pattern, expressed as a fraction a/b, for each of the four radiations in turn; and by using the fraction-coincidence method of calculation. A coincidence calculator scale is supplied for the easy determination of gauge errors, from fractional differences.

11.16 Twyman-Green prism and lens interferometers

The following account by Mr F. Twyman, F.R.S., in his book *Prism and Lens Making* gives the background to the interferometers which are now famous throughout the world. More recent developments are outlined on p. 338.

HISTORICAL INTRODUCTION

My justification for the rather full account I propose to give of Hilger interferometers is that their introduction at the Works of Adam Hilger Ltd established an epoch in the development of the firm. Not only so, but their importance in the control of optical production was early recognized by some of the foremost optical firms in the world.

These interferometers received recognition in three unusual ways. The author received on their account, in 1927, the Duddell Medal, awarded by the Physical Society for 'meritorious work on scientific instruments and materials'; in 1926 the John Price Wetherill Medal awarded by the Franklin Institute of Washington for 'the great scientific value of his interferometer for the testing of optical parts', and was, in 1931, with the co-inventor, Alfred Green, granted a ten-year extension of the two principal British patents. Once before only had so long an extension been granted for British letters patent.

W. Taylor, Governing Director of Taylor, Taylor & Hobson of Leicester, said of the camera lens interferometer in *The Times* of 24 April 1929:

> We have had one of these interferometers in use in our works upwards of four years and it has been one of a number of things which have enabled us to make advances in the design and construction of our optical systems.

This opinion supplemented that of A. Warmisham, Optical Director of the same firm, who said in a letter to me dated 9 May 1925:

We see in the Hilger photographic lens interferometer a new and most powerful means of revolutionizing the method of optical design. The present state of the art of photographic lens design is determined by the absence of specific means of dealing with the spherical aberrations of oblique periods. In spite of all the mathematical work that has been done on this subject there exists no means at present of forming a definite judgment, based on algebraic processes of the quality of definition given by a lens throughout a semi-field of 26°. We suppose that a fairly complete idea of the quality of definition could be obtained by a sufficient amount of trigonometrical computation, but an enormous amount of labour would be involved, especially to trace the effect of small departures from the spherical in any surface, for it would be necessary to compute a large number of rays skew to the axis. The lens interferometer will provide quick and certain means of determining what departures from sphericity are required, and in what surfaces they are best applied, in order to reduce to a minimum, at any selected angle, the outstanding oblique spherical aberrations.

Ross Ltd had a like experience and informed me (May 1925) that they found that with its aid they could definitely determine the cause of bad definition in optical instruments and, in any doubtful case, examination on the interferometer showed at once the location and the extent of the fault, and gave them the means of correcting these faults by local re-touching of one or two surfaces.

Wilfred Taylor, of Cooke Troughton & Simms, Buckingham Works, York, wrote (1 Oct. 1930):

We have now used one of your interferometers for a number of years, and regard it as a most remarkable weapon in the hands of the optician, particularly, as regards our own work, in connection with the figuring of large prisms. If it were possible to obtain consistently large pieces of perfect optical glass, the correct angling and working of the surfaces could be ensured by various tests, but, as is well known, imperfections in the material itself may mar the performance of a large and expensive prism, even though the surfaces and angles are free from error.

For example, Dr G. Hansen of the Zeiss-Opton Optischewerke, Oberkochen, Württenburg, sends me the following particulars of the use made of interferometers supplied by Adam Hilger Ltd. (Translated from Dr Hansen's letter of 17 April 1950.)

The interferometers were used in Jena to an ever-increasing degree for the testing of optical parts. In particular, since the beginning of 1930, such parts for all glass and quartz spectographs in my department were so tested. Further, particular lens systems were frequently examined interferometrically when it was desired to obtain quickly an exact determination of the degree of correction. Later the instruments supplied by your firm no longer sufficed and we then ourselves made similar interferometers for our testing. The large interferometer for photographic objectives was used in the correction of the prisms of the tower spectograph of the Einstein Tower in Potsdam.

The firm of Carl Zeiss, Jena, purchased in 1929 an entire series of these interferometers (prism and lens, camera lens, and microscope objective interferometers). Many years afterwards evidence was afforded that use was made of the instrument to test the performance of *non-spherical* lens systems.

These interferometers express the aberrations of optical systems in terms of departure from sphericity or planeness of wavefront arising from passage through the systems. None-the-less they are capable of being interpreted so as to give the aberration in any of the customary forms, namely longitudinal, lateral or rectangular (Perry, 1923).

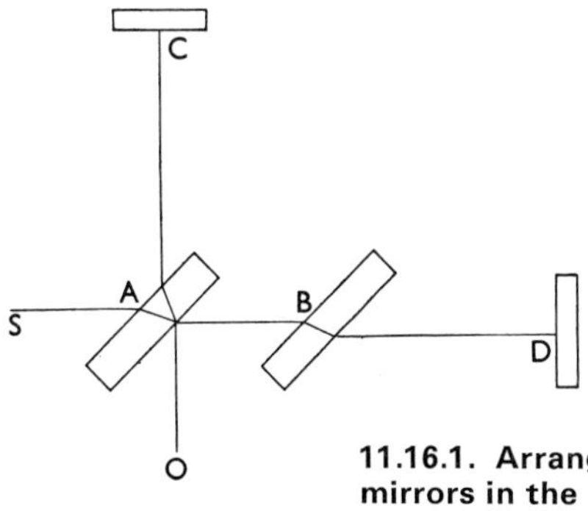

11.16.1. Arrangement of mirrors in the Michelson interferometer.

All these interferometers derive from that of Michelson (1907), illustrated in Fig. 11.16.1, as will be described later. This, by-the-by, was not the form originally described by him (Michelson, 1881) which had the disadvantage that the two comparison beams were side by side. Laurent used a form of Fizeau apparatus in which a lens focused the rays from the source on the eye. By this simple modification it was ensured that those rays which reached the eye of the observer had passed parallel through the plates.

11.16.2. The first Hilger interferometer.

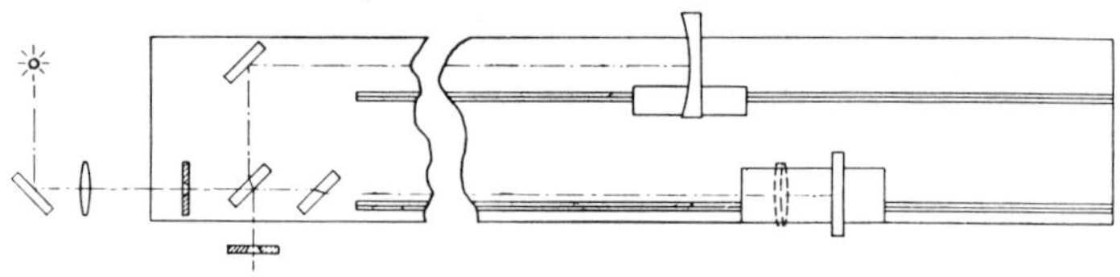

This was, in effect, the modification to the Michelson interferometer that transformed it into the Hilger prism interferometer now to be described, although in the development of the principle many other new devices were employed.

The first types were described in British Patent 103,832 (Twyman & Green, January 1916), of which one of the forms (Fig. 11.16.2) was appropriate for testing concave mirrors and telescope objectives only.

The most generally useful of the instruments built on the Twyman and Green principle is the Hilger prism and lens interferometer, made by Hilger & Watts Ltd [Rank Precision Industries] for testing prisms and lenses. This instrument is suitable for workshop use in the operation of retouching.

Numerous methods of testing telescope or camera objectives have been devised with a view to the control of retouching.

The opinions of Schroeder, Grubb, Czapski and Alvan Clark are cited in a résumé by H. Fassbender (1913) of the then known methods of testing object-glasses. It omits, however, an ingenious method of Dr Chalmers (1912).

The more recent methods of Waetzmann (Bratke, 1924) (founded on the Jamin refractometer), Ronchi (1926), and Lenouvel (1924) (these two derived from Foucault's test), are of interest and capable—in experienced hands—of yielding useful results. Of none of them, however, can it be said—as it can of the Twyman and Green forms of interferometer—that unskilled boys or girls can in a week or so be taught not only to test prisms and lenses and state precisely the nature of their defects, but to correct the optical performance by retouching the surfaces.

It must be added that the phase contrast test developed by Burch from that introduced by Zernicke is also capable of yielding very direct information concerning distortions of wavefronts caused by concave mirrors or lenses.

The Mach-Zehnder interferometer

The Mach-Zehnder interferometer may be mentioned for the sake of historical completeness. In application to the examination of plane parallel plates this interferometer presents appearances almost identical with those of the Hilger Interferometer; and, indeed, but for the fact that it does not lend itself very well to the examination either of prisms or of lenses, there would be little to choose between it and the latter instrument. Doubtless it could be modified to overcome these limitations, but it would then form an extremely cumbrous piece of apparatus.

It was described in 1891 by Zehnder and shortly afterwards modified by Mach, and has since received many modifications.

A very complete description of the instrument in its various forms is given in an article by Kinder (Kinder, W., 1946, *Optik*, **1**, 413). The paper has a very complete bibliography and a plate of beautiful coloured photographs of the interference fringes, which are like those obtained on the Hilger interferometer.

THE HILGER INTERFEROMETERS

(Figs 11.16.3 and 11.16.4.)

The Hilger interferometers here described produce a series of interference rings which may be regarded as a 'contour map' of the imperfections. This contour map can for practical purposes be considered as located at any of the optical surfaces involved and, in the case of the control of retouching, the observer may, if he likes, draw this map upon the surface under treatment. He is then in a position, without further preliminaries, to remove the superfluous material from the prominences by polishing with pads of suitable size and shape, the 'contour map' giving all that is necessary for him to know both as to the location and magnitude of the sources of the imperfections.

General construction principles

This instrument in its simplest form resembles the well-known Michelson interferometer, the main essential optical differences being that the light is collimated and the two interfering beams of light are brought to a focus at the eye of the observer.

Optical elements or combinations suitable for examination on it may almost all be classed in two

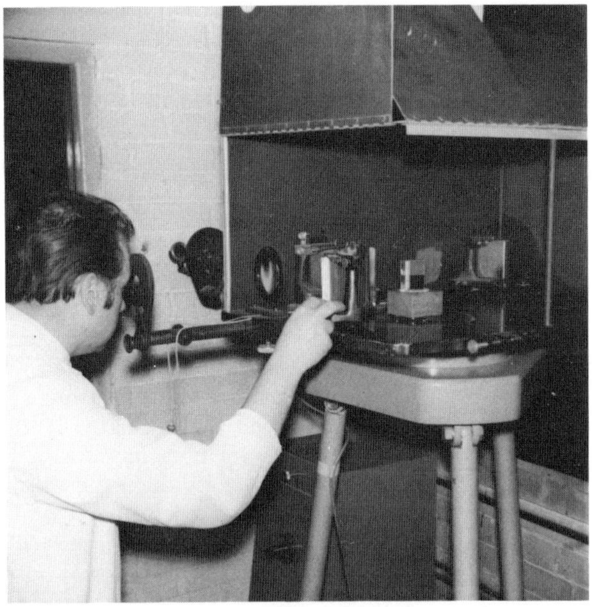

(*a*)

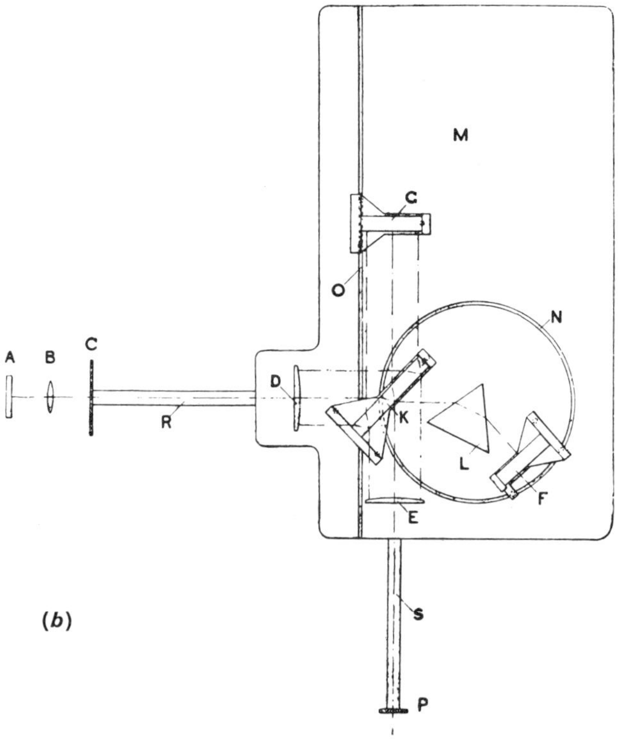

(*b*)

11.16.3. Three-inch aperture Twyman-Green interferometer of the original design.

categories. Into the one category fall those combinations which are required to receive a beam of light which has a plane wavefront and deliver it again after transmission with a plane wavefront, and into the other fall those the object of which is to impart spherical wavefronts to beams which are incident on them with plane wavefronts. The two corresponding arrangements will be referred to as the prism interferometer and the lens interferometer respectively.

The prism interferometer

The prism interferometer is shown in the diagram (Fig. 11.16.3) as arranged for the correction of a 60° prism, such as is used for spectroscopy.

The light used must consist of very homogeneous radiation. Such light may be obtained from a low-pressure mercury-vapour lamp with a glass tube, such as the Hewittic.

The light from the source is reflected by the adjustable mirror *A* through the condensing lens *B*, by means of which it is condensed on the aperture of the diaphragm *C*.

The diverging beam of light is collimated by a lens *D*, and falls as a parallel beam on a plane parallel plate *K*, the second surface of which is silvered (or aluminized) lightly so that a part of the light is transmitted and part reflected. The major part should be reflected. One part passes through

11.16.4. Seven-inch aperture Twyman-Green interferometer of the original design.

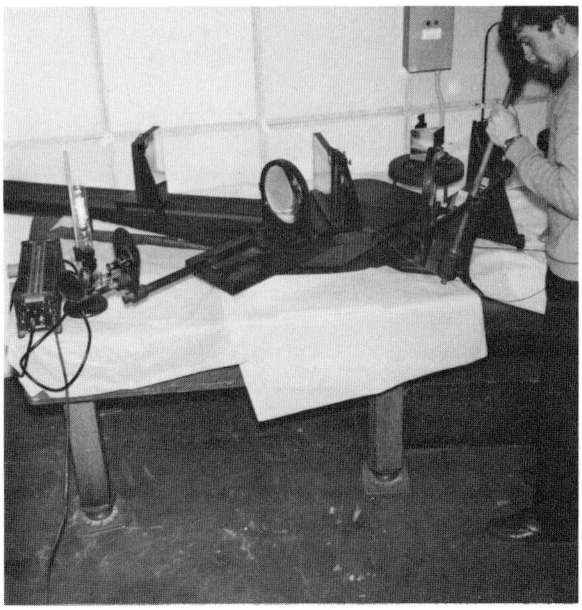

11.16.5. Diagram of interference pattern.

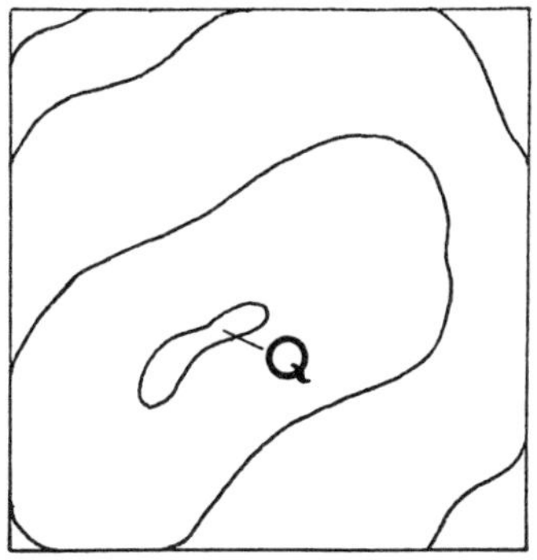

the prism L in the same way as in actual use, and, being reflected by the mirror F, passes back through the prism to the plate K. The other part of the light is reflected to the mirror G and back again to the plate K. Here the separated beams recombine, and passing through the lens E each forms on the eye, placed somewhat beyond the aperture in the diaphragm P, an image of the hole in the diaphragm C.

When the mirrors are adjusted, interference bands are seen which form a contour map of the glass requiring to be removed from the prism face in order to make its performance perfect.

Fig. 11.16.5 represents in diagram a typical map, where Q represents the highest point of a 'hill'. The procedure in such a case would be to mark out the contour lines on the surface of the prism with a paint brush and then to polish first on the region Q, subsequently extending the area of polishing, at first partly, then wholly, to the next contour line; and so on. The marking out of the prism surface can be done while one is observing.

It should be noted that variations in the contour lines are obtained by a tilt of the plane of reference. Thus a slight adjustment of mirror F (Fig 11.16.3b) might change a contour map from that shown in Fig. 11.16.6 to that shown in Fig. 16.11.7. The form of surface is in each case the same (see the sectional diagrams at the top of the figures), but correction can be carried out according to whichever plane of reference is the most favourable from the point of view of the operator. In order to find whether Q is a hill or a valley, the cast-iron table M can be bent with the fingers so as to tilt the mirror F in such a way as to lengthen the ray path. If the contour line at Q expands to enclose a larger area, a hill is indicated, and *vice versa*. Although the words 'hill' and 'valley' are convenient to use, it must not be supposed that the imperfections necessarily result from want of flatness either of one or of both surfaces of the prism. The contour map gives the total effect on the wavefront produced by double passage through the prism, and shows in wavelengths the departure from planeness of the resulting wave surface.

Increased illumination

One can get a greater contrast and considerably more light in the fringes by the use of zinc sulphide coating for the semi-reflecting surface instead of silver. In spite of the better effect obtained, however, we have not found it worth while to use it, since zinc sulphide coating is considerably more

11.16.6. Diagram of interference pattern for a lens.

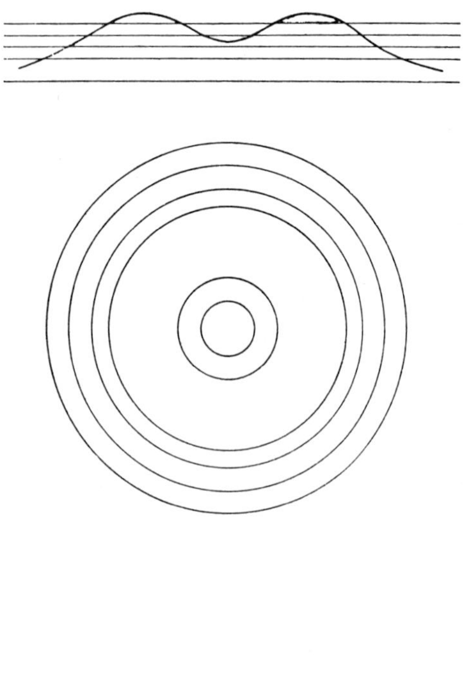

11.16.7. Diagram of interference pattern for a tilted lens.

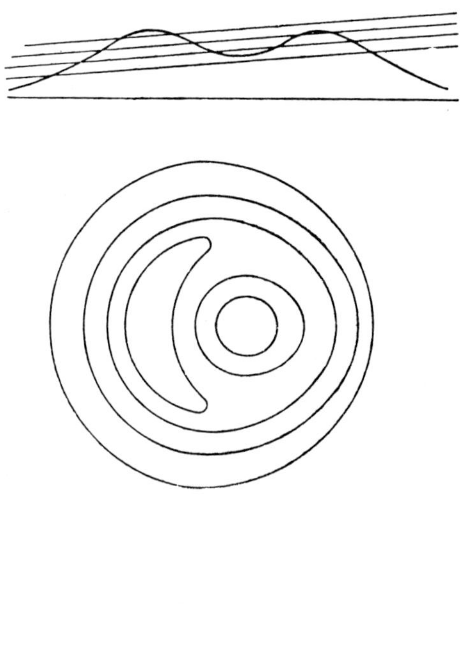

complicated than coating with silver or (as we now prefer) aluminium; and, although the zinc sulphide gives a better result than silver and the silver than the aluminium, the last named is on the whole less troublesome because of its durability.

It is quite easy to get photographs of the interference patterns by placing a camera at *P* in Fig. 11.16.3(b); using a high-pressure mercury-vapour lamp, good photographs have been obtained with 1/25th second exposure and still shorter exposures could be used with suitable selection of the optimum conditions.

A very complete set of interferometer patterns due to the primary aberrations are illustrated by Kingslake (1925); interferograms obtained by calculation and by photography on the Hilger lens interferometer are shown side by side.

Compensated interferometers; white light fringes

For some purposes, the interferometer must be 'compensated'.

Referring to Fig. 11.16.3(b) and supposing the prism to be removed and the mirror *F* adjusted perpendicular to the incident rays, a mirror identical with *K* is placed between *K* and *F* and parallel to the former.

By adjustment of the mirror *G* the paths of the two interfering beams can then be made exactly equal, when white light interference fringes are seen with a black central band which indicates the positions of exact equality of path.

Thermal uniformity in polishing

In the final stages of polishing large prisms it is essential that before testing the prisms the temperature should be allowed to settle down. It used to be our practice to stand the prism for this purpose on three projections of non-conducting material, such as ebonite, to allow free access of the air all round the prism.

Although by this means an approximate equalization of temperature throughout the prism is acquired fairly rapidly, yet until the prism has acquired the temperature of the air the equalization is not good enough for the purpose of a critical test. The method that has been adopted, therefore, for a number of years is to stand the prism on a metal plate (which should not be too thick, so that it can rapidly accommodate itself to the temperature of the room, and should be nicely flat, so that it can rapidly convey that temperature to the prism) and to place over the prism a metallic cover nicely fitting the metal plate at the bottom and rough blacked on the outside so that it, also, rapidly acquires the temperature of the room.

In these circumstances, half an hour is sufficient for a 60° glass prism 2 inches high and $2\frac{1}{2}$ inches length of face to settle down appropriately for the most critical test. With increasing length of prism the length of time required increases rapidly; for example, a prism 2 inches high by 3 inches length of face would require three-quarters of an hour.

It is scarcely necessary to add that for very large work a constant temperature room, in which the temperature can be held within 1/10°C, must be used.

The interferometer is very useful, also, for testing angles to a high precision.

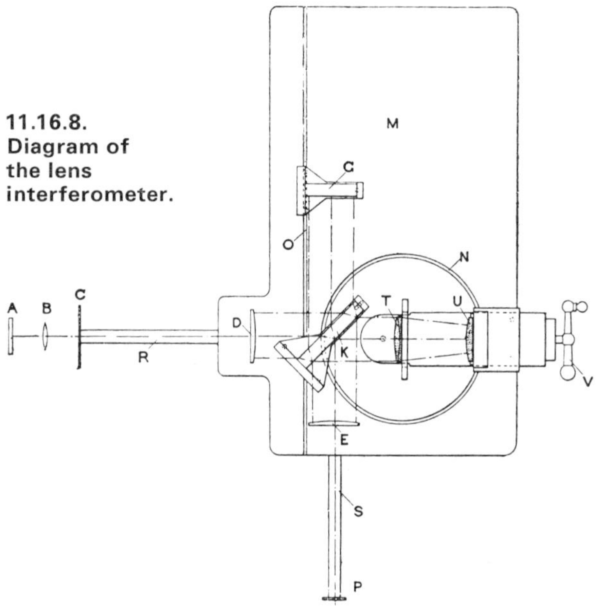

11.16.8. Diagram of the lens interferometer.

Lens interferometer

In the lens interferometer all parts are left as in the prism arrangement, except that the mirror *F* is removed and replaced by the lens attachment shown in Fig. 11.16.8. *T* represents the lens under test, *U* a convex mirror in such a position that it reflects back along their own paths the rays received from *T*. The mirror *U* can be moved by a screw motion actuated by the handle *V*, so that its distance from *T* can be varied at will. It will be seen that when the adjustment of this part of the apparatus is correct, the whole lens addition will, if the lens *T* be perfect, receive the beam of plane wavefront and deliver it back again with a plane wavefront. If it does not do so, the departures from planeness of the wavefront so delivered will give rise to a contour map of the corrections which have to be applied to the lens in order to make its performance, when in actual use, perfect.

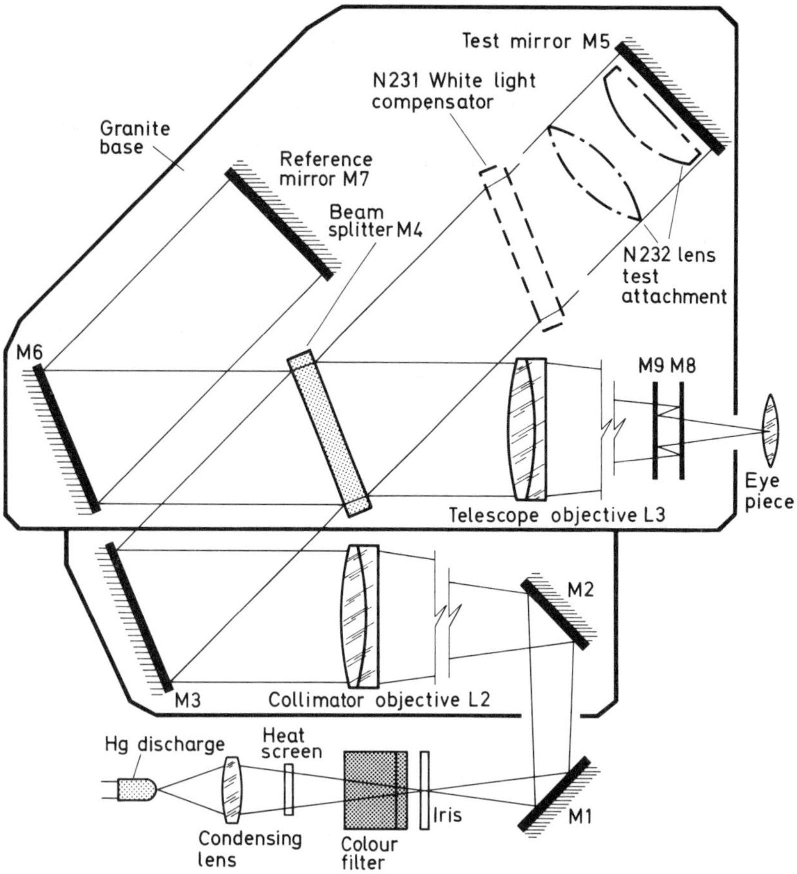

11.16.9. Three-inch aperture Twyman-Green interferometer of modern design. The optical system is shown above.

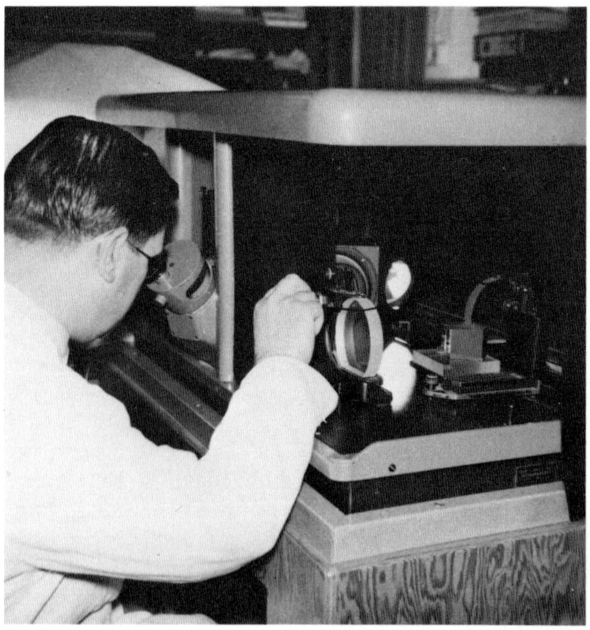

FURTHER DEVELOPMENTS

Mr Twyman's account is brought up to date by the following short account of further developments.

In 1966, the Twyman–Green interferometers were completely redesigned by Hilger & Watts Ltd (Fig. 11.16.9). The light path was folded to reduce external dimensions and the assembly mounted on a solid granite base to provide firmness and rigidity. For optimum performance the instrument must be sited in a draught-free position with the temperature controlled to 20°C ± 0·5°C.

The viewing aperture and quarter-plate camera form a combined observation and recording system, changeover being simply effected by a knob-rotated mirror. The light source is a low-pressure mercury lamp, and at a wavelength of 5461 Å the complete instrument is accurate to 1/20 wavelength.

11.16.10. **Hilger microscope interferometer of the original design.**

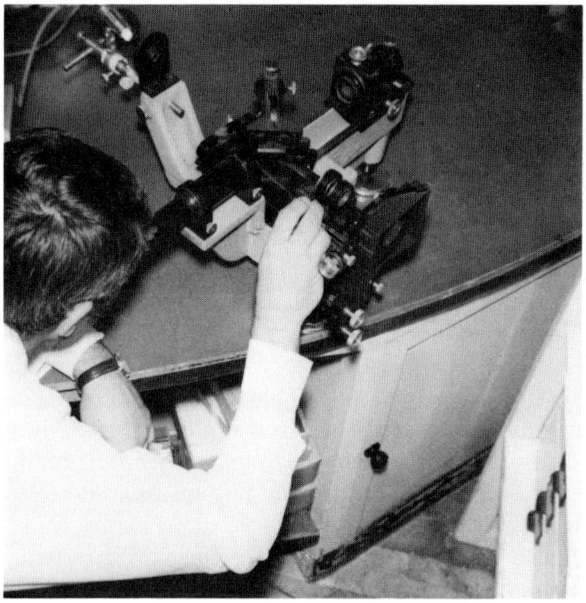

Hilger microscope interferometer

A Hilger microscope interferometer (Twyman, 1923) was developed from the lens interferometer to exhibit the aberrations possessed by a microscope objective when the latter is focused to produce its real image at infinity. (Fig. 11.16.10.)

11.17 Shearing interferometer for testing astronomical telescope mirrors

The primary mirror of a telescope which may be 400-cm diameter and weigh 22 000 kg (50 000 lb) has to be ground and polished to an accuracy better than 1/8 wavelength.

Usually interferometer components have to be about the same size as the optical system under test.

The shearing interferometer (Bates, 1947) demonstrates the possibility of testing an optical wavefront or surface against an identical sight of itself, the technique being to obtain two identical views of the same wavefront or surface and by superposition to produce interference between them. (Fig. 11.17.1.) If the two wavefronts exactly

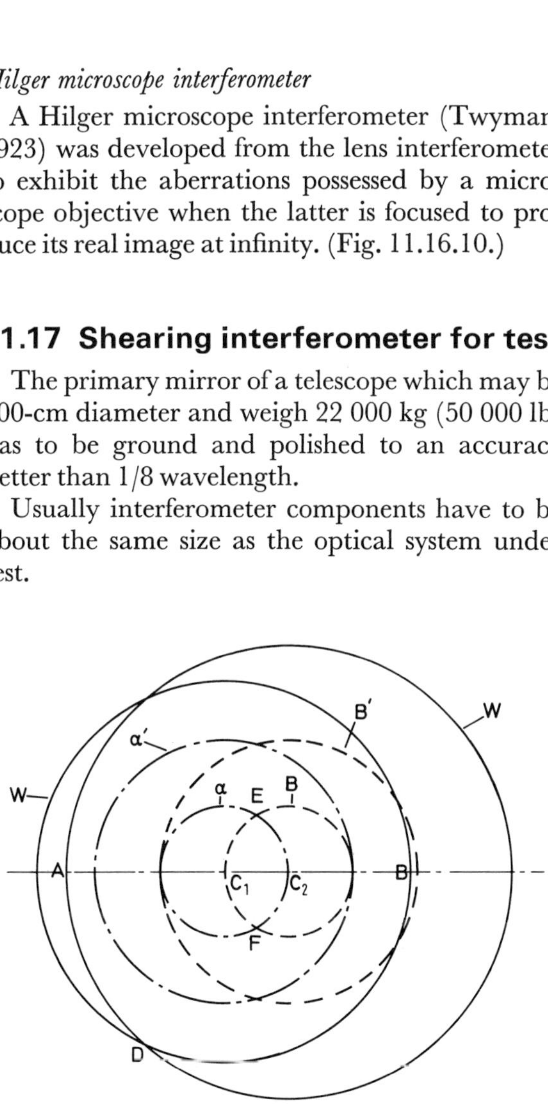

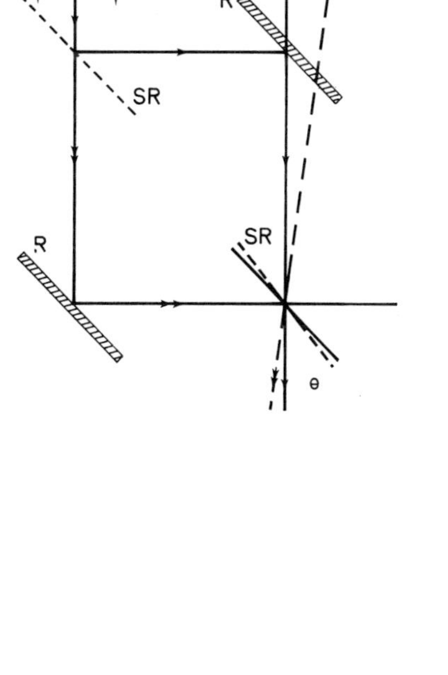

11.17.1. The Bates shearing interferometer. R is a reflecting surface; SR is a semi-reflecting and transmitting surface. The diagram left shows the form of the wavefront in the instrument.

overlap, then a single bright fringe covers the whole field of view. If, however, they are displaced laterally with respect to one another, the errors on the one no longer coincide in space with errors on the other, and the interference which one obtains in the overlap region is indicative of the errors involved.

Apart from the question of interpretation of shearing interferograms, the main obstacle in the path of the intending user of the shearing technique has been the interferometer itself. The instrument described by Bates is extremely flexible, but is not suitable for routine testing work. A simpler design of Drew is much easier to use but it cannot be relied on to remain in adjustment, with the degree of accuracy required for the testing of wide-angle cones, over long periods of time.

The desirable features in a shearing interferometer for workshop use are:

(1) Mechanical and optical reliability.
(2) Simplicity of operation.
(3) Reasonably small size.
(4) Independent, continuously variable controls.

The cemented units described by Brown meet the first three of these requirements but since both shear and tilt are fixed the range of testing for which a single unit may be used is limited.

A small interferometer developed by Grubb Parsons measuring only 58 mm × 30 mm × 46 mm can be used at the focus of any optical system capable of forming a real image of a slit source.

The instrument shown in Fig. 11.17.2 consists of a normal Jamin interferometer used in convergent light; into it is introduced a prism *P* and a compensating plane parallel *Q*, the prism being situated at the focus of the wavefront under test. Tilt between the sheared and unsheared wavefronts may be introduced either by a slight rotation of *Q* about the axis *A–A′* or by rotation of the second interferometer plate *S* about the axis *B–B′*. In the latter case the fringe pattern will have a slight lack of compensation similar to that found in the Drew instrument. The second method has the advantage that *P* and *Q* may be fixed to a single metal frame which may be made detachable so that prisms giving different shears may be used. For correct compensation, the faces of *P* and *Q* facing *S* should lie in the same plane.

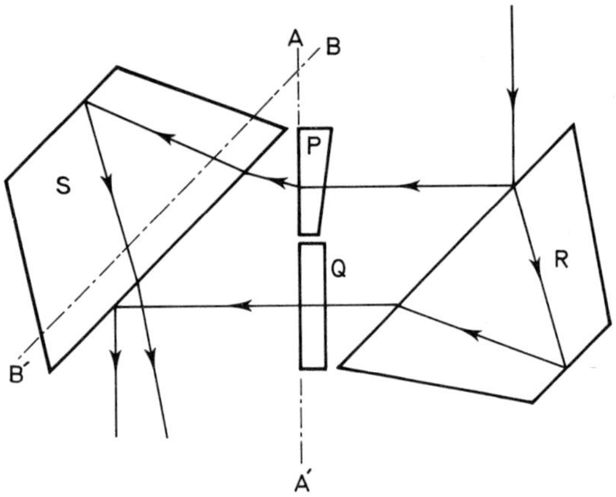

11.17.2. Optical arrangement of the Grubb Parsons shearing interferometer, refracting type.

The use of a prism to introduce shear is open to objection on two accounts. Firstly, the variation of shear with wavelength; secondly, the variation of shear across the field of view. Neither of these defects is very troublesome unless highly aspheric wavefronts are to be investigated.

The interferometer operates on the 'wavefront shearing' principle and is a small instrument that can be used at the focus of any optical system capable of forming a real image of a slit source. The interferometer divides the incoming light into two identical wavefronts which are recombined with shear and tilt to give an interferogram which can be inspected visually or photographed.

By suitable choice of shear and tilt the sensitivity of the interferometer may be made suitable for any testing requirement from the very highest (e.g. astronomical optics, for which the instrument was developed) to the lowest. Components of any size may be tested, the only limitations being to the focal ratio of the incident light which can be accepted (*f*/3) and a minimum distance from optical system to focus of $1\frac{1}{2}$ inches. Systems faster than *f*/3 or of focal length less than $1\frac{1}{2}$ inches can be tested with the aid of suitable auxiliary optical systems which are usually of a very simple nature.

The interference patterns produced by the primary aberrations are easily recognized, and aberration coefficients are obtainable with a minimum number of measured points on an interferogram. By a simple summation method the interferogram measurements may be converted to wavefront profile if desired.

There remains the problem of interpreting the interferogram into clear instructions on how much polishing is necessary and where, in order to achieve the accuracy required in the shortest possible time.

Grubb Parsons have developed (Brown) a computer programme to aid the problem of interpreting the interferograms taken from the interferometer.

In the case of astronomical tests there are problems of flexure of the mirror during test, the essential need for constant temperature control and the special difficulties of size and focal length.

A testing tower has been built about 100 feet high with seven floors (28 feet to the first floor) spaced at 10-foot intervals.

These floors are used according to the focal length of the mirror being tested and they are all constructed separately from the external structure to avoid vibrations which would upset the interferometer readings. The tower is 20 feet square at the base and in this space the telescope mirror is moved, mounted in the same way as it will be in the completed telescope. The tower has a double skin to ensure temperature control, as the maximum change permitted over 24 hours is 0·5°C and the change from one level to another must not exceed 0·2°C over 10 feet. Thermostats and heaters are fitted at every 10 feet to avoid a vertical temperature gradient.

The problem of avoiding distortion of the mirror itself by temperature changes has been greatly reduced by the availability of Cervit material which has a zero coefficient of expansion, and this has now completely replaced Pyrex glass and silica as materials for astronomical mirrors.

11.18 The Foucault test

This is a highly informative test for lenses and mirrors. (Fig. 11.18.1.)

The test of Foucault (1858) is described by him as follows:

> The mirror is so placed that it produces an image of a small hole in an opaque screen brilliantly illuminated by artificial light. To make the test this image is almost entirely masked by a second opaque screen with a straight edge. The rays which just pass this edge are received by the eye and the surface of the mirror is then seen in light and shade and one sees, in strong relief, all the reflections which do not participate in the exact focusing of rays. One can thus recognize the parts needing correction.

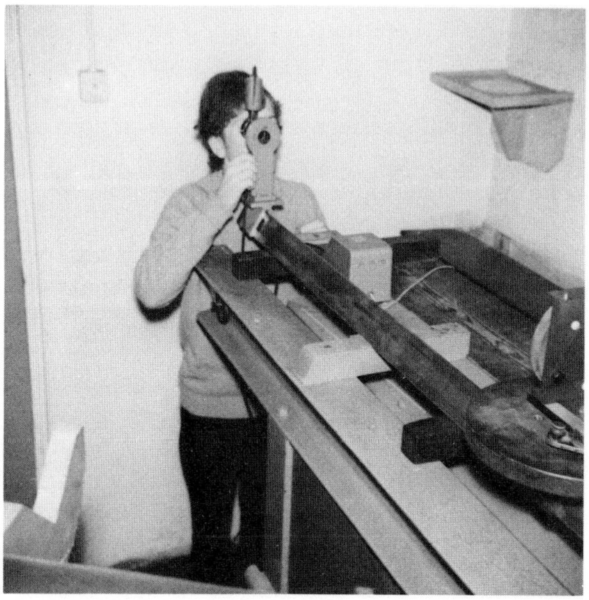

11.18.1. The Foucault test of a concave mirror, which is required with an aspheric surface. This knife-edge test is applied to a concave mirror by placing both knife and source at the centre of curvature.

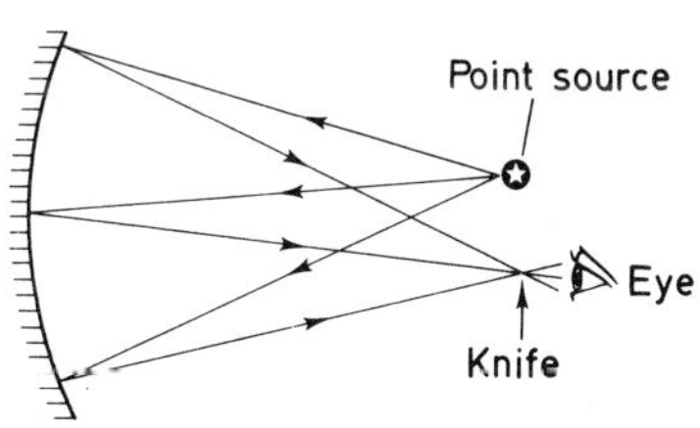

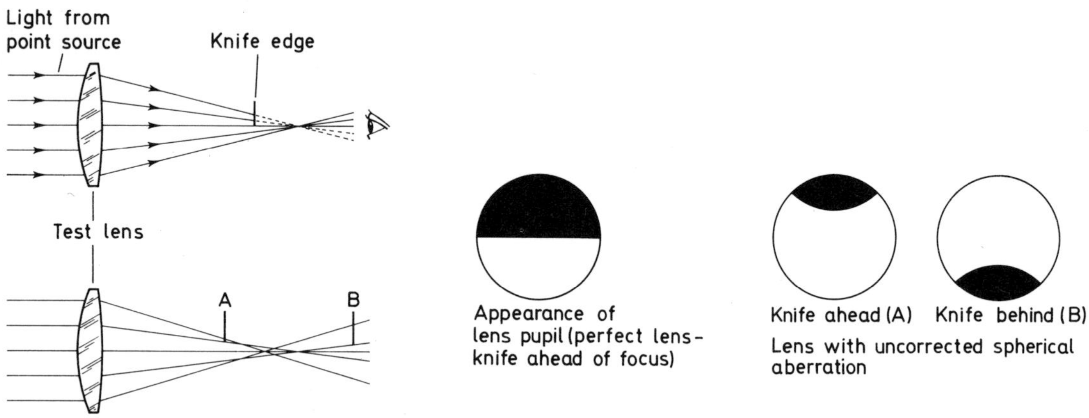

11.18.2. Foucault test of a lens.

In testing a lens, the object-glass is examined without any eyepiece. With one eye in the focus of an artificial star one sees the object-glass uniformly illuminated. If a knife edge is passed slowly across the focus from right to left, this uniform illumination disappears over the whole object glass simultaneously if, and only if, the rays all pass accurately through the focus. If they do not, the object-glass first darkens over regions from which rays pass to right of the focus. (Fig. 11.18.2.)

When the surface to be tested is an aspheric, the desired foci for the various zones are computed from the design data and the measurements are compared with the calculated values. Focus differences are then converted into errors in the surface contour and so determine the zones of the lens or mirror which require further working.

Although experienced optical workers will recognize the characteristics of parabolic or other aspheric surface, the test can be made quantitative by measuring the focus for successive zones by the use of diaphragms. (Fig. 11.18.3.)

When the knife-edge intersects the beams either nearer or farther than the focus, the two patches exposed by the diaphragm will not darken simultaneously. If the pinhole and the knife-edge move on the same carrier, a pointer can be made to read the relative positions of the centres of curvature of successive zones.

This test is often used for testing the main and secondary mirrors of reflecting telescopes. (See § 15.3.)

Sometimes the system is used for compressed air flow studies, where it is often called the Schlieren system. In most applications, only qualitative information is required which is easily obtained by the photographic record of an image and another after the knife-edge has been rotated through 90°.

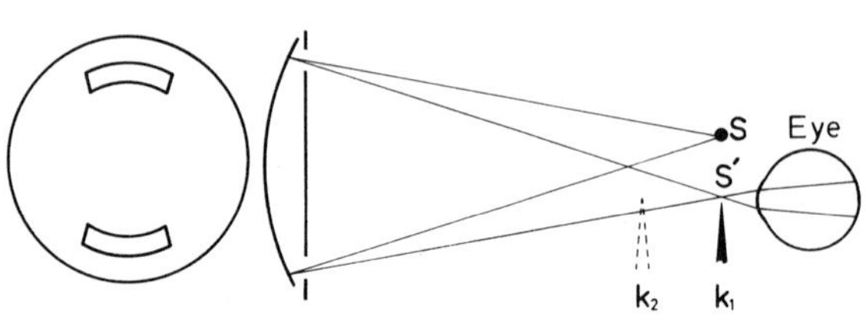

11.18.3. Zonal diaphragm for the Foucault test.

11.19 Angle Dekkors and Microptic Autocollimators

Hilger & Watts Angle Dekkors and Microptic Autocollimators are among the most versatile of optical instruments ever produced for the needs of engineering metrology. They are used for measuring small angular displacements, and for checking alignment, squareness, straightness, and parallelism.

General principles

When a parallel beam of light is projected on a reflecting surface which is not square to the axis of the beam, the angle through which the return beam of light is deflected is twice the angular tilt of the reflector. In this fact lies the principle of autocollimation.

An autocollimator is fundamentally a telescope having a target graticule, a built-in illuminating system, and means for measuring angular tilts of a reflector attached to, or in contact with, the part or the surface to be checked.

The instrument determines angular changes of the position of the reflector—relative to a datum setting—from corresponding changes in the return angle of a beam of collimated light emerging from the objective of the telescope. These changes are indicated by displacements of the return image of the target graticule, which are measured and directly related to the angular tilt of the reflector and, therefore, to the angular displacement of the part or surface being checked.

Any movement that can be made to tilt a reflecting surface can be measured. Irregularities in flatness and straightness are determined by measuring angular tilts of a reflector when positioned at successive stations along the surface or the straight-edge.

The Angle Dekkor

The Angle Dekkor is an autocollimating telescope which can simultaneously measure vertical and horizontal angular displacement of not more than 60 × 40 minutes. With very few exceptions, the Angle Dekkor must be used as a comparator. Small linear displacements, micrometer faces, set squares, vee blocks, small castings and tapers can be checked with this instrument. It is also most useful for setting jigs and workpieces for machining, for testing deflection under operating forces and for checking the true location of parts in motion, such as slides and spindles, for example.

Readings from two illuminated, figured scales in the focal plane of the objective are taken direct to 1 minute of arc without the use of a micrometer. The instrument measures by comparing the reading obtained from a standard with that from the work under test. It requires very little setting-up time, and once set up to a standard, will indicate at a glance the error in angle of the work.

The Angle Dekkor can be mounted on its own precision surface table-stand. This is most useful for locating small workpieces, gauges, vee-blocks, die blocks and micrometer anvils, and for repetition work. The instrument tube can be rotated through 360°, and can be swivelled in one plane about its horizontal axis from an arm clamp. The slotted arm allows for height adjustment of 5½ inches (140 mm).

A magnetic stand, for attaching the instrument to an upright, and a tripod stand, which can be adjusted to various heights, are also available. The tripod is especially useful for machine tool work or surface plate testing.

The basic unit (TA 105) consists of a steel body tube with an objective lens mounted at one end, and a clamp for holding the eyepiece at the other. There is a supporting bracket which can be clamped in any position along the body tube or removed so that the tube can be fitted to special attachments.

The standard Angle Dekkor with the general purpose eyepiece TA 104 is the instrument most commonly used in workshops. Maximum working distance from the objective lens is about 4 to 5 feet (1·5 m). Steel or glass plate reflectors can be used. The working distance can be greatly increased by using the long-range eyepiece (TA 103) which extends it to 30 feet (9 m). However, as the working distance increases, the reflected scale will be reduced in length from each end approximately in proportion to the distance of the work from the reflecting surface. This reduces the total angle of difference measurable but does not affect the accuracy of reading. The brightness of the scale is lower than with the general purpose eyepiece, and steel or metallised reflectors must be used.

With the projection model of the Angle Dekkor

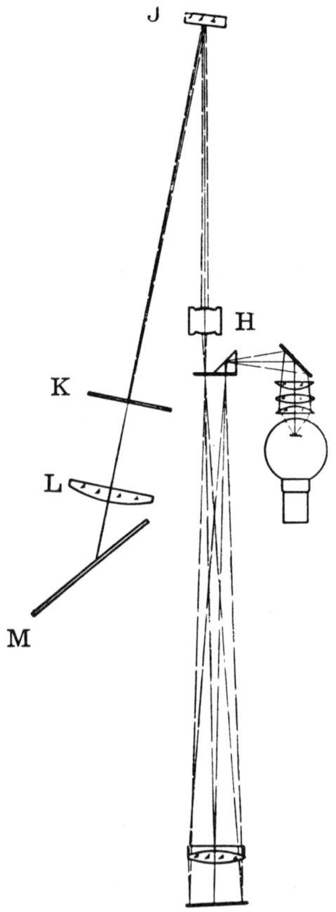

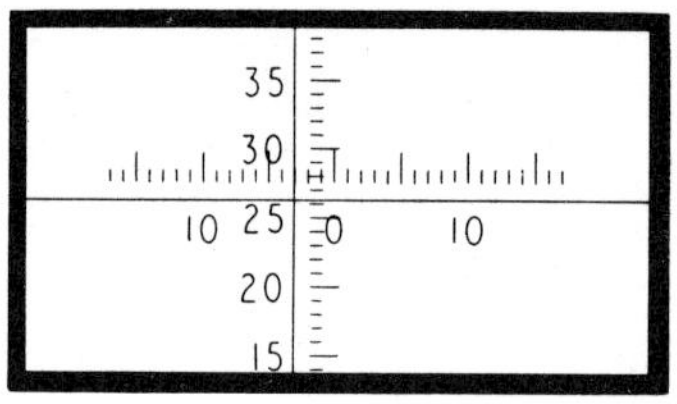

◁ **11.19.1. Optical system of the Projection Angle Dekkor. The projection lens H throws an image on the grey-glass screen K by way of the mirror J. The image is magnified by the reading lens L and viewed in the hinged mirror M, which can be tilted for vertical reading. The appearance of the scale is shown above.**

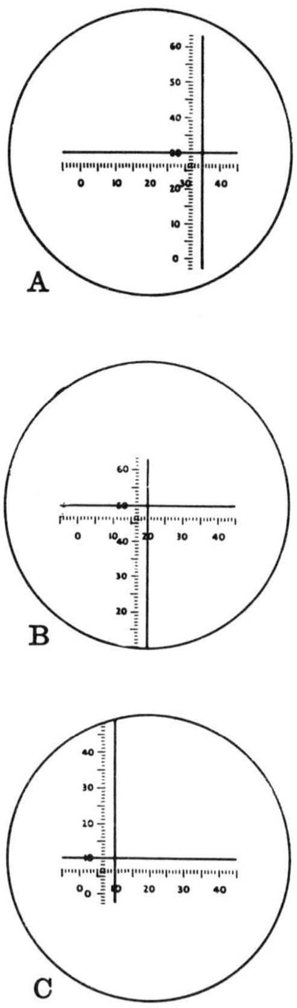

11.19.2. ▷
Scales as seen in the standard Angle Dekkor. (a) shows a horizontal displacement, (b) shows a vertical displacement, and (c) shows a combination of both. The horizontal scale is fixed. The vertical scale, which is brightly illuminated, is the reflected scale.

(TA 110) the indicating scale is seen in an aperture at the front of the casing, which is approximately 3 × $1\frac{1}{2}$ inches (76 × 38 mm) in size and has a hinged mirror attached to it. As this obviously minimizes eyestrain, this model is eminently suitable for repetition work. Moreover, setting up is easier because a bright beam of parallel light, which appears on the surface of the object to which it is directed, is projected from the objective lens of the instrument. The light reflected back to the objective is also clearly visible, and the instrument can easily be adjusted in relation to the beam or light returned.

When the instrument is in the horizontal position, the scale can be observed directly through an aperture in the bottom of the casing, so that it can be clearly seen at any reasonable distance in a normally lighted room. The instrument is strongly made. It is a little more bulky than the standard Angle Dekkor, but nevertheless easily transportable. (Fig. 11.19.1.)

On looking into the eyepiece of the Angle Dekkor a black fixed scale will be seen, figured from 0–40. Each division of this scale represents 1 minute of arc. When the Angle Dekkor is set up and a reflection of the light is obtained from the surface under test, another brightly illuminated scale, divided 0–60, will be seen crossing the dark fixed scale at right angles. The reading where the scales cross indicates the vertical and horizontal displacement of the working face under test. (Fig. 11.19.2.)

Alternatively, an angle gauge or optical flat may be set beside the test object so that both receive part of the light from the Angle Dekkor, and then one bright scale will appear in the eyepiece from the angle gauge and one from the test object. The separation between the two images will represent the error in minutes of arc. This method of use is much to be preferred when the work cannot be located on the Angle Dekkor surface plate, as a small movement of the instrument is then of no account.

Fig. 11.19.3 illustrates an Angle Dekkor in use

 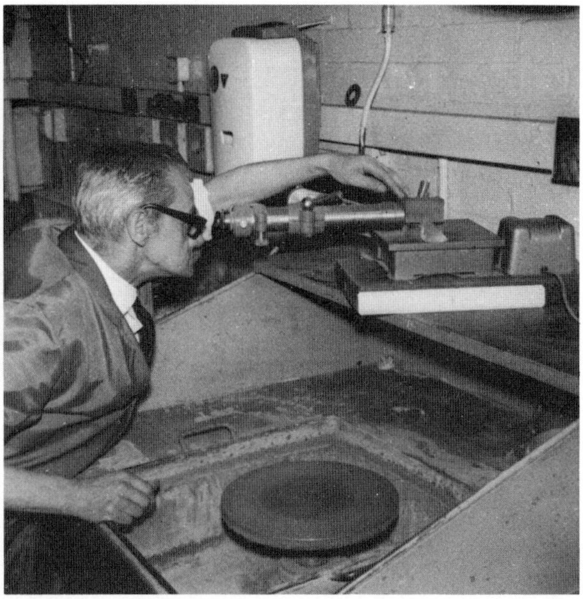

△ **11.19.3. Comparison between (*left*) a 45-degree reference gauge and (*right*) a 90-degree reference gauge with the buffed face of the adjacent side of a prism before truing.**

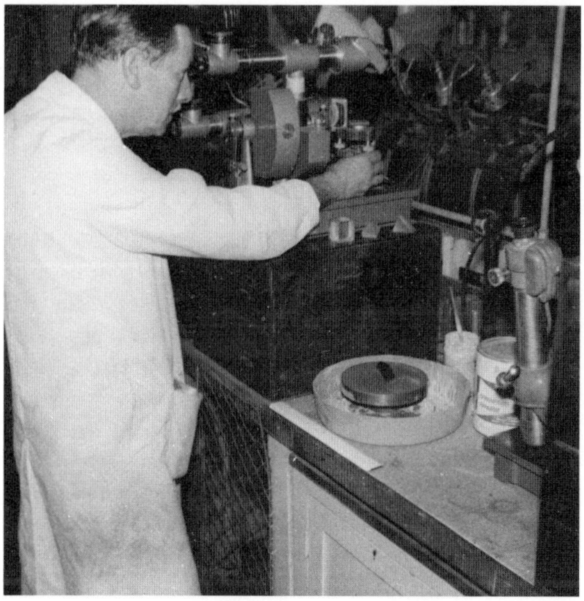

11.19.4. ▷
Using two Angle Dekkors to check the angle on a stack of prisms before truing the base.

to compare the angle of a prism with a reference gauge of known angle.

The reference gauge and the part under test are laid on a surface plate and the reflections of both viewed in the telescope. This operation is known as collimating, and is then followed by truing the prism with emery on the adjacent smoothing spindle.

It should be noted that the Angle Dekkor can be used on a fine grey surface under certain conditions, but usually the glass surface on the prism being tested has been buffed in order to give a bright reflected beam of light.

Fig. 11.19.4 illustrates the use of two Angle Dekkors assembled on a fixture for checking the angle along a stack of prisms.

The Microptic autocollimator

The Microptic autocollimator is the ideal optical tool—combining very high accuracy with speed and ease of operation—for solving many metrology problems in the laboratory, standards room, inspection department, and the toolroom. It has been chosen by well known firms as providing the fastest, most accurate, and easiest means of carrying out measuring operations such as calibrating large surface plates, checking the straightness of table movements, or the alignment of centres and spindle axes, the testing of dividing equipment, checking the diameter of very fine wire, aligning machine beds, and checking gear hobs.

The autocollimator consists of a telescope with the means of illuminating a graticule placed in the focal plane of the objective. Parallel or collimated light emerges from the objective and is directed at a plane surface, set up at right angles to the optical axis of the telescope. The rays from the illuminated graticule are reflected back through the objective and form an image. The displacement of this reflected image from the graticule is measured by a micrometer eyepiece reading direct to 0·1 second of arc over a measuring range of 10 minutes of arc.

The telescope tube, which is mounted in a cradle, can be rotated through 90° about its optical axis, so that measurements can be taken in both the horizontal and vertical plane. The instrument is finished in grey enamel, and the underside of the base is ground to a very high degree of flatness. Along one side of the base a locating pad is fitted which can be used for clamping if necessary. Three adjustable feet are provided for levelling the instrument on an uneven surface, and a spirit bubble is built into the base.

When taking a measurement, the cross-wires of the target are ignored; only the variation in the position of the reflected image is measured. The vertical parallel setting lines, shown already straddling the reflected image, are moved into position by the micrometer screw, revolutions of which are counted on the coarse scale. The horizontal setting wire serves to confirm that the axis of rotation of the tilting surface is at right angles to the plane of measurement. (Fig. 11.19.5.)

The micrometer drum is divided into 300 equal divisions, alternately light and dark; each one represents 0·1 seconds of arc. Every second of arc is figured. The index line has the same thickness as the drum divisions. One revolution of the drum signifies an angular displacement of 30 seconds of arc.

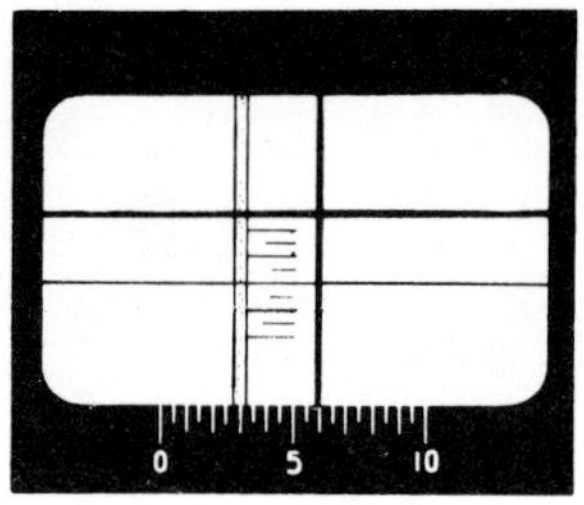

11.19.5. The field of view in an autocollimator.

Photoelectric autocollimator

The photoelectric version of the Watts Microptic autocollimator provides an easier, quicker, and more accurate method of setting than the visual instrument. It achieves a setting accuracy of 0·1 second at a single setting, and reduces operator fatigue caused by repeated exacting observations through the eyepiece.

The instrument is ideally suited for short-range work, especially when using standard polygons for checking angular indexing or for calibrating other polygons. When used together with accessories, it provides, like its visual counterpart, the means of determining accurately straightness, flatness, and small linear displacements.

The photoelectric setting system makes it unnecessary to view through the eyepiece when making exact settings.

The normal setting lines are seen through the eyepiece of the instrument and are used for preliminary adjustment.

Signals from the photoelectric detector pass through an amplifier and are displayed on a meter. The photoelectric system embodies a vibrating slit, the basic idea of which is attributable to the Standards Division of the National Physical Laboratory who assisted materially in the development of the instrument.

The system is of the null-setting type; exact setting is obtained by bringing the needle of the meter to the central zero. Readings are then taken from the micrometer drum of the instrument as usual.

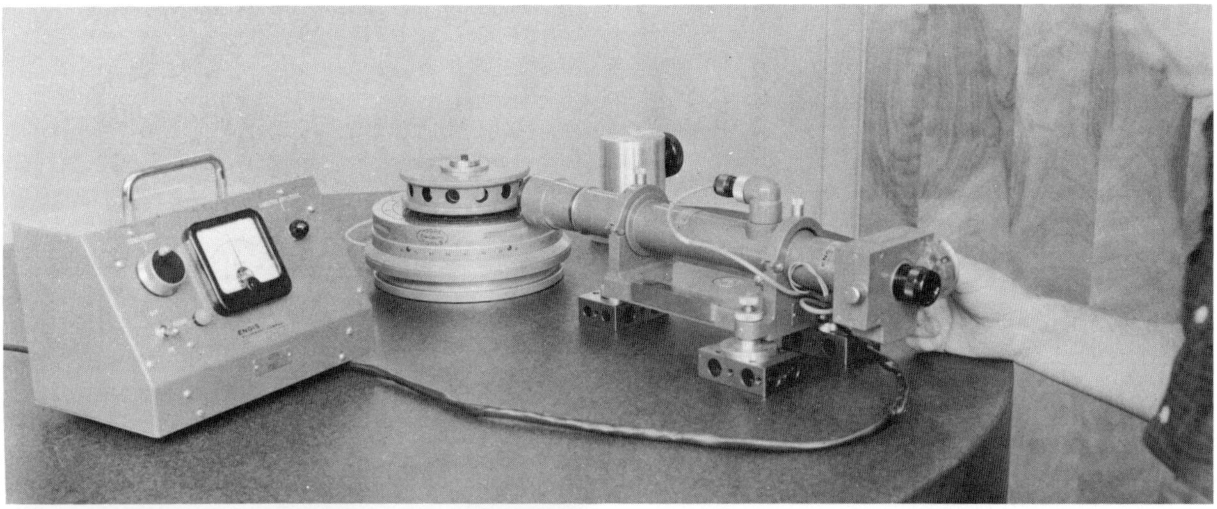

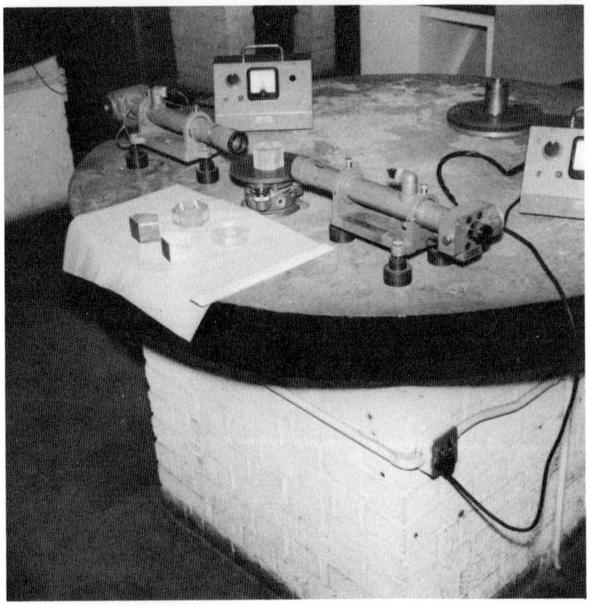

△ **11.19.6. (*a*) A photoelectric autocollimator in use for direct inspection of the 30-degree spacing of a 12-sided optical polygon.**

◁ **(*b*) Two photoelectric autocollimators set up at 180 degrees for testing the faces of a six-sided polygon.**

The sensitivity of the meter can also be adjusted to a value so high that it is unnecessary to set the meter needle to zero, because errors in the micrometer setting would be smaller than could be read on the micrometer drum.

The sensitivity of the meter reading can be adjusted by trial so that the divisions on the meter scale represent a definite angular value (as read from the micrometer drum). A small range of angular errors can then be read direct from the meter. The accuracy of such readings is considerably reduced, but the method is useful when a lower accuracy over a limited range is acceptable.

Fig. 11.19.6 illustrates the use of two photoelectric autocollimators set up for testing the reflecting faces on a six-sided polygon. By this method a reliability of plus or minus 1 second of arc at 95 per cent confidence level can be achieved.

The method becomes more involved when polygons having a greater number of faces have to be calibrated, although up to 72 peripheral reflecting faces have been tested and made true to within plus or minus 10 seconds of arc. This accuracy was achieved by repeatedly checking the angular position of each face with a photoelectric autocollimator.

11.19.7. The Moore 1440 Precision Index (mechanical polygon) inspecting an angle gauge block. Note that the gauge block has been specially prepared to ensure reflectivity of optical quality. The polygon is supplied with a certificate of accuracy that records the results of a calibrating test.

Moore Special Tool Co., Inc.

Certificate of Accuracy

MOORE 1440 PRECISION INDEX

± 0.1 Sec. of Arc Guarantee

This is to certify that calibration of Moore 1440 Precision Index Serial No. 8-1440-57

was performed for PTB*** on Feb. 2, 1969

CALIBRATION CHART

Angle	Calibration in Arc Seconds	Angle	Calibration in Arc Seconds
0°-30°	0	0°-210°	+0,06
0°-60°	+0,03	0°-240°	+0,09
0°-90°	+0,05	0°-270°	0
0°-120°	+0,06	0°-300°	-0,01
0°-150°	+0,02	0°-330°	+0,02
0°-180°	+0,01	0°-360°	0

The above calibrations of 30° angle segments have been determined on the basis of 144 individual readings. To obtain the readings, 1440 Precision Index Serial No. 8-1440-57 and 1440 Precision Index Serial No. 8-1440-59 have been compared.

See "Moore 1440 Precision Index" brochure.

Traceability–1440 Precision Index Serial No. 8-1440-5 which has been calibrated by the NBS*, 1440 Precision Index Serial No. 8-1440-10 which has been calibrated by the NPL**, and 1440 Precision Index Serial No. 8-1440-15 which has been calibrated by the PTB*** are reference masters which guarantee Traceability.

The calibration procedure followed by Moore is identical to that employed by the above-mentioned bureaus and is sanctioned by them.

Due to the design nature of the 1440 Precision Index, wherein all 1440 serrations are in engagement at each angular setting, it is considered that the above calibration method sufficiently represents the error curve of the 1440 Precision Index at all 1440 indexed positions.

*NBS–U.S. National Bureau of Standards (Washington, D.C.)
**NPL–National Physical Laboratory (Teddington, England)
***PTB–Physikalisch-Technische Bundesanstalt (Braunschweig, West Germany)

Date of Calibration: 2-2-69 Calibrated by: Approved by:
Chief Inspector

Comments:

Certificate No. 8-1440-57-A

Use with precision index table

The Moore Special Tool Co. Inc. have developed a very accurate precision index table, which, in conjunction with a Hilger & Watts photoelectric autocollimator, will measure angles very rapidly to within one-tenth of a second of arc. (Fig. 11.19.7.) This 1440 Index is discussed further in the next section.

11.20 Moore 1440 small-angle divider

Prior to serrated tables, the most common circular calibration standards were the optical polygons. They were stable, accurate, and readily available in various numbers of sides. Polygons were used for testing rotary tables, but the testing of the polygons themselves was carried out with the use of autocollimators.

By 1963 the Moore Special Tool Co. Inc. had developed the 1440 serrated-type precision index which gave an accuracy of plus or minus 0·1 second over $\frac{1}{4}^\circ$ increments, but the problem of checking smaller increments than $\frac{1}{4}^\circ$ still remained.

A recent development has been the addition to the 1440 Precision Index of a tangent-screw device which in turn, divides each $\frac{1}{4}^\circ$ step into 0·1 second increments. The heart of the tangent device is a Moore leadscrew and the high magnification is achieved by using a 120-pitch screw and an accurately graduated 4-inch-diameter dial with vernier for the small-angle settings.

The Moore 1/10th Second Small Angle Divider is constructed as an integral assembly to the Moore 1440 Precision Index.

The 1440 Index divides the circle into 1440 parts ($4 \times 360^\circ$ or $\frac{1}{4}^\circ$ increments). The Second Divider is responsible for dividing each $\frac{1}{4}^\circ$ segment of the 1440 Index into an infinite number of steps.

The 1/10th Second Divider reads to minutes and seconds on the dial and to 1/10th second on the vernier.

Using the 1440 Index and the 1/10th Second Divider together, then, the circle can be divided and *read directly* to 12 960 000 parts.

The advantage of being able to index (with the 1440) to as small as a $\frac{1}{4}^\circ$ increment, is that the maximum length of arc—0·025 inches on a 32-inch circumference—that the 1/10th Second Divider is required to generate can be considered a straight line movement; the difference in length of the arc and a straight line can be considered negligible.

Used in conjunction with an autocollimator and an optically-flat mirror, the Small Angle Divider can be mounted on rotary tables and other indexing devices to inspect them for any point in a circle, or parts such as polygons or angle gauges may be mounted on the Small Angle Divider for calibration.

Fig. 11.20.1 illustrates a Moore Small Angle Divider inspecting a prism and a Hilger & Watts photoelectric autocollimator being used as a nulling reference.

11.20.1. The Moore 1440 Small Angle Divider inspecting optical prism with an autocollimator as null reference.

11.20.2. The Moore Master 1440 Precision Index, accurate to ±0·1 second, undertaking a complete inspection of a rotary table worm and gear for errors including periodic errors. The table is indexed with a Slo-Syn preset indexer. The nulling meter and strip recorder feedback use a Hilger & Watts TA-5 autocollimator. The meter is sensitive to ±0·1 second and the accuracy of the rotary table is guaranteed to ±2·0 seconds.

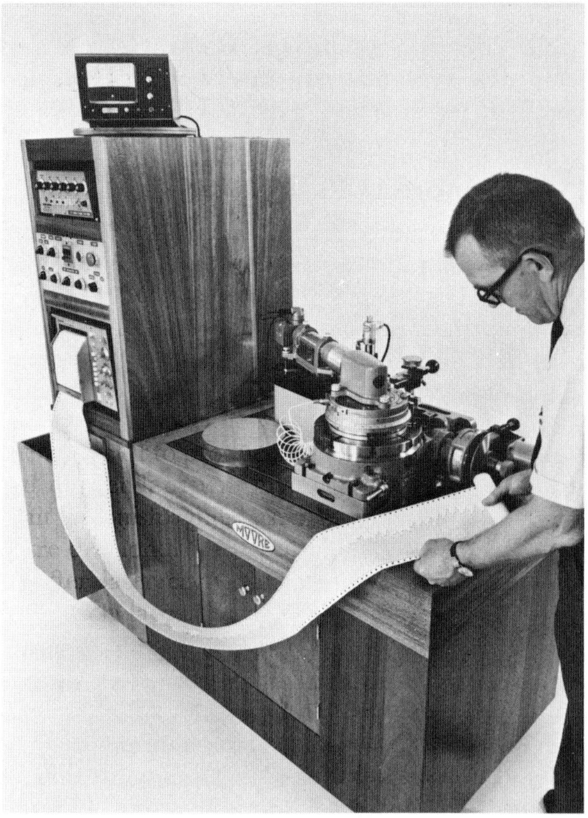

Fig 11.20.2 illustrates the method used for testing a rotary table by means of a Moore Master 1440 Precision Index and a Hilger & Watts TA 5 photo-electric autocollimator, with strip recorder.

11.21 Precise measurement of angles and refractive indices

Every optical factory making high-quality components should be provided with an accurate goniometer capable of measuring refractive indices with certainty to one unit in the fifth decimal place. Such accuracy is essential in the manufacture of high-quality camera and television lenses.

The Guild–Watts spectrometer

Fig. 11.21.1 illustrates a Guild–Watts Spectrometer being used to test a penta prism (optical square) to a tolerance of one second of arc on the right angle.

This is a precision instrument designed specifically for research work in refractometry and for the highest class of goniometric work. Provided all reasonable precautions are taken and a correct experimental technique is adopted a refractive index accuracy of one unit in the sixth decimal place can be obtained. By utilizing one accurately divided circle for reading both the prism table and telescope positions, both can be ascertained to the same accuracy. Both sides of the glass circle are read simultaneously so that the mean value is immediately seen through a single eyepiece and all centring errors are corrected. The circle is divided at 5-minute intervals and the micrometer scale is divided at 0·1-second intervals and figured at 5-second intervals.

Two telescopes are provided, both fitted with micrometers at the eyepiece ends. The symmetrical slit can be rotated about the axis of the collimator. Special, cemented, triple objectives are fitted, corrected to have minimum chromatic variation of spherical aberration rather than apochromatism.

The research spectrometer

Fig. 11.21.2 illustrates a research spectrometer used to test the right angle of a prism. Its components are shown in Fig. 11.21.3.

The research spectrometer, eminently suitable

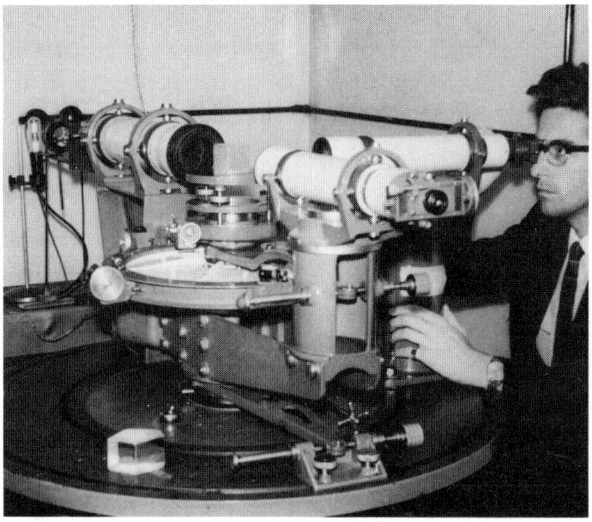

11.21.1. A Guild-Watts spectrometer testing the right angle of a pentaprism that must be accurate to one second of arc.

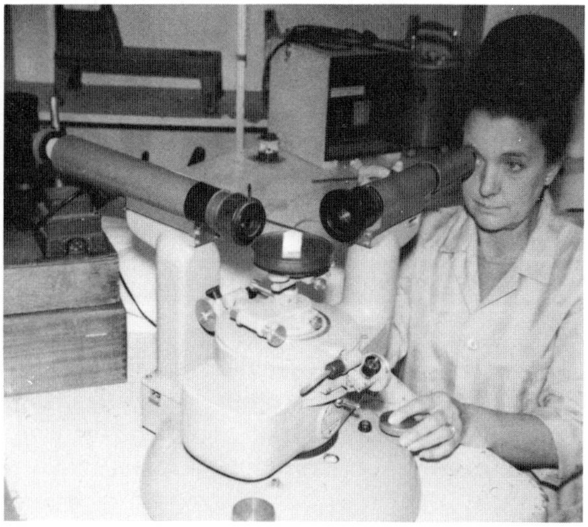

11.21.2. A research spectrometer testing the right angle dimensions of a prism.

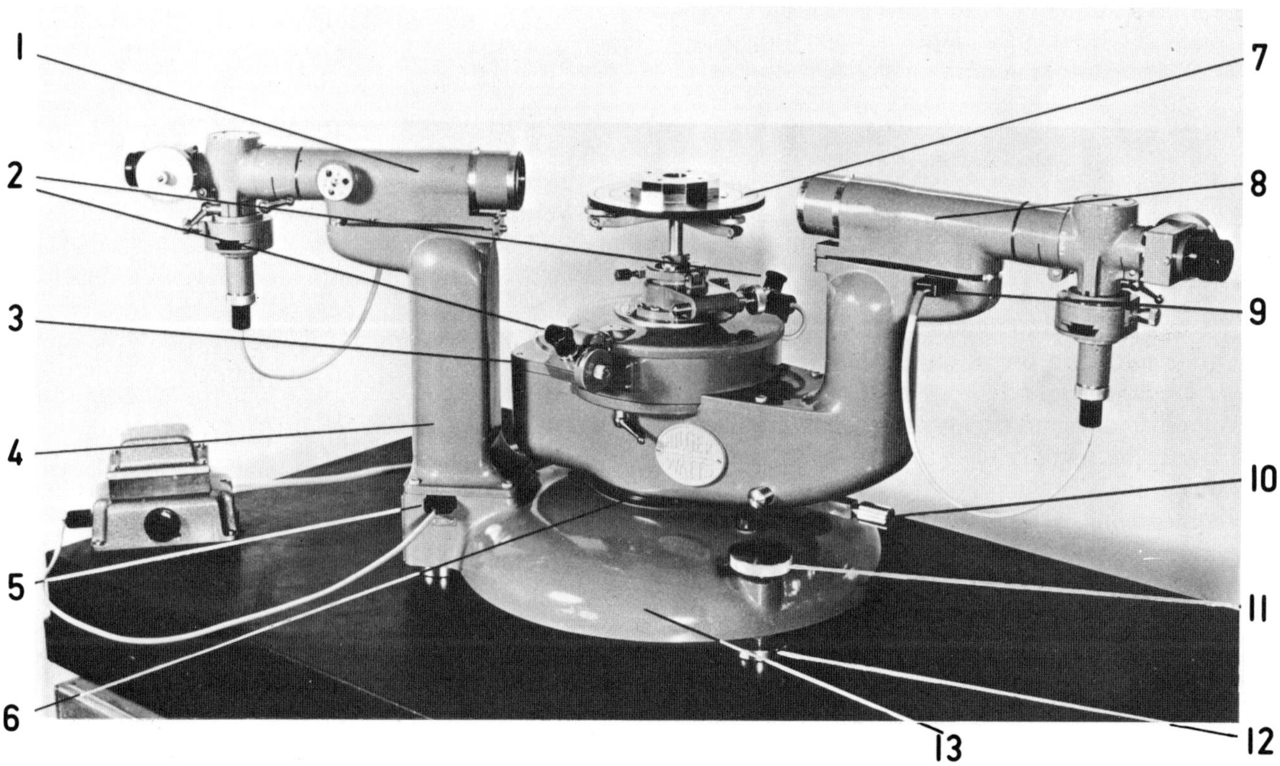

11.21.3. The basic parts of a research spectrometer. 1, sighting unit (collimator); 2, diametrically opposite circle-reading microscopes; 3, lower bearing counterbalance; 4, fixed vertical column; 5, 4-volt input socket; 6, 4-volt supply slip rings; 7, levelling table; 8, sighting unit (telescope); 9, 4-volt output socket; 10, lower bearing clamp; 11, levelling screw; 12, Lewis plate; 13, base casting.

for precision work, has a divided circle of glass with two opposite micrometer microscopes for observation of both telescope and table rotation. The circle graduations are at 10-minute intervals, straddled by the two fine parallel settings lines of the micrometer. One revolution of the drum takes the setting lines from one graduation to the next, and 120 divisions on the drum give direct reading to 5 seconds with estimation to 1 second.

11.22 Lens coating hardness testing

That quality of an anti-reflection film which is commonly called its hardness is mainly determined by two properties, resistance to surface scratching and adherence to the base on which the film is deposited. An abrader which tends to scratch the film and rub it off the base is therefore a suitable instrument for testing the quality of films in regular production.

In 1947, as a result of an investigation by Dr K. M. Greenland, Ph.D., B.Sc., an abrasive rubber was developed for use in an anti-reflection Hardness Testing Machine. This machine and abrasive rubber has, until 1970, been the standard method of testing the hardness of coatings.

The composition of the abrasive rubber is:

Smoked sheet	50	Zinc oxide	2·5
Brown substitute	25	Sulphur	3·0
China clay	70	Mineral oil	1·0
B.A.O. emery		M.B.T.S.	1·5
grade 302	200	T.M.T.	1·5

The rubber was made into cylindrical rods ¼-inch diameter and of a length such that 0·08 inch projected when the rubber was in its socket.

Under a load of 200 grams the rubber was reciprocated to and fro about 20 times and according to the severity of the scratch, so the coating was graded. Standards of hardness vary from one company to another and over the years the rubber hardness has changed and so the results were not consistent.

In 1970 a much simpler hand-operated hardness tester came on the market, manufactured by Summers Laboratories Inc., complying with the American Military Specification MIL–E–12397.

Directions for use. Press down on the shaft of the tester until the 2½-lb pressure mark appears. Rub the coated surface with strokes of about 1 inch (shorter for smaller lenses) for twenty complete strokes. All strokes should be made on one path. Wash the lens with acetone, and inspect for deterioration of the coating.

In testing a lens with an extreme radius, some attempt should be made to keep the axis of the tester perpendicular to the lens surface.

Care of the instrument. The eraser is pressure fitted in the tester. Before inserting a new eraser, remove the chuck and clean the chuck and plunger threads thoroughly. (Caution: Do not push plunger without chuck in place, since this will change the spring tolerance.) Water may be used as a lubricant for the new plug.

As the eraser wears, more can be exposed by holding the cap at the end of the indicating rod and turning the chuck.

The tester is lubricated during assembly. If slight binding occurs after long use, a small amount of any good lubricant should be applied to the indicating rod and plunger.

If the tester is used frequently, it should be returned for recalibration after a year or so. With normal care, and occasional spring replacement, the hardness tester will last indefinitely.

The anti-oxidant rubber abrasive used with this test does not damage the coated surface under the prescribed test conditions.

As the test is non-destructive, it may be applied to components directly, as well as to monitor plates, with the aid of the special loaded test pencils.

11.23 Reflectance and transmittance testing

Testing antireflection coatings

A Sira Institute development of a reflectometer described by Shaw and Blevin (*J. Opt. Soc. Am.*, 1964, **54**, 334) is a satisfactory instrument for measuring low reflectivity in the wavelength range from 1900 Å to 7000 Å.

The instrument consists of a monochromator, silica prism and drums, ultra-violet lamphouse,

11.23.1. Reflectometer for testing anti-reflection coatings. In the photograph, the monitor plateholder is in the withdrawn position to show the sample being tested. The diagram below shows the principles.

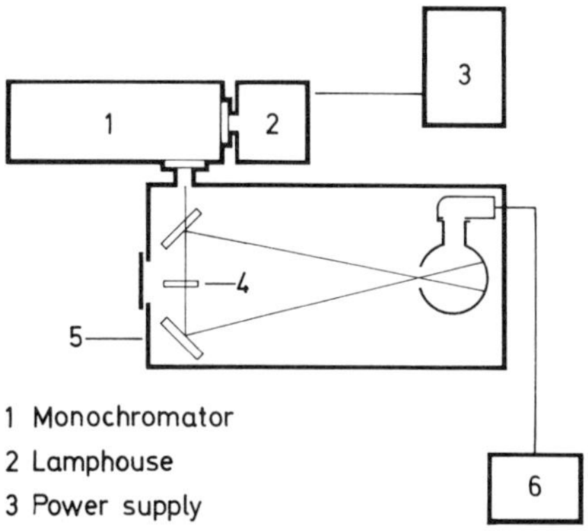

1 Monochromator
2 Lamphouse
3 Power supply
4 Sample
5 Photometer
6 Amplifier and display

11.23.2. ▷
Instrument for measuring the amount of light transmitted and reflected from a beam-splitting cube.

deuterium discharge lamp, stabilized power units, amplifiers, photomultipliers, and integrating sphere internally coated with magnesium oxide smoke. A microswitch is provided to cut off the HT to the photomultiplier as the sample is withdrawn from the box. (Fig. 11.23.1.)

The reflectometer is ideally suited to measurements of low-reflectance films and may also be used for high reflectance providing an absolute accuracy of 1–2 per cent is sufficient. It is not suitable for measuring multi-layer high-efficiency films.

The wavelength range is limited by the transmission of air at low wavelengths and the sensitivity of the photomultiplier at the high end of the wavelength scale.

With a 50-Å band width at 5000 Å, the absolute accuracy of low reflectance films is well under 0·1 per cent.

The specimens which the instrument will test are plane glass sheets, so a monitor plate is necessary for including in each vacuum coating load which is required for reflectance measurement.

Testing coatings of beam-splitting cubes

A suitable instrument consists of a collimated light source, a turret with filter glasses to control the wavelength of light to be used in the test, and an integrating sphere on a swivel arm graduated for angular settings. (Fig. 11.23.2.)

A meter reads the output from the photomultiplier in the integrating sphere.

By adjusting the position of the integrating sphere it is possible to measure transmission and reflection from the beam splitter at any chosen wavelength or angle.

Testing dichroic optics at various wavelengths and angles

The 50–601–010 Universal Reflectance and Transmittance Accessory adapts the Cary 14 to reflectance and transmittance measurements on a wide variety of solid samples, such as prisms, lenses, and mirrors. Such measurements are not possible with the standard in-line (source and sample in a straight line) sample compartments.

When measuring the transmittance of solid samples with an in-line sample compartment, the increased optical path length (due to the high refractive index of most solids) results in changes in the image formation and a consequent measurement error. While corrective measures are possible for any given measurement problem, where the sample length and refractive index are held constant, a general solution for measuring transmittance of high refractive index materials of considerable and variable length poses a difficult problem.

A further consequence of the in-line requirements in the cell space is that the transmittance of optical devices (which change the direction of the transmitted light by some mechanism) cannot be measured without redesign of the cell compartment. While systems can be designed permitting some flexibility in the geometry of the optical path through the sample, a general solution for devices such as beam splitters, prisms and lenses would be exceedingly complex and costly.

As shown in Fig. 11.23.3, a $2\frac{1}{4}$-inch diameter aluminium cylinder is driven by a synchronous motor at 1800 r.p.m. A 2-inch (5·1 cm) diameter sphere is generated inside the cylinder. A slot 1 cm wide by 90° is cut into the sphere; masks limit the viewing angle to 72°. As the sphere rotates, it alternately views standard sources A and B of spectral irradiance. As a result of the sphere rotation, an optical signal having a trapezoidal waveform with a duration of approximately 13 milliseconds at the base, at a frequency of 30 Hz, is generated from the standard sources.

An exit port $\frac{5}{8}$ inch (1·6 cm) in diameter is cut normal to the axis of rotation of the sphere. The sphere is coated with MgO deposited by burning Mg ribbon and collecting smoke. The sphere is sealed with quartz windows to prevent deterioration of the MgO surface by dirt. Dry nitrogen gas fills the sphere.

Light from the sphere is collected by a quartz lens M which provides for the imaging of slit and aperture. The slit is imaged on the sphere wall. A square aperture image exists at the exit port.

Radiation passes through the slit of the spectrophotometer and proceeds through the Cary 14 optical system, grating, prisms, collimator, and detectors in the normal manner. A 1P28 detector is used in the ultra-violet and VIS region, a PbS cell in the infra-red region.

Because the rotating sphere functions as a chopper, and because no switching of the beam is required, the signal generator is used only to generate timing signals for the relays. The signal generator is therefore located outside of the optical train and may be mounted on a convenient shelf.

To provide for a large dynamic range, stainless-steel screens are mounted on shift rods. The shift rods are operated from the end of the sphere housing. Positive detents are provided to permit reproduction of screen position. On the sample source side, four screen combinations having attenuations of 10, 100, 1000, and 2000 are provided. On the working standard source side, four screen combinations are provided with attenuations of 2, 10, 100, 1000. The screens are calibrated by actual measurements since the attenuation factors are approximate.

The working standard lamp is the same type used by the Bureau of Standards as a standard source of spectral irradiance. It is a General Electric 1000 W DXW quartz envelope, bromide-filled lamp. For long life, the current is standardized at 8·3 amperes a.c. or d.c. About 110 volts is required. Lamp current should be regulated for accurate results. The lamp mount is designed to permit reproduction of lamp position. Positive indexing is accomplished by securing the lamp with a fixed ball contact at one end and a spring-loaded ball contact at the other, and by registration of the press of the lamp against a stainless steel post.

The power supply unit will operate two 1000-watt halogen-filled lamps and consists of a 2kVA

11.23.3. The Cary recording spectrophotometer for testing dielectric coatings for colour television camera beam-splitters. The diagram below shows the optical system with a universal %T-reflectance accessory.

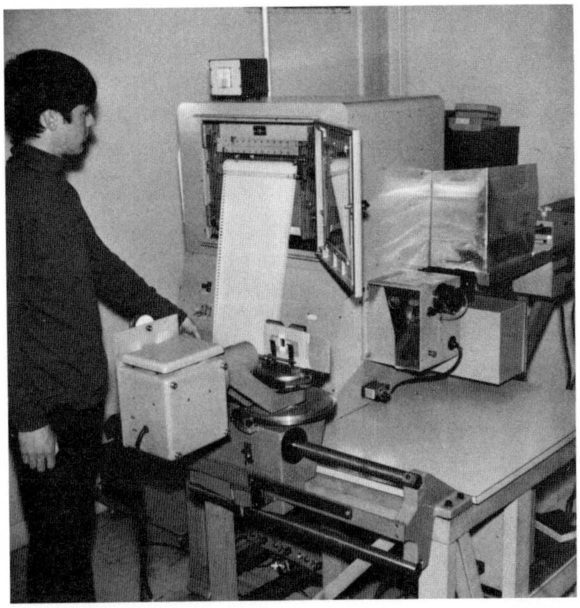

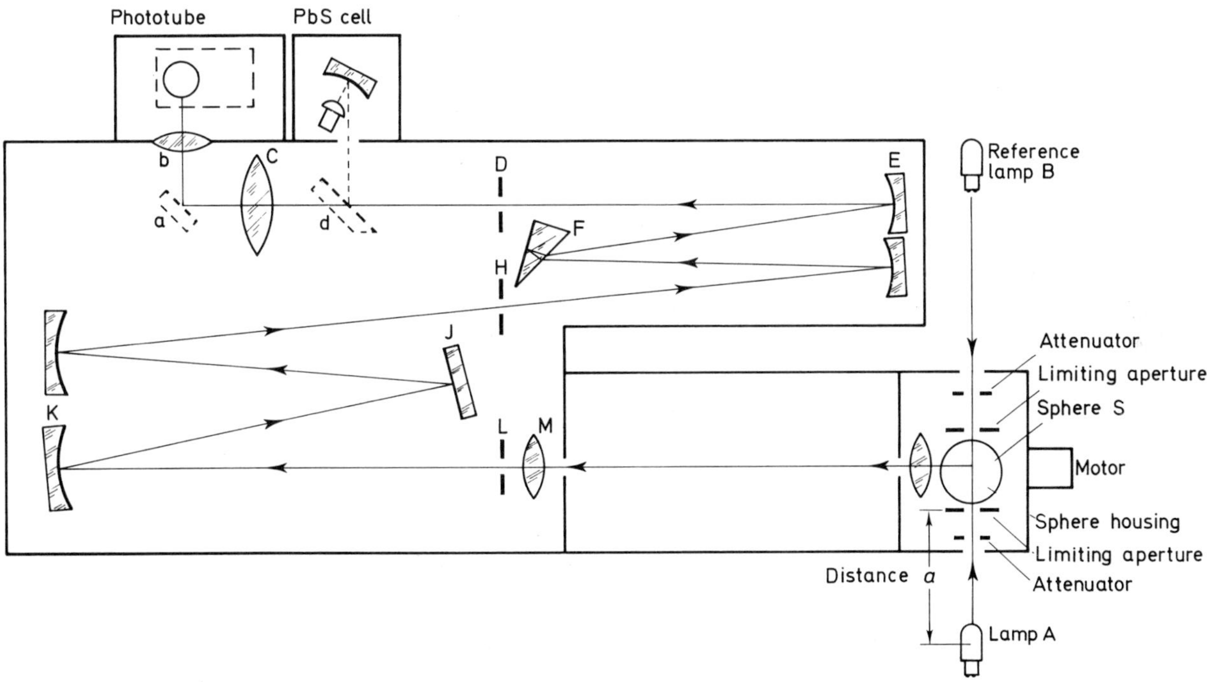

Sola regulating transformer along with a variable transformer for adjusting current. These are mounted on a small base with castors for ease in moving.

If sources A and B (Fig. 11.23.3) are lamps of the same type and are put in series with a regulated voltage or current source, a recording of unity ratio or 100 per cent T over the wavelength range can be established by means of the Cary 14 balance control and multipots, if the lamps are approximately equidistant from the aperture. This recording will be referred to as the baseline.

11.23.4. ▷
Optical system for measurement of lens transmittance.

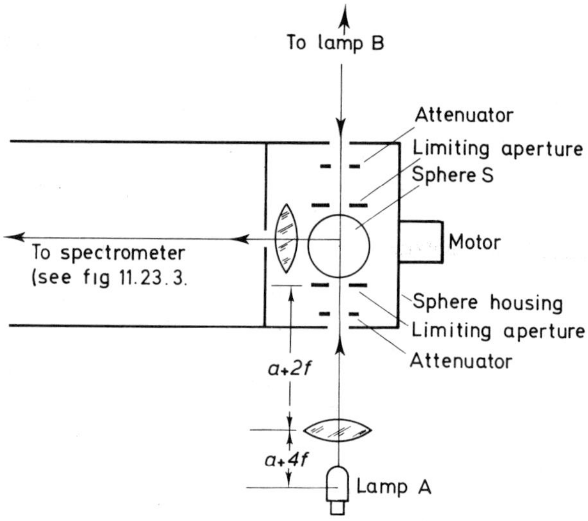

11.23.5. ▷
Optical system for measuring specular reflectance of a plane mirror.

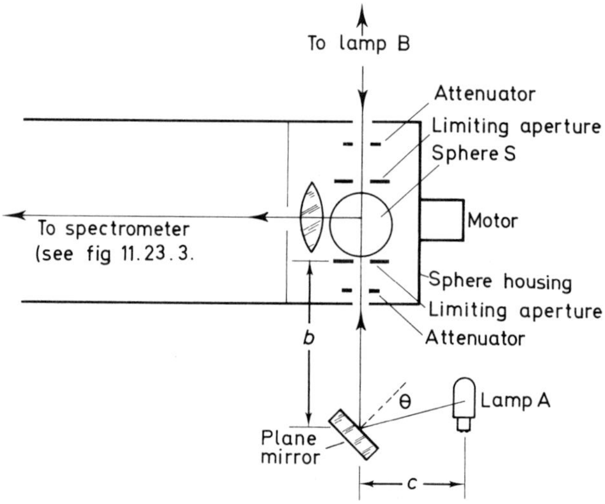

For transmittance measurements on spherical surfaces, a baseline is recorded, with no lens present as in Fig. 11.23.3, with source A at a distance a from S. Then the lens is placed at a distance $a+2f$ from S, and the source A at $a+4f$ as shown in Fig. 11.23.4. (The focal length of the lens is f.) Under these conditions, an inverted image of the source will be formed at the distance a whose radiance will be the same as that of the source at the distance a, if the lens transmittance is 100 per cent. The Cary 14 will then record the correct transmittance of the lens. The accuracy of the measurement will depend on the degree to which the rays are paraxial.

For reflectance measurements on plane surfaces, a baseline is recorded as before with the arrangement shown in Fig. 11.23.3.

If a plane mirror is centred on the optical axis of A and B (see Fig. 11.23.5.), the angle θ and source position can be varied such that the distance $b+c = a$, and the source A radiance after striking the mirror is directed along the same axis as when the baseline was drawn. The recorder will indicate the specular reflectance of the mirror at the angle θ.

Testing ultra-violet transmission of crystals

Crystals such as lithium fluoride are grown for use as ultra-violet transmitting components and therefore have to be tested for their transmitting qualities at low wavelengths. Coated mirrors and gratings, suitable for use in ultra-violet monochromators must also be tested at their working environment and at correct operating wavelength.

A 1-metre vacuum grating monochromator was modified for these tests and with a nitrogen-

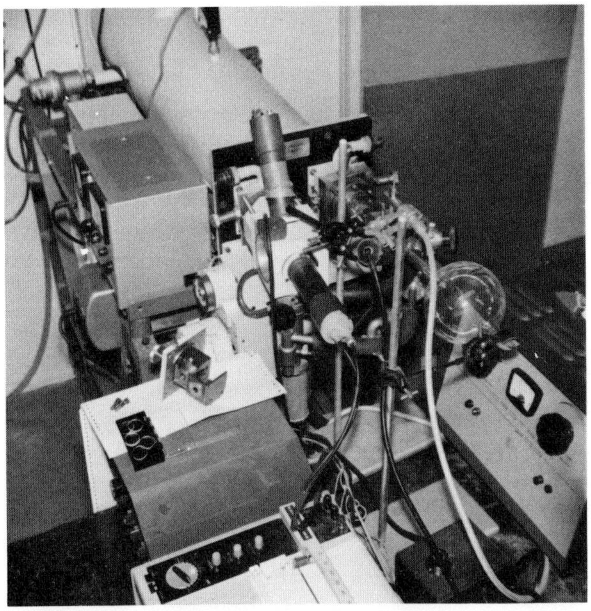

11.23.6. Ultra-violet test equipment for crystal transmission. The monitor holder for reflection tests is illustrated with sample monitor glasses. The glass bulb on the right side is the nitrogen-filled microwave lamp.

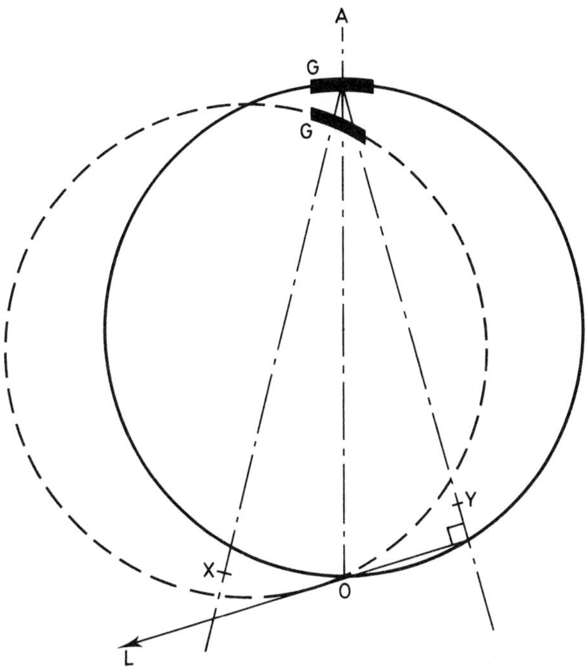

11.23.7. Pivoting Rowland circle.

filled microwave lamp source gives a range of 1150–2000 Å in the test chamber (Lyman alpha, 1216 Å).

The instrument used was based on a design by Professor R. W. Ditchburn, F.R.S., to provide a scanning monochromator in the vacuum ultra-violet (Fig. 11.23.6) with a fixed distance between the entrance and exit slits.

As shown in Fig. 11.23.7, the underlying principle of the design is the pivoting of the Rowland circle about a point O on its circumference, mid-way between the two slits.

The entrance slit Y and the exit slit X are respectively placed approximately 2 mm in front of and behind the Rowland circle when the instrument is set for the zero order.

The grating G and the two slits are all on the Rowland circle at 1 wavelength only. When scanning in the direction of increasing wavelengths the grating approaches the point O along the straight line AO and, at the same time, is rotated the required amount.

The grating is translated along the line AO by a carriage on a slide and is rotated through the medium of a bar which is attached to the grating mount at one end and to the leadscrew carriage at the other. Rotation of the leadscrew moves the carriage along the line OL.

The instrument consists of a vacuum chamber 24 cm diameter × 103 cm long (internal) made from non-magnetic materials, wavelength drive, revolution, counter, and micrometer scale. The drive moves a 576 lines per mm diffraction grating, coated for maximum reflection in the vacuum ultra-violet. Two separate high-vacuum pumping systems are required, one for the main chamber and the other for the sample chamber. In order to achieve the low wavelengths, a windowless lamp is used, excited by microwaves.

The sample chamber is normally used for samples of crystal for transmission tests, but an accessory will convert the sample chamber to take 2-inch square plates, which can be moved to various angles for reflectivity tests.

11.24.1. Talystep thickness measuring instrument with fibre-optic illuminating device. The close-ups show the viewing microscope, the stylus, and the fibre-optic illumination.

11.24 Measurement of ultrafine surface finishes on glass

The Talystep (Fig. 11.24.1) is an instrument designed primarily for the thickness measurement of thin deposits formed by evaporation, but it is also suited to the surface measurement of glass, as the diamond stylus at 1 000 000 × magnification is accurate to about 25 angstroms with a vertical resolution better than 5 angstroms.

When being used at these very high magnifications it is essential that the instrument be in a temperature controlled room and on a very stable concrete base with anti-vibration mountings.

Optical flats are used in the testing of the flatness of gauges and other flats either by contact or by non-contact with an instrument such as the Fizeau interferoscope. This method is economical,

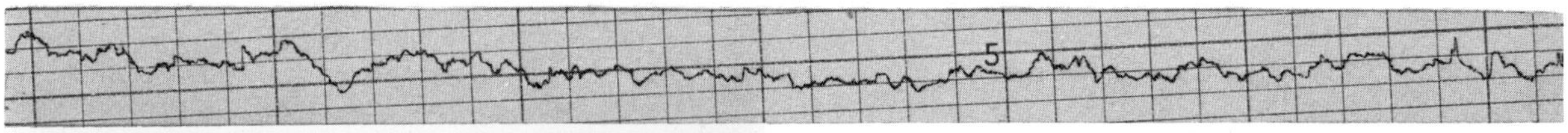

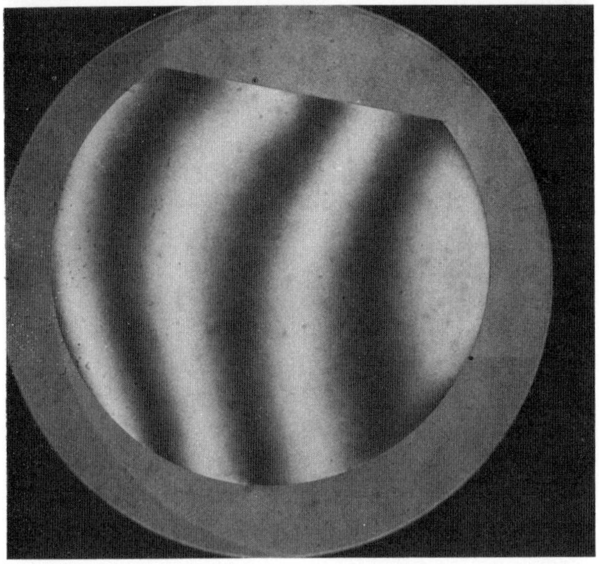

11.24.2. A Talystep graph of glass surface finish normally considered satisfactory for graticules. The glass was smoothed for 3 minutes with diamond pellets followed by 2 hours of plastic polishing. Vertical scale: 2·5 cm=250 angstroms (1 inch=1 μinch) Horizontal scale: 2·5 cm=12·5 μm (1 inch=0·0005 inch) Stylus loading: 0·6 mg. The photograph of fringes shows the figure of the testplate illustrated by the Talystep measurements.

and is accurate up to a millionth of an inch (0·025 μm), but to obtain the highest accuracy the reference surfaces must be finished to a flatness of 1 micro-inch and a surface finish with no irregularities greater than 0·2 micro-inch, i.e. 50 angstroms.

However, well-finished optical flats do have irregularities in the surface and, when investigating various methods of polishing, it is necessary to be able to measure the results of experiments. This requires very sensitive equipment and at 1 000 000 magnification, the Talystep trace of 0·2 micro-inch irregularities appear as 0·2-inch irregularities on the graphs. (Fig. 11.27.2.) It is therefore possible to detect surface differences as small as 0·02 micro-inch or 5 angstroms over a limited area.

There are various problems to be considered when carrying out such accurate tests:

(*a*) The diamond stylus which tracks over the glass surface has a definite width and it is the highest peaks or plains on the surface which cause the vertical stylus movement. If there is a noticeable peak near the edge of the stylus, two runs over the same track may cause a peak to disappear or a new peak to appear. A movement sideways of the stylus of a micrometre would be sufficient to cause this error.

(*b*) The stylus tip, however well ground, has two slightly different angles or opposite faces. Any hole may therefore have its contours altered by the stylus.

(*c*) There is always some dust on a glass surface and this can fill up holes or give different results from one run to the next.

(*d*) The drive roller on the chart recorder may have some slip and this can change the length between two peaks. A nominal error of 5 per cent should be allowed between two traces.

(*e*) Small changes in temperature, e.g. due to handling the instrument without gloves, can upset the trace. Draughts must be excluded.

(*f*) If traces are to be believed they must be repeated several times, and this leads to peaks wearing down.

(*g*) Pulses from neighbouring electronic equipment sometimes register on the chart recorder trace and give misleading information.

The eye can often distinguish between two samples of glass, the one fully polished and the other requiring further polishing. The Talystep will therefore supplement present methods of glass surface inspection and, in particular, provide a means of investigating the change in surface finish of glass which takes place some time after polishing is completed.

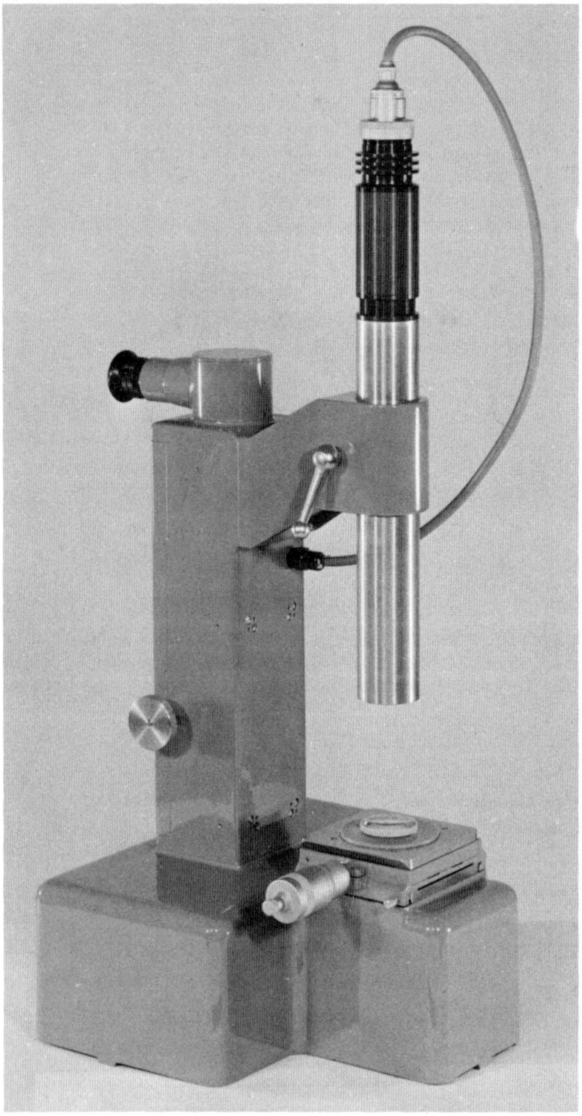

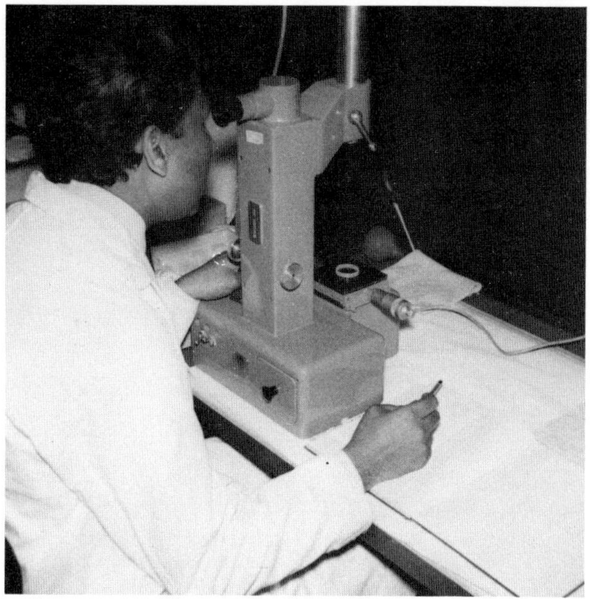

11.25.1. Focal collimator for measuring equivalent focal length. The operator is measuring the focal length of a convex lens.

11.25 The focal collimator

An important performance characteristic of a lens, whether it is a simple uncorrected component, or a complex mounted combination, is its equivalent focal length. Variations of this property from the design value will give an indication of errors in refractive indices of materials, surface curvatures, and thickness or separations of components. The Hilger & Watts Collimator has been developed to measure equivalent focal lengths accurately and rapidly. It is, therefore, a most valuable instrument for quality control of precision lenses. Measurements can be made on converging lenses having equivalent focal lengths up to 125 cm (50 inches) and on diverging lenses having equivalent focal lengths up to 50 cm (20 inches). It will accept lenses up to 20 cm (8 inches) in diameter, and 20 cm (8 inches) in axial length.

Precision optical benches, with or without nodal slides, can be used for a similar purpose but need more highly trained personnel to carry out the tests, which take a longer time for the same degree of accuracy.

The operating principle of the Hilger & Watts Focal Collimator is well known, and has been described in many textbooks. Light from a collimator is passed through the lens under test, and the image it produces is examined by a low-power microscope. A graticule placed in the focal plane of the collimator carries sets of special stadia marks that are separated by $\frac{1}{50}$th and $\frac{1}{25}$th of the collimator focal length. The image of these marks produced in the focal plane of the test lens therefore subtends $\frac{1}{50}$th or $\frac{1}{25}$th of its equivalent focal length. Consequently, this can be determined by measuring their separation with a scale graticule in the eyepiece focal plane of the microscope.

An alternative method to move the test lens laterally, setting the stadia marks in turn against a fiducial mark in the microscope field of view, and measuring the displacement by a micrometer screw. Multiplying the difference between the two readings by 50 or 25, as the case may be, again gives the equivalent focal length.

The photograph (Fig. 11.25.1) shows how the instrument embodies the following four important units, which are supported by a single casting.

(1) The stadia collimator, which can be unclamped and displaced vertically in order to provide clearance for long-lens systems. Illumination is provided by a small mercury-discharge lamp which, with a green filter, provides monochromatic green light. An electrical supply unit is built into the base of the main casting.

(2) The test lens support stage, with screw micrometer displacement, lies in a horizontal plane so that lenses may quickly and easily be placed into correct position.

(3) The focusing system, of very special design, refocuses the focal plane of the test lens, wherever it may be, on the object plane of the low-power microscope, *with a constant magnification.* An external coarse scale shows the position of focus for lenses varying from -50 cm through zero to $+125$ cm in focal length.

(4) The reading microscope carries a scale graticule in its eyepiece focal plane, which is always correctly calibrated owing to the constant-magnification property of the special focusing system.

11.26 Testing lenses after assembly

(See also §§ 11.27 and 13.3.)

(*a*) *Finite conjugate bench*

Fig. 11.26.1 illustrates a finite conjugate test bench for testing process and enlarging lenses.

Process lenses are employed in the main for copying work where the original and the copy are of the same order of size. It is, therefore, usual to test them under the conditions of one-to-one copying because there is bound to be some change in characteristics when any lens is used over a differing range of magnifications.

If it is known that a particular lens is needed to be used at some magnification other than one-to-one, this should be notified at the time of ordering so that the lens can be set to give of its best under these conditions.

All Wray process lenses undergo systematic tests before being passed for despatch, and the main tests ensure:

(1) that the image is sharp over the whole angle of the copy,

(2) that the image is the same size (within a very close tolerance) for all wavelengths of the visible spectrum,

11.26.1. Finite conjugate bench for testing the performance of process and enlarging lenses. A 20:1 reduction Wray 10-inch *f*/4 microcircuit lens is shown under test.

(3) that the image is free from distortion to close limits, and

(4) that the image is symmetrical; that is to say, no changes occur in the image when the lens is rotated about its axis.

Frequently also it is important that the focal length is known to great accuracy so that distances from the principal planes of the lens to object and image planes can be determined for a range of magnifications. This is of particular importance where cameras have automatic focusing and can be set for photographing at different magnifications by means of tapes.

To ensure that these tests and measurements can be carried out as accurately and efficiently as possible, Wray have recently installed an optical test bench, specially designed and built for this work.

The bench comprises a planed cast iron bed with a fixed image scale at one end, and carrying the lens under test in the middle, and the object scale at the other end. Both the object and image scales are 5 foot 6 inches long and the lens and object scale can be moved along the bed on rollers. Provision is made for measurement of their position using a scale which may be read to within 0·01 mm.

The maximum distance between the object and image scales is over 17 feet. This bed length and also the object and image scale length can be extended to cater for very long-focal-length lenses.

The lens may be rotated at suitable speeds by a reversible electric motor as a check on any local variations within the lens, representing performance variation between different directions in the field.

The object scale has divided and calibrated clear crosses every 10 mm in the background. The background is opaque, apart from pin-holes to act as artificial stars for checks on image quality by comparison with theoretical spot diagrams obtained by computer. Illumination is by means of a fluorescent tube behind the scale.

The image scale also has divided crosses similar to those on the object scale but these are dark on a clear ground. This scale is viewed with a microscope which travels along precision straight-edges and enables any error in reading between the image of the object scale and the image scale to be observed.

Micrometer movement of the lens and object-scale trolleys is afforded by clamping them to sliding bars moving along the length of the bench and operated by handwheels at the image end which control special backlash-free drives. One of the uses of this bench is to provide a means for the reliable calibration of lenses for scale purposes.

For process lenses designed to work in the region of 1:1 the object and lens are positioned to give an image of a chosen suitable size with a magnification ratio of 2:1. The settings of the lens and object trolley are noted and, keeping the object trolley fixed, the lens is moved to give a magnification ratio of 1:2. The distance moved by the lens between these two positions is recorded; it is equal to $3f/2$ where f = focal length of the lens, which is thus deduced. If d is the distance between the nodes of the lens, the object-to-image distance is $9f/2+d$, from which d is deduced. The absolute positions of the nodal points relative to the flange are also deduced after measurement of the flange position on the longitudinal scale. As a check on the measurements, the object trolley is moved a distance of $f/2$ towards the image scale, when the magnification should be unity.

This method is independent of individual choice of best focus and it also allows for any minute variations in image size due to distortion variation over the chosen range, as the measurements are made under conditions which simulate the actual conditions of use. For special cases other magnification ranges and object and image sizes may be chosen. Such checks on positional accuracy of the image are essential for cameras required for microcircuit mask production or mapping, but are clearly important in any work where strict register is required.

(b) Measurement of focal length using a collimating nodal slide test bench

An optical lens bench consists of a collimator, which produces an infinitely distant image of a test target, a holder for the lens under test and a microscope for examination of the image formed by the system.

The collimator consists of a well-corrected objective and an illuminated target, which is usually a pinhole for star tests, but could be a resolution target or calibrated scale according to the tests required.

11.26.2. **A 12-inch nodal slide, showing a Wray 6-inch wide-angle *f*/5·6 aerial survey lens under test. This lens has a maximum permitted distortion of 10 μm at 45° semi-angle.**

The lens-holder is usually mounted on a nodal slide of high accuracy.

The measuring microscope will have at least one micrometer slide for precision movements. The lens under test is placed in the path of the collimated light and is capable of:

(*a*) Rotation in a vertical plane about its axis.

(*b*) Rotation in a horizontal plane around the axis of the nodal slide by means of a graduated drum.

(*c*) A fore-and-aft movement of the nodal slide along the bed of the instrument.

The image of the spot of light made by the lens can be examined by means of the microscope:

(1) on its axis, by having the graduated drum at zero.

(2) At any suitable angle to the axis, by rotating the drum and refocusing the image made by the lens in the microscope.

One of the important tests with this instrument is the measurement of equivalent focal length. (Fig. 11.26.2.)

11.27 Optical transfer function tests

INTRODUCTION

A traditional method of judging definition is to determine the finest detail that an instrument can resolve when it is used to observe test charts marked with groups of black and white stripes. The result of such a test is usually expressed in lines per millimetre or lines per degree according to the type of instrument.

Owing to diffraction, even a perfect system has a limit to its resolving power.

A more satisfactory expression of performance is a statement of the imperfections suffered by all grades of detail and the optical transfer function is such a statement. It is based on the fact that a test-object whose brightness distribution obeys a sine law will always form a sinusoidal image.

Diffraction, aberrations, and scattered light will reduce the contrast of the image, and may shift its position, but the character of the image and its spacing (or spatial frequency) will be unaffected.

Thus the image-forming properties of an optical system can be represented by two graphs, one showing the variation of image position, and the other the variation of image contrast, with spatial frequency.

The first curve is termed the phase response of the system and the second the amplitude response function or the modulation transfer function. Both curves together form the optical transfer functions.

There are several advantages in obtaining the performance of a lens in terms of a transfer function.

Firstly, the transfer function of an optical system can be calculated from the design data; an accurate comparison of design and practical

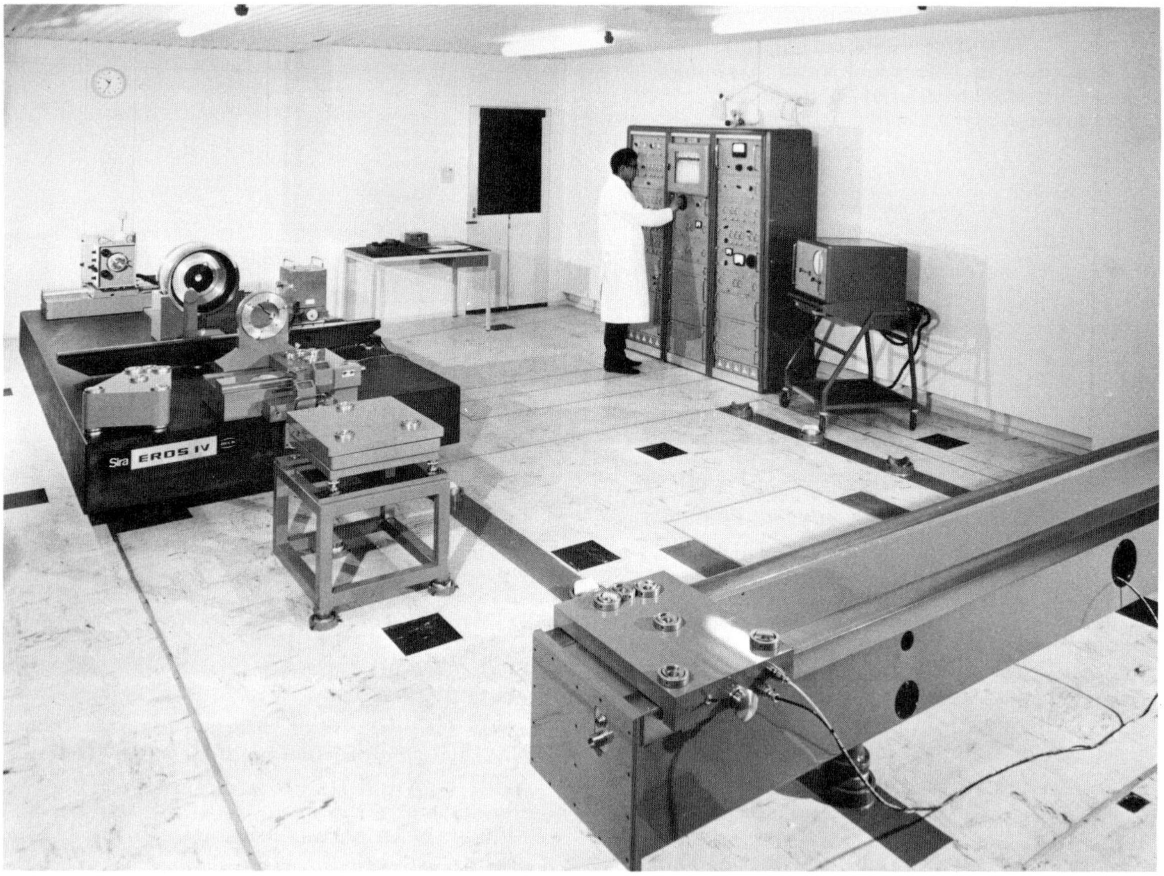

11.27.1. The Eros IV mounted on a universal image-analysis table. The extra long field slide is shown in the foreground. A lens of medium focal length is being tested at fixed conjugates. (*Left*) The equipment is shown with a reflecting collimator on its own stand in the background. (*Bottom left*) Lens systems up to 14 inches (35 cm) in diameter can be accommodated.

performance is possible. Secondly, the information contained in the OTF can be used to obtain the form of images other than sine waves. Finally the OTF of an optical system can be cascaded with the transfer function of a detecting system to obtain the image-forming properties.

OTF alone does not fully describe the quality of an optical image. Veiling glare due to scattered light and to multiple reflections from polished surfaces may have as serious an effect as have aberrations and must be measured separately.

As a result of research and development at the Sira Institute, R. & J. Beck Ltd have produced for Rank Precision Industries Ltd an Eros IV OTF system to measure modulation transfer function to 1 per cent and phase to 5 per cent, and which are therefore suitable for testing high-class camera and television zoom lenses. (Fig. 11.27.1.)

The rest of this section is extracted from the handbook 'Beck Optical Transfer Function Systems', by permission of Ealing Beck Ltd.

OPTICAL TRANSFER FUNCTIONS

Optical transfer functions are useful figures of merit for the performance of optical elements or systems of elements. These functions evaluate lenses or systems in terms of efficiency, over their entire operational range in an environment which is directly related to that of the final or intended use.

Classically, lens evaluation has depended upon the use of resolution targets. The higher the spatial frequency (e.g., lines per mm in a square wave bar target) which a lens was able to resolve, the better the lens was felt to be. This technique is analogous to determining the quality of an audio amplifier by measuring its high frequency cut-off. The information is undeniably useful, and as exponents of the test target measurement still argue, the experimental setup is straightforward and inexpensive. However, if the target contrast can not be varied, nothing will be learned about the system's performance at any other contrast level.

An obvious example of this problem is seen with lens systems which are coupled to vidicons or image orthicons. Typically, the tube has a scanning raster which is coarse in terms of optical resolution, and this resolution (or spatial frequency) is the upper limit to the system. One

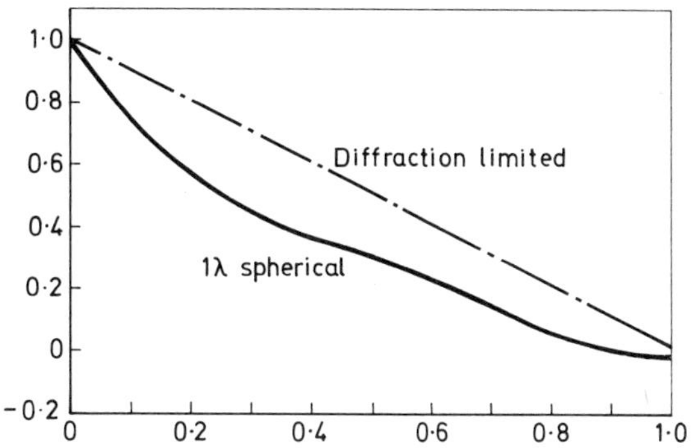

11.27.2. Modulation transfer function of a diffraction limited lens and of the same lens with one wave of spherical aberration.

wants, however, the optics feeding the tube to produce images of high contrast at this spatial frequency so that the best transfer of information may occur.

Perfect (i.e., diffraction limited) lenses transmit increasing spatial frequencies at ever-decreasing contrast levels; the higher the number of lines per mm, the lower the ratio of dark to light. Fig. 11.27.2 illustrates the MTF of a diffraction limited lens and the MTF of the same lens with one wave of spherical aberration.

In the vidicon problem, using standard resolution target evaluations, the only safe procedure is to pick a resolution limit for the optics substantially above the tube's resolution limit and to assume that this will assure adequate contrast at lower spatial frequencies.

An immediate alternative which comes to mind is to measure photoelectrically the contrast of the square bar resolution target at these lower and intermediate spatial frequencies. This approach is, however, substantially more complex than it would first appear. Square bar gratings are, when viewed as an inverse Fourier series, the summation of a great number of sinusoidal gratings of ever-increasing frequency. Instead of measuring simply the square bar contrast variations, we are, in fact measuring the contrast variations of a large number of individual, but superimposed, sinusoidal contrasts whose individual contributions are varying in a complicated manner. Furthermore, the boundaries of a three-bar target are so confined that they in turn have an influence on the results.

Optical elements are, in general, passive linear operators; i.e., they transmit a sinusoidal distribution of intensity in space as exactly a sinusoidal distribution without distortion. Changes in spatial frequency between the object and image planes are a function of the system's magnification. Changes of contrast (the ratio of light to dark) and phase (the positional relationship) are a function of the final diffraction limitation and the aberrations, scatter, and design or fabrication errors remaining in the optics.

An ideal lens or system testing technique would then entail the production of an object with known sinusoidal contrast and phase over an extended spatial frequency range. The Optical Transfer Function (OTF) states the ability of the lens to image this object. This function consists of both the *Modulation Transfer Function*, which is a measure of the contrast reduction from object to image over an extended spatial frequency range, and the *Phase Transfer Function*, which is a measure of the positional variation in a one to one relation between the object and the image over an extended spatial frequency range.

The Modulation Transfer Function (MTF) gives an entirely satisfactory evaluation of a system's performance; however, the Phase Transfer Function (PTF) is additionally required if adequate insight into the failure of a system to perform to its designed specification is necessary. Further, the PTF offers a rapid evaluation technique for sorting out elements which are entirely unsatisfactory. For instance, a phase reversal (the rendering of dark as light and vice versa) in the mid-spatial frequency regions would be completely unacceptable in most systems.

The instrumental difficulties encountered in measuring the PTF are clearly understood when it is pointed out that a 1° phase shift at 100 lines per mm is equivalent to a 300 Å positional variation in the image plane.

Instrumental techniques

At the present time only two general instrumental techniques for ascertaining the Optical Transfer Function appear to be commercially viable.

One of these general techniques involves the presentation of an object with a large and continuous variety of ever-present spatial frequencies with known contrast and phase (such as produced by a slit or pinhole), to the system under test. A Fourier analysis is then performed on the image. This is done by placing a varying spatial frequency filter of known contrast and phase in the image plane and measuring the resulting image content. Since the object spatial frequency contrast is known, the computations may be automatically handled and the resulting MTF displayed.

This technique suffers from several instrumental drawbacks:

(1) Undistorted sinusoidal spatial filters (of either variable area or variable density) are very difficult to make with high density or high spatial frequency. Reduction of the former degrades the signal/noise, and reduction of the latter limits the usefulness.

(2) The spatial frequency amplitude content of a line or point image decreases with increasing frequency. Signal/noise must therefore decrease at higher spatial frequencies.

(3) Because of metrological problems, phase is virtually impossible to measure. The PTF may, therefore, never be determined.

The second general technique for measuring the OTF involves the generation of one sinusoid spatial frequency at a time and presenting it as the object to the lens under test.

If the amplitude and phase of the smoothly or incrementally varying spatial frequency of the object are known, the resulting MTF and PTF may be measured throughout the frequency range of the object as scaled in image space by the system's magnification.

Difficulties encountered with this technique are:

(1) Again the problem of producing a purely sinusoidal target of adequate contrast over a substantial spatial frequency range. The problem is eased, however, by the fact that, generally, the system magnification is working in one's favour and very high spatial frequencies do not have to be generated.

(2) The metrological problems of measuring phase are still severe and, unless a technique

which does not depend on a knowledge of mechanical space is used, it will still be difficult to measure the PTF.

The Eros equipment described on the following pages utilizes the second technique for measuring the OTF. Smoothly varying spatial frequencies are generated in the object plane and the resulting transferred contrast is measured in the image plane.

This equipment was developed by Dr L. R. Baker at the SIRA Institute, and is an outgrowth of the work of Dr H. H. Hopkins and his associates at the Imperial College, University of London.

The instrumental techniques used solve the basic problems of the second method of measurement mentioned above in a straightforward yet ingenious manner.

A sinusoidal target is generated by time filtering an optically square target pattern. The extended square target pattern is the summation of a series of discrete frequency sinusoidal targets. The series consists of a strong fundamental frequency sine wave, whose period corresponds in time and space to the square wave, and other sine waves of ever decreasing amplitude at 1, 3, 5, . . . $(2n+1)$ times the fundamental frequency. In this instrument the time frequency remains constant, while the spatial frequency varies. A narrow-band electrical time filter will therefore remove all of the higher harmonics leaving only the fundamental sine wave component.

The square pattern requires care in manufacture but it may be made with certainty and exactness. The variation of spatial frequency is achieved mechanically, and the results are reliable and intuitively certain. This method of generating sinusoidal frequencies avoids the complex difficulties of photographic target preparation and the mechanical difficulties of moiré target generation.

The PTF measurement problem is equally well handled. Phase shifts are determined electronically as a time shift and the mechanical positioning problems are thereby avoided.

In a consideration of this technique, the following points should be borne in mind:

(1) Only at zero cycles per mm of spatial frequency is it possible to establish an absolute or normalized amplitude level.

(2) Failure to preserve the phase transfer function along with the modulation transfer function may lead to a very real ambiguity in the results.

(3) Calibration of optical transfer function systems with high contrast targets or calculated masks is helpful, but does not assure accuracy at any other contrast level.

(4) Equipment based upon masks or photographically produced targets is only as good as the reproduction technique used to make and calibrate the mask.

Operational theory

The operational theory of the Beck–SIRA systems is simplified if one assumes that the object slit and the image slit are infinitely narrow. Fig. 11.27.3 illustrates the arrangement of these slits.

Assume that the grating consists of parallel opaque lines separated by perfectly transmitting spaces. The clear and opaque areas are also assumed to be equally spaced.

The departure from these conditions which occurs in practice does not significantly affect the performance of the apparatus.

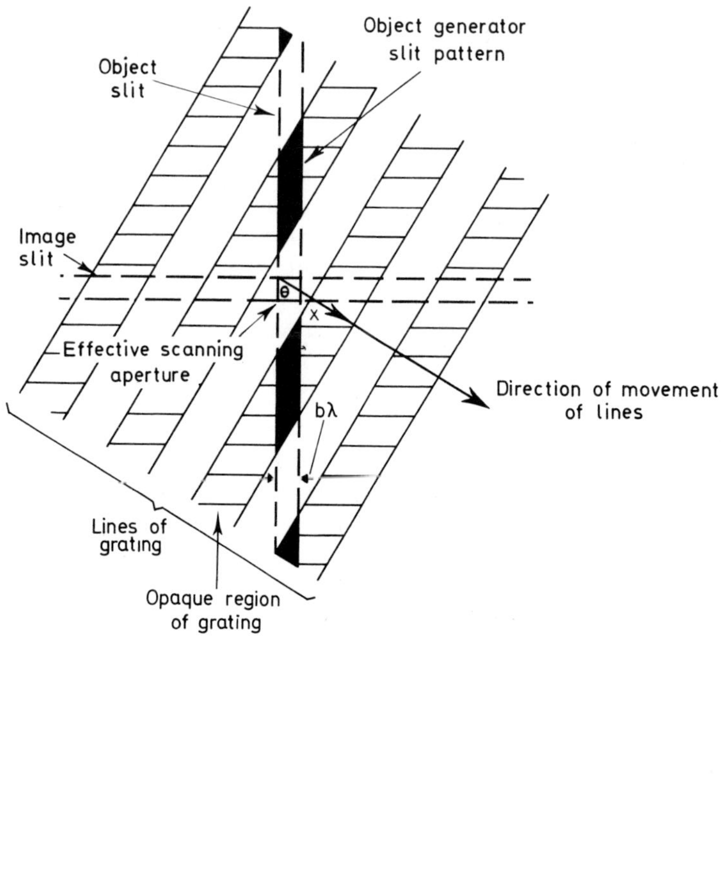

11.27.3. Arrangement of object and image slits.

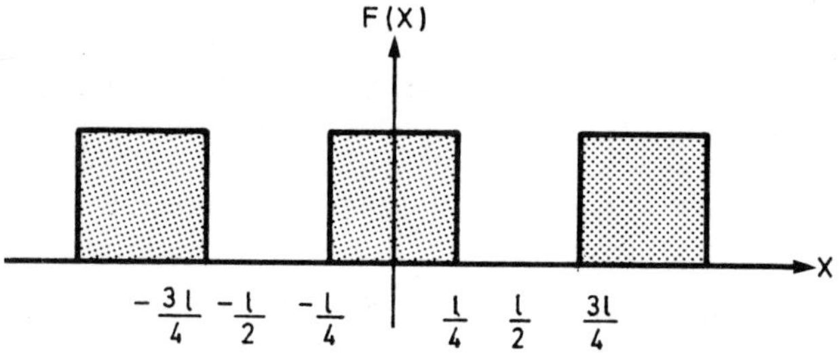

11.27.4. Intensity distribution along the line object slit.

The intensity distribution $F(x)$ along the line object slit will be of the form shown in Fig. 11.27.4.

The distribution is a square wave of amplitude A and period l (spatial frequency is $1/l$). By changing the angle θ between the object slit and the direction of rotation of the grating it is possible to vary the spatial frequency $1/l$. The relation between the grating period λ and the spatial frequency $1/l$ is given by

$$\frac{1}{l} = \frac{1}{\lambda} \cos \theta$$

$$F(x) = \frac{A}{2} + \frac{2A}{\pi}\left[\cos\left(\frac{2\pi}{l}x\right) + \frac{1}{3}\cos\left(\frac{6\pi}{l}x\right) + \frac{1}{5}\cos\left(\frac{10\pi}{l}x\right) + \dots\right]$$

The intensity distribution $F(x)$ along the object slit can be represented by the Fourier series where the first cosine term has the same fundamental frequency as that of the square wave. The amplitudes of the components of this series are modified by the optical system under test, and all the harmonic terms are rejected electronically so that only the fundamental frequency remains.

The frequency of the electronic signal from the image analyser is constant because the rate of passage of the image pattern past the image analyser slit is inversely proportional to the spatial frequency of the pattern. Only one electronic filter is required to ensure that only the fundamental frequency is passed under all test conditions. The use of electronic filtering produces results which are equivalent to those obtained from systems using a sinusoidal object pattern.

In actual fact both slits have a finite width and the function $F(x)$ does not follow a square wave distribution. Provided the slit widths do not exceed $\lambda/4$ the function $F(x)$ can be represented by the more general series,

$$F(x) = \frac{A}{2} + \frac{2A'}{\pi}\left[\alpha_1 \cos\left(\frac{2\pi}{l}x\right) + \alpha_2 \cos\left(\frac{4\pi}{l}x\right) + \alpha_3 \cos\left(\frac{6\pi}{l}x\right) + \dots\right]$$

which contains all harmonics of the fundamental frequency. The electronic filter is designed to operate at the fundamental frequency of the a.c. signal, 1000 Hz, and its bandwidth is about 100 Hz. The first harmonic will have a frequency of 2000 Hz. The attenuation of the filter at this frequency is about 40 dB and is even greater for higher harmonics. Thus, the filter effectively rejects the harmonics and passes only the fundamental sinusoidal signal.

It can be shown that the amplitude α_1 of the fundamental at the object generator slit does not vary significantly with the angle θ if one restricts the slit widths to less than $\lambda/4$. The degree of modulation of the amplitude α_1 by the optical system under test is, in fact, a measure of the transfer function of the system.

THE EROS SYSTEMS

There are three Eros systems. These are the Eros 100, the Eros III and the Eros IV. The operational principles are identical for all three systems. The intended uses are, however, quite different; and there are, therefore, marked differences of capability and convenience. Each system incorporates all of the necessary features to measure the modulus of the optical transfer function with a high degree of accuracy.

The Eros 100 measures modulus only, to an accuracy of 5 per cent. It is intended for production testing and laboratory use.

The Eros III measures modulus only to an accuracy of 3 per cent. It is intended for research and laboratory use as well as production testing of more demanding lens systems.

By the addition of certain separate accessories, this unit may be upgraded to an Eros IV at a later date.

The Eros IV measures both modulus and phase; modulus accuracy is 1 per cent and phase accuracy is 5°. It is used for research and laboratory measurements of the most demanding rigour.

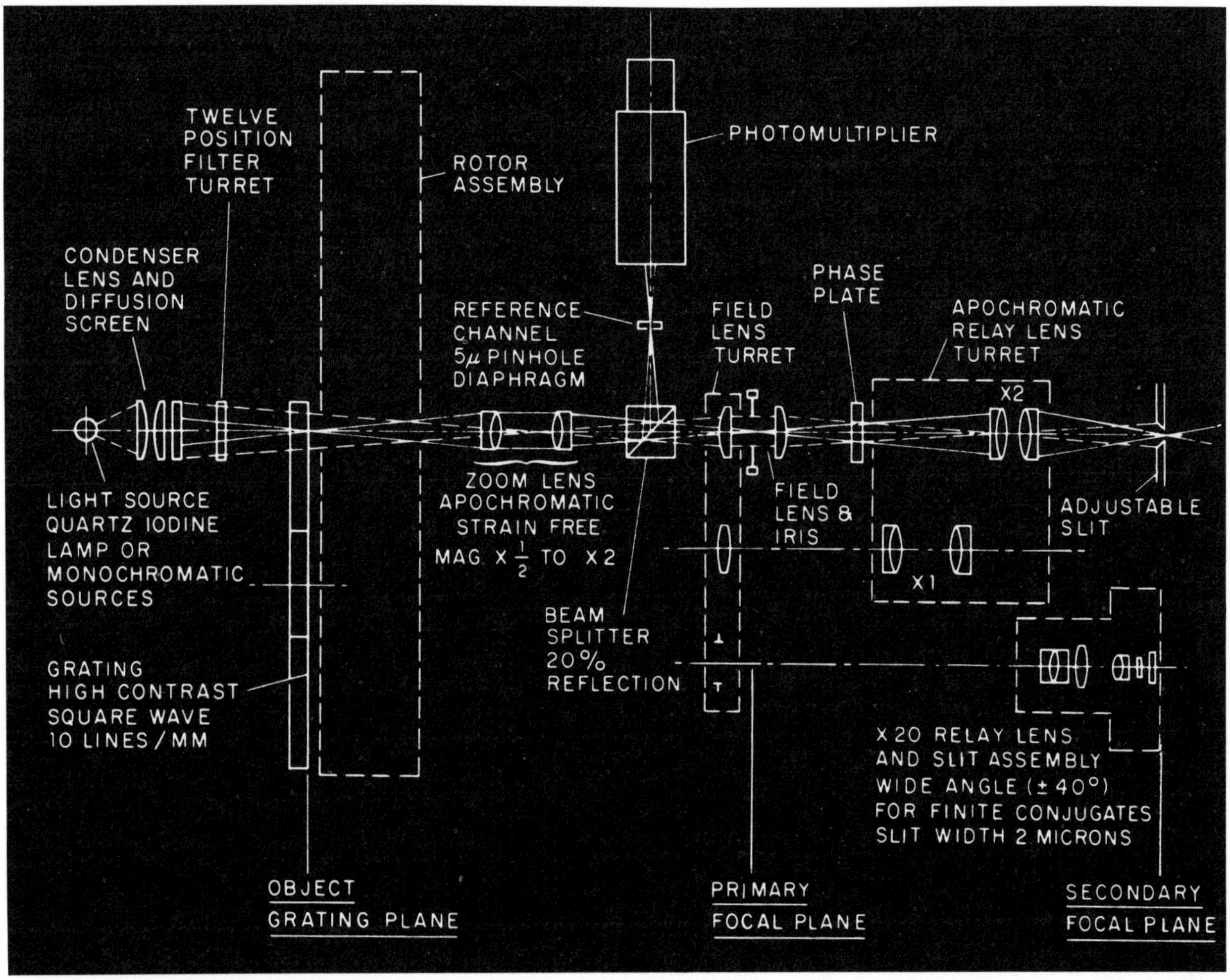

11.27.5. The Eros IV equipment.

Each Eros system consists of three basic elements:

(1) An object generator.
(2) An image analyser.
(3) An electronic control, analysis, and read-out unit.

While these elements differ from system to system, their essential design and function remain the same.

The object generator

The object generator produces an object of constant modulation or contrast (visibility in the Michelson sense) but variable spatial frequency for presentation to the lens system under test. The test object is formed by imaging part of a high contrast, radial, square-wave grating on to a slit. The grating frequency is 10 cycles per mm. The length of the individual lines on the grating itself is 15 mm. The radial lines consist of a hard chromium film deposited as an annular ring on a flat glass disc of 170-mm diameter.

The grating is rotated about its centre at a constant speed by means of a small synchronous motor. This rotation causes the line pattern focused on the object slit to drift over the slit at a constant rate, chosen to be 1 kHz. At the same time the whole grating and motor assembly is rotated at a much slower rate about a point in the centre of the grating annulus. This point is located on the principal optical axis.

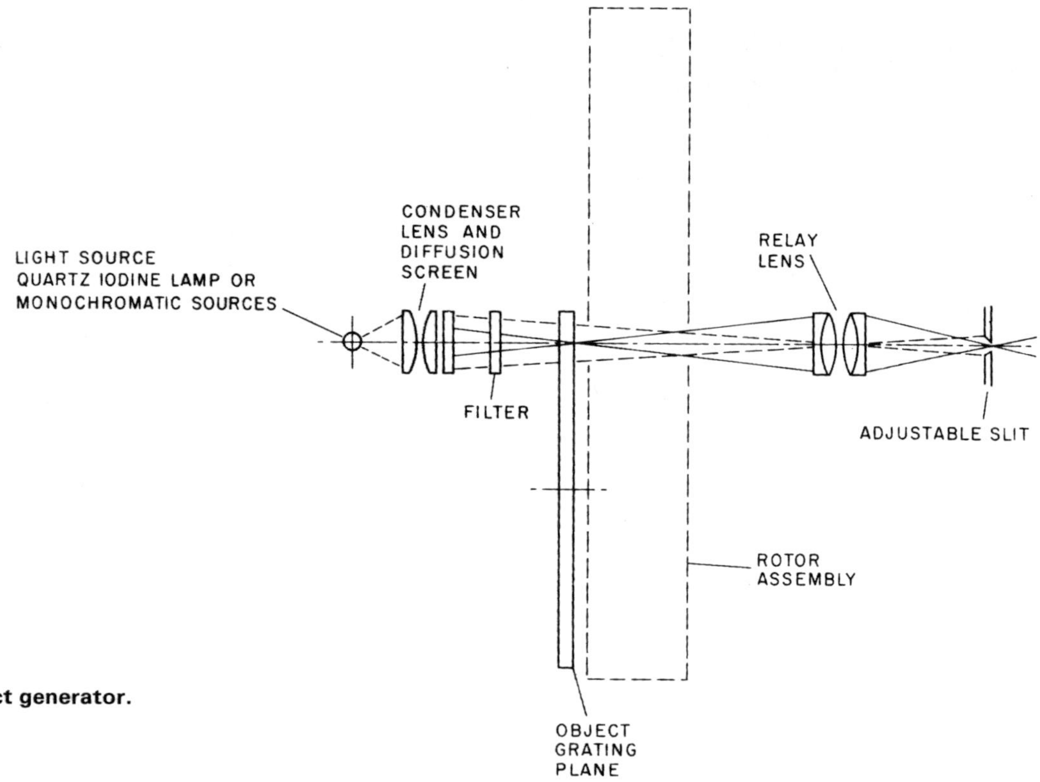

11.27.6. The object generator.

When the grating lines are instantaneously oriented parallel to the slit, a uniform distribution of light is produced along the length of the slit. This light is modulated at the 1 kHz carrier frequency by the drift of the lines across the slit. By rotating the grating and motor assembly about the principal optical axis, the orientation of the grating lines with respect to the object slit is changed. As a result, the spatial frequency of the line pattern seen at the object slit is increased from zero cycles per mm to the maximum value determined by the particular optical relay used to image the grating on to the object slit. The maximum spatial frequency is achieved when the lines of the grating are at right angles to the slit.

This square-wave object, limited in domain by the slit, is presented to the lens system under test. The signal received by the image analyser and the associated electronics is handled in such a way that only the fundamental frequency component of the square-wave pattern is accepted. It is immaterial to the system under test as to whether this filtering is performed before or after presentation; the result is that the lens system is evaluated for its sine-wave response, *not* its square-wave response.

The grating, with its square-wave pattern, rotates on its own axis at a constant time rate. This produces a constant square-wave modulation as a function of time which is used as an electronic carrier frequency. It is this signal which is filtered in the electronics section to remove the square-wave frequency components. The higher spatial frequency components of the square wave correspond to higher time domain harmonics.

This type of filtering is straightforward and unambiguous. The carrier frequency is a constant 1 kHz. The line to space ratio is very constant and the even harmonics are therefore nearly absent. The harmonics to be removed are therefore the third, fifth, and higher odd harmonics. A simple, three-section RCL filter is peaked at 1 kHz and designed to be flat for 100 Hz to either side (to allow for any possible line frequency drift causing a speed variation of the synchronous drive). At a frequency of 2 kHz, the filter attenuation is 40 dB.

The spatial frequency varies, as has been shown, by changing the orientation of the grating pattern while viewing it through a slit. The square wave components of the grating will be removed electronically and we may think of it as a sine-wave grating. When the object slit is parallel to the grating the spatial frequency is zero; when it is perpendicular the spatial frequency is a maximum. The specific value of this frequency is just the magnification of the optics imaging the grating on to the object slit times the grating frequency (10 lines per mm). The spatial frequency range at the object slit may, in practice, be set from 0–5 cycles per mm to 0–400 cycles per mm. These are the spatial frequencies in the *object plane*, not the image plane, of the system under test.

If the object slit were infinitely narrow, it would be clear that the system under test after time filtering would view a pure spatial sinusoid of varying frequency. However, energy, and therefore signal-to-noise considerations, require the slit to be as wide as possible. A rigorous analysis shows that a slit width of one-quarter of the projected frequency is an adequate approximation of zero slit width. The resulting harmonic and amplitude distortion due to a finite slit width are never greater than 0·1 per cent.

This approximation problem, common to all existing OTF systems employing slits, is of less concern for the Eros systems because the generation of spatial frequencies is in the object plane. Since maximum spatial frequencies of interest are nearly always far coarser in the object plane than in the image plane, the mechanical requirements of the solution are far less rigorous.

Setting of the slits on both the object generator and the image analyser is straightforward. The slit width is determined electronically, rather than mechanically, by observing the first signal maximum (slit width equals half of one cycle) as the slits are opened, and then reducing this width to half that value (slit width equals one-quarter of one cycle).

Rigorous requirements are not placed on the optics which relay the grating on to the object slit. The relay located in the object generator is operating at low spatial frequencies and the required numerical aperture is not demanding. Each relay system is well-corrected to produce high contrast images free from astigmatism. Effects of residual scatter are eliminated since the output is normalized at zero cycles per mm.

The Eros 100 relay optics are interchanged by unscrewing and replacing the entire relay lens. Spatial frequency ranges of 0–5 cycles per mm, 0–10 cycles per mm, 0–20 cycles per mm, 0–40 cycles per mm, and 0–100 cycles per mm are available as standard options.

The Eros III and IV relay optics are built into the unit and are interchanged by knobs on the control panel. A 4-power zoom system for continuous variation of the range of spatial frequencies is standard on the Eros III and the Eros IV.

Partially coherent light has a non-random phase relation, and interference effects will occur, to some degree, whenever the beam is acted upon. In OTF systems this results in a marked non-linear variation of the system frequency response. When coherence lengths approach the diameter of the Airy disc for a system, the effect on the frequency response curve becomes very serious.

Some degree of coherence exists in all light sources, and passing this light through a small aperture tends to increase the degree of coherence.

OTF systems utilizing line spread function analysis or Fourier analysis and spatial filtering in the image plane are most susceptible to this problem. Uncertainty levels within the readings become large and there is evidence that the results tend to be in error throughout the spatial frequency range; not simply in the higher spatial frequencies.

The Eros systems generate the spatial frequency pattern in the object plane, where apertures may be kept large with respect to the diffraction pattern.

Care has been taken in the illumination system design of the Eros instruments to assure the greatest possible degree of incoherence in the illumination.

The overall accuracy achieved in measuring the MTF by the Eros method is in no way dependent upon the idiosyncracies of masks of variable area or intensity, nor can there be any question of the proper normalization of the MTF curve since the instrument is normalized at true zero cycles per mm.

11.27.7. ▷
The recorded modulation transfer function of a 10-inch *f*/5·6 commercial collimator lens on axis. The record illustrates the variation of performance as a function of chromatic variation. The theoretical diffraction limit is indicated for reference.

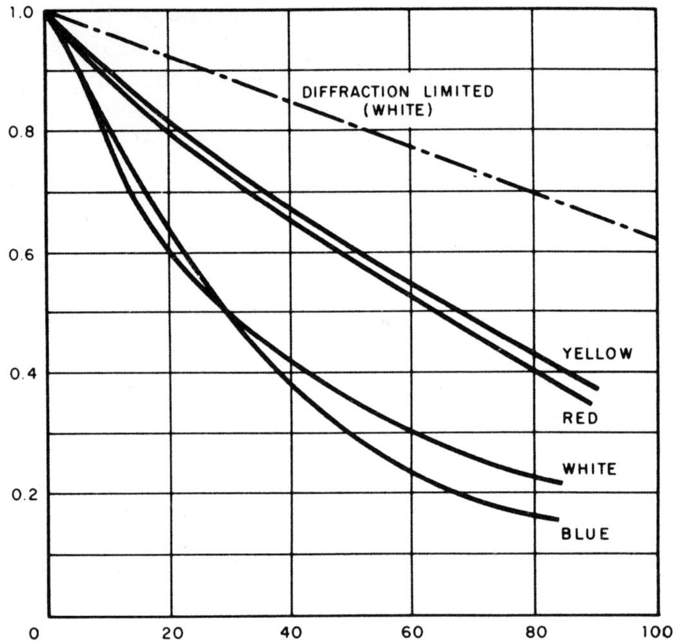

11.27.8. ▷
The recorded optical transfer function of a 3-inch *f*/2·8 projection lens. The MTF is indicated on axis and off axis at 5° and 8° in sagittal and tangential planes. The PTF is shown only for the 5° off-axis sagittal measurement.

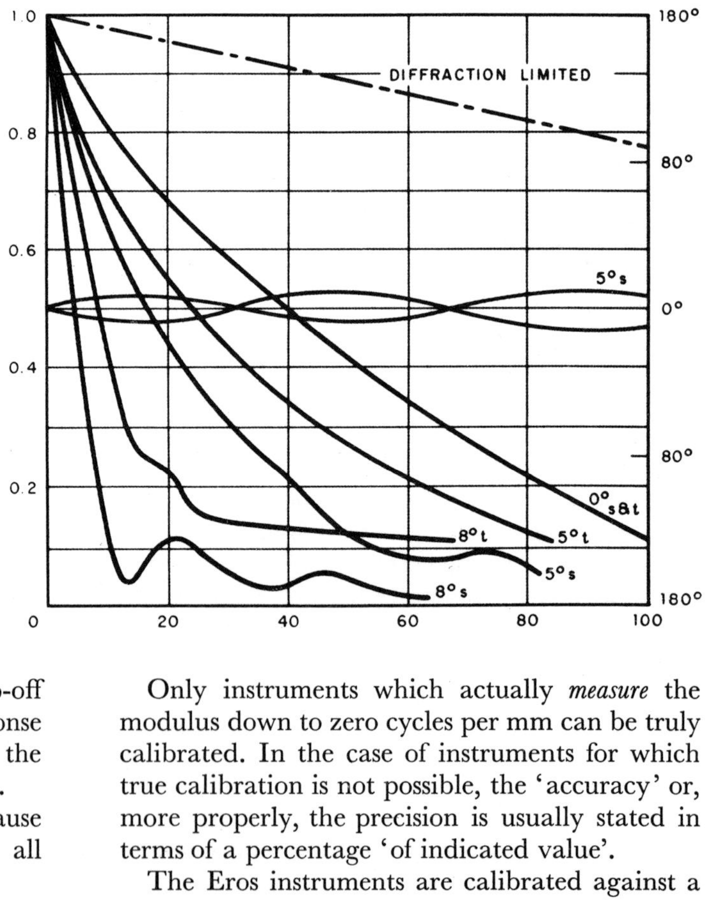

Lens designs which call for a contrast drop-off at very low frequencies, to allow a flat response through the mid-frequency ranges, illustrate the need for normalization at zero cycles per mm.

Failure to normalize the curve at *zero* will cause a substantial scaling error of the curve at all points.

Only instruments which actually *measure* the modulus down to zero cycles per mm can be truly calibrated. In the case of instruments for which true calibration is not possible, the 'accuracy' or, more properly, the precision is usually stated in terms of a percentage 'of indicated value'.

The Eros instruments are calibrated against a

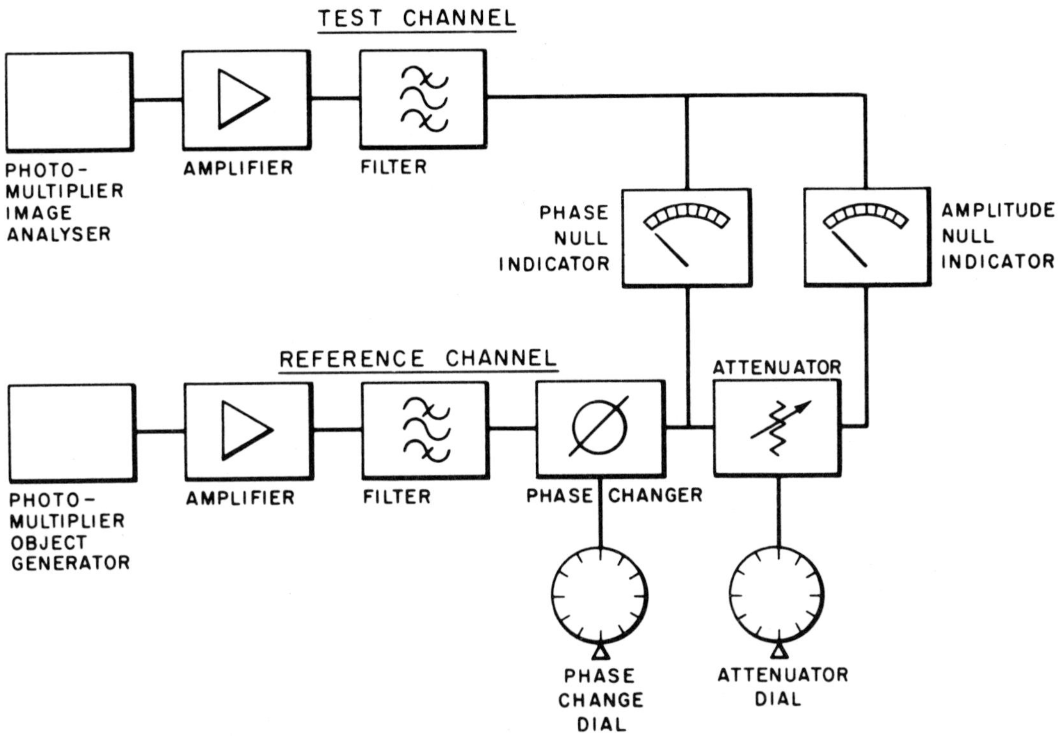

11.27.9. The electronics of the Eros IV equipment.

set of rectangular apertures whose dimensions are precisely known. The transfer functions of these apertures are directly calculable. The measured MTF and the theoretical MTF can therefore be directly compared. Typical sets of curves are shown in Figs 11.27.7 and 8. The accuracy of the Eros 100 is 5 per cent, that of the Eros III is 3 per cent and that of the Eros IV is 1 per cent.

The PTF is substantially more difficult, instrumentally, to determine than the MTF. Only the Eros IV system is equipped to measure the PTF, although the Eros III may be converted by a factory retro-fit to measure the PTF in addition to the MTF.

Image analyser

The image analyser in the Eros system is simply a high-precision microphotometer of compatible design. When testing a lens, the pattern produced by the object generator is focused on to the entrance slit of the image analyser. This slit is set at right angles to the object slit. Collecting optics behind the image slit relay the light on to the photomultiplier. The photomultiplier supplied with the Eros 100 has an S-20 response.

Eros III and IV are supplied with S-20 response photomultipliers, and a cooled S-1 response system is available as an accessory.

Electronics

The electronics units for each Eros system are entirely different in appearance and complexity. (See Fig. 11.27.9.) Each unit, however, provides both cathode-ray and X–Y recorder outputs.

The circuits are mainly solid state and of a modular design, built on plug-in printed circuit boards. Each circuit board has front-panel or easy-access test-points for rapid testing and service. Conservative design values have been used throughout.

Mounting systems

The mounting system for the Eros III and the Eros IV provides on-axis capabilities at both infinite and finite conjugate distances. (Figs 11.27.10 and 11.27.11.)

The major component of the system is an Ealing research optical bench. The specific bench must be selected on a length basis. The length chosen should correspond to the anticipated range of focal lengths and finite conjugate separations. The granite research benches provide the greatest length capabilities, but regardless of the bench selected, it will be necessary to mount the object generator for either the Eros III or the Eros IV on a base separated from the bench. This is required so that the optical centre line of accessories mounted on the bench carriers will correspond with the object generator centre-line. This separate base may be a simple poured concrete pier or a special granite mounting plinth.

The Eros III or Eros IV image analyser is

11.27.10. The Eros III system with image analyser on its slide unit and the accessory Nodal Slide mounted on an Ealing Precision Research Optical Bench. The object generator is mounted external to the bench on a poured concrete pillar.

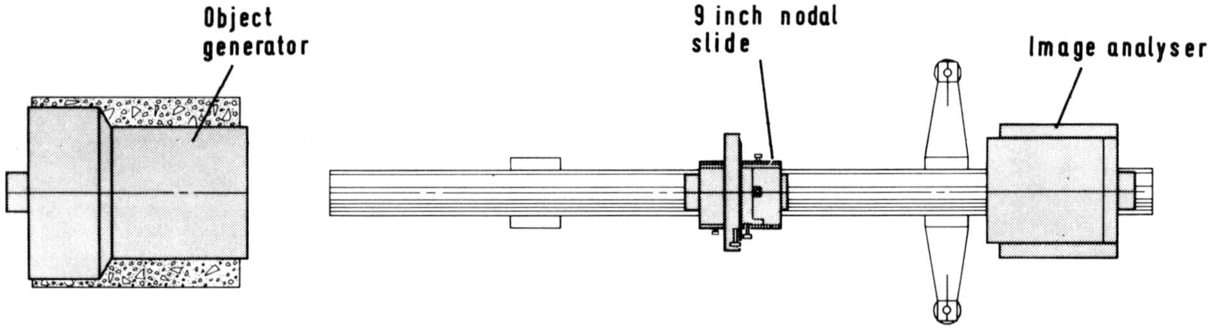

11.27.11. Eros IV system mounted on the image analysis table.

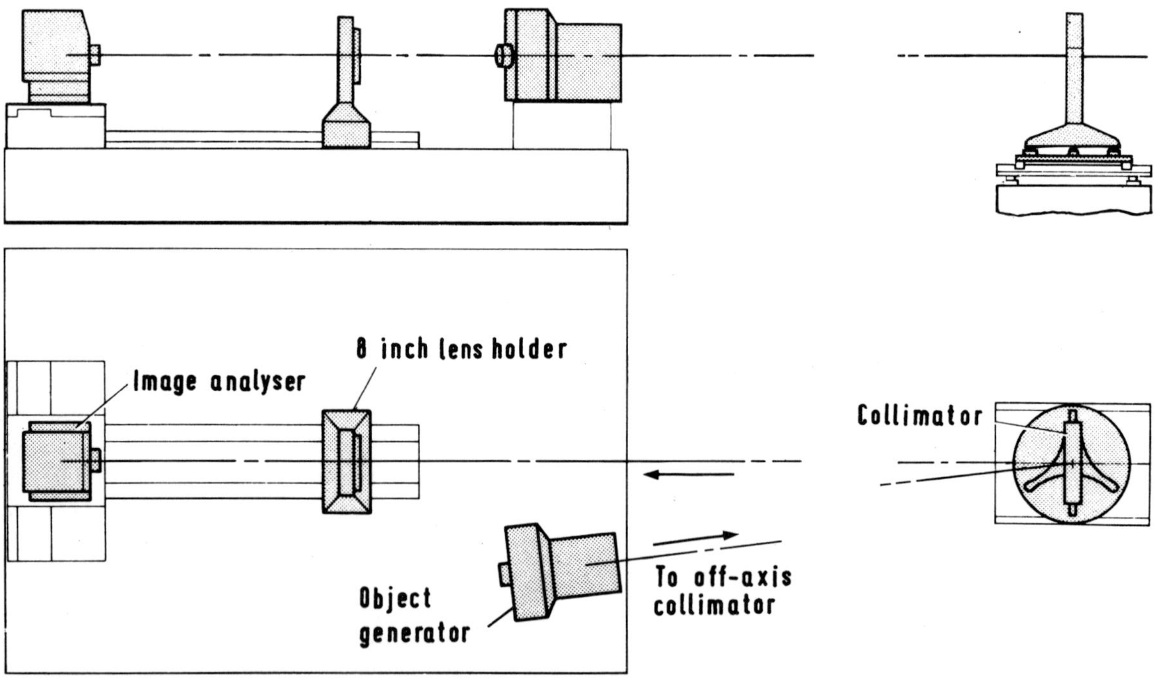

kinematically mounted on top of the field/focal slide coaxially with the lens under test. Fig. 11.27.11. indicates the general layout of the components.

The field/focal slide unit is located by a single 2-inch diameter spherical bearing set into the middle of the table. During setting the field/focal slide unit is lifted a few thousandths of an inch above the table surface by built-in air bearings. The resulting low friction allows the unit to be rotated up to $\pm 45°$ around the central ball pivot with great ease.

A unique system of time-sequenced air shut-off allows the field/focal slide unit to settle on the table surface with no sideways drift. A projected, luminous scale appears on a screen on the end of the field/focal slide unit, indicating the rotation in tenths of a degree.

In a similar fashion both the focal slide, the field slide, and the lens holder are supported by air bearings. The focal slide (with the field slide and image analyser mounted on top of it) may be moved easily over several feet. A hydraulic clamp, included in each unit, firmly locks the sliding unit in position after a setting has been made. A projection system relays an image of the appropriate stainless steel scale and associated vernier on to a screen on the side of each unit. Positions may be read to 0·1 mm.

The field slide is also mounted on air bearings with equivalent scale readouts.

The lens holder has an internal clear aperture of 8 inches. Its outside diameter is 12 inches. The lens holder mounts on the focal slide and is provided with a projected scale readout of position on the focal slide similar to those described above.

The lens holder is free to rotate and is radially supported by air bearings. Total deviation from true alignment and concentricity does not exceed 0·0002 inch. Three clamps firmly hold the unit in the desired position after setting.

The granite table is mounted on three steel jacking feet. Pneumatic vibration isolators *may not be used* directly under this system due to the external location of the collimating mirror. Care in installation should be taken to insure a temperature and humidity stabilized environment with a solid, vibration-free footing.

The diagrams in Figs 11.27.12 to 11.27.14 will more clearly illustrate the arrangement of the Eros III or Eros IV components for various test conditions.

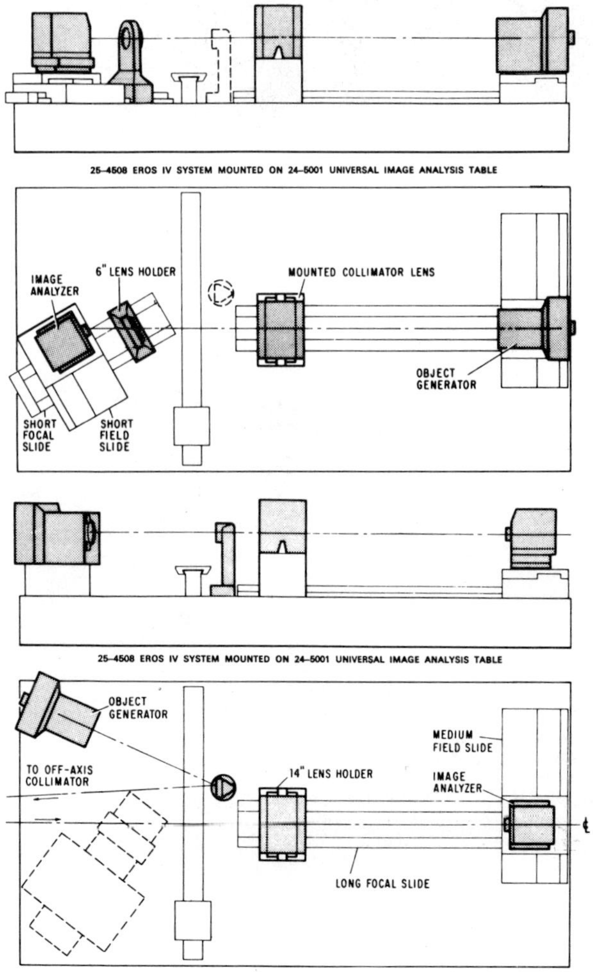

11.27.12. Testing (*top*) a short-focus lens off axis and (*bottom*) a long-focus lens on axis, both at one infinite conjugate.

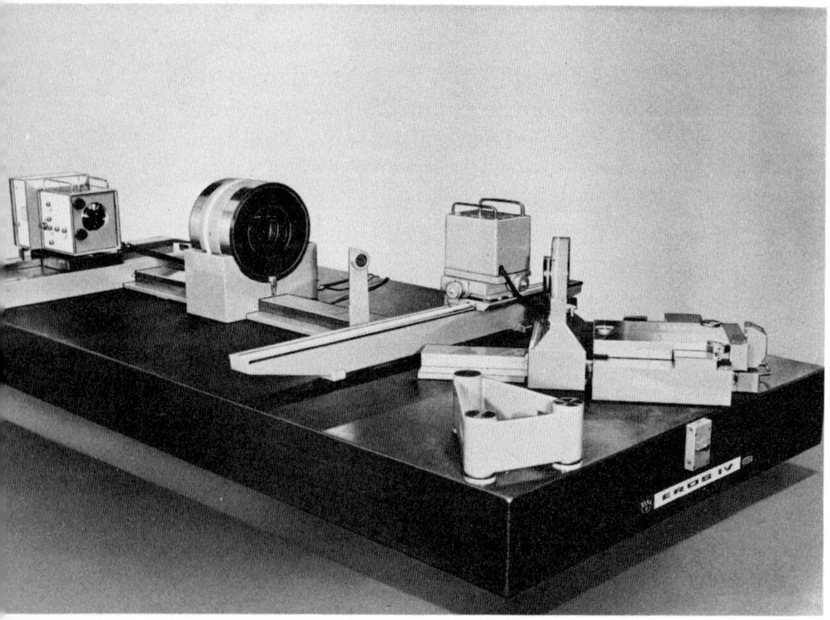

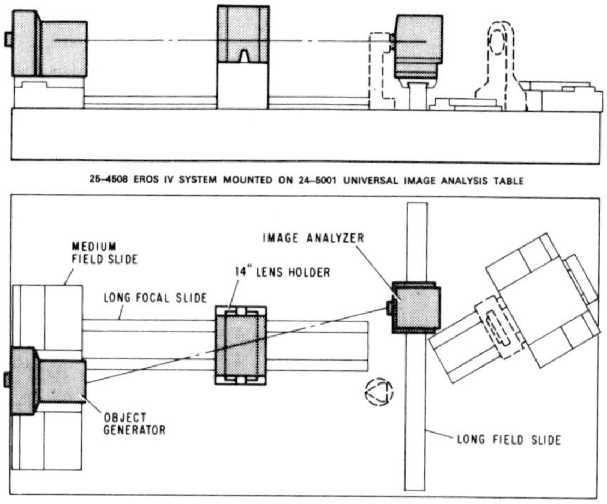

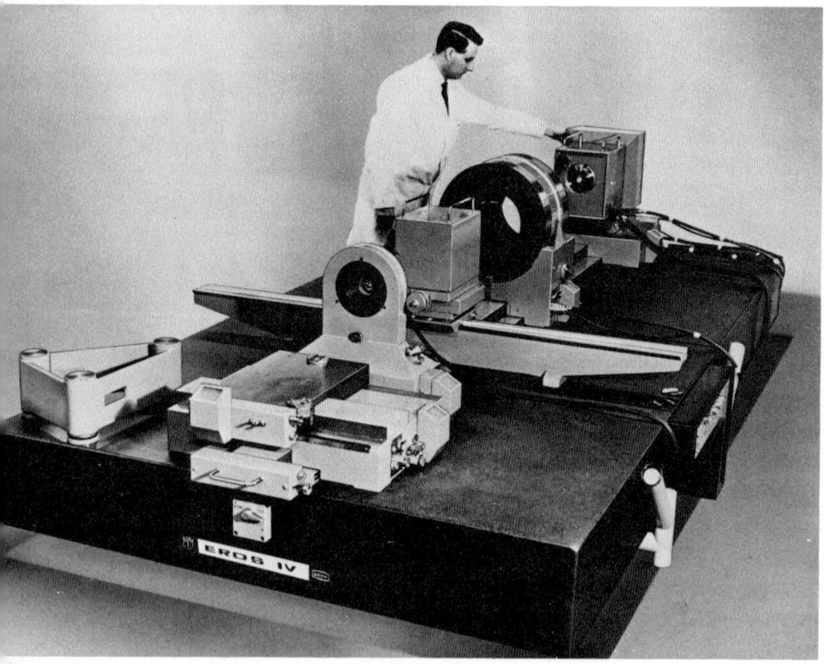

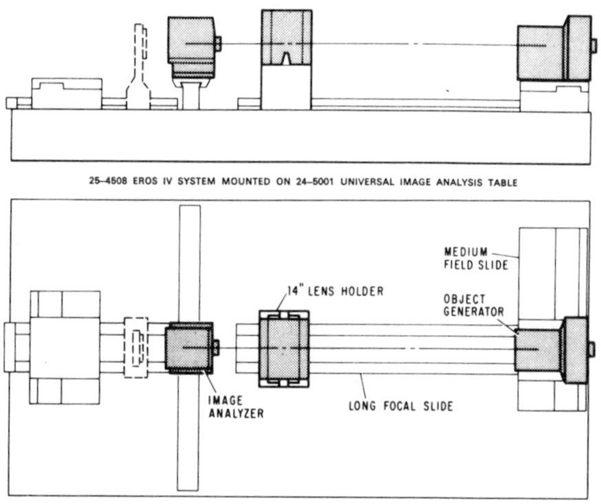

11.27.13. Testing (*top*) a long-focus lens off axis and (*bottom*) a long-focus lens on axis, both at two finite conjugates.

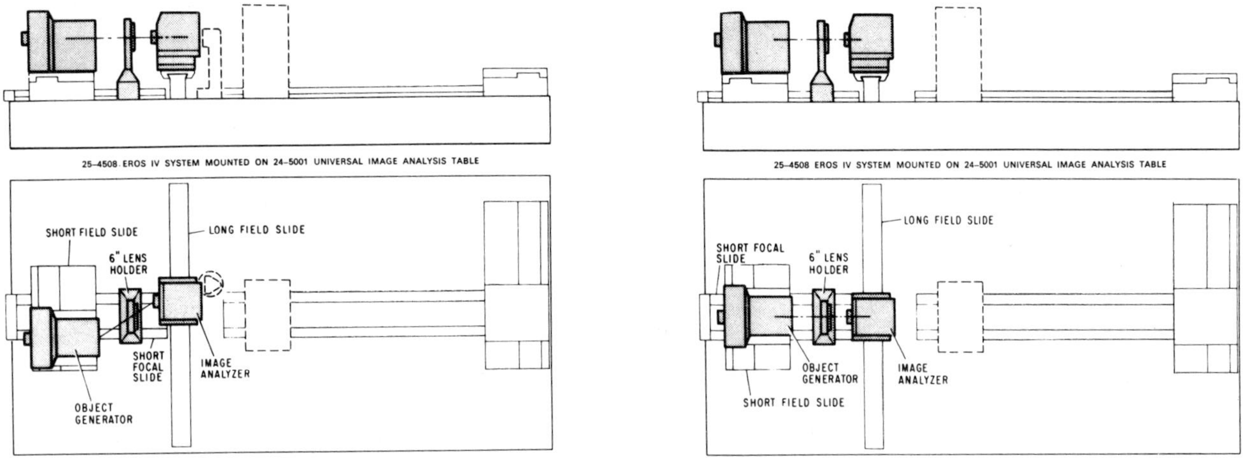

11.27.14. Testing a short-focus lens (*left*) off axis and (*right*) on axis, both at two finite conjugates.

Testing infra-red systems

OTF measurements of infra-red image converter and image intensifier systems may be made with the Eros 110 equipment which is now available both as a complete unit or as a conversion for the Eros 100 equipment. (Fig. 11.27.15.)

Since the basic Eros system is operationally reversible, the photomultiplier in the image analyser and the light source may be interchanged. No transmission optics need now be placed between the source and converter, and an infra-red source, to the air cut-off, may be used. The object generator (now working in the visible) will perform as a Fourier analyser.

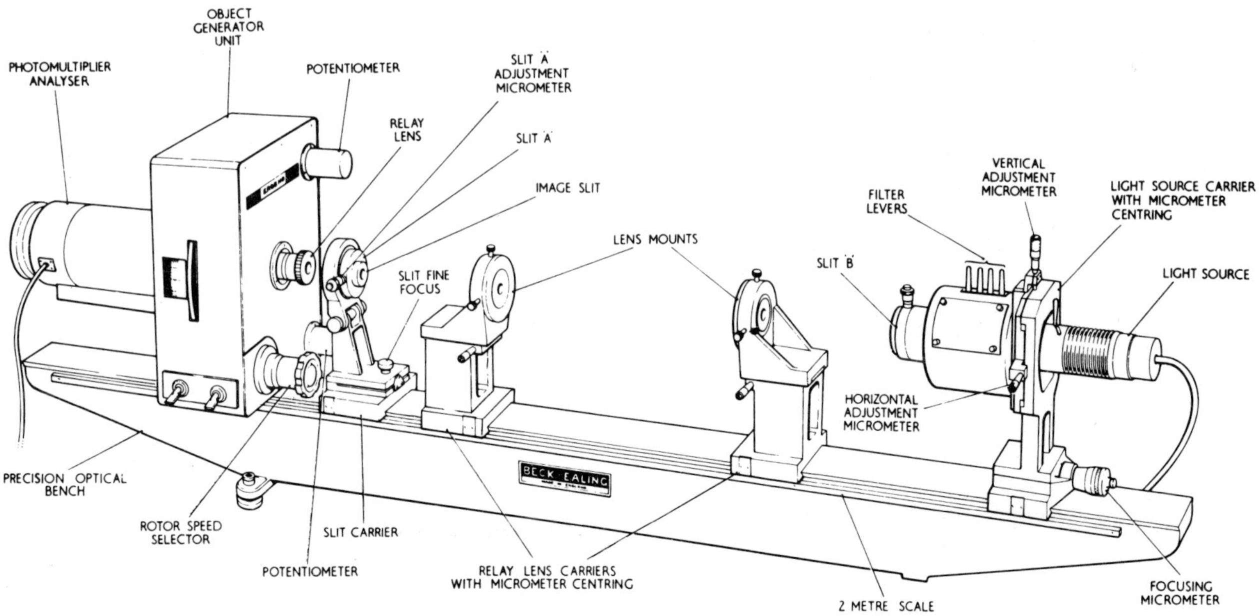

11.27.15. The Eros 100 equipment.

A GLOSSARY OF TERMS AND DEFINITIONS USED IN DISCUSSING IMAGE EVALUATION SYSTEMS

The following is a list of terms which are frequently encountered in image evaluation studies. Their definitions are still open to some dispute.

This list has been abstracted from the listing and definitions proposed by the International Commission of Optics.

Aberration

An aberration of a lens (or other optical system) is a property of the lens that causes the image formed by it to have a shape or position which differs from that produced on the basis of diffraction at the lens aperture.

Aberration of light distribution usually results in a distribution other than the Airy pattern formed by pure diffraction. Aberration of position (distortion and field curvature) results in a displacement from the Gaussian position of the image of a point.

Aperture limited

An aperture limited lens is one which is without aberration, so that the images formed by it can be predicted solely on the basis of diffraction at the aperture.

Acutance

The acutance of a photographic edge is the mean squared density gradient divided by the difference maximum and minimum density,

$$\frac{\overline{(\delta D/\delta X)^2}}{\Delta D}$$

and is the physical evaluation of subjective edge sharpness. Refer to Fig. 11.27.16 for graphic clarification of symbols. Also see 'sharpness' and 'edge response'.

Bandwidth

The bandwidth of a photo-optical system is sometimes referred to as a range of spatial frequencies which have transfer factors above zero or above some specified value.

Cascaded system

The tandem or series arrangement of photo-optical components in a reconnaissance system,

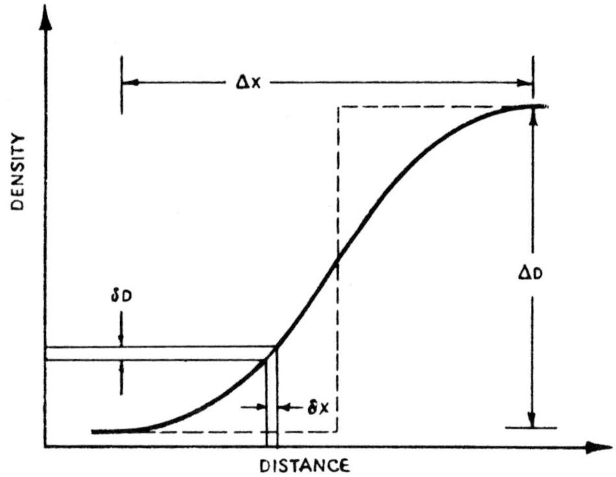

11.27.16. Density as a function of distance across a photographically recorded 'perfect' edge.

where the output of the preceding component directly becomes the input of the following component, is a cascaded system.

Depending on the nature of the coupling between the components, a cascaded system can in some cases be treated mathematically to give a combined function of those of the components. When the output of one component goes to the next as an intensity input, then the optical transfer functions can be multiplied. A cascade of several lenses gives their output-input as amplitudes, and the transfer functions cannot be combined in a simple way (unless amplitude functions are dealt with).

Contrast

Contrast is defined in several different ways. The following are most common:

(1) The contrast, C, is the ratio of the luminance of the object to its background, or $C = I_{max}/I_{min}$.

(2) In photography, since transmission, T, or density, D, is used, contrast is defined as $C = T_{min}/T_{max}$, or is given in densities as $\log C = D_{max} - D_{min}$.

(3) When using concepts involving the transfer function, contrast is sometimes defined in the same manner as modulation, i.e., $C = (I_{max} - I_{min})/(I_{max} + I_{min})$.

Contrast transfer function

(1) Contrast transfer function is a synonym for modulation transfer function, used before 1961 and still used in Germany (Kontrastübertragungsfunktion) and to some extent in France. It definitely should not be used in English but should be written 'modulation transfer function'.

(2) The contrast listed as 1 and 2 under the entry 'contrast' can be given as functions of spatial frequency, but this does not give a transfer function in the Fourier sense, and such a presentation should be avoided.

Convolution

The convolution of one function, $f_1(x)$, with a second function, $f_2(x)$, is a mathematical operation which produces a third function, $f_3(x)$, given by

$$f_3(x) = \int_{-\infty}^{+\infty} f_1(\epsilon)\, f_2(x-\epsilon)\, d\epsilon$$

The convolution of an object intensity distribution with a lens spread function produces the image intensity distribution.

Cut-off frequency

The cut-off frequency, N_0, of an optical system is that spatial frequency determined by the diffraction limit for the aperture at which the transfer function falls to zero. (Fig. 11.27.17.)

The cut-off frequency for a diffraction limited lens focused for infinity is given by $N_0 = 1/(\lambda f_{\#})$, where λ is the wavelength of light and $f_{\#}$ is the focal length divided by the aperture diameter.

Dynamic range

The ratio of the maximum signal to the minimum signal of a photo-optical system or component is defined as the dynamic range.

The limits I_{max} and I_{min} of the dynamic range of a photographic system are established by specifying a minimum discernible intensity difference, ΔI, which is a function of noise and sampling area.

Edge trace

The distribution of intensity or of density along the image of an edge is called an 'intensity edge trace' or a 'density edge trace'. The derivative of the intensity edge trace is the line spread function.

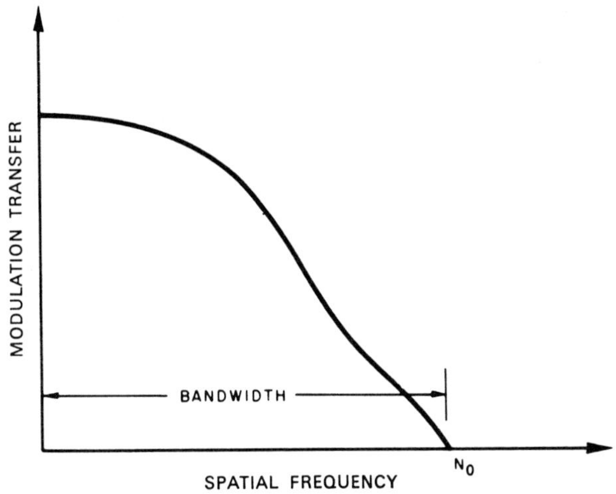

11.27.17. The transfer function of an optical system.

Encircled energy

Encircled energy is a function of a radius, r_0, in the point spread function which gives the fraction of the total flux in the image contained within the circle of the stated radius.

Frequency response

This is another name used by electronic engineers for the concept that here is defined as the 'modulation transfer factor'.

Frequency response function

See previous entry. This term corresponds to 'modulation transfer function'.

Graininess

The subjective impression or evaluation of granularity is graininess.

Granularity

Granularity is the spatial variation in the transmission and/or reflecting properties, due to the granular or particle-like structure of the developed photographic image.

Harmonic distortion

Harmonic distortion of a signal by a transfer means that not all of the harmonics in the output signal have the same relative amplitude and phase as in the input signal. It results from a nonlinear relation of the output signal to the input.

Line spread function

See 'spread function.' The line spread function is the illuminance distribution $A(x)$ in the image given by an element of a photo-optical system when the energy is incident as an infinitely narrow line.

The line spread function in a direction, x, corresponding to a point spread function is obtained from the relation

$$A(x) = \int_{-\infty}^{+\infty} a(x,y)\,\mathrm{d}y$$

A point spread function which is not symmetrical has different line spread functions in different azimuths.

Microcontrast, microdensities

When details in a photographic image are so small that the density values as defined under the entry 'contrast', 2, are affected by such factors as adjacency effects and the characteristics of the spread function, then the measured quantities are sometimes referred to as microcontrast or microdensity.

Modulation

Modulation (see Fig. 11.27.18.) is defined by

$$M = \frac{I_{max} - I_{min}}{I_{max} + I_{min}}$$

where I_{max} and I_{min} are the maximum and minimum values in a sinusoidal intensity distribution.

Modulation, M, and contrast, C, as defined under 'contrast', 1, are related by

$$M = \frac{C-1}{C+1}$$

Modulation transfer function

'Modulation transfer function' is the term recommended by the ICO Subcommittee for Image Assessment Problems (1961) for the modulus of the Fourier transform of a line spread function.

A transfer function $T(N)$ is the ratio of modulation at the output, M_{out}, of a photo-optical system or component to modulation at the input, M_{in}, as a function of spatial frequency, N:

$$T(N) = \frac{M_{out}\,N}{M_{in}\,N}$$

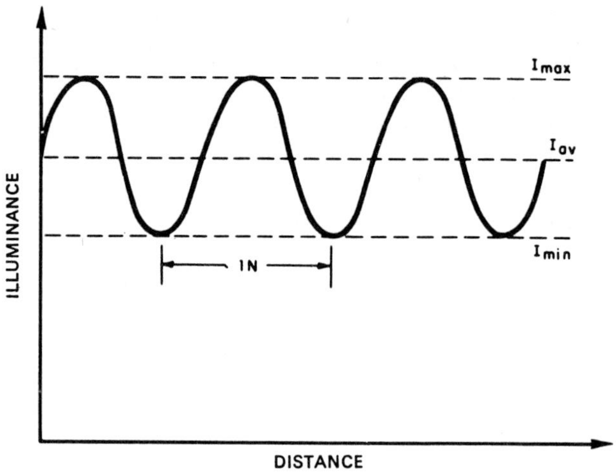

11.27.18. Sinusoidal periodic function of illuminance with distance.

A transfer factor, T, is the ratio of modulation at the output, M_{out}, of a photo-optical system or component to modulation at the input M_{in}, for a particular spatial frequency:

$$T = \frac{M_{out}}{M_{in}}$$

Noise

Noise is the ensemble of statistical variations in the energy of a system.

The type of noise that is usually relevant is the statistical background of a transmitted signal. Although granularity is not properly noise itself, it is amendable to the same sort of mathematical treatment and it gives rise to noise in transmission systems involving the photographic emulsion.

Noise power spectrum

Noise power spectrum is better defined as the Wiener spectrum of granularity (see Wiener spectrum).

Optical transfer function

'Optical transfer function' is the term recommended by the ICO Subcommittee for Image Assessment Problems (1961) for the complex Fourier transform of a line spread function.

The optical transfer function is complex; its modulus is called the 'modulation transfer function', and its phase is called the 'phase transfer function'.

Point spread function

The point spread function is the mathematical description, $a(x, y)$, of the light distribution in an element of a photo-optical system when the energy is incident as a point. Refer to 'spread function' and 'line spread function'.

Power spectrum

In optics, power spectrum, a concept introduced in electronics, is better referred to as Wiener spectrum [of . . .] because it might be related to a concept other than power. In optics, the Wiener spectrum is understood to be one-dimensional. Another equivalent term is 'spatial frequency content' [of . . .].

Rayleigh limit

When an aperture limited lens images two point objects, such that the central maximum of one image coincides with the first minimum of the other, the reciprocal of the separation between the images is the Rayleigh limit of resolving power.

The mathematical formula is

$$z = \frac{1{\cdot}22 f\lambda}{d}$$

where z is the geometrical separation of the point images, λ is the wavelength of the light, f is the focal length of the system (on the assumption that the object is at infinity; otherwise it is the image distance), and d is the linear diameter of the exit pupil of the system.

Although two points or lines closer together than this can usually be resolved, this criterion has the practical advantage of conservatism and the weight of usage, and it has the theoretical advantage that for an aperture limited lens, an infinite number of lines separated by this amount results in a uniform illumination in the image.

Resolving power

Resolving power, R, is a measure of the ability of a photo-optical system or component to produce an image of recognizable configuration at minimum size. It is measured in photographic practice by forming an image of a test object consisting of a few parallel bars on a dense background, the spaces between the bars having the common width of the bars themselves, and examining the image through a microscope under adequate magnification to determine the finest pattern in which the bars can be distinguished separately when the rounding of their ends is disregarded. The numerical value of resolving power is expressed in terms of lines per millimetre.

Compare also 'Rayleigh limit', which, however, is defined for aerial images of point sources.

Sharpness

Sharpness is the subjective impression associated with the density gradient and density difference across the image of an edge. Refer to Fig. 11.27.16, 'acutance', and 'edge response'.

Signal

A signal, S, is a photographic, optical, or electronic indication used to convey information. It consists of a spatial (density, illuminance) or temporal (voltage) distribution independent of noise.

Signal-to-noise ratio

The ratio of the strength of a wanted signal, S, to that of the interfering noise, N, is called the 'signal-to-noise ratio' (S/N). Refer to 'signal', 'noise', and 'detection'.

Selwyn granularity

Selwyn granularity, G_S, is a measure of granularity for a specific density given by

$$G_S = 1{\cdot}25\ \sigma_D \times d$$

where d is the diameter of the scanning aperture in microns and σ_D is the standard in density for that small aperture.

Sine wave

Sine wave should not be used as an adjective to express the sinusoidal nature of a function of time or space. The word 'wave' is misleading insofar as the functions it describes do not carry energy, as physical waves do.

Sine-wave response (function)

Sine-wave response and sine-wave response function are formerly used synonyms for modulation transfer function. These synonyms are no longer recommended.

Spatial frequency

Spatial frequency, N, is numerically the reciprocal of the distance between corresponding points on adjacent elements of a repetitive pattern, and it is expressed in cycles per unit length (usually cycles per millimetre). Refer to Fig. 11.27.18.

Spatial frequency content

Spatial frequency content is a synonym for the Wiener spectrum.

Spread function

The spread function is the illuminance distribution in an image corresponding to an infinitely narrow object, a point, or a line. For a lens, the spread function is a result of diffraction, aberrations, and defocus. For a film, it is due mainly to light scattering by the silver halide grains. Refer to 'point spread function' and 'line spread function'.

Square wave or bar pattern

'Square wave' is used adjectivally in the same way as 'sine wave' and is objectionable for the same reason. A good substitute would be 'bar pattern'. A square wave is an intensity distribution in space which alternately assumes two fixed values for equal distances, the distance of transition being negligible in comparison with the length of each fixed value.

In testing photo-optical systems, the number of bars in the target pattern and their length-to-width ratio should be specified.

Square wave (or bar pattern) response

If the input modulation at a given spatial frequency is that of a bar pattern, the output will be an attenuated and degraded bar modulation, and the ratio of the two is defined as the bar-pattern response.

The bar-pattern response is not a transfer function in the Fourier sense, and it is not combinable. From a given (sinusoidal) transfer function, the bar-pattern response can be calculated. The opposite, however, is not true; thus, for an experimentally obtained bar-pattern response curve, the transfer function cannot be unambiguously obtained.

Strehl criterion

The Strehl criterion, S, is a figure of merit for the image quality of an optical system. It is the ratio of the peak intensity of the spread function to the peak intensity of the diffraction pattern of equivalent aperture. Thus $0 < S < 1$.

Three-bar target

A three-bar target consists of a pattern of three parallel bars of equal width and separation, and of length equal to five times the width.

Wiener spectrum

The one-dimensional Wiener spectrum is the Fourier transform of the autocovariance (also called autocorrelation) function of any image contributing variable, such as intensity, amplitude, or granularity. If this variable is called $M(\xi)$, the autocorrelation function, $C(x)$, and the Wiener spectrum, $P(N)$, are related as follows:

$$C(x) = \lim_{x=\infty} \int_{-x/2}^{+x/2} M(\xi)M(\xi+x)\,\mathrm{d}\xi$$

$$P(N) = \int_{-\infty}^{+\infty} C(x) \exp(-\mathrm{i}2\pi Nx)\,\mathrm{d}x$$

or, because they are symmetrical,

$$P(N) = \int_{-\infty}^{+\infty} C(x) \cos 2\pi Nx\,\mathrm{d}x$$

A synonym for 'Wiener spectrum' is 'spatial frequency content'.

12 Surface Coating of Glass

12.1 Introduction

Anti-reflection coatings

By far the largest number of thin films manufactured are anti-reflection coatings, which have made complex optical systems possible by reducing light absorption and improving the image contrast.

Reflected light in an optical system appears as a veiling glare which causes a reduction in the image contrast, by illuminating regions of the image which should be dark. Lens surfaces which were planned as refractors may also behave as plane, convex, or concave mirrors and, in an optical system composed of many lenses, ghost images may be formed. Uncoated glass reflects from a little over 4 per cent per surface for a refractive index of 1·52 to over 5 per cent per surface for glass with higher refractive indices.

Although anti-reflection coatings can be of one-, two-, three-, or four-layer design, according to the characteristics required, probably 95 per cent of lenses have single-layer coatings.

Magnesium fluoride (refractive index 1·38) evaporated in high vacuum is the only material of low refractive index, which is mechanically robust and chemically stable, from which films can be made. A single coat of magnesium fluoride one quarter wavelength in thickness will reduce reflection as illustrated in Fig. 12.1.1.

A ray of light is shown moving from inside the lens through the coating and out into the air. It is partly reflected by each of the two surfaces it passes through. The momentary phase relationships are indicated. The phases of the unreflected ray are shown by a continuous line. Those of the

12.1.1. Quarter-wave coating. The reflected rays cancel each other, and thus the energy has no path by which it can leave the original wave.

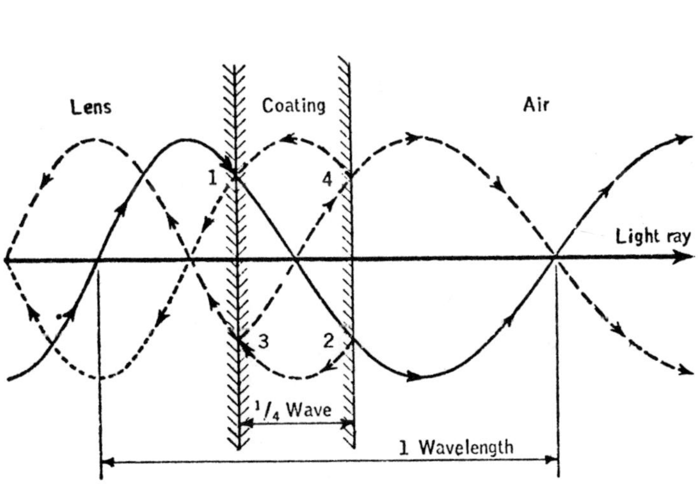

first reflection at point 1 are shown by a dotted line, and those for the remaining reflections are shown by dashed lines.

The ray that is reflected at point 2 and enters the lens at point 3 is exactly out of phase with the reflection from point 1. The resulting reflected wave in the lens is therefore very small, as it is nearly cancelled out. It will contain only the *difference* of the energy of the two. The reflected wave at point 3, which remains in the coating, is partly reflected at point 4, and, at point 1, it reinforces the first reflection or joins the original wave to repeat the cycle. If the refractive indices are properly balanced, the total energy of the two waves which move backward into the lens is zero. Then absolutely no reflection occurs; moreover, no energy is removed from the original wave.

At point 4 none of the energy can leave by moving into the air because the reflection is out of phase with the original wave.

Accordingly, under ideal conditions, none of the reflections shown can occur because each is cancelled by another reflection which is out of phase. The only wave that can exist is the original wave shown in black. Since no reflections occur, no energy is lost and the ray moves into the air without loss of energy and without creating any reflections.

Single-layer anti-reflection coatings have an advantage over multi-layer coatings in that glasses of different refractive indices can be coated simultaneously, with the same charge, and even a wrongly chosen thickness of film will reduce the reflectance. The coated reflectance will never be greater than uncoated glass. Thickness can be controlled to suit best optical results. For visual optics, such as binoculars and microscopes, the best anti-reflection property is achieved by choosing a minimum in the green, i.e. 5500 angstroms.

White light incident on the treated surface will be reflected with a purple to blue-violet colour and this is usually considered a good-quality colour. However, photographic objectives of high refractive index glass which absorb light in the short wave region (yellow-coloured glasses) require a coating with minimum reflectance at say 4500 angstroms, and this is necessary to maintain photographic colour balance. A straw-colour coating will be more efficient for this purpose.

Double-layer coatings are capable of giving a single wavelength zero reflectance and in multi-surface optical systems this refinement over the single layer coating is an important advantage. In the wavelength area adjacent to the zero of reflectance, the gain by using a double layer is lost, owing to high reflection at the edge of the curve. The advantage of zero reflection is greatest with a relatively narrow spectral region, such as green light, which is often used for illuminating optical instruments. If low reflectance over a broad band of wavelengths is required, then a three- or four-coat process is necessary with greatly increased costs and higher production risks.

An error in thickness on a single-layer film only means a shift in the minimum reflectance to another wavelength. An error on any of the coats of a multi-coat system usually implies that no minimum occurs, and possibly a reflecting surface instead of reduced reflectance is a result. This is disastrous, as coatings cannot be removed without repolishing and this is a very costly operation. Multi-coat films therefore demand better plant with adequate monitoring devices, and preferably with automatic controls to avoid the errors which can easily occur in a manually operated vacuum plant during complicated series of operations extending over 2 hours or more depending on the coating process. Some problems which prevent steady progress with the development of broad-brand anti-reflection coatings are:

(1) Glasses with high refractive index (and therefore high normal reflection at air-to-glass surfaces) require high refractive index coating materials which are not readily available.

(2) Accurate film thicknesses are essential and these are very difficult and costly to achieve on surfaces with steep curves or large areas.

Beam-splitters

Although any angle between the transmitted light and reflected light may be required, it is usual to have 45° angles with varying proportions of transmitted and reflected light according to the application. The diagonal planes of Michelson or Hilger interferometers usually have an aluminium semi-transparent surface produced by a vacuum process.

There are two physical methods by which metal

reflecting films are deposited on glass, and in both the metal is volatilized and condensed on the polished glass surface. The volatilization may be induced by bombarding the metal with gaseous ions in a moderate vacuum. In this process, called 'sputtering', the metal source remains comparatively cool. The other method is to heat the metal in a high vacuum to a temperature sufficient to make it evaporate freely.

To avoid double images, a cemented cube, instead of a simple plate, is used for beam-splitting components in optical instruments if high image-quality is required. Using a cement with the same refractive index as the glass avoids the unwanted reflections associated with metal films on glass plates.

Since beam-splitters always cause some loss of light, non-absorbing films have, to a large extent, replaced metal films (such as chromium), which may absorb half of the light. However, silver films are often used in cemented beam splitters, but apart from the 20 per cent light absorption, the coated surfaces are often fragile and easily scratched.

Non-absorbing metallic oxides are much more durable. It is possible, using a single layer of titanium dioxide, to achieve almost equal reflection and transmission with less than 2 per cent loss by absorption. If lower reflectivity and higher transmission is required, as for many types of interferometer, the thickness of the coating can be adjusted to give the reflectivity required.

Mirrors

Nearly all mirrors in general use are back-surface mirrors coated by chemical deposition of a silver film on one side of a plane glass substrate. However, unwanted reflections from the front surface make them unsuitable for image-forming systems. No such problem arises if prisms are used, and frequently a back-surface mirror is provided on the prism surface. Silver is the usual coating material as it has the highest reflectance in the visible region of the spectrum.

For wavelengths below 4000 angstroms, the reflectance of silver decreases, and aluminium must be used as a mirror coating for use in ultra-violet light. Aluminium-coated quartz dispersion prisms are used in monochromators. Front-surface mirrors must be used if good image quality is required, and the thickness of the glass substrate can be chosen to avoid distortion during polishing and figuring.

Aluminium is the most important metal for coating mirrors as it has good reflection properties from the ultra-violet to the infra-red, and also has good chemical stability. The reflectance of aluminium in the short-wave ultra-violet is dependent on manufacturing procedure, and a very high vacuum coupled with very fast evaporation of aluminium, followed by a coat of magnesium fluoride, are essential conditions to prevent the growth of an absorbing oxide film.

Front-coated aluminium mirrors are often given a protective film of silicon monoxide during the vacuum coating operation. The resulting coating has at least 80 per cent reflectivity in the visible range and, if immediately after coating the mirrors are irradiated with ultra-violet light from a high-pressure mercury lamp (without glass window) for 20 hours, the reflectivity down to 2600 angstroms wavelength will be increased to 65 per cent by making the silicon monoxide coating transparent.

By coating aluminium with dielectric films it is possible to enhance the visible reflectance to about 96 per cent and some specialized requirements justify the increased production costs.

High-intensity carbon-arc lamps and xenon lamps cause a considerable heat concentration in a cinematograph projector and it is important that as much light and as little heat as possible reach the film in the gate. 'Cold mirrors' solve this problem, as they reflect over 90 per cent visible light and transmit nearly 90 per cent infra-red, and so provide the means for dissipating heat without undesirable coloration of reflected light. Metals cannot be used for this purpose as they are good reflectors of infra-red. The Pyrex glass cold mirror is usually coated with a series of interference filters to achieve a high reflectance with an adequately wide bandwidth in the visible range.

Since 1950 a large number of books and scientific papers have been published on the application of thin films to optical components. One of the most comprehensive books on the subject is *The Vacuum Deposition of Thin Films*, by L. Holland (Chapman & Hall, London, 1956).

During this period, new designs of high-vacuum coating machines have been developed, making it commercially possible to reproduce the complex multiple thin-film coatings, which research establishments have devised for improving the efficiency of optical systems.

Interference filters

This is a very specialized subject and reference should be made to the book *Thin Film Optical Filters* by H. A. Macleod (Adam Hilger Ltd).

12.2 High vacuum coating machines and materials

Basic techniques

The material to be deposited is placed in a tungsten or molybdenum filament of suitable shape mounted on the base of a vacuum chamber, and the optical components are arranged at a certain distance around so that the surfaces to be coated face the filament. (Fig. 12.2.1.)

The vacuum chamber is lowered and pumped out until a vacuum between 10^{-5} and 10^{-6} torr is reached according to the requirement of the process. The filament is then heated by an electric current until the material evaporates. The resistance-heated filament is supplied from a low-voltage high-current a.c. step-down transformer and, as the amount of current required is high, the transformer must be designed to take the heat load resulting from continuous duty at 10 volts 400 amps or 20 volts 200 amps. A variable transformer is needed to contol the input to the power transformer primary. Examination of an ammeter in the line between the variable and power transformer enables the operator to warm up a source slowly and so prevent blasting the coating material in molten globs from the filament. The vapour molecules spread from the source in all directions in straight lines, as there are practically no gas molecules in the way with which they could collide. The vapour condenses on any 'cold' surface and a film is formed, the thickness of which depends on the duration and rate of evaporation, the geometry of the source, and the distance between the source and the coated surface.

In order to produce good adhesion between the film and the surface on which the film is to be deposited it is necessary to free the surface from molecular layers of gas, water, or grease before the film is deposited. This is usually done by chemical cleaning of the components before they are placed in the vacuum chamber and then, after pumping the vacuum, exposing the surfaces to the ionic bombardment of a high-tension glow discharge. Vacuum pressures have traditionally been measured in millimetres of mercury, but a new unit—the torr—is now in use, defined as 1/760 of a standard atmosphere. For all practical purposes, 1 mm mercury is equal to 1 torr. Another common pressure unit is the micron, which equals 10^{-3} torr.

The relationship between pressure and volume in vacuum systems is given by Charles' Law, $PV = nRT$. The number of molecules in 1 cc of gas at normal pressure is enormous and even at pressures considered very good vacuums, the number of molecules remaining in 1 cc is still large. The mean free path is the average distance a molecule travels before it collides with another molecule, and this quantity depends on temperature, pressure, and the molecule size.

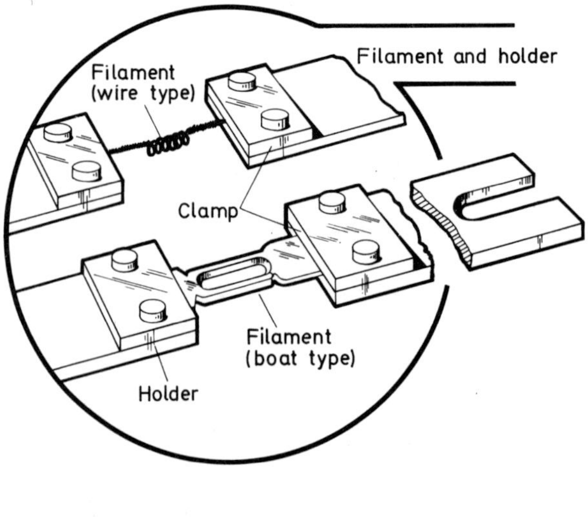

12.2.1. Comparison of the filament and boat-type holders as they are clamped in place in the vacuum chamber.

For air at 20°C the mean free path is given by:

$$\begin{aligned} L &= 6{\cdot}6 \times 10^{-5}\ \text{cm} && \text{at } 760 \text{ torr} \\ L &= 5 \times 10^{-3}\ \text{cm} && \text{at } 1 \text{ torr} \\ L &= 5\ \text{cm} && \text{at } 10^{-3} \text{ torr} \\ L &= 50\ \text{cm} && \text{at } 10^{-4} \text{ torr} \\ L &= 500\ \text{cm} && \text{at } 10^{-5} \text{ torr} \end{aligned}$$

It is sufficient to notice that the mean free path is inversely proportional to the pressure and decreases for increasing molecular diameter. On the other hand, as the gas pressure is reduced, molecules travel greater distances without colliding with other molecules.

Any material when exposed to a vacuum outgasses either from its surface, its interior, or both. Because the rate of outgassing increases with temperature, vacuum chambers are often heated to hasten gas removal. Since a finite amount of gas is contained within the material, the increased rate of outgassing due to heating permits the achievement of a given vacuum in a much shorter time than if no heat is applied. Materials used in the construction of vacuum plants must be capable of being cleaned and must not corrode or be permeable to gases. Different portions of a vacuum system may be at different pressures. In the low-vacuum areas, brass or copper is often used, whilst stainless steel and glass are used for the high-vacuum areas because their surfaces can be polished and easily cleaned. Special oils and greases have been developed for high-vacuum applications and they have very low vapour pressures at ordinary temperatures.

Any vacuum system approaches some lowest or ultimate pressure beyond which it will not pump because of leaks, diffusion of gas through the walls of the system or outgassing of materials within the vacuum chamber. The lower the pressure required, the greater is the volume of gas to be removed and this determines the design and capacity of the vacuum pump.

Two different sorts of pump are required: a mechanical pump, which is usually a rotary oil-sealed pump, and a diffusion pump in which a liquid produces a vapour stream which imparts motion to the gas molecules in the region of the pump. When in use, the liquid (usually silicone oil) is heated to boiling temperature and vapour rises up a chimney and then emerges downward. The high-speed vapour molecules collide with gas molecules which have diffused in from the vacuum chamber and this interaction gives a velocity to the gas molecules and drives them towards the bottom of the diffusion pump where they are removed by the mechanical pump. The silicone oil vapour strikes the cold walls of the diffusion pump, condenses, and returns to the reservoir.

In all diffusion pumps, small quantities of silicone vapour attempt to travel back from the pump to the vacuum chamber. This is known as back streaming and can contribute to contamination of the vacuum chamber. Baffles and traps are inserted between the diffusion pump and vacuum chamber to stop this contamination.

A variety of methods are available to measure 'pressure' in a vacuum chamber and the most important of these depend on gas ionization phenomena. Instruments for this purpose include thermionic and cold-cathode ionization gauges for total pressure measurement, as well as mass spectrometer devices for the analysis of individual components of a gas mixture. These instruments depend on the measurement of ionization currents, resulting from ionizing electron collisions with the gas, so that they really measure molecular gas density rather than pressure.

Complications arise because in practical vacuum-coating systems there is a continuous flow of gases due to evolution of gas from surfaces in the system and their removal or trapping by pumping devices. Also there are unequal temperature distributions because surfaces are heated for processing or outgassing, or cooled in cold traps.

The measurement of molecular density or equivalent pressure will therefore vary with the position of the gauge head in relation to gas flow. Both the position and orientation of vacuum gauge heads are of great importance in achieving significant measurements. The two main types of ionization gauge normally used are:

(*a*) Cold cathode, sometimes called Phillips or Penning gauges.

(*b*) Hot filament, which includes the thermocouple and the Pirani gauge.

Operation of these gauges is based on collisions between molecules and electrons with sufficient energy to ionize the molecules. If pressure is below 10^{-3} torr ion current is a direct measure of pressure. Cold-cathode ionization gauges cannot be degassed as rapidly or conveniently as hot-filament gauges, but they are less expensive, as no electronic amplifiers are required and there are no filaments to burn out.

Hot-filament gauges utilize the dissipation of heat by a sensing element placed in the vacuum chamber. Heat loss is by conduction and radiation and, as conduction depends on the pressure of the gas, the gauges can be calibrated to indicate pressure. The Pirani gauge uses a Wheatstone-bridge circuit with one leg of the bridge, usually a heated tungsten filament, enclosed in the vacuum system. Any change of pressure causes a change in filament temperature and wire resistance, so that the imbalance is a measure of the pressure in the vacuum chamber.

In order to cover the pressure range of 1 to 10^{-8} torr, it is usual to fit an ionization gauge for the higher pressure range $1–10^{-3}$ torr, and a Pirani gauge for the lower range $10^{-2}–10^{-8}$ torr. As the Pirani gauge will be monitoring the actual coating cycle usually at a pressure not less than 10^{-6} torr, it is frequently connected with a strip-chart recorder (which may have limit switches) to establish details of the vacuum cycle.

Controlled deposition of optical films

Except for single-surface blooming (anti-reflection coating) where exact uniform coatings are not so essential, it is usual to use vacuum plants designed with rotary work holders permitting the uniform sequential or simultaneous deposition of evaporated or sputtered deposits of any combination of optical materials.

The film thickness is either monitored optically by transmitted or reflected light using a photometer, or by a film-thickness monitor which measures thickness and rate of deposition using an oscillating quartz crystal. For the accurate control of individual film thickness on multi-layer interference systems, a modulated beam photometer, such as that made by Edwards High Vacuum International, is essential.

The Speedivac modulated beam photometer provides a method of controlling the optical thickness of films deposited by evaporation or sputtering by indicating the changing optical characteristics of the films as their thickness increases. The instrument measures the reflection from or the transmission through coated glass surfaces as a function of wavelength in the spectral range 4000 Å to 10 000 Å. Both these quantities can be measured alternately if two light sensing elements are used, selection being made by a simple change-over switch.

The photometer head illustrated in Fig. 12.2.2 shows photocell assemblies positioned to measure both reflectance and transmittance.

Light from a krypton-filled tungsten filament lamp passes through a collimating system and is modulated at $33\frac{1}{3}$ Hz by an interrupter disk driven by a synchronous motor. After passing through a beam-dividing mirror (where half of the light is lost, being reflected on to a black velvet absorbing surface) the modulated beam passes through a sealed glass window and into the chamber. Inside the chamber it is directed by beam guide tubes and mirrors to fall normally on to the work surface. The transmitted light continues on to be measured at the transmission photocell, and the reflected fraction passes back into the photometer head where it is reflected on to the reflection photocell by the beam-dividing mirror.

Wavelength selection is accomplished by a control filter, which has a suitable transmission band, interposed between the beam divider and the photocell.

The use of modulated light eliminates interference from daylight and by radiation from the evaporation source. The sensitivity of this instrument covers the range of minimum light levels specified for anti-reflection coatings on glass surfaces to maximum light levels as in the measurement of 100 per cent transmittance. The use of two photocells arranged to accept reflected and transmitted light respectively, permits measurement in rapid sequence of both reflectance, within an accuracy of 0·1 per cent, and transmittance, within 1·0 per cent.

For control of rate of deposition of optical films, or measurement of thickness when optical transmission is not critical, a quartz crystal film-thickness monitor is essential. The Edwards film-thickness monitor measures, on separate meters, the thickness and rate of deposition of thin films, using an oscillating quartz crystal positioned inside the coating chamber, and a reference crystal outside. As deposition proceeds, the frequency of the crystal in the coating chamber changes as the result of evaporant deposited upon it, whereas the frequency of the external crystal remains constant. The measurements taken are those of the frequency differences between the two crystals.

The thickness range and other parameters of the film-thickness monitor are dependent on the density of the evaporant, chamber temperature, monitor-crystal position, and other variable conditions, so that most data must be given in terms of frequency shift or meter-scale divisions per second. For extreme accuracy, calibration of frequency shift against film thickness by an independent absolute method, such as optical interferometry, is required for each crystal, but a graph supplied with the instruction manual can be used for most practical coating applications.

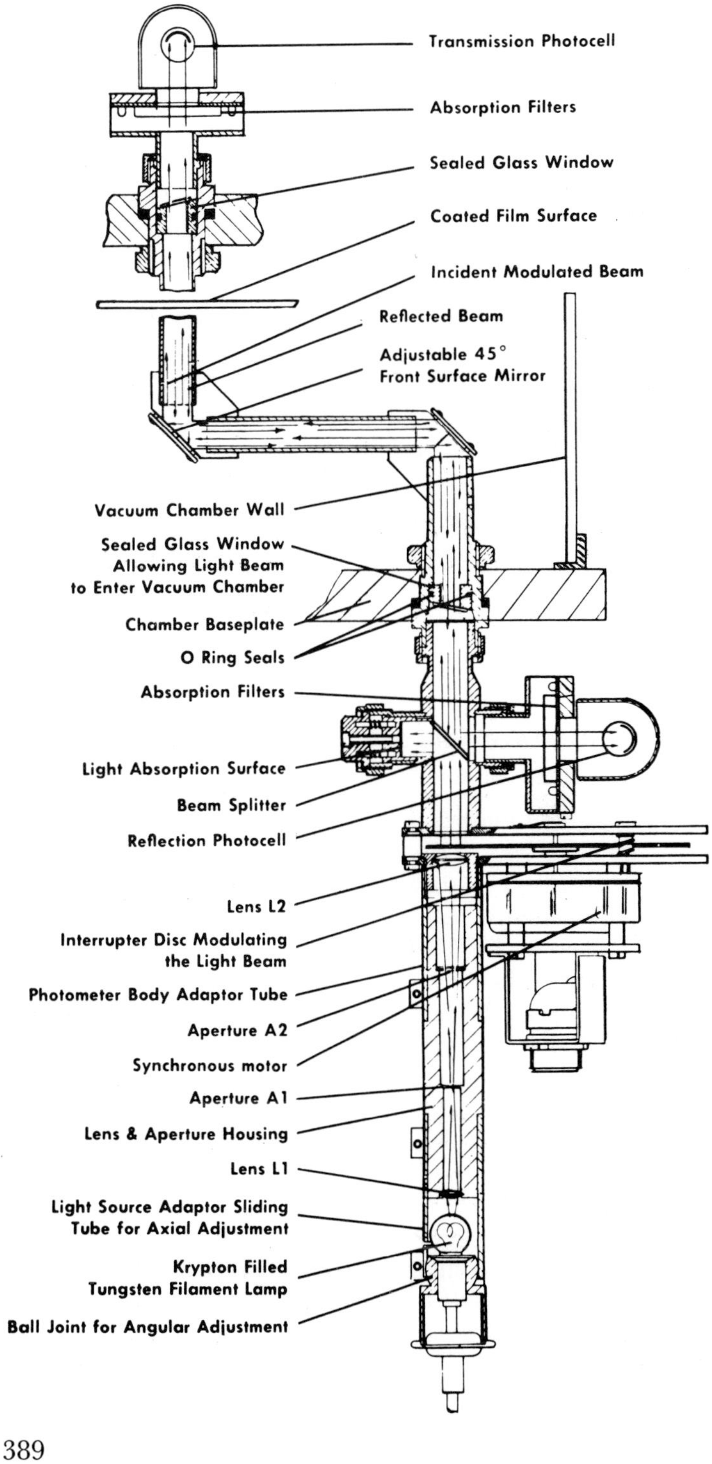

12.2.2. The photometer head of the Speedivac.

12.3 Cleanliness of vacuum coating machines and equipment

To ensure the continuous efficient operation of a coating plant, it is necessary to keep the following rules:

(*a*) Never allow air to enter the vacuum chamber through the pumps, but always through the inlet needle valve directly into the chamber.

(*b*) Never leave the plant standing idle overnight or for any other long period unless it is pumped down to a vacuum.

(*c*) At least once a month, the vacuum chamber should be removed from the plant and cleaned with FF emery cloth down to the bare metal. The metal should therefore have a polished appearance.

(*d*) Each month, the work-holders and electrode assemblies should be completely dismantled and cleaned to a polished appearance with FF emery cloth. A spare electrode assembly should be available to save machine time during the cleaning operation.

(*e*) During cleaning, a cover should be placed over the diffusion pump and paper held over the pressure gauge heads with rubber bands to avoid contamination.

(*f*) When coating is in progress, there should be no vacant positions in the coating jigs, or magnesium fluoride (or other coating material) will deposit on the radiant heater assembly and subsequently be re-evaporated. Spare positions in the coating jigs must be covered by blank plates.

(*g*) Before being used for coating, the vacuum chamber and work-holders must be cleaned by pumping down on the rotary pump to 0·1 torr, followed by a glow-discharge clean on the diffusion pump. If the discharge does not look magenta coloured within 45 minutes, the baffle should be fully opened and the radiant heater switched on. The electric input to the radiant heater should be controlled to keep the pressure below 0·3 micrometres on the Philips gauge until the temperature reaches 250°C. The glow discharge should then be continued for another 20 minutes.

(*h*) The shutter should be kept clean from loose powder or it may drop back into the boat and cause excessive spitting when the powder is heated.

(*i*) The thermocouple should be attached to the workholder and dummy runs carried out to measure lens temperatures with various workholder assemblies.

12.4 Materials used in vacuum coating departments

Tungsten boats

These tungsten boats and/or strip for evaporation sources are supplied by:

Balzers Aktiengesellschaft, FL 9496 Balzers, Principality of Liechtenstein. (Balzers High Vacuum Ltd, Berkhamsted, Herts).

Metallwerk Plansee GMBH, Reutte, Tirol, Austria. (Elvants Ltd, London, W13).

The Tungsten Manufacturing Co. Ltd, Portslade, Sussex.

The R. D. Mathis Company, 1345 Gaylord Street, Long Beach, California, 90813.

Molybdenum boats

These are supplied by:

Balzers Aktiengesellschaft, FL 9496 Balzers, Principality of Liechtenstein.

Murex Ltd, Rainham, Essex.

The R. D. Mathis Company, 1345 Gaylord Street, Long Beach, California, 90813.

Dielectric evaporative materials

These materials are supplied by:

Balzers Aktiengesellschaft, FL 9496 Balzers, Principality of Liechtenstein.

B.D.H. Laboratory Chemicals Division, Poole, Dorset.

Potter & Clarke, Ltd, Barking, Essex.

Levey West Laboratories Ltd, Harlow, Essex.

Union Carbide Ltd, Alloys Division, 8 Grafton Street, London, W1.

Metal evaporative materials

These are supplied by:

Johnson Matthey & Co. Ltd, 73–83 Hatton Garden, London, EC 1.

Imperial Metal Industries Ltd, PO Box 216, Witton, Birmingham 6.

Hopkins & Williams Ltd, Chadwell Heath, Essex.

Cleaning materials

(See also § 4.19.) Cleaning materials are obtainable as follows:

Quadralene Laboratory Detergent from Quadralene Chemical Products, Liversage Street, Derby.

DECON 75 detergent from:
Medical & Pharmaceutical Developments Ltd, Shoreham-by-Sea, Sussex.

Iso-propyl alcohol from:
Hopkin & Williams Ltd, Chadwell Heath, Essex.

12.5 Cleaning glass surfaces before coating

Preliminary cleaning processes are described in § 4.19, but, before coating, it is essential that optical components be chemically clean and, as far as possible, without any dust particles on the surface.

The stages may be:

(*a*) Clean in detergent.
(*b*) Clean in *iso*-propyl alcohol.
(*c*) Mount components in vacuum chamber and remove dust with static eliminator.
(*d*) Discharge clean before evaporation.

(a) Cleaning in detergent

Stage (*a*) has been described in § 4.20, and should always be followed by *iso*-propyl alcohol treatment if good adhesion of the coating is required.

(b) Cleaning in iso-propyl alcohol

The optical components should be wiped before cleaning in *iso*-propyl alcohol, and placed in a jig designed to hold them during treatment in the ultrasonic and vapour tanks. (Fig. 12.5.1.)

Mr T. Putner of Edwards High Vacuum International Reseach Laboratories, carried out some experiments in 1959 on the cleaning efficiency of carbon tetrachloride, trichlorethylene, and *iso*-propyl alcohol. After washing in detergent the specimens were suspended in solvent vapour for periods ranging from 15 seconds to 15 minutes. Maximum cleaning was achieved after an immersion period of 2 minutes, which generally corresponded to the time taken for the condensing vapour to heat the glass to the temperature of the saturated vapour so that condensation ceased. Obviously the greater the thermal capacity of the glass the longer would be the time for which condensing vapour washes the exposed surface. If the specimen had a small thermal capacity and a large area then condensation could cease before the surface was clean and it would be necessary to immerse the specimen in the vapour a number of times.

The highest value for the coefficient of friction obtained by vapour degreasing was 0·6 using *iso*-propyl alcohol as the cleaning agent.

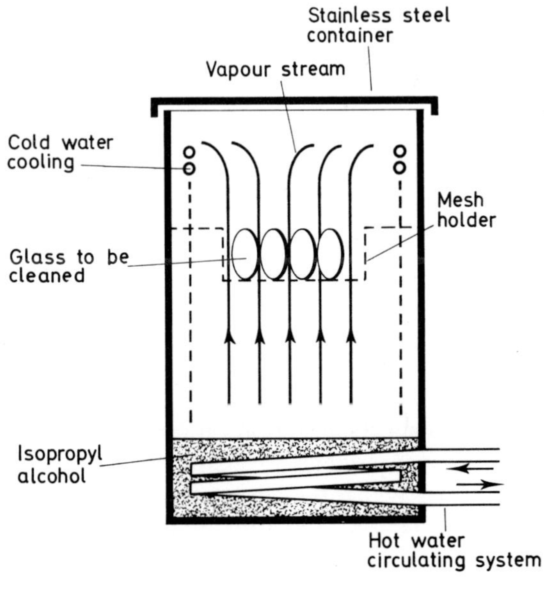

12.5.1. Vapour degreasing apparatus for cleaning glass.

12.5.2. On the right of the photograph is a jig loaded with lenses in *iso*-propyl alcohol liquid contained in a 25 kHz ultrasonic tank. In the centre is an *iso*-propyl alcohol vapour cleaning tank with its cover open. On the left, the loaded jig on the worktop has passed the liquid and vapour cleaning stages and is being blown with clean ionized air to remove static and so avoid the attraction of dust.

The *iso*-propyl alcohol vapour cleaning tank (Fig. 12.5.2) should be constructed from stainless-steel sheet and nickel-plated copper pipe. Tank dimensions suitable for most optical components are an overall size of 500 mm long, 300 mm wide, and 800 mm height including the heater chamber.

Although hot water in coiled pipes can be used for heating the *iso*-propyl alcohol, it is possible to use an electric heater with thermostat controls. At least five coils of cold-water pipe for vapour cooling are required and, of course, the water inlet must be at the top and outlet at the bottom of the cooling coils. A hinged cover must be fitted over the top of the tank when it is not in use and an extract duct with fan is needed to take away fumes.

(c) Mounting in vacuum chamber and removing dust with static eliminator

Glass which has been cleaned by *iso*-propyl alcohol vapour or other means usually has a static charge which attracts small dust particles. These do not matter on anti-reflecting coatings, but masks or films for precision ruling may be useless if they have pinholes caused by dust on the surface during the evaporation process.

To achieve pinhole-free coatings, the dust must be removed whilst the components are in the jig before the coating chamber is lowered. This is very difficult to carry out if the glass is charged with static electricity. One method is to use a radioactive brush (see § 6.11), but another and more satisfactory solution is to use a static eliminator, such as the Neutrostat air gun made by Meech Static Eliminators Ltd, 150 Clapham Manor Street, London, SW 4 (in association with Simco Company Inc., U.S.A.).

The function of the Meech Static Eliminator is to provide ionized air wherever it is necessary to remove a build up of static electricity. The air supplied to the gun must be filtered, clean, and dry. The air pressure must not be too high or the air friction on the glass will build up fresh electric charges. Generally 15 lb per square inch (1 kg per square cm) and an operating distance of 100 mm are satisfactory.

(d) Discharge cleaning before evaporation

Discharge cleaning conditions vary from one plant to another, but generally the glow-discharge electrodes are made of aluminium because this metal has a low sputtering rate. As a guide the following conditions will apply to an 18–20-inch (450–500 mm) diameter vacuum chamber.

H.T. voltage	1600 volts
Current	200 milliamps
Air pressure	0·02 mm of mercury
Time	5 to 10 minutes

Air should be flowing through the chamber

during the whole discharge. When glass is in a glow discharge, its surface is bombarded by electrons, positive ions, chemically active atoms and molecules, neutral molecules of high thermal energy, and radiations covering a range of wavelengths. There is no simple explanation of this cleaning process.

The results given in Table 12.5.1 were obtained by Mr T. Putner, of Edwards High Vacuum International Research Laboratories:

Table 12.5.1 Coefficients of static and kinetic friction of glass on glass after solvent and detergent cleaning (*after Putner*, 1959)

Method of cleaning	*Coefficient of friction*	
	Static	*Kinetic*
Vapour degreased in *iso*-propyl alcohol	0·5–0·64	0·4–0·62
Vapour degreased in trichlorethylene	0·39	0·31
Vapour degreased in carbon tetrachloride	0·35	0·28
Teepol washed and cloth dried	0·07	0·04
Teepol washed and glow discharge cleaned	0·8	0·6
Teepol washed, cleaned with alcohol and chalk, wiped with cotton wool	0·33	0·27
Teepol washed, cleaned with alcohol, wiped with cotton wool and glow discharge cleaned	0·8	—
Vapour degreased in *iso*-propyl alcohol and glow discharge cleaned	0·8	0·6
Teepol washed, cleaned with alcohol and chalk, wiped with cotton wool and flamed with gas flame	0·41	—

The value given for each cleaning method is a mean result for several tests.

12.6 Single-coat anti-reflection surfaces

A thin layer of magnesium fluoride, MgF_2, with thickness of one-quarter wavelength should be evaporated on to the optical surfaces, and as a result the reflectivity will be reduced as follows:

Refractive index of glass	*Reflectivity (percentages)*	
	Untreated surface	*Bloomed surface*
1·52	4·3	1·24
1·62	5·6	0·64
1·75	7·4	0·17

During the pump down of the plant and until after discharge cleaning (see § 12.5) the radiant heater (preferably a 3-kW sheathed-element radiant heater with temperature controlled by a simmerstat) should be heating the blanks to a temperature high enough to produce a hard durable coating, but not so high as to damage the figure on the surface or change the refractive index of the glass.

As a general rule, crown glass can, if necessary, be heated to 450°C, and flint to 320°C without damage but, as there may always be a mixed load, the blank temperatures should not normally exceed 320°C. As temperature measurements are taken from the jig surface and not the glass surface it is necessary to calibrate (see § 12.3) the thermocouple and jig assembly. The blanks may be at a higher temperature than the jig, due to the greater absorption rate of the glass to radiant heat.

Direct experiments with different size optical components will establish the correct simmerstat readings for the required blank temperatures. When the correct temperature has been reached and the vacuum is better than 2×10^{-5} torr, the evaporation of the magnesium fluoride can commence. As the magnesium fluoride is deposited, the reflected light from a tungsten-lamp filament goes through a series of colour changes when a monitor plate is viewed at or near normal incidence.

Each colour change corresponds to the correct coating for a particular wavelength or colour of light for which the optical component will be used. As soon as the correct reflected colour is achieved the shutter must be covered over the filament and, after coating, air can be let into the coating chamber, and the components removed. It is

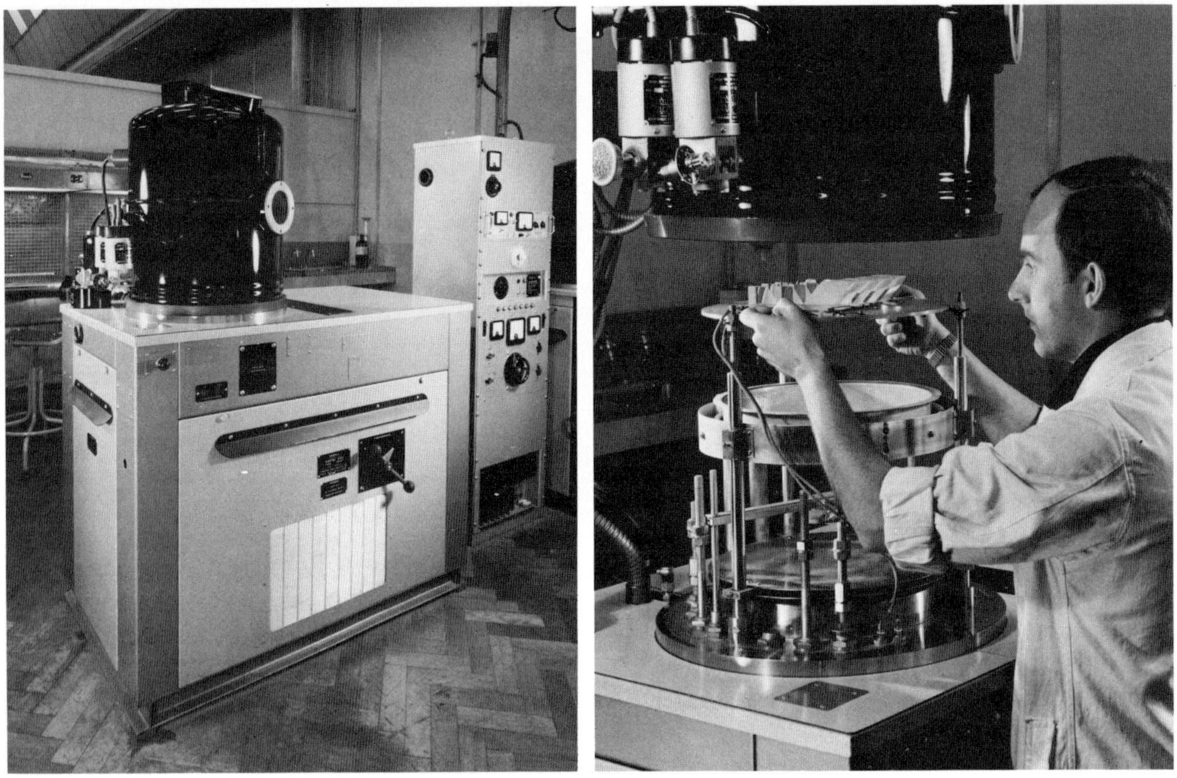

12.6.1. Edwards 19-inch plant with automatic pump down and a simmerstat for controlling the temperature of a radiant heater set-up for single-coat blooming. The operator is loading the vacuum chamber with prisms for single-coat blooming with magnesium fluoride.

essential that sufficient time be allowed for cooling before letting in air to the chamber or the glass components will crack. This cooling time for large lenses and prisms may be several hours and this risk of cracking, during heating and cooling, is one disadvantage of a hot process.

The colours associated with peak efficiency at various wavelengths are given opposite.

The monitor plate mounted with the work can subsequently be used in an abrasion test to establish the hardness of the coating.

Table 12.6.1 Monitor plate colour changes

Colour reflected	*Best efficiency* *Wavelength*	*Colour*
Very pale straw	3000–4000 Å	Near ultra-violet
Pale straw	3500–4500 Å	Violet
Straw	4000–5000 Å	Blue
Reddish purple	4500–5500 Å	Blue-green
Purple	5000–6000 Å	Green
Bluish purple	5500–6500 Å	Yellow
Blue	6500–8500 Å	Red
Blue-green	0·8–1 m	Near infra-red
Yellow-green	0·9–1·1 m	Near infra-red

12.7 Two-layer anti-reflection coatings

This process can be varied to give a minimum reflectance to suit a required wavelength. The vacuum plant must have sputtering electrodes, a rotary table for producing uniform films and a modulated beam photometer for control measurements. A plain glass monitor plate is required in the rotating jig of the same refractive index as the components to be coated and situated above the

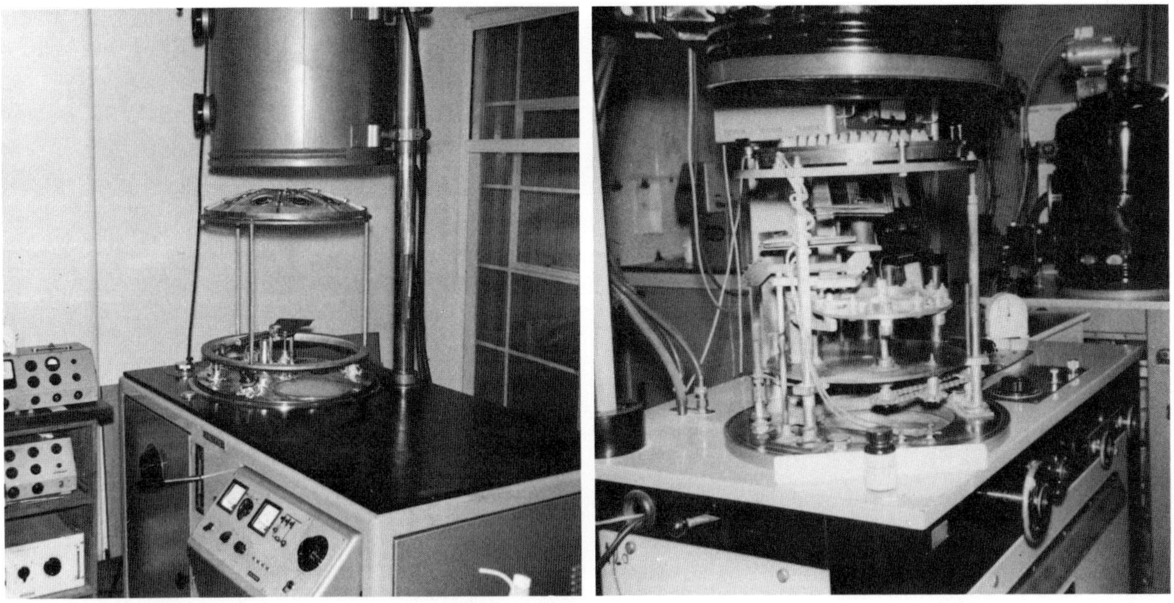

12.7.1. (*Left*) Balzers BA 710 machine with a 700-mm chamber set up for anti-reflection coatings. The pumping system is semi-automatic and single-lever operated. (*Right*) The Edwards 18-inch vacuum chamber loaded with prisms for two-layer anti-reflection coating of sputtered bismuth oxide and evaporated magnesium fluoride.

photometer. After loading the vacuum chamber with components, placing bismuth oxide powder on the sputtering cathode and magnesium fluoride on the molybdenum boat, the vacuum chamber should be evacuated to a pressure of better than 10^{-4} torr.

The work-holder should be rotated at 100 r.p.m. and, with the photometer set to read 8 per cent reflection, oxygen is leaked into the chamber through the needle valve to raise the pressure to 0·3 torr. Sputter bismuth oxide until the reflectivity increases from 8 per cent (the normal 4 per cent each side) to 12 per cent. The glow-discharge dark space should be approximately 25 mm above the cathode surface. Stop sputtering (after about 4–5 minutes) and close the needle valve. Pump the vacuum down to better than a pressure of 10^{-4} torr and preheat the work by switching on the radiant heater with the simmerstat set to a temperature of 250°C.

Set the photometer to 60 per cent reflection and evaporate the magnesium fluoride. The meter reading will rise initially and fall to a minimum point when the coating process must be stopped. The substrates must be allowed to cool before air can be admitted into the vacuum chamber.

Permission has been given by Dr L. Holland to publish extracts from a paper written in 1958 entitled 'Two-layer Anti-reflection Coating for Low Refractive Index Glass', by L. Holland and T. Putner of the Central Research Laboratory, Edwards High Vacuum International, Crawley, Sussex.

Introduction

To reduce the reflectance of a glass of low refractive index (approximately 1·5) to zero at a given wavelength with a $\lambda/4$-film requires a film material with an index of about 1·23, whereas dielectric materials which are capable of forming durable evaporated coatings, such as magnesium fluoride, have indices above 1·3. Consequently, it has not been practical to efficiently reduce the reflectance of a low index glass with a single-layer film. Several types of multilayer interference systems have been proposed for use on low index glass requiring the deposition of high and low index materials. However, such systems have not completely overcome the problem because it has proved difficult to evaporate high index films which are both durable and free from optical absorption. Zinc sulphide films are currently used as high index layers in multi-layer filters but they are not sufficiently durable for use in anti-reflectance layers on the exposed surfaces of optical elements. Vogt and later Hass have prepared durable films of high index by evaporating cerium dioxide, but this substance requires very high temperatures for its volatilization and elaborate pre-treatment before it can be evaporated at a controlled rate.

The authors have attempted to overcome these problems by using high index films of bismuth oxide prepared by sputtering bismuth in oxygen, because such films are reasonably durable and their deposition rate is easy to control. Holland has described a vacuum plant with which it is possible to sputter and evaporate thin films on to the same work-holder in sequence and this apparatus has been adapted for preparing two-layer anti-reflection coatings of sputtered bismuth oxide and evaporated magnesium fluoride. Given below is an account of the preparation and properties of these anti-reflection coatings using the modified apparatus.

Experimental apparatus

The apparatus used for depositing the two-layer anti-reflection coatings is shown in Fig. 12.7.2. The glass to be coated is mounted on a rotary work-holder above a pair of sector-shaped cathodes for sputtering and a molybdenum boat for evaporation.

The sputtering electrodes, shown in detail in Fig. 12.7.3, are of sector shape and made from $\frac{1}{8}$-inch thick copper sheet on which bismuth has been spread whilst in the molten state. Copper piping is brazed to the underside of the electrodes for water cooling to prevent the bismuth melting during sputtering. The cooling water for each sector electrode is taken in and out of the chamber by a pair of concentric tubes which pass through a vacuum sealed and insulated lead-in electrode in the base-plate. The cooling pipes on each sector cathode are connected inside the vacuum chamber to the water supply tube by a union with a rubber O-ring seal.

The sector electrodes are connected to the output terminals of a high tension a.c. transformer and operate as cathodes which sputter on to the work-holder on alternate cycles. The underside of the sector cathodes and the outside of the water-cooling tubes are screened with earthed plates mounted within the cathode dark space to prevent sputtering of the copper fittings.

When the sector cathodes are mounted on the same plane the glow discharge at the tips of the electrodes is partially suppressed, owing to the electrons in the cathode dark space being obstructed by the electrode serving as the anode on alternate cycles. This decreases the sputtering rate at the centre of the electrode so that the deposited film is thinner at the centre. The effect is avoided by mounting the two electrodes on separate levels, as shown in Fig. 12.7.2, so that each electrode does not hinder the cathode dark space of the other. The sector electrodes are arranged so that they do not unduly impede the vapour stream arriving at the rotary holder from the vapour source arranged under the edge of the holder. If the electrodes are mounted near to the work-holder the centre of the evaporated deposit shows a series of concentric rings of different interference colours where the electrodes cast shadows on the holder.

It has been found that sputtered bismuth oxide films are free from light absorption, providing they are sputtered at a high current density with oxygen continuously flowing through the coating vessel. Satisfactory conditions were obtained with a current density of 10 mA/cm^2 of electrode area and an applied voltage of 3 kV. Using the dimensions shown in Figs 12.7.2 and 12.7.3, the oxide was deposited at a rate of 1 Å/s.

The thickness of the deposits was controlled by measuring their reflectance during deposition, using a modulated beam photometer capable of reading to an accuracy of $\pm 0{\cdot}1$ per cent. The reflectance of the resultant films was also measured with the same photometer as a function of wavelength using a series of Wratten filters. The glass specimens were green plate glass with a refractive index of 1·5. They were cleaned with a detergent and by vapour degreasing in *iso*-propyl alcohol vapour.

Optical properties

A simple anti-reflection system for use on low index glasses has been proposed by Osterberg, Kashdan, and Pride. The method consists in adjusting the optical thicknesses of the layers so

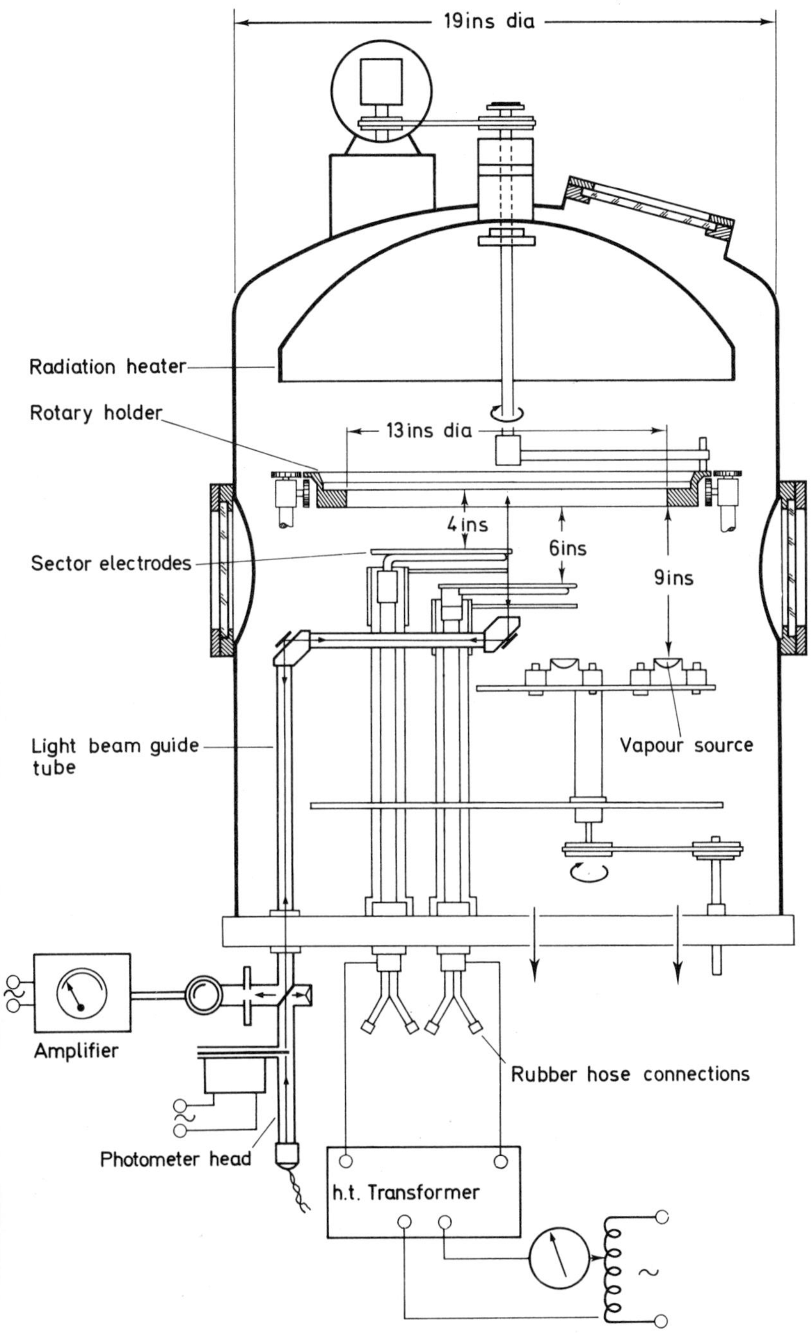

12.7.2 General arrangement of combined sputtering and evaporation plant for two-layer anti-reflection coating.

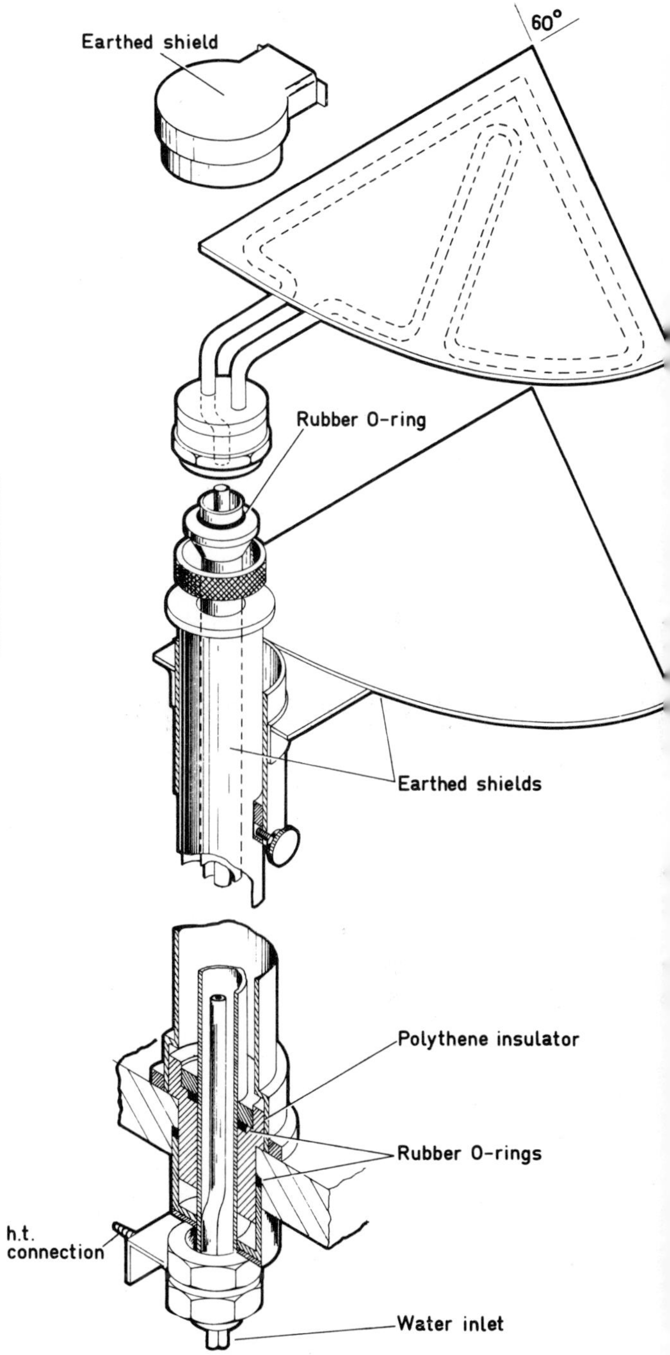

12.7.3. Sputtering electrode showing earthed shields and water-cooling connections.

that their reflection amplitudes sum to zero at a given wavelength. Sawaki and Kubota have compared the efficiency of this system with that of a single magnesium fluoride film and the 'chromatic system' proposed by Turner. They found that the coating described by Osterberg and others was more selective in its response at different wavelengths than a single magnesium fluoride film or the chromatic system, but that it had the lowest luminance factor for white light. For this reason the system proposed by Osterberg and others has been used by the present authors.

The refractive indices of sputtered bismuth oxide and evaporated magnesium fluoride are about 2·45 and 1·38 respectively. From these values the optical thicknesses of the films in the two-layer anti-reflection coating can be calculated for zero reflectance at a given wavelength. King and Lockhart have given a general expression for the reflectivity of a two-layer system and Cox, Hass, and Rowntree have calculated the optical thicknesses required for a range of film indices.

Using the materials discussed here the following values are obtained for their optical thicknesses: 0·049 λ bismuth oxide, and 0·33 λ magnesium fluoride. There is also a second solution for the optical thicknesses at which the reflectance of the combined layers is zero at 1 wavelength; i.e. 0·451 λ bismuth oxide, and 0·17 λ magnesium fluoride. Sawaki and Kubota have shown that this solution has a high luminance factor for white light.

For the purposes of controlling the optical thickness of the bismuth oxide layer it is necessary to know the reflectance of the initial film before the magnesium fluoride is deposited. Calculation shows that the reflectance of the bismuth oxide film should be about 8 per cent so that the reflectance of the glass surface being coated must be raised by 4 per cent.

Initial experiments determined the uniformity of the reflectance of the sputtered bismuth oxide layer when measured across the diameter of the rotary holder. It was found that the coating on the central area of the 13-inch diameter holder had a uniform reflectance of $7\frac{1}{2} \pm \frac{1}{2}$ per cent within a 10-inch diameter circle. The fall in the reflectance at the edges of the rotary holder was due to a decrease in the film thickness. The molecules which are sputtered and diffuse through the gas from the outer edges of the sector cathodes have a greater probability of capture on the chamber wall than the edge of the holder, because the chamber is nearer to the sector cathodes in that region. It has been demonstrated elsewhere that a condensing surface placed near the receiving plane influences the distribution of the sputtered film although it does not optically shield the receiver from the cathode. When the magnesium fluoride was deposited on the oxide-coated glass the reflectance was effectively reduced within an 8-inch diameter zone on the work holder.

Fig. 12.7.4 gives the measured reflectance, as a function of wavelength in the visible region, of a series of two-layer coatings with the bismuth oxide and magnesium fluoride films deposited to obtain a minimum reflectance at different wavelengths. The reflectance of a single-layer magnesium fluoride film has also been plotted for comparison purposes. The transmittance has been measured of a pile of plates with coated and uncoated surfaces, and the values obtained for up to six glasses agreed with those calculated using the reflectance values given in Fig. 12.7.4. From this it was deduced that the absorption due to the coatings was negligible.

Critical adjustment of the wavelength of minimum reflectance is made difficult when depositing the magnesium fluoride because the reflectance measured by the photometer changes slowly with the film thickness in the region of the minimum. This tends to make the operator wait, with the result that the minimum is displaced towards longer wavelengths. Also visual monitoring, as generally used with mono-layers, has not proved an accurate form of control. The reflectance curve of the two-layer coating is not symmetrical on either side of the minimum, as it is with mono-layers, but rises steeply on the lower wavelength side. Thus when the minimum is displaced towards the longer wavelength by some 150 Å above 5300 Å, the reflectance of the coating for blue light is greater than that of the uncoated glass. When a number of surfaces are incorrectly coated the resultant transmitted light is strongly yellow tinted. Elaborate apparatus has been devised for accurately locating the maximum or minimum reflectance of optical interference films used in multilayer filters, but for controlling broad band anti-reflection films the problem may be overcome by

12.7.4. Reflectance of two-layer anti-reflection coating as a function of wavelength with the minima adjusted for different wavelengths. A curve of single-layer magnesium fluoride film is included for comparison.

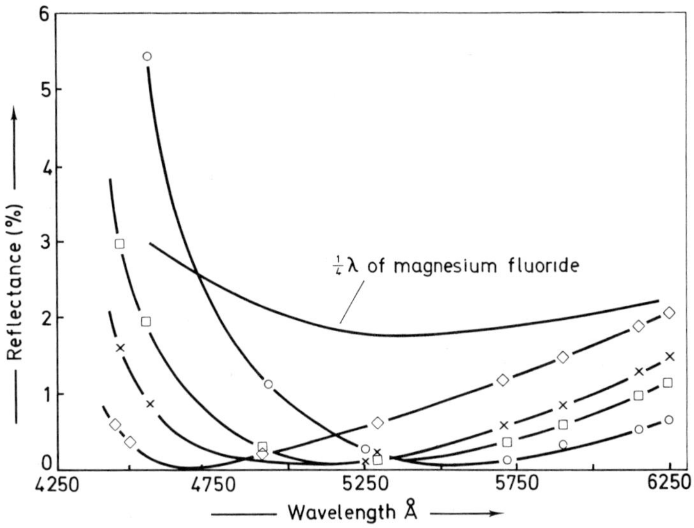

12.7.5. Calibration curve for adjusting wavelength of reflectance minima in terms of reflectance at the monitoring wavelength of 4550 angstroms.

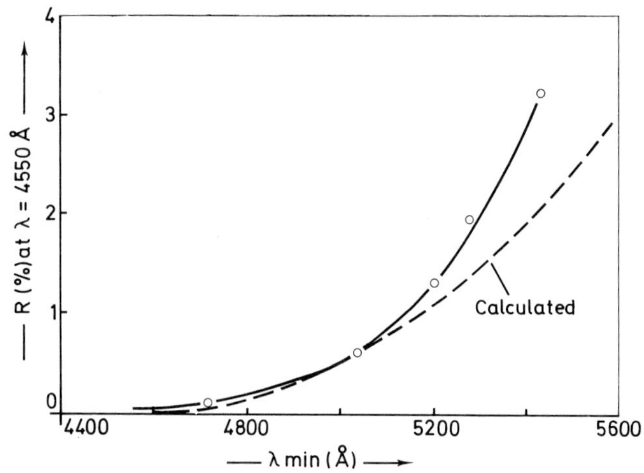

using a monitoring wavelength lower than the desired value and ceasing deposition when the photometer reading has passed through a minimum and reached a specified value.

With this method the authors have prepared films with a minimum at λ = 5300 Å using a blue Wratten filter No. 50 which, combined with the photometer, had a maximum sensitivity at 4550 Å. The optical thickness and corresponding reflectance of the bismuth oxide film required at λ = 4550 Å to obtain the optimum value at 5300 Å are 0·058 λ and $9\frac{1}{2}$ per cent respectively.

Using the values for the reflectance at λ = 4550 Å of the two-layer anti-reflection coatings given in Fig. 12.7.4, it is possible to draw a calibration curve relating the reflectance obtained at the monitoring wavelength after passing through a minimum to the wavelength of zero reflectance. The resultant curve is given in Fig. 12.7.5, together with the values obtained by approximate calculation using a vector diagram based on the Fresnel coefficients and neglecting multiple reflections within the layers. The optical thickness of the bismuth oxide was adjusted at each specified wavelength to provide minimum reflectance when the magnesium fluoride was deposited. This required varying the bismuth oxide reflectance, measured at the monitoring wavelength, from 8 per cent to obtain a minimum at a wavelength of 4550 Å, to 9½ per cent to obtain a minimum at 5300 Å.

Wear resistance

The wear resistance of the two-layer films has been determined by abrading their surfaces using a reciprocating machine of the type described by Holland and Van Dam for testing magnesium fluoride films. The abrasive head of the machine was fitted with a rubber impregnated with emery powder grade 302 supplied by the SIRA Institute. The test specimen was mounted on a support which rotated through one revolution for every three single strokes of the abrading head, so that the rubber traced out a series of flat spirals on the film. The rubber was ¼ inch in diameter and under a load of 112 lb/in^2. The length of the abrasive stroke was ¾ inch so that the overall diameter D of the abraded area was 1 inch. The abrasion of the coating gradually removed the deposit from a central region, which could be discerned by the reflection of the uncoated glass. A figure of merit for the wear resistance of a film could be obtained by measuring the diameter d of the removed coating and expressing its area as a percentage of the total area abraded, i.e. $w = (d/D)^2 \times 100$ per cent.

Average values of w obtained in each case for three samples of two-layer coatings prepared at different base temperatures and tested at 200 and 1000 single strokes of the abrading head are as follows. The figures given in brackets are corresponding values of w for hard $\lambda/4$-films of magnesium fluoride deposited on glass which has been cleaned in a glow discharge and baked at 300°C.

No. of rubs	w 20°C	w 200°C	w 250°C
200	6 (15)	3	3 (5)
1000	34	15	5 (6)

The glass specimens were not cleaned in a glow discharge before sputtering the bismuth oxide as is essential for producing adherent single-layer magnesium fluoride films, but their wear resistance is of a high order. This is because of the cleaning action of the glow discharge used for sputtering and the high adherence of the bismuth oxide to the glass base.

12.8 Four-layer anti-reflection coating

Fig. 12.8.1 compares the antireflection properties of typical one-, two-, and four-layer coatings.

This section describes the development of a fully automatic plant capable of producing four-layer coatings. A hard, broad-band anti-reflection coating process was required, suitable for large optical components. The process was to be 'cold', in that radiant heating would not be necessary.

At the request of the author (D. F. Horne), Edwards High Vacuum Ltd developed a fully automatic coating plant to produce anti-reflection coatings; a band width of at least 3000 angstroms (4250 Å to 7250 Å) at less than 0·5 per cent reflectance per surface was achieved by use of this plant.

Permission has been given by Dr L. Holland to publish extracts from a paper entitled 'Programmed Control of Vacuum-Deposited Multilayers', by J. English, T. Putner, and L. Holland of the Central Research Laboratory, Edwards High Vacuum International, Crawley, Sussex.

Introduction

A major requirement in the production of thin-film devices is repeatability of performance. An obvious variable when manually controlling the deposition conditions is operator error. Repeatability of film structure, adhesion, and related physical properties are determined by the operator's skill in controlling the film environmental and growth conditions and termination of deposition. When complex film systems are prepared whose properties critically depend on the composition, structure, and thickness of each layer the reject rate due to operator failure can be high. The use of thin films in the microelectronics field has stimulated the development of instruments for measuring growth rate, electrical resistivity, etc. However, in the optical field multi-layer systems are often prepared by manual control with the sole assistance of an optical measuring instrument. The use of multi-layer dielectric films for optical interference systems is rapidly increasing. The effect

12.8.1. The anti-reflection properties of one-layer, two-layer, and four-layer coatings.

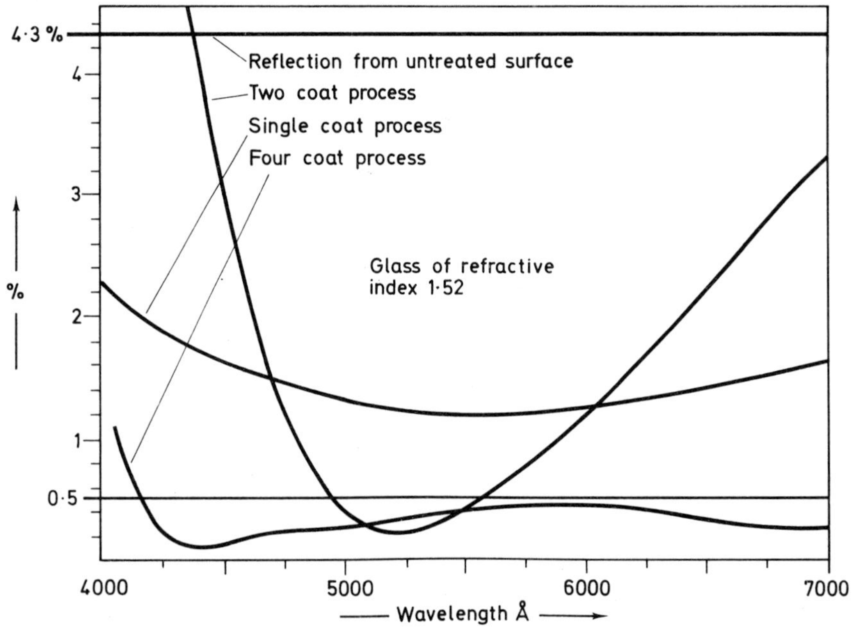

of the growth conditions on the optical properties of dielectric films has been well studied and instruments now exist for determining their growth and physical properties. Thus it is possible to remove human error from the deposition cycle by using direct instrument control and this paper describes an optical deposition system which can be programmed to operate automatically.

Environmental conditions

It is well known that for a given film/substrate pair the structure and composition of the deposit are chiefly influenced by its rate of growth, the nature and pressure of the gas atmosphere, and substrate temperature. The physical properties of a film depend upon its structure and composition. Thus when metal oxides are volatilized the evaporant may be reduced by source reactions or dissociated at the low oxygen pressure. Reduction of a metal oxide usually produces optical absorption and change in refractive index. Fortunately, oxygen may be restored to the film by sorption from the gas atmosphere during growth. Thus, it is a requirement of an automatic system to control the deposition rate, gas composition, and pressure during metal oxide deposition to produce films of consistent optical properties.

Optical characteristics

An optical film can be characterized by its reflectance R, transmittance T, and absorptance A, where $R+T+A = 100$ per cent. In the case of a dielectric film, interference characteristics are exhibited as the optical thickness is progressively increased, and both R and T, which are complementary (i.e. $R+T = 100$ per cent) tend to vary sinusoidally.

The optical thickness of a layer may be set by ceasing deposition at a given maximum or minimum point. Alternatively, if one chooses another monitoring wavelength the R (or T) values for the same optical thickness may occur in a region where dR/dt is greatest. One may also control in the high-slope region, because the optical thicknesses are not multiples of a quarter of the monitoring wavelength and therefore do not coincide with simple maximum and minimum points.

Growth and environmental control techniques

Experiments by the writers have shown that metal-oxide films of consistent structure and composition can be prepared if the evaporant growth rate and pressure of the oxidizing gas (e.g. O_2 and H_2O) are maintained constant. If the deposition rate of an active compound varies, then the change in the sorption rate alters the pressure of the oxidizing gas. Although the gas pressure can be adjusted to provide a constant value this will not necessarily produce a constant oxygen/evaporant molecule impingement ratio at the substrate. It is essential first to maintain a constant film growth rate and therefore sorption rate. This can be done with a crystal micro-balance monitoring the deposition rate and supplying a rate correction signal to a resistance-heated source via a silicon-controlled rectifier which otherwise stabilizes the source current (Bath *et al.*, 1966).

Some workers have maintained constant chamber pressure during reactive evaporation with a servo-controlled needle valve. However, this refinement is unnecessary if the system degassing rate is low compared with the inlet gas rate and the diffusion pump has a stable performance. The pressure can be pre-set with a needle valve and will remain constant if the deposition rate (and corresponding evaporation conditions) are maintained constant.

We have found measurement of total pressure adequate when admitting oxygen to the vacuum vessel, but the use of a gas analyser is being investigated where water vapour desorption is appreciable.

Film thickness determination

Termination of optical film deposition is best done by control from the optical characteristics as discussed above. A modulated beam photometer (Steckelmacher and English, 1961) has been used to monitor R (or T) at a specified wavelength. A voltage level switch was employed to control the termination point and with the aid of logic selection the instrument was made to discriminate between recurring R (or T) values when the optical signal was rising or falling.

The use of a differentiating amplifier was investigated for operating a termination switch as the optical characteristic passed through a maximum or a minimum ($dR/dt = 0$) with the growth rate constant with time. Although the instrument was effected by fluctuations of the photometer reading due to monitor vibration, etc., satisfactory results were obtained when preparing $\lambda/4$ films in deposition times less than 5 minutes. It was possible to select any integral number of quarter wavelengths for a layer.

Automatic system

Arising from the foregoing experimental experience a programmed apparatus was built for preparing broad-band interference filters and anti-reflection combinations using actively evaporated films. The deposition system and associated automatic apparatus is shown in Fig. 12.8.2. The chamber is initially evacuated using a conventional automatic pumping system with the operations of roughing, glow-discharge cleaning, radiant heating of the substrate, and high-vacuum pumping controlled by time switches. When the substrate is at temperature and the required pressure obtained the coating cycle can commence.

The deposition apparatus consists of a rotary plane work-holder with offset evaporation sources to provide uniform coating as described by Steckelmacher *et al.* (1959). For evaporating large quantities of material in sequence two pairs of sources are used. Each pair is provided with a water-cooled monitor crystal and an electromagnetically operated shutter. In the closed position the shutters prevent evaporant from reaching the substrates but leave the crystal exposed to allow pre-deposition control.

The vapour source current is stabilized by a thyristor control circuit programmed to give a current which rises gradually to a constant value at which the temperature of the vapour source is in the required evaporation region. At the end of the pre-heat period control is transferred to the quartz crystal monitor which adjusts the source current to give the required deposition rate. The shutter is then opened and deposition on the substrate commences. The growth rate is maintained constant until the required optical characteristic measured by the modulated beam photometer is reached. The shutter is then closed and source power removed.

The preheat current and time, deposition rate, current limit, and stability (time constant) are set independently to suit the characteristics of each vapour source and material. The process controller incorporates overriding manual controls which may be used to make experimental variations on the process or to carry out functional tests on individual sections. The stage counter

12.8.2. Edwards automatic coating machine capable of a seven-coat cycle, fitted with a Pirani-Penning pressure gauge, roughing and high-vacuum programme unit, modulated-beam photometer, mass monitor, and ratemeter, as well as electronic sequence controller. (*Above*) The plant is set up for four-layer anti-reflection coating. The vacuum chamber has a rotary workholder. On the worktable is a jig loaded with lenses ready for coating.

records each complete deposition stage or layer, while the operation selector controls each operation within a stage. An alarm system stops the process and provides a warning signal when a fault condition is detected.

The large number of switching operations involved makes electromechanical (relay) selection unattractive due to size, cost, and unreliability of the large number of contacts. Fluid logic was not considered here as the inputs and outputs are electrical, mostly inherently so. Resistance transistor logic was adopted for its low cost and high noise immunity. This immunity stems from its low speed by electronic standards (about 10 kHz) and high voltage, particularly in the discrete component form as opposed to solid state integrated circuits. Outputs from the logic system were taken electronically or via the reed relays for low-level signals, and via relays for power switching.

Application of the automatic controller

A two-layer anti-reflection coating consisting of reactively evaporated TiO_2 and SiO_2 is used to demonstrate the automatic system.

When the 'start' button is pressed and the system has been pumped to 3×10^{-5} torr, the first evaporation stage commences with selection of the TiO_2 source. The quartz crystal monitor reading is automatically set to zero, an adjusted leak valve connected to an oxygen supply is opened, and the pre-heat current starts to rise. The pre-heat current reaches 180 A within 30 seconds and remains at this value until a total time of $1\frac{1}{2}$ minutes has elapsed. Within this time the pressure has reached a steady value of about 10^{-4} torr. The desired rate control is quickly achieved and the shutter is opened. Deposition ensues at a rate of 10 Hz/sec (representing <1 Å/sec for TiO_2) until R for the coated surface rises to 8 per cent for $\lambda = 4750$ Å. The source shutter is then closed, the gas leak stopped, and the source power removed.

When the chamber pressure has fallen to 3×10^{-5} torr the second stage commences and the SiO_2 source is selected. The quartz crystal monitor reading is automatically reduced to zero, the desired oxygen leak valve is opened and the source pre-heat started. The current rise time is 30 seconds and the pre-heat value for SiO_2 is 160 A for 2 minutes. When the selected deposition rate of 5 Hz/sec (for SiO_2) has been reached, the source shutter opens. During this stage, reflectance first rises to a maximum, then falls to a minimum. Termination occurs as the reading falls to R = 0·25 per cent by closure of the source shutter. The leak valve is closed and power removed from the source. As this is the last stage, a 'process-complete' lamp is illuminated and the operator is able to admit air and unload the chamber.

The foregoing procedure can be used for repetitive deposition.

12.9 Anti-reflection coatings applied to steeply curved surfaces

The lenses illustrated in Fig. 6.11.31 provide examples of steeply curved concave and convex surfaces where hard even coats are difficult to achieve without special fixtures and coating methods. (See § 6.11.) During the development of these lenses, Mr D. J. Day and Mr D. G. Monk of Wray (Optical Works) Ltd, with the advice of Mr M. J. Shadbolt of SIRA Institute, designed apparatus which enabled the desired even coatings to be applied.

Concave surfaces

A flat surface of magnesium fluoride in a boat emits vapour in an upward direction distributed as the cosine of the angle made with the vertical (Fig. 12.9.1).

The effective area of the substrate is

$$(\text{actual area}) \times \cos\theta$$

The distance from source to substrate is $2R\cos\theta$.

Therefore amount of substance deposited in unit time on substrate is proportional to

$$\frac{\cos\theta\cos\theta}{4R^2\cos^2\theta} = \frac{1}{4R^2}$$

As R is a constant, the magnesium fluoride will be distributed evenly over the surface.

If the source were placed at a long distance from the lens the angle of incidence at any point on the lens would be 2θ, but by placing it on the extended circumference, this angle is reduced to θ, which results in a harder deposit, especially at the edge of the lens.

The method has been used on the deep concave surfaces of the 6-inch $f/5{\cdot}6$ survey lens to produce a hard even coating. On smaller lenses, the geometry of the system makes it more difficult to produce an even film, since small errors in setting up are magnified, and the source has a larger area in proportion. However, by using a smaller filament these difficulties can be largely overcome.

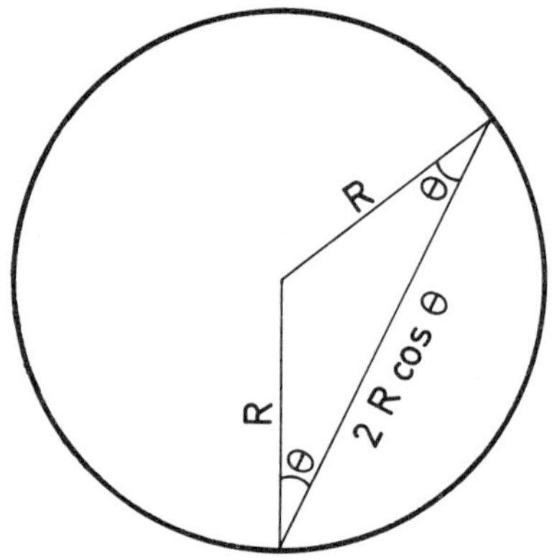

12.9.1. Geometry of evaporation coating.

If large quantities of lenses have to be coated, the arrangement could be duplicated in various parts of the vacuum chamber, and the separate filaments fired one after another. Satisfactory hardness of coating was achieved without radiant heating due to the relatively short distance of the filament to the lens.

Convex surfaces

If very-deep-curve convex lenses up to hemispheres have to be coated, it is essential that they be individually rotated on their polar axis, and the spinning axis inclined during the discharge cleaning and evaporation operations to ensure even coating.

Apparatus was constructed which would fit

into a vacuum chamber normally used for single coat blooming, but with provision for an external motor for driving a shaft in the chamber for rotating the lenses at about 100 r.p.m. The filament was placed relatively near to the lenses to improve efficiency, and baffles between the filament and each lens ensured that the vapour stream contacted the lens normally to the surface.

The aluminium discharge cleaning electrodes were shaped to bring them near to the lenses and this geometry is very important if good coatings are to be achieved. Radiant heating can be used but, in practice, the closeness of the filaments provided sufficient heat to ensure reasonably hard coatings. The evaporation process was generally as explained in § 12.6.

12.10 Aluminium-coated front-surface mirrors

(See § 15.13.)

New aluminium coatings have a reflectivity of about 90 per cent in the visible region, but lose some of their reflectivity witn aging. The aluminium surface is soft, so it is usual to coat it with silicon monoxide which reduces the immediate reflectivity to not less than 80 per cent but retains this value for a considerable period of time; this coating provides a tough surface which will withstand cleaning.

It is essential to use a coating machine which is in good order and will give a vacuum of at least 5×10^{-5} torr. The work-holder can be static or rotating, the latter giving a more uniform film.

The aluminium must be 99·99 per cent pure and wetted well on a triple-strand 0·5 mm tungsten-wire filament. The silicon monoxide should be on a molybdenum boat-shaped (Fig. 12.2.1) filament. After assembling the cleaned mirror blanks in the vacuum chamber, pumping down, and exposing to a glow-discharge cleaning, the process can commence.

At a vacuum pressure of 5×10^{-5} torr, the power to the aluminium filament is switched on with the shutter closed. When the aluminium has melted, the shutter is opened and the substrates exposed to the vapour. When the aluminium has evaporated, the shutter is closed.

The modulated beam photometer should then be switched on, the scale set at 100 per cent, and the silicon monoxide evaporated to a thickness giving maximum reflectivity. With the shutter closed, the silicon monoxide film should be $\frac{1}{2}$ wavelength thick, which results in maximum reflectivity. The substrates (mirror blanks) should then be removed from the vacuum chamber and baked in an oven for 3 hours at 200°C and allowed to cool slowly whilst in the oven. This heat treatment

12.10.1. Edwards vacuum machine set up for coating polygons with aluminium and silicon monoxide. The polygons are rotated during the coating operation. See also Fig. 8.7.7.

removes strain and hardens the film so that it can be rubbed with a clean polishing cloth (Selvyt) without marking the surface.

If the mirrors are required for use with wavelengths down to 2600 angstroms then, after coating and baking they should be irradiated with ultraviolet light from a high-pressure mercury lamp (without glass window) for 20 hours in order to make the silicon monoxide layer transparent, and this will improve the reflectivity below the visible region. For use at wavelengths below 2600 angstroms, the aluminium must be coated with very thin magnesium fluoride (instead of silicon monoxide) and, in order to be successful, the pressure must be lower than 10^{-6} torr during the evaporation of aluminium, which is followed immediately by the magnesium fluoride coating to prevent oxidization. Special precautions are necessary with this difficult process if good ultraviolet reflection properties are to be achieved.

12.11 Chemical silvering of prisms and mirrors

The chemical silvering process in common use is substantially that published by Mr Brashear in the *English Mechanic* in 1893, but modified to suit modern conditions.

A mirror may be silvered either face upward or face down as required. Small work is preferably silvered face down, but large mirrors are more easily handled face up. The dish for the bath should be of glass or porcelain, but large baths may be of wood or sheet metal thickly coated with paraffin wax and, for economy, should be of nearly the same size as the mirror to be silvered. In the case of very large mirrors, it is most economical and convenient to make the mirror itself form the bottom of the bath. Cleaning is an important operation. All dust must be removed from the glass and old silver cleaned off with strong nitric acid, using a swab of cotton wool. Wash with water, and swab with nitric acid, with ordinary water followed by distilled water, and then leave the mirror immersed in distilled water until ready for silvering.

Prisms or mirrors ready for silvering should be mounted in special jigs to cover up all surfaces except those to be silvered. A silvering solution temperature of about 20°C will give the best results. If the temperature is too high reduction will be rapid and the resulting silver film soft; if too low, action will be very slow and the film too thin. During the process, sediment will be formed, which must be prevented from settling on the mirror surface, so the solution must be constantly in motion. Baths are often mounted on a rocking mechanism to ensure that the sediment does not settle.

Before silvering, immerse jigs for 1 minute in stannous chloride solution (96 g of stannous chloride to 4000 c.c. distilled water) and then swab under running tapwater. Store under distilled water.

Bulk solutions of the following four chemicals should be prepared:

(A) Silver nitrate 214 g. Distilled water 4548 c.c.

(B) Ammonia (0·880) 454 c.c. Distilled water 4094 c.c.

(C) Potassium hydroxide (caustic potash) 298 g. Distilled water 4548 c.c.

(D) Sugar 908 g. Glucose (dextrose monohydrate) 454 g. Distilled water 4548 c.c.

From the bulk solutions a silvering bath is made with the following proportions by slowly adding B to A and then C to the mixture of A and B.

Solutions A, B, and C	100 c.c. each.
Distilled water	300 c.c.

Remove the jigs (with prisms and mirrors) from the distilled water storage and place them in the silvering bath. Add the reducing solution D in the proportion of 50 c.c. and agitate the bath until the colour has changed to black. Allow the chemical process to continue for about 20 minutes. Remove the jigs from the silvering bath and place them in distilled water. Prepare a second silvering bath, repeat the chemical process, and place the jigs in distilled water.

For back-surface mirrors, the jigs should be removed from the distilled water, wired up, and placed in a copper plating tank with standard copper sulphate solution for a neutral tank at pH 7 to 8. Apply a current of approximately 0·75 amps per square foot of silvered surface and continue plating for 10 to 20 minutes, after which the jigs should be removed from the plating tank and washed in distilled water, and the surfaces blown dry with filtered compressed air. Hand-paint the copper-coated silver surfaces with flake shellac and methylated spirit.

When the shellac is dry, remove the prisms or mirrors from the jigs and mount them in Plasticine. Paint the back of the silvered surface with a black mirror backing paint (such as Trebax 38998C supplied by Pearl Paints Ltd) and bake in an oven at 100°C for 1 hour. When cool, a second coat of paint should be applied and baked. The prisms or mirrors can now be removed from the Plasticine and any surplus silver or paint removed with a razor blade before cleaning, inspection, and wrapping.

A warning must be given of the explosion risk if, by accident, the addition of ammonia to silver nitrate becomes dry and forms silver nitride (silver fulminate). If ammonia (solution B) is added to silver nitrate (solution A) in an ordinary flask, there is a tendency for the substance to dry in the neck of the flask and if this happens, it should be washed down IMMEDIATELY. Silver fulminate is apt to explode with great violence if allowed to dry.

13 Mounting of Optical Components

13.1 Introduction

Optical designers now have the use of powerful computers, which can be programmed to optimize lens designs capable of a performance far better than was considered possible a few years ago. However, this performance can be achieved only by the use of optical components of greater accuracy and the careful design and manufacture of mounts to ensure accurate alignment.

Mount designs should, where possible, follow the basic principles of kinematics which control the six degrees of freedom; three translation movements along the rectangular coordinate axes and rotation about these three axes. In practice, owing to the soft materials such as brass and aluminium used in mounts, it is impracticable to have point contacts, and small area or line contacts have to be used instead.

Optical lens elements are usually mounted in close fitting cells. Fig. 13.1.1 illustrates alternatives on the four most popular methods of retaining the lens—by spring, threaded lock ring, spinning, or

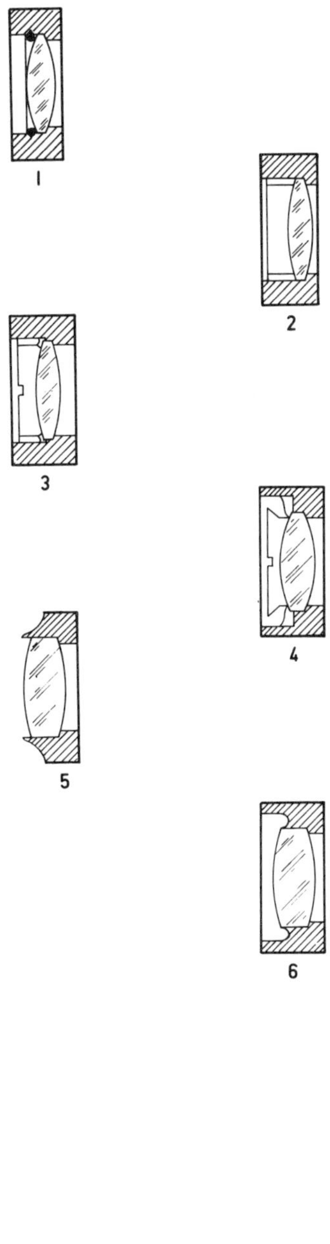

13.1.1. Methods of mounting lenses in cells.
(1) Wire spring ring in V groove.
(2) Flat spring ring.
(3) and (4) Threaded lock ring.
(5) Lens spun in mount.
(6) Lens cemented in mount (with trough for overflow).

by the use of an adhesive such as silicone rubber. Of these, the threaded lock ring and spinning are used in most designs of precision optics.

When the mounted lens is finished we are only interested in the alignment of the optical components, but during manufacture we have to consider:

(*a*) The grinding of the edge of the lens in relation to its optical axis.

(*b*) The tolerance on the grinding of the lens.

(*c*) The tolerance on the boring of the cell for the lens.

(*d*) The alignment of the cell in relation to the optical axis of the lens system.

(*e*) The tolerance necessary during assembly and the resulting clearances between lens and cell.

13.2 Typical examples of lens and mirror mountings

Binocular lenses

Fig. 13.2.1 illustrates a typical binocular lens where the aluminium alloy mount rim is designed to be spun over. The spinning operation provides not only a firm retainer but also centralizes the lens in the cell. In this example, both lens and cell diameters have a 0·002 inch (0·05 mm) tolerance with a minimum clearance of 0·001 inch (0·025 mm). It is clear that at the maximum tolerances, it would be possible to have a clearance of 0·005 inch (0·125 mm) and this would not be acceptable without the spinning action helping to centralize the lens. The chucking thread is only used during the spinning operation.

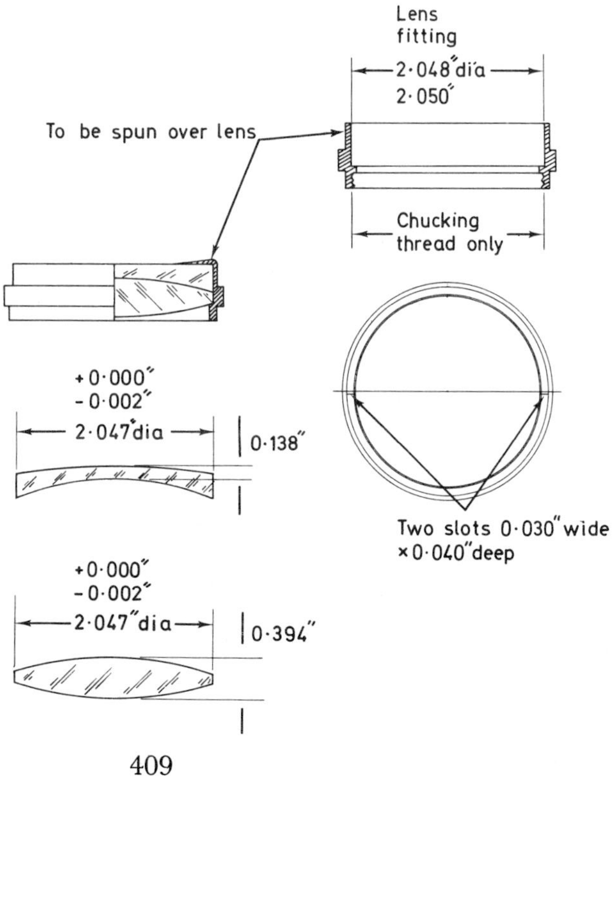

13.2.1. Tolerances on a good-quality lens with the lens spun into an aluminium mount.

13.2.2. Focusing eyepiece in focusing binoculars. The doublet lens and the graticule are spun into their mounts.

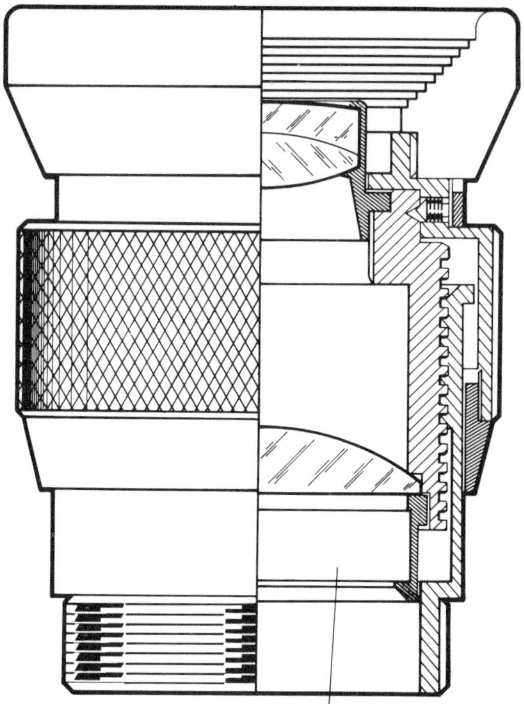

Focusing eyepiece

In the focusing eyepiece (Fig. 13.2.2) a spun cell is illustrated and also a graticule spun into its mount. The mount is made from moulded plastic and brass, with a multi-start thread for controlling the focusing movement.

Copying lenses

High-quality copying or enlarging lenses must be made with great precision or they will not give the required performance.

A typical one-to-one copying lens is illustrated in Fig. 13.2.3, and this is designed for zero vignetting over 8·75 inches (223 mm) diameter, although, when used in a copying machine, a slit 0·75 inch wide (19 mm) is the active surface. Very accurate components are needed, and both lens and cell diameters have a 0·001 inch (0·025 mm) tolerance, with a minimum clearance of 0·001 inch and a maximum of 0·003 inch (0·75 mm). (See § 6.13.)

The concave lens will centralize itself by the 45° chamfer, and the fit of the convex lens is aided by the black paint on the edge. A lens of this type must have its components centred in the mount to an accuracy better than 0·002 inch (0·05 mm). The 0·002-inch maximum tolerance on optical alignment must include all other manufacturing tolerances.

Lenses requiring very accurate alignment

If the maximum tolerance on optical alignment must be better than 0·001 inch, then it is not feasible by existing types of machinery to premachine the cells to finished size because the close tolerances would make production uneconomic. The method adopted in these very accurate lenses (Fig. 13.2.4) is to give large tolerances on the metal cells and, providing optical centring is very good, large tolerances on the diameter of the lenses.

The cell is screwed on a side lathe by the chucking thread provided and the cell bored to suit the diameter of the lens. During this operation the lens is used as a plug gauge and can be held either by vacuum or Plasticine. This is a skilled operation, and care will give a better result than any other method.

13.2.3. Assembly tolerances in a four-element one-to-one copying lens system, designed for zero vignetting over 8·75 inches diameter. The image-to-object distance is 17·25 ± 0·25 inches at *f*/5·6. The minimum resolution is 5 lines/mm, and the fixed stop (created by the chamfer of lens No. 2) is 16·2 mm (0·637 inch) in diameter. All internal glass-to-air surfaces are bloomed normal purple.

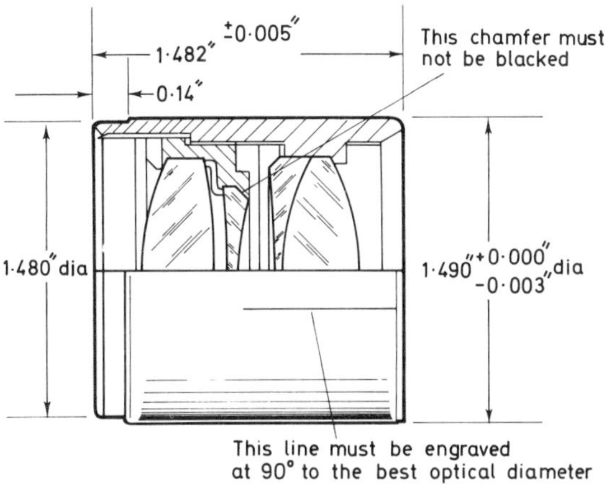

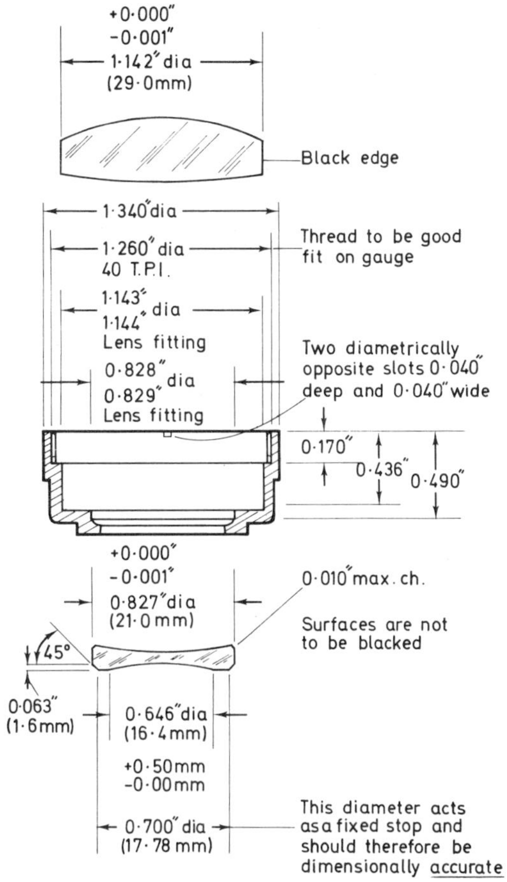

◁**13.2.4. Tolerances on very accurate lenses for micro-circuits, copying, or wide-angle profile projection, where the final cell machining is on assembly to fit lenses.**

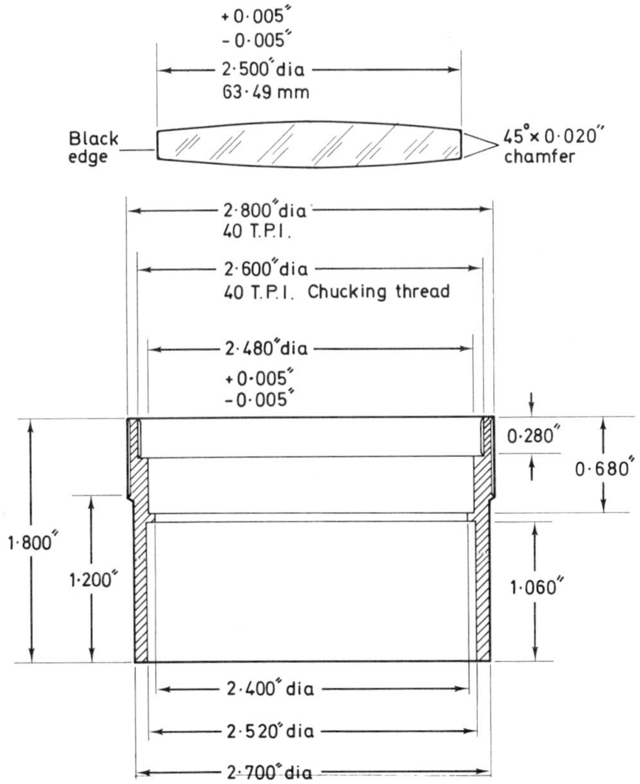

Lenses subject to heat

The mounting of condenser lenses near to a projection lamp is always a problem because of the intense heat. The system illustrated in Fig. 13.2.5 is suitable for a profile projector. The two meniscus lenses nearest to the lamp are located by a spiral spring to allow for movement due to expansion and contraction. The semi-reflecting mirror, under the ×10 projection lens, is held in position by sheet metal clips.

35-mm slide projector mounts

The optical components of a 35-mm slide projector must be held by a low-cost mount. (Fig. 13.2.6.) Usually light sheet-metal fittings are satisfactory as high-accuracy spacing or alignment is unnecessary, and some movement is essential to allow for the heat from the lamp. The projection lens is often made of plastic with plastic threads and spacers. Lens spacing or centring of high accuracy is not essential, but low cost is the

13.2.5. Collimating system for profile projector. The lens spacing springs provide for movement due to heat from the lamp.

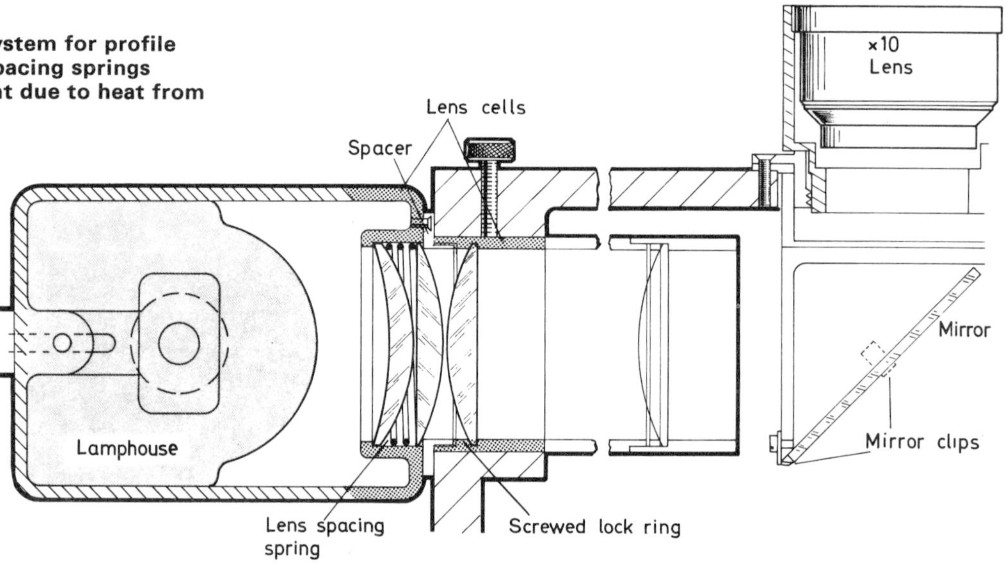

13.2.6. Slide projector optical system including mirror, quartz-iodine lamp, flame-polished aspheric lens, heat filter, condenser lens, and plastic-mounted projection lens.

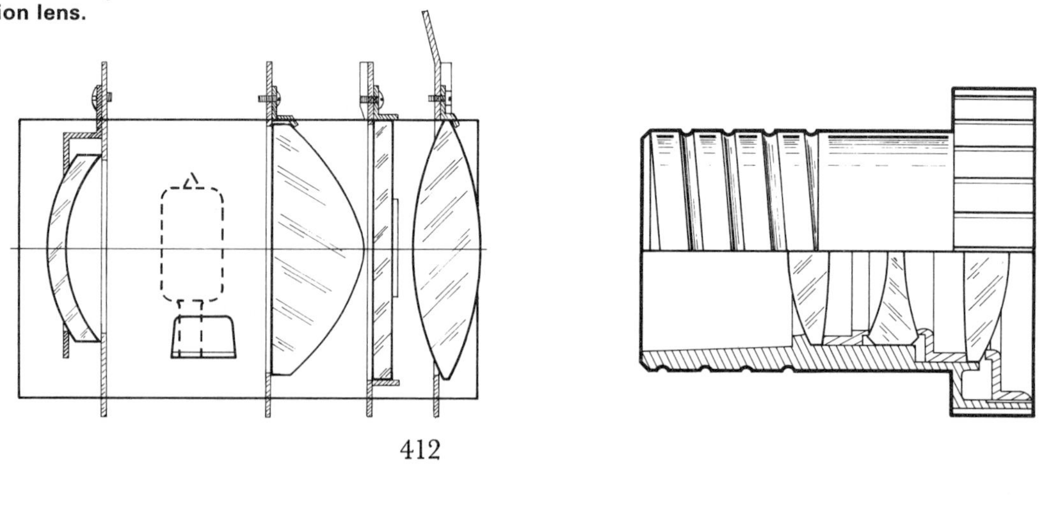

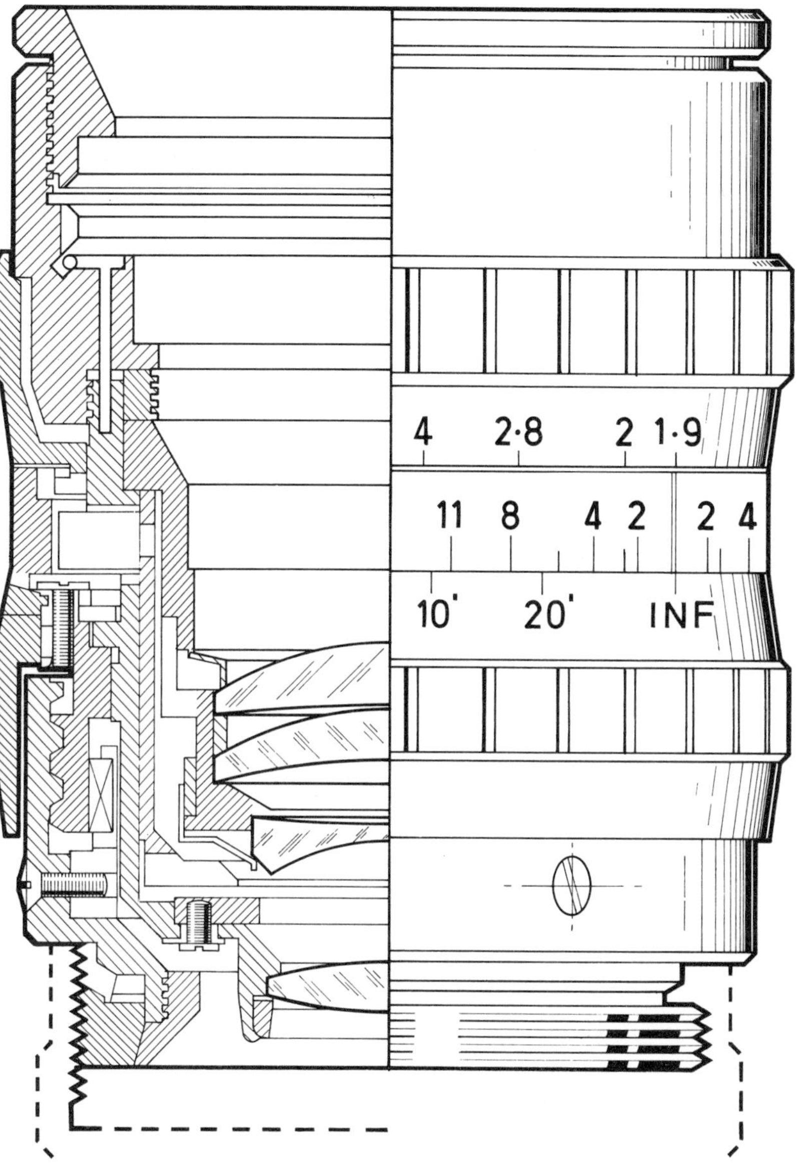

◁ **13.2.7. Taylor-Hobson 1-inch $f/1{\cdot}9$ lens in a focusing mount for 16-mm film.**

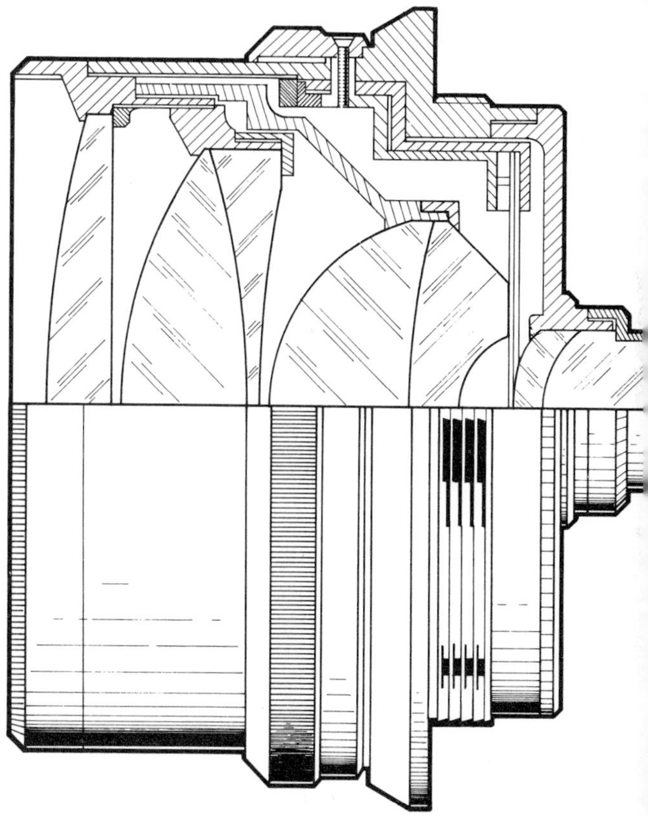

▽ **13.2.8. Wray 64-mm $f/0{\cdot}71$ fixed-focus lens for 16-mm ciné radiography.**

all important issue, within the limits of a generally satisfactory optical performance and attractive appearance.

Ciné camera and TV lenses

16-mm and 35-mm ciné cameras lenses and television lenses must be made very accurately to give the high performance demanded by professional photographers. Fig. 13.2.7 illustrates a 16-mm ciné focusing lens which was made in considerable quantities and justified the cost of tools. In addition to the traditional method of threaded lock rings the meniscus lens is held in position by a metal spring clamp spot welded to the mount. The front of the lens has a retaining ring which can be removed for fitting a filter glass.

Ciné radiography lenses

The 16-mm ciné fixed-focus radiography lens (Fig. 13.2.8) is of interest because of the high aperture, $f/0{\cdot}71$, and the very deeply curved lenses, which are not only difficult to make accurately but are also a problem to coat uniformly. (See § 12.9.)

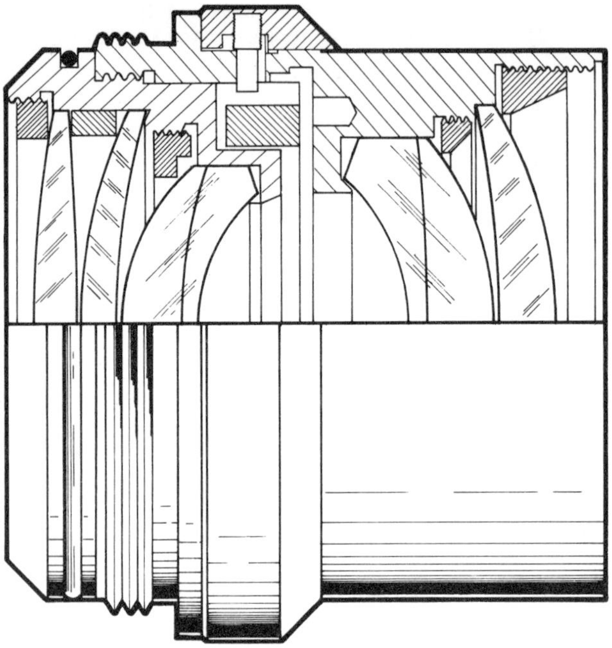

△13.2.9. **Cooke Speed Panchro 40-mm *f*/2 lens for 35-mm film (prebored construction).**

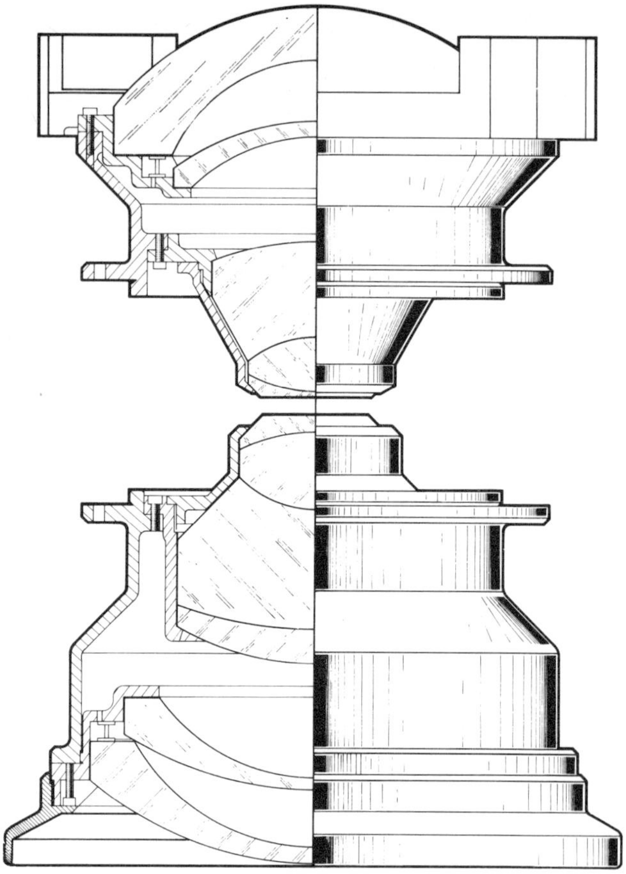

13.2.10.▷ Six-inch *f*/5·6 air survey lens.

Cooke Speed Panchro ciné lens

The 35-mm ciné Cooke Speed Panchro lens (Fig. 13.2.9) is an example of good lens design from the point of view of both glass and metalwork. The line contact between lens, spacer, lock ring, and cell face is clearly illustrated. Thus, providing the bore is true, the lenses will be accurately aligned in the cell.

Air survey lens

An air survey lens (Fig. 13.2.10) is a good example in which very close tolerances are an essential requirement of optical performance. Because the lens has to be used in an aircraft, there is a need for weight reduction, and therefore the mount metalwork must be as thin as possible consistent with the rigidity required for high accuracy. Lenses of this type are necessarily required in small quantities, and therefore each component is carefully made with individual attention. This is a case where the mount is finish-bored to suit the diameter of the lens with a minimum of clearance to ensure accurate centring. Great care is taken over the centring of the individual optical components on a chuck before edging. (See § 6.11.)

Zoom lenses

Zoom lenses are a fairly recent development and, in the cinematograph and television field, have to a large extent replaced a turret of accessory lenses with fixed focal lengths.

A typical 16-mm ciné lens (Fig. 13.2.11) made by Bell & Howell, Chicago, with eleven glass elements, illustrates the problem of maintaining the accurate alignment of moving components, which is the essential requirement for good optical performance.

This problem is more acute with 35-mm and television zoom lenses (Fig. 13.2.12). A typical Cooke Varotal zoom lens made by Rank Taylor Hobson has twenty-one glass elements, and fourteen have to move in precise alignment with the optical system, to achieve the high-quality zoom performance specified. Control over the refractive index of the glass (see § 5.2) accurate machining of metal work (§ 6.13), reliable and comprehensive testing of performance (§ 11.26), and good design are all essential for high-quality products.

Anamorphic Systems

Before 1953, 35-mm ciné films had an aspect ratio of about 1:3:1 for camera, projector, and screen, but in order to improve presentation, the 20th Century Fox Film Corporation decided that a wide picture was needed.

In September 1953, the first Cinemascope picture, *The Robe* (Fig. 13.2.13), was given a premiere in New York with a screen width of twice the normal size—the actual dimensions being 65 feet × 25 feet. The method used was to optically squeeze the picture in the horizontal direction by a cylindrical lens during photography and then expand the picture during projection.

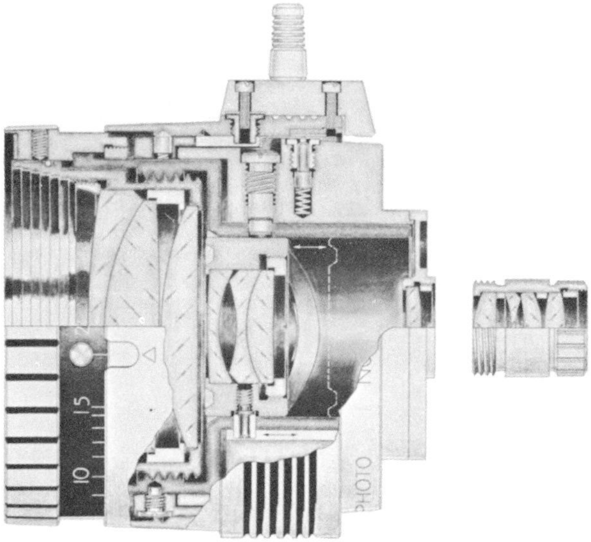

13.2.11. Bell & Howell ciné zoom lens for 16-mm film.

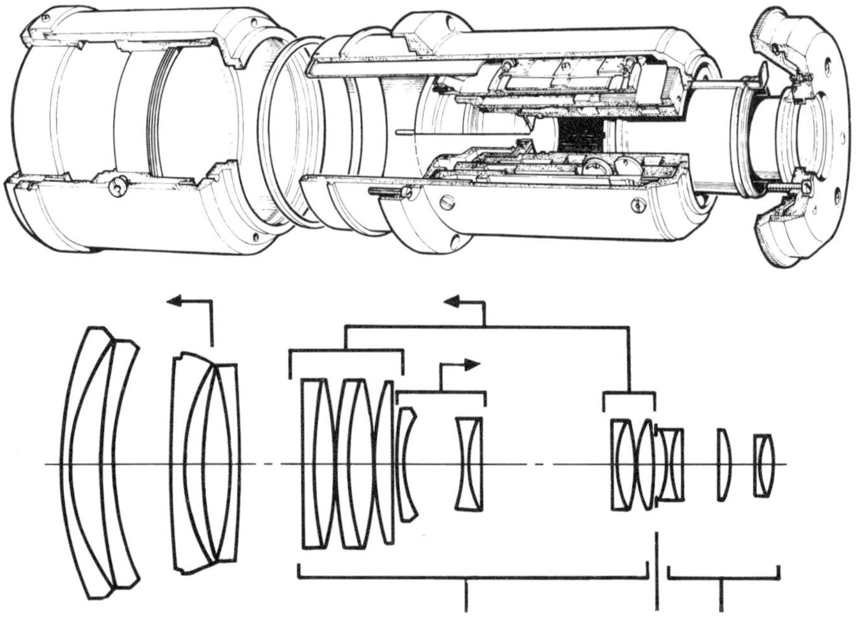

◁13.2.12. Cooke Varotal 35-mm film and television zoom lens system.

13.2.13.▷ A frame from the first 35-mm Cinemascope picture, *The Robe*, produced in 1953. The images are long and thin on the film. When projected, they are expanded 2:1 in the horizontal meridian by the anamorphic lens.

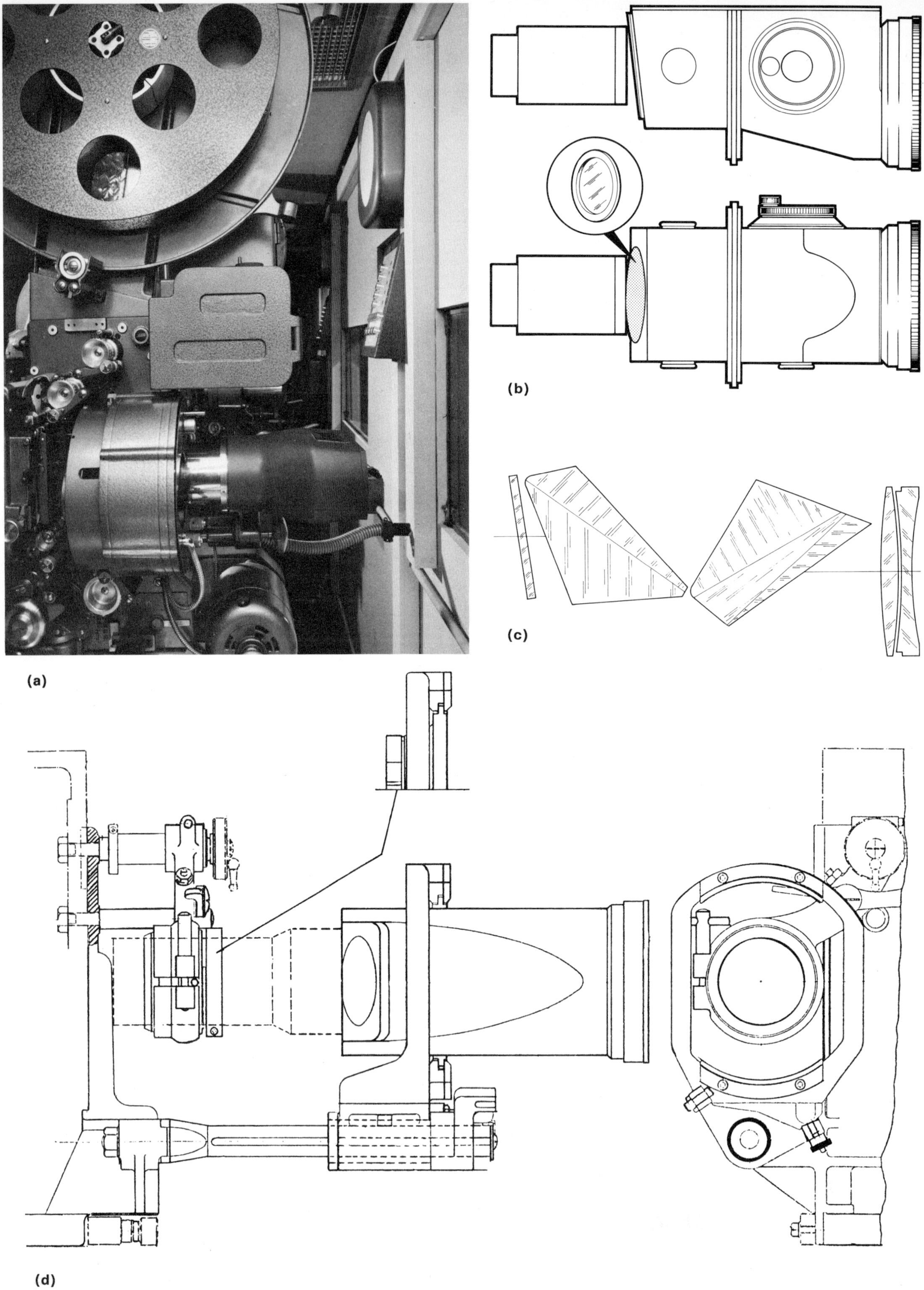

(a) (b) (c) (d)

◁13.2.14. **Prismatic anamorphic optical system.**
(a) **The anamorphic attachment is shown fitted to a 35-mm projector for Cinemascope films.**
(b) **The top and side elevation of the 'Varamorph' variable prismatic lens, showing how it is positioned as close as possible to the backing lens J. Flange A is for fixing the lens to the Gaumont-Kale anamorphic bracket. Knob B is for adjusting the aspect ratio of the prisms, and C is a locking nut. The focus of the front lens is adjusted (by scale D) to suit the projection throw.**
(c) **The optical components of the 'Varamorph'. On the left is the cover glass, and on the right the focusing lens. The additional sections of the prisms are for colour and distortion correction.**
(d) **The mechanical construction of the attachment.**

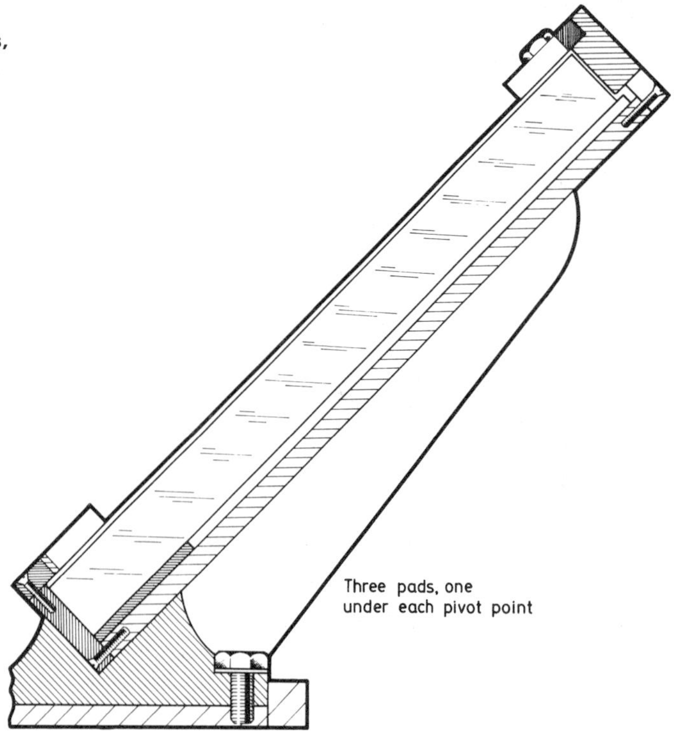

13.2.15. Mounting a mirror in a profile projector. The diagram illustrates an $8\frac{3}{4}\times 7\times \frac{3}{4}$ inch mirror.

The anamorphic optical system, which made this system possible, is one which has a different magnification in one principal meridian from that in the other. The desired result can be achieved by the use of either cylindrical lenses or prisms. (Fig. 13.2.14.) A typical anamorphic projection system consists of a prismatic or cylindrical lens adaptation in front of an ordinary spherical objective lens, but the combination shortens the focal length of the prime lens, and thus widens its field of view for a given image size.

In order to show Cinemascope films, it was necessary to have an anamorphic adaptation, a new high quality projection lens of longer focal length and a new wide screen and frame. The prismatic projection system is more flexible than the cylindrical system as it is easy to adjust the expansion ratio by moving the prisms. A refracting prism has different sized exit and entrance beams and thus produces a magnification in the meridian in which it produces a deviation. In order to eliminate the angular deviation, two prisms are arranged to cancel the deviation but combine the magnifications. A focusing lens is needed to correct axial astigmatism.

Front-surface mirrors

Mirrors can be mounted in a variety of ways depending on the shape and accuracy of the plane surface. (Fig. 13.2.15).

Front-surface mirrors of high orders of flatness, such as those used in a profile projector or interferometer, must necessarily be thick to ensure stability and hold the surface figure. Usually they can be mounted in accordance with the basic principles of kinematics, using a three-point pivot.

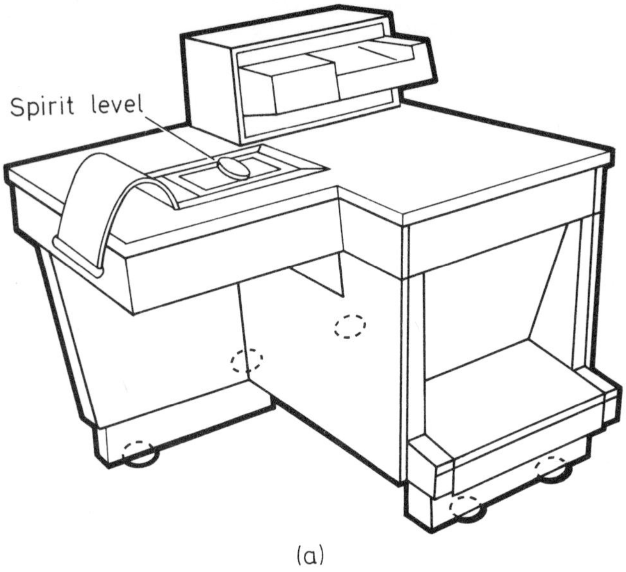

13.2.16.
(a) The Xerographic 914 copier showing method of levelling.
(b) Optical system.
(c) Optical path.

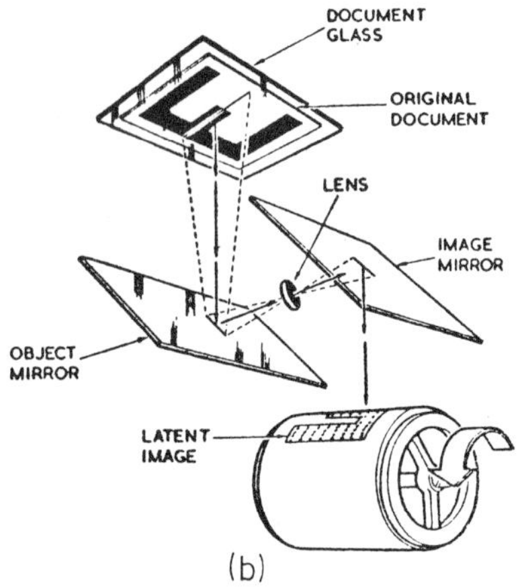

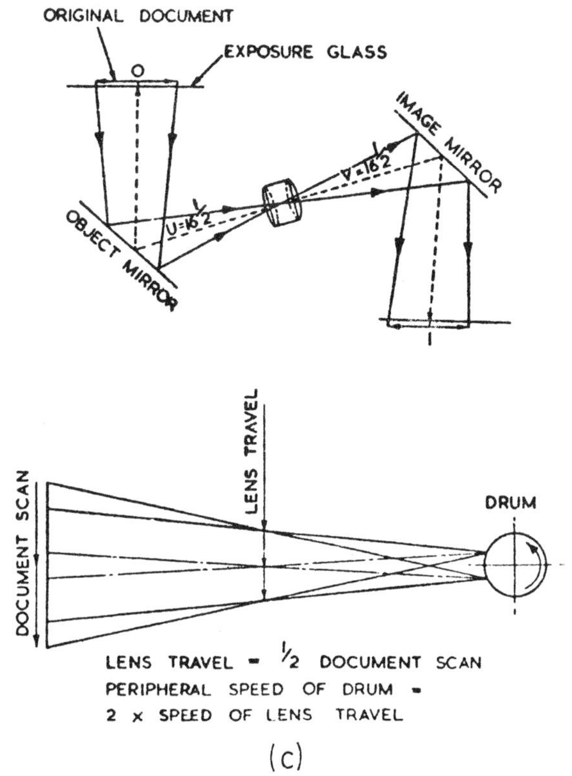

Xerographic copying machine

A different mounting problem arises with the optical system on a Xerographic copying machine (Fig. 13.2.16.) The optical components are arranged in a periscopic relationship. The document exposure glass is illuminated by twin fluorescent tubes mounted on a carriage under the exposure glass. The image of the document is reflected at an angle by a surface-silvered mirror mounted at 40° to the horizontal below the exposure glass (the object mirror); the reflection passes through a lens to another mirror (image mirror) mounted above the drum, also mounted at 40° to the horizontal but facing the opposite direction to the first mirror. From the second mirror, the image is reflected through the exposure housing on to the surface of the drum, as shown in Fig. 13.2.16 (b).

The lens used in the 914 Copier is an 8·25-inch focal length *f*/4·5 lens set at the factory to give a magnification of $1{\cdot}005 \pm 0{\cdot}3$ per cent. The reason for this very small magnification is to prevent the edge of the original being reproduced as a black line around the edge of the copy.

The length of the optical path (i.e. from original to image) is $16\frac{1}{2} + 16\frac{1}{2} = 33$ inches for same size reproduction. In order to obtain this length of path in the 914 Copier, the optical path is bent by means of the object mirror and the image mirror.

In the optical system used in the 914 Copier, the document original remains stationary and the exposure lamps and lens scan the document in synchronism with the rotating drum. The lens travel is half that of the lamps and its speed is half of the periphery of the drum. During each scan, the u/v distances vary but the magnification always remains at approximately one to one because the relationship between them remains the same.

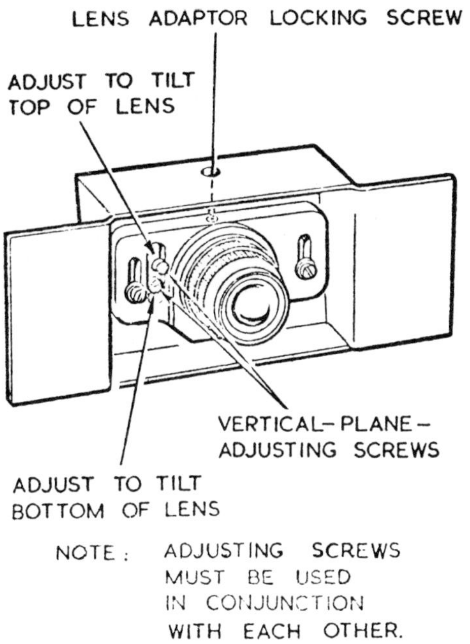

13.2.17. Optical mountings of the Xerographic 914 copier.

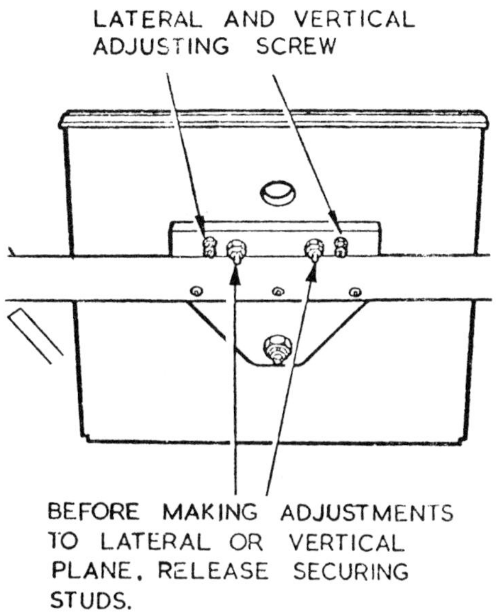

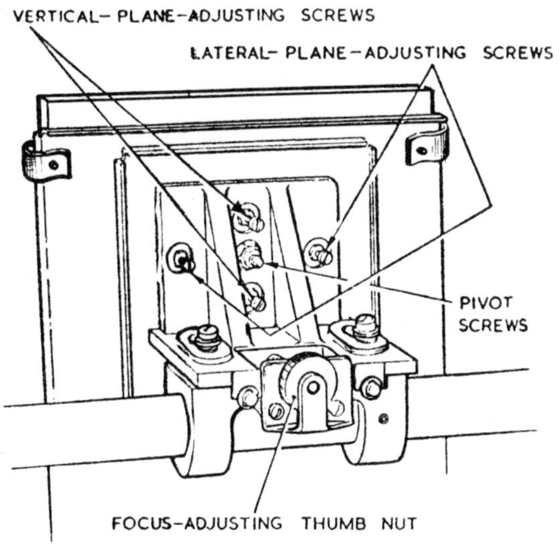

13.3 Optical alignment by image sharpness instruments

Chapter 11 gives details of traditional optical benches and the new technique of measurement of lenses for optical transfer functions.

The choice of best focus for a lens is generally that of the most attractive spread function taking into account all residual aberrations. When searching by eye for the plane of sharpest image one looks for a sharply defined edge, but usually the edge has a colour fringe and, for consecutive measurements, the eye has to come back to the same shade of colour in the fringe; thus one has to rely on memory in order to achieve accurate results. The settings are therefore not likely to be the same from one operator to another.

This is a very serious problem where high volume photographic equipment is concerned, as the visual setting of focus is a tedious job and sample testing with photographic film is necessary to maintain standards.

The first presentation of a new method to determine the plane of best focus was made by Dwin R. Craig of LogEtronics Inc. at the Society of Motion Picture and Television Engineers Conference in 1959.

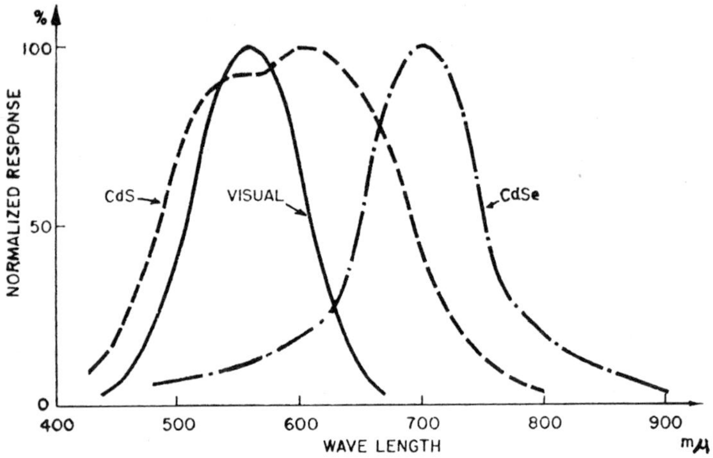

13.3.1. Typical spectral response curve for cadmium sulphide and cadmium selenide.

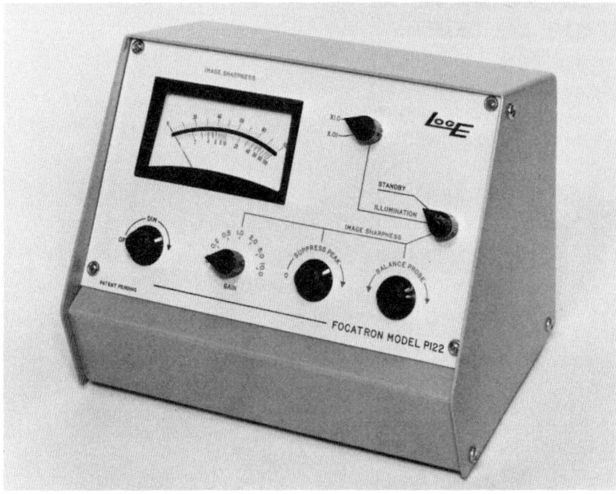

13.3.2. The Focatron Model P-122.

It is well known that the conductance of a photoconductive cell using cadmium sulphide (CdS) or cadmium selenide (CdSe) varies with the amount of incident light (Fig. 13.3.1), but an important fact is that the conductance also varies when the distribution of light changes, even if the total incident light is constant. This property of the photoconductive cell is not exhibited by photovoltaic or photoemissive cells. The local conductance in each part of the photoconductive layer is directly proportional to the local illumination. Many characteristics of a lens can be evaluated using a method based on changes in the light distribution which takes place in the image plane as the image of a high contrast target is moved through focus. A photoconductive surface in the image plane will measure this information in the image, without resort to the human eye or photographic techniques. (Figs 13.3.2 and 13.3.3.)

The basic components of the system are a Ronchi target with metal rulings on a glass substrate and mark/space ratio of 1:1, the image-forming lens, and the probe containing photoconductive cells. The probe must be connected to either a Focatron Model P-22 or Model P-122R readout unit. The standard probe covers a spectral range of 4000 to 7000 angstroms wavelength and can be made to fit cameras, microfilm cameras, precision enlargers, optical film printers, or Xerographic Copying machines. The photo-

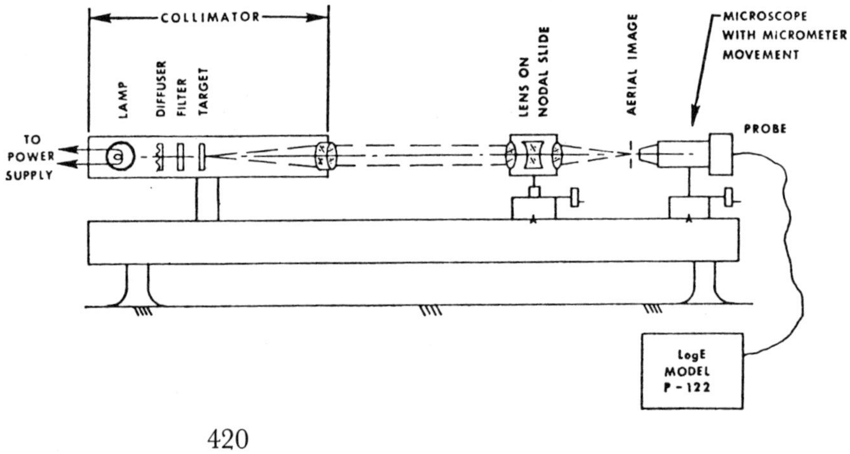

13.3.3. Typical optical bench arrangement.

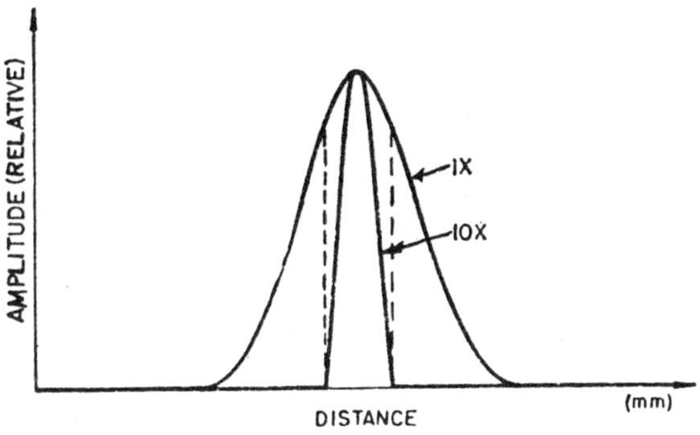

13.3.4. Probe output, with and without suppression.

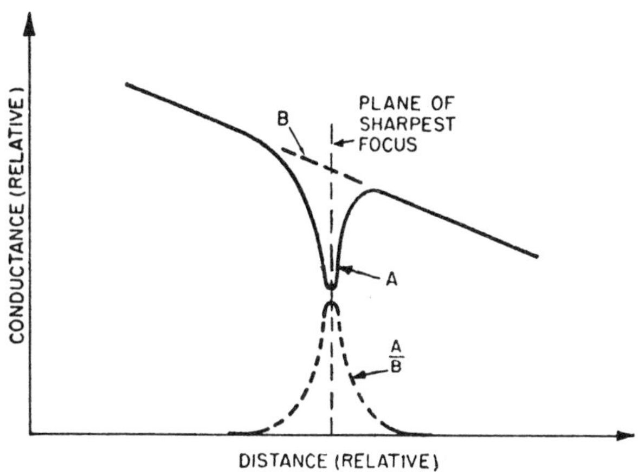

13.3.5. Basic focus-indicating curve.

conductive cells in the probe can resolve to about 250 lines per mm. As the output of the Focatron is electrical, not only can it drive meters and recording equipment, but it can be made to show when the image is defocused and to detect the direction of travel to the plane of best focus, thus making automatic focus control practicable.

When measuring a particular lens on the optical bench for the plane of sharpest image, one projects a line target through a collimator and the lens being tested, then via a microscope objective on to the probe. (At finite magnifications, one projects the target through the lens directly on to the probe.) The microscope containing the probe is then moved back and forth along the optical axis while the meter is observed. While the image is grossly out of focus, no change in the meter deflection occurs. As the image forms, the meter needle moves up-scale. Finally, when the image is in sharpest focus, the meter needle is at its peak position. As the probe is moved further, the meter needle falls until the image is again grossly out of focus. (Fig.13.3.4).

At the top of the curve (Fig. 13.3.4, curve 1 ×), the rate of change is fairly low; therefore, for such a curve it would be hard to detect accurately the peak indicating the plane of sharpest image. By increasing the gain of the system, we proportionately increase this rate of change. However, the deflection then occurs off-scale. Using a suppressing feature, the peak of the signal is brought back onto the scale. In this mode of operation (Fig. 13.3.4, curve 10 ×), the plane of sharpest image can be determined much more accurately.

If we plot cell conductance as a function of distance from the plane of best focus for the clear image, curve *A* in Fig. 13.3.5 results. Curve *B* represents the change in conductance which takes place in the cell with the diffused image. Thus, this second cell is used to compensate for changes in the total flux present at the probe, i.e., for the absolute light level.

The ratio of conductances of these two cells is displayed on a meter indicating by a peak reading the position of best focus. The display then looks like curve *A*/*B* in Fig. 13.3.5.

PRODUCTION ALIGNMENT IN XEROX COPIERS

Permission has been given by Mr William L. Kacin, Vice-President and General Manager of Log E Photomechanisms Inc, 15, Stepar Place, New York 11746, to publish the following article, 'Production Alignment of Optical Systems', by Donald A. McTarnaghan and Robert W. Morris of the Xerox Corporation, Rochester, N.Y. This article was originally published in Vol 2, Issue 3, of *Optical Spectra*, May–June 1968.

Introduction

Alignment of optical systems has always been a particularly difficult task both in the laboratory and on the assembly line. Conventional practice has been to establish alignment visually—specifically, determine the condition of sharpest focus in the image plane with the human eye. Depending thus on subjective judgement, there is the constant problem of doubtful accuracy and poor repeatability, whether for one individual or from person to person. At the same time, satisfactory or not, visual alignment is a time-consuming process.

Until relatively recently, visual methods of alignment were the only means available. Many devices employing optical systems are still aligned visually, often with the aid of complex, costly fixtures. If human errors are made early in assembly, they are caught in final testing and the system realigned. Even though visual alignment takes a good deal of time and painstaking adjustment, it is tolerated mainly because of tradition and existing investment in personnel, fixtures and standard procedures. However, with high-volume production of electrostatic copiers, Xerox and other firms have felt the need to improve alignment procedures by setting accuracy standards, narrowing the acceptable tolerances, and reducing the overall time.

When production of the Xerox copiers first began back in 1960, it was standard procedure to align all optical systems with a visual target. A position of sharpest focus was established by counting the lines on a standard NBS target, a procedure still widely used in industry today. The optimum position is that at which the viewer can count the most lines in the target. As is well known to anyone engaged in aligning optical systems, however, people often disagree on just where this position is. Therefore, about five years ago, we began to study the application of a new type of instrument that promised the capacity to establish a plane of sharpest focus without depending directly on the human eye. These Focatron image sharpness instruments, manufactured by Photomechanisms Division, Inc. of Huntington Station, N.Y., a subsidiary of LogEtronics Inc., have since been established as part of the standard optical alignment procedures for the Xerox Model 2400 and Model 660 copiers.

Aligning the 2400 optical system

The schematic drawing in Fig. 13.3.6 shows the optical system of the 2400 copier. The object surface is the curved platen where the document to be copied is located. The optical axis is reflected from the pivot mirror, through the lens to a fixed mirror, and then through an exposure slot to the surface of the selenium drum. The total length from the curved platen to the selenium drum is about 38 inches.

Since the focal lengths of lenses inherently vary somewhat from lens to lens due to manufacturing variations, it is necessary in alignment to match exactly the equipment path length to the particular lens being used. In aligning the 2400, this is accomplished by pivoting the frame supporting the fixed mirror in angle θ. When the condition of sharpest focus at the drum surface has been attained, it is locked in position. The image plane must be located within $\pm 0{\cdot}005$ inch from the actual drum surface. By using this adjustment,

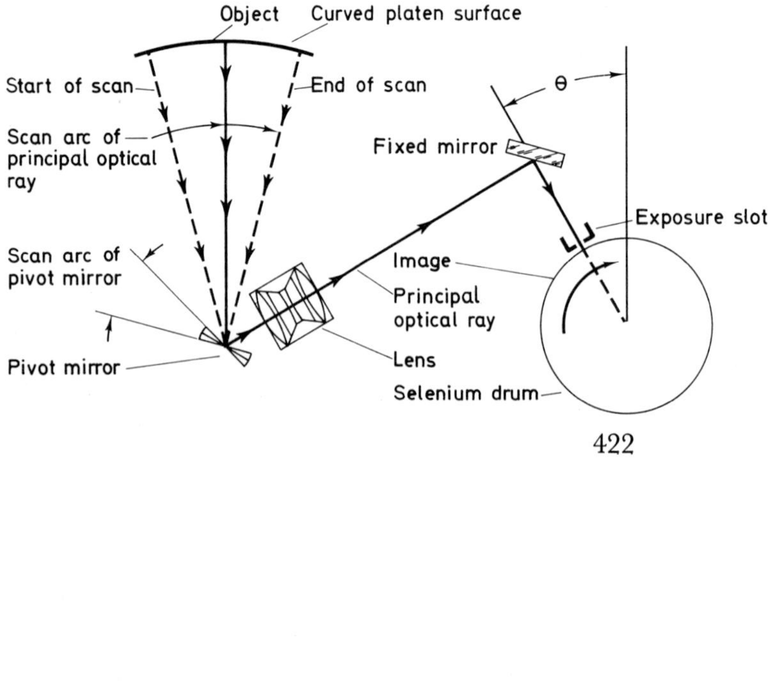

13.3.6. Schematic diagram of the optical system in the Xerox Model 2400 copier.

13.3.7. The optical system is placed in the alignment fixture at the right. The electrical output of the probe is indicated on the meter or recorded on the chart of the P-222R at the left.

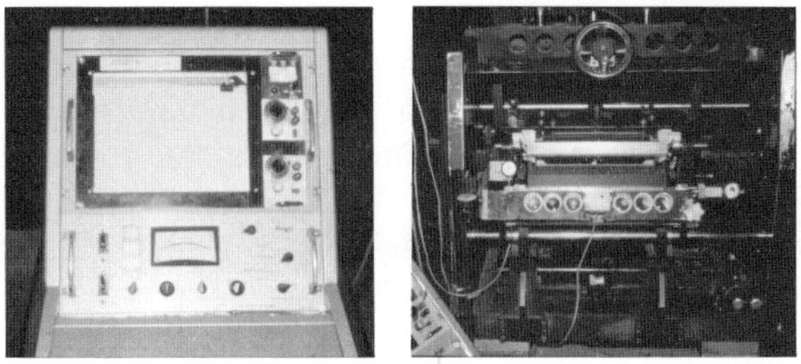

the normal anticipated build-up of manufacturing variations is cancelled and the drum surface will be located within the depth of focus of the lens.

In the original alignment procedure, a 1952 NBS visual resolution target was placed in the object plane and the image viewed on ground glass in the image plane at the nominal position of the surface of the selenium drum. The angle θ was then varied by moving the fixed mirror until the maximum number of lines was observed in the image plane.

With the Focatron® instruments, the significant difference in method is that the optimum angle (θ), which ranges between 28° and 30°, is set on the basis of a peak meter reading rather than visual observation and judgement. The instrument, whether an indicating or recording/indicating type, consists basically of a meter and a light sensing probe. The probe is a compact, solid-state device capable of producing an electrical output corresponding to the sharpness of the image falling on its photosensitive surface. The probe is placed in an image plane fixture located at the nominal position of the drum surface. A back-illuminated Ronchi ruling target with 200 lines/inch is placed in the object plane at the curved platen. At the position of maximum sharpness, the electrical output of the probe will be maximum, and the pointer on the meter will reach its peak deflection.

The alignment fixtures

Having established that the instrumentation would provide greater repeatability than we were able to attain with visual alignment, it was necessary, of course, to design fixtures that would incorporate the Ronchi target in the object plane and a probe in the image plane and allow convenient, fast preparation of each optical system for alignment.

The optical system is relatively large and heavy even though it does not yet have the selenium image drum or curved object platen in place (although the platen-retaining frame has been assembled). Therefore, it was necessary to develop the fixture shown (right) in Fig. 13.3.7, that would accept the optical assembly for alignment while still mounted on a pallet. With this arrangement, an optical system is easily pushed along a roller conveyor into the back of the fixture and locked into position for alignment. Once locked in, the Ronchi ruling target in the fixture is automatically located precisely at the centre of the object plane, and the probe is located at the nominal position of the selenium drum surface. The probe holder can be moved with precision through the nominal drum surface position along the optical axis, with its deflection shown on a dial indicator.

The electrical output of the probe bridge is connected both to a meter and the Y-axis of a recording pen on the console of a P-122R recording/indicating instrument. The horizontal deflection signal for the recording pen is taken from a transducer which parallels the dial indicator and represents the physical position of the probe in and near the image plane.

After first setting the position of the lens for proper centring and magnification, the operator makes a final adjustment of the position of the fixed mirror (Fig. 13.3.6), by observing the point of maximum needle deflection on the indicator. The position of the fixed mirror is then locked in and a trace is made on the graph of the probe output (Fig. 13.3.7), as the probe is moved through the drum surface position, zero reading on the dial indicator. These traces (two or more may be made to attain a high confidence level) are shown at right in the chart record.

The operator then sets the dial indicator at 0·005 inch above and below the nominal drum surface and traces a vertical line at each position. As shown in Fig. 13.3.8, an acceptable optical alignment has been attained when the peaks of the traces appear between the vertical lines representing the ±0·005 inch tolerance limits. When the optical alignment has been completed, the angle of the fixed mirror and serial number of the optical system are marked on the chart. The optical system is then removed from the alignment fixture and rolled along the conveyor to the alignment gauge at the next station in the optical assembly area. (Fig. 13.3.9.)

The alignment gauge is identical in construction to the alignment fixture and also is arranged so that the recorder traces the image sharpness peak as a function of probe position at the drum surface. However, the operator at this station cannot make any adjustments in the optical system. He simply places the chart traced at the alignment fixture into the recorder at the alignment gauge and makes several confirming traces on the same sheet (peaks

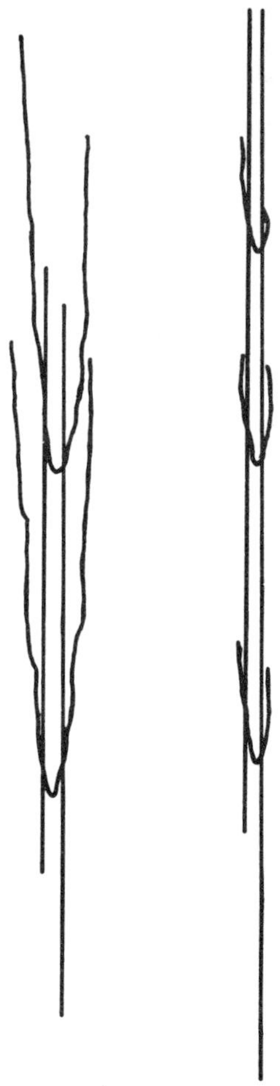

△13.3.8. **Traces made on the P-122R show peaks of focus recorded in the alignment fixture (*left*) and the alignment gauge (*right*). In each case, several peaks are traced as a check. The vertical lines represent the ±0·005-inch allowable range of peak position at the selenium drum surface.**

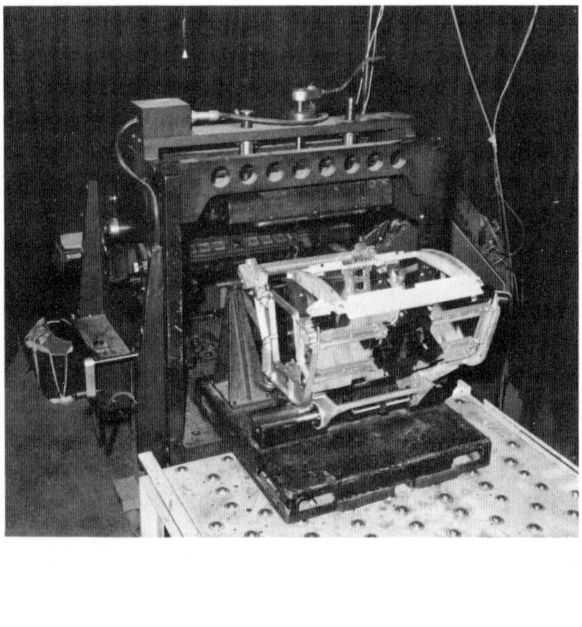

◁13.3.9. **Alignment fixture from the rear side, showing the optical system of the Model 2400 copier on its conveyor ready for sliding into the fixture.**

on the left in Fig. 13.3.8; but note that the horizontal gain of the recorder is slightly lower, so that the $\pm 0{\cdot}005$ inch lines are closer together). If the alignment is confirmed the gauge operator, who is actually a quality control inspector, approves the optical system and it goes on to the next assembly operation. If the alignment is rejected by the inspector at the gauge station, the optical system is sent back to the assembly operator at the alignment fixture. However, the requirement for realignment within this subassembly area has been cut to less than 1 per cent. The optical alignment can later be rejected in final inspection, at which point an inspector reviews quality by studying a resolution document. However, there are virtually no rejects for optical misalignment in final inspection.

The total time required for the alignment and inspection cycles has been reduced to about half of that for previous methods.

Other alignment tasks

The alignment operation at the assembly station also includes, prior to the adjustment of the fixed mirror, a check on magnification by observing whether two tee-shaped target lines located at the edges of the optical field fall within reference reticles located on the ends of the probe holder. Intermediate visual resolution targets are located in several places across the optical field and ground glass images of the targets in the probe holder are visually checked. Of course, these visual checks are critical only when there is some doubt that the sharp focus indicated at the centre by the probe is not also acceptable at other points across the optical fields.

Four additional probe positions are available in the probe holder in the event that it becomes necessary to check image sharpness at intermediate points between the centre and edges of the optical field. The lenses in the optical systems have been of such high quality, however, that alignment on the basis of the reading at the central probe position only has resulted in acceptable image sharpness all the way across the field. Squareness and angular relationships are established mechanically by the fixture. Lead screw-actuated mechanical movements create a smooth motion of the probe through the image plane.

There are nine work stations along the conveyor in the optical subassembly area, including two alignment fixtures and one alignment gauge. The instruments have proved to be extremely rugged production tools.

Alignment of the 660 copier

Alignment of the optical system of the 660 copier also takes place in separate assembly and inspection stations. The optical assembly of the copier consists of a sheet metal frame supporting a lens and two mirrors. With the optical components not yet tightened into final positions, the optical assembly is placed in the alignment fixture in which the lens, mirrors and frame are independently located. Adjustments are made in the fixture to bring the image produced by the lens within dimensional limits and to locate the focal plane reference surface at the proper position. An indicating instrument like that shown in Fig. 13.3.2 is used to find the position of peak sharpness in a manner similar to that of the recording instrument described above.

The operator's first steps in alignment are to set the object mirror position to the lens total conjugate as indicated by a colour code on the lens. He then tightens the lens slightly so it can be positioned smoothly. To obtain maximum resolution, he then adjusts the position of the image mirror until a peak is indicated on the meter. Finally, he checks the lens position for proper magnification as indicated by lines on targets and tightens the lens.

The 660 optical assembly, which is much smaller than the optical system of the 2400, is placed on a cart with other assemblies for transfer to a quality control station to check the alignment.

The optical assembly is placed into a gauge that is identical to the alignment fixture and represents the position of the optical assembly in the final product. The position of the probe in the gauging station is moved through the selenium drum plane (zero on dial indicator). As in alignment of the 2400, the position of sharpest focus, indicated by peak deflection of the pointer on the P-122, must be within $\pm 0{\cdot}005$ inch of the drum surface.

There are two alignment fixtures and one alignment gauge assigned to 660 production. The times for the alignment and gauging operation cycles have been reduced substantially over previous alignment methods.

Summary

Experience with both the Xerox 2400 and 660 has indicated the value of Focatron instrumentation in aligning optical systems in both engineering and production. The reduced alignment time, of course, has produced a more than satisfactory return on investment in the instruments and fixtures. More important, however, the much superior accuracy and repeatability of instrumental over visual methods contributes substantially in manufacturing and assembling optical systems of higher quality in large volumes.

13.4 Mounting a large telescope mirror

A very large mounting problem arises with the design of supports for an astronomical telescope primary mirror—such as the 98-inch Isaac Newton telescope, which is fully described in § 15.9. This mirror weights 9000 lb and must be supported within its cell so that flexure of the glass due to its own weight at any attitude does not cause distortions greater than the polishing limits, which are measured in millionths of an inch.

An air-bag axial support system has been used and 182 axial spacers are arranged in a symmetrical circular pattern over a three-section air-bag. The air pressure is 1·3 lb per sq. inch, which is sufficient to float the mirror. (See Fig. 13.4.1.)

Positive location in the axial direction is provided by three fixed pads in contact with the base of the mirror, and careful adjustment of the air pressure is necessary to avoid strain.

Support of the mirror in the radial direction is by a 36-lug mechanical system, together with four fixed constraining pads, for location of the mirror centrally to within a few thousandths of an inch throughout the range of working temperature. These pads perform a difficult function as any significant restraint in the axial direction would produce distortion in the mirror.

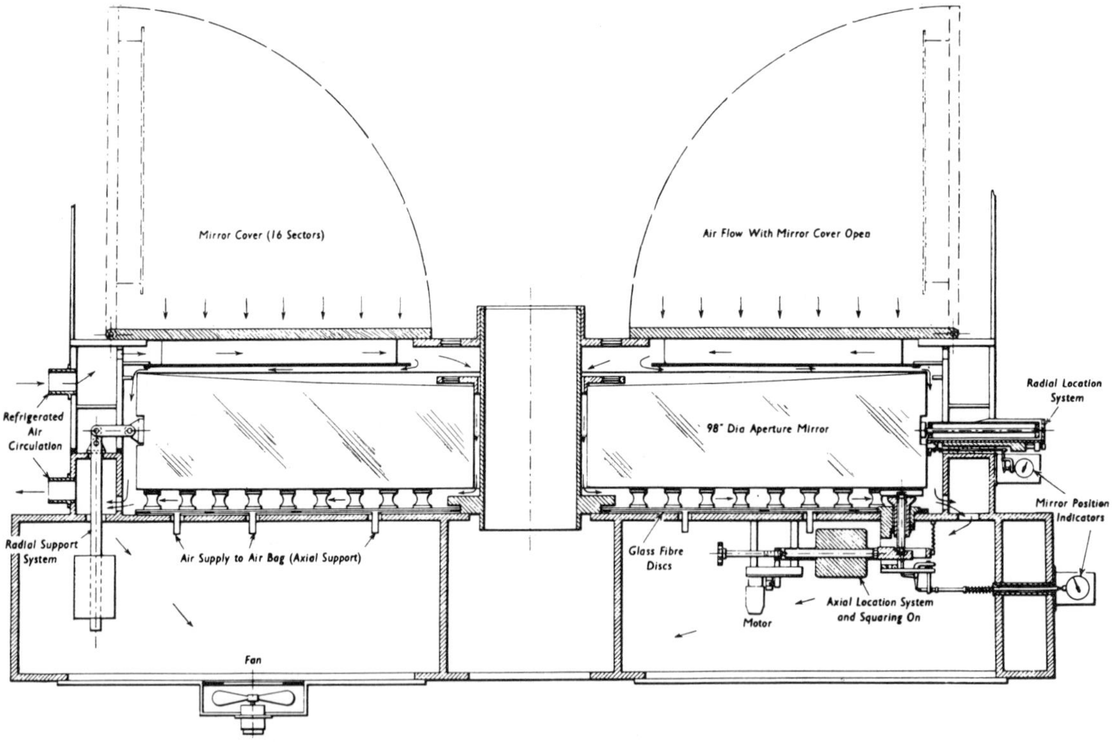

13.4.1. Mounting of the primary mirror in the 98-inch Isaac Newton telescope.

14 Production of Optical Glass

14.1 Introduction

Glass is formed by mixing silica sand (SiO_2) with carefully controlled quantities of various inorganic substances, and a proportion of scrap glass (cullet), followed by heating the mixture to about 1500°C and cooling the melt at a rate designed to prevent crystallization. This is followed by an annealing process. Full details of glass compositions are given in Chapter 5.

Glass-melting furnaces are constructed from special refractory materials which do not damage the glass. There are two main types of glass-melting furnace. The traditional pot furnace heats one or more fireclay melting pots at a time, each pot holding up to 1600 lb (725 kg) of glass depending on the type being made.

The modern tank furnace melts the raw materials required for a continuous flow of glass which is needed for automatic forming machinery. It is more economical in its use of fuel than the pot furnace and is used in all large-scale glass production. A tank furnace has a capacity of up to 2000 tons (2032 tonnes) of glass, and is in use night and day for a period of over two years, after which time it has to be rebuilt.

Glass-forming processes are all based on the fact that the viscosity of glass increases as the temperature falls. Mouth blowing was, for nearly 2000 years, the main forming method, and the quality of product achieved depended on the viscous property of glass.

Sheet glass is drawn upwards continuously from a tank—the ribbon being started on a metal former known as a bait. The rate of drawing determines the thickness, and the width is kept constant by passing the edges of the ribbon through cooled rollers which chill the edges of the glass. The chilled edges then act as supports for the less viscous glass suspended in between them.

Molten glass may also flow continuously from a tank between water-cooled rollers which control the thickness and may give a surface pattern to the glass. Alternatively, parallel surfaces may be ground and polished in the same continuous process to give polished plate-glass.

A recent development is the float process in which a continuous ribbon of glass, about 11 ft (3·35 m) wide, flows from the tank to float on the surface of molten tin at a controlled temperature in a controlled atmosphere. This ribbon of glass becomes very flat and parallel, giving undistorted vision, and does not require a subsequent grinding or polishing process.

Sheet and float glass are used in making optical components such as mirrors, graticules, theodolite circles, etc., where controlled refractive index is not required but reflection or transmission, and not refraction, are the essential properties.

For optical purposes, specially refined molten glass is extruded from a tank, in the form of blanks, which can afterwards be reheated and moulded into any required shape. Similarly, for optical purposes, molten glass can be poured out from the pot and allowed to set in a block, after which it may be cracked or sawn into small pieces, reheated, and moulded into any required shape.

14.2 Sheet glass

It is known that by the fourteenth century B.C. there was a well-established glass-manufacturing centre in Egypt and also in the Aegean. Some of the glass produced at that time was transparent and colourless.

The earliest use of glass for spectacles appears to have been in the thirteenth century A.D., and it was the seventeenth century A.D. before plate-glass for mirrors was made in England. In 1826, William and Richard Pilkington, with three other local partners, entered the flat-glass-making industry in St Helens, Lancashire, by forming the St Helens Crown Glass Company. The site was chosen because of the local availability of raw materials—sand, soda ash, dolomite, limestone, and coal for the furnaces.

The present company, Pilkington Brothers Limited, St Helens, Lancashire, have given permission for publication of the details of glass manufacture in the remainder of this section.

RAW MATERIALS

Glass is a soda-lime silicate, that is, it is made from silica (sand) soda and lime. If sand is mixed with soda in the form of soda ash or salt cake and heated, the soda melts and the sand dissolves in the soda. If there is an excess of soda, on cooling, a thick syrupy liquid is obtained, but if more sand is added, a hard, transparent, glassy substance results. This substance, unfortunately, is soluble in water and is known as water glass. By the addition of lime, the solubility of the sodium silicate is reduced, and with a sufficient quantity of lime, a durable glass is obtained which will stand up to the weather and to all strong acids except hydrofluoric.

Any process of manufacture is governed first by the article which it is required to make and secondly by the physical properties of the material employed. In the manufacture of glass, the principal physical properties which influence the processes are:

Viscosity. Glass has no definite melting point. If it is heated, it first softens so that it can be bent. As the temperature rises it reaches a point when it becomes a thick syrupy liquid—a state in which it can be 'worked', Finally, at higher temperatures it becomes a thin watery liquid.

Devitrification. Although weathering properties can be assured by a lime content, there is always the danger of crystallization, or even devitrification occurring. Above a certain temperature, known as the devitrification temperature, glass may be kept in a liquid condition without any change occurring, but if the glass is kept below that temperature for any length of time, crystallization or devitrification occurs. It is, therefore, essential in any process that the time taken to complete the operation shall not allow devitrification. The tendency to devitrify can be reduced by decreasing the amount of lime and increasing the soda, but this can only be done at the expense of the weathering properties.

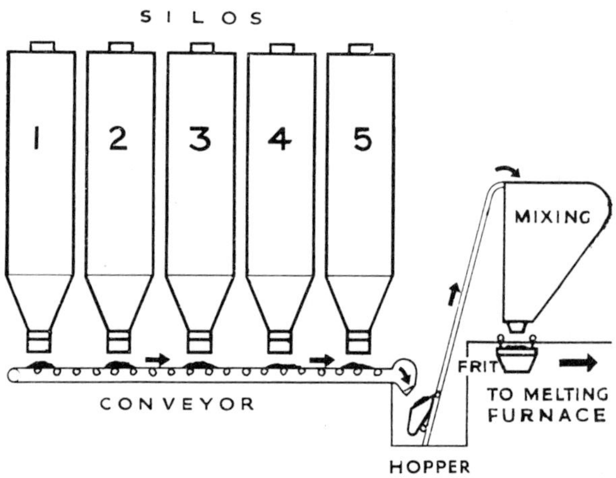

14.2.1. Central mixing plant for sheet glass manufacture, showing how the raw materials are 'weighed off' from the silos.

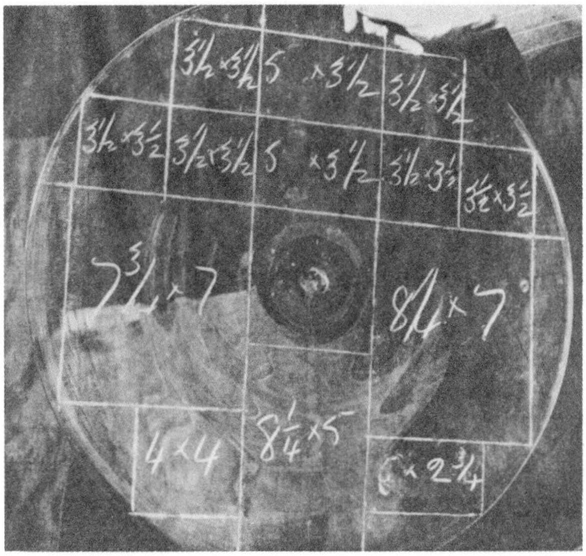

14.2.2. A crown of glass marked for cutting into square panes.

MANUFACTURE

The raw materials—mainly sand, soda, dolomite, and limestone—having been mixed together in a fine powder (which is known as 'frit'), are fed into the filling end of a furnace or tank and melted. (Fig. 14.2.1.)

The melting process takes place in three stages:

(1) The initial melting—that is the chemical reaction between the three ingredients. This results in a sticky mass full of bubbles.

(2) The next stage is the fining operation, which consists simply of raising the temperature, so that the glass loses its viscous nature and becomes quite watery, thus allowing the gases forming the bubbles to rise to the surface. At this stage the glass is so thin it is quite unworkable.

(3) The third stage consists of cooling the glass down until it is of the correct consistency for the particular process desired.

A sheet of glass left to cool naturally will break, and in order to obtain whole sheets of glass the sheet must be annealed, that is cooled down gradually in what is known as a lehr.

DEVELOPMENT OF MANUFACTURING PROCESSES

Crown

The first method of manufacture of sheets of glass through which clear vision could be obtained was the old crown glass process. In this process, the glass was gathered at the end of a blowpipe and blown into a form which was approximately pear-shaped, the pipe being at the thin end. An iron rod known as a punty, on the end of which was a blob of hot glass, was pressed against the thick end of the pear, to which it stuck. The neck of the pear, together with the pipe, was then broken off. The remaining portion was re-heated, and by spinning was opened out into the form of a disk.

This method resulted in relatively small usable sizes of glass and determined the size of the window of that time. It was also the source of the bull's eye or bullion. Although in the original process of manufacture this was practically waste and only used for the cheapest form of glazing, that is, cottage windows, today these bullions are made and used to give an antique air to modern homes and inns.

Cylinder processes

The next step in progress was the cylinder process in which the glass was blown in the form of a cylinder, split lengthwise and then flattened. Then followed the cylinder drawn process.

In this process the glass was ladled from the tank into a double-sided crucible or pot contained in a kiln heated by producer gas. A pipe with a re-entrant lip was lowered into the glass which welled over the edge of the lip and solidified, forming a solid ring by which the cylinder was drawn up from the pot. Air was supplied through the pipe to maintain the required diameter of the cylinder. In this way a cylinder 30 inches in diameter and 40 feet high was drawn. At the end of the draw the cylinder was cut off from the pot and lowered on to a cradle. There it was cut into lengths by means of an electrically heated wire which, in this process, takes the place of the hot thread of glass used by the hand-blower. The pot from which the cylinder was drawn was turned over so that the remains of the cylinder were melted out in the hot kiln, thus presenting a clean side of the pot for the next draw.

As the cylinders drawn in this way were too wide to be flattened in one piece, they were generally split lengthwise into two, three, or four sections or 'shawls'.

All the cylinder-made glass had to be flattened. The cylinder was usually placed on a carriage and introduced into the back of a flattening kiln where it was warmed up until nearly soft. It was then lifted from the carriage on to the flattening stone, opened out, and rubbed down with a piece of wood known as the Polissoir, which was kept wet, so that when the sheet was rubbed, the hot glass produced a buffer of steam between itself and the wood, which prevented the sheet from being scratched. After flattening, the sheet travelled down an annealing lehr, and was dipped in warm water in order to remove the scum which was generally formed from the sulphur in the gases in the lehr.

Flat drawn process

All the blown processes were indirect and for many years attempts were made to draw a flat sheet in the first instance. The difficulty in doing this was to maintain the width of the sheet. If a 'bait' in the form of a sheet of metal is dipped into molten glass, the glass adheres to the bait, and as the bait is drawn up the glass is drawn with it in the form of a sheet, but this sheet gradually narrows until finally only a thread is left. There are three successful methods of overcoming this difficulty: the Fourcault process, the Libby-Owens process, and the modern flat draw process.

In the *Fourcault process*, width is maintained by forcing the glass under hydrostatic pressure through a narrow slit in a fire-clay float or debiteuse. The debiteuse floats on the surface of the glass in the forehearth or kiln, and is depressed so that the slit is below the surface of the molten glass. Consequently, the glass wells up through the slit and is drawn away as fast as it is formed. Water coolers, i.e., metal boxes through which cold water is circulated, are situated on either side of the sheet so as to solidify it as quickly as possible. As soon as the sheet is solid, it passes between asbestos-covered rollers up an annealing tower, and emerges at the top of the tower, cool and annealed, where it is cut into lengths as required.

This process was invented at the beginning of the century, but it is only comparatively recently that more satisfactory glass has been made in this way, owing to the fact that as the glass has to be solidified quickly by means of the water boxes, the temperature at the top of the float soon drops to below devitrification temperature, and devitrification occurs at the edge of the slit. This causes the characteristic lines of most Fourcault glass. In order to overcome this difficulty, changes were made in the composition of the glass to prevent devitrification, but this led to further disaster and the glass was not durable. Not only did the glass 'weather' very rapidly, but in many instances when boxes of the glass were opened, it was found that the sheets were stuck together in a solid mass. This difficulty has now been mainly overcome by alterations in the composition of the glass, magnesia and alumina being used to prevent devitrification and maintain durability, but devitrification still occurs.

At the beginning of the operation, really good glass is drawn, but as time goes on devitrification begins and this gets worse and worse until finally, after some days, the process has to be stopped and the whole kiln re-heated so as to melt away the devitrified glass.

In the *Libby-Owens process* the width of the sheet

is maintained by pairs of small rollers which grip the edges of the sheet just above the level of the metal in the tank. The sheet is solidified quickly by means of water boxes, as in the Fourcault process. It is then re-heated and bent horizontally over a roller and travels along a roller lehr for annealing and at the end of the lehr is cut into pieces as desired.

This process eliminates devitrification troubles with the slit in the Fourcault process, but the surface of the glass is not so good owing to the re-heating and bending over the rollers. In the early days of the process, this trouble was much greater than it is now, and glass made by the Libby-Owens process could always be detected by the surface which has been in contact with the roller, and which was always slightly spoiled. This difficulty has been largely overcome by the use of highly polished rollers made from untarnishable metals, but the surface is still never as brilliant as the fire-polished surface of the old crown glass.

The difficulties of the previous flat draw methods were overcome by the introduction of the *Pittsburg process*, which has been modified and improved to become the *modern flat draw process*.

The melting in this modern method of sheet glass manufacture is carried out in a large tank. The raw materials, which consist of the actual ingredients and broken glass—known as 'cullet'—are fed into the filling pocket of the tank.

A glass tank, which may be as large as 120 feet long by 36 feet wide and 5 feet in depth, has sides and bottom made of clay blocks and the roof of silica bricks, and may contain anything up to 1200 tons of molten glass, with temperatures varying from 1200°C to 1530°C in different parts of its length. There is no end to the variations which may be made to the shape of the tank so as to melt the frit at one end and produce seamless and homogeneous glass at the other. It is comparatively easy to forecast the convective currents in a beaker in a laboratory, but the convective currents in a glass tank of such large dimensions are more difficult, and to understand them is to know how to produce good glass. Actually, the amount of glass flowing down the

14.2.3. The flat-drawn process. At the end of the furnace, the molten glass (metal) is drawn in a thin sheet up a 30-feet-high annealing tower. At the top of the tower, the glass is cut into large plates and removed with suction pads.

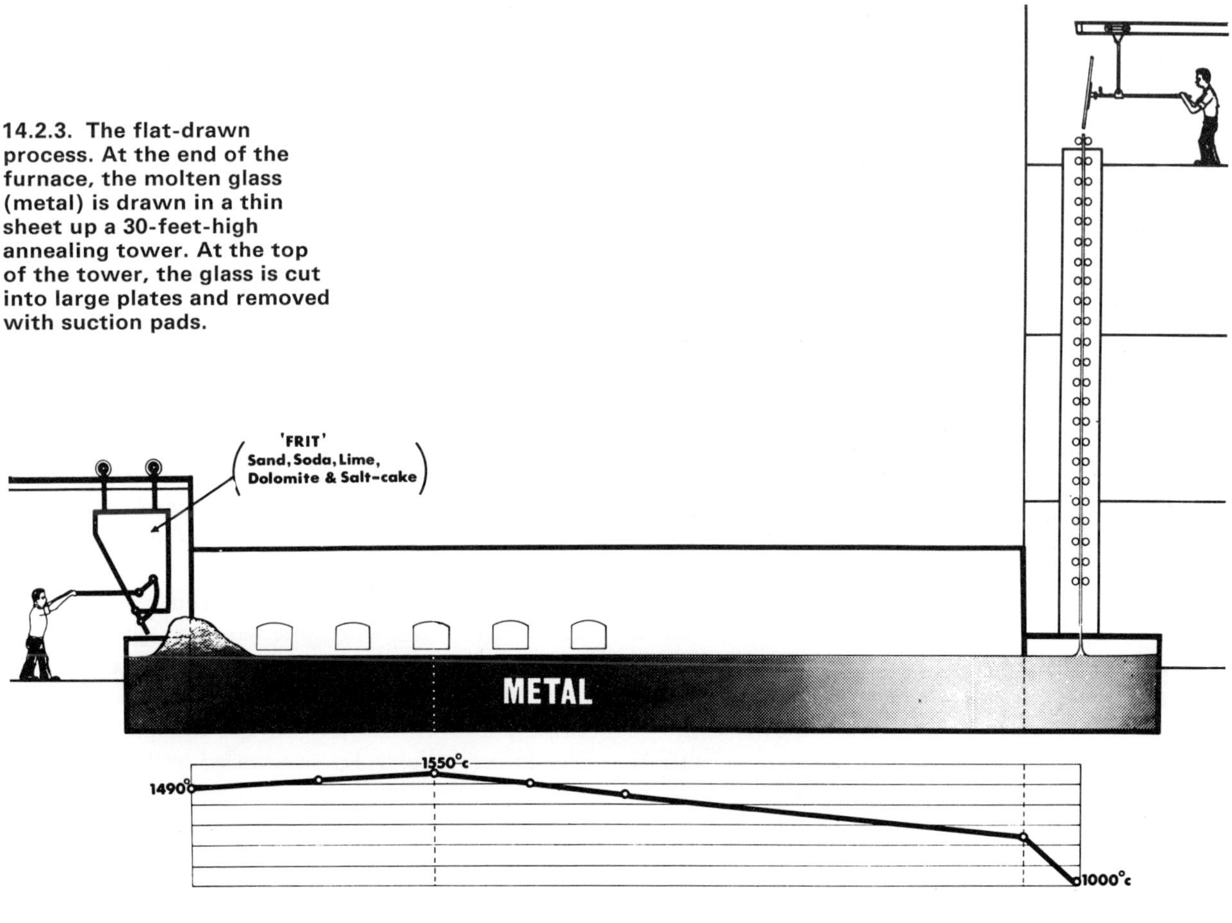

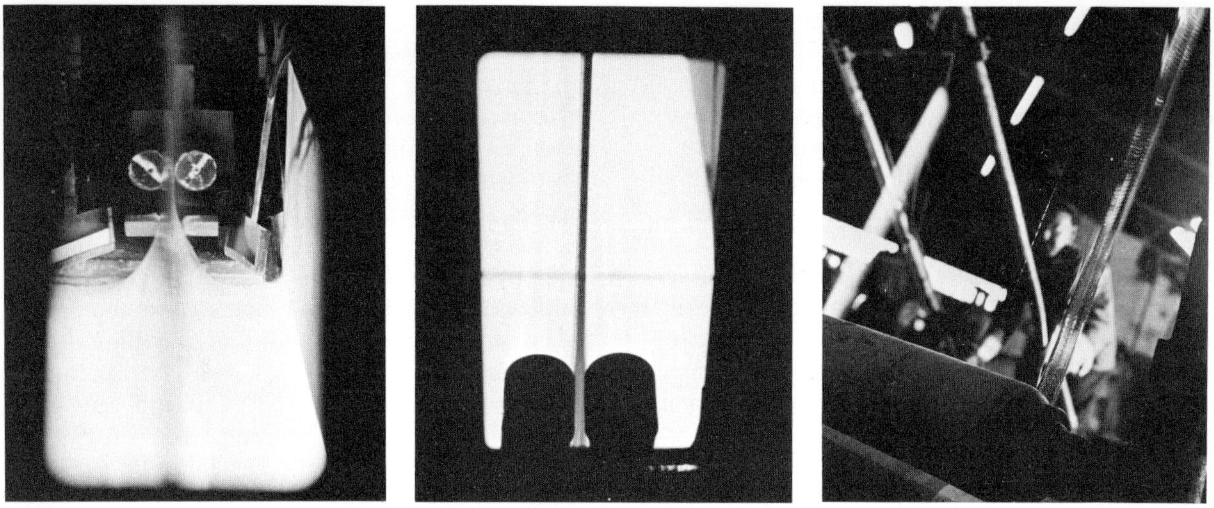

14.2.4. The flat-drawn process. ***(Left)*** **The sheet glass being drawn vertically up the 30-foot tower from the end of the furnace.** ***(Centre)*** **The glass part of the way up the tower.** ***(Right)*** **The sheet glass, with its perfect fire finish, before it is cut off at the top of the tower. Only the edges gripped by rollers are marked.**

middle of the tank, due to convective currents, is about twenty times as much as that being withdrawn at the working end.

To form the glass into a sheet, it first passes from the tank into a drawing kiln, a relatively small extension to the tank and separated from it above the level of the glass surface by a tweel and shut-off. The tweel is a slab of refractory material suspended from one edge and lowered until it rests on the shut-off and completes the seal between the tank atmosphere and that of the kiln. The shut-off is a block of refractory material which floats on the surface of the glass. There are usually four or five drawing kilns to each melting tank.

After entering the drawing kiln underneath the shut-off, the glass flows round either side of the submerged clay block, which is of special design, and known as the draw bar, and is drawn up in the form of a sheet from the surface above a series of electrically driven asbestos-covered rollers which are mounted in pairs in a cast-iron tower, situated above and parallel to the length of the draw-bar.

To start the process, an iron grille known as the 'bait' is lowered between the tower rollers into the glass in the kiln. When the bait has remained there for a short period the molten glass sticks to the iron, and the bait is slowly lifted, drawing behind it a sheet of glass. When the leading edge of the sheet has passed between the first few tower rollers, the bait can be cracked off from the glass, and the rollers engaging the sheet of glass which has followed the bait draw up a continuous strip or sheet into the annealing tower. (Fig. 14.2.4.)

The success of the process lies in the provision of devices for maintaining the width of the ribbon of glass being drawn; since it is in a plastic condition there is a marked tendency for the glass to 'waist'. This would occur progressively until the sheet ceased to draw at all. The usual mechanism employed for this purpose comprises two parts—a form and a pair of knurled air-cooled rollers. The form—a slightly curved steel plate with a machined slot in it—is placed just above the level of the glass in the kiln, so that the edge of the sheet draws through the slot. The knurled rollers are of steel and are placed a few inches above the form, engaging an inch or so of the edge of the sheet of glass. The two rollers are pressed towards each other so as to grip the glass firmly, and are usually driven independently of the tower rollers. These knurled rollers cool the edge of the glass sheet and thereby prevent subsequent 'waisting'.

14.2.5. The flat-drawn process. When the newly formed glass emerges, an auto-cutter enables it to be snapped clean and removed by an operator using hydraulically controlled suckers.

Facing the sheet at a position just above the level of the glass in the kiln are water-cooled steel boxes. The purpose of these is to assist in solidifying the sheet as soon as it has been formed. Once the sheet has been formed, it may be drawn continuously over lengthy periods, the flow of the continuous sheet only being interrupted when major repairs are necessary.

As the glass is drawn up the tower, which acts as an annealing lehr, it gradually cools off, and at the height of 30 feet above the drawing kiln, is sufficiently cool to be cut off. This cutting off is accomplished by putting a wheel cut on the back of the sheet, the cut being produced by a mechanically driven cutter, operated automatically, when the sheet reaches a certain predetermined height. The glass is then pulled away from the cut by means of an automatic suction pad, which enables the sheet to be removed and trimmed. (Fig. 14.2.5.) Trimming consists of the removal of the edges of the glass which bear the marks of the knurled rollers. The glass lost by this edge trimming is returned to the tank and re-melted.

From the trimming stage, the glass is sent into the warehouse where it is examined, sorted, and cut up into saleable sizes ready for despatch.

Flat drawn sheet glass can be drawn in standard thicknesses ranging from 2–6 mm—the thickness of the sheet being determined by the speed of drawing, the skill of the operator and, of course, the accuracy of the machine.

Flat drawn sheet glass is used where through vision is required but some degree of distortion is tolerable in the interests of economy; that is, on housing estates and in factories and for horticultural purposes. For better quality, distortion-free glazing float or clear plate glass should be specified.

Rolled and cast glass

The most obvious and simple way to make a sheet of glass from the molten metal is to cast it or roll it. This is the method employed in the manufacture of rolled and rough cast, cathedral, figured rolled, wired, and also polished plate glass. (Fig. 14.2.6.)

In making *rolled and rough cast glass*, molten metal was taken from a melting tank in a ladle, poured on to a cast-iron table and rolled out into a sheet by a roller which travelled on the table, the thickness of the sheet being determined by metal strips between the rollers and the table.

This system has now given place to a continuous method of casting in which the raw materials are fed into one end of a gas-heated tank, melted, refined, and cooled and then drawn from the other or working end of the tank in the form of a horizontal continuous ribbon and not vertically as described for flat drawn sheet glass. The ribbon passes on rollers into an annealing lehr, at the end of which it is examined and cut into the required sizes.

Rolled and rough cast glasses are used where clear vision is not required; for instance in factory roofs or vertical glazing.

Cathedral and *figured rolled glasses* nowadays are made by a method similar to that described above for rolled and rough cast glasses, but in addition a pattern or texture is impressed usually on one surface by a patterned roller. These glasses are also made in a series of standard tints for lead-light making, etc. This group of glasses possesses a varying and progressive degree of light diffusion and obscuration, and are used for internal partitions, windows of rooms where partial or total privacy is desired, and also in factory glazing where direct vision is not essential.

14.2.6. The rolled-glass process. A ribbon of molten glass flows from the melting tank and passes between rollers before being annealed.

About the middle of the nineteenth century when the use of cathedral glasses extended to roofs and canopies, the danger of objects falling on to the glass and causing resultant damage was realized. The question of fire also arose, because ordinary glass would break and allow the flames free passage through a building so glazed.

These difficulties led to the introduction of *wired glass*, that is, a rolled glass into which wire netting is inserted during the process of manufacture. The first patent for such a glass was taken out in 1865, the hexagonal type of wire being used.

14.3 Polished plate and float glass

On 13 February 1969, Sir Alastair Pilkington, F.R.S., gave a lecture to the Royal Society, London, on 'The Float Glass Process'. This section consists of extracts from that lecture, which is published with his kind permission.

HISTORICAL INTRODUCTION

The importance of the float process can be placed in perspective by taking a look at the processes used for making flat glass before and at the time of its invention.

There have, through the ages, been two basic methods of forming flat glass: the window glass processes and the plate glass processes.

Window glass processes have all depended on forming a sheet by stretching a lump of molten glass. They all have the characteristic of brilliant fire finish. Three processes, crown, cylinder, and finally drawn sheet, have been used to make windows. (See § 14.2.)

Sheet or window glass production was first mechanized on a large scale in the early part of this century. In 1903, the American Window Glass Company developed a method for the mechanical blowing of cylinders many times larger than hand blown cylinders. This drawn cylinder process drew cylinders up to 13·4 m long and 1 m in diameter from a molten pool of glass. The quality of the glass, however, was inconsistent.

The problem with the cylinder process was that it was discontinuous and the cylinders, however large, had to be split and flattened, which was costly and harmful to the surface. The logical evolution of this process was to draw a flat continuous sheet of glass from the molten pool. The modern sheet glass process was first developed by Fourcault in about 1914. (See § 14.2.)

The surface of glass made in this way has what is called 'fire-finish' (Fig. 14.2.4). The fire-finish surface is achieved by letting the glass cool down on its own without touching anything solid while it is still soft.

The major problem with all the glass made by these window processes was that it was bedevilled by distortion. All the manufacturing methods involved stretching the molten glass whether by spinning, blowing, or pulling it and this stretching converted inhomogeneities into distortion. The processes also made only a comparatively thin glass.

Neither of these properties was acceptable for mirrors or coach windows, or, later, for car windows or for the large distortion-free shop windows which came into vogue from 1850 onwards.

THE PLATE PROCESS

The plate process was developed to meet these requirements. To create distortion-free glass the molten metal was cast on to a table, rolled into a plate and, after annealing, ground flat and then polished. Grinding involved several stages, using finer and finer sand, and polishing was done with rouge.

The principles and, as we shall see later, the disadvantages of this process remained with the industry until the 1950s. Financial requirements were, from the beginning, quite enormous. The ordinary small glasshouse for making crown or cylinder glass could be set up fairly cheaply. But much more was required for plate manufacture. It needed a big melting pot furnace, a casting table, cuvettes or cisterns, cranes, grinding and polishing machinery, and extensive warehousing facilities. All this required unprecedented organization and greatly increased capital.

Some indication of the scale of the process is given by the factory which made Britain's first plate glass in 1773. Built by British Plate Glass Company at Ravenhead, St. Helens, the casting hall was 103 m long by 46 m wide, the largest industrial building of the period.

From 1688 until the First World War, plate glass manufacture moved ahead only very slowly. Progress was keyed to the logical development of greater mechanization, with the process always embodying the principles already described. It was in the 1920s that this classic example of evolutionary progress accelerated in leaps and bounds mainly due to the brilliance of development engineers.

Just after the First World War the Bicheroux process was introduced. Glass was still melted in pots but it was then rolled into a sheet between mechanical rollers rather than being cast on to a table and then rolled. This made a smoother sheet with a consequent saving in time and material in the grinding process.

Simultaneously, however, the search was on for a method of making the process continuous.

The first breakthrough came from Ford in America, where it was shown that glass could be rolled continuously.

As a result of Pilkington cooperation with Ford this discovery was successfully exploited. And after Ford's initial work the major developments in plate glass manufacture came from Britain.

We first developed a process which successfully combined a continuous melting furnace with the continuous rolling of a ribbon of glass. This was then sold back to America.

In 1923, the next advance towards continuity of plate production was made when Pilkington introduced the first continuous grinding and polishing machine. It took our plates of glass on to a series of tables which moved through the grinders and polishers. At the end of the process the table dropped into a tunnel and returned to accept another plate of glass.

The next logical step was a machine which could grind the ribbon of glass on both sides simultaneously as it came out of the annealing lehr and *before* it was cut into plates. Pilkington developed such a machine during the early 1930s and put it into service at their Doncaster works in 1935.

The machine, called the twin grinder, is acknowledged as the final and most remarkable development in the long history of plate glass manufacture. It must rank as one of the finest examples of large scale precision engineering and it gave Britain world technological leadership in the manufacture of high quality flat glass.

In the machine a continuous ribbon of glass about 300 m long was ground on both surfaces at the same time with enormous grinding wheels fed with progressively finer sand. The machine was driven by 1·5 MW and this power was expended in grinding a slender ribbon of brittle material. Even more remarkable was the fact that the bottom grinding wheels were kept perfectly flat and level while they were actually wearing away.

THE FLOAT PROCESS

The polished plate process was used in Britain until the 1960s. It met all the demand for thick and thin distortion-free windows. Why then has it been superseded?

Glass wastage amounting to 20 per cent of production and high capital and operating costs were its drawbacks. Its sister process, sheet, could not replace it because it could not be used to make the high quality products free from distortion which were required by the market. It could, however, make glass which retained its natural brilliance without the need for grinding and polishing.

Many minds had dreamed of combining the best features of both processes. They wanted to make glass with the fire polish and inexpensiveness of sheet and with the distortion-free quality of polished plate. The first time this was successfully achieved was in 1959 when the development of float was announced.

In the float process, a continuous ribbon of glass moves out of the melting furnace and floats along the surface of an enclosed bath of molten tin (Fig. 14.3.1). The ribbon is held in a chemically controlled atmosphere at a high enough temperature for a long enough time for the irregularities to melt out and for the surfaces to become flat and parallel. Because the surface of the molten tin is dead flat, the glass also becomes flat.

The ribbon is then cooled down while still advancing across the molten tin until the surfaces are hard enough for it to be taken out of the bath without the rollers marking the bottom surface; so a ribbon is produced with uniform thickness and bright fire polished surfaces without any need for grinding and polishing. (Fig. 14.3.2.)

14.3.1. In the float-glass process, raw materials (1) are fed into a regenerative furnace for melting. The molten glass then passes (2) into the float bath. Here, in the controlled atmosphere (3) the ribbon of glass (4) floats on the molten tin (5) and acquires its perfect flatness. The glass then passes out over rollers (6) for automatic cutting and dispatch (7).

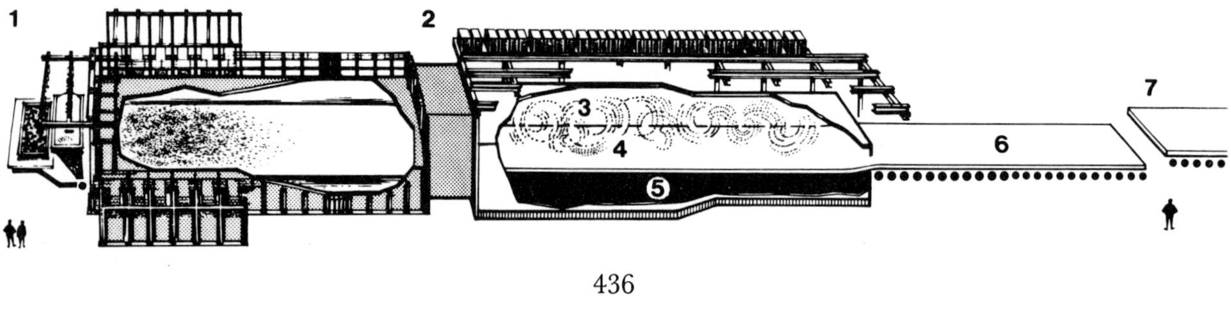

14.3.2. Float glass leaving the production line at the Pilkington Cowley Hill works, St Helens. Grinding and polishing are eliminated by the process.

Advantages of the float process

It has been said that float approaches close to a theoretically perfect process. It certainly has many built-in advantages.

First, in such a floating system the liquid surfaces become naturally flat and parallel. While the glass floats on the surface of the molten metal, and is held at a high enough temperature, the forces of gravity and surface tension combine to produce a good planimetry. There is little tendency for the forces which form the ribbon to convert any thermal or chemical inhomogeneity into distortion.

Secondly, we have not yet found a limitation on the output from the float bath. The speed at which the ribbon is formed is determined by the melting capacity of the furnace. We have already run at speeds in excess of 14 m a minute to make 3-mm-thick glass and there seems to be no reason why we should not increase this speed.

Thirdly, the variable costs for converting molten glass into a finished ribbon are remarkably low and only a small fraction of the equivalent part of the polished plate process.

Fourthly, the width of the ribbon can be easily altered at any time within the limits of the float bath and lehr; even the dimensions of the bath itself can be changed without much difficulty. This is important in reducing warehouse loss by matching the ribbon to the orders being cut.

Fifthly, the process produces glass with a very bright finish, and surface damage is much lower than with any polished plate glass we ever made.

Sixthly, because it is a horizontal process, annealing the glass is relatively simple.

Seventhly, it produces a finished ribbon at the

end of the annealing lehr, which encourages the use of an automatic warehouse.

And finally, the process lends itself to long trouble free runs. We have in fact run a plant for as long as 24 months without a major stop.

Thickness control

After the announcement in 1959 that we could make 6·5 mm float glass, we aimed to make it thicker and thinner. The same forces which made the glass flat were tending to determine the overall thickness of the ribbon at approximately 6 mm. This was fortunate since 50 per cent of our market for polished plate glass had been for this thickness.

But we had the other 50 per cent to think about. The full commercial potential of float could not have been realized without mastery of ribbon thickness. Just two years after float was announced we had learned to make a product half the thickness of our original float. The principle was to stretch the glass but in a gentle and controlled way so that none of the distortions arising in sheet glass processes could occur.

In the next three years we managed to make float thicker. The principle is simple in that we arrest the spread of molten glass in the float bath and allow it to build up into thicknesses which could be as great as 5 cm. The commercial range now available is from 3 to 15 mm ($\frac{1}{8}$ to $\frac{5}{8}$ inch).

Warehousing development

Another problem we had been conscious of during all stages of development was the need to produce a warehousing facility capable of handling the flow of good glass that the process promised to produce and cutting it into the myriad sizes our customers require. Now we have developed fully mechanized warehouses for cutting and handling this glass. This year they will cut and handle 16 000 km of glass ribbon.

The warehouses make full use of computers and stand by themselves as a major advance in flat glass technology. The computers are used to match orders for a very complex range of glass sizes to the continuous ribbon of glass. They in fact instruct the cutting machines in such a way that optimum usage of glass is obtained.

The electro-float process

In addition to warehousing advances, the float process has opened the way to other radical developments in flat glass manufacture. One such example is the electro-float process, which was announced in 1967, and which in itself is a major invention.

The process enables the surface of the clear float ribbon to be modified while it passes over the bath of tin. It exploits the unique features of the float process to effect rapid ion replacement between metals and glass. Although the full potential of this process remains to be realized, it has already proved capable of producing, economically, small quantities of heat rejecting glass on a mass production line. This first product, called 'Spectrafloat', is already being used in buildings and in prototype glass roofed cars.

14.4 Ophthalmic glass

This section describes the manufacturing processes in producing the raw material required for spectacle lenses. (See Chapter 7.) It consists of an article, 'The Manufacture of Ophthalmic Glasses', by Mr A. B. Scrivener, B.Sc., Technical Sales Service Manager of Chance-Pilkington (Pilkington Optical Division). It is published with his kind permission.

Ophthalmic properties

The supply of prescription spectacles involves the services of a number of industries and professions in order that each individual wearer of spectacles can be assured that the performance of these spectacles is adequate to his needs. In this article we are concerned with the early stages of this manufacturing process, which is the supply of spectacle glass from which prescription lenses can subsequently be ground and polished and fitted into frames according to various styles and personal tastes. Plastic materials are also used in the construction of ophthalmic lenses and have particular advantages in certain conditions, but it is with the glass lens and its manufacture that this article is concerned as this material is used in the manufacture of the greater proportion of spectacle lenses.

For simple visual corrective work, the material

chosen to form the spectacle lens should have no appreciable colour or light absorption and should have a controlled refractive index such that the curvature control of the lens is the only control necessary in order to provide the range of powers necessary to give visual correction. The optical power of a thin lens is directly related to the value of the refractive index and the curvature on the two major faces of the lens by the following relationship:

$$D = (N-1)\left(\frac{1}{R_1}+\frac{1}{R_2}\right)$$

where D is the power in dioptres, N is the refractive index of the lens and R_1 and R_2 are the radii of curvatures of the two major surfaces measured in metres.

The control of the optical properties of spectacle glasses is the responsibility of the glass manufacturer. The nominal value for refractive index of spectacle crowns is set at 1·523 with a tolerance of $\pm$0·001. These values are specified in the appropriate British Standard (B.S. 3062) together with other values for different glasses. Where prescriptions call for multi-focal lenses the glass manufacturer can meet this either by supplying crown pressings which have different curvatures over different parts of one of the faces, or alternatively by supplying one or more flint glasses of higher refractive index to combine with the crown base in the construction of these lenses. The actual fusing operation involved is carried out at a later stage in the sequence of processes, but the glass manufacturer in supplying raw material to this operation has to ensure that not only are the optical values of the material controlled to appropriate standards but also that the thermal characteristics of these glasses enable the components to be fused together at temperatures in the region of 600°C–700°C and then cooled to room temperatures without serious stresses being developed which could constitute a risk of the fracturing of the glass.

So far consideration has only been given to the clear or white spectacle glasses. Circumstances can arise where protection against excessive glare has to be given. This is achieved by using coloured glasses and a whole range of tints can be made by additions of colouring agents to the basic composition of the crown glass. As with the other physical properties of the glass it is the glass manufacturer's responsibility to control the variation of these tinted glasses from the point of view of colour and transmission, so that no objectionable differences in tint are apparent in the finished spectacle lenses. The specifications in this respect are obtained by mutual agreement between both the glass manufacturer and the lens processor, and in the majority of cases this control is effected by the visual comparison of production samples with limit samples that have been established. Whilst it is possible to determine the colour of the glass in numerical terms, it has so far been found impracticable to rely entirely on this method of specifying colours.

Raw material control

In its normal use the prescription lens has to permit vision undistorted by the presence of chemical inhomogeneities within the glass material or inclusions of other kinds. In addition, similar faults which, although not affecting the performance of the lens, are visible when viewed in other directions, and therefore appear to be rejectable on aesthetic grounds, also have to be avoided. Here again, it is the glass manufacturer's responsibility to ensure that the quality of the material he is supplying for subsequent conversion into prescription lenses is of such high standard that these kinds of defects do not present a serious problem to the subsequent processers.

The physical properties referred to in the previous paragraph depend largely, though not entirely, on the basic chemical composition of the glass. In order therefore to have adequate control over the physical properties, it is necessary that suitable control be exercised over the quality of raw material supplied to the glass manufacturer. For the white crown and flint glasses, some dozen raw materials are involved and an even larger number are concerned when the colouring materials are considered which are used in the various tinted crown glasses.

Some of these materials are as follows: silica, soda ash, potash, limestone, nitre, magnesium carbonate, barium carbonate, barium nitrate, borax, lead oxide, antimony, arsenic.

Colouring materials include ceria, manganese, cobalt, didymium, nickel, iron, vanadium.

It is easily seen that the presence of colouring

materials must be avoided at all costs in those materials which will be used for the clear glasses. Iron is present as a contaminant in most of our raw materials, and particularly in the silica, which is obtained from natural deposits of sand. It is normal for the remaining materials to be supplied by the chemical industry and again material specifications are agreed between the individual suppliers and the glass manufacturer. Sand, however, since it is a natural material, is usually processed further by the glass manufacturer by drying and magneting to obtain a purer form of silica.

The production process

The present process for manufacture of ophthalmic glasses is a continuous type in which raw materials are mixed together according to basic composition and then fed regularly to a melting furnace. Here the raw materials break down and fuse together into a glassy substance which at this stage is far from uniform in chemical composition. The glass moves to a further part of the furnace where temperatures are raised so that complete solution of the more refractory of the raw materials takes place and various gases involved during the fusion process can escape. The glass is cooled and stirred in order that a chemical and temperature homogeneity can be established before the glass is converted by mechanical means into lens blanks.

The raw materials required for a particular glass composition are initially transferred to hoppers situated above the weighing machine. This is illustrated in Fig. 14.4.1. By remote control the correct weights of each material are transferred from these hoppers to the weighing machine and their weights are recorded and can subsequently be compared with the standard batch specification to ensure that the correct composition has been obtained. The weighed materials are then transferred to mixing machines and from there, with the aid of batch boxes, the material is transferred to the melting units. Sufficient material is contained in each of these batch boxes to last several hours production time.

To obtain glass of the correct uniformity of composition and quality it is necessary that the whole temperature regime of the furnace from melting to stirring operations be critically controlled. After the stirring operations, glass is fed through a delivery feeder which controls the overall flow rate of the melting unit. The facility with which glass can be controlled depends upon the relationship between the viscosity of the glass and its temperature. As the temperature of glass changes from a high level to a lower level, a gradual stiffening of its form occurs and therefore it is possible to reduce the flow of glass through a feeder by reducing

14.4.1. Weighing raw materials for glass production.

14.4.2. Spectacle pressing set-up. Three stations are allowed for the glass to cool before it is lifted off the press table.

14.4.3. Spectacle pressings emerging from the annealing lehr.

the temperature of the glass in that region. At the end of the delivery feeder the continuous flow of glass is sheared into controlled weights of glass.

The shearing action is synchronized to the operation of the multi-stage press on which the glass gobs are formed. As the press table rotates discontinuously from one station to the next, the glass gob in turn is pressed between moulding tools which control the curvatures of the two major surfaces of the subsequent lens blank, and then sufficient cooling is allowed to take place so that the lens blank can be lifted from the press table and transferred to the annealing lehr without distortion. A typical view of a spectacle pressing set-up is shown in Fig. 14.4.2.

The annealing operation carried out by the lehr is one of controlled cooling so that temperature variations within the formed spectacle blank can be equalized at that temperature where stresses cannot be supported within the glass because the viscosity is sufficiently low. At the end of the cooling process the stresses are uniformly distributed within the spectacle blanks and low in magnitude. When examined in a polarized light system the residual birefringence caused by stress in such pressings can be seen and measured. By this method it is possible to limit the birefringence to a maximum of 75 nm/cm.

An important part of the continuous process is the lehr end sorting operation. Typically the defective level produced in ware by the continuous automatic processes runs at approximately 10 per cent, and in order to reduce this to a more acceptable 2 per cent level the production has to be examined visually and sorted accordingly. (Fig. 14.4.3.)

Quality control

The stability of optical properties and the thermal characteristics of ophthalmic glasses are assessed from samples which are taken of the process at regular intervals and annealed on controlled schedules. Earlier, it was said that the chemical composition of glass controlled the physical properties of the glass almost entirely but not completely. The distinction here is that it is possible for example to alter the refractive index of glass by a considerable amount, say 0·001, by simply altering the rate at which the glass is annealed during its manufacturing process. Therefore in order that the property charts which are maintained throughout each production run can represent the true variation in composition it is essential that all samples receive identical thermal treatment to eliminate this particular variable. The annealing rate specified in B.S. 3062 is 360° per hour.

Examination of the control charts enables the short term fluctuation in properties to be distinguished from the long-term fluctuation in

14.4.4. Statistical sample examination.

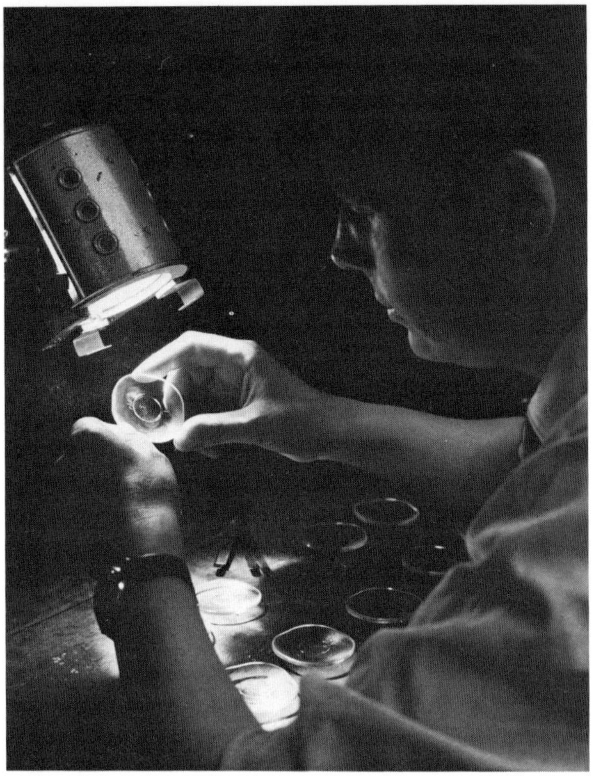

properties. The essential difference is that, in the first case, corrective action on the part of batch control is not required whereas, for a long term fluctuation, control can be exercised by carefully changing the composition of the basic glass. Where tinted glasses are involved the controls are more extensive because not only do the normal ophthalmic characteristics have to be controlled but also the colour variation. Here again it may be necessary to make minor changes to the batch composition in order to keep the colour of the glass within its prescribed tolerance.

Quality control implies far more than simply inspecting the production in order to eliminate that which does not comply with the required specification. Its purpose is primarily to prevent a fault rather than eliminate its effect and secondly to gain a feedback of information so that it is possible to determine the quality level produced by a process.

It is possible to examine pressings immediately after the pressed stage before they have been through the annealing operation. This position is an essential point of inspection since here it is possible to confirm that the overall dimensions and marking of any spectacle blank are correct. If any such aspect is found to be out of tolerance then action can be taken to correct this particular fault before the glass has to be annealed, an operation which can take at least one hour and would mean something like 3000 faulty pressings before the corrective action was taken. One of the more important dimensions concerned in the control of ophthalmic lens manufacture is the curvature of the two principal surfaces. It is insufficiently accurate to determine these by use of a dioptre gauge and instead one has to use a sweep gauge or a template in order that the overall departure of the curve from the required true curve can be estimated. In the sweep gauge a dial gauge is attached to a radius arm, which is swept across the curve. Variations in radius are then read directly on the dial gauge. Alternatively, a simpler depth gauge may be used. This has a reference plane formed by a ring of the same diameter as the glass blank, and a dial gauge is fitted to measure the size of the spherical cap height or compare that of a production sample with that of a true curve.

At the lehr it is possible to take further samples of glass and critically examine them for dimensional and internal quality faults. This operation is illustrated in Fig. 14.4.4. Over a period of time the sample size taken can be used with statistical significance to estimate the production performance for this period. Comparison of the theoretical and actual figures can on occasions lead to corrective actions where for example the difference between the two figures suggests that more acceptable glass is being rejected than should be.

After passing through the lehr end sorting operation, the defective level in the production should be approximately 2 per cent. In order to ensure that this is not exceeded a check inspection function is applied, in which batches of glass are statistically sampled according to plans prepared by Dodge and Romig which ensure that the average outgoing quality limit remains at 2 per cent. For each batch examined there is a specified limit for the number of defectives that can be permitted within that batch. Should the defective level exceed this limit then the whole batch is returned for re-sorting.

Repressing process

In the foregoing we have considered only the continuous manufacture of ophthalmic pressings since this constitutes the larger proportion of current ophthalmic processes. However, for certain purposes, particularly where small quantity orders are concerned, it is more convenient for the spectacle lens to be produced by a repressing process. The raw material for this process is taken from a rolled strip produced continuously. Basically the same conditions apply for melting, property and quality control, but the final form is a rolled sheet of glass. This sheet is subsequently cut into pieces of known weight which are now reheated until sufficiently soft to be formed between mould tools. The process does not yield such a clean pressing as can be obtained by direct methods, but has the advantage of greater flexibility where smaller numbers of pressings are involved.

14.5 Properties and production of optical glass

This section consists of an article, 'Optical Glass', by H. Mackenzie, M.A., and A. B. Scrivener, B.Sc., of Chance-Pilkington (Pilkington Optical Division), which is reproduced with permission.

CHARACTERISTICS OF OPTICAL GLASS

Optical glass is a material which modifies the direction of the propagation of light and the relative spectral energy distribution of that light be it ultra-violet, visible or infra-red.

Whereas these characteristics are not critically controlled in the normal window glass or beer bottle glass, they are the essence of optical glass.

Furthermore, optical glass must be virtually free from internal defects and strain, with a high standard of homogeneity. It must be produced and reproduced within relatively tight limits with respect to all its characteristics.

Optical glass characteristics are inevitably a compromise between the ideal required by the lens designer burning the midnight oil beside his computer and what the glassmaker can technically achieve sweating in front of his furnaces. In between, in no man's land, sit the research and development men of the optical glass world who are constantly trying to equate the possible with the ideal.

Designers look for certain optical constants as a basic requirement for their work. Two essential properties are the refractive index n_e and the constringence V_e.

Fundamentally the refractive index for an optical medium is the ratio of the velocity of light in a vacuum to the velocity of light in the medium. In practice it is calculated from the measured angular deviation of a light beam as it passes through a refractometer prism.

Constringence is related to refractive index. No medium has a constant refractive index for all colours of the spectrum. The difference between the refractive indices determined for two coloured sources, one at the blue and the other at the red end of the spectrum, is referred to as the dispersion.

Today, three specific colours are used as references, corresponding to red, green, and blue emissions for cadmium, mercury, and cadmium-discharge lamps respectively. The refractive index is measured for all three sources and the constringence calculated from:

$$V = \frac{n\,(\mathrm{gr}) - 1}{n\,(\mathrm{bl}) - n\,(\mathrm{r})}$$

These terms form the basis for categorizing glass compositions in terms of their essential optical characteristics and are frequently plotted as illustrated in Fig. 14.5.1, taken from the Chance-Pilkington catalogue, which illustrates differences between crown and flint glasses.

The early optical glass types were of the basic

soda lime composition, hard crowns and the lead-containing dense flint glasses which permitted simple colour corrected lenses.

Developments in Germany towards dense barium crowns brought a significant change in approach. The U.S.A. initiated the early special barium crowns and flints containing rare earth elements. Some more recent extreme glasses become dangerous to manufacture, containing, for example, thorium which is radio-active or tellurium which is poisonous. Those glasses containing germanium are extremely expensive.

There are now hundreds of optical glasses on the market, all with differing n_e and V_e values. Further developments away from the traditional are increasingly apparent. This is because larger aperture lenses, or lenses operating in low illumination conditions, demand combinations of glass where the designer will aim at dramatic differences in V value, together with a desire for a high refractive index.

Transmission

There is little point in having a lens system if very little light comes through. In particular with TV zoom lenses with a large number of glass elements of fairly substantial size, the transmission—particularly of the flint glass elements, traditionally with lower light transmission than crowns—becomes critical.

Again in the case of glass for fibre optics, where the raw glass is drawn into long fibres to be used to transmit light, users seem to want performances in excess of 100 per cent. Whole new families of high transmission flint glasses are being developed. Instead of a transmission figure it is becoming standard practice to show an extinction or absorption coefficient.

Secondary spectrum

The day has gone when the lens designer only looked at the V value. Additionally, he will want to know the partial dispersions.

For example in collimator lenses for optical-transfer-function benches with a white light source, the width of the transmitted spectral range is important and the lens should ideally be corrected for colour throughout. In these cases combinations of weird glasses with abnormal secondary spectrum characteristics are in demand.

Infra-red transmission

Infra-red missile-tracking devices have received much attention in the past few years. Glasses are now required where the designer is primarily, but not exclusively, concerned with the infra-red throughput of the lens system. In night-vision equipment for the armed forces, glass which will transmit only infra-red, and will positively exclude the transmission of light in the visible, is required.

Laser devices

High-purity glasses have been doped with neodymium to produce pulses of radiation at specific wavelengths in the 'micron range'. These glasses resemble solidified methylated spirit.

Size and homogeneity

It is little good if your spectral control is in hand and you come unstuck on other problems. For aerial cameras, 800-mm diameter lenses are often considered. Once it is decided that it is possible to manufacture the size of lens required, the homogeneity of the glass must be guaranteed and often glass with homogeneity of the order of a unit in the sixth decimal place with respect to variation of refractive index is required.

Index control in large batches

In the case of high-quality copying lenses the distance between object and lens is highly critical and a small change in track length can lead to huge differences in magnification. Copying machine makers must inevitably allow for mechanical tolerances and errors. They will not want to retool at frequent intervals to alter glass curvatures in an attempt to correct cumulative errors. Consequently, the glassmaker is virtually obliged to keep his sights pretty firmly on three or four units in the fourth decimal place over thousands of blanks.

Strain

A major requirement is to produce glass as strain-free as possible. This means very careful annealing.

Selection of glasses

Figs 14.5.1 and 14.5.2 are pages reproduced from the Chance-Pilkington catalogue, and show

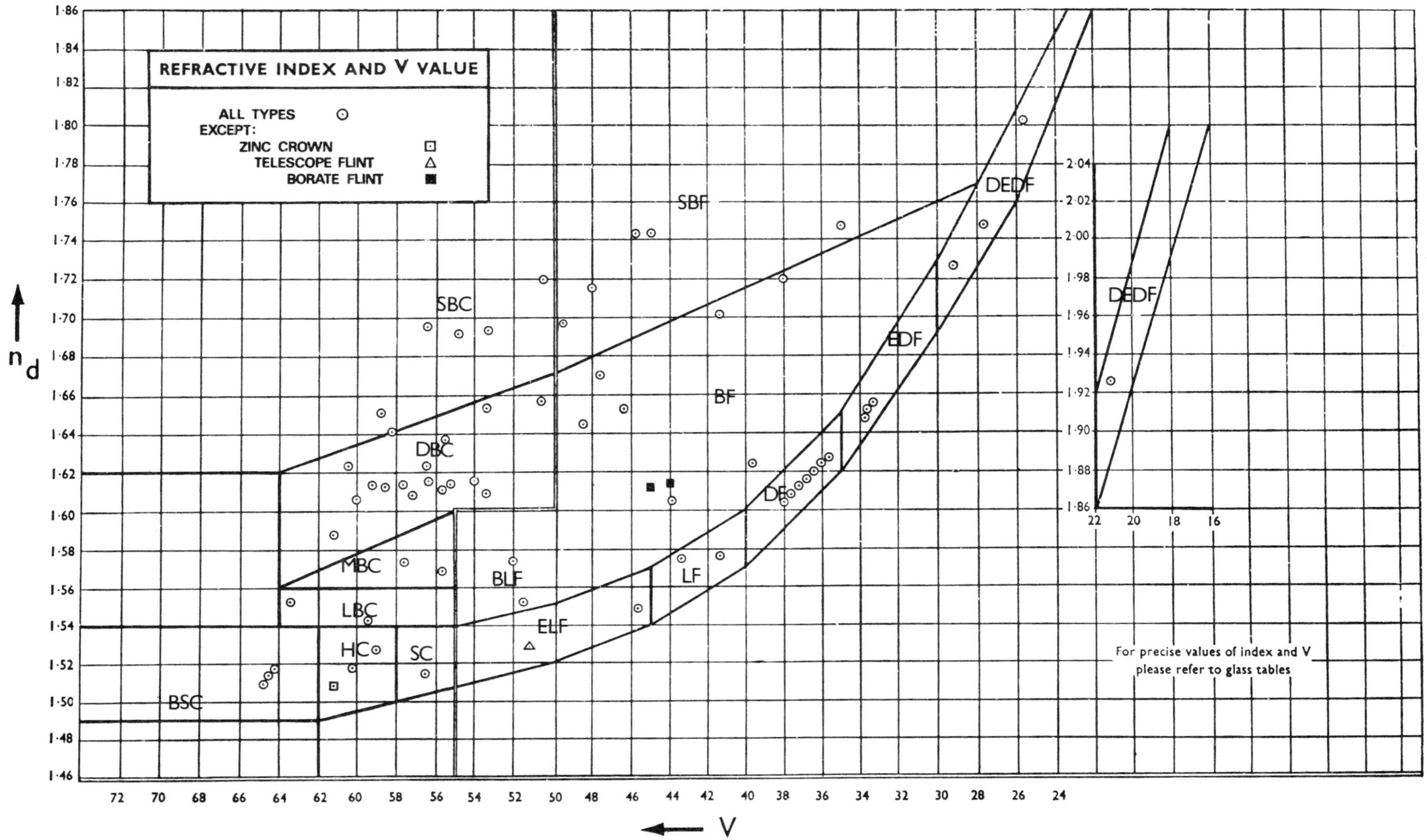

14.5.1. The relationship of the refractive index and *V* value for various types of glass. (From the Chance-Pilkington catalogue.)

	Wavelength λ	Refractive Index n	$\frac{n_\lambda \sim n_e}{n_{F'} - n_{C'}}$	Spectral Line	Temperature Coefficient of Refractive Index X 10^5 °C^{-1}
ULTRA-VIOLET	3126	1.54093	3.6846		
	3342	1.53516	2.9598		
	3650	1.52882	2.1638		0.49
	3663	1.52859	2.1349		
	3906	1.52477	1.6552		
VISIBLE	4047	1.52290	1.4202	h	0.37
	4358	1.51941	0.9817	g	
	4678	1.51656	0.6238		
	4800	1.51562	0.5064	F′	0.26
	4861	1.51518	0.4507	F	
	5086	1.51369	0.2634		
	5461	1.51159	0.0000	e	0.21
	5876	1.50970	0.2370	d	
	6438	1.50766	0.4936	C′	0.22
	6563	1.50727	0.5425	C	
	6678	1.50692	0.5858		
	7065	1.50586	0.7186	b	
	7699	1.50440	0.9026	A	
INFRA-RED	7800	1.50419	0.9289		0.25
	8521	1.50284	1.0983		
	8944	1.50214	1.1859		
	10140	1.50041	1.4037		0.33
	11287	1.49895	1.5873		
	13951	1.49580	1.9820		
	15295	1.49421	2.1814		
	18131	1.49064	2.6298		0.52
	19701	1.48848	2.9010		
	20581	1.48720	3.0616		
	23254	1.48297	3.5930		
	25628	1.47873	4.1251		0.66
	26173	1.47769	4.2557		
	26300	1.47745	4.2866		

REFRACTIVE PROPERTIES

n_d	$n_F - n_C$	Vd	n_e	$n_{F'} - n_{C'}$	Ve
1.50970	0.00791	64.44	1.51159	0.00796	64.27

THERMAL PROPERTIES

T_A°C	T_F°C	T_S°C	αX10^7 °C^{-1}	ΔT	Index Annealing Factor, F
559	688	750	77	9.2	8.4

EXTINCTION COEFFICIENTS

Wavelength	Extinction Coefficient (cm^{-1})
3200	0.90
3600	0.035
4000	0.009
4500	0.005
5000	0.003
5500	0.002
6000	0.004
6500	0.006
7000	0.006
8000	0.007
9000	0.008
10,000	0.010

DURABILITY

Resistance to atmospheric attack	Resistance to acid attack
S.1	1

DENSITY

g.cm^{-3}	oz.in^{-3}
2.49	1.44

Bubble classification—3

Wavelengths in Ångstrom units

All Refractive Indices for 20 C

Glass Type
Reference Number
BSC 510644

REFRACTION AT LASER WAVELENGTHS

LASER SOURCE	Wave-length	Refractive Index	$\frac{n_\lambda \sim n_e}{n_{F'} - n_{C'}}$
Neodymium in glass	10600	1.49981	1.4794
Neodymium in glass overtone	5300	1.51243	0.1064
Neodymium in Calcium Tungstate	10650	1.49974	1.4874
Helium-Neon	6328	1.50802	0.4479
Helium-Neon	11530	1.49865	1.6244
Ruby	6943	1.50618	0.6786

14.5.2. Properties of various types of glass. (From the Chance-Pilkington catalogue.)

many of the measurements that have to be made.

Reference to optical glass catalogues shows several hundred different glass types. It may be in the interest of designers to have a wide choice but it does pose problems of capacity and flexibility for the glassmaker. If this spread is reduced it can lead to rational production of bigger quantities of fewer types, easing delivery problems for the user. The major glass companies have tried to condense their ranges. Designers certainly recognize that with automatic design techniques some discipline is possible in the number of glasses which are selected for a lens programme.

METHODS OF MANUFACTURE

The methods of manufacture of optical glasses are governed largely by their application. In the examples to be considered, there is need for large quantities of small unit sizes (100 000 pieces at 5–50 g) as well as small quantities of large unit size (one or two at 50 kg). At the same time, these have to be produced from approximately fifty regularly used types taken from over 100 practical possibilities.

All this indicates that a very flexible process is required. An early and important stage in glass manufacture is the selection and control of raw materials. The presence of contaminants such as iron or chromium can give objectionable colour if present in amounts exceeding 10×10^{-6} and 2×10^{-6} respectively. The proportions in which materials are mixed before melting have the greatest influence on the optical characteristics. Control is necessary to the extent that variation from the true composition of as little as 0·1 per cent of the gross weight can put the optical properties out of specification.

Two scales of melting are available with present-day methods of manufacture:

To supply relatively large volumes of glass a continuous process, in which mixed raw material is melted, refined, stirred and delivered in a hot stream to a forming operation, was developed in the 1950s. The refining operation removes gaseous and solid inclusions, but the stirring operation is a continuation of the classic technique developed in the early nineteenth century to improve the homogeneity of the melted glass by mechanical means.

Glass coming from this process is in a plastic condition; and is formed by at least five different methods of interest to the optical world.

(a) Cutting into controlled weight gobs which are then pressed between mould plungers to form a lens or prism blank.

(b) Passing between rollers to form a strip of controlled cross-section.

(c) Extruding through rectangular or prism section dies to give a bar of controlled section.

(d) Casting into a mould to give a relatively large mass.

(e) Drawing from an orifice to give a fire-polished rod.

To supply relatively small volumes of glass the classical pot process is still used, although much improved in operational detail. Here all the glass making stages described for the continuous process take place in a pot. Because of the different scale of operation, the hot glass is formed by one of two methods: (*a*) casting into a mould to form a block; (*b*) casting into an open-ended die to form a bar.

14.5.3. Bars of extruded optical glass as produced.

One essential difference between today's practice and that of the nineteenth and early twentieth centuries is the change in material from which pots are made, to avoid erosion and contamination caused by the much more reactive glass compositions required to give optical designers the glasses they require.

Both scales of melting give glass in a form which can be further processed. Bar, strip, prism, and block forms can be sawn or trepanned with diamond tools. Bar and strip forms can be scribed and mechanically snapped into smaller pieces which, after reheating to make them plastic, are pressed in moulds to form simple or complex product shapes.

LIMITATIONS OF OPTICAL GLASSES

Whilst some optical crown glasses are simply much purer forms of commercially available material, most of the compositions for the higher-refractive-index glasses do possess some limiting characteristics.

Colour contamination will be exaggerated in these glasses and can limit the colour performance of some optical instruments.

Glasses are not of a crystalline structure, but certain components in these special compositions tend to devitrify more easily and appear as solid crystals within the glass. Reducing the risk of this occurrence imposes a limit on the size of product and its 'remouldability'.

Whilst composition is of major importance in determining optical properties, heat treatment after melting critically modifies these values. Rates of cooling can control homogeneity and stress levels within individual pieces, which can give rise to optical aberrations.

Compositions for the more extreme optical values generally give glasses which may stain during polishing or may weather when exposed to humid conditions. These compositions require extra care in processing and may need durable protective coatings applied by vacuum deposition.

One limitation of the continuous process is that there are small variations of composition with time so that for any quantity of glass produced there may be variation in the refractive index. It is possible to predict limits of this variation relative to quantities produced, but the relationship is dependent on the ease of making the various glass types.

For the non-continuous process the main limitation is the maximum quantity available for each cast. Due to the melting procedure there will be virtually no refractive index variation through each melt.

14.6 Moulding and remoulding

Optical glass mouldings for lenses or prisms (Fig. 14.6.1) could be produced on line, immediately after extruding from the tank of molten glass, if the quantities were large enough. In fact, some binocular prisms (Fig. 5.2.4.) and ophthalmic lenses (Fig. 14.4.3) are produced in this way.

However, the majority of optical orders are for very small quantities and, as moulding on line would be uneconomic, a remoulding process is necessary. The type of remoulding process depends on the weight of the moulding and the time it will take to heat up to plastic temperature. For

14.6.1. Glass mouldings for lenses and prisms.

small prisms and lenses, where the heating-up time can be only a few minutes, the flat type of furnace is used, in which mouldings are heated on the floor of the furnace and, if necessary, patted roughly to shape. (Fig. 14.6.2.)

The plastic glass is then transferred to a heated mould and pressed to final shape (Fig. 14.6.3). Larger mouldings need more time to heat up, and it is therefore convenient to pass them through a Lehr remoulding furnace with a gradually increasing temperature. (Fig. 14.6.4.) It is more important to pat the moulding whilst it is on the furnace floor and less desirable to use force from a

△14.6.2. Press operation with flat Seige furnace for moulding small lenses and prisms.

△14.6.3. Moulding press, from left-hand side of the flat Seige furnace, producing small lens mouldings. The close up shows the moulding tool.

◁14.6.4. Lehr remoulding furnace for larger lens and prism mouldings requiring lengthy heating.

△14.6.5. Lehr remoulding furnace.

△14.6.6. Close-up of tank prism moulding after being remoulded in a lehr furnace. The moulding has been knocked out of the inverted tool.

press to achieve the required shape. (Fig. 14.6.5.) Because of the weight of the plastic glass, it is more likely to take the form of the mould without any other external force. (Fig. 14.6.6.) Once again, because of the larger weight of glass, the rate of temperature drop must be slower in order to avoid cracking the glass or introducing unacceptable strain.

All mouldings must be annealed before use. (See § 5.2.)

14.7 Optical glass pot melting process

Jena^{er} Glaswerk Schott and Gen was founded at Jena in 1884, when Dr Otto Schott started a new era in optical glass making. Because, for Otto Schott, the science of glassmaking was only part of pyrochemistry, he realized that glass could be produced from more and different kinds of earth and metal oxides. Glass varied from other melt fluxes only in that it solidified and remained isotropic and transparent. His systematic melting tests established glassmaking as an applied science and made possible the development of new types of optical glass with their own specific chemical and physical properties.

The original plant at Jena was dismantled in 1945 and, since 1952, Mainz has been the site of the glassworks. The classical pot process consists of melting together the batch materials, with broken glass of a similar composition from a previous melt, in a fireclay vessel which may be up to 6 feet (1·83 m) in diameter. After melting is complete and the glass considered free from bubble, it is stirred with a refractory rod. (Fig. 14.7.9.) Each melt has a liquid sample of glass taken whilst the glass is in the pot and it is annealed by a rapid process so that, within 3 hours, it can be checked for refractive index by an Abbe refractometer.

If there is an error, further materials can be added to the melt and the glass must be stirred again. When melting is completed, the pot is removed from the furnace and transferred to a cooling arch. (Fig. 14.7.1). After cooling, the pot is broken open (Fig. 14.7.2) and the pieces of optical

14.7.1. A pot melt (*left*) after completion of stirring and ready for removal from the furnace and (*right*) after removal from the furnace and with the cooled raw optical glass in the pot.

14.7.2. Breaking open a pot of raw optical glass.

glass remoulded into the sizes and shapes required. (Fig. 14.7.3.) The pot may be used for only one melt, after which it is scrapped. The classical pot process was used until about 1930, but the yield was low and there were too few large pieces of optical glass.

14.7.3. The raw optical glass is reheated in fireclay moulds to form blocks.

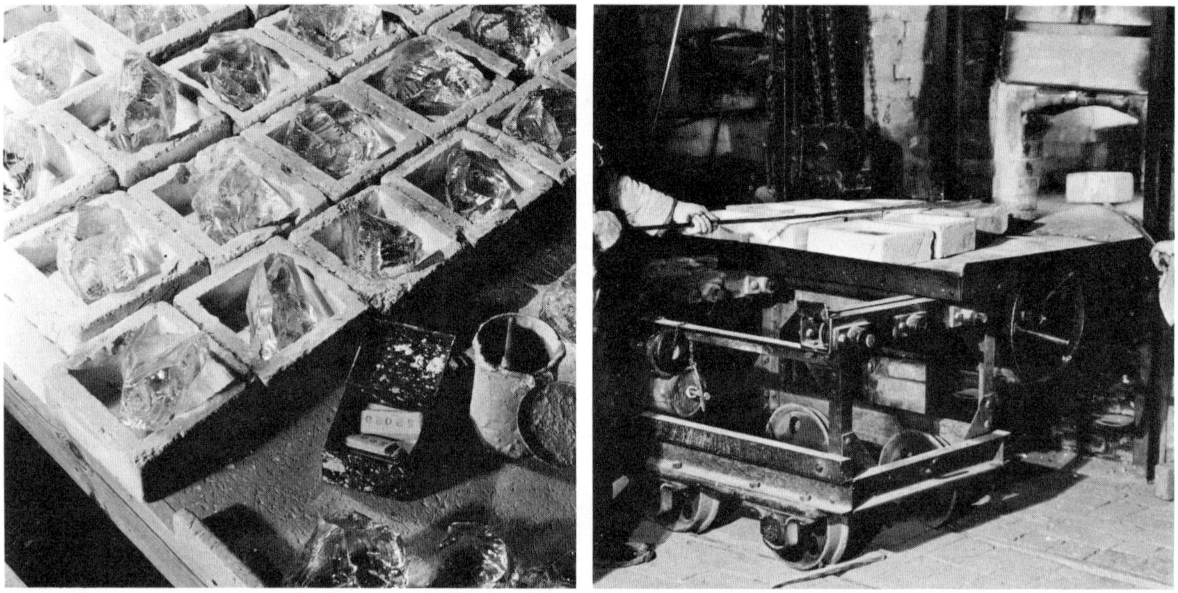

14.7.4. The mould for a pot is made from acid plastic beaten into a cloth over a metal former. *(Right)* Pulling the core mould out of the soft pot, which is held together by the plastic-coated fabric. The external metal former, which is clamped together, is then removed so that the pot can dry out.

14.7.5. ▷
Pouring the 'basic' clay mixture into the mould.

14.7.6. ▷
The pots are placed in a room for drying and storage until required for melting optical glass.

The production of fireclay pots

The production of fireclay pots for melting glass is an exacting task. 'Acid' pots are made out of plastic clay and beaten by hand into a fabric material. (Fig. 14.7.4.) The more important basic pots are made from a mixture of burnt and unburnt clays, as received from the pit, which have been dried and fine ground into a powder. Water and alkali (soda, waterglass, or both) are added to the powder and, after thorough mixing by stirring, the slurry is poured into a gypsum mould. This is removed after about a week, when the pot is sufficiently stiff to retain its shape. (Fig. 14.7.5.)

The pot is then dried for about three weeks at room temperature. (Fig. 14.7.6.) The temperature is then increased to 35°C for four weeks, followed by 70°C for four weeks, after which the pot is transferred to the electrically heated tempering furnace. The pot is then slowly heated up to 900°C when it is quickly pushed into the glass melting furnace (Schott 'Harbour' furnace) to try to avoid a heat shock from the higher temperature which may cause cracking of the fireclay pot. Altogether, the pot takes about 12 weeks to manufacture.

The poured pot process currently used by Schott at Mainz

The real production of glass starts in the mixing machine where often more than ten materials are thoroughly mixed together in the proportions

14.7.7. **After the raw materials and chemical powders for glassmaking have been mixed, they are drawn off from a chute and loaded in a bulk-transport wagon.**

14.7.8. **Ladle for depositing the bulk quantities of raw materials in the melting pots.**

specified. The mixture is drawn off from a chute to the transport waggon. (Fig. 14.7.7.) The mixture is taken to the melting furnace where it is ladled into the hot fireclay pot. (Fig. 14.7.8.)

The pot furnace is heated by the regenerating process, invented by the Siemans brothers in the nineteenth century, which makes the best possible use of available heat energy. The regenerating chambers on each side of the furnace contain a refractory bar system and are heated by the hot exhaust gases. The incoming air, necessary for burning the oil, is passed over hot bars on the way to the furnace. The intake air and exhaust gas automatically alternate from one side of the furnace to the other side every few minutes. Ladling the glass raw material mixture into the pot takes several hours according to the type of glass—the slower the mixture melts, the slower it is ladled into the pot. With some glasses, which melt with difficulty, it may take forty hours to fill the pot. The hot flames of the furnace burner cause the mixture to shrink together and melt at a temperature of about 1200°–1400°C into a viscous liquid. Bubbles are formed by the chemical reaction of the mixture ingredients and also by the gases which are leaving the mixture, and these bubbles leave very slowly. The procedure for removing bubbles is to increase the temperature from 1300°C to 1500°C so that the clearing ingredients, such as arsenic, antimony oxide or halogens which have been added to the mixture, can remove the gas. This process takes several hours.

When the glass is clear of bubbles, the temperature is reduced until the desired viscosity for stirring is reached. At the end of stirring to achieve the required homogeneity, the temperature is gradually decreased so that the glass gets even more viscous. (Fig. 14.7.9.) Immediately after stirring is completed, the pot is taken out of the furnace with the help of pot tongs and a special cart, and it is bound with iron straps to avoid the risk of bursting. (Fig. 14.7.10.)

A special pouring crane takes the pot from the cart and tips the glass into an iron mould which has been preheated to about 200°C. (Fig. 14.7.11.) Immediately the pouring is finished, the iron mould, filled with molten glass, is taken to a preheated 'cooling' furnace. In this furnace, the glass is cooled down to room temperature over

△14.7.9. Stirring the optical glass molten in the pot. The arm of the stirrer is water cooled. The rod in the liquid glass is made from refractory material.

▽14.7.10. Removing the pot of molten optical glass from the furnace.

△14.7.11. Pouring the viscous optical glass from the pot into preheated iron moulds.

▽14.7.12. Grinding and polishing opposite sides of a raw optical glass block before inspecting for striae, bubbles and other defects.

several weeks or even months in order to avoid strains. The quality of the glass is decisively influenced by this process as only a carefully cooled-down glass shows a consistent homogeneity without cracks. When cool, the glass block is freed from the iron mould and opposite faces are ground and polished for examination. (Fig. 14.7.12.) The

14.7.13. Chipping a raw glass block along the line where a break is required and breaking it with a hydraulic press.

14.7.14. Polishing flat surfaces on the glass blocks, which are set in a circular plaster block, ready for inspection and testing.

14.7.15. Sawing up slabs of optical glass with diamond cutters, using paraffin as a lubricant.

glass block is then broken under a hydraulic press. (Fig. 14.7.13.)

As it is not possible to identify fine faults in the glass from the 'inspection windows' in the block, it is necessary to make some polished test-pieces. (Fig. 14.7.14.) In a darkened room, the glass test-pieces are examined for striae, bubbles, and other defects, such as stones. Defective pieces are cut off the block with a diamond saw, using paraffin as a lubricant. (Fig. 14.7.15.) The remaining good glass can either be worked as a large piece or be cut up into many small pieces for lenses or prisms. (Figs 14.7.16 and 14.7.17.)

It takes about twenty weeks from preparation of the pot for the furnace to the despatch of the completed glass.

△14.7.16. An assortment of optical glass, including plates, cut prisms, round slices, pressed mouldings.

14.7.17.▷ Production of trepanned glass rod drilled from slab, a process ensuring consistent refractive index in lens blanks cut from the rods after centreless grinding. Such blanks are highly suitable for hard-on blocking ready for diamond generating.

14.8 The platinum crucible process

When rare element optical glasses were being developed, one of the problems encountered was that of attack of the glass on the fireclay pot. This attack caused striation and bubbles so that the conventional pot technique was quite unsuitable. The only material which will withstand the corrosive action of these glasses is platinum, and therefore platinum crucibles had to be used. With these it is possible to obtain almost 100 per cent yield of colourless bubble-free glass, and slabs of a considerable size can be cast from which blanks can be cut or moulded. This method is limited to glasses which do not involve a melting temperature more than about 1400°C; for beyond that temperature the mechanical strength of platinum fails. This means that the bulk of ordinary-type glass will continue to be made in refractory pots or tanks.

The latest advance in the art of glass-melting is the continuous manufacture of optical glass in a platinum tank. This new process, developed at the Schott glassworks at Mainz, has proved to be a success with large-size optical blanks such as bubble-chamber windows. The glass is completely free from striae and almost without bubbles, so that fully automatic pressing of lens blanks for cameras and other optical instruments is a practicable production method.

15 Large Object Glasses and Mirrors

15.1 Introduction

Advances in astrophysics demand new optical telescopes of ever increasing size. A large astronomical telescope is a complex precision mechanical structure, supporting mirrors which must be aligned with a celestial object to an accuracy better than a second of arc.

Images of distant stars upon a photographic plate may be not more than 20 micrometres in diameter and may require an exposure time of several hours. During this time, the telescope must follow the object, allowing for the rotation of the earth, so that the image on the plate does not move out of alignment by more than a few micrometres.

Objective mirrors now (1971) being made are 150 inches (381 cm) in diameter and, in order to give the required resolution, must be polished to an accuracy better than one-eighth of a wavelength. In other words, surface errors on the mirror must not exceed two-millionths of an inch (0·05 micrometres). Early mirrors were made from a copper and tin alloy, then later from plate-glass, followed by Pyrex glass, which had better thermal stability. The performance demanded from modern optical systems would be impossible without the development of transparent glass ceramics such as Cervit (Owens-Illinois) and Zerodur (Schott and Gen, Mainz) which have a zero coefficient of expansion. (See § 5.5)

The grinding, polishing, and figuring of a large objective mirror may take $1\frac{1}{2}$ to 2 years. Much patience, and dedication to a tedious job, is needed.

The techniques of testing vary and, although the method of tilting the polishing table 90° during testing is well established, it seems highly desirable for large objectives weighing many tons to be tested in the same attitude as they are polished. (See §§ 8.10 and 10.4) This implies the use of a high tower with stable temperature conditions. (See § 11.17.) Associated with the requirements of space optics, Kanigen-coated beryllium mirrors have been developed in recent years and have given high stability over a temperature range of +50°C to −40°C.

Generally, glass lenses and mirrors more than 3·5-inch (89 mm) diameter should be worked as single surfaces. Where the diameter exceeds 4·5 inch (114 mm), problems arise due to flexure which are not dealt with in Chapter 3.

Sections 2 to 8 of this chapter are adapted from Mr Twyman's *Prism and Lens Making*. They trace the development of making and testing large objectives, principally through abstracts of historically important papers by Draper, Grubb and Ritchey. An account of the 200-inch Mt Palomar telescope is given, and the work of Texereau on the effects of minor irregularities in objectives is described. The material has been supplemented by accounts of contemporary projects and of modern techniques in the casting, grinding and polishing, and aluminizing of large objectives.

15.2 The development of astronomical telescopes

Sir Harold Spencer-Jones, the Astronomer Royal, in his Presidential address to Section A of the British Association in 1949, gave a sketch of the development of astronomical telescopes (Spencer-Jones, 1949).

The following passage is quoted, with permission, from this address.

Because of the invention of the achromatic object glass and because also of the difficulties of casting and figuring specula, the refractor for a time almost completely superseded the reflector. In 1828 Lord Rosse wrote, 'Many practical men whom I have spoken to seem to think that since Fraunhofer's discoveries the refractor has entirely superseded the reflector and that all attempts to improve the latter instrument are useless.' Larger and larger objectives were made by Merz and Mahler, the successors of Fraunhofer in Germany; by Cauchoix in France; by Grubb and Cooke in England; and by Alvan Clark in America. The largest objective in 1824 was the $9\frac{1}{2}$ inch made by Fraunhofer for Dorpat; in 1839, the 15 inch, made by Merz and Mahler for Poulkova; in 1871, the $24\frac{1}{2}$ inch made by Cooke for Newall, now at Cambridge; in 1878 the 26 inch made by Grubb for Vienna; culminating in 1888 with the 36 inch by Alvan Clark for the Lick Observatory, and in 1897 with the 40 inch, also by Alvan Clark, for Yerkes Observatory. One or two attempts to make larger objectives have not been successful.

The development of the silver-on-glass mirror, by avoiding the difficulties inherent in specula, began to bring the reflecting telescope back into favour in the seventies and eighties of the last century. Lassell's 47-inch reflector, made in 1860, was the last large reflector with a speculum mirror to have been used, as far as I have been able to ascertain; the 48-inch speculum reflector made in 1867 by Grubb for Melbourne was never used. The idea of making mirrors of glass was not new. Short in 1734 had made some glass mirrors, silvered on the *back*, which Maclaurin declared to be excellent. In 1827, Airy had proposed that a layer of silver should be deposited on the *surface* of a figured glass disk. But the method did not become practicable until Liebig discovered a simple chemical process for depositing a thin film of silver on a glass surface. The methods of figuring and testing glass disks were due primarily to Foucault who, in 1857, presented his first telescope with a silvered mirror to the French Academy. His knife-edge test is still a valuable tool to the optician. Further developments in technique were made by Draper in America, by Common in England, and by Ritchey in America.

In recent years, aluminium has almost entirely replaced silver as the reflecting coating on a figured glass surface.

No material has yet been found which is more suitable than glass as a medium for supporting the reflecting film. But its relatively large coefficient of expansion and poor thermal conductivity are disadvantages. Assuming the temperature to be uniform throughout the disk, the position of the focus of the mirror will depend on the temperature; for a given change of temperature, the displacement of the focus will be greater the larger the coefficient of expansion. If the temperature is not uniform throughout the disk, distortion of the figure of its surface will occur, which will impair the quality of the images; the effect will be greater the thicker the disk and the lower its thermal diffusivity. Pyrex glass is now used in preference to plate glass because of its smaller coefficient of expansion and its larger thermal diffusivity. The thinner the disk, the more rapidly it will reach a state of temperature equilibrium, but then also the more flexible it is, and the more elaborate the support system must be. The requirements are therefore to some extent self contradictory and the choice of thickness is a matter of compromise.

15.3 The work of Draper

Draper's work was directed towards making mirrors of up to $15\frac{1}{2}$-inches diameter for astronomical telescopes. The following material is taken from his paper on the construction of a silvered glass telescope.

EXTRACTS FROM DRAPER (1864)

In the summer of 1857, I visited Lord Rosse's great reflector, at Parsonstown, and, in addition to an inspection of the machinery for grinding and polishing, had an opportunity of seeing several celestial objects through it. On returning home, in 1858, I determined to construct a similar smaller instrument; which, however, should be larger than any in America, and be especially adapted for photography. Accordingly, in September of that year, a 15-inch speculum was cast, and a machine to work it made. In 1860, the observatory was built, by the village carpenter, from my own design, at my father's country seat, and the telescope with its metal speculum mounted. This latter was, however, soon after abandoned, and silvered glass adopted. During 1861, the difficulties of grinding and polishing that are detailed in this account were met with, and the remedies for many of them ascertained. The experiments were conducted by the aid of three $15\frac{1}{2}$-inch disks of glass, together with a variety of smaller pieces. Three mirrors of the same focal length and aperture are almost essential, for it not infrequently happens that two in succession will be so similar, that a third is required for attempting an advance beyond them. One of these was made to acquire a parabolic figure. . . .

During the winter of 1862, the art of local corrections was acquired, and two $15\frac{1}{2}$-inch mirrors, as well as two of 9 inches for the photographic enlarging apparatus, were completed. The greater part of 1863 has been occupied by lunar and planetary photography, and the enlargement of the small negatives obtained at the focus of the great reflector.

Experiments on a metal speculum

My first 15-inch speculum was an alloy of copper and tin, in the proportions given by Lord Rosse. His general directions were closely followed and the casting was very fine, free from pores, and of silvery whiteness. It was 2 inches thick, weighed 110 lb, and was intended to be of 12 feet focal length. The grinding and polishing were conducted with the Rosse machine. Although a great amount of time was spent in various trials, extending over more than a year, a fine figure was never obtained—the principal obstacle to success being a tendency to polish in rings of different focal length. It must, however, be borne in mind that Lord Rosse had so thoroughly mastered the peculiarities of his machine as to produce with it the largest specula ever made and of very fine figure.

During these experiments there was occasion to grind out some imperfections, 8/100 of an inch deep, from the face of the metal. This operation was greatly assisted by stopping up the defects with a thick alcoholic solution of Canada balsam, and, having made a rim of wax around the edge of the mirror, pouring on nitro-hydrochloric acid, which quickly corroded away the uncovered spaces. Subsequently an increase in focal length of 15 inches was accomplished, by attacking the edge zones of the surface with the acid in graduated depths.

An attempt also was made to assist the tedious grinding operation by including the grinder and mirror in a Voltaic circuit, making the speculum the positive pole. By decomposing acidulated water between it and the grinder, and thereby oxidizing the tin and copper of the speculum, the operation was much facilitated, but the battery surface required was too great for common use. If a sufficient intensity was given to the current, speculum metal was transferred without oxidation to the grinder, and deposited in thin layers upon it. It was proposed at one time to make use of this fact, and coat a mirror of brass with a layer of speculum metal by electrotyping. The gain in lightness would be considerable.

Of the speculum metal for mirrors Grubb says:

The composition of metallic mirrors of the present day differs very little from that used by Sir Isaac Newton. Many and different alloys have been suggested, some including silver or nickel or arsenic; but there is little doubt that the best alloy, taking all things into account, is made with 4 atoms of copper, and 1 of tin, which gives the following proportions by weight: copper, 252, tin, 117·8.

Silvering glass

At Sir John Herschel's suggestion (given on the occasion of a visit that my father paid him in 1860), experiments were next commenced with silvered glass specula. These were described as possessing great capabilities for astronomical purposes. They reflect more than 90 per cent of the light that falls upon them, and only weigh one-eighth as much as specula of metal of equal aperture. . . .

In order to guard against tarnishing, experiments were at first made in gilding silver films, but were abandoned when found to be unnecessary. A partial conversion of the silver film into a golden one, when it will resist sulphuretted hydrogen, can be accomplished as follows: Take three grains of hyposulphite of soda, and dissolve it in an ounce of water. Add to it slowly a solution in water of one grain of chloride of gold. A lemon yellow liquid results, which eventually becomes clear. Immerse the silvered glass in it for twenty-four hours. An exchange will take place, and the film become yellowish. I have a piece of glass prepared in this way which remains unhurt in a box, where other pieces of plain silvered glass have changed some to yellow, some to blue, from exposure to coal gas. . . .

Grinding and polishing glass

Some of the facts stated in the following paragraphs, the result of numerous experiments, may not be new to practical opticians. I have had, however, to polish with my own hands more than a hundred mirrors of various sizes, from 19 inches to $\frac{1}{4}$ inch in diameter, and to experience very frequent failures for three years, before succeeding in producing large surfaces with certainty and quickly. It is well nigh impossible to obtain from opticians the practical minutiae which are essential, and which they conceal even from each other. The long continued researches of Lord Rosse, Mr Lassell, and M. Foucault are full of the most valuable facts, and have been of continual use.

It is generally supposed that glass is possessed of the power of resistance to compression and rigidity in a very marked manner. In the course of these experiments it has appeared that a sheet of it, even when very thick, can with difficulty be set on edge without bending so much as to be optically worthless. Fortunately in every disk of glass that I have tried, there is one diameter on either end of which it may stand without harm.

In examining lately various works on astronomy and optics, it appears that the same difficulty has been found not only in glass but also in speculum metal. Short used always to mark on the edge of the large mirrors of his Gregorian telescopes the point which should be placed uppermost, in case they were removed from their cells. In achromatics the image is very sensibly changed in sharpness if the flint and crown are not in the best positions; and Mr Airy, in mounting the Northumberland telescope, had to arrange the means for turning the lenses on their common axis, until the finest image was attained. In no account, however, have I found a critical statement of the exact nature of the deformation, the observers merely remarking that in some positions of the object glass there was a sharper image than in others.

DRAPER'S METHODS OF RETOUCHING

Draper corrected his mirrors by local retouching. He used two distinct modes of examination: first, observing with an eye-piece of power 20 the image of an illuminated pin-hole at the focus, and the cone of rays inside and outside that plane: second, Foucault's test.

He also seems to have used Foucault's method of controlling the retouching, namely, as Draper says, first to bring the mirror to a spherical surface, and then 'by moving the luminous pin-hole toward the mirror, and correspondingly retracting the eye-piece or opaque screen, to carry the surface —avoiding aberration continually by polishing— through a series of ellipsoidal curvatures, advancing step by step towards the paraboloid of revolution'.

15.4 The work of Sir Howard Grubb

Sir Howard Grubb's article (Grubb, 1886) is chiefly concerned with the working of achromatic object glasses, which he carried up to a diameter of $27\frac{1}{2}$ inches. No one before seems to have given so much attention to the effect of flexure both in the production of optical surfaces and on the performance of the finished article.

Speaking of the spherometer in its then customary form with three legs, he says:

I do not find the points satisfactory for regular work. They are apt to get injured or worn, and for ground surfaces are a little uncertain, as one or other of the feet may find its way into a deep pit. This particular spherometer has 3 feet, of about $\frac{1}{2}$ inch long, which are hardened steel knife-edges forming three portions of an entire circle. In using this it is laid on the surface to be measured, and the screw with micrometer head is turned till the point is felt to touch the surface of glass. This scale and head can then be read off. The screw in this instrument has fifty threads to the inch, and the head is divided into 100 parts, so that each division is equal to 1/5000th of an inch. With a little practice it is easy to get determinate measures to 1/10th of this, or 1/50 000th of an inch, and by adopting special precautions even more delicate measures can be taken, as far probably as 1/100 000th or 1/150 000th of an inch, which I have found to be practically the limit of accuracy of mechanical contact.

To give an idea of the delicacy of the instrument, I bring the screw firstly into contact with the glass. Now the screw is in good contact; but there is so much weight still on the three feet, that, if I attempt to turn it round, the friction on the feet oppose me, and it will not stir except I apply such force as will cause the whole instrument to slide bodily on the glass. Now, however, I raise the whole instrument, taking care that my hands touch none of the metal-work, and that the screw is not disturbed. I lay my hands for a moment on part of the glass where the centre screw stood, and thus raised its temperature slightly, and on laying the spherometer back in the same place, you now see that it spins on centre screw, showing how easily it detects what to it is a large lump, caused by expansion of the glass from the momentary contact of my hand.

Flexure

One of the greatest difficulties to be contended with in the polishing of large lenses is that of flexure during the process.

It may appear strange that in disks of glass of such considerable thickness as are used for objectives, any such difficulty should occur; but a simple experiment will demonstrate the ease with which such pieces of glass can be bent, even under such slight strain as their own weight.

We again take our spherometer and set it upon a polished surface of a disk of glass of about $7\frac{1}{2}$ inches diameter and $\frac{3}{4}$ inch thick. I set the micrometer head as in the former experiment to bear on the glass, but not sufficiently tight to allow the instrument to spin round. This has now been done while the glass, as you see, is supported on three blocks near its periphery. I now place one block under the centre disk and remove the others thus, and you see the instrument now spins round on centre screw.

It is thus evident that not only is this strong plate of glass bending under its own weight, but it is bending a quantity easily measurable by this instrument, which, as I shall presently show, is quite too coarse to measure such quantities as we have to deal with in figuring objectives.

After this experiment no surprise will be felt when I say that it is necessary to take very special precautions in the supporting of disks during the process of polishing to prevent danger of flexure; of course if the disks are polished while in a state of flexure, the resulting surface will not be true when the cause of flexure is removed.

For small-sized lenses no very special precautions are necessary, but for all sizes over 4 inches in diameter I use the equilibrated levers devised by my father, and utilised for the first time on a large scale in supporting the 6-foot mirror of Lord Rosse's telescope.

I have also sometimes polished lenses while floating on mercury. This gives a very beautiful support, but it is not so convenient, as it is difficult to keep the disk sufficiently steady while the polishing operation is in progress without introducing other chances of strain.

So far I have spoken of strain or flexure during the process of working the surface; but even if the surface be finished absolutely perfectly, it is evident from the experiment I showed you that very large lenses when placed in their cells must suffer considerable flexure from their own weight alone, as they cannot then be supported anywhere except round the edge.

To meet this I proposed many years ago to have the means of hermetically sealing the tube, and introducing air at slight pressure to form an elastic support for the objective, the pressure to be regulated by an automatic arrangement according to the altitude. My attention was directed to this matter very pointedly a few years ago from being obliged to use for the Vienna 27-inch objective a crown lens which was, according to ordinary rules, much too thin.

I had waited some years for this disk, and none thicker could be obtained at the time. This disk was very pure and homogeneous, but so thin that, if offered to me in the first instance, I would certainly have rejected it. Great care was taken to avoid flexure in the working, but to my great surprise, I found no difficulty whatever with it in this respect. This led me to investigate the matter, with the following curious results.

If we call f the flexure for any given thickness t, and f_1 the flexure for any other thickness t_1, then $f/f_1 = t^2/t_1^2$ for any given load or weight approximately. But as the weight increases directly as the thickness, the flexure of the disks due to their own weight, which is what we want to know, may be expressed as $f/f_1 = t/t_1$.

Let us now consider the effect of this flexure on the image. In any lens bent by its own weight, whatever part of its surface is made more or less convex or concave by the bending has a corresponding part bent in the opposite direction on the other surface, which tends to correct the error produced by the first surface. This is one reason why reflectors which have not this second correcting surface are so much more liable to show strain than refractors. If the lens were infinitely thin, moderate flexure would have no effect on the image. The effect increases directly as the thickness. If then the flexure, as I have shown, decreases directly as the thickness, and the effect of that flexure increases directly as the thickness, it is clear that the effect of flexure of any lens due to its own weight will be the same for all thicknesses; in other words no advantage is gained by additional thickness.

This has reference, of course, only to flexure of the lens in its cell after it has been duly perfected, and has nothing to do with the extra difficulty of supporting a thin lens during the grinding and polishing processes.

Figuring and testing

By the figuring process I mean the process of correcting local errors in the surfaces, and the bringing of the surfaces to that form, whatever it may be, which will cause the rays falling on any part to be refracted in the right direction. When an objective has undergone all the processes I have described, and many more which are not so important, and with which I have not had time to deal, and when the objective is centred and placed in its cell, it is, to look at, as perfect as it will ever be, but to look through and use as an objective it may be useless. The fact is that when an objective has gone through all the processes described, and is in appearance a finished instrument, I look upon it as about one-fourth finished. Three-fourths of the work has probably to be done yet. True, sometimes this is by no means the case, and I have had instances of objectives which were perfect on the first trial; but this is, I am sorry to say, the exception and not the rule.

This part of the process naturally divides itself into two distinct heads:

(1) The detection and localization of faults —what may, in fact, be termed the diagnosis of the objective.

(2) The altering of the figures of the different surfaces to cure these faults. This may be called the remedial part.

It may be well here to try to convey some idea of the quantities we have to deal with, otherwise I may be misunderstood in talking of great and small errors.

I have before mentioned that it is possible to measure with the spherometer quantities not exceeding 1/50 000th of an inch or with special precautions much less even than that; but useful as this instrument is for giving us information as to the general curves of the surface, it is utterly useless in the figuring process; that is, an error which would be beyond the power of the spherometer to detect, would make all the difference between a good and a bad objective.

Take actual numbers and this will be evident. Take the case of a 27-inch objective of 34 feet focus; say there is an error in the centre of one surface of about 6 inches diameter, which causes the focus of that part to be 1/10th of an inch shorter than the rest.

For simplicity's sake, say that its surface is generally flat; the centre 6 inches of the surface therefore, instead of being flat, must be convex and over 1 000 000 inches radius. The versed sine of this curve, as measured by spherometer, would be only about 1/250 000th, 4 millionths of an inch, a quantity mechanically unmeasurable, in my opinion.

If that error was spread over 3 inches only instead of 6 inches, the versed sine would only be about 1/1 000 000th. Probably the effect on the image of this 3-inch portion of 1/10th inch shorter focus would not be appreciable on account of the slight vergency of the rays, but a similar error near the edge of objective certainly would be appreciable. Until, therefore, some means be devised of measuring with certainty quantities of 1 millionth of an inch and less, it is useless to hope for any help from mechanical measurement in this part of the process.

For concave surfaces, Foucault's test is useful. I shall not trespass on your time to explain this in detail, as it is described very fully in many works, in none better than in Dr Draper's account of the working of his own reflecting telescope.

If an objective have but one single fault, its detection is easy; but it generally happens that there are many faults superposed, so to speak. There may be faults of achromatism, and faults of figure in one or all of the surfaces; faults of adjustment, and perhaps want of symmetry from some strain or flexure; and the skill of the artist is often severely taxed to distinguish one fault from the rest and localize it properly, particularly if, as is generally the case, there be also disturbances in the atmosphere itself, which mask the faults in the objective, and permit of their detection only by long and weary watching for favourable moments of observation.

It would be impossible to enumerate all the various devices that are practised for the localization of errors, but a few may be mentioned, some of which have never before been made public.

For detection of faults of symmetry, it is usual to revolve one lens on another and watch the image. In this way it can generally be ascertained whether it is in the flint or crown lens.

With some kinds of glass the curves necessary for satisfying the conditions of achromatism and spherical aberration are such that the crown becomes equi-convex and the flint a nearly plano-concave of same radius on inside curve as either side of the crown. This form is a most convenient one for the localization of surface errors in this manner.

The lenses are first placed in juxtaposition and tested. Certain faults of figure are detected. Now, calling the surfaces *A*,*B*,*C*,*D* in the order in which the rays pass through them, place them again together with Canada balsam or castor-oil between the surfaces *B* and *C*, forming what is called a cemented objective. If the fault be in either *A* or *D* surface, no improvement is seen; if in *B* or *C*, the fault will be much reduced or modified. Now reverse the crown lens, cementing surfaces *A* and *C* together. If same fault still shows, it must be in either *B* or *D*. If it does not show, it will be in either *A* or *C*. From these two experiments the fault can be localized.

It often happens that a slight error is suspected, but its amount is so slight that it appears problematical whether an alteration would really improve matters or not. Or the observer may not be able to make up his mind as to the exact position of the zone he suspects to be too high or too low, and he fears to go to work and perhaps do harm to an objective on which he has spent months of labour and which is almost perfect. In many such cases I have wished for some means by which I could temporarily alter the surface and see it so altered before actually proceeding to abrade and perhaps spoil it.

During my trials with the great objective of Vienna, I thought of a very simple expedient, which effects this without any chance at all of injuring the surface. If I suspect a certain zone of an objective is too low, and that the surface might be improved by lowering the rest of it, I simply pass my hand, which is always warmer than the glass, some six or eight times round that particular zone. The effect of this in raising the surface is immediately apparent, and is generally too much at first, but the observer at the eye end can then quietly watch the image as the effect goes off, and very often most useful information is thus obtained. The reverse operation, that of lowering any required part of the surface, is equally simple. I take a bottle of sulphuric ether and a camel's-hair brush, and pass the brush two or three times round the part to be lowered, blowing on it slightly at the same time; the effect is immediately perceived, and can always be overdone if required.

So far then for the diagnosis. Now for the remedy. When the fault has been localized, the lens is again put upon the machine and the polisher applied as before, the stroke of the machine and the size of the pitch patches being so arranged as to produce, or tend to produce, a slightly greater action on those parts that have been found to be too high (as before described while treating of the polishing processes).

The regulation of the stroke, eccentricity, speed, and general action of the machine, as well as the size and proportion of the pitch squares, and the duration of the period during which the action is to be continued, are all matters the correct determination of which depends upon the skill and experience of the operator, and concerning which it would be impossible to formulate any very definite rules. All thanks are due to the late Lord Rosse and Mr Lassell, and also to Dr de la Rue for having published all particulars of the process which they found capable of communication; but it is a notable fact that, as far as it is possible to ascertain, every one who has succeeded in this line has done so, not by following written or communicated instructions, but by striking out a new line for himself; and I think I am correct in saying that there is hardly to be found any case of a person attaining notable success in the art of figuring optical surfaces by rigidly following directions or instructions given or bequeathed by others.

There is one process of figuring which is said to be used with success among Continental workers. I refer to the method called the process of local touch. In this process those parts, and those parts only, which are found to be high, are acted upon by a small polisher.

This action is of course much more severe; and if only it were possible to know exactly what was required, it ought to be much quicker; but I have found it a very dangerous process. I have sometimes succeeded in removing a large lump or ring in this way (by large I mean 3 or 4 millionths of an inch), but I have also and much oftener succeeded in spoiling a surface by its use. I look upon the method of local touch as useful in removing gross quantities, but for the final perfecting of the surface I would not think of employing it.

In small-sized objectives the remedial process is the most troublesome, but in large-sized objectives the diagnosis becomes the more difficult, partly on account of the rare occurrence of a sufficiently steady atmosphere. In working at the Vienna objective it often happened when the figure was nearly perfect that it was dangerous to carry on the polishing process for more than ten minutes between each trial, and we had then sometimes a week to wait before the atmosphere was steady enough to allow of an observation sufficiently critical to determine whether that 10 minutes' working had done harm or good. It must not be supposed either that the process is one in which improvement follows improvement step by step till all is finished. On the contrary, sometimes everything goes well for two or three weeks, and then from some unknown cause, a hard patch of pitch perhaps or sudden change of temperature, everything goes wrong. At each step, instead of improvement there is disimprovement, and in a few days the work of weeks or months perhaps is all undone. Truly anyone who attempts to figure an objective requires to have the gift of patience highly developed.

In view of the extraordinary difficulty in the diagnostic part of the process with large objectives, it is my intention to make provision which I hope may reduce the trouble in the working of the new 28-inch objective for the Royal Observatory Greenwich.

Two of the greatest difficulties we have to contend with are: (1) the want of homogeneity in the atmosphere, through which we have to look in trials of the objective, due to varying hydrometric and thermometric states of various portions; and (2) sudden changes of temperature in the polishing-room. The polisher must always be made of a hardness corresponding to the existing temperature. It takes about a day to form a polisher of large size, and if before the next day the temperature changes 10° or 15°, as it often does, that polisher is useless, and a new one has to be made, and perhaps before it is completed another change of temperature occurs. To grapple with these two difficulties I propose to have the polishing-chamber under ground, and leading from it, a long tunnel formed of highly glazed sewer-pipes about 350 feet long, at the end of which is placed an artificial star illuminated by electric light; on the other side of the polishing-chamber is a shorter tunnel, forming the tube of the telescope, terminating in a small chamber for eyepieces and observer. About half-way in the long tunnel there will be a branch pipe connected to the air-shaft of the fan, which is used regularly for blowing the blacksmith's fire, and through this, when desired, a current of air can be sent to 'wash it out' and mix up all currents of varying temperature and density. By this arrangement I hope to be able to have trials whenever required, instead of having to wait hours and days for a favourable moment.

There is a general idea that the working of a plane mirror or one of very long radius is a more difficult operation than those of more ordinary radii. This is not exactly the case. There is no greater difficulty in figuring a low curve than a deep one, but the difficulty in the case of absolutely plane mirrors consists simply in the fact that in their figuring there is one additional condition to be fulfilled, viz. that the general radius of curvature must be made accurate within very narrow limits. In figuring a plane mirror to use, for instance, in front of even a small objective, say 4-inch aperture, an error in radius which would cause a difference of focus of 1/100th of an inch would seriously injure the performance. This would be about equivalent to saying that the radius of curvature of the mirror was about 8 miles, the versed sine of which, with the 6-inch spherometer, would be about 1/50 000th of an inch. Now what I mean to convey is this: that it would be just as difficult to figure a convex or concave lens of moderate curvature as a flat lens of the same size if it were necessary to keep the radius accurate to that same limit, i.e. one-tenth of a division of this spherometer.

15.5 The work of Ritchey

Ritchey's paper (Ritchey, 1904) describes the methods employed by him at the Yerkes Observatory in making and testing spherical, plane, paraboloidal, and (convex) hyperboloidal mirrors up to 5 feet in diameter. Ritchey's own wording is used throughout except for the substitution of oblique for direct narration.

He used disks made at the glass-works of St Gobain, near Paris, of sizes from 8 inches in diameter and $1\frac{1}{2}$ inches thick, to the great one which is 5 feet in diameter and 8 inches thick, and which weighs a ton; in every case it was specified that great care should be given to thorough stirring and annealing.

For mirrors of 24 and 30 inches in diameter, a thickness of one-sixth of the diameter was preferred. In the cases of the large paraboloidal mirror of a reflecting telescope, and the large plane mirror of a coelostat or heliostat, which should always be supported at the back to prevent flexure, the thickness should be one-seventh of the diameter.

All mirrors should be polished (not figured) and silvered on the back as well as on the face, in order that both sides shall be similarly affected by temperature changes when in use in the telescope; for the same reason the method of supporting the large mirror at the back, in its cell, should be such that the back is as fully exposed to the air as possible. . . .

The grinding and polishing machines were similar in principle to Dr Draper's, but were more elaborate.

The turntable upon which the glass rests consists of a vertical shaft 5 inches in diameter, carrying at its upper end a heavy triangular casting, upon which, in turn, is supported the circular plate upon which the glass lies. This plate is of cast-iron, weighs 1800 pounds, is 61 inches in diameter, is heavily ribbed on its lower surface, and is connected to its supporting triangle by means of three large levelling screws. The surface of the large plate was turned and then ground approximately flat; two thicknesses of Brussels carpet are laid upon this, and the glass, with its lower surface previously ground flat, rests upon the carpet.

The entire turntable, with the heavy frame of wood and metal which supports it, can be turned through 90° about a horizontal axis, thus enabling the optician to turn the glass quickly from the horizontal position which it occupies during grinding and polishing, to a vertical position for testing.

A long transverse slide on the secondary arm allows the grinding and polishing tools to be placed so as to act on any desired zone of the glass, from the centre to the edge; and this setting can be changed as desired while the machine is running. The secondary crank, which turns at the same speed as the large one which drives the main arm, enables the optician to change as desired the width of the (approximately) elliptical stroke or path of the tool with reference to the length of this stroke; this change is especially desirable when figuring the glass; it is, of course, impossible when only one driving-stroke is used.

To recapitulate briefly: this method of connecting the grinding and polishing tools allows them to be controlled in all of the following ways simultaneously: (1) the stroke of the tool is given by the motion of the main arm; (2) the slow rotation of the tool is rigorously controlled by the belting; (3) the tool is allowed to rock or tip freely by means of the universal coupling, in order that it may follow the curvature of the glass; (4) the tool rises and falls freely by means of the sliding of the $1\frac{3}{8}$-inch vertical shaft in its bearings, in order that it may follow the curvature of the glass; (5) the tool is counterpoised by means of the lever on the main arm, through the medium of the same vertical shaft and universal coupling.

An important question concerns the size of grinding tools—should they be of the same diameter as the mirror? For mirrors up to 24 or 30 inches in diameter full-size tools are generally used. For concave mirrors larger than 30 inches in diameter Ritchey used grinding tools whose diameter is slightly more than half that of the glass, i.e., a 16-inch tool for a 30-inch glass; a 32-inch tool for a 60-inch glass. Full-size tools are, of course, much more expensive and difficult to make; they are many times heavier than half-size tools of equal stiffness; and they require a much stronger grinding machine to counterpoise them properly; grinding can be done with them, however, more quickly than with the smaller tools. Half-size tools are economical and are quickly prepared; they are easily counterpoised; and a much greater variety of stroke can be used with them, so that with a well-designed grinding machine he found it easier to produce fine-ground surfaces, entirely free from zones, with half-size than with full-size tools. If temperature conditions and uniform rotation of the glass are carefully attended to, the surface of revolution produced by the smaller tools is fully as perfect as that given by the larger ones.

The curvature of the tools and of the glass is measured by means of a large spherometer. The spherometer is of the usual three-leg form; the legs terminate in knife-edges, the lines of which are parts of the circumference of a 10-inch circle. The screw is of $\frac{1}{2}$-millimetre pitch, and the head, which is 4 inches in diameter, is graduated to 400 divisions. On fine-ground surfaces settings can be made to one-half or one-third of a division, corresponding to a depth of 1/40 000th or 1/60 000th of an inch approximately.

The tools adopted after trying various kinds consisted of a wooden disk or basis covered on one side with squares of rosin faced with a thin layer of beeswax. The wooden disk may be replaced, in the case of small polishing tools up to 12 or 15 inches in diameter, by a ribbed cast-iron plate so designed as to be extremely light and rigid; the bases of larger tools may be made of cast aluminium alloy; such a basis for a 30-inch polishing tool weighs about 60 lb. It is possible that a metal basis possesses an advantage over a wooden one in that its surface is less yielding. Ritchey's full-size iron tools for a 24-inch mirror weigh about 150 lb, or $\frac{1}{3}$ lb for each square inch of area. This weight, or even $\frac{1}{2}$ lb to the square inch, is not objectionable with emeries down to 5-minute or 10-minute washed; but when this weight is allowed with finer emeries, scratches are liable to occur. The pressure on the glass is therefore decreased, by counterpoising the tool, to approximately $\frac{1}{5}$ lb to the square inch for 12- to 20-minute emeries, $\frac{1}{8}$ lb per square inch for 30- to 60-minute emeries, and about $\frac{1}{2}$ lb per square inch for 120- to 240-minute emeries. *This rule is followed, approximately, in all fine-grindings, whether of back or face.* This obviates, to a great extent, the danger of scratches in grinding, provided that thorough cleanliness is practised in the preparation and use of fine emeries. (See § 4.10.)

In fine-grinding a 24-inch glass, the 2-minute and 5-minute emeries are used for $\frac{3}{4}$ hour each; the 12- and 30-minute emeries for $1\frac{1}{2}$ hours each, and the 60-, 120-, and 240-minute emeries for $1\frac{1}{2}$ hours each. The fine-ground surface resulting takes a full polish very readily.

Polishing

The preparation of polishing tools has already been described. The polishing rouge which Ritchey used was of the quality which is used in large quantities commercially in polishing plate-glass.

The pressure per square inch used by Ritchey in polishing is of interest. His 24-inch polishing tool, with its wooden basis, weighs about 25 lb; this is not heavy enough for the best action in polishing; so about 50 per cent additional weight is put on in the form of twelve lead blocks which are distributed uniformly and screwed to the back of the tool. This gives a weight of about $\frac{1}{12}$ lb for each square inch of area, which is found to work well for all large tools. For tools 18 inches or less in diameter somewhat greater pressure per square inch of area may be used. A 36-inch tool, with wooden basis $3\frac{3}{4}$ or 4 inches thick, weighs 75 or 80 lb, and needs no additional weighting.

Testing and figuring spherical and plane mirrors

Ritchey, like Draper and Grubb, used Foucault's Test which was originally described in Vol. V of the *Annals of the Paris Observatory.*

The making of large plane mirrors of fine figure is usually regarded as much more difficult than that of large concave mirrors. The difficulty has been, in the past, largely one of testing. With a satisfactory method of testing the large plane surface *as a whole,* in a rigorous and direct manner, the problem is greatly simplified. So far as Ritchey was aware, no such test had hitherto been fully developed. In *Monthly Notices,* **48**, p. 105, Mr Conumon suggests, very briefly, the testing of plane mirrors in combination with a finished spherical mirror, and gives a diagram in illustration. This method was developed and used by Ritchey in testing plane mirrors up to 30 inches in diameter. When this test is used, the difficulty of making a 24-inch plane mirror which shall not deviate from perfect flatness by an amount greater than 1/500 000 inch is neither greater nor less than that of making a good spherical mirror of 2 feet aperture and 50 feet radius of curvature, when it is required that the radius of curvature shall not differ from 50 feet by a quantity greater than 1/100 inch.

Testing and figuring paraboloidal mirrors

The work of changing a spherical mirror to a paraboloidal was accomplished entirely by the use of polishing tools, by shortening the radii of curvature of the inner zones, instead of by increasing or lengthening those of the outer zones.

Testing can be done at the centre of curvature, by determining there the foci or the radii of curvature of successive zones of the mirror; it may be done at the focus of the paraboloid, by the aid of a finished mirror which should be at least as large as the paraboloidal one; and it may be done directly on a star. The first two methods named have the very great advantage that they may be conducted without interruption, under the practically perfect atmospheric and temperature conditions of the optical laboratory.

A knowledge of the properties of the parabola enables the optician to compute the positions of the centres of curvature of successive, definite, narrow zones of the mirror, and the surfaces must be so figured that the radius of curvature of each zone agrees with the computed value. In testing, each zone in succession is exposed by means of a suitable diaphragm, all of the rest of the surface being covered. In practice, two different formulae are used, depending upon the position of the illuminated pinhole.

The diaphragms which Ritchey used do not expose entire zones, but only pairs of arcs on the right and left sides of the mirror.

This ends the summary of Ritchey (1904).

The method of testing a paraboloid at its focus was briefly described by Ritchey in the *Astrophysical Journal,* November 1901. It is more simple, direct, and rigorous than the test at the centre of curvature. A well-figured plane mirror, which should not be smaller than the paraboloidal one, is necessary in order that the testing may be done in the optical laboratory.

Changing a spherical surface to a paraboloid

This is accomplished by shortening the radii of curvature of all of the inner zones of the surface, leaving the outermost zone unchanged. There are two distinct methods of accomplishing this: (1) by the use of full-size polishing tools, the surfaces of which are cut away in such a manner as to give a large excess of polishing surface near the central parts of the tool; (2) by the use of small polishing or figuring tools worked chiefly upon the central parts of the mirror, and less and less upon the zones towards the edge.

(1) *Parabolizing with full-size tools.* The rosin surface can be trimmed in a variety of ways to give a great excess of action on the central parts of the mirror. A form of tool used for this purpose was very similar to that shown in Fig. 10.2.9 (Chapter 10). The form of the polisher areas can be altered as desired, and thus the amount of action on any zone can be in some measure controlled. Length of stroke and amount of side-throw are also very important factors in controlling the figure of the mirror. Tools of this kind serve admirably in parabolizing mirrors up to 36 or 40 inches in diameter, when the angular aperture is not very great.

(2) *Parabolizing with one-third-size and smaller tools.* In the case of very large mirrors, when full-size tools are almost unmanageably heavy, and in

the case of mirrors of great angular aperture, in which the departure from a spherical surface is great and is effected with difficulty with full-size tools, one-third-size and smaller figuring tools may be used. The machine should invariably be employed in this work, the transverse slide being used to place the tool in succession upon the various zones. In order to preserve the surface of revolution the setting of the transverse slide should be changed only at the end of one or more complete revolutions of the glass.

The mirror should be tested very often, and the utmost care taken to keep the apparent curve of the surface, as seen with the knife-edge test, a smooth one, i.e. free from small zonal irregularities, at all stages of the parabolizing; this is not extremely difficult when the optician has become experienced in the use of the transverse slide.

15.6 Hartmann's testing of large objectives

In the testing of telescope objectives, a large degree of direct knowledge of the course of the rays is to be had from the method of Hartmann (1900 and 1904). In this, a diaphragm is pierced with rings of holes at the various zones. The diameter of the holes is between 1/200 and 1/400 of the focal length of the objective. This diaphragm is placed close to the object glass, and the image of a star photographed within and without focus. Such photographs consist of dots, each of which corresponds to one of the holes in the diaphragm; and from the distance apart of corresponding dots in the two photographs the course taken by the rays from the corresponding parts of the object glass can be deduced. Highly accurate results have been obtained, but the method is laborious.

Kingslake (1926) points out that the Hartmann test may be complicated when applied to the measurement of the aberrations of oblique pencils and shows how to deal with the determination of coma and astigmatism.

An account was published by Professor Hartmann (1908) of the use of his method of test for the correction of the 80-cm objective of the Potsdam objective, then the largest outside America. The method has been fully described by Hartmann and others and its utility widely accepted.

It should be remembered that the Hartmann test is the only one which determines the exact course of the rays and so yields a quantitative statement of the true residual aberrations of optical systems.

It was concluded from the tests that the zonal aberrations could be reduced by local retouching; in such local polishing only those zones are worked which have too great a curvature, the work being usefully carried out on the last surface.

As Kingslake (1928) points out the assumption that is made in the ordinary method of performing the Hartmann test that all the rays cross the optical axis is only true if both surfaces and homogeneity of the glass are perfect. 'Technical' aberrations which arise from the manufacture of the glass or in the working or figuring of the surfaces may result in rays which do not do so. This is especially the case in the large astronomical objectives and mirrors to which the Hartmann test is most generally applied. Kingslake's paper and the discussion stress the limitation of the Hartmann test to the actual parts of the lens or mirror which happen to fall under the holes in the diaphragm.

This limitation was appreciated by Professor Hartmann. The measurements on the 40-inch (102 cm) objective at the Yerkes Observatory (Philip Fox, 1908) were made at Professor Hartmann's suggestion and the results of the measurements communicated to him. He said in reply, 'So far as one can see from the few points observed the objective is excellent. . . .' Hartmann, however, considered the investigation incomplete because so few zones and points were included and therefore advocated the desirability of applying to the objective, in addition to his own test, the modification of the Foucault test which he had recently applied to the 80-cm objective at Potsdam. The improvement consisted merely in replacing the eye by a camera, the objective of which throws upon a photographic plate a sharp image of the objective to be investigated, with the knife edge in a suitable position.

This application of the Foucault test revealed an astonishing amount of detail in the surface which, of course, was not revealed by the Hartmann test. It was even possible to see traces of the epicycloidal motion of the polishing tool.

15.7 The 200-inch Mt Palomar objective

The following passage is quoted, with permission, from Sir Harold Spencer-Jones' Presidential Address to Section A of the British Association (Spencer-Jones, 1949).

For the 200-inch Hale telescope at Mount Palomar it was decided to reduce the weight of the disk by constructing it with a thin face supported on a honeycomb structure, consisting of 36 cylindrical pockets, lying on five concentric circles, each pocket being connected to each of the adjacent pockets by a glass rib, thereby providing stiffness. These pockets were designed to form part of the support system. Each was ground accurately cylindrical and fitted with a steel sleeve, gaskets being inserted between the bearing faces and the glass. Each sleeve is connected to the cell-frame by an integral two-component system of balance weights and levers. Double gimbal bearings on the steel sleeve and on the support fixed to the cell frame prevent constraints. Three of the lever systems are tied down by circumferential springs to constrain the disk against rotation about the axis, while three others are tied down as axial defining points. A central tube passes through the 40-inch diameter central hole in the mirror; four central radial jacks, made of invar and steel to compensate for expansion of the central steel tube, define the radial location of the mirror, no edge band being required. These centre jacks and 12 edge radial squeeze arcs, to remove gravitational astigmatism, are provided with ball-bearing faces to allow freedom in the plane normal to the applied reaction. A number of fans have been mounted in the back of the mirror cell to draw air from the front to the back of the mirror in order to secure better temperature equilibrium.

Work on the mirror

Through the kindness of Mr J. V. Thomson, of Messrs Cox, Hargreaves & Thomson Ltd, Mr Twyman was able to give the following notes about the work on the mirror.

Work on the 200-inch mirror began in 1936 in the specially built optical shop at the California Institute of Technology, Pasadena.

After the front surface of the 20-ton disk had been ground flat the disk was turned over and the back surface dealt with in the same manner, the pockets and cavities being filled with wooden plugs and plaster of Paris to prevent chipping of the ribs. The 36 pockets were then accurately ground to receive the steel sleeves of the support mechanism.

The work of grinding out the curvature of the mirror to a depth of $3\frac{3}{4}$ inches at the centre occupied three months, during which operation 10 tons of coarse abrasive grit were used, removing a little over 5 tons of glass. A grinding tool 100 inches in diameter was used for this part of the work. When the approximate depth had been reached, a full-sized 200-inch tool was constructed for grinding the surface spherical and establishing a true surface of revolution. This tool, of welded sheet-steel construction, was surfaced with about 1950 Pyrex glass blocks and the abrasive grits were fed to the surface of the mirror through funnels in the upper surface of the hollow tool.

After the last grade of fine abrasive had left the surface ready for polishing, $3\frac{1}{2}$ months were spent in cleaning up the optical shop and removing all trace of grit from the grinding machine, in order to reduce the danger of scratching the optical surface during the polishing and figuring.

Altogether 31 tons of abrasives, including rouge, were used in the work on the mirror, which, first to last, took 180 000 man-hours of labour.

The full-sized tool was used during most of the initial polishing and for this purpose squares of a special pitch mixture were cemented to the Pyrex glass blocks, each square being cut and channelled into smaller squares, so that the 200-inch polisher comprised a mosaic of some 8000 pitch facets which distributed the rouge equally over the whole surface of the mirror. The pitch used was a mixture of rosin, paraffin wax and cylinder oil, later replaced by pine tar. The temperature in the shop ranged from 65° to 85° over the year, making it necessary to vary the proportion of the ingredients according to the existing temperature. Air conditioning and temperature control plant had been installed but its action disturbed the air more than enough to make optical testing impossible, and its use was therefore abandoned. After the pitch had been melted it was poured into specially constructed wooden trays. When cool the trays were taken apart and the pitch cut into uniform squares ready for cementing to the under surface of the polishing tools. After a polisher had been made up in this way it was first suspended over gas stoves to soften the pitch facets, and then lowered on to a thick cream of rouge and water on the mirror's surface, additional weight being then applied to press all the facets into perfect contact with the optical surface. The tool was then lifted off and the facets trimmed and bevelled. The smallest polisher used on the 200-inch mirror measured 8 inches in diameter.

It had originally been planned to test the 200 inch with parallel light, employing a 120-inch plane mirror, offset laterally, for this purpose. Although this mirror had been fine ground ready for polishing it was decided to abandon the original plan for certain reasons. It was estimated that far less time would be occupied in parabolizing the 200 inch and testing it by other methods, than would be the case if it had first to be polished and figured truly spherical in order to test the 120-inch flat mirror, during the figuring of which no further work could be done on the 200 inch itself. The work of parabolizing the mirror was therefore begun as soon as it had received its initial polish, and as the surface at the centre had to be deepened by the relatively large amount of 1/200 inch it was accomplished by alternately grinding and polishing until a stage was reached when the final figuring could be done by polishing alone.

The mirror, tipped up on its turntable into a vertical position, was tested at its radius of curvature 110 feet away. A Foucault screen, dividing each radius of a diameter into thirteen 6-inch apertures, was wheeled in front of the mirror, and the measured zonal aberrations compared with the calculated parabolic aberrations. From these differences a graph would be drawn indicating in wavelengths the differences in height between the actual surface and the required parabolic surface, showing clearly where the next spell of figuring should be done.

For the very final figuring the Strong and Gaviola system was used.

The 160-ton grinding and polishing machine consisted of three main parts; the horizontal beam, the spindle section and the turntable. The heavy horizontal beam, called the bridge, moved back and forth on wheels above the surface of the mirror. This bridge supported a carriage, also on wheels, which held the spindle, at the lower end of which were attached the grinding and polishing tools. The directions of motion of the bridge and of the spindle being at right-angles to each other, a variety of different strokes could be obtained. The turntable was designed to play the dual role of turntable and mirror cell, the 36 support mechanisms being secured in their pockets and supporting the mirror during the optical work. When the mirror was finished the turntable, carrying the mirror, was detached from the machine and transported as a complete unit to the observatory on Palomar Mountain, and bolted to the base, of the tube after the mirror had been aluminized.

A further period of figuring, confined to the outermost 18 inches of the mirror, was carried out at the observatory during 1949, since when the performance of the telescope has exceeded the hopes of all concerned.

At the time of writing (July 1950) the most recent report concerning the performance of the telescope (Bowen, 1950) describes extensive tests carried out to disentangle the complex effects of temperature changes and flexure due to gravity and to correct them. As a result a system of fans has been installed to accelerate adjustment to temperature changes of the central portion, combined with extra insulation to retard the effect of temperature changes around the rim.

The criterion of performance was the proportion of light from a point source which was concentrated at the focus within a circle of specified diameter. On the three best nights the average figure was 84 per cent within a circle of 0·05-mm diameter.

15.8 Texereau's study of the effects of minor irregularities in objectives

Texereau (1949) studied and analysed the faults arising from various polishing techniques. The remainder of this section is abstracted from Texereau's paper on the subject.

THE SEARCH FOR A CRITERION OF 'DEFINITION'

To characterize by a single figure the quality of an objective is a convenient and serviceable simplification, but in many cases it is insufficient.

The celebrated quarter-wavelength rule of Lord Rayleigh has been useful because it is founded on a physical property and not on a simple and more or less artificial mathematical treatment. It may be formulated thus:

> The wavefront emerging from a good objective is contained completely between two spheres whose radii differ by a maximum of a quarter wavelength.

Without wishing to detract from the practical comprehensiveness of this rule, which is often sufficient, one must issue a warning against the widely relied upon supposition that it is a criterion of unsurpassable quality.

Actually, an optical surface (and consequently the emergent wavefront) of an objective is a very complex physical entity, and there can be no question of predicting the full extent of the influences of its imperfections on the image in terms of a single parameter.

Before attempting a more detailed analysis some facts may be called to mind which may reduce our enthusiasm for the quarter wavelength limit.

First of all, there is a factor which does not come within the optician's province, but may well spoil the result of his work—atmospheric disturbances. Following upon a programme of investigations into the quality of imagery in different places, A. Danjon (*Réunions Inst. d'Optique*, 1933, *2nd*, 20) has formulated an important law: To the faults of an objective must be added those atmospheric disturbances which change the incident wavefront (and also the emergent wavefront if there are thermal heterogeneities in the air within the tube of the instrument). Naturally, if the objective has faults of its own approaching the quarter wavelength limit, the total effect frequently exceeds this limit. Such an objective is therefore much more sensitive to atmospheric agitation.

The smallest perceptible alteration of the focal diffraction pattern of a star, which gave Rayleigh the basis for formulating his rule, is not the most sensitive test for determining the faults of an objective and their effect upon the images. The effect on the diffraction patterns near to the focal plane becomes apparent for a phase difference of $\lambda/10$.

Astronomical observers often speak of small contrasts as being better or less well seen with different objectives of comparable powers. This effect goes so far that, in special circumstances, observers can recognize the objectives used merely by inspecting the image of Mars (see for example *Annales des Observations Jarry–Desloges*, 1907, p. 115; 1913, p. 296).

M. Françon has shown convincingly that for the faintest perceptible contrasts ($\lambda = 0{\cdot}03$), the efficiency of an instrument corrected to $1/4\lambda$ falls to 0·62, while it remains at 0·92 if the residual imperfections do not exceed $1/16\lambda$.

The most satisfactory criterion is, without doubt, that of A. Danjon and A. Couder (*Lunettes et telescopes*, p. 522), which combines the rule of Lord Rayleigh with the requirement that the transverse aberrations shall be smaller than the radius of the diffraction pattern.

A. Couder has for a long time been observing and describing (*l'A.*, **50**, February 1936, p. 65) particular defects of the wavefront, often very small, but nevertheless sometimes serious because of their slope or periodicity. In 1944 he summarized many fruitful ideas in a note (*Cahiers de Physique*, **26**, 27).

F. Zernicke (1935) in describing the method of phase-contrast, drew attention to the 'phase-mosaic', which is a wavefront disturbed by small periodic undulations (*M.N.*, 1934, p. 377).

The study of the diffracted light at a large distance from the central image has more recently led B. Lyot (*Comptes Rendues*, **222**, 765) to evolve a phase-contrast method of another form, that reveals a whole class of defects which were comparatively unknown before, but which appear to deserve attention.

CLASSIFICATION OF WAVEFRONT IMPERFECTIONS

It is not intended to consider here the faults caused by refraction but only those of the optical surfaces themselves.

In Fig. 15.8.1, let us call a the height of a fault that represents the phase-difference, and b the width of the element of the inclined wave; hence $p = a/b$ is the corresponding slope.

With these three parameters we can lay down a satisfactory classification of the imperfections. This is more important to the optician than to the user, because every form of imperfection generally originates in the method of working employed.

15.8.1. Wavefront imperfections.

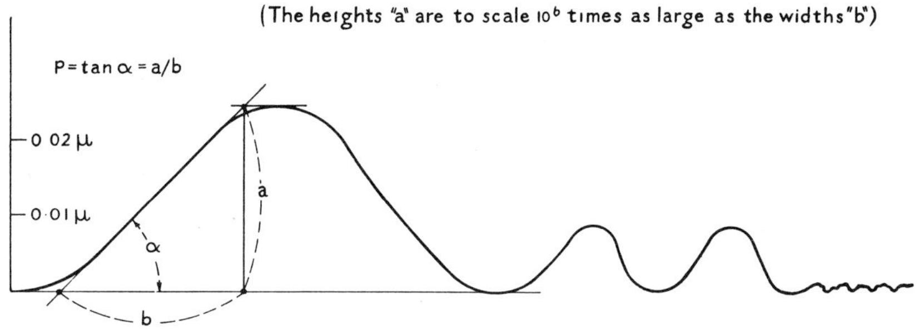

The conditions which make elimination of the faults possible are not always compatible and one has to decide on the compromise to be aimed at for each particular case.

We shall at once distinguish between two completely different and clearly defined classes of faults. The 'goose-flesh' (or 'orange peel') effect on cloth polished surfaces would form an intermediate class, but it would be of no interest for the optics of astronomy.

(A) *Surface faults*: *a* and *b* are usually small and of the same order, the slope *p* is therefore considerable and of the order of unity.

(B) *Faults of shape*: In the case of finished surfaces *a* is restricted to a maximum of $\lambda/4$, but may be perceptible when as small as 1/1000 of a wavelength; *b* also may vary within wide limits, from the radius of the surface down to 1/10 mm. The slope of these imperfections is always very small; in a good surface it is of the order of 10^{-6} to 10^{-5}.

The amount of light scattered can be estimated quantitatively by the use of simplifying assumptions which are more or less unsatisfactory. It will be enough to mention the general effect for the main categories, with the reminder that a direct investigation into the energy distribution away from the main image is always advisable.

Surfacing faults

(i) In the case of spots of 'grey' *a* and *b* are of the order of one micron. The stray light is diffracted within a solid angle of considerable extent, of the order of several tens of degrees. There are also micro-spots which are only visible with a powerful ultra-microscope, and which diffract very little light away from the main image.

(ii) In scratches, *a* and *b* are of the order of 1/10 mm. In order to realize their seriousness, other parameters are necessary. If a scratch is straight and sufficiently long, a diffraction flare will be visible with a sufficiently bright star, but if the scratch is curved, the effect will be scattered and mostly not noticeable.

(iii) With sleeks, *a* and *b* are of the order of one micron, and the length of the order of one centimetre. These very fine marks may occur sporadically (only a few over the whole surface), and then the effect will mostly be negligible, or they may be present in great numbers, in which case the effect will be similar to that of greyness. One could also mention micro-sleeks, which are invisible by ordinary means, and which may occur in very great numbers on a surface polished with little pressure, but whose optical effect is in most cases negligible.

Faults of shape

(i) Faults of large amplitude in which *a* may be up to $\lambda/4$ and *b* may be of the order of several centimetres, the slope being 10^{-5} to 10^{-6} at most, are the classical faults, the most serious ones, because of their direct bearing on the nucleus of the diffraction pattern and its immediate neighbourhood. It is also necessary to distinguish between zonal and astigmatic defects. A test report including the analysis of these defects is given with every well-made astronomical objective.

(ii) Faults of medium amplitude comprise those in which *a* is several tenths of a wavelength and *b* of the order of a centimetre. Local faults, such as superficial veins in the glass (the slope can exceed 10^{-5} even in a good mirror) or traces left by local polishing, are generally of negligible effect.

(iii) The result of more or less periodic faults of fairly great number (primary ripples) can be serious, in extreme cases (e.g. not properly tended machine work) the difference in the diffraction pattern is clearly visible. The diffraction pattern rings from the second and third outward are disturbed and appear as a luminous haze. A similar specially disastrous effect is produced by turbulence, which can break the wavefront into minute elements of greater inclination. Contrast in the telescopic images of planets is reduced and the image is blurred.

(iv) Faults of small amplitude (micro-ripples) in which a is of the order of 1/1000 of a wavelength and b is of the order of a millimetre, the slope being about 10^{-6} to 10^{-5}, cannot be considered as local imperfections, but rather as a general condition of the surface, there being millions of them on a surface. It is not astonishing that, even if the distribution of these faults is not very regular, a considerable part of the mirror can be well away from the main image in phase and produce a diffraction spectrum. The diffracted light, especially up to 1° off axis, can in bad cases ($a \approx 30$Å, $b \approx 1$ mm) be as much as several thousandths of the total energy at 2° off axis. It shows up specially clearly if one moves a bright object out of the field, and it affects the detection of the solar corona with a Lyot coronograph much more seriously than do many defects of surfacing.

Finally there are the elementary faults which could be investigated at grazing incidence by the electron microscope, but information is not available to permit the classification or even detection of these imperfections.

15.9 The Isaac Newton 98-inch telescope

The preceding material from Mr Twyman's book will now be supplemented by accounts of two contemporary projects, the Isaac Newton telescope at Herstmonceux, and the Anglo-Australian telescope at Siding Springs.

Permission has been given by Mr George Sisson, O.B.E., and Mr G. E. Manville, both of Sir Howard Grubb, Parsons & Company Ltd, to publish the following article on the Isaac Newton telescope.

Introduction

On 4 December 1959, the Admiralty placed an order with Grubb Parsons, of Walkergate, Newcastle-upon-Tyne, for the design and construction of a reflecting telescope having a main mirror of 98 in aperture. The reason for this particular size lay in the possession by the Admiralty of a Pyrex disk generously provided in 1949 for this project by the MacGregor Trust of Michigan, U.S.A. For various reasons the project suffered delay in its early stages. If the construction of such a telescope were started to-day, a Cervit disk would be used in preference to the low expansion glass.

The telescope, known as the Isaac Newton telescope, was installed at the Royal Greenwich Observatory at Herstmonceux, Sussex, and will be the largest instrument in Europe. (Fig. 15.9.1.)

The immense strides made in radio astronomy have tended to over-shadow the valuable contribution which can still be made to the science by the optical telescope. Actually the observations by radio techniques are complementary to those obtained from visual astronomy, with the result that today several countries are contemplating optical telescopes of large size. In Europe, there are plans for three telescopes of about 140-inch aperture and in the U.S.A. for two telescopes of 150-inch aperture. Other projects for instruments of similar size are in hand in Canada and in Australia, and Britain will collaborate with these countries. The possession of a large telescope in the U.K. forms a valuable training ground for astronomers who will then be competent to use the even larger instruments located in more favourable climatic conditions.

General problems

The large astronomical telescope is a mechanical structure supporting optical elements which have to be pointed with great accuracy at a chosen celestial object. The usual mechanical arrangement comprises a polar axis of rotation parallel to that of the earth's axis and a second axis at right angles to it called the declination axis. By turning the polar axis at a uniform rate in the opposite direction to that of the rotation of the

15.9.1. The completed 98-inch Isaac Newton telescope in the Grubb-Parsons factory. The telescope is now installed at the Royal Greenwich Observatory, Herstmonceux, Sussex. The shutters over the primary mirror (Fig. 13.4.1) can be seen in the closed position for protection of the mirror surface.

earth the telescope tube, containing the optical elements, can be held pointing steadily at a chosen object in space. The mechanical and optical problems combine in such a way that deflections normally considered insignificant by engineering standards do in fact carry an unusual influence, so that the large fabricated steel sections and solid pieces of glass may be considered as behaving with the characteristics of rubber. The main mirror surface must retain its curvature measured in terms of millionths of an inch and a total moving weight of some 87 tons must be pointed into space with an accuracy better than 1 second of arc.

The optical system of the Grubb Parsons telescope—like most large telescopes—is arranged so that three different focal positions are available, these being known as the prime focus, cassegrain focus and coudé focus. Their relative positions are shown in the optical diagrams. (Fig. 15.9.2.) Each focal station has its individual characteristics suitable for a particular type of astronomical work. For example, the coudé focus is at a fixed position and is used where it is required to feed the incoming starlight into heavy equipment such as a large spectrograph which has to be kept in a thermally controlled environment.

The cassegrain focus will be used for a lighter type of spectrograph than is the coudé, so that advantage can be taken of the fact that the light will only have been reflected by two mirrors in the telescope system. It will be particularly useful for photometric work. The prime focus will be used with a small spectrograph for work on very faint objects, as well as for photography when the correcting lens is required to give a useful field of good definition.

The telescope is housed within a dome which not only provides protection but retains within itself an environment approximating to ambient temperatures at night. Since the dome must be opened for observation, it is important that the temperature within it should be held as near night temperature as possible during the day, otherwise convective air currents would disturb the imaging of the telescope. This situation means that the telescope itself must function satisfactorily over a wide range of temperatures corresponding to the annual variations in night temperature.

Probably the most difficult problem posed by the optical system of the Isaac Newton telescope is maintenance of the alignment of the prime focus correcting lens relative to the main mirror. The alignment must be correctly established with great precision in the first place and thereafter maintained to very close limits. Support of the main mirror was in itself an interesting problem and the general mechanical design of the mounting presented difficulties for which engineering solutions had to be devised.

Manoeuvring the telescope and its use demand a wide range of speeds, some constant and some variable, all of which can conveniently be obtained by the arrangements provided. In the initial stages, setting up of the telescope can be carried out by a night assistant at the central control console on the observing floor; during actual use on a star, fine adjustments and corrections are continually made by the observer at the focal station in use. For instance, an astronomer working in the prime focus cage may be taking a photograph of a particular region of the sky. Under the best observation conditions images of the stars upon the photographic plate will turn out to be about 20 micrometres diameter for an exposure lasting, perhaps, several hours. During all this time, the telescope follows a moving object and this motion must be monitored by the observer so that the image does not shift on the photographic plate by more than a few micrometres. Naturally, the work of a large telescope such as this will be devoted to the study of faint objects not readily observed by smaller instruments—including a new class of mysterious objects recently identified by both radio and visual astronomers and given the name quasars. The existence of quasars may well eventually disprove several existing notions on the formation of the universe. They radiate enormous quantities of energy, but are so distant that the quantity so far registered on earth barely amounts to a calory.

The observing floor in the observatory is at a height of 48 feet and at this level there is a control desk from which all the motions of the telescope at any attitude can be read. The control panel has a display showing digital indication of sidereal time and dials indicating the attitude of the telescope to an accuracy within 1 min of arc. The dials receive their information from coarse and fine synchros directly geared to large spur wheels on the declination axis and polar disk.

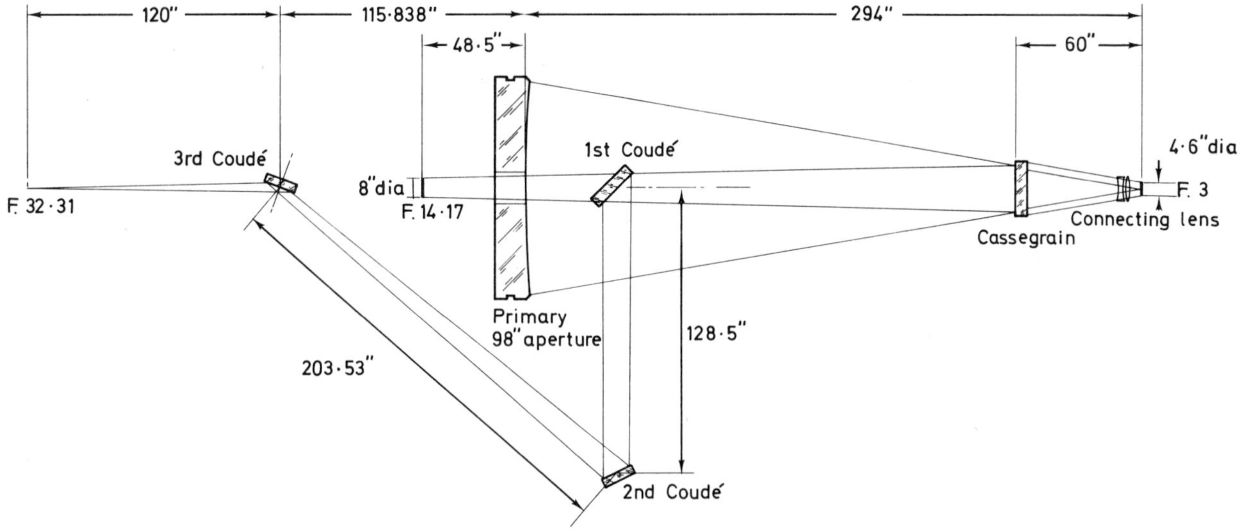

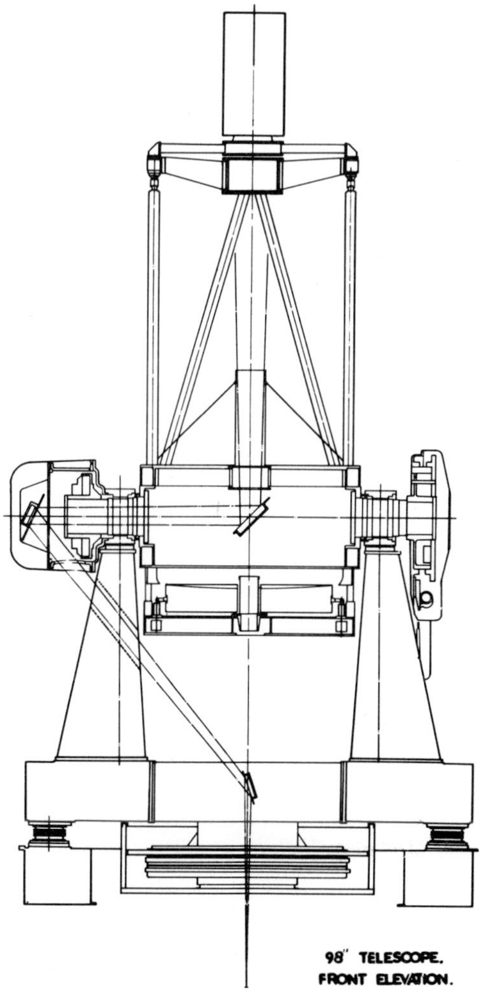

15.9.2. Optical arrangement and front elevation of the Isaac Newton telescope, showing the location of optical components in the mounting.

The mirror cell

The Pyrex mirror (Fig. 15.9.3) is 98·2 inch diameter, 16·1 inch thick at the edge, and has a central hole 13·3 inch diameter. It weighs 9000 lb. The front surface is concave to a depth of 2 inch at the centre, and a reflecting coat of aluminium, evaporated on to the surface, reflects the starlight. The glass surface is polished to an accuracy measured in millionths of an inch and, to maintain usefulness in service, the mirror must be supported within its cell so that flexure of the glass due to its own weight at any attitude does not cause distortions greater than the polishing limits. The need for this extreme accuracy can best be appreciated if it is realized that the beam of parallel light falling on the mirror surface must be reflected to converge at the prime focus so that on test 90 per cent of the light is contained within an image of 5 micrometres diameter.

To support the mirror at changing attitudes of the telescope, forces must be applied to it axially and radially to balance gravity, but in such a way as to leave the shape of the mirror undistorted. Moreover, the mirror must also be positioned accurately within its cell both in respect of tilt and centring. Various multiple support devices have been employed in the past, but the Isaac Newton telescope exploits a completely new system.

For most of the time, the mirror will be required to work with its axis at less than 45° depression so that more weight must be supported axially than in the radial direction. This situation is fortunate since the new air bag axial support system that has been used approaches closer to the ideal than any known radial support system. Since the concave surface reduces the depth of glass to be carried progressively towards the centre and, therefore, as a slightly different supporting pressure is required at the centre than at the edge, the air bag has been made with three separate annular sections. For thermal considerations it is desirable that the back surface of the mirror shall not be in direct contact with the bag and distance pieces have been interposed between the two. Provided they are in sufficient number the actual arrangement of these spacers is not greatly significant—they simply transfer the uniform upward pressure of the bag to the mirror. (Fig. 13.4.1.)

The main body of the cell is a fabricated steel structure within which the lower surface is machined flat. A glass fibre centring plate $\frac{3}{16}$ inch thick rests on this face and is capable of radial adjustment in any direction. The air bag is buttoned to the plate for location and is supported by it when the telescope is turned to the horizontal. The air bag itself is built up of rubberized fabric 0·012 inch thick with vulcanized joints and internal divisions at 29-inch and 49-inch radius. The centre hole within the air bag is 17-inches diameter and the outside diameter 100 inches, compared with corresponding diameters of $13\frac{1}{4}$ inches and 98 inches of the Pyrex disk. The nominal outside thickness of the bag in its working condition is $\frac{1}{4}$ inch and it lies below a second glass fibre sheet to which the spacers are glued. When the telescope is nearly horizontal, this sheet is 'on its edge' and is supported by a system of multiple spokes which provide restraint in the radial direction.

The axial spacers, numbering 182, are arranged in a symmetrical circular pattern which results in a distance of approximately 7 inches between each support; the flexure of the glass across this distance can be disregarded. The upper faces of the spacers are covered with a thin layer of synthetic rubber and, when differential expansion takes place between the glass disk and the support system, the spacers become slightly inclined without exerting an undesirable force radially on the lower surface of the mirror. The nominal air pressure required within the three sections of the bag is only 1·3 lb per sq. inch, and while this arrangement is sufficient to float the glass comfortably it offers no positive location in the axial direction.

This location is provided by three fixed pads in contact with the glass and disposed in D-shaped cut-outs around the circumference of the air bag. Besides positioning the mirror these pads also support the weight of the glass not carried by the air bags at these points.

After allowing for the load carried by the pads, residual weight which must be borne by the air bag amounts to 8550 lb and the pressure must be adjusted to maintain this value within quite close limits: moreover the pressure must vary sinusoidally with the angle of depression of the telescope axis. Tests indicate that, if the forces applied to the mirror by the locating pads differ by more

15.9.3. The 98-inch primary mirror of the Isaac Newton telescope on the polishing machine and after aluminizing. The coating machine for aluminizing is reflected in the mirror.

than 17 lb then perceptible triangular distortion is introduced at the image. Two obvious sources of error are apparent: decentring of the air bag, causing the resultant lift to be displaced relative to the centre of gravity of the mirror, and a pressure error within the air bag itself. Accurate location of the bag on the centring plate can be accomplished empirically once and for all to an accuracy of about one-hundredth of an inch, but continuous control of air pressure presents a problem.

The arrangement employed consists quite simply of a piston floating within a cylinder. Air lubrication from an external supply at 20 lb per sq inch is applied between the piston and cylinder wall to eliminate mechanical friction, and the piston floats freely at all angles down to horizontal. A secondary air supply at $1\frac{1}{2}$ lb per sq. inch is applied beneath the piston and raises it to the level of a port through which surplus air escapes. Thus the pressure maintained below the piston is always exactly that required to carry its weight and the port acts as a relief valve for the bag supply. Three such controllers are employed, one for each compartment of the bag, the whole device being mounted on the mirror cell so that it is always at the same angle as the mirror, thereby ensuring that the bag pressure is just sufficient to support the mirror.

The action of this controller in service is such that, when the mirror is lying with its axis vertical, the pressure within each compartment of the air bag is reliably controlled and the total support offered by the bag is a nominal figure of 8550 lb $\pm$ 1 lb.

Support of the mirror in the radial direction is provided by a mechanical system consisting of 36 steel lugs cemented on to the periphery of the mirror at uniform spacing. Each of these lugs is pivoted to a link extending radially outwards and, at the outer end of each link, there is a weighted lever which pivots tangentially to the disk. When the mirror lies horizontally the links exert no force upon it, but, as it turns on its edge, all those links above the horizontal centreline exert a radial force upwards which increases progressively to reach a peak at the top of the mirror. Below the horizontal centreline the thrust bears radially inwards in a similar manner.

Radial location of the mirror is also essential; it is provided by fixed constraining pads at four points around the circumference. These pads perform a difficult function since they must hold the mirror central to within a few thousandths of an inch throughout the range of work temperature. Any significant restraint in the axial direction would produce distortion.

The telescope tube

The mechanical function of the tube is to unite the optical elements of the instrument so that they are correctly located with respect to each other and can be pointed in the desired direction by the mechanical mounting.

The tube has an axis of rotation, the declination axis, passing through its centre of gravity and provided by two hollow trunnions. These trunnions are mounted on the 'centrepiece', a ribbed structure of fabricated steel beneath which the main mirror cell is supported with its surface 42 inches from the axis. The appropriate convex secondary mirror is carried at the upper end of the tube together with the focusing gear, or alternatively prime focus equipment comprising plate holder, guiders, focusing gear and correcting lens. Using the prime focus, the observer sits within a small capsule supported independently of the photographic equipment. The apparatus at the upper end of the tube is held in position by four radial strips fixed to form a mild steel box section ring at the top of the tube. Electrically operated, sliding trimming weights at either side of the tube provide a balance for any auxiliary apparatus in use.

For optical reasons it is essential that the axis of the main mirror shall pass centrally through the upper end of the tube at all attitudes of the telescope. Thus deflection of the main mirror and the tube top must be equal and relative tilt between them must be eliminated.

The mechanical structure used for the tube consists of four V-frames sprung from a massive central section within the centrepiece. The upper V-frames are made of steel tubes of 6-inch diameter and contain, at each end, a short section of solid steel rod which acts as a circular spring flexure element. The attachment on the centrepiece is extremely rigid; it is devised so that the forces from the upper V-frames are transmitted directly through the central section to the lower V-frames. The mirror cell supporting frames are of $3\frac{1}{2}$-inch diameter tubes and are also fitted with flexure elements at both ends. It is estimated that the main mirror and the tube top will each sag by 0·016 inch when the telescope is in the horizontal position. The arrangement is such that, theoretically, the angular error between top and mirror should be zero at all angles. The combined effect of tilt and inequality of flexure must not displace the axis of the main mirror by more than 0·02 inch at the centre of the top tube, or this would interfere with the correct operation of the prime focus correcting lens.

On the outside of the declination bearings one of the trunnions carries a driving gear, with both quick and slow motions for the axis, while the other trunnion carries a balance weight. For each trunnion, the load carried outside the bearing is approximately equal to the weight of the telescope acting inside the bearing, with the result that torque exerted by the trunnion on the centrepiece side wall is very small. The hollow passages within the two declination axis trunnions are occupied on one side by electric cables and on the other side by the coudé focus light beam.

The declination axis motor drive transmits quick motion to the axis through a final engagement spur gear 74 inches diameter, at speeds of $22\frac{1}{2}°$ per minute or 90° per minute. A double-acting clamp is mounted on the trunnion in a way that permits it to be locked solidly either to the quick motion spur gear or to a tangent arm assembly which is free to turn upon the axis. A variety of slow motion speeds can thus be imparted to the tube under the control of a servomotor.

The polar disk

A conventional form of equatorial mounting, much used in the past, is to carry the telescope tube between the tines of a fork, the stem of the fork being the polar axis. This basic design has been chosen for the Isaac Newton telescope but an entirely novel arrangement has been substituted for the conventional polar axis. In attempting to minimize the flexure of the conventional fork mounting, most of the difficulties are encountered, not so much in the tines themselves, but in the cross-piece which joins their roots to the central axis and is subjected to varying types of bending and torsion. These difficulties can be eliminated by the replacement of the axis with a disk of rugged construction from which stout tines can be sprung. The polar disk is fabricated from mild steel and has an outer circumference of 22-feet diameter and a thickness of 3 feet. It weighs 40 tons complete and is made up of three sections, with the mating surfaces hand scraped to very fine limits and held together by multiple internal bolts. This hand facing operation had to be carried out before the final machining.

Beneath the disk is a large fabricated bedplate that is set upon a concrete pier which supports the whole instrument. Interposed between the bedplate and the disk is a system of externally pressurized oil pads; this is a system which has been used for telescope mountings before, though not for a disk as large as this. Its function is to float 87 tons upon a film of oil without mechanical contact so that precise motion, free from frictional uncertainties, can be imparted to the telescope.

Radial location of the disk is provided by two pads, each with a load bearing surface of 24-inches diameter, disposed at 80° to each other. All the pads follow the same general design with the working area divided into quadrants, each with a pressure pool working at 100 lb per sq. inch fed through control jets from a supply pressure of 200 lb per sq. inch. The oil escapes from the pressure pool into an annular drainage passage around the pad and then back into the oil circulation system. A nominal working clearance of 0·004 inch is maintained and monitored by four microswitches in each pad to give warning if at any point clearance is reduced by half.

Axial support is provided by pads generally similar to those already described. At the lowest point, a pad 18 inches in diameter, carrying a load of about 11 tons, acts as a locating pad, and under the upper half of the disk two other pads, each $9\frac{1}{2}$ inches in diameter and carrying $3\frac{1}{4}$ tons, complete the location system. In addition to these pads, there are four other weight-relief pads on bellows.

The polar disk drive

In the event of a pump pressure failure, drive to the disk is automatically interrupted while oil pressure to the pads is maintained by a nitrogen bottle for at least 10 seconds to allow the disk time to come to a standstill. When the telescope is not in use there is no oil circulation, and direct contact takes place between the disk and the pads, when the disk is completely locked. A system of jacks has been incorporated as a maintenance measure that would support the disk to allow removal of any pad if required.

Tests for circularity of the disk conducted with a radius arm rig show the circumference to be within $\pm 0{\cdot}002$ inch of a true circle. Flatness of the underside was more difficult to check but optical siting on a target indicated that switchback error of the working surface did not exceed $\pm 0{\cdot}003$ inch from a true plane, and the quality of the surface finish was within 35 micro-inches.

Checking the accuracy of rotation has revealed that the total radial wander of the centre of rotation is less than 0·004 inch in a complete revolution; the maximum error of tilt in any position does not exceed 7 seconds of arc; and the error is smoothly accumulated over a travel of 90° rotation.

In addition to manœuvring motions it is, of course, necessary to impart to the polar disk a very smooth continuous fine motion corresponding to that of the earth's rotation. This motion is imparted by a worm wheel and is known as the sidereal drive.

15.10 The 150-inch Anglo-Australian telescope

A new telescope is to be erected at the Siding Springs Observatory of the Australian National University, near Coonabarabran, some 300 miles from Canberra. Design and construction of the telescope is being carried out under the direction of the Project Manager, Mr W. A. Goodsell, and the Anglo-Australian Telescope Board. Funds are being provided equally by the Science Research Council, U.K., and the Department of Education and Science in Canberra.

Because of the long construction period, and the rapid pace of astronomy, it is not possible to forecast the objectives to be achieved by the telescope, but it will occupy a central position in the advance of astronomy in the Southern Hemisphere during the late 1970s and 1980s.

The contract for grinding and polishing the optical components was received by Sir Howard Grubb Parsons and Company Ltd, Newcastle-upon-Tyne, and the mirror blanks were made from Cervit glass ceramic by Owens-Illinois, Toledo, Ohio, who delivered them to Newcastle-upon-Tyne, England, before the end of 1969.

The telescope mounting will be made by Mitsubishi (MELCO) (Tokyo) and the building will be constructed by Leighton Contractors Ltd (Sydney) who have subcontracted the dome to Evans Deakin Industries Ltd (Carringbak, N.S.W.).

The weight of the primary mirror blank was 33 500 lb (15 000 kg). It cost about £250 000. The grinding and polishing are expected to cost more than £100 000.

The telescope optics consist of a Ritchey-Chrétien system and the general dimensions of the Cervit mirror blanks are as follows:

Primary mirror	155-inch diameter (393 cm)
f/8 convex	58-inch diameter (147 cm)
f/15 convex	35-inch diameter (89 cm)
Elliptical flat	37-inch × 27 inch (94 cm × 68 cm)
Second coudé flat	28-inch diameter (71 cm)

Also four flats of 14-inch (35 cm) diameter for use on the coudé system.

DESIGN OF THE TELESCOPE

Permission has been given by Mr E. L. Freedman, Executive Secretary of the Australian and New Zealand Association for the Advancement of Science, to reprint the following article by Professor S. C. B. Gascoigne on the Anglo-Australian 150-inch telescope originally published in the *Australian Journal of Science*, *32*, No. 12, June 1970.

Introduction

Following discussions which extended over several years, and in which a prominent part was taken by the Australian Academy of Science and the Royal Society, agreement between the British and Australian Governments to build the telescope was finally reached in April 1967. Costs were to be shared equally between the two governments, the telescope was to located at the Siding Spring station of Mt Stromlo Observatory, near Coonabarabran, N.S.W., and the design was to be based on that of the 150-inch being built at the

Kitt Peak National Observatory, U.S.A. At a cost expected to exceed $10 000 000, the telescope will be the largest purely scientific project yet undertaken in this country; and only two other telescopes in the world will be substantially bigger, the Palomar 200-inch and a 238-inch currently being built in Russia. Progress to date has been good. The 20-ton blank for the primary mirror has been successfully manufactured and delivered, most of the main contracts have either been let or are pending, and estimates that on-site assembly will begin early in 1973, and the telescope should be operational by the end of 1974, still seem reasonable.

The optical system

The optical arrangement of the telescope is shown in Fig. 15.10.1. There will be four focal positions, a primary focus, *f*/8 and *f*/15 cassegrain foci, and a coudé position. The relevant parameters are set out in Table 15.10.1. The principal observing station is expected to be the cassegrain, the coudé being used for heavy instrumentation, notably a coudé spectrograph, and the primary for direct photography. Observers at the primary and cassegrain foci will ride with the telescope, the former in a small cage at the upper end.

The mirrors will be of Cervit, a new zero-expansion glass-like substance developed by the firm of Owens-Illinois. Chemically it is similar to the felspars, basically SiO_2 with some of the Si replaced by Al or Li. Its low thermal expansion follows from its micro-crystalline structure, which is induced by nucleating agents and controlled by a heat cycle. Because of the low expansion, prolonged annealing is not necessary, and our blank was in fact ready for rough machining about a month after pouring. Our mirror will have a clear aperture of 154 inches, be 24 inches thick and, when finished will weigh about 16 tons (it is solid). It will have a thermal relaxation time of about fourteen days. If made of normally expansible material, a mirror of this size would be very subject to thermal deformation arising from its immediate past thermal history; the passage of a severe cold front can distort a big glass mirror for days. One of the major advantages of Cervit should be its freedom from this sort of deformation.

The optical design will be of the now conventional Ritchey–Chrétien type, in which the primary is a hyperboloid (of eccentricity 1·0832) and the secondaries also hyperboloids, chosen to form point images at the cassegrain foci. The *f*/8 focus will present an essentially aberration-free field of 14 inches (40 minutes arc) in diameter, and so be advantageous for direct photography. Because it will be hyperboloid, the primary mirror alone will produce an on-axis image 6 seconds arc diameter at the prime focus, and this, of course, must be corrected before any observations, especially direct photography, can be carried out there. But the design of such correctors is by now highly developed, and is in fact much facilitated by the hyperbolic form of the primary. Our primary focus will have available corrected fields of up to a degree (9·2 minutes) diameter, across which the image spread should not exceed half a second arc diameter; such a field will be a major

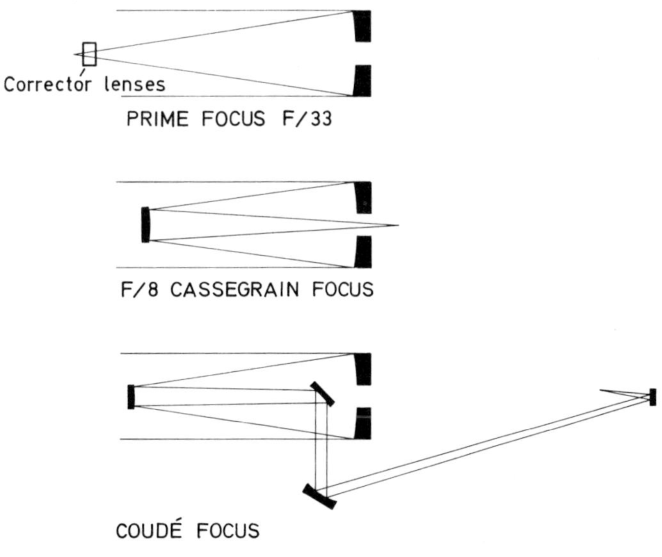

15.10.1. Optical configurations of the 150-inch telescope.

Table 15.10.1 150-inch telescope optical data

Focal position	*Focal ratio*	*Focal length*	*Scale* Per inch	*Scale* Per mm.	*Field diameter*
Primary	f/3·3	500 in.	6′·9	16″·2	1° = 9·2 in.
Cassegrain	f/8	1225 in.	2′·8	6″·7	39′ = 14 in.
Cassegrain	f/15	2290 in.	1′·5	3″·5	15′ = 9·8 in.
Coudé	f/36	5540 in.	37″	1″·5	5′ = 8 in.

advance on anything achieved hitherto with large telescopes.

A mirror of these dimensions, simply supported, will deform under its own weight many times in excess of any allowable tolerance. Its weight must be relieved to near 0·1 per cent, and this calls for a carefully designed support system, which must be effective at any mirror aspect, and in which particular attention must be paid to friction and hysteresis. Adapting an idea due originally to the Canadians, the AAT mirror will be supported from behind by thirty-three air pistons and three locating points arranged in two concentric rings. The air pressure in the pistons will be made to vary as the cosine of the zenith distance, while the diameters of the rings will be chosen to minimize flexure. Side supports will be twenty-four counterweights working through a push-pull mechanical lever system. The secondaries, being with one exception used upside down, will be supported from the back by partial vacua, and from the side by annular bags filled with mercury. To change the optical configuration, for instance from *f*/3·3 to *f*/8, the telescope must be pointed vertically upwards and the upper ends of the main tube interchanged. These weigh four tons each, and the operation is expected to take 30 minutes to complete.

The contract for the mirror figuring, the mirror supports and the tube has been let to the well-known firm of Grubb Parsons, of Newcastle-upon-Tyne, England. The primary was delivered late last year. Preliminary work is well advanced and the actual figuring should begin shortly, with the expectation that it will take about two years. The maximum departure from the nearest sphere will be 66 wave-lengths. We might add a word here on optical performance. A large mirror which would concentrate 80 per cent of the incident light into an image 0·5 second arc diameter would be accepted, not to say welcomed, by most astronomers. Better performance is possible, and given enough money, especially for testing, one could go a fair way towards achieving the 0·03 second arc images theoretically attainable by a diffraction-limited 150-inch. But the improvement would be lost in the residual flexure still allowed by the mirror support system, in imperfect optical collimation, in local 'seeing' effects created by the temperature differences which will often exist between the mirror surface and the ambient air, and in irregularities in the telescope drive, each of which can contribute 0·2 second or more to the image size. Superimposed on these are the highly variable effects of atmospheric seeing, that is, the extent to which the stellar image is degraded by its passage through turbulent elements in the earth's atmosphere. All in all, a large telescope which can consistently produce images smaller than 1-second arc diameter is not only performing very well, but also is located at a very good site.

Mounting and control system

The telescope follows the main lines of the 200-inch—the observer rides with the tube, the tube is a Serrurier truss, designed to bend an equal distance at each end under gravitational loading and so preserve optical collimation, and the main load is carried by a horseshoe bearing, in this case of 40 feet outside diameter, moving on oilpad bearings. A change from the 200-inch is that the horseshoe has been shifted to coincide with the declination axis. This enables the declination axis to be used as a stiffening member, at the cost of heavier loading on the bearings, and some problems in keeping them aligned as the horseshoe twists. The mounting, complete, will weigh 316 tons.

If the mechanical design is fairly conventional, the drive and control system will break substantial new ground. The diurnal drive, in which to counteract the earth's rotation the telescope is driven about the polar axis at an accurately controlled rate of 15 seconds arc per second of time, will be provided not by the traditional worm and wheel, but by three spur gears, a wheel and two pinions, which can now be ground to the very high precision this task demands. Spur gears make it easier to eliminate backlash, one pinion providing the drive, the other the preload; they facilitate slewing—rapid movement from one part of the sky to another; they are less vulnerable to damage, and they lend themselves better to servo control. The positional encoders will be driven by separate pinions, two for each axis, and similar drives will be used for both axes. A computer will be incorporated from the outset as an integral part of the control system (though it will be possible to operate the telescope without it). The computer should simplify and speed up normal telescope operation, make for more accurate setting and driving, facilitate the incorporation of devices like automatic guiders, and have wide application to the control of instrumentation and to on-line data reduction.

In this connection it is important to remember that our telescope (with others like it) will be used not by an experienced resident staff, but by a series of visitors who will have only a nodding familiarity with it. All astronomers are aware of the time it takes to master a large telescope: much of the first night can be wasted, and several nights may pass before one feels reasonably at home, let alone begins to appreciate the real potentialities of the instrument. Strenuous efforts are being made to by-pass these difficulties. These will appear in the carefully designed cassegrain head on which all the cassegrain instruments will be mounted, on the consistent use of automatic guiding and automatic star acquisition, and on the use of television cameras and image intensifiers to facilitate the identification of faint objects. With the general exception of radio sources, it is not usual for the positions of objects of astro-physical interest to be known to better than 1 or 2 minutes arc; they must be identified from charts, at times a maddening and time-consuming process, especially on an unfamiliar telescope, and television, which can now be made more sensitive than the eye, should help substantially.

Considerable attention is being paid to instrumentation, for which a substantial sum has been set aside. At the cassegrain focus there will be a low-dispersion spectrograph designed to go as faint as possible, a medium-dispersion spectrograph, photometers, a single-channel photoelectric spectrum scanner, and means for direct photography. Direct photography will also be carried out at the prime focus, where there will be a choice of correctors, and planning for a coudé spectrograph has gone some distance. Four coudé stations will be available, one for a spectrograph with a 6-inch beam, one for a possible large (20-inch) spectrograph, and two others for future heavy instruments, as yet unspecified.

Dome, building and site

The building and dome (Fig. 15.10.2) will have a common outside diameter of 120 feet, with the top of the dome 161 feet above ground level. The dome is expected to cost more than $1 000 000, and the building more than $2 000 000, and the two are on the critical path on the PERT chart, so that it is on their completion more than on any other factor that the time scale of the whole enterprise depends. This follows because the telescope cannot be assembled without the dome, the dome cannot be erected without the building, and the design of neither dome nor building could really get under way until the main telescope dimensions had been settled.

The dome will be a steel structure weighing about 500 tons, rotatable around a horizontal track; the slit through which the telescope will work will be closed by an up-and-over shutter. It will carry a 40-ton hoist, to be used in the first instance for erecting the telescope, thereafter for heavy maintenance. The building will be erected around a 60-foot high pier with a hammerhead or flat cantilevered top which will carry the telescope and coudé spectrographs. Long-term stability and freedom from vibration are, of course, important here. Besides immediate telescope ancillaries, the building will house the aluminizing plant, photographic dark-rooms and plate store, and electronic laboratory, library, observers' offices, a visitors' gallery and finally an extensive ventilation plant. Considerable attention has been paid to active air control of the dome, partly to avoid large daytime temperature fluctuations, but more particularly to keep the dome air and telescope near ambient temperature from the time of beginning observations to the end of the night. This will be achieved by a thorough flushing of the dome with ambient filtered air, beginning perhaps 2 hours before observations commence. The air will be supplied by large fans moving up to 110 000 cubic feet/minute, along ducts and up through grilles in the observing floor.

There have been extensive developments to the site, now either complete or well in hand. These have included the construction of power and telephone lines, the running of an 11-mile pipeline from the Coonabarabran water supply reservoir, and major improvements to the road, the latter necessitated by the need to move some quite massive telescope components up the mountain. Five houses have been built for permanent maintenance staff, the observers' lodge has been approximately doubled in size to accommodate visiting astronomers, and a combined workshop

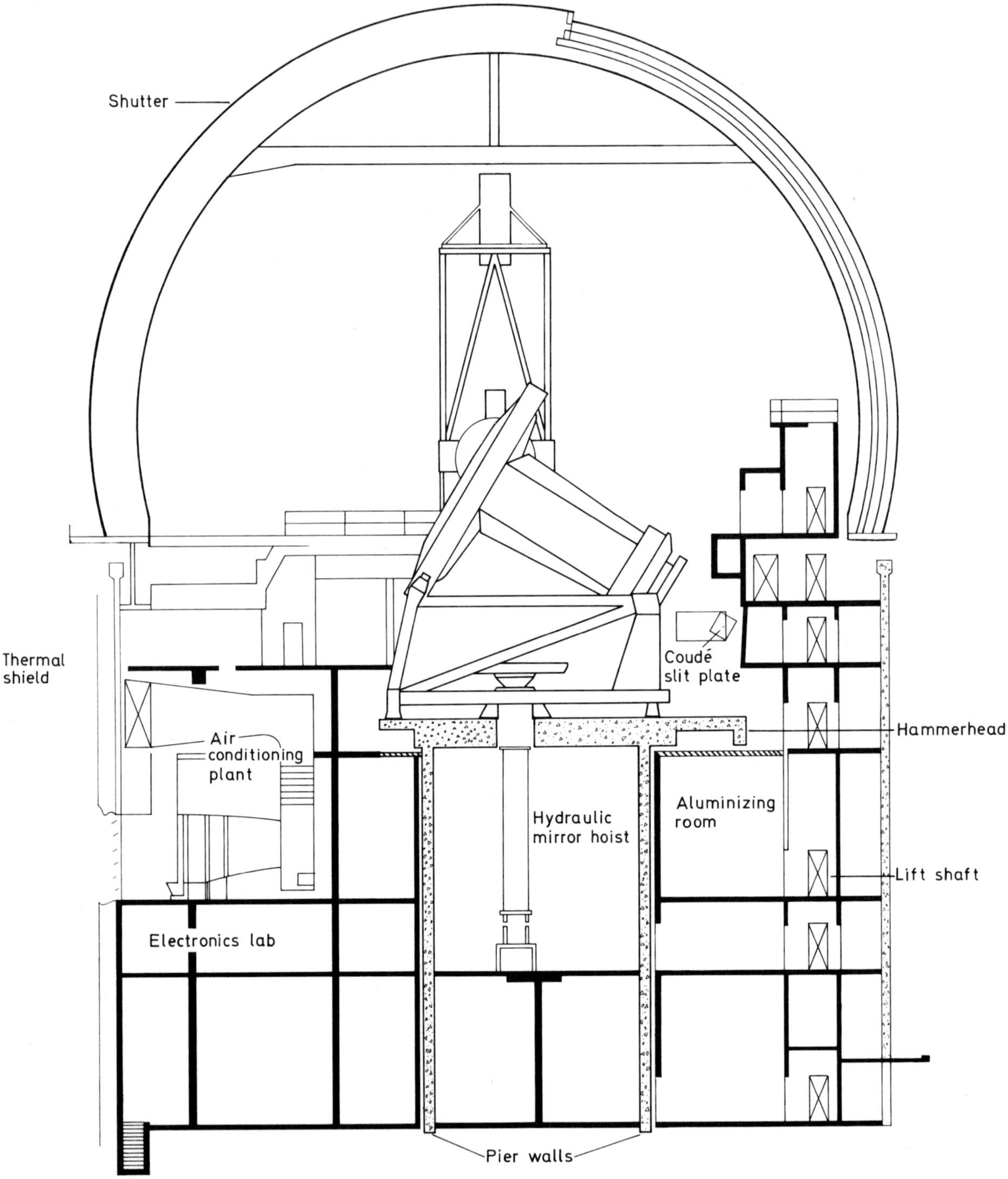

15.10.2. The 150-inch telescope and its dome.

and maintenance depot should be completed shortly.

The Anglo-Australian telescope promises to be a first-class instrument. It is well abreast of current practice, which in some respects it leads; an immense amount of care and ingenuity has gone into the design; it should be highly efficient, and readily adaptable to almost any type of astronomical observation; and the instrumentation should be as good as any which has been built. One can confidently predict that the telescope will have a long and productive life. That it will provide a major stimulus over the whole range of Australian astronomy goes without saying.

15.11 Making Cervit blanks for large objectives

The stages of manufacture of a Cervit mirror blank are as follows:

(1) Pouring the Cervit into a mould (Fig. 15.11.1).

(2) Curing and crystallizing the material for several weeks in a kiln (Fig. 15.11.2).

(3) Removing the blank from the kiln (Fig. 15.11.3).

(4) Removing the mould from the blank (Fig. 15.11.4).

(5) Grinding to drawing dimensions (Fig. 15.11.5).

(6) Crating for shipment (Fig. 15.11.6).

△15.11.1. After the disk for the Anglo-Australian telescope had been poured, a temperature-controlling lid was lowered over the 25 tons of white-hot glass. Without this heat shield there would have been a risk of melting the metal roof rafters.

15.11.2. ▷ The casting after being poured from the furnace (*top left*). The blank was cured and crystallized for several weeks in the kiln (*lower right-hand edge of the photograph*)

15.11.3. The Anglo-Australian mirror blank in its mould.

15.11.4. The Anglo-Australian blank after its removal from the mould. It is being moved to the rough-grinding machine for shaping to its final dimensions before shipment to Grubb-Parsons for smoothing and polishing.

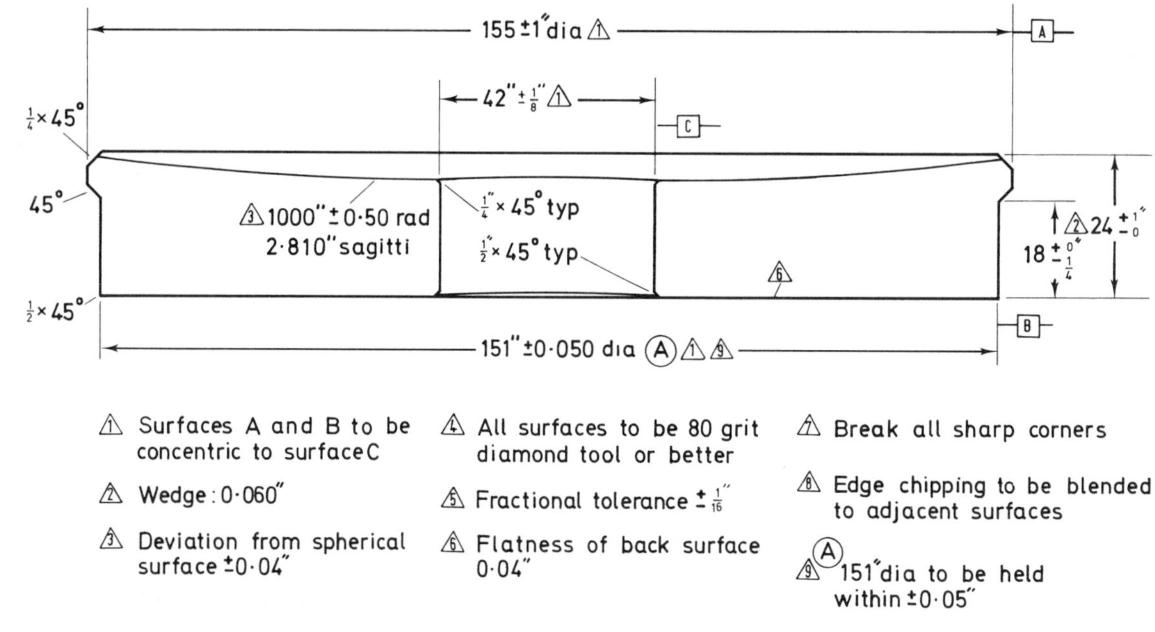

15.11.5. The Cervit primary mirror blank for the 150-inch Anglo-Australian telescope.

15.11.6. The Anglo-Australian mirror blank, after rough grinding, being crated before dispatch to Grubb-Parsons in England from Owens, Illinois.

Pouring large Cervit blanks

Permission has been given by Owens–Illinois to publish the following report on the pouring of the primary blanks for the Cerro Tololo telescope and the Anglo-Australian telescope.

In the largest single pouring of glass attempted in the western world, 50 000 pounds of Cervit were cast on 25 June 1967 for the 158-inch mirror of Cerro Tololo Inter-American Observatory in Chile (see *Sky and Telescope*, February 1968, p. 72).

Another similar disk had been poured on 9 April 1969 at the Owens–Illinois Development Center in Toledo, Ohio, for the 150-inch Anglo-Australian reflector to be installed at Siding Spring Observatory in Australia. A third giant blank of this material, which has practically zero expansion coefficient, will be made for the 144-inch telescope being built by the French government for operation in southern France, perhaps in the Pyrenees Mountains.

Although similar to conventional disk-making, the new process is far less time-consuming. A Cervit mirror begins with weighing and mixing nine ingredients—principally the oxides of lithium, aluminium, and silicon. For the Cerro Tololo project, about 110 000 pounds was melted in a furnace of 140 000-pound capacity (enough for a solid 200-inch mirror). Only ten days were required to charge, melt, and refine the mixture, whereas in 1934 it took a month to prepare 65 tons of Corning Pyrex borosilicate glass for the honeycomb 200-inch blank for the Hale reflector.

The necessary high degree of homogeneity of the Cervit was accomplished with a battery of five mechanical stirrers, which churned the viscous glass at 3000° Fahrenheit (1489°C).

On pouring day, a fifteen-man crew handled the casting, one man monitoring the glass stream and the mould surface with an optical pyrometer. There was also a safety team and an emergency squad ready for the critical operations of removing and replacing the ceramic plug at the bottom of the furnace. Below floor level a large volume of water was ready to quench the molten glass in case of a failure.

The Cervit was then transferred into the 169-inch diameter titanium-brick mould for the 158-inch blank. The stream flowed at 3800 lb per minute for $13\frac{1}{2}$ minutes. The plugging of the furnace to cut off the flow was decided by visual inspection when the mould was filled to within 2 inches of the top, a glass depth of 28 inches. Then men in silver thermal suits signalled for a fork-lift truck to pull the mould out from under the furnace. For safety the mould cart was also connected to an overhead crane.

Compare this one-step process with that for the 200-inch, where the filling of the mould was laboriously done with huge ladles that scooped 750 lb of molten glass at a time, but poured only 450 into the mould. Over a hundred such transfers were required for the 20 tons of glass in the Hale telescope's ribbed blank.

At Toledo on 25 June, the freshly poured Cervit glowed intensely at 2900°, the 25-ton mass containing about 25 million British thermal units (Btu) of heat energy. To prevent the radiation from cooling the top surface too rapidly, a heat shield was lowered over it, while the cooling rate of the bottom of the mould was enhanced with forced cold air.

Water-cooled pipes attached to the heat shield supported an array of temperature-sensing elements inserted into the centre of the glass. In only 14 hours, the blank's temperature fell to about 1500°F (745°C), so the mould could be placed in the kiln for slow cooling and the vital heat-treating process. (At this point the 200-inch blank began an 11-month slow annealing in which its temperature was reduced about a degree a day.)

Gas-fired in a 700°—2000°F range, the kiln can be heated or cooled at rates of less than a degree per hour. This permits altering the nucleation and crystal growth of the Cervit mix to determine its final properties.

When removed from the kiln after several weeks of heat treatment the blank was ready for smoothing and edge-grinding to the specified dimensions. Trimming 10 000 pounds off the oversized 162-inch casting for the Anglo-Australian telescope required only about forty days.

When inspected at the beginning of August 1969, the blank for the Siding Spring Observatory had been finished to near its final dimensions of 24 inches thick and 155 inches in diameter. Still to be done at Toledo were the finish grinding of the surfaces and edge, coring of the 42-inch central hole (for the Cassegrain focus), and hewing out of a radius of curvature of 1000 inches on the front or mirror surface.

The Anglo-Australian blank was scheduled for final inspection and shipment on 15 October 1969. The finished piece was cradled in close-fitting steel banding and trunnion-mounted in a large crate. The mounts carry large vertical loads and isolate the mirror blank from outside shocks. The journey to England was made by rail to the Toledo portside, thence by St Lawrence Seaway and the North Atlantic.

15.12 Grinding and polishing large mirror blanks

The following notes have been prepared with the help of Mr G. E. Manville, General Manager of Sir Howard Grubb Parsons & Co. Ltd. who ground and polished the blank for the Anglo-Australian telescope.

After receipt of the rough ground blank, the early milling operations were carried out with an 8-inch diameter 80–100 grit diamond cup or peripheral wheel. This was followed by a grinding operation. (Fig. 15.12.1.)

The smoothing tool was 155-inch diameter, to match the blank, and was made of aluminium with the surface turned to a 1000-inch radius of curvature. Pitch pads, which were rolled to constant thickness, were stuck on the aluminium face with the aid of Bunsen burners. (Fig. 15.12.2.) Ceramic tiles were then stuck onto the pitch pads. These tiles, which are the grinding material in contact with the mirror, must not chip, and they must be softer than the glass—but not too soft, or they will wear away without grinding the mirror. Aloxite abrasive is used in several grades to suit the particular smoothing stage.

A polishing tool, also 155-inch diameter, has squares of polishing pitch stuck on it in such a pattern as to generate the aspheric surface required on the mirror. (Figs 15.12.3 to 15.12.4.)

The accuracies required in optical working of large mirrors are such that the surfaces are milled to about 0·001 inch (25 micrometres); they are then lapped to remove the machine marks and, by using finer and finer grades of abrasive, a surface accurate to about 0·0001 inch (2·5 micrometres) on a spherometer of appropriate size is achieved. Controlled polishing is then used and measurements made optically.

In general if the aspheric surface departs from the sphere by more than 0·0005 inch (12·5 micro-

△15.12.1. The Anglo-Australian 150-inch mirror being ground prior to polishing at Grubb-Parsons optical factory, Newcastle upon Tyne.

▽15.12.2. Making the 155-inch smoothing tool. Pitch pads of constant thickness have been stuck on the aluminium tool. Ceramic tiles are being stuck on the pitch pads with the aid of bunsen burners.

△15.12.3. Grinding and smoothing. The tool is in position on top of the mirror blank. Note the protective plastic covers over the clamp screws to prevent grinding compound from coming into contact with the screw-threads.

▽15.12.4. This plan view of the smoothing tool shows the holes through which grinding compound can be passed to the ceramic tiles in contact with the mirror blank surface.

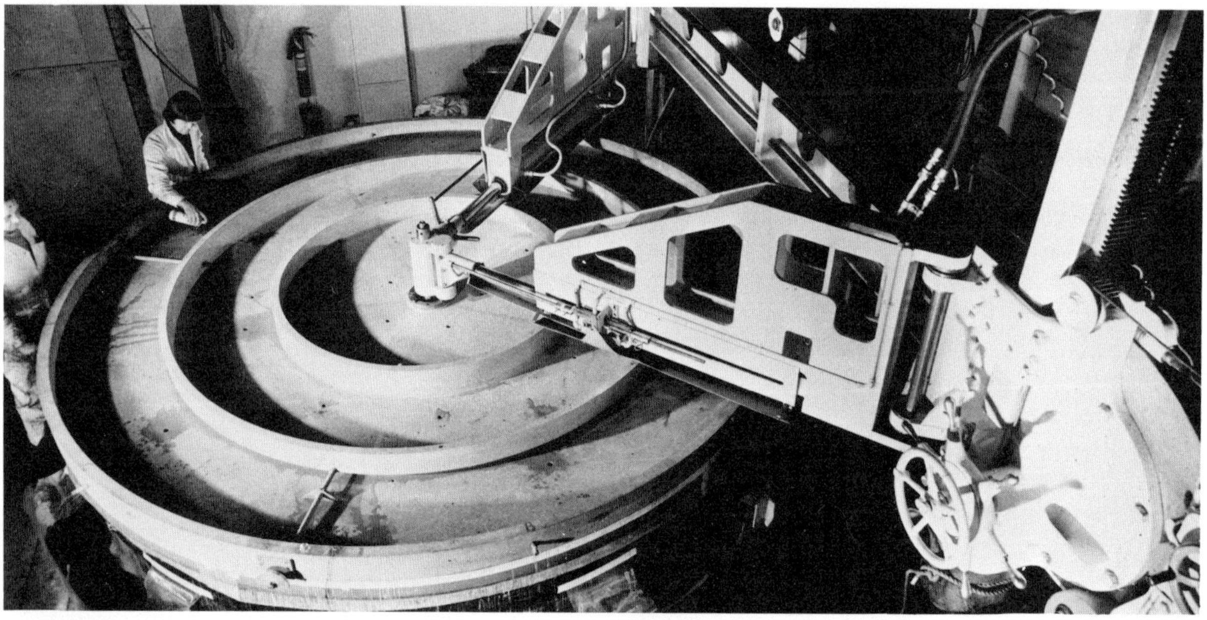

▽15.12.5. An early stage in polishing the 150-inch primary mirror.

metres) it is usual to grind, whereas below this, polishing methods are used.

The distribution of pitch, the polisher movements, and the length of run are calculated on an electronic computer, which provides an accurate and logical assessment of the work necessary.

The use of ceramic tools in the grinding process is general for work of 12 inches or more in diameter. Below this, cast iron tools are frequently used.

15.13 Aluminizing astronomical mirrors

Because the primary mirrors of astronomical mirrors must always have a high reflectivity, it is usual to recoat them every 6 to 12 months and, therefore, vacuum-coating equipment must be available at the observatory.

Primary mirrors may weigh 2 tons for a 76-inch diameter and up to 16 tons for a 150-inch diameter, and therefore the design of the coating chamber and handling equipment demands considerable care. Not only does a 150-inch mirror weigh 16 tons, but the air pressure on the chamber ends will exceed 130 tons under vacuum conditions.

The coating machine must be designed with gimbals on the chamber so that it can be loaded in the horizontal position and then turned to the vertical position for coating. This is necessary to avoid bits of aluminium or dust particles that might fall on the mirror surface. (Fig. 15.13.1.)

The chamber must be able to be split so that the

15.13.1. An Edwards High Vacuum coating plant with a chamber 84 inches in diameter for the 76-inch mirror supplied to the Helwan Observatory, United Arab Republic. Loading is carried out with the chamber horizontal, whereas coating must be effected with the mirror vertical.

15.13.2. In these two views of the coating plant can be seen the supports for the 2-ton mirror and the two strong pads for the mirror when in the vertical position for coating. (*Right*) The high-tension electrodes at the top are for discharge cleaning before aluminizing by evaporation from the thirty-two sources.

mirror mounts are in one section (the gimbal mounting section) and the discharge cleaning electrodes, evaporation sources, etc., are in the other section. A large number of sources are needed to give a uniform coating distribution, and a high-capacity power transformer is needed to provide adequate current for heating the sources. (Fig. 15.13.2.)

Inspection windows are provided in each section of the coating chamber. As mirror sizes get larger, the mean free path increases and therefore a lower vacuum is necessary; thus more efficient equipment is needed to avoid molecular collisions which would give an uneven coating.

Whereas a vacuum pressure of 10^{-5} torr may have been adequate for a 76-inch diameter mirror, it will be necessary to have a pressure of less than 5×10^{-6} torr during the evaporation of a 150-inch-diameter mirror to achieve a uniform coating. The usual opaque coating of aluminium for a mirror surface is about 1000 angstroms in thickness.

Design factors in coating plants

The following details have been supplied by Edwards High Vacuum Plant Ltd.

The main factors to be considered in the production of an aluminized mirror are:

(1) The reflectance performance over a specified wavelength region.
(2) The quality of the film.
(3) The distribution of the film.

With reference to item (1) the reflectivity of the aluminium film, which is of prime importance for a light-gathering component, the pressure, rate of condensation, evaporation distance, and angle of vapour incidence are the four variables which determine the optical performance of the mirror.

Burridge has shown that, as the product *P.t* increases, so the reflectance decreases, where P = system pressure and t = the condensation rate. It can be shown that at a pressure of 3×10^{-5} torr about 2×10^{16} molecules arrive at 1 cm²/second, and for an evaporation rate of 100 Å/second about 2×10^{17} atoms arrive at 1 cm²/second; therefore the ratio of gas to metal molecules is about 1:10. Thus it can be seen that an aluminium film can be contaminated with gas even under relatively good vacuum and high rates of evaporation. Fortunately, in normal-size coating systems, the first aluminium molecules that condense take up the active gases by gettering action. In a large vacuum chamber with a limited pumping system, the gettering action may be swamped by gas, i.e. mainly H_2O vapour, and reaction with gas could continue through the film.

The above figures are quoted for a system where the mean free path L is much greater than the dimensions of the coating system. However, where L is approaching the dimensions of the system the condensation rate will be further influenced, as discussed by Holland.

It is shown that the fraction of evaporated molecules ($\mathcal{N}$) which traverse a distance in the vacuum system without collision with residual gas molecules is determined by the mean free path and is given by the expression

$$\mathcal{N} = \mathcal{N}_0 \exp - (l/L)$$

where l is the distance traversed, L the mean free path (m.f.p.), and $\mathcal{N}_0$ the initial emitted number of atoms.

Assuming a condensation rate of 100 Å/second, we can calculate and compare the percentage of molecules arriving at the mirror surface for two pressures at the greatest source-to-mirror separation.

First Case: Pressure = 1×10^{-4} torr, L (m.f.p.) = 50 cm, source to mirror separation 300 cm diagonal distance. The percentage of the emitted aluminium atoms arriving without collision will be approximately one-quarter.

Second Case: Pressure = 1×10^{-5} torr, L (m.f.p.) = 500 cm, source to mirror separation 300 cm as above. In this case, the percentage of atoms arriving without collision will now be approximately 50 per cent.

It is shown again that, with a limited pumping performance, where the pressure could rise to 1×10^{-4} torr during evaporation one is working dangerously close to maximum collision conditions.

Also, in relationship to gas contamination of the film, one must consider the influence of angle of vapour incidence on gas absorption by the film.

Holland has shown that as the angle of vapour incidence increases, then so does the quantity of gas absorbed by the film.

With a plant of the dimensions specified for a large diameter mirror, one is concerned with high angle vapour incidence; thus low pressures during evaporation are of prime importance.

Finally, one must consider the spectral range over which the mirror is expected to reflect. In order to maintain high reflectance at the ultraviolet end of the spectrum, it has been shown by Hass and ourselves that a low *P.t* value is required.

16 Fibre Optics

16.1 The human eye

The most complex fibre-optic bundles are undoubtedly those connecting the human eye to the brain (Fig. 16.1.1). Light entering the eye passes through a crystalline lens which is of variable focal length and whose thickness and curvature are controlled by ciliary muscles. The lens focuses an image on the retina, a light-sensitive nerve plate. If a bright object is viewed, the retina gives a signal to the brain which makes the iris close and so reduces the light falling on the retina.

The retina consists of a mosaic of light-sensitive cells operating on a photochemical principle, and these are in two groups—'rods' and 'cones'. The cones are sensitive to colour and are the receptors for day vision, with a maximum efficiency at 5500 angstroms wavelength. (Fig. 16.1.2.)

The rods are insensitive to colour and are the receptors for night vision, with a maximum efficiency at 5100 angstroms wavelength.

In the region of the fovea, the point on the retina cut by the axis of the eye lens, the receptors are crowded densely together and consist almost entirely of cones. Each of the foveal cones has a separate optic nerve joining it to the brain. The density of receptors is less in the outer regions of the retina where the receptors include a proportion of rods which increase in number with increasing distance from the fovea. In the outer

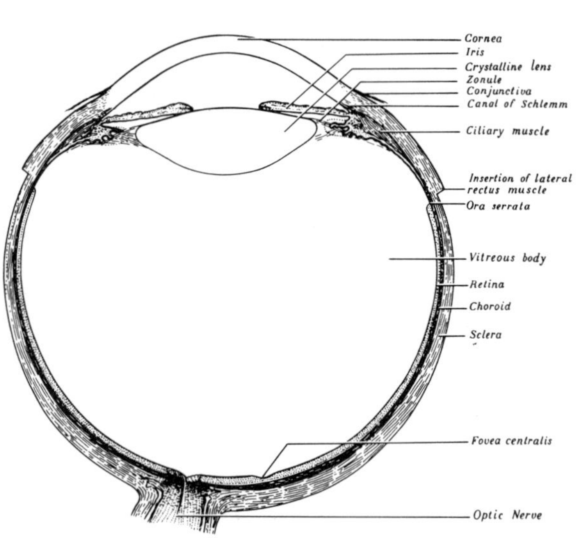

◁16.1.1. **A horizontal section through the human eye.**

▽16.1.2. **Day-vision colour sensitivity.**

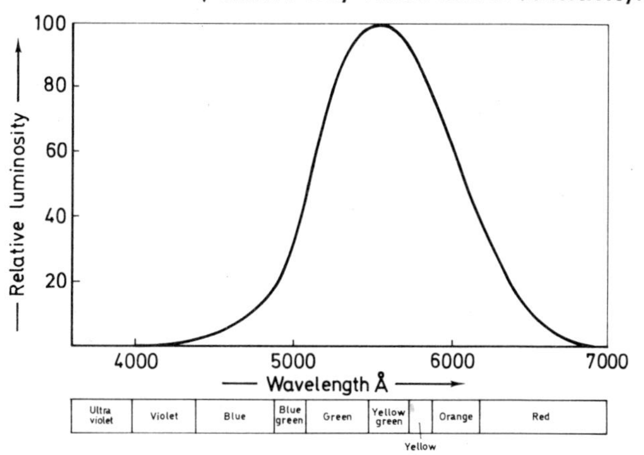

regions, receptors do not have a separate fibre each, but are joined to the brain in groups. At the point where the optic nerve joins the retina, there are no receptors, and this causes the blind spot, which does not normally affect vision as it is situated out of the normal visual field.

The human eye responds to a range of wavelengths slightly greater than 4000–7000 angstroms, but the retina does not respond equally to light of different wavelengths. A yellow surface appears to have a higher luminance than a blue surface reflecting the same amount of energy because the eye is more sensitive to yellow light than to blue.

The mechanism of colour vision is not fully understood, but it would seem that there are three elements in each cone giving three sets of colour perceptors, each element being active over a different range of wavelength representing the blue, green, and red regions of the spectrum. Light of a particular wavelength causes a reaction in the three colour perceptors, and a related signal is transmitted to the brain, which produces a sensation depending on the relative proportions of the signals and so enables an observer to recognize colours.

In the retina of each human eye is a highly organized fibre-optics bundle of about 130 000 000 fibres of which about 120 000 000 are rods and 10 000 000 are cones.

The visual receptive cells of all vertebrates are fibre-optic elements. The diameters of the individual receptors approach the wavelength of light and so these photosensitive cells exhibit waveguide properties, but an analogy with a fibre-optics bundle whose termination is viewed or photographed is only partially valid.

16.2 Glass fibres and fibre bundles

The earliest attempt to manufacture fibre optics would appear to have been about 3500 years ago, when the Egyptians produced coloured glass fibres for decoration and ornamental purposes. Very-small-diameter glass and quartz fibres were made by Faraday in 1832, and since then glass, quartz and plastic fibres have been made for optical purposes.

Methods of fibre production are by (*a*) extrusion, (*b*) hot drawing from molten bulk material through an orifice or (*c*) drawing of uncoated, coated, and multiple fibres from rods or tubes through a hollow cylindrical furnace. Uncoated quartz fibres down to 1 micrometre diameter or less can be produced and their strength compares favourably with the strongest of materials.

Quartz fibre, for optical purposes, has the advantage of transmission in the ultra-violet and infra-red regions of the spectrum; but unfortunately its very high softening point, small thermal working range, and lack of thermal compatibility with most low-index coating materials, make it unsuitable for fibre optics.

Glass is a satisfactory material for fibre optics. A lot of information is available on different types of glass, with their physical characteristics, and the technology of glass fibre drawing is already well established for the production of fibre glass yarn (Fig. 16.2.1).

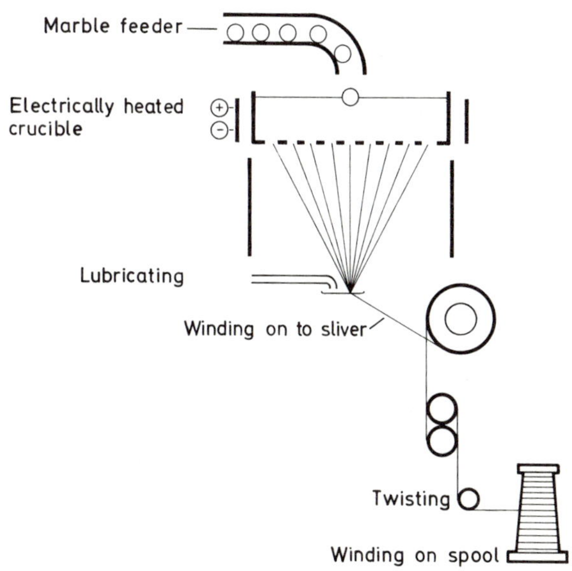

16.2.1. Making glass fibre by the continuous process. Glass marbles are fed into an electrically heated platinum crucible with a perforated base through which the molten glass drips. The fibres are drawn off on a revolving drum and then spun.

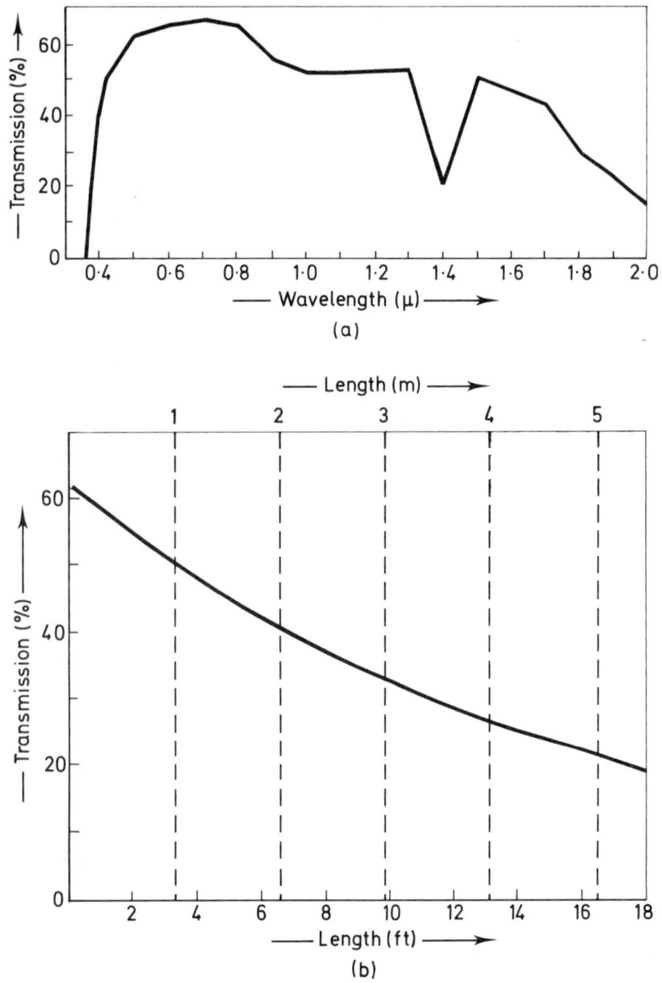

16.2.2. (*a*) Typical spectral transmission curve for glass fibre. (*b*) Transmission curve for 3-mm light guides.

16.2.3. A fibre optic drawing machine. In the lower picture, the hollow cylindrical furnace is in the top left corner. From it, the fibre passes down to the thickness-measuring instrument and the drum, the circumference of which determines the length of the fibre bundle.

Glass used in the production of long lengths of fibre must be free from bubbles, seeds, and inhomogeneities. The fibre optics in which we are now interested, depend on not only the individual fibre core but also the material of lower refractive index used for coating the core. (Fig. 16.2.2.) This coating has three purposes:

(*a*) It protects the total internal reflecting surface of the core from contamination and damage.

(*b*) The lower refractive index coating provides optical insulation between core fibres.

(*c*) The coating material can be used to make vacuum-tight optic assemblies by suitable fusing processes.

Manufacture of fibre optics

A high refractive index glass rod about 15 mm to 30 mm diameter is inserted in a tube of compatible low index glass; the assembly is placed in a hollow cylindrical furnace and drawn into fibres down to about 10 micrometres diameter.

The tubing forms the coating around the fibre. (Figs 16.2.3 and 16.2.4.)

The lower end is drawn around a drum after passing through a thickness-measuring instrument. For most applications, the cross-section area of the fibre bundle is important and the feedback from the measuring instrument controls the number of turns on the drum before it racks across for the next bundle.

The fibres are wound on top of paper on the drum, which is of a diameter to suit the particular length of bundle required, the length of the bundle being the circumference of the drum. When winding is completed, the drum is removed, the bundles are cut, and the ends are then potted, ground, and polished by conventional grinding and polishing machines.

Coherent and non-coherent bundles

There are two different types of fibre-optic bundle, known as coherent and non-coherent bundles. (Figs 16.2.5 and 16.2.6.)

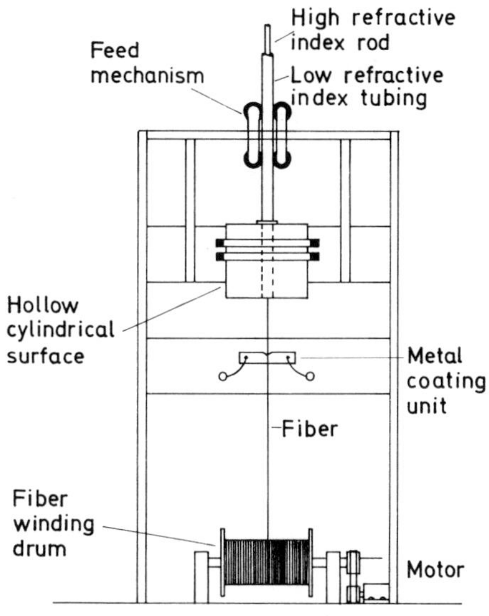

16.2.4. Machine for drawing single and multiple glass fibres.

16.2.5. Rank Kershaw Fibrox fibre optic components.

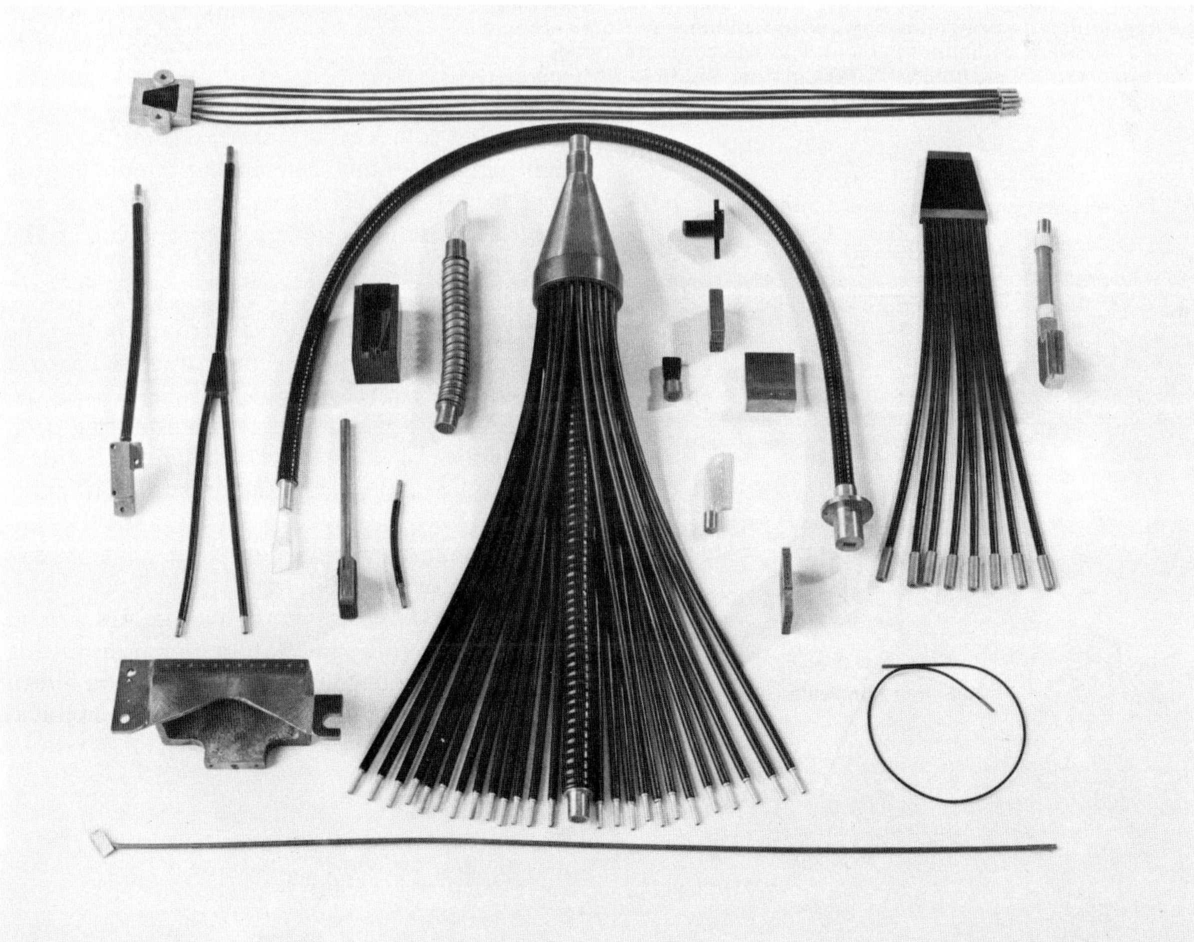

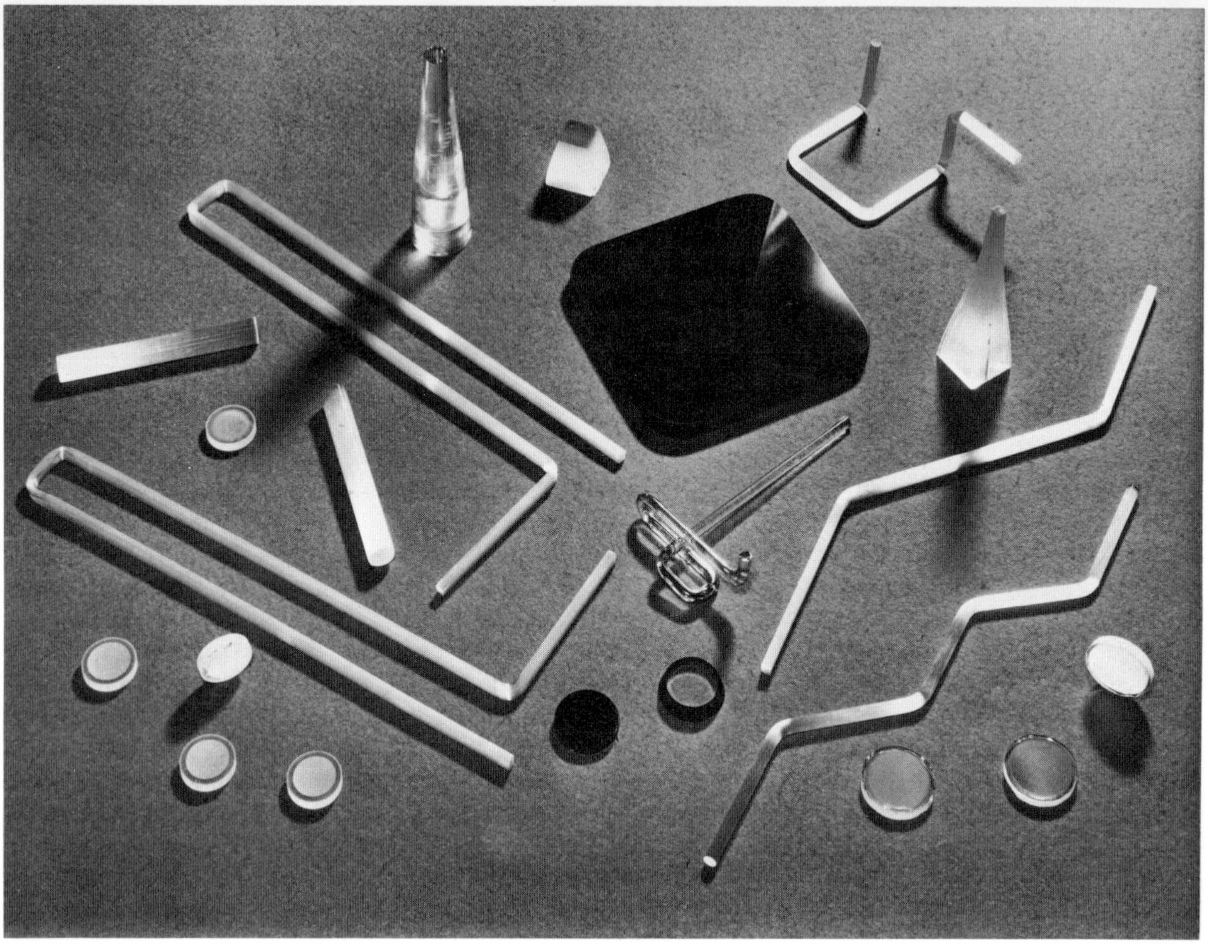

16.2.6. Two types of coherent fibre optic: (a) solid coherents, where the fibres are aligned and fused together to form rigid components, and (b) flexible coherents, where the fibres are aligned and consolidated at the ends only, with flexible free fibres along the length. Coherent components include image conduits, image inverters, magnifiers, minifiers, face plates, and field flatteners.

16.2.7. Coherent and non-coherent bundles.

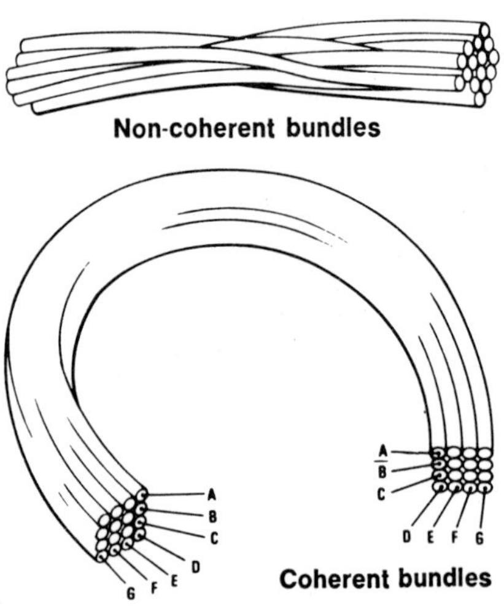

The non-coherent bundle transmits light for illumination uses but the coherent bundle has the fibres in a logical order at both ends and can therefore transmit an image from one end to the other end of the fibre optic. (Fig. 16.2.7.) A face plate is a coherent bundle in which the end viewing area is considerably greater than the bundle area, and the individual fibres are fused into a solid plate. (Fig. 16.2.8.)

Face plates are short, image-transmitting optical systems for large surfaces. The individual fibres are fused together exactly parallel to allow images to be transmitted point by point. Depending on the surface of the face plate, plane images can be bent, or curved images flattened.

The main applications of face plates are as front plates of cathode-ray tubes for transmitting the fluorescent image on to the surface of a screen and as flatteners of image fields in classical optical systems.

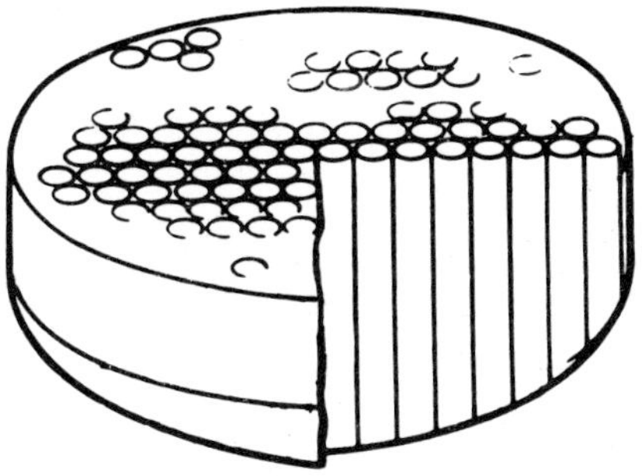

16.2.8. Fibre optic face plates having plano-convex or plano-concave surfaces can be used to flatten an image on a curved surface.

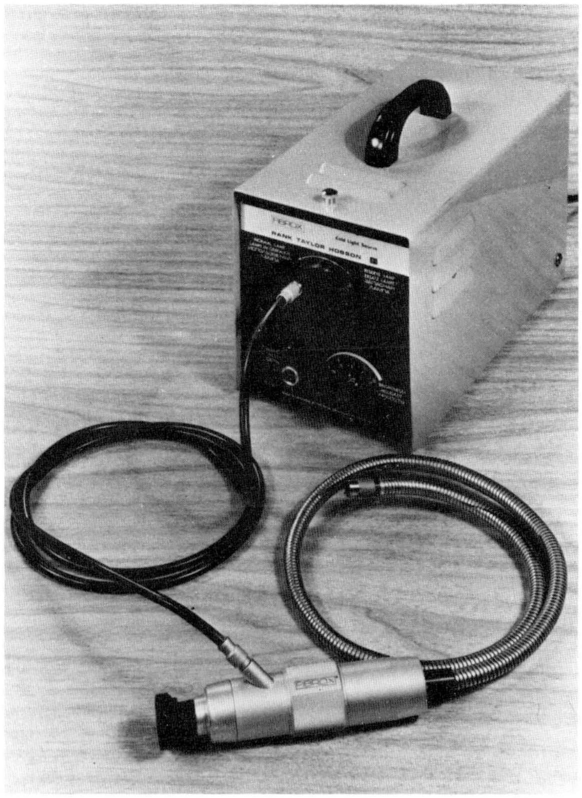

16.2.9. ▷
Fibrescope remote-viewing instrument consisting of an objective lens, a coherent fibre-optic bundle, an eyepiece, and an integral light path.

If a coherent bundle is fitted with an objective lens and an eyepiece, it becomes a flexible form of periscope or fibrescope. (Fibroxscope is the name given by Rank Kershaw, and Flexiscope by Bausch and Lomb.) Applications include inspection of hard-to-reach areas, including medical endoscopes for examination of internal body cavities. (Fig. 16.2.9.)

The remainder of this chapter consists of two reports on fibre optics, by researchers at Schott of Mainz, and Barr & Stroud Ltd.

16.3 New developments in fibre optics

With the permission of Jena^{er} Glaswerk Schott and Gen. is published 'New Developments in Fibre-Optics', by Alfred Jacobsen, Dipl.-Phys., Mainz.

In the early 1960s Schott, of Mainz, commenced the development and manufacture of light-conducting fibres. The starting point came from the optical glasses which were melted in our optical works and have been further developed for this application ever since. Up till then optical glasses were almost without exception cast in large blocks. These blocks were then sawn into sections, and through grinding and polishing of the new surfaces the glasses were prepared for their applications.

Light-conducting technology calls for large numbers of optical glasses with a particularly clean and flawless surface, which can be achieved by using the so-called fire-polishing process. For some years now various optical glasses have been manufactured in bar shape in such a way that the surface formed during the cooling process remains smooth and clean.

It is a well-known fact that light-conducting fibres consist of very fine fibres of highly refractive optical glasses which are embedded in a thin cladding of another glass with a lower refractive index.

Total reflection of the light entering the front face of the fibre takes place at the sealing zone of both glasses. The light is then reflected and re-reflected thousands of times as it passes through the fibre until it reaches the exit end. It is therefore understandable that optimum light transmission can only be guaranteed when the glasses and the manufacturing process satisfy severe demands. The surfaces of the glasses have to be very clean and smooth, with no foreign matter or air bubbles melted into them; otherwise the light is dispersed and is lost for transmission. On the other hand it is essential that the optical core glass should possess an unusual degree of light transmission so that sufficient light can also be led through fibres of great length. It is just as essential to keep a constant check on possible improvements in existing manufacturing methods, since even the most minute irregularity multiplies many times and may lead to rejection of the product.

Light-conducting fibres have secured a definite place in scientific and technological applications. Not only do they improve and simplify existing methods, but they also offer new solutions to problems which have arisen within the framework of modern science and automation.

Flexible light conductors consist of several thousand light-conducting fibres of the same overall length whose ends are tightly secured in a metal sleeve offering the best possible packing density. They are housed in a flexible metal or plastic hose for added protection against damage. Up to now, it was only possible to manufacture light conductors of not more than 3·6 m. But improvements in light transmission and the overall manufacturing process of the fibres now allow production of flexible light conductors of great lengths. The present limit lies around 7 m length, but in special cases it is also possible to produce longer light conductors, for instance, up to 14·5 m.

In the past, light conductors of such length were produced by joining several shorter flexible conductors. This resulted in considerable loss of reflection at the various contact points, which could be reduced by applying immersions; these, however, frequently introduced other difficulties. The new long light conductors are also made of type B fibres which possess maximum optical transmission characteristics. The angle of aperture on these fibres is $2\alpha = 67°$ (n.a. = 0·65). Light conductors of any diameter and cross-section, or with several branches if required, can now be produced.

Fig. 16.3.1 shows transmission as a function of light-conductor length. A logarithmic scale for the length was chosen for the sake of clarity. Transmission was measured at a light wavelength of $2\lambda = 546$ nm and was calculated as a mean value with the dispersion range of several hundred measured light conductors of various lengths. The light aperture angle in the measuring equipment was $2\alpha = 11°$.

Messrs. Schott of Mainz have developed a new method which will make possible the manufacture of much less expensive light conductors in future. The desired number of fibres, of 70 μm thickness, are pulled together out of the oven and are bundled into a plastic foil with a thickness of a few thousandths of a mm. Afterwards they are housed

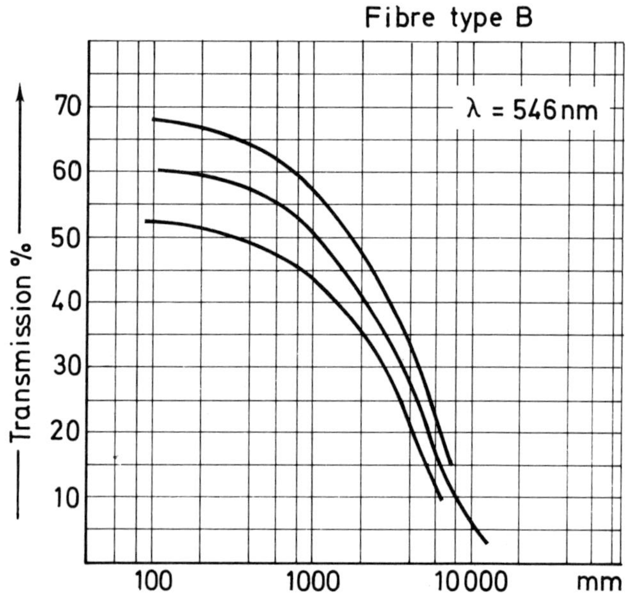

16.3.1. Transmission as a function of light-conductor length.

in a closely fitting black PVC sheath. This permits the manufacture of light-conducting cables in one efficient and continuous operation. Cables of more than 1000-m length can be wound on to drums, and correspond in their construction to the copper-stranded wire used in the electrical industry. At the same time their ends can be stripped with suitable tweezers, in order to glue the fibres together and confine them in a small section of shrunk-on hose. These new flexible light conductors are produced in two standard thicknesses. One comprises approximately 250 individual fibres of 70-μm thickness with an overall bundle diameter of 1·25 mm; the other consists of about 1000 fibres of 70-μm thickness with a bundle diameter of 2·5 mm.

16.4 Fibre optics: methods of manufacture, and new applications

This section is a reprint from *Science Journal* published with the permission of Iliffe Science and Technology Publications Ltd and the Director of Barr & Stroud Ltd. The article, 'Fibre Optics', is by J. M. Ballantine and W. B. Allan who, at the time, were responsible for fibre-optic research and development at Barr & Stroud Ltd.

Historical introduction

John Logie Baird, while experimenting in 1926 with his television system, considered a variety of mechanical and optical methods for transferring an image. One idea was to assemble a bundle of small diameter glass fibres in a parallel manner so that each fibre would transmit a minute portion of the image which would be reproduced in total at the far end of the fibres. Baird thus became the first man to patent a device incorporating the principles of fibre optics. At the time he little realized that these principles would lie dormant for nearly three decades but would then be revived and developed into a completely new branch of technology.

Between the mid-1920s and early 1950s fibre optics was regarded with little enthusiasm. Sending light down fibres of glass could not be accomplished without great difficulty; the light seemed to issue everywhere except where it was meant to go. It was a very expensive process also, and commercial possibilities seemed meagre. The few workers who remained interested produced no major advance, and commercial security restricted their publication of reports, which makes it difficult to compile a comprehensive history of fibre-optics development.

Nevertheless a 'breakthrough' came in the 1950s. Fibre optics can now control light in ways impossible by other means and, when it challenges a known method, can frequently do the job better. The subject has now found applications in fields which range from astronomy to medicine and from engineering to electronics. Nor has the technology by any means reached its zenith. Every newcomer to the field seems to be able to find a new application to suit his own needs.

In the past full-scale development has been prevented primarily by the high cost of the materials involved. For this reason large programmes of research and development are now under way concerned principally with the problems of cheap manufacture. Successful conclusion of these programmes should, in the near future, make the use of fibre optics economic in an even wider field of applications than at present.

The principle of fibre optics

The principle of fibre optics is exploited in the illuminated water fountain, where a submerged source of light illuminates the water jets with light trapped inside the jet by internal reflection at the water/air interface. Total internal reflection requires an angle of incidence above a critical value, dependent upon the difference in refractive index of the air and water, so that any imperfection in the surface, or reduction of angle of incidence, allows some of the light to escape and so give the jet a luminous appearance. This reflection process is much more efficient than the normal metal mirror reflection and, as a result, is made use of in high-quality optical systems such as the reflecting prisms of binoculars.

Light entering one end of a rod of glass is similarly trapped within the rod by internal reflection and is guided by multiple reflections along the length of the fibre to the other end. This effect is independent of rod diameter provided this is greater than a few wavelengths of the light being conducted. Thus light can be successfully guided along fibres with diameters as small as 1/10 000th inch—much finer than human hair.

Baird's original idea involved very advanced techniques which account, in part, for the long time delay before its potential was realized. Meanwhile many other light guiding systems were devised, mainly constructed of polished plastics, whose function was simply illumination.

The low refractive index medium was, in all cases, air. As in the water fountain, discontinuities and defects in the reflecting surface of the light guide affected the efficiency of the light transfer.

But although it is essential to have light loss in an illuminated fountain, such a loss is intolerable where maximum transmission is required. During the process of internal reflection light inevitably penetrates a small distance—about half a wavelength—into the lower-index surrounding medium, so that dust particles on the surface of the light guide cause scattering of the light with attendant loss. Dielectric light guide materials may attract dust particles electrostatically. The main problem with early fibres was light loss due to surface scratches and dust, light leakage at touching fibres, and finger grease causing a change of critical angle.

Nevertheless, although in the 1930s and 1940s interest in fibre optics was very low, early light guides were capable of high illumination efficiency and their defects were tolerated for some applications. One of the first fields of use of fibre optics devices was surgery, retractors for which were made using polished Perspex. Light introduced into the handle was carried to the other end, by internal reflection. It was found that the light guide could be bent without loss of efficiency providing that the radius of curvature was much larger than the diameter of the rod. This is easy to achieve with fine fibres, and one application of the latter was to convert the circular image of a star to a slit image for analysis by a spectrometer to improve the efficiency of utilization of the available light. In another application the fibres transmitted light from a hot source to an area which had to be illuminated but not heated. This is especially important in the application to medical inspection.

Sheathing

A major advance occurred in the early 1950s when A. G. S. van Heel in the Netherlands and B. O'Brien in the United States recognized the necessity for a protective insulation around the fibres, in the form of a sheathing of lower refractive index than the fibre cores, to avoid the difficulties at the core surface. Any practicable sheath material would automatically have a higher refractive index than air and, to maintain a suitable

16.4.1. Optical fibres consist of a core of glass of high refractive index surrounded by a sheath (shaded) of glass of lower index. A beam of light incident on one end passes down the core by a series of total internal reflections at the interface with the sheath, into which it penetrates about half a wavelength during the reflection. One of the key problems in manufacturing fibres is to eliminate scratches in the surface deep enough to come within half a wavelength of the core surface, since they allow light to escape from the fibre prematurely.

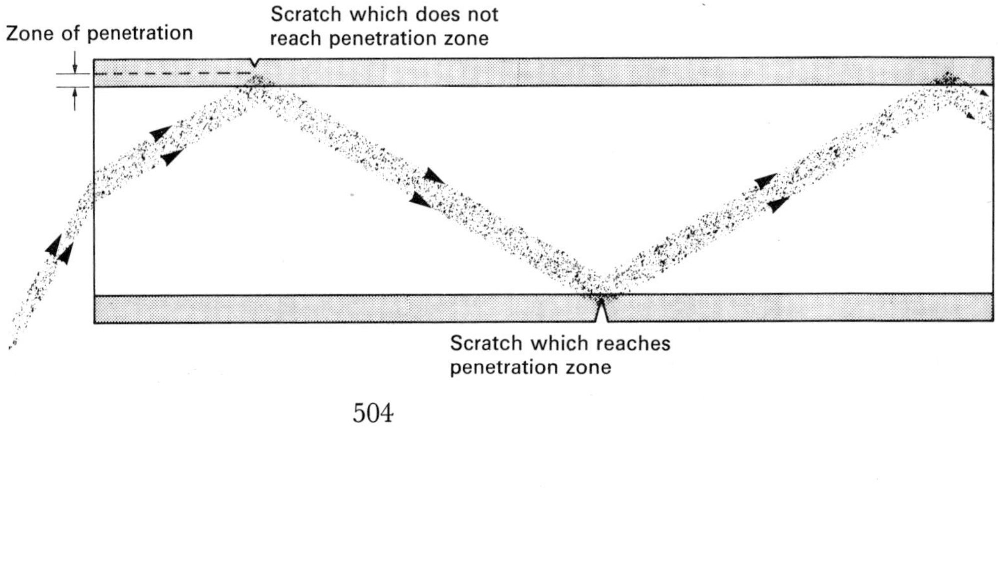

difference in refractive index between sheath and core, a core material of even higher index is necessary. (Fig. 16.4.1.)

Materials which possess properties suitable for both core and sheath are glasses with well-known characteristics. A benefit from using glass is that its plastic properties allow it to be pulled into fine fibres. Moreover, a glass rod of high refractive index sheathed with a tube of low refractive index can be pulled while in the plastic state as a single fibre to form a complete sheathed light guide.

Fibre bundles

There is no limit to the number of configurations of the fibres in a bundle. While one end of the bundle cross-section may be square the other end may be a slit, a ring, or a disk of the same illuminated area. The light guide can therefore transform a given light input configuration to any desired output configuration. Again, the bundle can be split into several smaller bundles illuminated either from a common source or from any of the small bundle ends.

Since the sheathing prevents light in one fibre from leaking into its neighbours the intensity of light emerging from the end of each fibre depends solely on the intensity of light entering the other end. In other words the sheathing optically insulates each fibre. An image formed on the end of a bundle of sheathed fibres is dissected into as many pieces as there are fibres in the bundle, and each 'bit' is transmitted through its fibre with uniform attenuation caused only by light absorption in the fibre core material. If the fibres occupy the same relative positions at each end, the image is reassembled with a definition dependent on the fibre size. Thus smaller fibres resolve finer detail.

Some of the first examples of applied fibre optics were flexible viewing systems for inspection. Among the pioneers in this field were H. H. Hopkins and N. S. Kapany in the Imperial College of Science and Technology, London, in the early 1950s. These devices aroused great interest. In particular the medical profession saw the technique as a means of replacing many of the rigid endoscopes for viewing inside the human body. The flexibility allowed greater penetration into the body with less discomfort to the patient. Engineers found a host of similar inspection uses.

The basic requirement in these bundles is strong sheathed optical fibres, and much research has gone into their production. The first stage in the process is to insert a polished rod of optical glass into a close-fitting polished tube of glass of low refractive index. The combination is fed down at a constant rate through a localized hot zone in a furnace where the sheathed rod reaches a plastic condition. A fibre can then be pulled from the lower end. This is done at a precisely chosen speed to give the required uniform fibre diameter. For a given rate of feed, the faster the fibre is pulled the smaller will be its resultant diameter.

In order to produce strong fibres by this process the glasses must have compatible properties. Their expansion coefficients must be matched; if possible, the sheath glass must have a lower expansion than the core so that, when the core and sheath are fused together and cool, the fibre sheath experiences a longitudinal surface compressive stress which improves its mechanical strength. In addition, the softening points of the two glasses must be similar.

The refractive indices of the known optical glasses yield a reasonable number of suitable pairs but, when the additional requirements are also considered, the number of suitable pairs is much reduced. Further, since the polishing of tubes from optical glasses is a prohibitively expensive procedure, the starting point for the sheath must instead be commercially available glass tubing as used in industry. In turn this further restricts the choice of core glasses.

Fibrescopes

Flexible light guides are made from strong fibres of as little as 0·002 inch diameter, but there is a limit to the minimum diameter of fibre which can be handled easily. The two factors controlling this limit are the difficulty of seeing fibres much thinner than 0·001-inch diameter and the fact such fibres have a small breaking strength. Fibres of minimum practical diameter are used for making flexible 'fibrescopes' because the smaller fibre diameter yields higher resolution. One method of manufacturing these fibrescopes is to wind a close helix from one drum on to a second in much the same way as one winds a reel of thread. Additional layers are then wound on until the desired size of bundle is produced. Cutting across the layers leaves a coherent bundle with flat ends

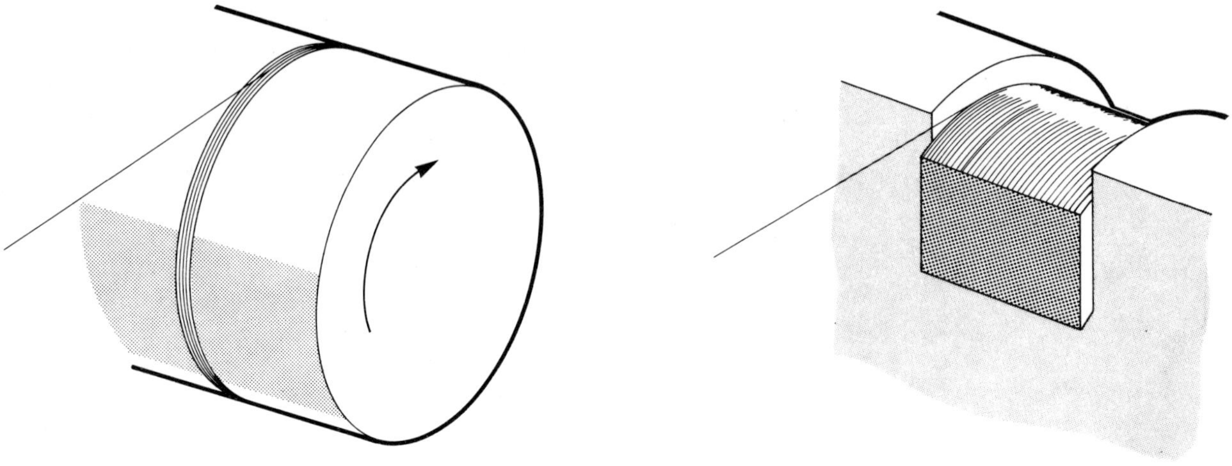

16.4.2. Fibrescope manufacture involves the winding of a close helical coil of sheather fibre into a bundle of the desired cross-section. This bundle is then sliced through and the ends of the fibres are joined together, often by potting them in a mass of suitable resin. After that, the ends are polished.

which may be sealed or potted in resin. Polishing each end yields a flexible image-transferring fibrescope. (Fig. 16.4.2.)

A strong fibre is obviously essential since it can be guided into its allotted position only under tension. Because of the limitation upon minimum fibre size, methods have been sought of overcoming this handling problem. One obvious solution is to wind several fibres at a time, making use of the increased strength of the group. For instance, 0·010-inch fibres laid into a coherent bundle can be drawn out in a similar manner to the single rod and tube combination. This yields a multiple fibre with the individual cores embedded in a matrix of sheath glass.

These multiple fibres can be produced with any desired cross-section. Square or hexagonal sections can be packed more closely than simple round fibres and eliminate dead space. A further advantage of square multiple fibres is that they do not twist during the winding operation.

An extension of this process is to increase the diameter of the multiple fibre to, say, one-eighth inch square to yield a rigid coherent bundle or image conduit. Because of the fineness of the individual cores rigid conduit can be bent in a gas flame to various shapes without impairing its image-transferring efficiency. In such a conduit there are typically 90 000 fibres (300×300) which allows a resolution of approximately 200 lines. A 600×600 fibre bundle would have about the same picture quality as television on the 405 line system. For most applications, this is adequate.

The advantages in using image conduit in preference to conventional optics are twofold. First, the greater light-gathering power of the fibres gives enhanced transmission. Second, the fibre system requires only two lenses, eyepiece and objective, so that lens aberrations are much reduced.

Fibre cones

One by-product of multiple fibre manufacture is the fibre cone, the tapered section formed during the reduction in diameter of the original bundle. In this cone each fibre is reduced in diameter by the same amount. An image incident on the large face is reduced in size on passing to the smaller face, in the ratio of the two end diameters. Conversely, magnification can be obtained by the reverse procedure. If the image formed by the lens system of a camera is focused on the large end of a tapered bundle, and the photographic film placed in contact with the smaller end, the intensification thus produced enables good photographs to be obtained at relatively low light levels. The larger face can be curved to correct the image curvature caused by the lens system, thus correcting this lens aberration.

Oscilloscope face plates

Increasing use of oscilloscopes to display high-speed transient events generated a need to record such rapidly changing displays. Recording oscilloscope traces by normal photographic techniques is very inefficient. The oscilloscope-screen phosphor emits light in all directions and, even with the best lens system, the proportion of this light reaching the photographic plate is meagre. The fast events now being investigated with oscilloscopes give phosphor traces with such low light output that they cannot be recorded even on specially sensitive films.

As fibres can accept wide angles of light an improvement in recording efficiency appeared possible. In the late 1950s, an oscilloscope face plate formed from a coherent fibre bundle was developed by N. S. Kapany—who by then was working in the United States—and independently by workers at the American Optical Company. In an ideal face plate, the glasses are chosen to give fibres which accept all light within a semi-angle of 90° to the axis; the phosphor is deposited on the inner face of the plate and all the light emitted in a forward direction by the phosphor is trapped in the fibres. The image of the trace is thus transferred to the outside face of the plate with little loss of light and a photographic film in contact with the outer face would record with high efficiency a sharp image of the trace.

Typically, oscilloscopes for such work use screens measuring 1 × 4 inches. To make fibre plates of these dimensions multiple fibres are stacked and fused together in a furnace; the resulting block is then cut into slices across the fibre axes. As the plates in an oscilloscope form part of a vacuum system the fusion process is critical and the plate must not leak. The glasses used in these plates must also possess thermal properties similar to those of the glass envelope of the oscilloscope tube. This restricts the choice of glass for fibre pairs and as a result the ideal case of 180° acceptance angle at the end of a fibre has not yet been realized.

The cone of light accepted by a fibre for transmission is determined by the difference in refractive index between its core and sheath. The sine of the semi-angle of this cone is called the numerical aperture (n.a.) of the rod or fibre. Thus a fibre with an n.a. of 1·0 will trap light within a cone having a semi-angle of 90°; in other words it will accept all light incident on the end of the rod. By suitably choosing the glass pair the acceptance angle, and therefore the exit cone angle, can be varied right up to the limit of 180°. Current plates yield acceptance angles of up to 120°—an n.a. of 0·86—and current research is directed to modifying the refractive indices of glasses available, to achieve the ideal case.

Electron guns of modern cathode ray oscilloscopes have been improved to give a trace using a spot one-thousandth of an inch in diameter. Necessarily the fibre optics face plates for such tubes are made from fibres less than 0·001 inch in diameter; in fact the fibres are normally one quarter of the spot size. Since there are some 70 million individual fibres in the typical 4 × 1 inch plate it is hardly surprising that some of these come to grief during manufacture and lose their ability to transmit light.

Breakage is not the cause of the trouble. The process of pulling creates large areas of new surface and makes the fibres electrically charged. As a result they attract dust from the surrounding atmosphere; even though this has no effect on a single-coated fibre, in later processes it forces the adjacent fibres to be unnecessarily distorted and so causes the light carried by them to escape. The larger the contaminating dust particles the greater the area of the resulting blemish, and great care must be taken to eliminate such contaminants from the atmosphere. In currently available plates, the blemish area is less than 1 per cent and blemish sizes are kept below 0·002 inch. But if an oscilloscope spot size is 0·001 inch, it is desirable to make the largest blemish smaller still, and current research is devoted to this end.

Similarly care must be exercised to ensure that the fibres maintain the same relative position on both faces of the plate. Present techniques produce plates in which the fibres are not displaced by more than half the spot size.

In a plate made from light conducting cores in a matrix of sheath material, not all the light imaged on the plate is conducted by the core. Some falls directly on the sheathing glass and some enters the cores at too large an angle to the axis, outside the n.a. of the fibres, and so escapes into the sheathing. Such light is not trapped by any of the cores and leaves the plate at a random point not necessarily opposite its entrance. Not only does this light not contribute to the image transfer but it degrades the resulting image by making the dark portion lighter and so decreasing the contrast and therefore the resolution.

An elegant method of avoiding this degradation is to place around each fibre a second sheath of a highly absorbing glass such as a black glass. The function of this sheathing is to absorb any light not guided by the cores. The clear sheath is however, still essential since the light conducted by the core penetrates the sheath at each reflection to a depth of about half a wavelength. Obviously were the whole sheath black the light carried by the core would be strongly attenuated. This black sheath is known in this country as black cladding and in the United States as extramural absorption (e.m.a.).

One important consequence of the introduction of black cladding was the realization that cathode-ray tubes, incorporating low n.a. plates with this cladding, could be viewed comfortably in bright ambient lighting whereas a normal tube, as in a television set, has poorer contrast and is preferably used in subdued lighting. The disadvantages of viewing in subdued lighting are that the viewer fatigues easily, requires time to become dark adapted and must lose his dark adaption on viewing a normally illuminated scene. These disadvantages become serious in the operation of aircraft and ship radar displays.

In a normal cathode-ray tube the ambient light falls freely on the granular phosphor of the screen and is scattered. Light emitted by the trace on the phosphor has low intensity and is easily swamped by this scattered ambient light. With a fibre plate most of the ambient light is absorbed by the black cladding and so is prevented from reaching the screen. It is true also that most of the light from the phosphor outside the n.a. is similarly absorbed. But this is no loss to the observer since his eyes operate well within the n.a. of the plates, using only a portion of the available unattenuated light from the plate when viewed directly.

As the size and availability of fibre plates increase they will find applications in other electro-optical devices. In particular high contrast plates may be used in portable television tubes to aid viewing in daylight conditions.

Much work also remains to be done in producing light pipes with high transmission over distances measurable in miles. If this could be achieved, communication networks might be set up in which light carried the information as electrical currents do today. It is possible that such a development could involve the coherent light from a laser.

Fibre lasers

When glasses were made which displayed 'lasing' action it was a natural development to pull fibres using them as the core, and the resulting fibres also 'lase'. Laser light amplification is dependent on the optical path length in the laser material. To give high amplification this path length is made long in a normal laser, using mirrors which cause multiple reflection of the light. In a fibre, it is possible to produce a path length of several miles without the need for multiple passes through the material, so the mirrors can be eliminated.

The fibre laser shows great potential as a simple light amplifier which may find application in high-speed computers. A pulse of light passing through it is amplified but leaves behind a region which is temporarily deactivated. Thus a second pulse following closely behind the first will receive much less amplification. This creation of a 'dead zone' immediately following the amplified pulse has been achieved electrically and has been employed in electronic circuits to create logic networks. It seems feasible that a fibre laser could be made to duplicate such networks optically.

17 Projection Screens

17.1 Introduction

Both back- and front-projection screens are important optical components which need care in design and manufacture if they are to give the best magnified image and light distribution for factory, studio, theatre, or lecture room. There are at least ten different types of screen in general use, each with its own optical features, designed to give the best possible illumination and definition of image to the viewer or audience of viewers.

The various types of screen may be classified as follows.

Back-projection screens

(*a*) Greyed-glass screen for wide-angle metrology projectors.
(*b*) Wax screens for high contrast and narrow angle on metrology instruments.
(*c*) Etched-glass screen.
(*d*) Emulsion-coated glass screen.
(*e*) Fresnel plastic screens for back projection of television or ciné films.
(*f*) Translucent screens for TV or film studios, or back-projection theatres.

Front-projection screens

(*a*) Glass-beaded screens.
(*b*) Lenticulated aluminium-coated plastic-on-cotton-fabric screens.
(*c*) Aluminium-painted welded plastic screen.
(*d*) Pearl-essence-painted welded plastic screen.

17.2 The production of back-projection screens

(a) Greyed-glass screens for wide-angle metrology projectors

The process of greying sheet glass for such screens is as follows.

The glass sheet to be greyed must have the edges either ground or polished smooth. The grinding operation is by hand rubbing the screen blank on a flat ground-glass surface using Aloxite 225 optical powder mixed with water. This grinding must be very even over the whole surface and the finished screen must be opaque. Any surplus grinding powder should be washed away with water and the glass thoroughly dried on both surfaces. A screen of size 250 mm × 200 mm should be about 6 mm thick to ensure an even grey finish.

(b) Wax screens for high contrast and narrow angle on metrology instruments

The two pieces of glass between which a thin wax film is required, must be of exactly equal dimensions and the surface to be in contact with the wax must be greyed as in the previous paragraph.

The greyed surfaces must be cleaned, by a ball of clean cotton wool soaked in paraffin, until the surface has been 'wetted' and the paraffin has disappeared. Possibly up to twelve cotton-wool balls will be needed for a screen 130 mm × 80 mm. Finish by rubbing with a pad of clean white muslin. During the cleaning process, the glass should be held in a wire handle-fixture to avoid accidental damage. To obtain a wax film of the required opacity and minimum scintillation or loss of transmission, the wax thickness should be about 0·004 inch (0·1 mm). This thickness is controlled by the use of four wire spacers each 0·004 inch (0·1 mm) thick and about 2 inches (50 mm) long.

The wires must be cleaned by holding them in tweezers and passing them through a methylated-spirit-lamp flame. The flame should also be allowed to clean the tweezers. The wire can now be bent around the glass and secured on the back surface with adhesive tape. Finally, rub the glass with a pad of clean muslin. A thermostatically controlled oven, at least 6 inches (150 mm) all round larger than the largest screen must be levelled and set for 135°C. A drip-tray (loaded with the jig) should be placed in the bottom of the oven. Place the first glass blank face upwards on the jig. Place the other blank (in its wire handle fixture) in a convenient position in the oven. The blank will be held to the wire handle fixture by adhesive tape and, since this must not come adrift when taking the weight of the hotplate, it is most important to have good-quality material. Speedfix tape No. M109A has been found to be strong enough to hold a 10-inch (250 mm) × 7-inch (175 mm) blank weighing 0·8 kg, and has good adhesion when it is hot.

Place a beaker of Amber Micro Crystalline Wax No. 3562 M.P. (obtainable from Poth Hill & Co. Ltd) in the oven and close the oven door until the temperature reaches 135°C. With the use of thick rubber gloves, open the oven door wide and pour a pool of molten wax on the first glass blank in the jig. Pick up the other glass by its handle and, starting from the back of the oven, lower this on to the wax-covered glass very slowly so as to exclude any trapped air. Ensure that the upper glass blank lies squarely on the lower one. Close the oven door and allow the temperature to return to 135°C. Switch off the oven and allow to cool very slowly to room temperature.

When the oven is cold, remove the jig and screen. Remove the wire handle fixture and adhesive tape from the back, scrape away surplus wax, cut off the surplus spacing wires level with the screen edge. Wipe marks with cotton wool damped with benzene, and remove any smears with acetone. Polish clear faces with Selvyt cloth. To ensure long life, the edges of the screen must be sealed by painting them with I.C.I. insulating varnish No. F268–585. After 24 hours for drying, the excess varnish can be removed with a razor blade. After polishing with a Selvyt cloth, the screen is ready for use.

Wax-screens from $\frac{1}{2}$-inch (12·5 mm) diameter upwards can be made by this method given sufficient care in the processing; if required for reading a scale, one of the blanks can be in the form of a plano-convex lens. The plano side should be greyed before processing.

(*c*) *Etched glass screens*

Because hydrofluoric acid is strong enough to dissolve glass, etching is a very dangerous process and any vessels used must be made of lead or polythene. Areas of glass which are required to be etched should be mounted on suitable polythene holders and rotated within 25 mm of the surface of neat 70 per cent concentrated hydrofluoric acid. The time for the fume process varies according to the depth of the etch required. The process must be carried out under controlled conditions in a plastic-coated fume cupboard designed for the process. Usually etched screens have a finer finish than greyed screens, but they also have a more pronounced 'hot spot'.

(*d*) *Emulsion coated glass*

Kodak day-view screens are made of a special diffusing material dispersed in gelatin and coated on glass. In many optical applications, they are much superior to ordinary ground glass because they produce a brighter image at wider viewing angles and with higher resolution than does ground glass. Thus, they can be used to advantage in applications where many people must view an image projected on a relatively small screen, such as a microfilm reader. Kodak day-view screens also produce a superior image to that of ground-glass screens when high ambient levels of illumination are present on the viewing side of the screen.

The curves in Fig. 17.2.1 show the relative brightness as a function of viewing angle for the four types of Kodak day-view screens. The curves also indicate the directional distribution of transmitted light relative to that which would be produced by a perfect diffuser under the same illuminating conditions. A directional distribution curve for a perfect diffuser is a straight horizontal line, i.e. the relative brightness of a perfect diffuser is

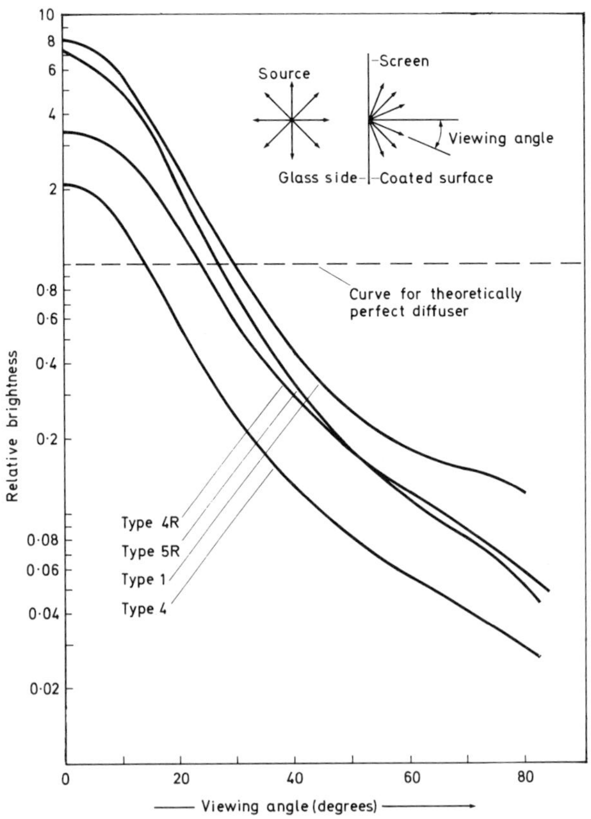

17.2.1. Goniophotometric curves for Kodak day-view screens.

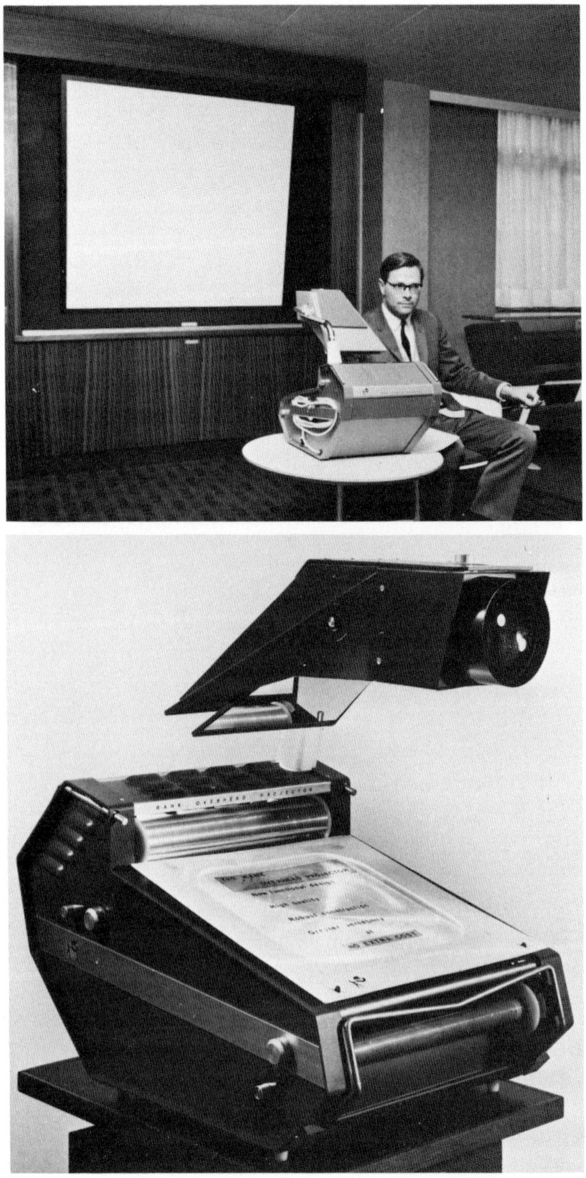

17.2.2. ▷
Rank Aldis overhead projector for use in lecture halls and conference rooms. The close-up shows the Fresnel screen lens, silicon-protected mirror, and three-element projection lens.

the same at every viewing angle. The *absolute* brightness of the four Kodak day-view screens, however, will depend on the intensity of the light projected on the screen.

(e) Fresnel screens

Screens of good image quality are made from moulded plastic. The screens are used on photographic enlargers and overhead projectors, but there is a limitation on size and tooling is expensive. (Fig. 17.2.2. See also § 5.8.)

(f) Translucent screens

Translucent screens are used in television and film studios as well as in some back-projection cinemas. The screen consists of a welded transparent plastic base, about 0·012 inch (0·3 mm) thick, which is sprayed with a diffusing emulsion to give a uniform illumination and to eliminate, as far as possible, the hot spot caused by direct light from the projector.

The diffusing emulsion can be made from a wide variety of materials but vinyl lacquer with a cellular particle additive, such as 'Santocel' powder (Monsanto Chemicals), will matt and diffuse the plastic base. The emulsion must be substantially transparent and must not include impurities, which would cause scintillation, grain, or marks.

The screen normally transmits about 50 per cent of the light from the projector. The emulsion formula depends on the requirements of the studio or auditorium and whether the image on the screen will be viewed by a camera or an audience. Usually television studios require the most uniformly illuminated picture, at the expense of some reduction of light. In this case, both sides of the PVC base should be sprayed with an emulsion. When used in studio lighting, this emulsion is sometimes tinted grey, to enrich the effect of colour television.

To help achieve a uniform picture, the projector is usually fitted with a long-focal-length lens, but the long throw inevitably requires additional floor space.

17.3 The production of front-projection screens

(a) Glass beaded screens

Beaded screens have been made for auditorium use since 1932 and at the time their advantage was a high forward gain on the axis of projection of about 2 foot lamberts per foot candle compared with matt white, whilst the fall-off, down to matt white reflective brightness, was not until about 25° off axis. This is a directional type of screen suitable for narrow halls and small screens up to a width which can be accommodated on a roller coating machine. (Fig. 17.3.1.) The screen base can be plastic or cotton and this is coated with PVC laquer on to which the glass beads (1·66 refractive index) are spread.

However, a development in 1963, with a studio front projection system, led to a demand for a highly directional process screen. The new development was a system of composite photography, consisting of projecting the background image from a point alongside the camera upon a special front-projection screen, the actors being situated between camera and screen (Fig 17.3.2).

When compared with a translucent screen (§ 17.2(*f*)), studio space is halved, setting-up time is reduced, the difficulty of alignment between camera and projector does not arise, there is no hot spot, and it is immaterial if the foreground lighting falls on the screen. It will be objected that if the background image is front-projected upon the screen, it will be projected equally upon the actors in front of that screen. However, the special process screen has such a high reflection capacity at a very small angle (approximately 2°) that the projection light falling on the actors is not visible in the photograph.

Another advantage of the system is that as the optical axis of the projector is made to coincide with that of the camera, by means of a semi-reflecting mirror, the actors cannot cast shadows on the screen which would be visible to the camera. (Fig. 17.3.3.)

The beaded screen is made from material 24 inches wide (609 mm) coated with microscopic beads of a very high refractive index (refractive index 1·927, specification DEDF927210) inlaid in a silver base. The material, which is supplied by Minnesota Mining and Manufacturing Co. (3 M's) under their trade name 'High Gain Scotchlite Type 7610', is cut into panels which are adhered to a perforated plastic base (which avoids air pockets) and made to a size required by an installation. As with any form of lenticulated system there is a problem of accurately matching one panel to another as slight variations show up under arc light.

The high intensity screen (Fig. 17.3.3) represents, on the face of it, a complete contradiction in terms. When viewed by stray light it is charcoal grey in colour; if a beam of light is shone upon it, it still appears almost black. Yet actually it reflects over 90 per cent of incident light, but this within an extremely narrow angle.

The screen is similar in construction to the glass-beaded screen which has for many years been used for film and still projection. In Fig. 17.3.3 is shown the construction of the screen, in which 1 is the tough plastic base; 2 is an intermediate layer upon which glass beads (3) are spread, and embedded in grey plastic (4).

These beads are almost microscopic in size and spherical in shape. As indicated in the second sketch, a ray of light falling upon the bead is not

◁17.3.1. **Coating a glass-beaded or white matt screen on a cotton cloth base. The width is limited by the size of the coating machine because joins are not practicable. The white plastic being coated can be seen as a bulge on the side of the 'doctor' plate.**

▽17.3.2. **A projector on the right with a camera and semi-reflecting mirror on the left for front projection in film studios.**

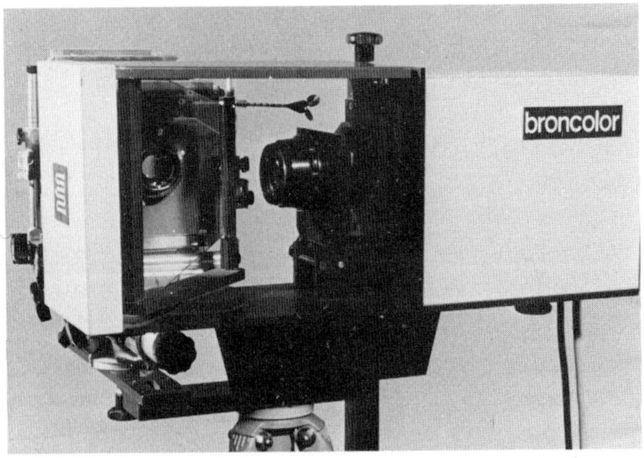

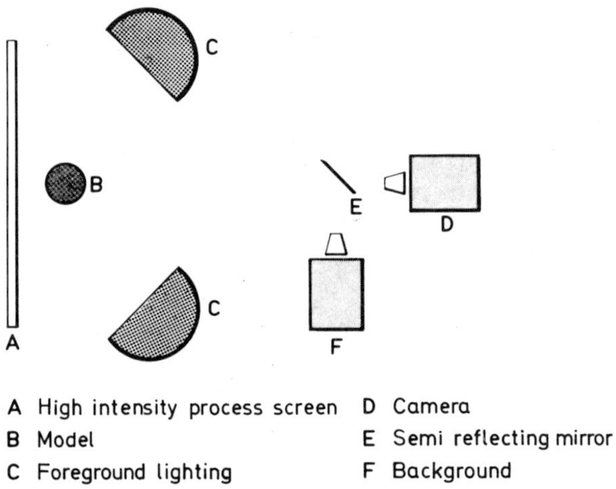

◁17.3.3. **The high-intensity screen. (A) the screen, (B) model, (C) foreground lighting (tungsten or flash), (D) camera, (E) semi-reflecting mirror, (F) background projector.**

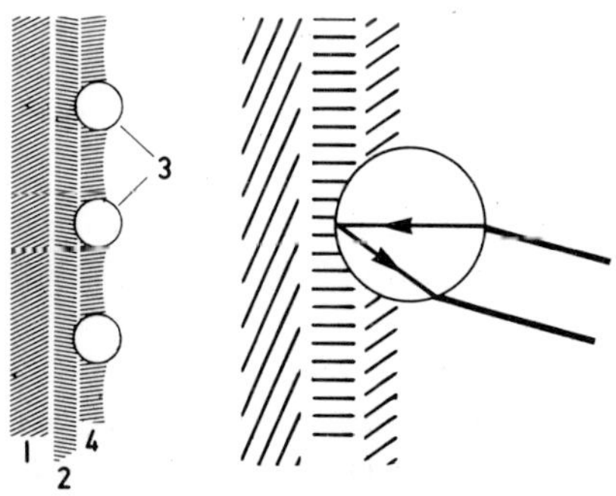

dispersed, but is reflected in its direction of origin. Thus from whatever direction the screen is illuminated light will be reflected back principally towards the light source. This process is known as catadioptric or reflex reflection.

The screen differs from all previous types of screen in two respects: (1) the beads are made of costly optical glass of high refractive index and (2) they are partially embedded in dark plastic. As a result the reflection from each bead is controlled to a very narrow angle, and diffuse reflection is practically eliminated. If an image is projected upon the screen, it will be practically invisible except from the vicinity of the optical axis of the projector.

Lenticulated aluminium-coated plastic on cotton fabric

When screens are viewed under arc light, the eye can see any small marks or discolorations which show up against a moving picture—particularly in light sky scenes. Lenticulated screens would give the best optical performance of any design if they could be made accurately in one piece. Embossed screen material must necessarily be narrow enough to pass through a roller-coating machine and perforator; therefore it cannot be much more than 54 inches wide (1370 mm). The material cannot be welded or complete rows of lenticles will be destroyed. There are problems in obtaining precision embossed, consistent, long lengths of material from the roller coating machine owing to the very sensitive nature of high reflectance material. Marks, variations in colour, etc., occur owing to problems of precise control during coating.

In 1953, when the 20th Century Fox Film Corporation launched CinemaScope, they realized that a high-efficiency wide screen would be needed for use with anamorphic lenses, and so they designed and developed the 'Miracle Mirror' screen. This was the most successful screen of its type, but it suffered from the problem of seams and variations between adjacent panels, which were visible to the eye and distracted from the picture on the screen.

The Miracle Mirror screen (Fig. 17.3.4) had two alternative patterns embossed on its surface. The 'head on' pattern was intended for level or small projection angles, whereas the 'tilted' pattern was better suited to steep projection angles, as it had a built-in tilt of 11°. The choice of pattern depended on the shape of theatre and position of projector. Further control of light distribution is obtained by tilting the top of the screen away from the projector.

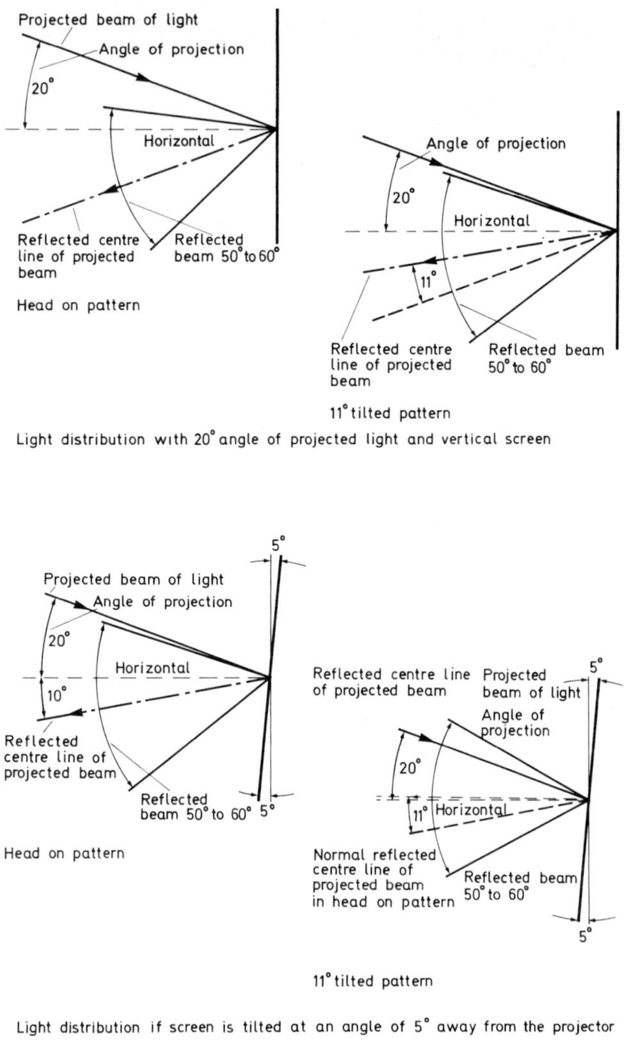

17.3.4. 'Head-on' and 'Tilted' types of the Miracle Mirror screen, together with the effects of tilting the screen itself.

To ensure the maximum picture brightness throughout the theatre, the screen should be mounted on a frame curved to a radius equal to the distance from the projector. (Fig. 17.3.5.) To permit undistorted sound to pass from the speakers

17.3.5. CinemaScope aspect ratio screen with curved frame.

behind the screen to the auditorium, perforations are needed. A square pattern of 1-millimetre diameter holes is the most suitable for this design to give the required acoustic performance and suppress moiré fringes on the screen surface. (Figs 17.3.6 and 17.3.9.)

The base material consisted of strong, non-stretch, Sea Island cotton cloth, 160 count, 0·006 inch (0·150 mm) thick, 54 inch (1370 mm) width, which shrinks to about 51 inch (1300 mm) width during coating. Base coatings are given to each side of the cloth and then one or two filler coats (which do not include aluminium). Three aluminium coats are then given, the second having a very fine embossing to keep the material flat and smooth, and the last coat, which is the thickest and most important, gives the final embossing for the mirror pattern. The finished material weighs about 4 oz per sq. foot. During manufacture the

17.3.6. Square pattern of holes for sound in a lenticulated screen.

17.3.7. **High gain with the Harkness Dayvue screen.**

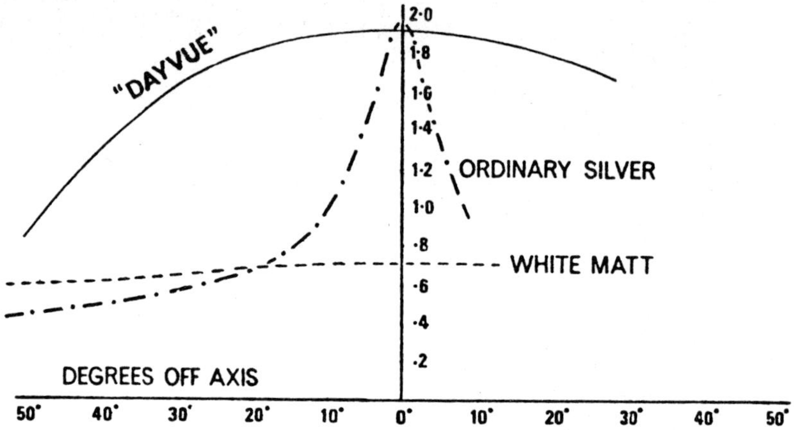

coats are controlled by a beta-ray thickness measuring machine.

This screen had 'invisible' seams made by partially cutting through the screen material from the front surface with a sharp blade, and sewing the panels with stitches in the cut of the material, the object being to conceal the row of stitches when the screen was stretched on its frame. This attempt to conceal the join was only partially successful and, in service, it was not long before the seam became a visible distraction and the screen unserviceable.

However, for small screens there is no doubt that the optical performance from a rolled plastic with a suitable pattern is good, and screens with one width of material are made in this way.

The Harkness 'Dayvue' screen is made from cotton-backed plastic with lenticulations at 47 lines per inch and, for an exceptionally wide angle, a high gain is achieved. (Fig. 17.3.7.)

Aluminium-painted welded plastic screens

Large cinema screens must necessarily be coated after manufacture if the seams are to be made invisible to the eye. If speakers for sound are to be placed behind the screen then it must be perforated with holes 1 mm diameter spaced every 4 mm across and 4·5 mm down, (a diamond shape pattern). (Fig. 17.3.8.) Providing there are no hole omissions and black serge or felt is hung behind the screen to prevent dust or shadow marking of the surface, then it will not be possible to notice the hundreds of thousands of perforations under normal projection conditions.

17.3.8. Diamond pattern of holes for sound in a painted plastic screen.

17.3.9. Perforating machine for punching the sound holes in a screen before welding the seams. The press reciprocates automatically up and down whilst the rollers rack the plastic through the press.

17.3.10. A Radyne r.f. plastic-welding machine being used to double seam-weld Harkness screens.

17.3.11. Automatic spray painting machine for aluminium or pearl finish. The spray gun travels up and down a distance of 30 feet and the screen slowly passes the track of the spray gun. The gun is photographed in the down position and the spray is just visible.

The most important operations are as follows:

(1) Perforate the 0·010-inch (0·25 mm) polyvinylchloride base (Fig. 17.3.9) and cut into panels.

(2) Double weld the screen panels with an h.f. electric welding machine to make up the required size of screen. (Fig. 17.3.10.)

(3) Sew a strong PVC plastic edging to the screen and press.

(4) Stretch the screen on a frame and paint with an automatic-spray painting machine which scans up and down and slowly across the screen. (Fig. 17.3.11.) The paint consists of aluminium powder, suspended in a PVC lacquer, to which is added a small quantity of Santocel powder consisting of cellular particles which matt and diffuse the lacquer. The effect is to coat the screen with a very large number of random 'reflectors' caused by the aluminium-coated cellular particles. By

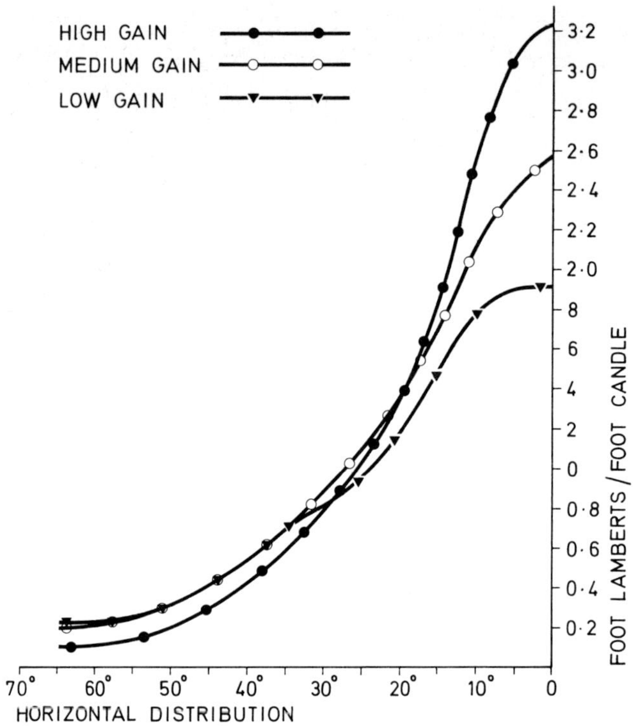

17.3.12. Different gains with aluminium-painted screens.

varying the mixture it is possible to vary the gain from 1·9 (relative brightness compared to a diffuse matt white screen) for wide halls with relatively short throw, to a 3·2 gain suitable for long narrow halls. All colours are reflected uniformly, so the screen is equally suitable for colour or black and white pictures. (Fig. 17.3.12.)

The PVC base is resistant to fungus and mildew corrosion and the elastic properties of the material make it suited for stretching on curved, tilted, or straight frames.

(5) After painting, the screen is hung up for the paint to dry thoroughly before it is rolled up for packing and despatch to the customer.

(d) Pearl-essence-painted welded plastic screens

For some situations, it is preferable to use a white instead of an aluminium screen. A Harkness 'Perlux' screen has a brightness equal to that of a high-gain aluminized screen and is suitable for all types of light distribution, including the 'Deep Curve' specification. (Fig. 17.3.13 shows a Deep Curve screen.)

The plastic base is made into a screen by the same methods as for aluminium-coated screens. However, instead of aluminium, the paint sprayed by the automatic machine consists of transparent plate-like crystals of basic lead carbonate suspended in a transparent PVC lacquer. The proportion is approximately 85 per cent lacquer to 15 per cent crystals, but this is varied to suit the required gain from the screen. The crystals have a high refractive index, and are hexagonal in shape, with a thickness of less than 1 micrometre. The coating consists of layers of crystals overlapping each other, some lying parallel to the base material and others at a variety of angles.

The result is a screen surface reflecting light from surfaces of the crystals, through refraction and multiple reflections within the crystals, and by diffuse reflections from the white matt base.

The 'Deep Curve' screen specification is similar except that fine mica powder is used, instead of basic lead carbonate crystals, suspended in matt PVC lacquer.

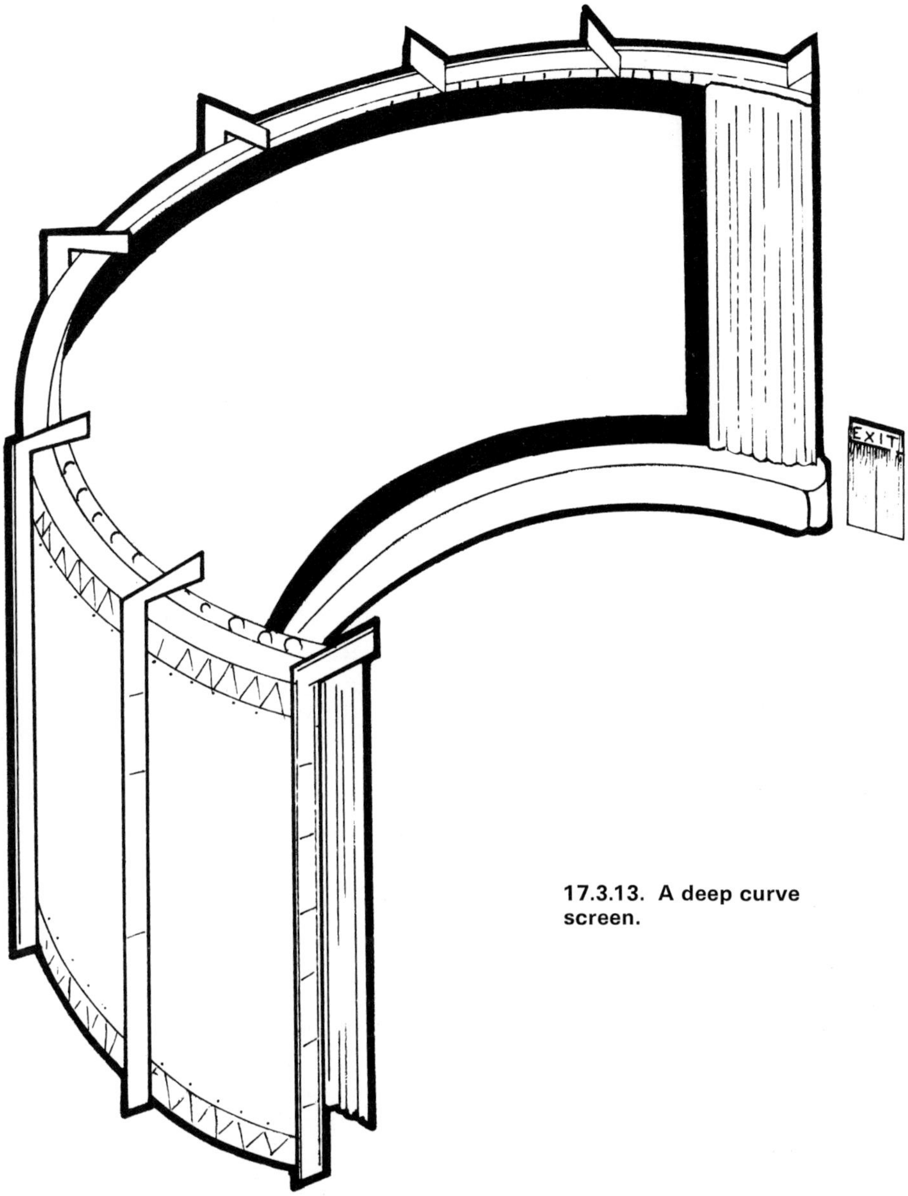

17.3.13. A deep curve screen.

17.4 Light distribution on screens

There is a limited amount of light available from a projector and, in most cases, the maximum possible is projected whilst avoiding the destruction of the film in the projector gate by the heat from the arc lamp. In view of this limitation on available light and the demand for the largest possible screen, it is essential that the light which does fall on the screen is reflected to the auditorium seats and not wasted on walls or ceiling. This is the background to the design of theatres, screens and frames, and seating layouts.

A perfect matt white surface reflects all the light falling on it and diffuses the light so that the density is the same in all directions. A simple test rig can be made consisting of a flat board 2 feet square (610 mm) which has in one corner a sample carrier for a 3-inch (75 mm) disk of screen material. Pivoted from the centre of the sample carrier is an arm at the end of which is fitted a 3-inch diameter circular selenium photo cell connected to a light meter. The board is marked in 5° steps, up to 60°, from the centre of

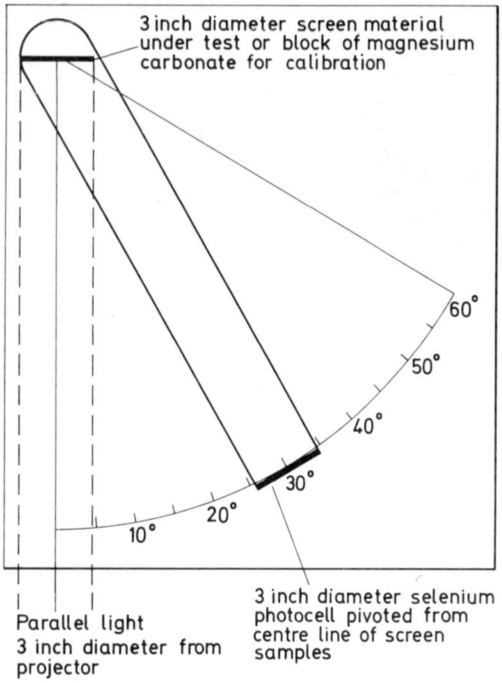

17.4.1. Test rig for measuring reflectance of screen material.

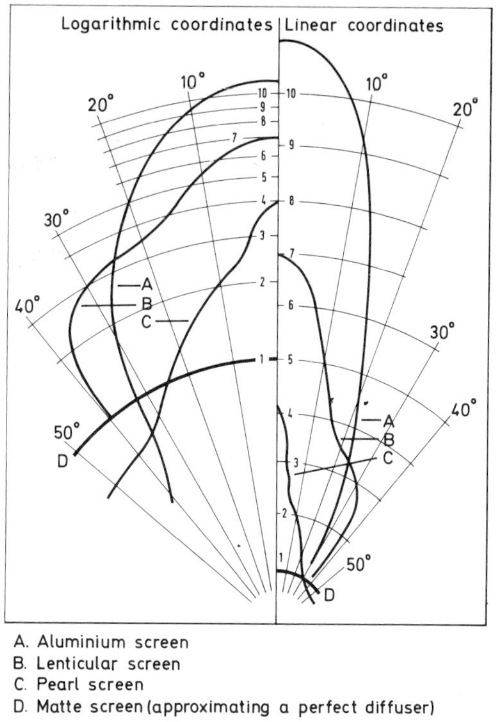

17.4.2. Logarithmic and linear co-ordinates for matt white, pearl, aluminium lenticular, and aluminium smooth screens.

the sample carrier. Light from a 300-watt projector, fitted with an aperture plate to give a 3-inch diameter disk of light, is directed at the screen sample, and photometer readings are taken at 5° intervals. The screen sample is then replaced by a 3-inch diameter magnesium carbonate block and the test readings repeated. The ratio between the two sets of readings from each angular setting are then plotted in chart form. (Fig. 17.4.1.)

If desired, a more elaborate Goniophotometer can be used and, instead of a photocell, light-sensitive paper is suitable for measuring the reflected light. The relative brightness of the screen material under test and the standard flat, freshly exposed, magnesium carbonate is converted into percentage gain.

The usual presentation of light distribution of reflection screens in the form of polar diagrams, with linear coordinates, does not give a result which corresponds to the luminance sensation received by the audience in a theatre. This is because the luminance of the screen, for a human observer, follows the Weber-Fechner law and varies with the logarithm of reflected light.

The selection of type of screen required should take into account that, for example, to double screen luminance, one should double on a linear scale the luminance plotted on a logarithmic scale. A comparison between linear and logarithmic coordinates is given in Fig. 17.4.2. In each case the value of 1 corresponds to the reflection distribution pattern of a perfectly diffuse reflector (such as a magnesium carbonate block or a matt white screen).

17.5 Screen and projection-lens sizes

To take full advantage of the available screen area, for a given throw, it is obviously necessary to use the correct focal length of lens with the largest aperture available. A table (pp. 521–2) of lens focal length sizes for 16-mm film and 35-mm slides is published, and also a chart (Fig. 17.5.1) for 35-mm lenses to be used together with an anamorphic attachment for the CinemaScope film format and photographic sound track (Fig. 13.2.4).

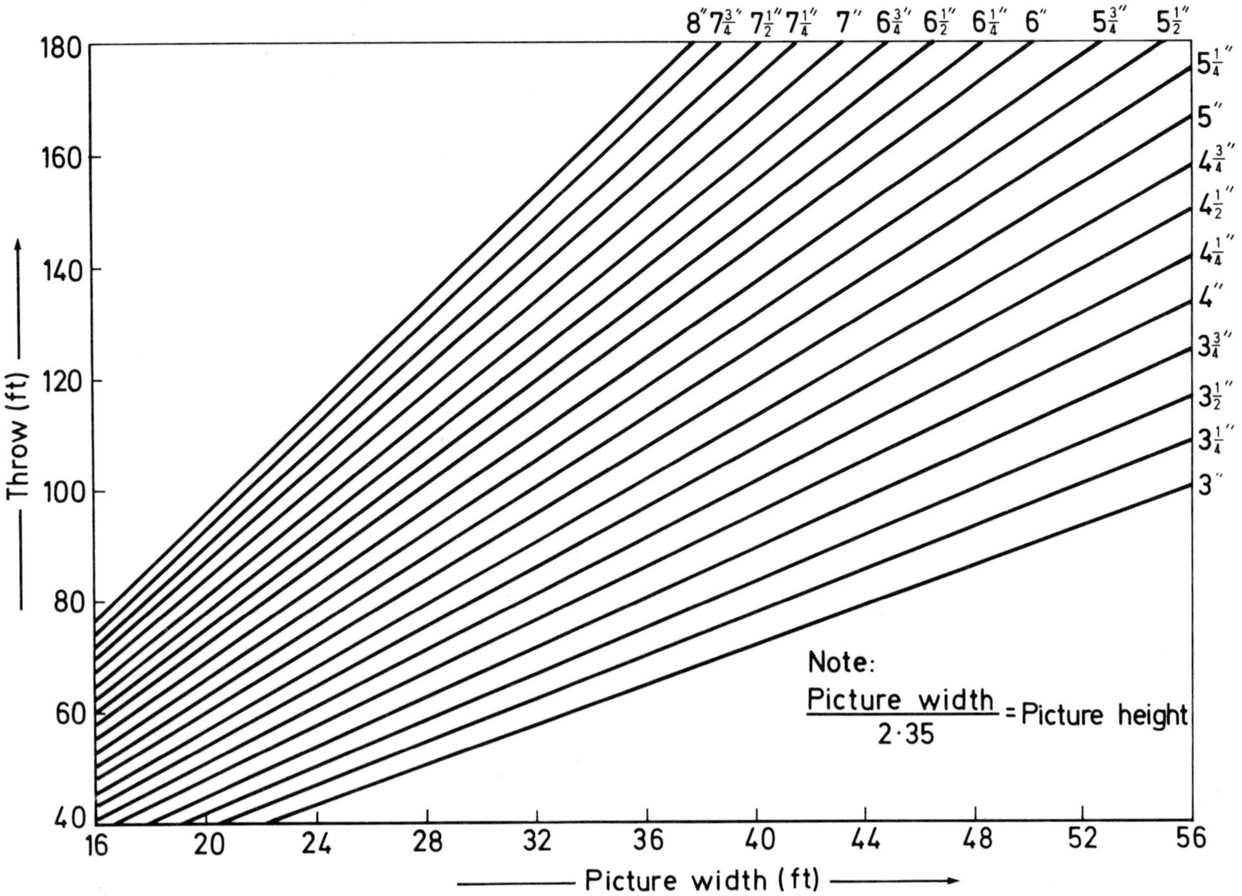

17.5.1. Width of a CinemaScope picture when projected at different distances by lenses of different focal lengths. The film aperture is 0·839×0·715 inch with an aspect ratio of 2·35:1 (2:1 anamorphic expansion).

Table 17.5.1. Projection chart for 16 mm projectors.

Distance from Lens to Screen	*FOCAL LENGTH OF LENS USED — SIZE OF PICTURE*															
	¾ inch				*1 inch*				*1½ inch*				*2 inch*			
	Width		*Height*		*Width*		*Height*		*Width*		*Height*		*Width*		*Height*	
Feet	*ft.*	*in.*	*ft.*	*in.*	*ft.*	*in.*	*ft.*	*in.*	*ft.*	*in.*	*ft.*	*in.*	*ft.*	*in.*	*ft.*	*in.*
3	1	6	1	2	1	0		9	0	9		7				
4	2	1	1	7	1	6	1	2	1	0		9	0	10		7
5	2	6	1	10	1	10	1	4	1	4	1	0	1	0		9
6	3	0	2	2	2	4	1	9	1	6	1	2	1	2		11
7	3	6	2	7	2	8	2	0	1	9	1	4	1	4	1	0
8	4	0	2	11	3	0	2	2	2	0	1	6	1	6	1	2
10	5	0	3	9	3	9	2	10	2	6	1	10	1	10	1	4
12	6	0	4	6	4	7	3	5	3	0	2	2	2	3	1	8
15	7	0	5	3	5	8	4	2	3	10	2	10	2	10	2	1
18	9	2	6	10	6	10	5	0	4	6	3	4	3	5	2	6
20	10	0	7	6	7	6	5	6	5	0	3	9	3	10	2	10
25	12	6	9	4	9	4	6	11	6	4	4	8	4	8	3	6
30	14	10	11	0	11	6	8	7	7	6	5	7	5	8	4	3
35	17	6	13	0	13	4	9	11	8	10	6	7	6	6	4	10
40	20	0	14	11	15	0	11	2	10	0	7	5	7	6	5	7
45	22	6	16	9	16	10	12	6	11	4	8	5	8	6	6	4
50	25	0	18	8	18	8	13	10	12	6	9	4	9	4	6	11
60					22	0	16	5	15	6	11	7	11	7	8	11
75									18	8	13	9	14	2	10	6
100									25	0	18	8	18	9	14	0

Based on projection aperture 0·284 × 0·380.
(Table continues overleaf.)

Table 17.5.1. (continued).

Distance from Lens to Screen	FOCAL LENGTH OF LENS USED — SIZE OF PICTURE							
	2½ inch		3 inch		3½ inch		4 inch	
Feet	Width ft. in.	Height ft. in.	Width ft. in.	Height ft. in.	Width ft. in.	Height ft. in.	Width ft. in.	Height ft. in.
6	0 10	7						
7	1 0	9						
8	1 2	10						
10	1 6	1 2	1 3	11	1 0	0 9		
12	1 9	1 4	1 5	1 0	1 3	0 11	1 1	0 10
15	2 3	1 8	1 10	1 4	1 6	1 2	1 4	1 0
18	2 10	2 1	2 2	1 7	1 10	1 4	1 7	1 2
20	3 0	2 2	2 6	1 10	2 1	1 6	1 10	1 4
25	3 10	2 10	3 2	2 4	2 8	2 0	2 4	1 9
30	4 6	3 4	3 8	2 9	3 4	2 6	2 10	2 1
35	5 2	3 10	4 4	3 3	3 10	2 10	3 2	2 4
40	6 0	4 6	5 0	3 9	4 4	3 3	3 10	2 10
45	6 9	5 0	5 8	4 2	4 10	3 7	4 2	3 1
50	7 6	5 6	6 4	4 8	5 5	4 0	4 8	3 6
60	9 6	7 0	8 0	6 0	6 6	4 10	5 10	4 4
75	11 4	8 6	9 6	7 0	8 2	6 0	7 2	5 4
100	15 2	11 4	12 8	9 5	10 10	8 1	9 6	7 0
125	19 8	17 7	15 7	11 7	13 4	10 0	11 8	8 8
150	22 5	18 0	18 8	13 11	16 0	12 0	14 0	10 0

Table 17.5.2. Projection chart for miniature camera slides.

Distance of lens from screen (feet)		6	8	10	12	15	20	25	30	40	50	60	75
		SIZE OF PICTURE											
Lens size mm		ft. in.	ft. in.	ft. in.	ft. in.	ft. in.	ft. in.	ft. in.	ft. in.	ft. in.	ft. in.	ft. in.	ft. in.
35	W	6 0	8 1	10 4	12 1	15 6	20 8	25 10	31 0	41 4	51 9	60 7	77 8
	H	4 0	5 4	6 9	8 1	10 2	13 7	17 0	20 5	27 3	34 1	40 6	51 3
40	W	5 3	7 1	8 10	10 7	13 3	17 9	22 3	26 7	35 6	44 6	52 11	66 10
	H	3 6	4 8	5 11	7 8	8 11	11 10	14 10	17 0	23 9	28 0	35 4	44 7
50	W	4 2	5 7	7 0	8 5	10 6	14 1	17 8	21 1	28 2	35 4	42 1	53 1
	H	2 9	3 9	4 8	5 7	7 1	9 5	11 10	14 2	18 11	23 8	27 11	35 7
75	W	2 9	3 8	4 8	5 6	7 0	9 4	11 8	14 0	18 9	23 6	27 8	35 2
	H	1 10	2 5	3 1	3 8	4 8	6 2	7 9	9 4	12 5	15 7	18 4	23 5
85	W	2 5	3 3	4 1	4 10	6 2	8 2	10 3	12 4	16 5	20 7	24 2	30 11
	H	1 7	2 2	2 8	3 2	4 1	5 5	6 10	8 3	11 0	13 9	16 2	20 8
105	W	1 11	2 7	3 5	3 10	5 2	6 11	8 8	10 5	13 11	17 4	19 5	26 0
	H	1 3	1 9	2 2	2 7	3 3	4 5	5 6	6 7	8 10	11 0	12 11	1 67
120	W	1 8	2 3	2 10	3 4	4 3	5 9	7 2	8 7	11 6	14 4	16 8	21 7
	H	1 1	1 6	1 11	2 3	2 10	3 10	4 9	5 9	8 7	9 7	11 3	14 5
135	W	1 5	2 0	2 6	2 11	3 9	5 1	6 4	7 7	10 2	12 9	14 9	19 2
	H	1 0	1 4	1 8	2 0	2 6	3 5	4 3	5 1	6 9	8 6	10 2	12 9
150	W	1 3	1 9	2 3	2 7	3 5	4 6	5 8	6 10	9 1	11 4	13 2	17 1
	H	10	1 2	1 6	1 9	2 3	3 0	3 9	4 7	6 1	7 7	8 10	11 5
165	W	1 2	1 7	2 0	2 4	3 1	4 1	5 1	6 2	8 3	10 3	11 11	15 4
	H	9	1 1	1 4	1 7	2 0	2 9	3 5	4 1	5 6	6 10	7 11	10 4
180	W	1 1	1 5	1 10	2 2	2 9	3 9	4 9	5 7	7 6	9 7	10 10	14 5
	H	8	1 0	1 3	1 5	1 10	2 6	3 1	3 9	5 0	6 3	7 2	9 5
200	W	11	1 3	1 8	1 11	2 6	3 4	4 2	5 0	6 8	8 5	9 8	12 8
	H	7	10	1 1	1 3	1 8	2 3	2 9	3 4	4 6	5 7	6 5	8 5

Based on miniature camera size 24 mm × 36 mm.

18 Electro-optics and Opto-electronics

18.1 Introduction

Lasers

In 1960 an entirely new source of light, the ruby laser, was demonstrated by the inventor Theodore H. Maiman. Since then many different kinds of laser have been made which have in common an active material (e.g. the ruby) to convert energy into laser light, a pumping source (e.g. a flash tube) to provide energy, and mirrors, one of which is semi-transparent, to make the beam of light traverse the active material many times and so become amplified.

The light from the laser is concentrated, powerful, and 'coherent', which means that it is highly monochromatic and all the light waves in the beam are exactly in phase with each other.

Using a laser, E. N. Leith and J. Upatnicks at the University of Michigan produced the first laser hologram in 1963. The technique of holography was published by D. Gabor of Imperial College, London, in 1949, but the rapid development in applications for holography has been almost entirely due to the availability of solid pulsed lasers and continuous wave gas lasers.

Some solid pulsed lasers are made from ruby crystals which consist of highly purified corundum (Al_2O_3) doped with Cr_2O_3. Variations in crystal quality such as refractive index, strain, bubbles, or the scattering of particles in the crystal, control optical effects and may cause changes in coherence, energy, or spectrum of the laser beam.

Yttrium aluminium garnet (YAG) is another versatile crystalline host material for solid lasers. Because of their numerous optical transitions and sharp line fluorescence character, the rare earths are preferred for laser impurities. YAG doped with neodymium has acquired increasing interest, and the addition of lutetium to the YAG:Nd enables higher power outputs to be obtained from pulsed as well as from 'Q'-switched laser operation; in many cases, 20–30 per cent more power can be achieved than with standard YAG crystal.

Electro-optic light modulators

Since 1960 a new range of single-crystal materials have been developed, many of which have optical properties that can be varied by the application of an electric field or current. Electro-optic light modulators will provide micro-second light pulses when an electric field is applied to them. They are made of materials such as ammonium dihydrogen phosphate (ADP), or potassium dihydrogen phosphate (KDP), or potassium tantalate-niobate (KTN).

KTN is particularly attractive because it is a low-voltage electro-optic material, and light can be modulated up to 10 MHz with the application of 100 volts or less and 150-volt bias. Light beam deflections in prisms made of KTN are caused by a change in refractive index of up to 0·5 per cent, when a voltage is applied.

Semi-conductor diodes

Semi-conductors form a class of electrically resistive materials which can change from being good conductors to being good insulators. Semi-conductor light-emitting diodes operate at low

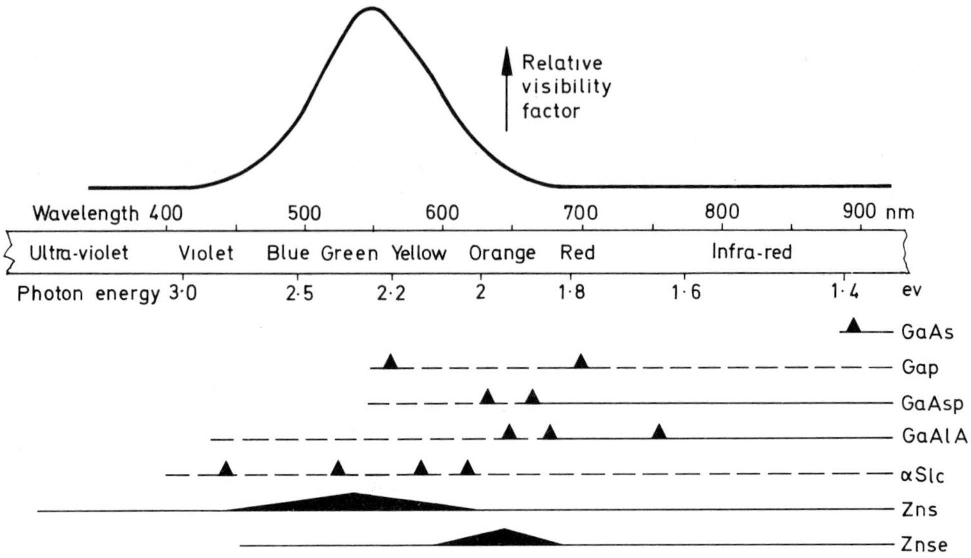

18.1.1. Emission wavelengths of some electroluminescent materials shown in relation to the spectral response of the eye.

voltage and relatively low current. They emit visible radiation which is reasonably monochromatic, have fast switching times like most semi-conductor devices, and they are small, rugged, and very reliable.

Semi-conductors such as cadmium selenide, cadmium sulphide, zinc sulphide, and zinc selenide are particularly strong luminescent materials (Fig. 18.1.1).

Gallium arsenide–phosphide *p–n* junction lamps emit radiation in the red and amber region of the spectrum, whilst gallium phosphide will emit red or green light. Gallium arsenide diode emitters are made in two distinct types:

(*a*) Diodes which produce spontaneous emission when forward biased.

(*b*) Diodes which, under certain drive conditions, act as lasers.

The two types have a common output wavelength as this is an inherent property of the semi-conductor material. Other characteristics are very different but can be summarized as follows:

Used as an emitter

(1) If duty factor must exceed 2 per cent.

(2) If pulse duration must exceed 2 microseconds.

(3) For continuous operation.

Used as a laser

(1) If high peak power is needed.

(2) For high repetition rates.

(3) When narrow beams of more power are needed.

(4) When fan shaped beams are needed.

Gallium arsenide emitter and laser radiation is invisible, but is readily detected by image converter tubes. In the future a full range of coloured light will be possible from photo-diodes, and opto-electronics will become a very important part of the optical and electronic industries.

18.2 Growing electro-optic crystals

BRIDGMAN AND CZOCHRALSKI TECHNIQUES

Crystals which can be grown in vacuum or in an inert or neutral atmosphere can be grown by the Bridgman technique. (Fig. 18.2.1.) The controlled method of inducing a temperature gradient allows the easy growth of oriented crystals from a seed, either in a graphite crucible or a soft mould of magnesium or aluminium oxide powder. The complete visibility of the crucible and melt allows continuous assessment of growth

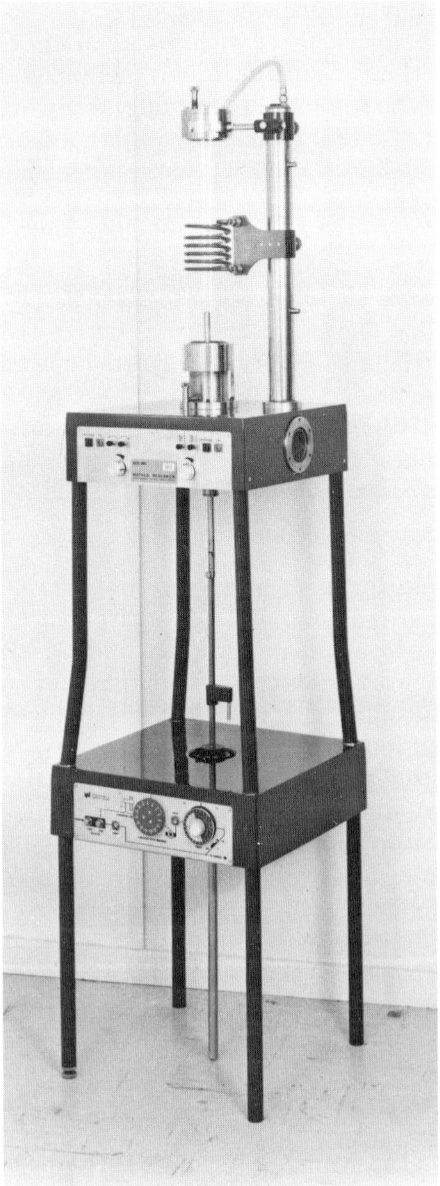

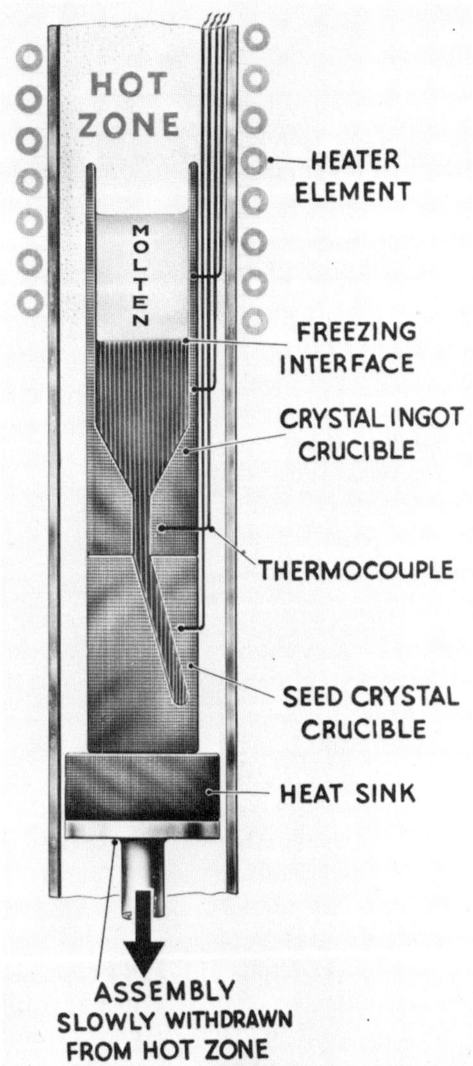

18.2.1. The Bridgeman BCG 265 crystal-growth unit for use with radio frequency induction heating. Single crystals can be grown up to 25 mm in diameter and 250 mm long.

conditions and avoids the use of expensive and complex temperature control equipment.

Ruby and YAG laser crystals, ADP, KDP, and semi-conductors such as gallium arsenide and gallium phosphide, are all grown in a Czochralski type of crystal puller. A large number of semi-conductor compounds decompose at the melting point. Usually one of the components is volatile and distils from the melt.

A major problem in the single-crystal growth of gallium phosphide is the high vapour pressure of phosphorus amounting to 35 atmospheres at the melting point of 1467°C. The usual method of growing large single crystals is by the Czochralski technique, with liquid encapsulation, and a high-pressure chamber for the crystal puller. In this method a melt of gallium phosphide, doped as required just above the melting point, is contained in a carbon crucible in an atmosphere of nitrogen (with low oxygen content) at a pressure of 50 atmospheres. A thin molten layer of inert boron trioxide, a transparent low softening-point glass, above the gallium phosphide prevents loss of phosphorus to cooler parts of the pressure chamber. An oriented seed of gallium phosphide is dipped into the melt and slowly withdrawn in the usual way as the crystal grows. (Fig. 18.2.2.)

Observation of the growth of the crystal is possible by a TV camera looking through a sapphire window in the furnace.

The R.R.E. crystal furnace

The Royal Radar Establishment, Malvern, Worcs, have taken a leading part in the development of production processes for these and other crystal materials and the following description is of a furnace developed by the R.R.E., and marketed by Metals Research Ltd, Melbourn, Royston, Herts. (Fig. 18.2.3).

The R.R.E. crystal furnace is a versatile apparatus designed for the laboratory or small-scale industrial production of the widest range of single-crystal metal, semi-conductor, and optical materials of the highest quality, generally by the 'pulling' (Czochralski) method. Accessories are, however, available to permit the use of the floating zone or Stockbarger techniques. The temperature range of the standard apparatus as set up for pulling extends upwards to include the growth of ruby and sapphire, and is limited by the availability of suitable crucibles rather than by the design. Crystals at least 15 cm long and up to about 2 cm in diameter are readily grown. The liquid encapsulation technique may be used in this apparatus for growing volatile materials. (J. B. Mullin *et al.*, *Phys. Chem. Solids*, 26, 782–784.)

It is now well-known that one of the difficulties in crystal growing lies in avoiding small-scale fluctuations of impurity distribution in deliberately 'doped' crystals, and also in avoiding precipitates in crystals grown from starting material which is less pure than one would wish. In both situations a suitable slow uniform growth rate is necessary; any transient change in growth rate caused by mechanical or thermal fluctuations can change the impurity distribution in the portion of crystal growing at that time. For these reasons extreme care has been taken to achieve the necessary mechanical and thermal stability.

An interesting mechanical feature of the apparatus is the arrangement whereby a variety of work chambers can be fitted. Removable top and bottom plates are provided, forming the work chambers ends, so that if other than the standard simple silica chamber is to be used, new sealing grooves can be cut, or a set of new plates made, by the simplest turning operation. Various other accessories are available, including a metal work chamber to work at pressures up to 2000 p.s.i.g., and a top plate for use with a closed lower-end envelope to facilitate the use of electromagnetic stirring techniques.

Generally the crucible is heated by r.f. induction, but resistance heating by a wirewound heater has been used below 1000°C.

The frame

The apparatus consists of a rigid main frame supporting near the lower end a counterbalanced cast bracket which moves vertically and which supports the work-chamber bottom plate. For loading or unloading, this bracket is lowered and the work chamber removed, thus exposing the crucible and the pull rod. The main frame also carries the rotation and lift motors with gear-boxes and the lead screw.

The work chamber

The working volume is designed to vacuum standards, although crystals are generally grown in a gas atmosphere. The two end-plates are water cooled and carry flat silicone rubber gaskets against which the silica chamber ends bear, with axial compression provided by a single, hinged, tie rod.

The lower plate carries a hollow spigot which locates the tubular refractory crucible support. The spigot is drilled to take a refractory thermocouple sheath or the sapphire rod of a radiation pyrometer if used, and is fitted with a demountable vacuum seal. The plate also has ports for gas, vacuum, or power lead entries, and a calibrated adjustable bracket arrangement to support the r.f. coil at known heights.

The standard silica chamber has a side arm, at 45° to the vertical axis, which terminates in a demountable water-cooled silica window. The window frame has ports which may be used to provide a gas curtain to keep the window free

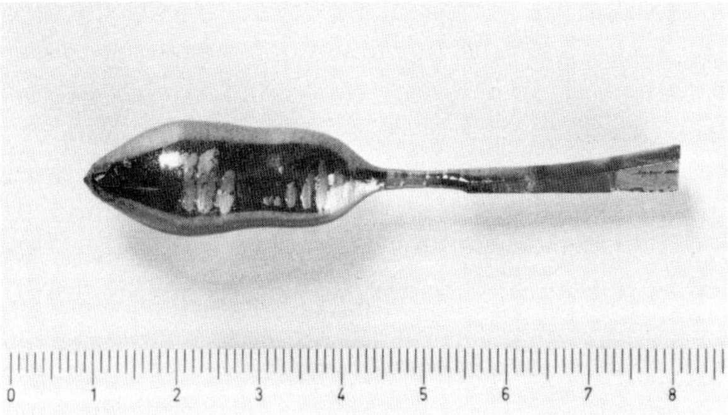

△18.2.2. Single crystal of gallium phosphide.

18.2.3. ▷
The Royal Radar Establishment Czochralski furnace for small-scale industrial production of very high-quality single crystals by the pulling or liquid-encapsulation method. The furnace can produce single-crystal metal, semiconductor, and optical materials. Work chambers for vacuum or high pressures can be used. Generally, the crucible is heated by r.f. induction. The crystal growth can be watched on a TV screen from observation by a camera through a sapphire window in the top of the furnace.

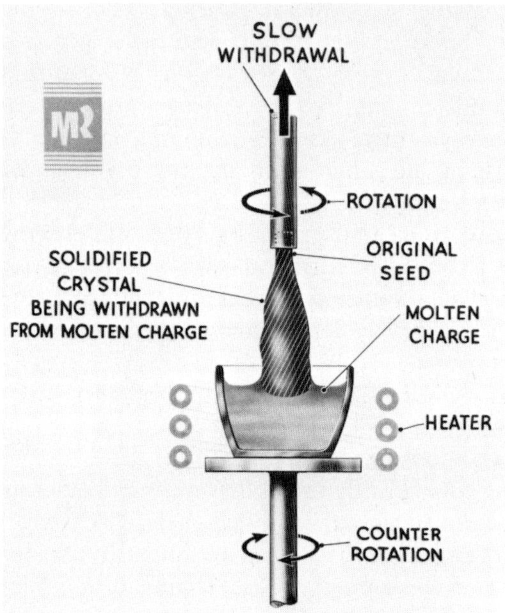

from condensed evaporated material when unencapsulated volatile substances are being melted; alternatively the ports may be used for introducing dope pellets or electrical leads.

The upper plate forms part of the head assembly and is easily removed for modification. Both plates and the window frame are rhodium plated on their inner faces to provide a non-contaminating surface and to permit easy cleaning.

The complete crucible assembly may be withdrawn through one alternative type of bottom plate without removing the envelope, while another version is fitted with a velodyne motor and gearbox for crucible rotation.

The head assembly

This carries the upper work-chamber plate and a long water-cooled portion through which the pull rod passes, which is one of the important and original design features of the apparatus; instead of a fixed pull-rod seal located in the head as is usual, the O-ring seal is mounted in a groove in the pull rod itself. With the usual arrangement, the dry pull rod, which may be covered with an abrasive evaporated deposit, is drawn through the seal. The latter then wears rapidly, and in our experience requires replacing after each run if true vacuum and/or gas-tightness is to be maintained. In the R.R.E. system any evaporated material remains on the pull rod, which is of reduced diameter below the O-ring. The O-ring seal thus slides in the water-cooled housing carrying its own oil supply, always works against a lubricated surface, and effectively forms a zero clearance bearing. The O-ring has an indefinitely long life when used in this manner. By selecting the neutral gear change position the lead screw can be rotated manually. Alternatively, the pull rod can be lifted manually and directly, for example to remove a crystal quickly from the melt. A peg is provided to hold the pull rod in the raised position.

The pull-rod drive mechanism

A precision 10TPI stainless-steel lead screw, running in a bronze nut, is driven by a Velodyne motor through a gearbox giving alternative overall ratios of 6862/1 and 650/1, with a neutral selector position. A shear pin limits the torque in the event of accidental jamming.

A similar motor and two-speed gearbox rotates the pull rod over the normal range of $\frac{1}{2}$ to 350 r.p.m. The Velodyne motor is an R.R.E. invention and consists of a reversible split-field d.c. motor with a d.c. generator on a common shaft. The generator voltage output is proportional to rotational speed, and is compared with an adjustable reference voltage. Any difference is amplified and applied to the motor field windings so as to minimize the difference by altering the motor speed. Thus, by varying the reference voltage the speed may be varied from zero up to 7000 r.p.m., the mechanical limit of the type-126 motor-generator, at which speed it develops $\frac{1}{8}$ h.p. At any set value, the speed varies by only a few r.p.m. from zero to full torque, and is virtually independent of fluctuations in supply voltages and ambient temperature. The motors are usually run at a few hundred r.p.m. only to minimize vibration, but full speed in each direction is available, by means of an over-riding key switch on the controller, for fast positioning of the pull rod, which thus can be used for positioning the seed crystal at the beginning of growth by pulling.

The heating system

The r.f. method of heating is convenient and versatile for heating the crucible, 18-kW output from a high kVA industrial heater at 500 kHz being adequate for the full range of temperatures used. Wire-wound heaters have, however, been used at crucible temperatures of 1000°C or lower.

The r.f. heater power is controlled by a saturable reactor, driven usually by a Honeywell–Brown 0–1 mV Electro–O–Volt instrument. It has been found that the best results are obtained by using the servo system to maintain constant r.f. coil current rather than by attempting to control the crucible temperature directly by means of a thermocouple. This method is also mechanically more convenient and probably the only practicable method at temperatures beyond the range of precious metal thermocouples. It is not generally appreciated that it is the rate of change of temperature rather than absolute temperature stability which is of importance in crystal growth.

The device for providing the coil current signal consists of a small pick-up loop mounted near the 'earthy' output lead of the r.f. heater. The signal is rectified by a semi-conductor diode or a vacuum thermojunction smoothed by an R.C. network and reduced to approximately 20 mV at working temperature. The d.c. signal is then compared with a voltage derived from a dry cell, and the difference fed into the Honeywell–Brown controller, or other control system.

Crucibles

The art of crystal pulling is intimately bound up with the choice of a suitable crucible material. Graphite, vitreous carbon, base metal, and precious metal crucibles have been used, with and without refractory liners. The choice must generally fulfil the following criteria: the crucible (or the molten charge) must be electrically conducting (for r.f. heating), refractory at the intended temperature, and must neither react with the melt nor introduce unwanted impurities.

Chucks and extensions

The end of the pull rod, which is alternatively of molybdenum or stainless steel, is tapped to receive the seed crystal chuck or an extension piece. The chuck supplied is of the simple split type and, like the extension pieces, is available in stainless steel or molybdenum; other types can be fitted as required. It is worth noting that once the metal work has been heated the threads must be lubricated with a low surface tension liquid such as ethyl alcohol before they are unscrewed. Failure to observe this precaution inevitably leads to seizure and stripped threads.

SYNTHESIS OF GALLIUM PHOSPHIDE

A special technique of manufacture is necessary for a material such as gallium phosphide, as one of the components of the compound has a very high vapour pressure.

Gallium phosphide, a pale orange solid, can be synthesized in bulk by either a direct combination of the elements or treatment of the gallium metal with phosphine.

The first method, by direct combination of the elements, is the more usual one. Gallium metal is placed in a horizontal carbon tube within a larger thick quartz tube. (Fig. 18.2.4.) The tube is locally

18.2.4. Synthesis of gallium phosphide by direct combination of the elements at a temperature of 1450°C.

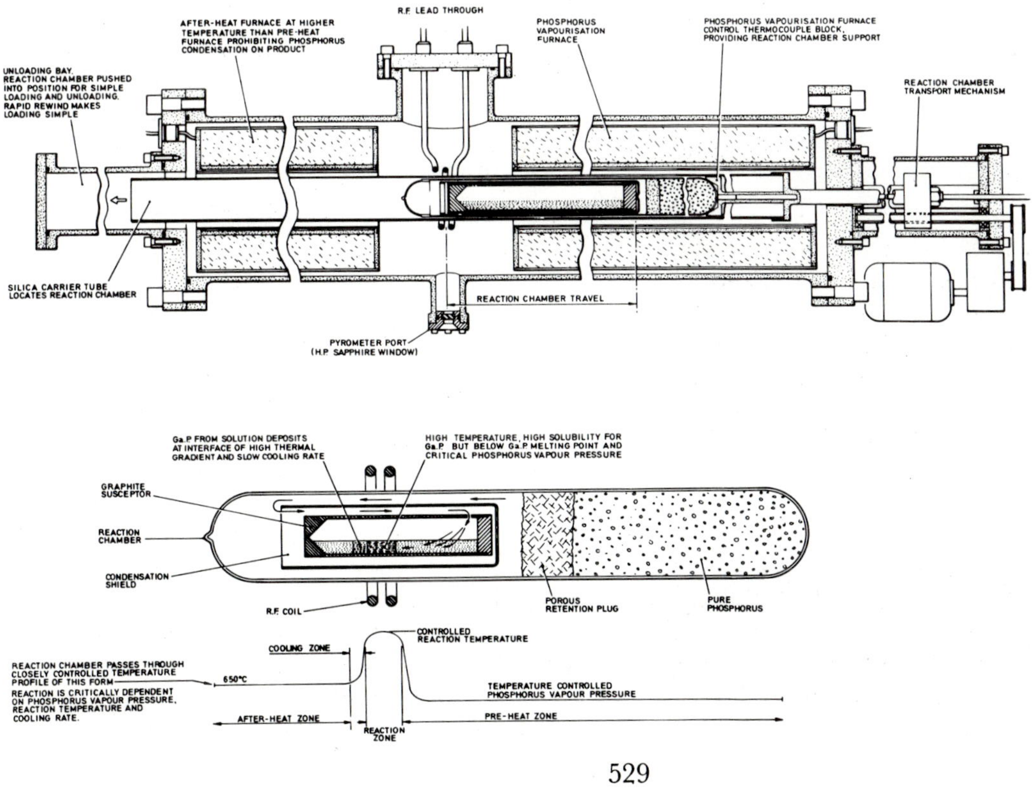

heated by an r.f. furnace to 1450°C at which temperature it reacts with phosphorus vapour to form a compound. The phosphorus is contained in a separate compartment of the reactor, and its vapour pressure held at 5–6 atmospheres by controlling the temperature of the phosphorus container to 470°–480°C. The hot zone is slowly passed through the gallium metal and, at the same time, the phosphorus vapourizes and reacts with it. Two passes of the zone will usually convert more than 90 per cent of the gallium into the phosphide in the form of a dense block, which forms the raw material for the crystal growing.

GALLIUM ARSENIDE-PHOSPHIDE

A planar technology has been established for making individual lamps, as an array, with the required number of evaporated lead-through wires. Aluminium wires are then ultrasonically bonded to the array mount.

Gallium arsenide-phosphide, suitably doped with selenium, is grown as an epitaxial layer on gallium arsenide substrates using the arsine-phosphine system. Planar techniques are then employed to fabricate the *p–n* junction by the diffusion of zinc into the epitaxial layer through a silicon dioxide mask.

18.3 The slow-speed sawing of brittle crystal materials

Most electro-optic crystals are at present very expensive, and usually the polished faces are required to be flat to an accuracy of from 1 wavelength to one-tenth of a wavelength, parallel to 15 seconds of arc, and at a correct angle to the crystal axis to within 1 minute of arc.

These are very stringent requirements, but necessary for the efficient operating of the device, and so errors are costly. Before cutting, the first essential is to take a back reflection X-ray photograph of the crystal to establish the orientation of the axis. (Fig. 18.3.1.) The crystal must be mounted on a goniometer head which can be transferred from the X-ray machine to the cutting machine without loss of accuracy, and after the necessary adjustment is calculated, the goniometer must be capable of precision movement in two planes for the correct setting on the cutting machine.

At the Royal Radar Establishment, Malvern, Worcs, research by Mr G. W. Fynn and Mr W. J. A. Powell has led to the production of three efficient cutting machines marketed by Metals Research Ltd, Melbourn, Royston, Herts. These machines are known as Macrotome 1 and 2, and Microslice. (Fig. 18.3.2 and 18.3.5.)

The Macrotomes

The Macrotomes are precision slicing machines working on a principle of a carefully counterbalanced 'see-saw' device. (Fig. 18.3.3.)

A precision slicing blade of diamond or carborundum is rotated on an accurately machined shaft at a relatively slow controlled speed of between 50 and 1200 r.p.m. dependent on the nature of the workpiece. The workpiece is fed into the wheel by the damped counterbalanced see-saw.

18.3.1. Laue back-reflection X-ray photograph of gallium phosphide, which has a cubic structure.

18.3.2. The Macrotome II will take diamond or carborundum saw blades up to 150 mm in diameter, either single or multiple. Cutting pressure is adjustable by a counterweight system. The close-up shows the cutting of a crystal ruby.

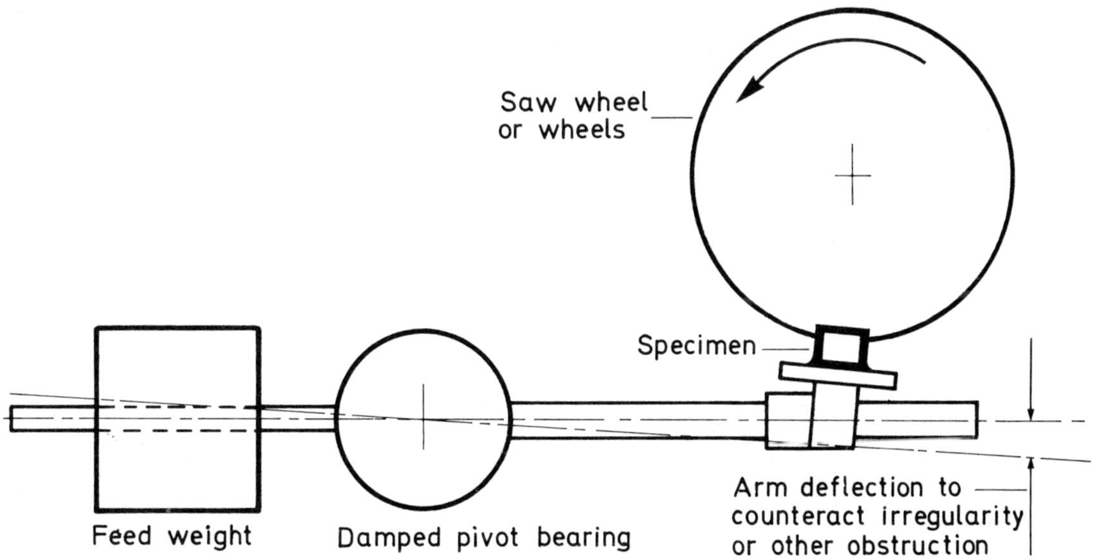

18.3.3. The use of the counterweight on the Macrotome II

The damping is arranged to give the system a low effective inertial mass and the counterweight is chosen to give the appropriate feed pressure. This principle allows the work to back away automatically if the blade encounters a hard spot or surface protrusion or if there is a peripheral variation in the saw blade. This avoids excessive mechanical and thermal stresses that such features could impose on the workpiece if the feed was uniform and unyielding.

Two versions of the Macrotome exist, the Macrotome I and the Macrotome II. Both operate on the same principle and have the same advantage in terms of performance. The Macrotome II, however, has provision for cutting longer specimens by the use of a longitudinal slide assembly and also provision for the fitment of stacks of up to fifty blades for multiple slicing and dicing.

The sample to be cut is mounted on the appropriate table which is then attached to the sawing arm by means of a universal clamp. The position of the work table is adjusted so that the table forms a tangent to the blade, i.e. the specimen will pass through the centre of the blade. Care must be taken while these adjustments are being carried out so that the blade is not damaged in any way.

The correct weight should now be selected (from the three provided) so that when it is positioned on the arm, with clamp D released, it just balances the specimen and table. If the weight is now moved away from the pivot the specimen will impart a load on to the blade. This load which can be varied by re-positioning the weight will depend upon the nature of the specimen and the desirable rate of cutting.

The Macrotome can now be switched on, the correct speed selected and the flow of lubricant regulated. The swing arm should be supported, and the sample allowed to come gently into contact with the blade. Once cutting has commenced the arm can be completely released and the blade will cut through the sample at its own rate.

If the rotary divided table is being used this is attached to the swing arm of the Macrotome by means of the clamp on its base. After making the first cut the table can be rotated through the required angle in increments of 1° and the next cut made.

The goniometer is supplied with attachments such that it can be transferred from an X-ray machine to the Macrotome, without loss of orientation of the monocrystal being cut.

The goniometer head consists of two worm drives at right angles such that any angular displacement of the work table can be made in two perpendicular planes. The scales marked on the goniometer are in degrees with a vernier scale attached to give 10-minute increments. Movement of the two arcs is by means of the socket screw driver provided.

For alignment of the goniometer normal to an X-ray beam, the goniometer axis should first be set to zero.

A vee block assembly is available for cutting long specimens which require supporting along their length.

Attachment of the vee block to the swing arm is by means of the universal clamp. Specimens are mounted on the vee block using the screw clips provided.

The sliding table is mounted on the machine in place of the swing arm attachment. Specimens are mounted on a special table which locates on the sliding tray. This positive fixing is by means of a ball fixing which locates into a groove. There are two grooves on the fixture on the sliding tray so that the table can be rotated accurately through 90°. The sliding carriage is mounted with low-friction bearings on to a track, and by attaching a suitable weight at the end it can be drawn along towards the blade but still maintaining a non-positive feed. (Fig. 18.3.4.)

It is possible with Macrotome II to gang several blades together on the main spindle with spacers in between, so that a series of cuts can be made in one operation. The standard Macrotome II will allow up to about six blades to be used at once, depending on the thickness of the spacing collars used. The Macrotome II model with the extended spindle will take up to fifty blades at once.

It has been found that ganging of silicon carbide disks gives a far more accurate cut than using sintered diamond or electrometallic blades. This is due to the fact that less edge chipping results from using these blades. The spacing collars can be made from ferrous or non-ferrous flat strip but care must be taken to ensure that there is no edge turn over or burring. The thickness of the spacing collars

18.3.4. The Macrotome II blade passing through the centre of a specimen.

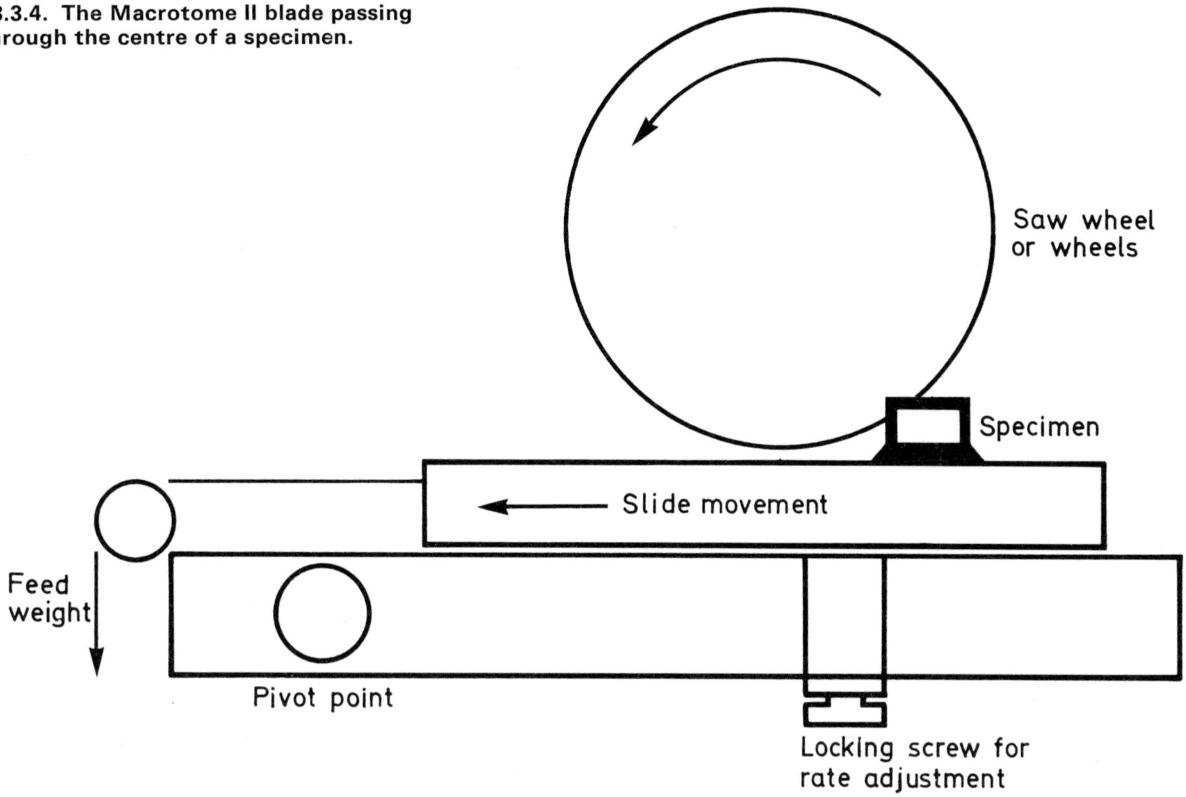

will depend on the required size of the cut sample. It is sometimes necessary to true the disks after mounting on the cutting spindle, because the bores vary slightly in diameter and this could result in an uneven saw load.

For simple cutting without using the longitudinal slide attachment the workpiece is attached to a piece of glass (e.g. a microscope slide) by a suitable compatible adhesive. The glass is then fastened to the work-table by double-sided adhesive tape or any other conventional adhesive. When very thin slices are to be cut it may be necessary to encapsulate the specimen completely to eliminate movement. Backing plates, cover slips, or thin glass slides can be used to give the necessary support to very delicate or multi-constituent materials.

Delicate cylindrical specimens may be first mounted in a plastic or glass cylinder of suitable diameter, and then the composite workpiece can be cut.

When using the longitudinal slide assembly, the specimen should be mounted more rigidly to prevent movement during long cuts.

In using the multiple blades for materials other than semi-conductors, conventional adhesion methods can be used. It is often advisable to cover the specimen with a layer of lacquer or wax to prevent chipping. When dicing semi-conductors, the specimen should be mounted on a soft but firm material (e.g. carbon, Perspex, Tufnol, or methyl methacrylate) and lacquer should be applied over the top of the specimen before making the second cut at 90°.

The quality of saw cut produced will always depend upon the quality of the blade, the correct loading, and correct speed. Of these three factors the quality of the blade is the most important and a damaged blade should never be used. Damage usually occurs in two ways.

(*a*) Chipping the blade through bad handling.

(*b*) Twisting the blade by exerting excess pressure.

Thus it is extremely important to prevent knocking the blade with the swing arm or specimen, or allowing the blade to seize up in the specimen due

Table 18.3.1. Examples of materials cut on the Macrotome

Material cut	*Type of blade*	*Diameter and width of blade (mm)*	*Thinnest section obtained*	*Grit size*	*Cutting (r.p.m.)*	*Pressure (g)*	*Depth of cut (mm)*	*Time (mins)*
Indium antimonide	Sintered and electrometallic	100 × 0·3		170 280 280	250–300 50	20–30	20	10
Indium antimonide	Sintered and electrometallic	75 × 0·25		400	100	5	0·05	2
Germanium	Sintered	100 × 0·3		170	250–300	20–30	6	5
Ruby/sapphire	Sintered and electrometallic	125 × 0·4 100 × 0·3	0·1	160 280	100–400	350	20	20
Calcium fluoride	Sintered and electrometallic	125 × 0·4 100 × 0·3		160 280	50–200	60	35	10
Calcium tungstate	Sintered	100 × 0·3		170	100	30	2	5
Neodymium doped glass	Sintered	100 × 0·3		170	200	30	12	2
Vanadium germanium alloy	Sintered	100 × 0·3		170	100	40	5	5
Uranium carbide	Electrometallic	75 × 0·25	0·1	400	200	10	3	3
Lanthanium bi-fluoride	Electrometallic	100 × 0·3		280	150	25	18	4
Yttrium-aluminium garnet	Sintered and electrometallic	100 × 0·3		170 280	100–400	250	10	4
Calcium molybdate	Electrometallic	100 × 0·3		280	150–250	25	10	5
KDP	Electrometallic	100 × 0·3		280	150–250	10	10	10
High alumina ceramics	Sintered	75 × 0·25	0·05	300	200	60	0·5	2
Tungsten	Sintered	100 × 0·3	0·25	170	150	250	6	20
Anthracene trans-stilbene	Sintered and electrometallic	100 × 0·3		170 280	100	10	12	5

to excess loading. Damage can easily be caused to specimens which are not rigidly mounted especially if the longitudinal sliding table assembly is being used.

It is often necessary to 'dress' the saw blade by lightly holding a piece of alumina or similar material against it, especially when cutting plastics which might clog the blade and lead to seizure.

The Microslice

The Microslice is a precision slicing device working on the same principle as the Macrotome. (Fig. 18.3.5.) A tensioned annular cutting blade rotates an accurately machined shaft at a relatively slow controlled speed of between 50 and 1200 r.p.m., dependent on the nature of the workpiece. The workpiece is fed into the wheel by the damped counterbalanced see-saw. The damping is arranged to give the system a low effective inertial mass and the counterweight is chosen to give the appropriate feed pressure. This principle allows the work to automatically back away if the blade encounters a hard spot or surface protrusion or if

Material cut	*Type of blade*	*Diameter and width of blade (mm)*	*Thinnest section obtained (mm)*	*Grit size*	*Cutting (r.p.m.)*	*Pressure (g)*	*Depth of cut (mm)*	*Time (mins)*
Quartz	Sintered	100 × 0·3		170	200	250	6	1
Strontium titanate	Sintered	100 × 0·3		170	200–250	200	12	10
Bone	Sintered	100 × 0·3	0·1	170	200	100	12	10
Plastics	Sintered	125 × 0·9	0·1	170	300	80	25	20
Granite	Sintered	75 × 0·25		300	200	60	0·5	2
Bone (encapsulated)	Sintered	100 × 0·3	0·1	220	500–600	100	25	60
Arteries (encapsulated)	Sintered	100 × 0·3	0·1	220	400–500	50	12	30
Teeth (encapsulated)	Sintered	100 × 0·3	0·25	220	100	50–75	25	60
Wood cores	Electrometallic	75 × 0·3	1·0	150	150–200	100	10 × 150	15
Silica glass plate	Sintered	100 × 0·3	0·25	220	600–700	150	8 × 125	20
Rock	Sintered	125 × 0·4	0·25	220	650	75	25	45
Glass fibre	Sintered	100 × 0·3	0·1	220	600–700	100	12	10
Carbon fibres in bonded medium	Sintered	75 × 0·25		220	50	50	50 × 3	10
Gallium phosphide	Sintered or carborundum	100 × 0·3 100 × 0·15	0·25	220	100	100	35	15
Gallium arsenide	Carborundum	75 × 0·15	0·17	220	50	20	8 × 5	10
Lithium ferrite	Sintered	75 × 0·25	0·25	220	250	40	10	2
Potassium lithium niobate	Sintered	75 × 0·25		220	75	30	5	1
Barium titanate	Sintered	75 × 0·25		220	75	30	5	1
Mild steel	Sintered	100 × 0·3	0·25	220	700	70	10	20
Ferrite	Sintered	75 × 0·25	0·1	220	400	40	10	15
Titanium alloy	Sintered	75 × 0·25	0·5	220	600	60	10	10
Boron carbide	Sintered	75 × 0·25	0·1	220	600	250	10	45
Silicon carbide	Sintered	75 × 0·25	0·1	220	600	250	10	45

18.3.5. The Microslice annular saw with electrometallic diamond cutting surface on the inside edge of the blade.

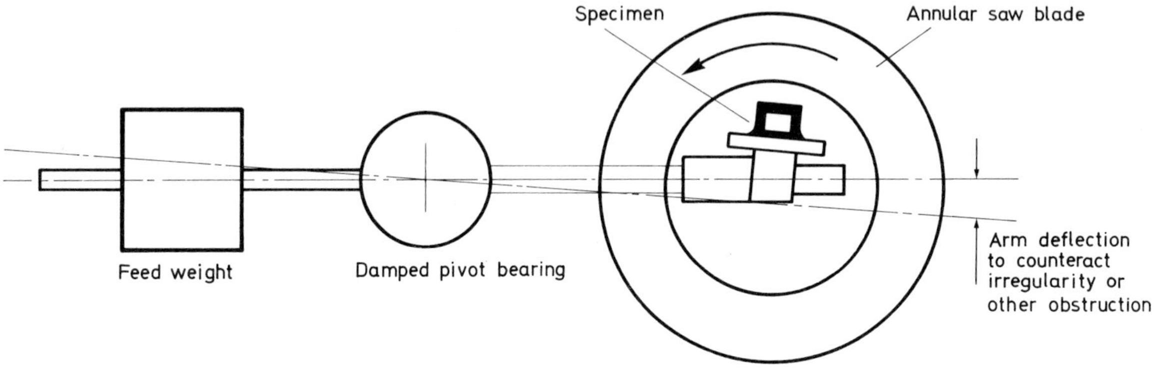

18.3.6. Cutting pressure is by a variable counterweight system. The thinnest slice that can be cut is 0·004 inch (0·1 mm) thick.

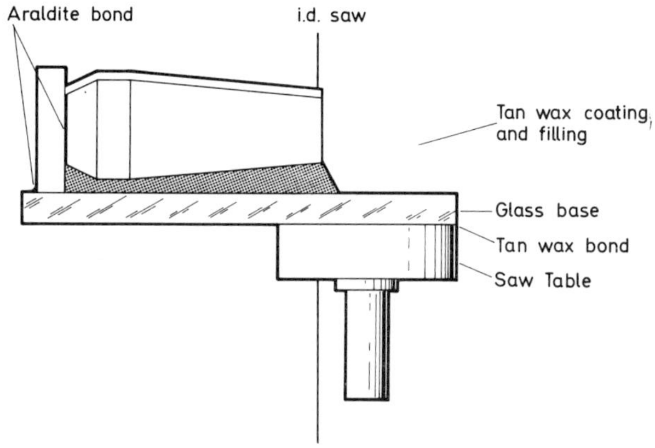

18.3.7. An ingot prepared for cutting.

there is a peripheral (internal) variation in the saw blade. This avoids excessive mechanical and thermal stresses that such features could impose on the workpiece if the feed was uniform and unyielding. (Fig. 18.3.6.)

Since the blade of the microslice is held in tension from its outside periphery it is possible to make it from much thinner material than a normal peripheral cutting disk. This means that the width of each cut is extremely small and the tensioned blade cannot 'wander' through the workpiece as it is cutting.

It is not possible to put a positive figure on the life of a blade since this depends on the material cut and the durability of the annular disk itself. Electro-metalled annular blades are not recommended for materials with a Moh's hardness figure > 8, though relatively short-term successes can be had at the slow r.p.m. used.

The smaller mesh diamond sizes (*circa* 300), though giving the finer finishes and lowest kerf losses, do not last as long as blades with 240 grit. When monitoring of the cut slices shows that these are beginning to exhibit undulations or wedging, then it is almost certainly the sign of an ageing blade. The fact that one blade has been found to cut hundreds of slices before renewal is not a safe guide to the life of the next identical blade, operating identically. The best indication is gained from sampling the slices cut.

The lubricant is recirculated on the microslice. If the lubricant is poured into the level of the cutting edge of the blade, it will be found that very little is lost from the system. A tray is provided to catch any slight splash or drips which may occur. It is advisable to replace the lubricant regularly since particles suspended in it will tend to force any of the slices being cut away from the main bulk of material and thus raise the lower limit of slice thickness.

The sample to be cut is mounted on the table provided (Fig. 18.3.7), which is then attached to the sawing arm by means of a universal clamp. The position of the work-table is adjusted so that the table forms a chord to the inside periphery of the blade. Care must be taken while these adjustments are being carried out so that the blade is not damaged in any way.

The correct weight should be selected (from the three provided) so that it is positioned on the arm; just balancing the specimen and the table, when the clamp is released. If the weight is now moved away from the pivot the specimen will impose a load on the table. This load, which can be varied by repositioning the weight, will depend upon the nature of the specimen and the desirable rate of cutting (consistent with the thickness of slice required).

The Microslice can now be switched on and the correct speed selected. The swing arm should be supported and the sample allowed to gently come into contact with the blade. Once cutting has commenced the arm can be completely released and the blade will cut through the sample at its own rate.

The general procedures for Macrotome sawing are applicable. The specimen is mounted on a glass strip, long enough to extend completely within the chuck cavity. Tan wax is used to encapsulate the boule and to stick it to the glass finger. Obviously, it is possible to attach extensions to the normal work-table and arrange that the latter never enters the chuck; but when material is being sawn with related sides, it is an advantage to use the rotational axis of the specimen table (or turntable) in order to avoid having to unfasten, realign, and refasten the boule on its extending finger. The standard practice is to set up the work so that the end sawn is remote from the machine—that is, the crystal is completely within the chuck cavity. This permits the operator to inspect a slice more conveniently. Since one has here the classical case of sawing off the branch on which one is cutting, an end stop is usually arranged to limit the travel of the counter-weight arm to the required arc. If the crystal is being sawn relative to a crystallographic axis, the best practice is to provide it with a reference face on an X-ray room saw. This axis is then re-established by aligning the prepared place surface on the front of the boule with the sawing plane. A convenient way of achieving this is to remove the chuck-facing screw from the shaft and, in its place, mount the dial test indicator (d.t.i.) accessory. The stylus is then traversed in two directions, approximately at 90°, one by swinging the d.t.i. on the saw spindle and the second by moving the work on the damping head spindle. The various

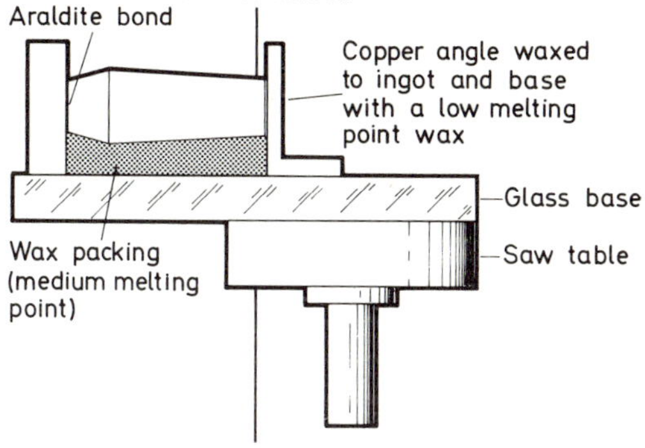

18.3.8. The ingot mounted for cutting with angle plate support.

degrees of freedom allowed for in the work-table construction are used in order to achieve the required orientation. A check on the uniformity of thickness of the first slice is a good test of the degree of alignment attained.

The boule mounting system is important and ideally should have the following features: (i) rigidity, (ii) a base which can be sawn without damage to the blade, (iii) wax encapsulation of the crystal to protect the edges, (iv) a wax that will not load the saw, (v) an arrangement which provides some space between the boule and the base, which is filled with wax. This last provision allows either the whole boule to be sawn and left *in situ*, or single slices to be removed in series as sawing progresses, with a hot-wire wax-cutter.

A system which satisfies these requirements is shown in Fig. 18.3.8. It is, in effect, a fabricated-glass angle plate, made from 6 mm thick plate and cemented together with an epoxy resin. At the first stage of the mounting process a small flat is hand-ground on the end of the boule remote from the oriented face. This flat faces the vertical of the glass angle and is positioned with an epoxy resin in the interspace. In order that the boule may be spaced from the base of the angle plate it is packed up with thickness of card. The join is now cured at 100°C. Before the assembly cools, the whole of the exterior of the boule is coated with tan wax, and the space left when the packing card is removed is allowed to fill, too. By these means the whole of a boule may be cut with each slice retained in place, firmly held in the bed of wax at the bottom and protected around its periphery. Alternatively, each slice may be removed by the hot wire cutter.

The techniques so far described are satisfactory for a wide range of delicate materials, but there are those that do not give repeatable results below 600 micrometres in thickness. Notable exceptions are lead telluride (PbTe) and indium arsenide (InAs), and in order to obtain slices down to 200 micrometres some face support (Fig. 18.3.8) should be given by using an 18 mm length of 18 × 12 × 12 mm nickel-plated copper angle plate waxed to the face of the boule. It is achieved by warming the angle section and melting on a liberal coating of a wax with a melting point *circa* 80°C. The hot-waxed angle is then pressed lightly against the exposed end of the boule and allowed to cool. This bonds the end of the boule to the glass base and a cut made for the required thickness is permitted to enter the glass. The angle plate is removed with the slice *in situ* by applying a pulse of heat from a small soldering iron which melts the wax locally. The slice itself can be allowed to slide off under its own weight when warmed in an oven or over a hot plate.

The vicinity of the boule end is quickly cleaned —for example, with an aerosol degreasing agent— and the angle plate rewaxed in its former position.

On the rare occasion when heat-sensitive materials have to be thin-sliced, temporary support is given to the specimen by use of a thick grease, or an embedding resin such as acrulite or a standard polyester glass or slate-filled resin. The slices are separated by soaking in dichloromethane to remove the polyester resin; the polymethylmethacrylate is best treated by immersion in agitated chloroform.

Table 18.3.2. Examples of materials cut on the Microslice

Material cut	*Typical blade sliced (r.p.m.)*	*Typical load (g/cm cut length)*	*Kerf loss (micrometres)*	*Blade type (grit size)*	*Surface finish (micro-inches c.l.a.)*
Gallium arsenide	180	50	175	280	12
Gallium phosphide	180	50	150	FX	12
Indium arsenide	150	30	150	FX	12
Indium phosphide	150	30	150	FX	12
Lead telluride	150	20	150	FX	6
Cadmium telluride	150	20	150	FX	12
Lithium niobate	200	50	175	280	8
Proustite	150	20	175	280	16
Tri-glycine sulphate	150	30	150	FX	12
Rutile	200	50	175	280	8
Spinel	200	50	175	280	8
Crown glass	200	50	150	FX	8
Quartz	200	50	175	280	8
KDP, ADP	150	30	150	FX	12
Yttrium aluminium garnet	200	50	175	280	8
Silicon	200	50	150	FX	8
Brass	200	40	150	FX	6

18.4 Cylindrical grinding of crystals

Raw crystals can vary from rectangular to roughly cylindrical in form and therefore cutting, sizing, and mounting are necessary before polishing. Most crystals can be cut on the machines described in § 18.3 above, but some very highly stressed crystals cannot be shaped, without danger of cleaving, even by the most delicate sawing. In these cases the entire surplus material must be ground or lapped away.

Crystals in the square-sawn form can be hand-lapped to an octagonal section by the use of suitable abrasives—such as 45-micrometre diamond for ruby and BA302 emery for softer materials—on a flat cast-iron plate. The octagonal crystal can be reduced to the circular section with two semi-circular cast iron laps, the internal diameter of which approximately equals the corner to corner distance of the octagon. (Fig. 18.4.1.)

The finished diameter is produced with two more half-laps slightly above the final size. Crystals can be finished to within ±0·0002 inch (0·005 mm) on a 2 inch (50 mm) ×0·25 inch (6 mm) diameter rod by this method.

Centre grinding can be very satisfactory for very long delicate crystals, and for this method Tufnol extension centres are cemented on the sawn ends of the crystal in a jig which registers them centrally and applies a clamping spring pressure. The adhesive used between the Tufnol and the crystal can be Araldite resin AY111 mixed one to one with hardener HY111. The whole assembly can be transferred to an oven and the adhesive cured for 3 hours at 60°C.

When the adhesive is cured, the crystal can be mounted in the centres of a small lathe and ground with a 120-grit Resinoid diamond wheel. The Tufnol centres can be removed from the crystal by immersion in dichloromethane for several hours or destructively ground away.

Centreless grinding, on a machine such as that made by Walter Bunter, S.A. Lausanne, Switzerland (Fig. 18.4.2), is the fastest and the most

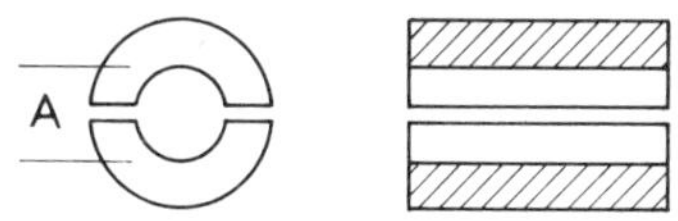

18.4.1.▷ Cast-iron half laps.

Diameter A:
One lap to fit corners of rough crystal
One lap to match finished crystal

18.4.2. Bunter centreless grinder, shown with the control wheel in the open and closed positions. The machine will grind cylinders from 3 to 8 mm in diameter. The control wheel is 100 mm diameter × 60 mm. The grinding wheel is 200 mm diameter × 60 mm.

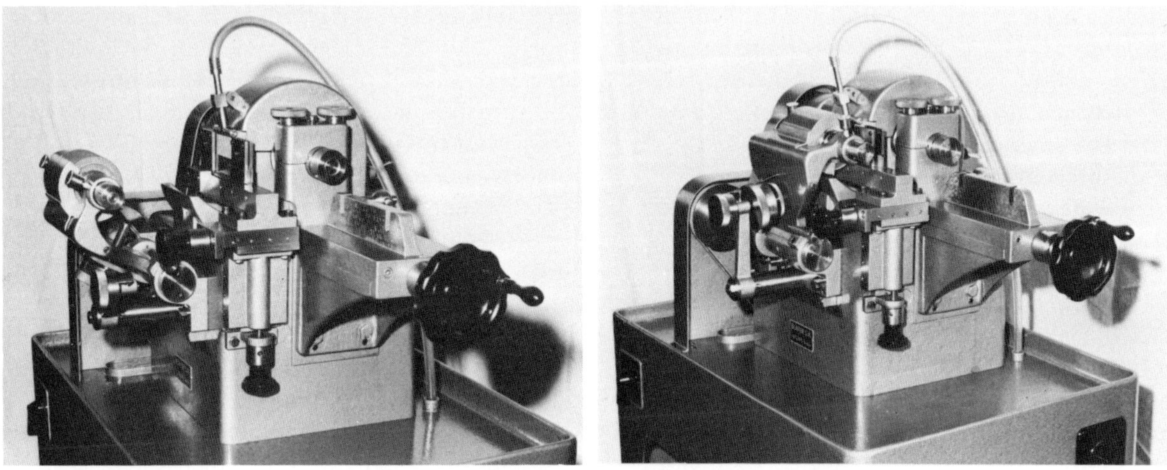

18.4.3. The Bunter centreless grinder in close-up.

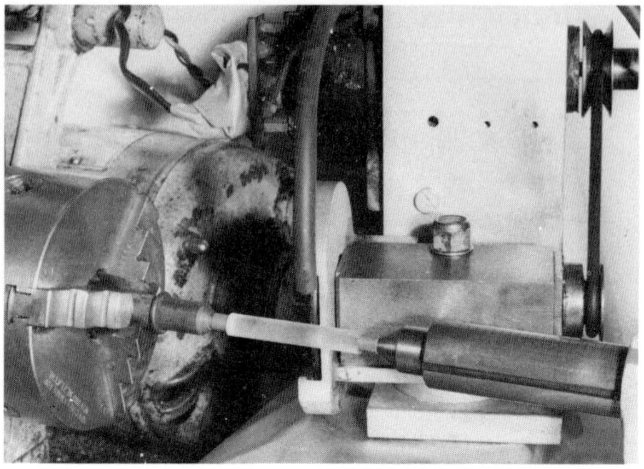

18.4.4. The grinding of a cylindrical crystal with Tufnol extensions.

convenient method but is mainly restricted to plunge-grinding crystals up to 2·25 inch (56 mm) long and varying in diameter from 0·08 inch (2·0 mm) to 0·05 inch (12 mm).

Fig. 18.4.3 shows the Bunter centreless grinder in close-up. The grinding of a cylindrical crystal with Tufnol extensions is carried out as shown in Fig. 18.4.4.

18.5 Polishing laser crystals

Owing to the toxic nature of many of the electro-optic crystals, it is necessary for all polishing machines to have an extractor fan for taking away poisonous gases. Separate water tanks and circulation systems are necessary because of the toxic by-products. A supply of filtered deionized water must be available.

The work-holding system for laser crystal polishing differs from that for lens polishing owing to the need to control parallelism to within a few seconds of arc. A pressure bias must be applied to the polisher to control the parallelism in a similar manner to the Hilger parallel-plate polishing machine. (See § 8.6.)

The methods of polishing in this section have been developed by Mr G. W. Fynn and Mr W. J. A. Powell of the Royal Radar Establishment, Malvern, Worcestershire, and published in 1969 as R.R.E. Tech. Note No. 709 (2nd edition), Very hard crystals are polished plane parallel singly or severally by the following methods. Single ruby rods are polished in a mild-steel lapping block approximately 1·25 inch (30 mm) in diameter (Fig. 18.5.1), which is faced at each end with either a ruby or sapphire ring or three separate disks as feet. It is important that these stabilizing feet are of the same crystal cut as the central laser rod. Araldite resin AY111 and HY111 is used to cement the ring or feet in place. It should be noted that there is no close support to the crystal edges.

When several uniform rods are to be polished, a mild-steel cylinder should be jig bored with one central and six surrounding holes, and axially shorter by 0·25 inch (6 mm) than the rods. The laser rods should be a close sliding fit in the block and protruding an equal amount from each end. They must be cemented in position. This arrangement results in fairly even wear on cast-iron laps with diamond abrasives and is satisfactory with hard crystals because mechanical length measurements can be taken on the optical faces of the six outer rods.

Softer crystals require close support to their periphery and the lapping blocks are more difficult to manufacture. The internal diameter of the packing washers must be worked to fairly close limits, to suit the rod, if the support is to be

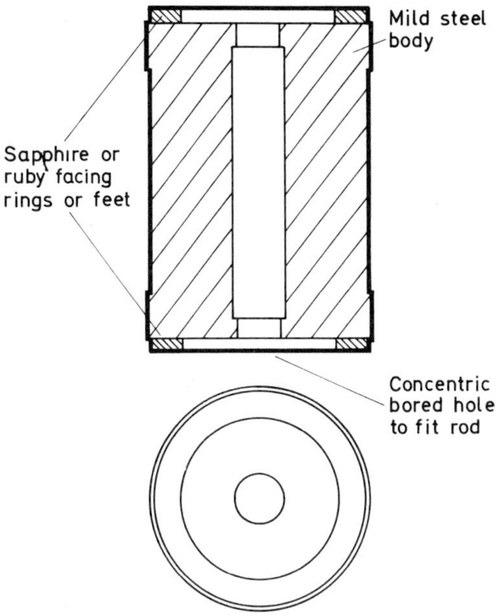

△**18.5.1. Section of a single ruby rod lapping block**

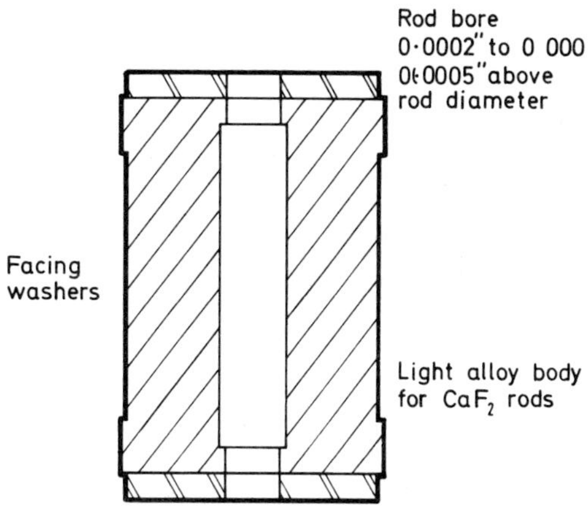

△**18.5.2. Section of a single rod lapping block.**

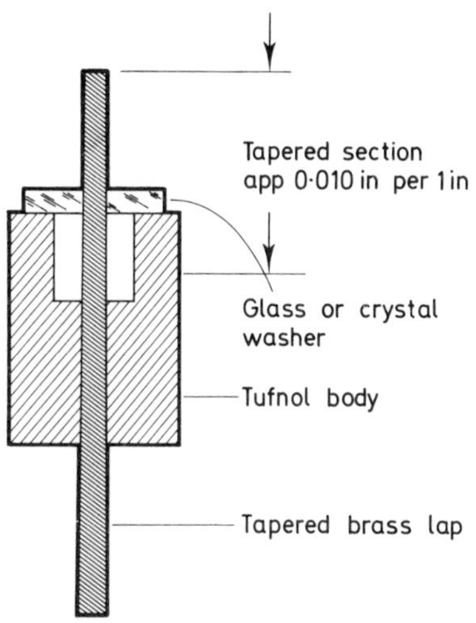

◁**18.5.3. Lapping jig for precise packing washers.**

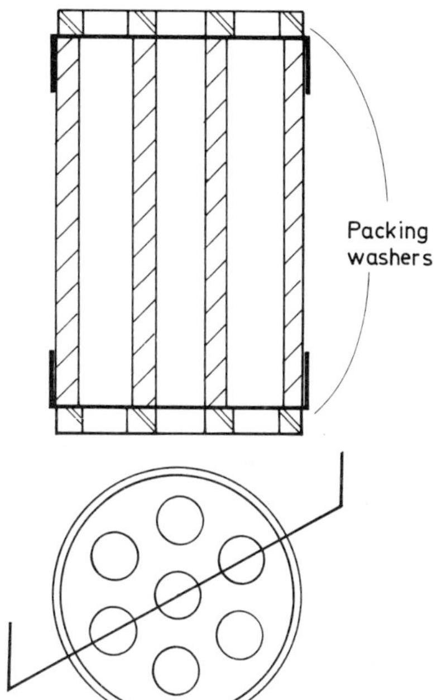

18.5.4.▷ **Section of a seven-rod lapping block for glass or soft crystal.**

of any real value in maintaining flatness to the extreme edge of the crystal—particularly during the later polishing stages. (See § 5.7.)

An aluminium alloy block should be turned cylindrical, bored concentrically, and its ends faced square to the axis of the hole. (Fig. 18.5.2.) Two glass washers are prepared by trepanning their central holes 0·005 inch (0·127 mm) below the laser rod size. The special lapping jig is required to bring the washer hole square and 0·001 inch (0·025 mm) below nominal rod size. The lapping operation consists of slowly rotating a brass-tapered plunger, coated with 25-micrometre diamond or BAO 303 abrasive, until its parallel section emerges on the far side of the washer. (Fig. 18.5.3.)

The register plug and hole are then coated with a release agent, such as wax polish, and the washer cemented to the block with Araldite. The register plug is then withdrawn and a tapered lapping plug used to size the washer hole finally slightly above the laser rod diameter. The second glass washer is registered, cemented, and lapped in the same way. For polishing quantities of identical soft crystals or glass rods, seven-way blocks can be made. (Fig. 18.5.4.)

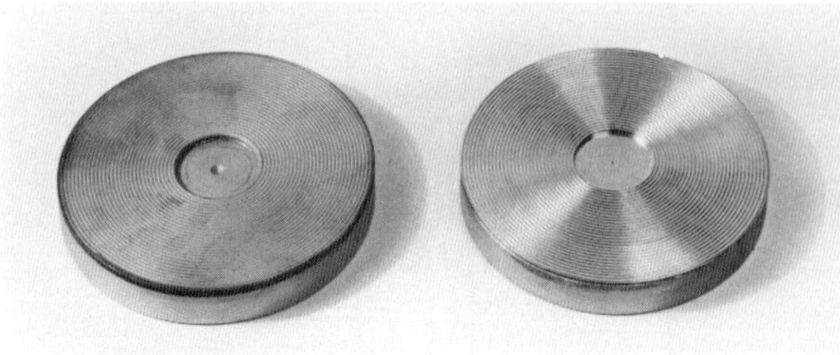

18.5.5. Pitch-faced and electro-deposited laps. Wax and pitch laps are for the final polishing of optical surfaces. Electro-deposited laps are used when the final polish is achieved with the help of an etching solution.

A washer of double the thickness required is prepared using the same technique as for a single block. It can then be sliced on a low-speed diamond saw to give two equal thickness washers which must be marked for identification. With care a concentrically turned housing will be made to within 0·0003 inch (0·007 mm) of parallel on a 1-inch (25 mm) measuring circle.

A final degree of parallelism to 0·00002 inch (0·0005 mm) is necessary and this is achieved by a machine, such as the Multipol polishing machine. When a lapping block is prepared the crystal rod and block are heated and optical stick wax worked into the extremely small gap between the rod and block to stop any chatter. As a general rule metal laps are used for polishing very hard crystals and pitch laps for the softer materials. Cast-iron Meehanite laps are used for lapping ruby rods with a diamond abrasive and, after turning and grinding, they are lapped on a Lapmaster machine to within one light band. No grooving is necessary for blocks of open rubies but with larger and continuous lapped surfaces three annular grooves in the surface are needed to interrupt the lubricant film.

With some specimens of ruby and sapphire it is often difficult to produce a good, sleek-free surface, and copper laps will give a noticeable improvement in finish when used with 8- and 3-micrometre diamond.

Softer laser crystal materials must be finished with pitch laps. (See §§ 3.2 and 4.7.) Wax laps can give surfaces freer from sleeks than would normally be possible with pitch, but they are more difficult to form and flatten. Wax laps must always be fed with a liberal supply of abrasive and must not dry to the point of damp dry which is sometimes considered ideal for pitch. (Fig. 18.5.5.)

The Multipol machine

The Royal Radar Establishment have developed a polishing machine suitable for crystals, and this is now marketed by Metals Research Ltd, Melbourn, Royston, Herts.

The Multipol polishing machine is basically of the Draper type, modified at the polishing head to allow an eccentric bias pressure to be applied to the crystal polishing block. This gives a fine control of parallelism during the lapping and polishing stages.

18.5.6. The Multipol precision polishing machine is constructed of stainless steel and P.T.F.E. for all parts liable to chemical attack. The lap capacity is 8 inches (20 cm) diameter, and it has an infinitely variable rotation speed up to 400 r.p.m. and stroke of ± 2 degrees.

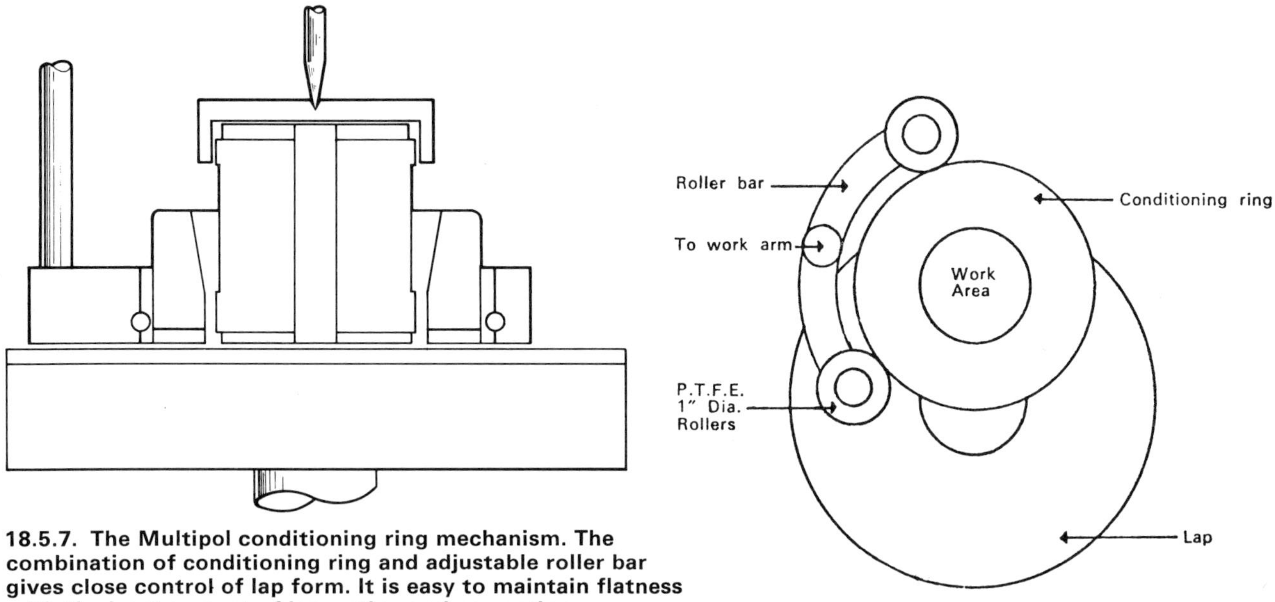

18.5.7. The Multipol conditioning ring mechanism. The combination of conditioning ring and adjustable roller bar gives close control of lap form. It is easy to maintain flatness with the least amount of inspection and correction.

A glass-faced conditioning ring, about one-half to two-thirds times the diameter of the lap, is used to maintain flatness of the lap. The ring runs between the two p.t.f.e. rollers at the base of a roller bar. The stroke is operated on one side only of the lap centre, and in this way any high or low zone formed by load differences between the work and the conditioning ring will be ironed out by the ring. (Fig. 18.5.6.) Thus instead of a lightly loaded specimen creating a high zone on the lap, the conditioning ring will cause considerable corrective wear to be imposed on this zone. Conversely, a relatively heavily loaded workpiece will not form a low zone since the conditioning ring will again compensate and wear will take place on the two bands on either side of the low zone.

The stroke of the sweep arm can be adjusted by first sliding back the cover door on the left-hand side of the machine case and manually turning the visible pulleys until the knurled screw is visible through the side aperture. If this screw is turned anticlockwise the stroke will be increased up to 2°, and vice versa.

The mean position of the sweep can be adjusted by turning the large knurled knob at the pivot end of the sweep arm. This control will govern the length of time the workpiece spends over the edge of the lap and must be carefully set to ensure that the lap is not being 'turned down' on the edge.

Laps are attached to the machine by screwing on the centre drive spindle. Waste lapping and polishing lubricants are released from the stainless bowl by removing the small plug situated on the right. The lubricants are released through a nozzle on the right-hand side of the machine, and it may be useful to attach a small length of pipe to this to direct the flow into some vessel.

To attach the push rod assembly to the machine, the free end of the sweep arm must be unclamped and tilted back. The post can then be located in the sweep arm and locked in position by the clamp. The end of the sweep arm can now be lowered back to the horizontal position so that the post is vertical and about 0·05 inch from the lap surface.

The spring-loaded piston assembly is then slid on to the post and its height adjusted to allow correct positioning of the push rod under the p.t.f.e. ring piston and pressing down on to the workpiece. Location of the push rod on to the workpiece is by a small dimple drilled in the back of the specimen.

When the push rod is being used it is the conditioning ring which retains the workpiece in position on the lap. (Figs 18.5.6 and 18.5.7.) The load can be applied to the work by screwing in a clockwise direction the large knurled screw on the piston assembly. This applies load by increasing the compression of the spring within the piston head.

To give bias loading on the workpiece the push-rod can be offset from the centre position on the workpiece. Thus, by drilling a series of dimples in an eccentric pattern on the back of the workpiece

18.5.8. Bias loading. Where the faces of the work must be made parallel, as in laser rods, the Multipol has a push-rod spring-arm attachment that allows bias loading of the specimen.

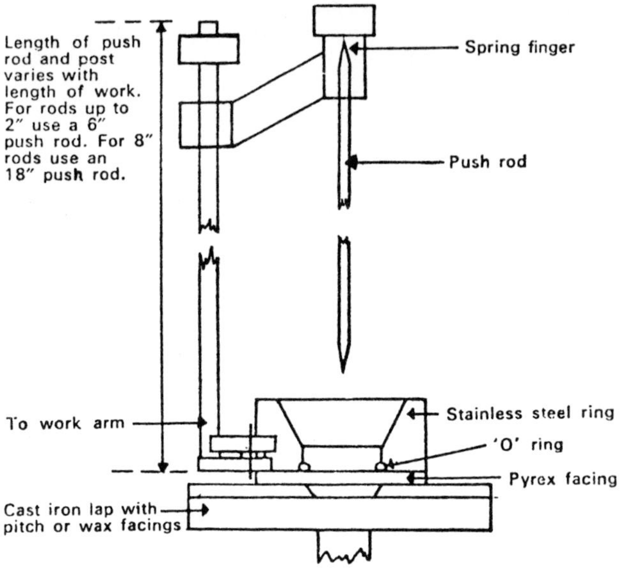

or cover cap the push-rod can be repositioned to give bias loading. This is a most desirable feature when it is necessary to polish to fine tolerance of parallelism. (Fig. 18.5.8.)

For general polishing of samples where extreme flatness and parallelism is not important, the ball-point attachment can be used. This is located on the sweep arm of the machine and clamped into position by means of the locking clamp. A small dimple should be drilled into the back of the specimen to be polished and the ball-point stem locates into this. Loading the specimen is done by placing weights on the vertical pillar on top of the ball-point stem. This attachment is very useful when it is desired to polish several specimens at once. The specimens can be stuck to a flat plate which has a small dimple drilled in the back of it. This will act as the location point for the ball-point end.

With the use of the precision polishing jig it is possible to produce specimens, either in disk or rod form, which are exactly flat and parallel. Fig. 18.5.9 shows a cut-away diagram of the polishing jig. The glass-faced conditioning ring in this case is attached to the bottom face of the jig and the complete jig runs between the p.t.f.e. rollers at the base of the push-rod post.

The specimen to be polished is waxed to the centre plate using an optical wax. By unscrewing the bottom section the distance from the glass face to the centre plate is increased, and the jig is designed to take a specimen of up to 1-inch depth. Long laser rods can be polished by inserting them up the hollow centre tube and waxing them in. Rods of up to $\frac{7}{16}$ inch can be accommodated by removing the three Allen screws which retain the centre polishing plate.

To load the specimen, disk or rod, the large knurled centre screw should be turned clockwise. This applies load by increasing the compression of the internal spring in the screw head.

To adjust the angular tilt of the centre plate holding the specimen there are three knurled screws positioned around the centre column. Screwing these up or down will alter the tilt of the specimen.

All bearing surfaces in the jig are protected from abrasives by a flexible diaphragm as shown in Fig. 18.5.9. Periodical lubrication of the main spindle bearing will also give longer life of the accurately machined parts. This is done through an oil filler on the central column, access to which is gained by rotating the small hole in the outer cylinder case until it corresponds with the oil filler.

Meehanite cast-iron laps are used for lapping and polishing ruby rods and for lapping glass crystal materials to the grey state. The working faces of the laps are machined and surface ground and then flattened to within one light band on a Lapmaster (registered trade mark) machine. The

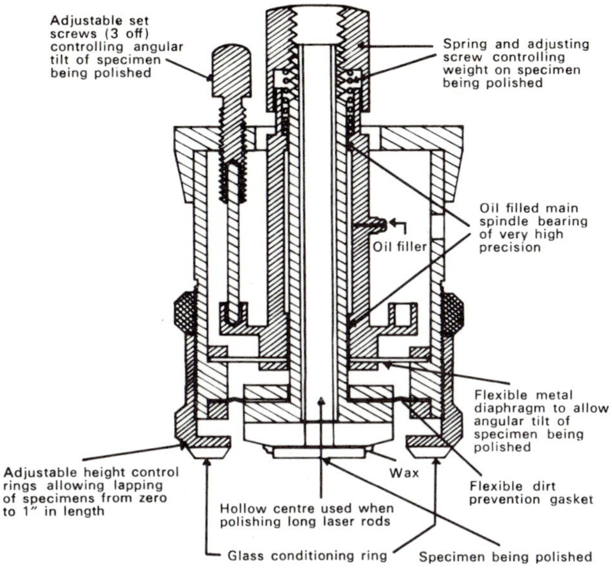

18.5.9. Precision polishing jig. With this attachment, it is possible to produce specimens either in disk or in rod form and exactly flat and parallel. The stand is used when the surfaces are being tested for flatness and parallelism.

cast-iron laps are used with diamond abrasives for lapping and polishing ruby, but various grades of emery or aloxite can be used for glass and crystal materials if the lap is very carefully cleaned between grades.

When resurfacing an old lap, the top 0·010 inch should be machined off before grinding and lapping so that all abrasive compounds are completely removed. For good sleek-free surfaces, when polishing some ruby and sapphire materials, it may be necessary to use the softer surface of a copper-faced lap. These laps maintain flatness well and can be loaded with 8- and 3-micrometre diamond abrasive.

Laps are usually made by electro-depositing soft materials, e.g. (copper, tin, indium, etc.) on to a stainless-steel master lap. Thus the surface of the final lap will have the properties of the deposited metal, but the great advantage of this type of lap is that resurfacing can be achieved very easily by stripping the electro-deposit from the stainless-steel master and re-electroplating.

The minimum fluid lap is used for etch polishing when it is found that benefit is gained by etching at the same time as polishing. The lap is made on a stainless-steel base and has a p.t.f.e. bowl attached to it to prevent the etchant flowing off the lap. The etchant is usually a weak (0·1 per cent–0·01 per cent) concentration of bromine in methanol or ethanol. A commercial mixture which is available is Lustrox, which is based on sodium hypochlorite with additions of sodium silicate and zirconium dioxide as a mechanical polishing agent. The etchant is used diluted with water in a ratio of 1:1 to 1:3. However there is a possibility of etch pits being formed when a material such as gallium arsenide is polished.

The table overleaf summarizes the polishing processes recommended for various materials on the Multipol machine.

Table 18.5.1. Materials polished on the Multipol machine

Material	Scratch hardness (*Moh's scale*)	*For grey finish* *Lap*	*Final abrasive*	*For polished finish* *Lap*	*Abrasive*
Ruby, sapphire	9	Cast iron	15μ diamond	Soft metal: electroformed copper or tin; or cast solder	8μ, 3μ, 1μ diamond
Spinel	8	Cast iron	Carborundum 600	1 mm pitch, wood loaded	Linde A
Quartz	7	Cast iron	Carb. 600	2 mm pitch, wood loaded	Cerium oxide
Silicon	7	Cast iron	303½ emery or 1200 carb.	(a) Cloths or (b) Wax	(a) 8μ, 3μ, 1μ diamond or (b) Linde A
Rutile Y.A.G., E.I.G.	6·5–7	Cast iron	Carb. 600	0·3 mm–0·5 mm wood loaded pitch	Linde A
Laser glasses	circa 6	Cast iron	Emery 303½ or 1200 carb.	2·5 mm pitch, wood loaded	Cerium oxide
Periclase (MgO)	5·5–6·0	Cast iron	Carb. 600	Wood loaded 3 mm pitch	Linde A
Germanium	5	Cast iron	Emery 303½	(a) Soft metal cloth or (b) Wax	(a) 2μ, 1μ, D.A. (b) Linde A
CaMo	5	Cast iron	Emery 303½	2·5 mm pitch	Linde A
Barium strontium niobate	5	Cast iron	Emery 303½	0·5 mm pitch* and beeswax	Linde A
$CaWo_3$	4·5	Cast iron	Emery 303½	(a) 2·5 mm pitch wax or (b) 4 mm pitch	Linde A colloidal Fibrils 100 A
Lithium niobate	4·5	Cast iron	Emery 303½	Solder or electrodeposited tin	3μ, 1μ diamond
CaF_2 CdF_2	4	Cast iron	Emery 303½	2·5 mm pitch	Linde A
LaF_3 CeF_3	4–4·2	Cast iron	Emery 303½	Wood loaded 2·5 mm pitch	Linde A
InSb	3–3·5	Cast iron	Emery 303½	(a) Cloths or (b) Wax	(a) 3μ, 1μ diamond or (b) Linde A
PbTe	3·0	Cast iron	Emery 303½	Wood loaded 5 mm pitch	Linde A
Calcite A:Se:glasses	3·0	Cast iron	Emery 303½	Wood loaded 4 mm pitch	Linde A
GaAs	2·5–3·0	Cast iron	Emery 303½	(a) D.P. cloth (b) Micro cloth	(a) 3μ, 1μ diamond (b) Linde A
KDP, ADP, KBr, KCl, NaCl	2–2·5	Cast iron	Aloxite 175, 125 and 50 if possible	Wax	Linde A finish on dry Selvyt cloth
Proustite	2·0–2·5	Cast iron	Emery 303½ (with oil)	(a) Wax or (b) Cast or electro-deposited indium	(a) Linde A (b) 3μ, 1μ diamond
Cd:Hg:Te	2	Cast iron	Emery 303½	Wax	Linde A
AgCl	1–1·5	Cast iron	Aloxite 175 to 50	Wax	Linde A
Electroless nickel and S/steels	5–8	Lapmaster cast iron plates	Lapmaster 1900 AL203	4 mm pure pitch	Linde A
LiS	2·5	Cast iron	Aloxite 175 to 50	Solder or electrodeposited soft metal	3μ diamond
Antimony sulphur iodide (SbSI)	2	(From sawn surfaces)		Wax	Linde A
Ferrites	circa 7	Cast iron	Carborundum 600	(a) Pitch or (b) Soft metal (solder)	(a) Linde A or (b) 3μ diamond
T.G.S.	circa 2	Cast iron	'Bramet' paper 600	Soft cloth ('Pellon K')	3μ diamond

Lubricant	*Approx load g.cm^{-2}*	*Typical polishing time (mins)*	*Remarks*
Hyprez fluid	100	60	Chamfer polish is critical for sleek free finish. Abrasive is fed in minute quantities every 2–3 mins.
Water 5% Na_2CO_3	100	30	Would be speeded up with 3μ, 1μ diamond on soft metal laps.
Water 5% Na_2CO_3	100	60	
Water 5% Na_2CO_3	150	30	Etch-polishing most common for damage-free surfaces. Otherwise wax laps for least edge turn-down.
Water 5% Na_2CO_3	100	15–30	All polish well on soft metal laps.
Water 5% Na_2CO_3	50	60	Generally, glass polishing techniques. Little benefit from finishing on wax.
Water 5% Na_2CO_3	250	180	
Water 5% Na_2CO_3	150	30	Etch-polish for final damage free surfaces. Wax laps for least turn-down.
Water 5% Na_2CO_3	75	45	Wax laps may remove final sleeks. Avoid thermal shock.
Water 5% Na_2CO_3	250	60	
Water 5% Na_2CO_3	50	45	Avoid thermal shock.
Hyprez fluid	150	15	
Water 5% Na_2CO_3	50	45	Would probably respond more quickly to metal-lap treatment. Avoid thermal shock.
Water 5% Na_2CO_3	50	45	
Water 5% Na_2CO_3	75	30	Etch-polishing can be advantageous at final stage.
Water 5% Na_2CO_3	75	30	
Water 5% Na_2CO_3	100	15	
Water 5% Na_2CO_3	50	45	Final etch-polish for 10 mins. on micro cloth and hypochlorite soin.
Ethane diol	100	30	Water soluble crystals.
Ethane diol or Hyprez fluid	75	60	Clean polished surface with chamois leather.
Water	60	30	Final etch polishing will improve surface condition.
Ethane diol	60	15	Sparingly soluble in H_2O
Water 5% Na_2CO_3	150	60	
Ethane diol	100	30	Water soluble.
Water Na_2CO_3	90	30	Can be polished on side or ends of needles.
Water or Hyprez fluid	200	60	
Hyprez fluid	100	5	Water soluble crystal. Can be mounted with 'raw' pitch and released with benzene or toluene.

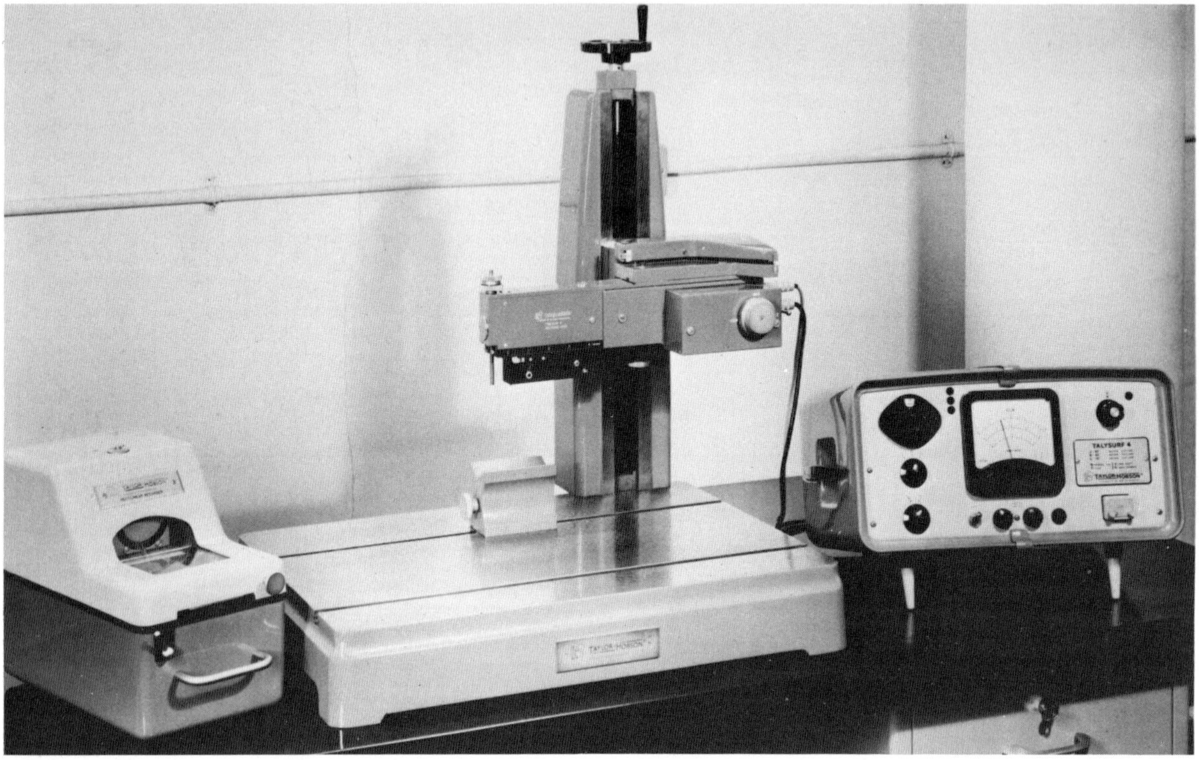

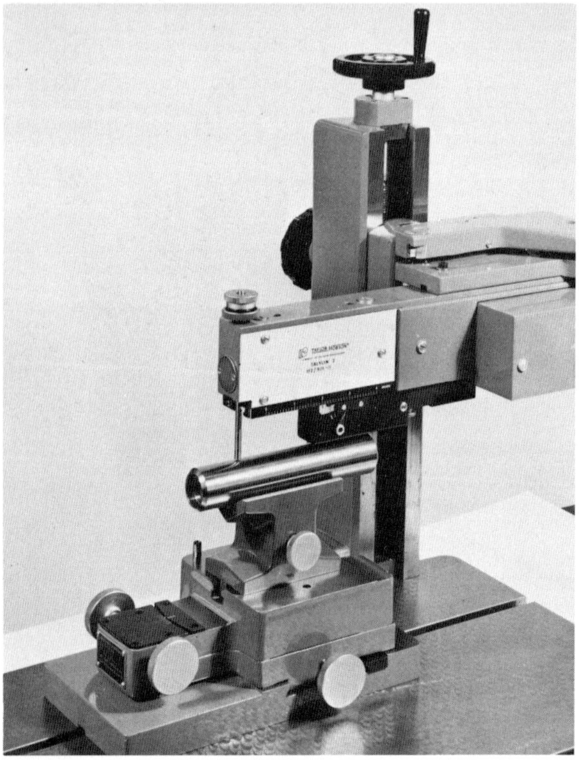

18.5.10. The Talylin measuring unit comprises an electrical displacement pick-up, drive carriage, motorized gearbox, and an accurate datum plate in the form of an optically polished straight-edge. The recorder provides graphs for which the ordinates are straight and the scale is linear. (*Left*) The pick-up is power-driven over a traverse length of up to 4 inches (100 mm) and draws a stylus across the surface to be measured. Movement caused by errors in the surface is amplified electronically.

Measuring parallelism

When laser crystals are at the final stage of polishing, and parallelism to 3 seconds of arc and flatness to one-tenth of a fringe are required, a process of alternate measurement and correction by the application of an eccentric lapping load during working is necessary.

A stand is provided to support the jig when it is required to check the flatness and parallelism of specimens. This may be done by the use of an autocollimator or alternatively a suitable instrument for the measurement of parallelism is the Taylor–Hobson Talylin straightness-measuring instrument, which has a vertical magnification of $\times$ 10 000. (Fig. 18.5.10.)

18.5.11.▷
Netting press for flattening and netting the pitch polisher.

18.5.12.▷
Typical fringe patterns on the faces of seven rubies.

Netting press

A netting press for flattening and netting the pitch polisher is shown in Fig. 18.5.11. A small square of wet net is inserted between the cell pitch and a water-warmed stainless steel optical flat. It is then stripped from the pitch as soon as the lap is removed from the hot plate and before cooling makes this impossible.

18.6 Polishing crystal slices

For the production polishing of silicon or gallium arsenide slices, similar polishing conditions are required with some detail changes in polishing materials.

A suitable machine for this purpose is the Engis Mark III, designed to allow both lapping and polishing with Hyprez diamond compounds and abrasive slurries; it is shown in Fig. 18.6.1. A spray bar washes away waste materials and slurries through a central sump to the waste outlet, and a flexible tube, with fine control valve, feeds distilled water and lubricating fluids to the lapping plates.

The driving plate has been gear-cut on the periphery to drive the automatic lapping attachment, which is used in conjunction with lift-off

18.6.1. The Engis Mark III lapping and polishing machine. Three heads are arranged for lapping and polishing semi-conductor slices, which are mounted on stainless steel work-holding blocks.

disks. The disks are carriers for self-adhesive grinding disks or polishing cloths.

Silicon slices can be lapped and polished in a two-stage operation, each lasting about 20 minutes, with the lapping and pre-polishing in the first stage, and the final polishing in the second stage.

As a carrier for the diamond compound a self-adhesive close-napped polishing cloth (Microcloth) is attached to a 12-inch diameter cast-iron lapping plate. (§§ 4.5 and 4.8.)

During the first lapping stage of 15 minutes, the weights are loaded on the silicon slice blocks and the polishing cloth is impregnated with 1·6 g of 1-micrometre particle size Hyprez diamond compound formula L (IL 05/36) and 20 ml of Hyprez lubricating fluid W.S. Pre-polishing should continue for 5 minutes without the weights in order to increase the degree of finish.

The second stage consists of two polishing cycles, the first of 10 minutes with weights on the work-holding blocks and with 1·6 g of $\frac{1}{4}$-micrometre particle size Hyprez diamond formula L and 20 ml of alkaline A.D.W.S. fluid.

The second polishing cycle is at a lapping plate speed of 165 r.p.m. (413 r.p.m. for the work-holding block without the weights) for 5 minutes. At the end of this final polish, the surface will be practically scratch free at × 500 magnification.

Gallium phosphide slices can be polished by use of the normal lapping technique and keeping the abrasive very fluid and at very low pressure. The first lapping stage should be at a lap speed of 200 r.p.m., with 6-micrometre diamond on Polimetal C polishing cloth, care being taken not to overheat the gallium phosphide specimens.

The second lapping stage should include polishing with, progressively, 3-micrometre, 1-micrometre and $\frac{1}{4}$-micrometre diamond on Polimetal C or Microcloth, Hyprez spray diamond lapping compound and Type W lubricating fluid. The Engis Mark II polishing machine has a fibreglass body which is not attacked by the caustic solutions used in chemical polishing.

Appendix I Bibliography

The reference books are listed in alphabetical order of author in each chapter to which they apply.

Chapter 1

BEDINI, SILVIO A., 'Lens Making for Scientific Instrumentation in the Seventeenth Century', *Appl. Opt.*, **5**, No. 5, May 1966.

Chapter 2

CORNISH, D. C., 'The Mechanism of Glass Polishing', SIRA Institute.

HOLLAND, L., 'The Properties of Glass Surfaces', Chapman and Hall, 1964.

ISHIDA, YASUHIRO, 'High-speed Polishing of Optical Glass', *Journal of Mechanical Laboratory of Japan*, **10**, No. 2, 1964.

TEXEREAU, JEAN, 'Les Principaux Défauts Réels des Surfaces Optiques Engendrées par Différentes Techniques de Polissage', extracts from the bulletin *Ciel et Terre* of the Societé Belge d'Astronomie de Météorologie et de Physique du Globe, March–April 1950.

Chapter 3

MOORE, WAYNE R., 'Foundations of Mechanical Accuracy', The Moore Special Tool Co., Connecticut.

STRONG, JOHN, 'Modern Laboratory Practice', Blackie & Son Ltd.

Chapter 4

BURRIDGE, A. D., 'Microscope on Performance', De Beers Diamond Research Laboratory.

HEPWORTH, M. A., and THOMPSON, R., 'Machining Glass and Ceramics', *Machine Shop and Engineering Manager*, Dec. 1968.

HUSH, J. S., 'Diamond Abrasives in the Optical Industry', *Manufacturing Optics International*, April 1970.

ISHIDA, YASUHIRO, 'High-speed Polishing of Optical Glass', *Journal of Mechanical Laboratory of Japan*, **10**, No. 2, 1964.

PROSSER, J., 'The Versatility of Electro-plated Diamond Wheels', *UNI Review*, **6**, 1970.

SHEKHTER, YU. N., *et al.*, 'Cutting Fluids for Grinding Glass', *Machines and Tooling*, **XL**, No. 2.

SAXENA, NIGAM, SENCUPTA, 'Fungal Attack of Optical Instruments and its Prevention', *Indian J. of Technology*, **1** (7), 283, 1963.

SWAN, R. J., 'The Role of Diamond Polishing in Optical Manufacture', *Manufacturing Optics International*, April 1970.

Chapter 5

BABCOCK, DR C. L., 'Transparent Zero-expansion Materials for Optical Uses', presented at the New England Section Meeting of the Optical Society of America in Boston, Mass., May 1966.

BRICE, J. C., 'The Growth of Crystals from the Melt', Vol. V, Selected Topics in Solid State Physics, North-Holland Publishing Company.

BUERGER, M. J., 'The Stuffed Derivatives of the Silica Structures', *Am. Mineral.*, **39**, 600, 1954.

COX, ROBERT E., 'Working a Mirror Blank of "Zero"-expansion Coefficient', *Sky and Telescope*, **32**, No. 3, 2, Sept. 1966.

FINCHAM, W. H. A., 'Optics', Hatton Press.

FYNN, G. W., and POWELL, W. J. A., 'Polishing Lithium Fluoride for High Transmission in the Ultraviolet', *Journal of Physics, E. Scientific Instruments*, **4**, 1971.

GILMAN, J. J., 'The Art and Science of Growing Crystals', John Wiley and Sons Inc., New York.

HOUSTON, R. A., 'Physical Optics', Blackie & Son Ltd.

LONGHURST, R. S., 'Geometrical and Physical Optics', Longmans, Green and Co., Ltd.

MCMILLAN, P. W., 'Glass-ceramics', Academic Press, New York, 1964.

NEILSON, GEORGE F., Jr., 'Small-angle X-ray Scattering Study of Nucleation and Devitrification in a Glass-ceramic Material', presented at 1966 Pittsburgh Diffraction Conference.

PINCUS, ALEXIS G., 'Melt-formed Ceramics', *Frontier* (published by I.I.T. Research Institute), Autumn 1966.

POWELL, L. S., 'Light', *Physics*, Vol. III, Pitman.

REDWINE, R. H., and CONRAD, M. A., 'Microstructures developed in Crystallized Glass-ceramics', presented at Third International Symposium on Materials, June 1966, at the Univ. of Calif., Berkeley. The proceedings will be published by John Wiley and Sons.

SHUBNIKOV, A. V., and SHEFTAL, N. N., 'Growth of Crystals', Vol. 3, Consultants Bureau.

SHAND, E. B., 'Glass Engineering Handbook', McGraw-Hill Book Company.

SMITH, WARREN J., 'Modern Optical Engineering', McGraw-Hill Book Company.

Chapter 6

Hopkins, H. H., and Tiziani, H. J., 'A Theoretical and Experimental Study of Lens Centring Errors and their Influence on Optical Image Quality', *Brit. J. appl. Physics*, **17**, 1966.

Martin, L. C., 'Technical Optics', Sir Isaac Pitman & Sons Ltd.

Twyman, F., 'Prism and Lens Making', Adam Hilger Ltd.

Chapter 7

Emsley and Swaine, 'Ophthalmic Lenses', Hatton Press, 1946.

Chapter 8

Dickinson, C. S., 'A Method of Flatness Control in Optical Polishing', *J. scient. Instrum.*, Series 2, **1**, 1968.

Garratt, G., 'Producing Optical Polygons', *Machinery and Production Engineering*, 13 November 1968.

Otte, G., 'An Improved Method for the Production of Optically Flat Surfaces', *J. scient. Instrum.*, **42**, 1965.

— 'The Use of Teflon Polishers for Precision Optical Flats', *J. scient. Instrum.*, Series 2, **2**, 1969.

Rumsey, N. J., 'The Effect of Eccentric Loading of Optical Flats during Grinding or Polishing', *J. scient. Instrum.*, **32**, 338–9, 1955.

Chapter 9

Radford, J. D., and Richardson, D. B., 'The Management of Production', Macmillan Press, London, 1968.

Shaw, Anne G., 'The Purpose and Practice of Motion Study', Harlequin Press.

Chapter 10

Bracey, F. J., 'The Technique of Optical Instrument Design', The English Universities Press Ltd.

Burch, J. M., *Nature*, **171**, 889, 1953.

Cox, A., and Royston, M. F., 'The Manufacture of Aspheric Surfaces', Proceeding of I.C.O. Conference, London, 1961, Bell & Howell Co., Chicago.

Hines, R. L., *J. Appl. Phys.*, **28**, 587, 1957.

Holland, L., 'Machining and Etching Targets by Impact Sputtering with Ion Beams', V. Jugoslovanski Vakuumski Kongres, Portoroz, May 1971.

James, W. E., and Waterworth, M. D., *Optica Acta*, **12**, 3, 223, 1965.

Kumanin, K. G., 'Generation of Optical Surfaces'; Ch. VIII, 'Methods of Generating Aspheric Surfaces', The Focal Library.

Larmer, J. W., and Goldstein, E., 'Some Comments upon Current Optical Shop Practices', *Appl. Opt.*, **5**, No. 5, May 1966.

Martin, L. C., 'Technical Optics', Vol. II, Sir Isaac Pitman and Sons Ltd.

Meinel, A. B., Baskin, S., and Loomis, D. A., *Appl. Opt.*, **4**, 1674, 1965.

Narodny, Leo H., 'Ion Beam Technology', Gordon & Breach, 1971.

Perry, W. H., 'Simple Aspherical Surface Generator', *Appl. Opt.*, **5**, No. 5, May 1966.

Random, George, and Wallerstein, Edward P., 'Air Gauge Measurement and Driven Lap Polishing in the Production of Aspheric Surfaces', *Appl. Opt.*, **5**, No. 5, May 1966.

Rawstron, G. O., 'Improvements in or relating to Cam Devices', Patent No. 93951.

— 'Improvement in or relating to the Manufacture of Optical Cam Devices', Patent No. 936952.

— 'Improvements in or relating to Polishing', Patent No. 940732.

— 'Improvements in or relating to Polishing', Patent No. 947176.

— 'Improvements in or relating to Polishing', Patent No. 947177.

Rawstron, G. O., and Reason, R. E., 'Improvements in or relating to the Generation of Shaped Surfaces', Patent No. 947174.

— — 'Improvements in or relating to the Generation of Aspheric Surfaces', Patent No. 947175.

Schmidt, F. J., 'Electroforming of Large Mirrors', *Appl. Opt.*, **5**, No. 5, May 1966.

Strong, J., 'Modern Physical Laboratory Practice',

Strong, J., and Gaviola, E., *J. opt. Soc. Am.*, **26**, 4, 153, 1936.

Williams, T. L., *Inst. Pract.*, **19**, 8, 773, 1965.

Chapter 11

BS 1134:1961, '"Centre-Line-Average" Height Method for the Assessment of Surface Texture', British Standards Institution.

BS 4301:1968, 'Recommendations for the Preparation of Drawings for Optical Elements and Systems', British Standards Institution.

DIN 3140, Deutscher Normenausschuss, Berlin W.15.

MIL-STD-34, United States Government Printing Office, Washington.

British Calibration Service, Document 0715, May 1969, 'Supplementary Criteria for Laboratory Measurement of the Optical Transfer Function of Optical Systems'.

Baker, L. R., 'Properties and Applications of the New Position-Sensitive Photocells', Part 3, *Control*, **5**, 81, 1962.

— 'Automatic Recording Instrument for Measuring Optical Transfer Functions', *Jap. J. appl. Phys.*, Suppl. 1, **4**, 146, 1965.

Baker, L. R., and Moss, T., 'Electro-Optical Methods of Image Evaluation', Electro-Optical Systems Design Conference, New York, Sept. 1969.

Baker, L. R., and Whyte, J. W., 'New Instrument for Assessing Lens Quality by Pupil-scanning (Spot Diagram Generation)', *Jap. J. appl. Phys.*, Suppl. 1, **4**, 121, 1965.

Barrell, H., and Marriner, R., 'A Liquid Surface Interferometer', *Br. Sci. News*, **2**, 130–2, 1949.

Barrell, H., and Preston, J. S., 'An Improved Beam Divider for Fizeau Interferometers', *Proc. Phys. Soc.* (*B*), **64**, 97–104, 1951.

Birch, K. G., 'A Survey of OTF-based Criteria as used in

the Specification of Image Quality', *Nat. Phys. Lab. Op. Met.*, 5 April, 1968.

BROCK, G. C., 'A Review of Current Image Evaluation Techniques', *J. Photogr. Sc.*, **16**, 241, 1968.

— 'Reflections on Thirty Years of Image Evaluation', *Photogr. Sci. Engng*, **11**, 356, 1967.

BROWN, D. S., 'The Application of Shearing Interferometry to Routine Optical Testing', *J. scient. Instrum.*, **32**, 1955.

— 'Radial Shear Interferometry', *J. scient. Instrum.*, **39**, 1962.

BROWN, RONALD, 'Lasers', Aldus Books, London, 1968.

CLAPHAM, P. B., and DEW, G. D., 'Surface-coated Reference Flats for Testing Fully-aluminised Surfaces by means of the Fizeau Interferometer', *J. scient. Instrum.*, **44**, 1967.

DE VELIS, J. B., and REYNOLDS, G. O., 'Theory and Applications of Holography', Addison-Wesley.

DEW, G. D., 'A Method for the Precise Evaluation of Interferograms', *J. scient. Instrum.*, **41**, 160–2, 1964.

— 'The Measurement of Optical Flatness', *J. scient. Instrum.*, **43**, 409–15, 1966.

— 'Systems of Minimum Deflection Supports for Optical Flats', *J. scient. Instrum.*, **43**, 809–11, 1966.

DYSON, J., 'The Rapid Measurement of Photographic Records of Interference Fringes', *Appl. Opt.* **2**, 487–9, 1963.

— 'Interferometry as a Measuring Tool', The Machinery Publishing Co. Ltd.

EMERY, JACKSON K., 'How to Measure Flatness', *American Machinist*, August 1966.

FIZEAU, H., *C.R.Ac. Sci. Paris*, **54**, 1237, 1862.

GATES, J. W., 'A Slow Motion Adjustment for Horizontal Interferometer Mirrors', *J. scient. Instrum.*, **30**, 484, 1953.

— 'The Evaluation of Interferograms by Displacement and Stereoscopic Methods', *B.J. appl. Phys.*, **5**, 133–5, 1954

HARPER, D. W., 'A Comparative Method of Checking Angles on Glass Samples', *J. scient. Instrum.*, **2**, No. 1, 1968.

HUME, K. J., 'Metrology with Autocollimators', Adam Hilger Ltd.

LONGHURST, R. S., 'Geometrical and Physical Optics', Longmans, Green and Co. Ltd.

LITTLE, R. E., 'Practical Considerations in Specifying Lens Performance', *Opt. Technol.*, May 1969.

MARTIN, L. C., 'Technical Optics', Vol. II, Sir Isaac Pitman & Sons.

MIYAMOTO, K., 'Progress in Optics', Vol. 1, ed. E. Wolf, North Holland Pub. Co., Amsterdam, 1961.

PRIMAK, W., 'The Determination of the Absolute Contours of Optical Flats', *Appl. Opt.*, **6**, No. 11, 1967.

RAYLEIGH, LORD, 'Interference Bands and their Applications', *Nature, Lond.*, **48**, 212–4, 1893.

ROESLER, F. L., and TRAUB, W., 'Precision Mapping of Pairs of Uncoated Optical Flats', *Appl. Opt.*, **5**, 463–8, 1966.

ROSENHAUER, K., and ROSENBRUCH, K. J., 'Reports on Progress in Physics, Vol. XXX, Part 1, The Measurement of the Optical Transfer Function', 1967.

SMITH, WARREN J., 'Modern Optical Engineering', McGraw-Hill Book Company.

SYDENHAM, P. H., 'Position-sensitive Photocells and their Application to Static and Dimensional Metrology', *Optica Acta*, **16**, 377, 1969.

TAYLOR, W. G. A., 'Spherical Aberration in the Fizeau Interferometer', *J. scient. Instrum.*, **34**, 399–402, 1957.

TIMOSHENKO, S., and WOINOWSKY-KRIEGER, S., 'Theory of Plates and Shells', p. 295, 2nd edn., New York, McGraw-Hill, 1959.

TOLANSKY, S., 'Multiple Beam Interferometry of Surfaces and Films', Oxford, Clarendon Press, 1948.

WILLIAMS, T. L., 'A Spot Diagram Generator for Lens Testing', *Optica Acta*, **15**, 553, 1968.

WILLIAMS, T. L., and ASHTON, A., 'The Use of Standard Test Lenses for Verifying the Accuracy of O.T.F. Equipment', *Appl. Opt.*, Oct. 1969.

Chapter 12

ANDERS, H., 'Thin Films in Optics', Focal Press.

BATH, H. H. A., ENGLISH, J., and STECKELMACHER, W., *Electronic Components*, 239–47, March 1966.

CANFIELD, L. R., HASS, G., and WAYLONIS, J. E., 'Further Studies on MgF_2 Overcoated Aluminium Mirrors with Highest Reflectance in the Vacuum Ultraviolet', *Appl. Opt.*, **5**, No. 1, Jan. 1966.

ENGLISH, J., PUTNER, T., and HOLLAND, L., 'Programmed Control of Vacuum-deposited Multilayers', Fourth International Vacuum Congress, 1968.

HASS, GEORG, 'Optical Film Materials and their Applications', *The Journal of Vacuum Science and Technology*, **4**, No. 2.

GLANG, R., and MAISEL, 'The Handbook of Thin Film Technology', 1970.

HOLLAND, L., 'Vacuum Deposition of Thin Films', Chapman and Hall, London, 1956.

— 'Thin Film Microelectronics', Chapman & Hall.

— 'Review of Vacuum Deposition Mechanisms', *Electronic Components*, February, March, April, May, 1970.

HOLLAND, L., HACKING, K., and PUTNER, T., 'Titanium Dioxide Beam-splitters of Large Surface Area', *Vacuum*, **III**, No. 2, April 1953.

HOLLAND, L., and PUTNER, T. 'Two-layer Anti-reflexion Coating for Low Refractive Index Glass', *J. scient. Instrum.*, **36**, 81–4, Feb. 1959.

LETHAM, DARYL L., 'High Vacuum', *Machine Design*, 4 Feb. 1965.

MACLEOD, H. A., 'Thin Film Optical Filters', Adam Hilger Ltd.

PERRY, D. L., 'Low-loss Multilayer Dielectric Mirrors', *Appl. Opt.*, **4**, No. 8, Aug. 1965.

STECKELMACHER, W., 'Can we really measure Low Pressures?' Edwards High Vacuum International Ltd.

STECKELMACHER, W., PARISOT, J. M., HOLLAND, L., and PUTNER, T., *Vacuum*, **9**, 3/4, 171–85, 1959.

STECKELMACHER, W., and ENGLISH, J., Trans. 8th Vac. Symp. and 2nd Internat. Congress, Pergamon Press, London, 1961.

Chapter 13

CRAIG, DWIN R., 'Image Sharpness Meter', *Photographic Science & Engineering*, **5**, No. 6, 1961.

HABELL & COX, 'Engineering Optics', Freeman, 1958.

JACOBS, 'Fundamentals of Optical Engineering', McGraw-Hill Book Company, 1943.

KINGSLAKE, R., 'The Development of the Zoom Lens', *Journal of the Society of Motion Picture and Television Engineers*, Aug. 1960.

— 'Applied Optics and Optical Engineering', Vol. III, Academic Press.

PARGAS, PAUL, 'Phenomena of Image Sharpness Recognition of CdS and CdSc Photoconductors', *J. opt. Soc. Am.*, **54**, No. 4.

— 'A Lens-measuring Method using Photoconductive Cells', *Journal of the S.M.P.T.E.*, **74**, June 1965.

SMITH, W. J., 'Modern Optical Engineering', McGraw-Hill Book Company, 1966.

STRONG, J., 'Concepts of Classical Optics', Freeman, 1958.

Chapter 14

MOLONEY, F. J. T., 'Glass in the Modern World', W. H. Allen & Co.

ROLLASON, E. C., 'Metallurgy for Engineers', Edward Arnold (Publishers) Ltd.

Chapter 15

'Amateur Telescope Making—Advanced', Munn & Co. Inc., U.S.A., 1946.

BURRIDGE, J. C., KUHN, H., and PERCY, A., *Proc. Phys. Soc. Lond.*, B.66, 1953, 963.

FLUGGE, S., 'Optical Instruments', Handbuch der Physik (Encyclopaedia of Physics) Vol. XXIX, Springer-Verlag.

GASCOIGNE, S. C. B., 'The Anglo-Australian 150-inch Telescope', *Aust. J. Sci.*, **32**, No. 12.

HOLLAND, L., 'Vacuum Deposition of Thin Films', Chapman & Hall, London, 1956.

NICOLSON, Iain, 'Astronomy', Hamlyn.

TEXEREAU, J., 'La Construction du Telescope d'Amateur', Societé de France, 1951.

WEIGERT, A., and ZIMMERMAN, H., 'ABC of Astronomy', Adam Hilger Ltd.

Chapter 16

BALLANTINE, J. M., and ALLAN, W. B., 'Fibre Optics', *Science Journal*, Sept. 1965.

HICKS, WILBUR J., and KIRITSY, PAUL, 'Fibre Optics', *The Glass Industry*, April and May 1962.

HOPKINS, H. H., and KAPANY, N. S., 'A Flexible Fibrescope using Static Scanning', *Nature*, **173**, 39, 1954.

KAPENY, N. S., 'Fiber Optics', Academic Press, New York.

— 'Fibre Optics; Part 1, Optical Properties of Certain Dielectric Cylinders', *J. Opt. Soc. Am.*, **47**, 413, May 1957.

KINGSLAKE, R., 'Applied Optics and Optical Engineering', Academic Press, New York.

Chapter 17

KINGSLAKE, R., 'Applied Optics and Optical Engineering', Vol. 1, Academic Press, New York.

Chapter 18

CASEY, H. C., and TRUMBORE, F. A., 'Single Crystal Electroluminescent Materials', Bell Telephone Labs. Inc.

DANCE, J. B., 'Photoelectronic Devices', Iliffe.

EPSTEIN, A. S., and HOLONYAK, N., 'Solid State Light', *Science Journal*, Jan. 1969.

FYNN, G. W., and POWELL, W. J. A., 'Machine Polishing of Laser and other Solid State Materials', R.R.E. TN 709, 2nd edn, Ministry of Technology, Reports Centre, St Mary Cray, Kent, 1969.

HASS, G., and THUN, R. E., 'Physics of Thin Films', Academic Press.

HORNE, D. F., 'Optical Component Manufacture', Professional Advancement Course Electro-Optical Systems Design Conference, 28 Feb. 1972. Published by Kiver Communications Ltd.

Metals Research Ltd., Melbourn, Royston, Herts., Instruction Manuals on Macrotome, Microslice and Multipol.

NISHIZAWA, JUN-ICHI, 'Optoelectronics goes Digital', *Electronics*, Dec. 1967.

ROBERTSON, D. S., and JONES, O., 'Stockbarger Growth of Cuprous Chloride', *Br. J. appl. Phys.*, **17**.

ROSTRON, A. J., 'Levitation Melting', *Science Journal*, July 1967.

RUNYAN, W. R., 'Silicon Semi-conductor Technology', McGraw-Hill Book Company.

Appendix II Glossary of Terms used in the Optical Industry

English	*French*	*German*	*Japanese*
abrasive	abrasif	Schleifmaterial	研磨砂
acetone	acétone	Aceton	アセトン
adhese. To join two pieces of glass by heat treatment without distortion; a method developed in the Research Department of Adam Hilger Ltd.	——	Schweissen ohne Erweichung	2 枚のガラスを面が変らない程度に温めて接着させる
alcohol	alcool	Alkohol	アルコール
Aloxite	Aloxite	Aloxite	アロキサイト
Alundum	Alundum	Alundum	アランダム
angling	——	Winkel, an einen Block genaue ausschleifen	ブロックの角度を正しく磨く
angle Dekkor	cales d'angle	Autokollimations fernrohr	アングル・デッコウ
angle gauges. Prisms of polished glass or metal of accurately known angle	cales étalons	Winkel-Normalien	角度ゲージ
anneal (to stress, relieve)	recuire	Umformungspunktes schnell zu erreichen Vorwärmen	アンニール；焼鈍
anneal (to normalize)	tremper	tempern	徐冷(正規状態化)
aspherize. To depart deliberately from spherosity in working a surface	faire des surfaces aspherique	deformieren	非球面化する
astig. Abbreviation of astigmatic; denotes the shape of a surface which has different curvatures, or the performance of a lens system which has different foci, in two axial planes at right angles to each other	cylindre ou tore	astigmatisch	非点的
beeswax	cire d'abeilles	Bienenwachs	蜜蠟
benzene	benzine	Benzol	ベンゾール
bevel gauge	fausse équerre	Winkellehre	角度定規
block; blocking	bloc, blocage	Block (in einen Gipsklotz); in einen Block zu setzen	ブロック；ブロックを作る
blocking tool, tool on which optical pieces are placed for making a plaster block	outil à bloquer	——	ブロック台皿
block holder	——	Tragkörper	ブロック台皿
bort (a form of diamond)	bort	Diamantbort	黒色ダイヤモンド
bottle, drop. Stoppered bottle for allowing a drop of liquid to fall at a time	——	Tropfflasche	滴壜

English	*French*	*German*	*Japanese*
boxwood sticks. Name erroneously applied in the Hilger workshops to the thin orangewood sticks recommended for cleaning the slits of spectrographs and as vehicles for swabs of cotton wool	fusain	Orangenbaumstab	橙の木の箸
breath on, to	(contrôler) à la buée	anhauchen	呼気をかけてきれいにする
bruiser (for breaking down and spreading abrasive)	brisoir	Schleifmittel, Werkzeug zum Zerkleinern und Verteilen	研磨砂を潰し且つ広げる道具
brush	——	Pinsel, Wischer	ブラシ
bubbles in glass brought to the surface in grinding or polishing	points crevés	Blasen	気泡
bubbles, air balls in glass	bulles	Blasen	気泡
bubble, natural, within a crystal	vacuole	Lufteinschluss	結晶体内の気泡
buff, to; to buff up. To polish free from grey but without regard to surface flatness	——	blankpolieren, ohne Rücksicht auf Flächengüte	表面の平坦に関係なくバフ磨きを行う
caliper with scale but no vernier used in measuring thickness. Also double-ended caliper used in conjunction with a gauge for the same purpose	nonius	——	カリパス
cave. Abbreviation of concave, synonymous with shallow. Used to indicate that a surface, if concave, is more, or, if, convex, less so than it should be	concave	——	凹面の略語に使用することあり
carborundum	carborundum	Karborundum	カーボランダム
cement	ciment	Kitt	接着剤
Cenco Sealstix (formerly called Khotinski cement)	Silastic	Cenco Sealstix	センコ・シールスティックス
centre, to. To set a lens for edging so that the optical axis shall be central with the edge	centrer	zentrieren	心合わせを行う
cerium oxide	rose à polir oxyde de cérium	Ceroxyd	酸化セリウム
chamfer, to. To grind away sharp edges	chanfreiner	Kanten brechen; facettieren; fasen	面取を行う
chamois leather	peau de chamois	Putzleder	鹿皮
chromic acid	acid chromique	Chromsäure	クローム酸
chromium oxide	oxyde de chrome	Chromoxyd	酸化クローム
chuck	mandrin	Bohrfutter	チャック
clean	nettoyer	sauber	きれいな
cleave, to	cliver	spalten	劈開面に沿つて剝がす
colophony; rosin (the solid residue after distilation of oil turpentine from crude turpentine)	colophane	Colophonium (or Kolophonium)	コロフォニウム
colour, to put down in. To clean and lay two surfaces together so that they show Newton's fringes	mettre aux couleurs	auf Passe prüfen	2 つの面を合わせてニュートン・リングを見る
contact. To put surfaces in optical contact, viz. so cleanly that they adhere together without reflection at the interface; a process first used by Adam Hilger Ltd.	adhérer	ansprengen	2 つの面を光学的に接触させる
contrivance, apparatus	——	Vorrichtung	考案，仕掛
convex, convex surfaces	boss	Konvex, Konvexflächen, konvexe Flächen	凸の，凸面
correct, to	corriger	korrigieren	補正する
corundum	corindon	Korund	コランダム
cotton wool	ouate	Baumwolle Watte	脱脂綿
creosote (in U.S.A. creosote oil or liquid pitch oil)		Kreosotol	クレオソート

English	*French*	*German*	*Japanese*
crown (of a lens)	couronne	Linsenscheitel	レンズの頂点
crystal, cloud defects within	neige	Wolken	結晶体の中の曇
crystal, twin or other defect in; such as a small crystal embedded in the mass	macle	Einschluss	双晶その他の瑕
crystolon (an artificial corundum)	——	——	クリストロン
curvature, radius of	rayon de courbure	Krümmungsradius	曲率半径
curve, to	——	krümmen	曲線または曲面にする
curve; shaping cloth, felt, pitch saturated taffeta etc. to curvature for a polisher	gaufrage	——	布地，フェルト，ピッチの飽和したタフェータに研磨皿に必要な曲面を与える
cut, to	scier	schneiden	切断する
cut, severe scratch on a polished surface	——	Krätzer	引掻き瑕
cutter, sharp blade for trimming and grooving a polisher	tranchet	Messer	カッタ，刄物
dead metal. Small piece of opaque material in optical glass	——	Gallen, Stein	石
deep (used of a surface whose curvature is too great)	courte (surface)	(zu stark gekrümmt)	深面
depth gauge	jauge de profondeur	Tiefenmass	深さゲージ
Diamantin	——	Diamantine	ディアマンティン
diamond	diamant	Diamant, Diamantbort	ダイヤモンド
diamond, cutting. Glazier's diamond	——	Gläserdiamant	ガラス切り
diamond dust (generally in olive oil or vaseline)	egrisé	Diamantstaub	ダイヤモンド粉末
diamond, machine for cutting circles by	tournette	Brillenglasschneidemachine	丸切機
diamond, marking. Diamond chip set in metal for engraving or writing on glass	pointe à graver	Schreibdiamant	ダイヤモンド・カッター
digs on an optical surface	échignures	Grübchen	光学面の突き穴
disk, slice	disque, feuille de glace	Scheibe	円板
double image	——	Doppelbild	二重像
drill, to. See also trepanning. To bore small holes in glass	saccade, par perçage	bohren, bohren klein Löcher	孔あけ ガラスに小さな孔をあける
dry, to	sécher	trocknen	乾燥する
dust	poussière	Staub	塵埃；埃
edge, to	déborder	rundieren	縁摺する
edging with a peripheral groove to take a special mounting, etc.	rainer	Nut einschleifen	縁に溝を切ること
elutriation (of emery, etc.)	minutage	Schlämmen	清洗（金剛砂などの）
emery	——	Schmirgel	金剛砂
emery stone, natural; Turkey stone	pierre de Levant	Schleifstein	金剛砂砥石
emery, working down; increase of fineness of abrasive during working	raffinage	——	加工中金剛砂が細かくなる
errors, imperfections	erreurs	Fehler	誤差
eyeglass; eyeglasses	——	Augenglas; Augengläser	眼鏡用レンズ；眼鏡
faults	défauts	Fehler	欠点，不良個所
fluor spar	——	Fluss-spat	方解石
feathers; feather shaped defects within transparent materials. The term is applied also to collections of bubbles or flecks	feuillets	Federn	霧
figuring. Altering the shape of a surface to improve definition	rétoucher	retouchieren, korrigieren	局部修整

English	*French*	*German*	*Japanese*
flake. A shallow chip broken out of an optical part	——	Muschle	ガラスの砕片
flonk. A percussion tool whereby the surface of a piece of glass is flaked roughly to curvature to save grinding	fioner (pince à)	——	衝撃用工具
focus	foyer	Brennpunkt	焦点
forceps, spring	brucelles	Federzange	鋏
form, to (sometimes called to press out)	mouler	formen	成形する
former. Tool for forming polisher (sometimes, misleadingly, called polishing tool)	——	Druckkörper	研磨皿の表面成形工具
Fuller's earth (used with rouge in polishing on cloth polisher)	terre à foulon	Fullererde	フーラース・アース
gasolene (in England petrol)	essence	Benzin	ガソリン
gas pitch (in U.S.A. coal tar pitch)	poix	Pech	ピッチ
gauge (for optical tools)	calibre	Schablone, Lehre	ゲージ
glass	verre	Glas	ガラス
glass, unworked. Glasses	verre brut, verres	Rohglas, Gläser	ガラスの生地
glass-cutting	——	Glasschneiden	ガラス切断
goniometer	goniometre	Goniometer	測角器
grain	grain	Korn	粒子
grains	grains	Körner	粒子
graticule	réticule	Glasteilung	刻線板（焦点鏡）
graze, see scratch	——	Schramme, Kratzer	掠り傷
grey (of a ground surface)	gris	grau, narbig	砂摺面の艶消
greyness of an insufficiently polished surface	chair	nicht auspoliert	砂目
grind	meuler ou ébauchage	schleifen	砂摺する
grinding	meulage	Schliff	砂摺
grinding bench	machine à meuler	Schleifbank	砂摺機
grinding material, grinding powder	abrasif	Schleifmittel, Schleifpulver	研磨砂
grinding tools	meules	Schleifschalen	砂摺用皿
groove, to	encocher	Rillen; zu erzeugen	溝を作る
grooving tool	outil à rainurer	Rillen in der Polierschale Werkzeug zum Erzeugen	溝を作る工具
gypsum	gypse	Gips	石膏
handle, wooden, on which a small lens is worked individually	molette	Holzheft	木の柄
hardness	dureté	Härte	硬度
holder for working a diamond	dopp	Diamanthalter	ダイヤモンド保持具
homogeneity	homogénéité	Gleichmässigkeit, Homogeneität	均斉
hone. Oilstone or other fine sharpening stone	pierre d'aiguiser	Abziehstein	油砥
hot plate	plaque chauffante	Heizplatte	熱板
hydrochloric acid	acid chlorhydrique	Salzsäure	塩酸
ice box	glacière	Eisschrank	冷蔵庫
Iceland spar	spath d'Islande	Kalkspat	方解石
imperfections, see errors	défauts	——	不良個所
impregnated diamond wheel	meule diamant	Diamantscheibe	ダイヤモンド填込円板
jigs	tamis	——	治具；ジグ
Khotinski cement, now sold as Cenco Seal-stix	Silastic	Khotinski	ホチンスキ・セメント
knocker off. Hardwood stick or mallet for detaching 'mallets' from lenses by giving a sharp blow near the junction of pitch and work piece	maillet	Schlegel	槌
knocking off	——	——	たたき剝がす

English	*French*	*German*	*Japanese*
knocking-off mallet	deglanter, maillet à	——	たたき離す槌
lens	lentille	Linse	レンズ
lens holder (small pencil shaped handle on which small lenses are cemented for working)	cotret	Kittheft (Linsenhalter)	レンズ保持器
mallet	glaud	Kittklumpen	やに玉
malletting (putting mallets on) lenses or disks as for blocking	glauter	aufkitten	やに玉でレンズまたは円板を固めブロックを作ること
magnetic chuck	mandrin magnétique	Magnetspannfutter	電磁チャック
magnifier	loupe	Lupe, Vergrösserungsglas	拡大鏡
marker, piece of glass cemented to a tool with pieces which are to be ground as an indication of the approach to the desired thickness	témoin	Richtpunkt	荒摺の際厚さの標準として皿に貼付けられるガラス片
methylated spirit	alcool à brûler	Methyalkohol	メチルアルコール
micrometer gauge, 'mike'	comparateur micrometre	Schraubenmikrometer	マイクロメータ・ゲージ
milling (by diamond wheel)	——	fräsen (mit Diamantwerkzeug)	(砥石車で)フライス削り
mirror, speculum metal	mirror	Spiegel, Spiegelmetal	鏡，鏡用金属
mortar; steel mortar	mortier; mortier d'acier	Morser; Stahlmorser	モルタル；鋼乳鉢
Neven wheel	——	——	ネーヴェンの輪
Newton's fringes, working to	travail au couleurs	auf passe arbeiten	ニュートン・リングを出す
Nicol prism	Nicol	Nicol	ニコルのプリズム
Nicol prism, jig in which Iceland spar is cut for	hirondelle	——	方解石を切断するために入れる治具
nitric acid	acid nitrique	Salpetersäure	硝酸
ochre, red	ocre rouge	Ocker, rot	弁柄
ochre, yellow	ocre jaune	Ocker, gelb	黄土
o.g., object glass	objectif	Objektiv	対物レンズ
oil	huile	Oel	油
oilstone or other fine sharpening stone	pierre d'aiguiser	Oelstein	油砥
opacity	opacité	Undurchsichtigkeit	不透明度
optical bench	banc optique	optische Bank	光学測定台
optical contact	contact optique	optische Kontakt	光学的接着
oven	four	Ofen	窯
packing (strips of cardboard, sheet metal, etc., cemented to a curved tool to modify its curvature for use as a polisher or block holder)	cuirasse	Packung	皿の曲率を変えるためにこれに貼付ける厚板または金属板
paper for polishing (as used in France)	Berzelius	Polierpapier	研磨用紙
paraffin wax	cire de paraffine	Paraffin wachs	パラフィン蠟；洋蠟
paraffin (in U.S.A. kerosene)	paraffine	Paraffin	パラフィン
parallel up, to	paralleler à	parallel zu machen	平行にする
petrol (in U.S.A. gasolene)	essence	Benzin	ガソリン
pimply or granular appearance of a polished surface under magnification. Greyness of an imperfectly polished surface	gris	Orangeschale Anschein	研磨面の凸凹
pitch, gas (in U.S.A. coal tar pitch)	poix de Suede	Pech	ピッチ
pitch, pitch polishing	poix, polissage à poix	Pech, pechpolitur	ピッチ，ピッチ磨き

English	*French*	*German*	*Japanese*
pitch pad (or block). Tool covered with a layer of pitch to act as backing for working pieces plane-parallel	——	Plankörper	
pitch, Swedish (in U.S.A. Stockholm pitch)	——	Schwedisches Schiffspech	瑞典ピッチ
pits (in incompletely polished surfaces)	piqûres de 'gris'	Grübehen	突き穴
plaster block	blocs de plâtre	Gipsvorrichtung	石膏で固めたブロック
press, to (e.g. to form a polisher by pressure of a warm tool)	mise en forme	abdrücken	研磨皿に温い皿を当てて成形する
plaster of Paris	Plâtre de Paris	Gips	焼石膏
plunger; steel plunger	piston d'acier	Stempel, Stahlstempel	プランジャ；鋼のプランジャ
polish, to produce the early stage of polishing	éclaircir	polieren anzufangen	研磨する
polisher	polissoir	Polierschale, Polierwerkzeug	研磨皿
polisher holder	——	——	研磨皿の金物の部分（台皿）
polishing machine	machine à polir	Polierbank	研磨機
polishing or grinding machine	machine à polir	——	研磨砂摺兼用機
polishing powder	poudre à polir	poliermittel	研磨剤
polishing tool, see also former	outil à polir	Polierschale, Polierwerkzeug	研磨皿
Portland cement	——	Portland Zement	ポートラント・セメント
pot (as, for instance, emery pot)	pot	Topf	壺
press out, to (see form)	former	——	成形する
proof plate for spherical surfaces	calibre	Probeglas	球面用ニュートン板
proof plate for plane surfaces	calibre	Probeplatte	平面用ニュートン板
protector or any piece used as a help in making anything	——	Hilfsstuck	保護捧
protractor	rapporteur	Transporteur	分度器
pumice powder	poudre de pierre ponce ou ponce	Bimsstein	軽石の粉末
putty powder	potée d'étain	Zinnoxyd, Zinnasche	パテの粉末
quartz, incorrectly matched prisms of; as a Cornu prism with the two components of similar rotation	macle	——	十字形水晶プリズム
red ochre, see ochre	ocre rouge	——	弁柄
retouching. Correction of defects in a lens or mirror by local polishing	retouche locale	retouchieren	局部修整
roof prism	prisme en toit	Dachkantprisma	屋根形プリズム
rosin (in U.S.A. colophony)	résine	Kolophonium	ロージン
rouge, jeweller's	rouge à polir, rouge Anglais colcotar	Polierrot	弁柄
roughing	dégrossissage	schruppen	荒摺
ruby powder	ébauchage	Rubinpulver	ルビー粉末
ring, brass strip or tube round the outside of a plaster block	——	Ring	環
roughing tool	platine	Schruppschale	荒摺皿
rub up, to correct a polished surface by rubbing a master (or truing) tool on it	retoucher le polissage	Schaben corregieren	研磨皿の面を整形するためにこする
ruler (with scale in inches or millimetres)	——	Lineal, Maßstab	物差
sand, or grit used for grinding, or occasionally the natural emery sands which occur in Asia Minor and Macedonia	grés	Sand	砂
saw (circular)	scie circulaire	Kreissäge	丸鋸

English	*French*	*German*	*Japanese*
saw, to (in general)	——	sagen	鋸引する
sawdust, willow	scier	Sagespane von Weide	柳の木の鋸屑
scratch (slight, between a sleek and a scratch)	frayure	Kratzer	瑕
sealing wax	cire à sceller	Siegellack	封蠟
seeds. Opaque specks of unmelted material in glass (not infrequent in glass-ware)	points	Steine	石
set, to. To solidify	——	erstarren	固体化する
shallow (of a surface; not up to curve)	jeune	flach	浅面
shanks	équarrir, pince à; mivoitier, pince à	Brockelzange	柄（ツカ）
shank, to	——	abbrockeln	柄を着ける
Sira abrasive	——	Siraschmirgel	シラ研磨砂
sleeks; sleeks, heavy	filandres; ragures	Wischer, Haarrisse	スリーク
slit, to	——	schneiden	切断する
slitting wheel (or blade)	disque à refendre ou scie	Trennscheibe	切断用円板
smoothing	doucissage	Feinschleifen	精摺；砂掛
smoothing and polishing	surfaçage	——	精摺および研磨
smoothing tool	outil à doucir	Schleifwerkzeug	精摺用皿；砂掛皿
soft	douci	weich	軟かい
spectacles (pair of)	lunettes	Brille	掛眼鏡；眼鏡
speculum metal, see mirror	métal à polir spéculaire	Spiegelmetal	鏡用合金
spherical	sphérique	sphärisch	球状の
sponge	éponge	Schwamme	石綿
springing; alteration in shape due to cementing, uncementing, etc., due to introduction or relief of strain	déformation dûe aux pressions	springen	盛り上がり
spring off, to. To cause a protector to become detached from the work by suddenly heating or chilling it	——	absprengen	離してしまう
squeegee, e.g. for pressing out cement or adhesive under paper polisher in fixing to a tool	colloir	Presse, Druckvorrichtung	ローラー
stack (of disks cemented for edging en masse)	carrotte	Stapel, Rolle, Paket	円板を縁摺のために貼合わせたもの
stack, pile of disks or plates	colonne	——	円板などを積重ねたもの
stamping gauge	——	Pragegaze	ピッチ皿用網
stones; dead metal	pierres	Gallen	石
strain	tension	Spannung	歪み
straining gauze; closely woven muslin through which pitch is poured to remove foreign particles	tarlatane	Filtriertuch	濾過用ガーゼ
stray light	——	Streulicht	散光
sulphuric acid	acide sulfurique	Schwefelsäure	硫酸
surfaces	surfaces	Flächen, Oberflächen	表面；面
Swedish pitch, see pitch	poix de suede	——	瑞典のピッチ
testing	contrôle	Prüfung	検査
test plate	calibre	Probeplatte	平面原器
thickness, cavity ground in spotting for	mouche	Richtpunkt	
tilting chuck	mandrin oscillant	——	傾けられるチヤック
tin-oxide, putty powder	potée d'étain	Zinnasche	酸化錫；パテの粉末
tissue	chiffon	Gewebe, Stoff	薄い織物
tool handle	manche ou parpin	Handgriff	光学用工具の柄

English	*French*	*German*	*Japanese*
tool, optical, double-sided with transferable handle for grinding both convex and concave tools	outils à ébaucher	——	皿；工具
tools, optical; grinding tools	paire d'outils	Schalen, Schleifschalen	皿；砂摺皿
tools, optical, concave (for generating convex surfaces)	bassins	Schalen, Hohlschalen	凹面皿
tools, optical, flat	plateaux	Scheiben, Planscheiben	平面皿
tools, optical, convex (for generating concave surfaces)	balles	erhabene Schalen	凸面皿
tools, optical, pair of	——	Schalenpaar	一対の皿
tools, regrinding to bring them to correct radius and contact	réunir	Schalen korrigieren	擦に合わせ用皿
tool, optical, with circular grooves for grinding small convex lenses	caillebotter, plateau à	Schleifwerkzeug mit Ringnuten	円形の溝を有する皿（小さな凸レンズ砂摺り用）
transparency	transparence	Durchsichtigkeit, Durchlässigkeit	透明度
trepanning tool	trépan	Kernbohrer	丸切用工具
Tripoli powder	tripoli	Tripel	トリポリの粉末
truing	ébauchage	schleifen	整形
Turkey stone, natural emery stone	pierre de Levant	Naturschleifstein	天然産金剛砂砥石
turpentine	térébenthine	Terpentin	テレピン油
tweezers, spring	brucelles	Kluppzange	鋏
twinning, association in one piece of crystals of the same mineral but differently oriented or of opposite optical rotation	macles	Zwilling	雙晶
veils, see feathers	——	Schleier, Wolken	霧
veins, sharply defined or diffused	fils, secou gras	Schlieren	脈理
Vienna lime	——	Wiener Kalk	ウインナの石灰
warm, to	réchauffer	erwärmen	温める
washing (of emery, etc.)	lavage	Schlämmen	洗別
washing, grading (emery, etc.)	minutage, tamisage	Schlammung, Ausgeschlammung	洗別
Water-of-Ayr stone	——	Graustein	ウォータ・オブ・エーヤの石
wax	cire	Wachs	蠟
wet, to give a; to polish a piece of work from the wet to the dry condition of the polisher	faire une séchée	trockenpolieren	濡れた状態から乾く所まで研磨する
Windolite	——	——	ウィンドライト
yellow ochre, see ochre	ocre jaune	——	黄土
Xylol	Xylol	Xylol	キシロール

SUBJECT INDEX

NAME AND AUTHOR INDEX